Nanotechnology in Plant Sciences

Volume 3

Series Editors

Jameel M. Al-Khayri, Department of Agricultural Biotechnology, King Faisal University, Al-Ahsa, Saudi Arabia

Shri Mohan Jain, Department of Agricultural Sciences, University of Helsinki, Helsinki, Finland

This book series provides a platform for researchers to update advances in nanotechnology research and applications relevant to various fields of plant sciences. This series is an excellent reference source of cutting-edge knowledge for agriculturists, breeders, biotechnologists, and molecular biologists interested in plant nanotechnology. It is essential for professionals, researchers, policymakers, and commercial entrepreneurs concerned with understanding the mechanisms and innovative applications of nanotechnology to achieve enhanced crop production improvement, agricultural sustainability, environment protection, and food and nutritional security.

Jameel M. Al-Khayri • T. R. Anju
Shri Mohan Jain
Editors

Nanotechnology Applications in Modern Agriculture

Springer

Editors
Jameel M. Al-Khayri
Department of Agricultural Biotechnology
King Faisal University
Al-Ahsa, Saudi Arabia

T. R. Anju
Department of Biotechnology
Newman College
Thodupuzha, Kerala, India

Shri Mohan Jain
Department of Agricultural Sciences
University of Helsinki
Helsinki, Finland

ISSN 3004-9334 ISSN 3004-9342 (electronic)
Nanotechnology in Plant Sciences
ISBN 978-3-031-90505-6 ISBN 978-3-031-90506-3 (eBook)
https://doi.org/10.1007/978-3-031-90506-3

This Springer imprint is published by the registered company Springer Nature Switzerland AG
The registered company address is: Gewerbestrasse 11, 6330 Cham, Switzerland

Preface

Nanotechnology is revolutionizing modern agriculture by enhancing crop productivity, improving soil health, and ensuring sustainable farming practices. The book, ***Nanotechnology Applications in Modern Agriculture***, serves as a comprehensive guide to the latest advancements, challenges, and future prospects of nanotechnology in the agricultural sector. The book begins with an introduction providing a fundamental understanding of the role of nanoscience in agriculture. Subsequent chapters delve into the synthesis and modification of nanoparticles for plant applications, highlighting innovative approaches for developing effective and sustainable agricultural solutions. A key focus of this book is the application of nanotechnology in enhancing agricultural inputs and plant disease management. The book also explores cutting-edge applications of nanotechnology in plant breeding and genetics. It explores the latest advancements in nanoparticle synthesis, nanoencapsulation, and nanosensors for precision agriculture, while also addressing biosafety, ethical considerations, and future trends. The book consists of 22 chapters, enriched with 27 tables and 63 color figures to ensure better understanding and clarity. Chapters are contributed by 79 recognized scientists representing nine countries after a rigorous review process to maintain quality and scientific accuracy. Each chapter begins with an introduction that offers fundamental background information, followed by a detailed examination of the topic enriched with high-quality color figures and tables. To inspire future research, every chapter concludes with insightful recommendations for further exploration. Through this book, researchers, scientists, agronomists, and students will gain a deeper understanding of how nanotechnology is reshaping modern agriculture. It serves as a valuable resource for exploring new frontiers in agricultural science, ensuring a sustainable and technologically advanced future for global food production.

Al-Ahsa, Saudi Arabia — Jameel M. Al-Khayri
Thodupuzha, Kerala, India — T. R. Anju
Helsinki, Finland — Shri Mohan Jain

Acknowledgment

The editors extend their appreciation to the support of the Deanship of Scientific Research, Vice Presidency for Graduate Studies and Scientific Research, King Faisal University, Saudi Arabia (Project No. KFU253023).

Contents

1 **Introduction to Nanotechnology Applications in Modern Agriculture** 1
Rahimol Ramesh, T. R. Anju, Jameel M. Al-Khayri, and Shri Mohan Jain

2 **Nanoparticle Synthesis and Modification for Plant Applications** 17
Panangattu Sankarji Jayalakshmi, Sanu Mathew Simon, and Aji Abraham

3 **Nanoencapsulation of Agricultural Inputs** 51
Gracesuganthi Jayaraj, Mohanapriya Balasubramaniam, and Kamaraj Raju

4 **Nanotechnology in Plant Disease Management** 81
I. K. Nishitha, Tintu Jose Manicketh, and Arya P. Mohan

5 **Nanosensors for Monitoring Plant Health** 101
Anju Susan, K. R. Raghi, and T. M. Archana

6 **Nanotechnology for Soil Improvement and Plant Nutrition** 133
Puthamohan Vinayaga Moorthi, Ravi Bavyataa, Kumar Vignesh, Divya Sri Kathiresan, and Rubadevi Balasubramani

7 **Nano Fertilizers for Precision Agriculture** 153
Akhila Sen, P. Faseela, Julie Jacob, and T. Siju Thomas

8 **Nanoparticle-Based Delivery Systems for Plant Nutrients and Stress Mitigation** 179
Jaya Mary Jacob, G. K. Shamnamol, Shynu Mary John, and P. Fathima Rini

9 Engineered Nanomaterials for Abiotic and Biotic Stress Tolerance in Plants 203
Mawuli K. Azameti, Muazu Issifu, Abdul-Wahab Imoro, Wadzani Palnam Dauda, and Kishor Prabhakar Panzade

10 Nanotechnology Application in Crop Improvement 237
Nisha Pallath, Tharakupeedikayil Abdul Majeed Sajeena, Aswathy Vijay, and Shalet Varghese

11 Nanoparticles for Crop Protection 271
K. R. Raghi, Anju Susan, and T. M. Archana

12 Nanotechnology in Crop Breeding and Genetics 307
Meril Sara Kurian, C. Akshaya Prakash, S. Sanjana, A. M. Shackira, Delse Parekkattil Sebastian, and Nair G. Sarath

13 Nanoscale Nutrient Delivery for Hydroponics and Aeroponics 327
Rashmi Singh and Kamaraj Raju

14 Nanotechnology Applications in Modern Agriculture 347
Mariat George, Amritha Krishnakumar, and T. R. Anju

15 Biosafety and Environmental Impact Assessment of Nanopesticides 381
Rashmi Singh and Kamaraj Raju

16 Sustainable Agriculture and the Role of Nanotechnology 407
Vidya Niranjan, Lavanya C, Anagha S Setlur, Chandrashekar Karunakaran, and Sinosh Skariyachan

17 Ethical and Environmental Considerations in Plant Nanotechnology 437
Jenifer Christy Yacob, Vidhya Ramachandran, Dhannia P. Narayanan, and Sumod George

18 Current Trends of Nanoparticles in Insect Pest Management 451
U. Pirithiraj, R. P. Soundararajan, M. Yuvaraj, Jaiby Cyriac, R. Sathya Priya, N. Jagathjothi, N. Suganthi, R. Sharmila, and M. Saranya

19 Nanotechnology Solutions for Post-harvest Challenges 479
Emad Hamdy Khedr and Nagwa Khedr

20 Nanomaterials for Enhancing Fruit Yield and Quality 505
Emad Hamdy Khedr and Nagwa Khedr

21 Nano Fertilizers Influence on Plant Physiological Activities 523
Ghulam Abbas, Muneeba, Muhammad Asif Jamal, Kaleemul Din, Ilkay Yavas, Safdar Ali, Asif Ameen, Saddam Hussain, and Mustafa Surmen

22 Future Trends and Challenges in Agricultural Nanotechnology 547
Viji Mary Varghese, O. Kavitha, Gigi Poulose, and Nithya Joseph

Index 585

Editors and Contributors

About the Editors

Jameel M. Al-Khayri is affiliated with the Department of Agricultural Biotechnology, King Faisal University, Saudi Arabia. He received B.S. Biology in 1984 from the University of Toledo, M.S. in Agronomy in 1988, and Ph.D. in Plant Science in 1991 from the University of Arkansas. He is the National Correspondent of the International Association of Plant Tissue Culture and Biotechnology. He has authored 145 research articles, 80 chapters, and edited several journal special issues. In addition, he edited 34 reference books on plant biotechnology, genetic resources, breeding, genomics, and nanotechnology. In addition to teaching, advising, and research, he held administrative responsibilities as the Assistant Director of Date Palm Research Center, Head of Department of Plant Biotechnology, and Vice Dean for Development and Quality Assurance. He served as a Member of Majlis Ash Shura (Saudi Legislative Council) for the 2009–2012 term. He is interested in agricultural innovations and sustainability.

T. R. Anju is an Assistant Professor in Biotechnology at Newman College, Kerala, and the Secretary of the Society for Biotechnologists, India. She completed a master's degree in biotechnology, and a doctoral degree at the Centre for Neuroscience, Cochin University of Science and Technology, Kochi, India. She completed her postdoctoral research on a neonatal hypoglycemia project funded by the Department of Biotechnology, Government of India. She has authored 44 publications in international peer-reviewed journals, 20 book chapters, and edited 2 books. She is the recipient of Kerala State Young Scientist Award, IBS Award in Medical Biotechnology, and Prof. Edathil Vijayan Award in Neuroscience. In addition to teaching and research, she acts as coordinator of institutional grants funded by the Government of India. Her research is mainly focused on molecular signaling, neurotransmitter receptor functions, and nanotechnology, especially in green nanomaterials and their impact on plants, the environment, and sustainable development.

Shri Mohan Jain (Adjunct) is currently associated with the Department of Agricultural Sciences, University of Helsinki, Helsinki, Finland. He received his M. Phil in 1973 and Ph.D. in 1978 from Jawaharlal Nehru University, New Delhi, India. He served as a Technical Officer, Plant Breeding, and Genetics, International Atomic Energy Agency (IAEA), Vienna, Austria, 1999–2005 when the agency was awarded Nobel Peace Prize, 2005. He is on the Editorial Board of *In Vitro* and *Propagation of Ornamental Plants* and a reviewer in *Plant Cell and Organ Culture*, *Plant Cell Reports*, and *Plants*. His publications include 225 peer-reviewed journals, book chapters, and conference proceedings, and 82 edited books; he is an invited speaker and has acted as a Chairperson in several international conferences worldwide. He is a former consultant to the IAEA, the European Union, the Government of Grenada, the Iranian Private Company, and the Egyptian government.

Contributors

Ghulam Abbas Department of Agronomy, University of Agriculture, Faisalabad, Pakistan
Agronomic Research Institute, Ayub Agricultural Research Institute, Faisalabad, Pakistan

Aji Abraham Department of Botany, Mar Athanasius College (Autonomous), Kothamangalam, Kerala, India

C. Akshaya Prakash Department of Botany, St. Joseph's College (Autonomous), Devagiri, Kozhikode, Kerala, India

Safdar Ali Department of Agronomy, University of Agriculture, Faisalabad, Pakistan

Jameel M. Al-Khayri Department of Agricultural Biotechnology, King Faisal University, Al-Ahsa, Saudi Arabia

Asif Ameen Agronomic Research Institute, Ayub Agricultural Research Institute, Faisalabad, Pakistan

T. R. Anju Department of Biotechnology, Newman College, Thodupuzha, Kerala, India

T. M. Archana Postgraduate and Research Department of Microbiology, Sree Sankara College, Kalady, Ernakulam, Kerala, India

Mawuli K. Azameti Department of Applied Biology, C. K. Tedam University of Technology and Applied Sciences, Navrongo, Ghana

Mohanapriya Balasubramaniam Department of Biotechnology, Rathinam College of Arts and Science, Coimbatore, Tamil Nadu, India

Rubadevi Balasubramani Department of Human Genetics and Molecular Biology, Bharathiar University, Coimbatore, Tamil Nadu, India

Ravi Bavyataa Department of Human Genetics and Molecular Biology, Bharathiar University, Coimbatore, Tamil Nadu, India

Chandrashekar Karunakaran Department of Biotechnology, RV College of Engineering, Bangalore, India

Jaiby Cyriac Department of Botany, Newman College, Thodupuzha, Kerala, India

WadzaniPalnam Dauda Crop Science Unit, Department of Agronomy, Federal University, Gashua, Nigeria

Kaleemul Din Department of Botany, University of Agriculture, Faisalabad, Pakistan

P. Faseela Department of Botany, Korambayil Ahamed Haji Memorial Unity Women's College, Manjeri, Kerala, India

P. Fathima Rini Department of Biotechnology and Biochemical Engineering, Sree Buddha College of Engineering, Pattoor, APJ Abdul Kalam Technological University, Thiruvananthapuram, Kerala, India

Mariat George Department of Botany, Newman College, Thodupuzha, Kerala, India

Sumod George Department of Chemistry, Nirmala College (Autonomous), Muvattupuzha, Ernakulam District, Kerala, India

Saddam Hussain Department of Agronomy, University of Agriculture, Faisalabad, Pakistan

Abdul-Wahab Imoro Department of Forestry and Forest Resources Management, University for Development Studies, Tamale, Ghana

Muazu Issifu Department of Agriculture Biotechnology, University for Development Studies, Nyankpala Campus, Tamale, Ghana

Jaya Mary Jacob Scientist, Kerala State Council for Science Technology and Environment (KSCSTE), Thiruvananthapuam, Kerala, India

Julie Jacob Department of Biochemistry, Mar Athanasius College (Autonomous), Kothamangalam, Kerala, India

N. Jagathjothi Tamil Nadu Agricultural University, Coimbatore, Tamil Nadu, India

Shri Mohan Jain Department of Agricultural Sciences, University of Helsinki, Helsinki, Finland

Muhammad Asif Jamal Department of Agronomy, University of Agriculture, Faisalabad, Pakistan

Panangattu Sankarji Jayalakshmi Department of Botany, Mar Athanasius College (Autonomous), Kothamangalam, Kerala, India

Gracesuganthi Jayaraj Department of Biotechnology, Rathinam Technical Campus, Coimbatore, Tamil Nadu, India

Shynu Mary John Department of Biotechnology and Biochemical Engineering, Sree Buddha College of Engineering, Pattoor, APJ Abdul Kalam Technological University, Thiruvananthapuram, Kerala, India

Nithya Joseph Polymer Scientist, Bioscience and Water filtration Division, Parker Hannifin Corporation, Oxnard, CA, USA

Divya Sri Kathiresan Department of Human Genetics and Molecular Biology, Bharathiar University, Coimbatore, Tamil Nadu, India

O. Kavitha Department of Biotechnology, St. Joseph's College (Autonomous), Irinjalakuda, Thrissur, Kerala, India

Emad Hamdy Khedr Department of Pomology, Faculty of Agriculture, Cairo University, Giza, Egypt

Nagwa Khedr Department of Pomology, Faculty of Agriculture, Cairo University, Giza, Egypt

Amritha Krishnakumar Department of Botany, Newman College, Thodupuzha, Kerala, India

Meril Sara Kurian Department of Botany, Mar Athanasius College (Autonomous), Kothamangalam, Kerala, India

Lavanya C Department of Biotechnology, RV College of Engineering, Bangalore, India

Tintu Jose Manicketh Department of Botany, St. Teresa's College (Autonomous), Ernakulam, Kerala, India

Arya P. Mohan Department of Botany, St. Teresa's College (Autonomous), Ernakulam, Kerala, India

Puthamohan Vinayaga Moorthi Department of Human Genetics and Molecular Biology, Bharathiar University, Coimbatore, Tamil Nadu, India

Muneeba Department of Agronomy, University of Agriculture, Faisalabad, Pakistan

Dhannia P. Narayanan Department of Botany, Mar Athanasius College (Autonomous), Kothamangalam, Kerala, India

Vidya Niranjan Department of Biotechnology, RV College of Engineering, Bangalore, India
Research and Collaborations, MIT Vishwaprayag University, Solapur, India

I. K. Nishitha Department of Botany, St. Teresa's College (Autonomous), Ernakulam, Kerala, India

Nisha Pallath Department of Biosciences, MES College Marampally, North Vazhakulam, Aluva, Ernakulam (Dist), Kerala, India

Kishor Prabhakar Panzade Department of Plant Biotechnology, SDMVM College of Agricultural Biotechnology, Georai Tanda, Maharashtra, India

U. Pirithiraj Centre for Plant Protection Studies, Tamil Nadu Agricultural University, Coimbatore, India

Gigi Poulose Department of Zoology, St. Joseph's College (Autonomous), Irinjalakuda, Thrissur, Kerala, India

R. Sathya Priya Tamil Nadu Agricultural University, Coimbatore, Tamil Nadu, India

K. R. Raghi Department of Integrated Chemistry, MES College Marampally, Aluva, Ernakulam, Kerala, India

Kamaraj Raju Department of Biotechnology, Rathinam College of Arts and Science, Coimbatore, Tamil Nadu, India

Vidhya Ramachandran Department of Biochemistry, Idhaya College of Arts and Science, Pakkamudayanpet, Pondicherry, India

Rahimol Ramesh Department of Biotechnology, Al Ameen College, Edathala, Aluva, Kerala, India
Post Graduate Department of Biosciences, M.E.S College, Marampally, Aluva, Kerala, India

Tharakupeedikayil Abdul Majeed Sajeena Department of Biosciences, MES College Marampally, North Vazhakulam, Aluva, Ernakulam (Dist), Kerala, India

S. Sanjana Department of Botany, Sir Syed College, Kannur University, Kannur, Kerala, India

M. Saranya Tamil Nadu Agricultural University, Coimbatore, Tamil Nadu, India

Nair G. Sarath Department of Botany, Mar Athanasius College (Autonomous), Kothamangalam, Kerala, India

Delse Parekkattil Sebastian Department of Botany, St. Joseph's College (Autonomous), Devagiri, Kozhikode, Kerala, India

Akhila Sen Department of Botany, Mar Athanasius College (Autonomous), Kothamangalam, Kerala, India

Anagha S. Setlur Department of Biotechnology, RV College of Engineering, Bangalore, India

A. M. Shackira Department of Botany, Sir Syed College, Kannur University, Kannur, Kerala, India

G. K. Shamnamol Department of Biotechnology and Biochemical Engineering, Sree Buddha College of Engineering, Pattoor, APJ Abdul Kalam Technological University, Thiruvananthapuram, Kerala, India

R. Sharmila Tamil Nadu Agricultural University, Coimbatore, Tamil Nadu, India

T. Siju Thomas Department of Botany, Mar Athanasius College (Autonomous), Kothamangalam, Kerala, India

Sanu Mathew Simon Department of Physics, Mar Athanasius College (Autonomous), Kothamangalam, Kerala, India

Rashmi Singh Department of Zoology, Dayanand Vedic College, Orai, Uttar Pradesh, India

Sinosh Skariyachan Department of Microbiology, St. Pius X College, Rajapuram, Kasaragod, Kerala, India

R. P. Soundararajan Centre for Plant Protection Studies, Tamil Nadu Agricultural University, Coimbatore, India

N. Suganthi Tamil Nadu Agricultural University, Coimbatore, Tamil Nadu, India

Mustafa Surmen Department of Field Crops, Faculty of Agriculture, University of Aydın Adnan Menderes, Aydın, Turkey

Anju Susan Department of Chemistry, Mar Athanasius College (Autonomous), Ernakulam, Kerala, India

Shalet Varghese Department of Biosciences, MES College Marampally, North Vazhakulam, Aluva, Ernakulam (Dist), Kerala, India

Viji Mary Varghese Department of Biotechnology, St. Joseph's College (Autonomous), Irinjalakuda, Thrissur, Kerala, India

K. Vignesh Department of Human Genetics and Molecular Biology, Bharathiar University, Coimbatore, Tamil Nadu, India

Aswathy Vijay Department of Biosciences, MES College Marampally, North Vazhakulam, Aluva, Ernakulam (Dist), Kerala, India

Jenifer Christy Yacob Peer Review Services Team (Wiley publishing company), Integra Software Services Pvt. Ltd., Pakkamudayanpet, Puducherry, India

Ilkay Yavas Koçarlı Vocational School, Department of Crop and Animal Production, University of Aydın Adnan Menderes, Aydın, Turkey

M. Yuvaraj Agricultural College and Research Institute, Vazhavachanure, Tamil Nadu, India

Chapter 1
Introduction to Nanotechnology Applications in Modern Agriculture

Rahimol Ramesh, T. R. Anju, Jameel M. Al-Khayri, and Shri Mohan Jain

Abstract Nanotechnology has rapidly evolved as a transformative approach in modern agriculture, offering groundbreaking solutions that address challenges related to crop productivity, soil quality, and environmental sustainability. By leveraging nanoparticles, researchers have developed methods to target specific cellular components within plants, functioning as a highly precise "magic bullet" for delivering critical molecules—such as genes, herbicides, and fertilizers—directly into plant organelles. This targeted delivery system not only optimizes the effectiveness of agricultural inputs but also significantly reduces the overuse of chemicals, minimizing the environmental impact associated with conventional farming practices. Moreover, recent advancements in nanoscience have provided deeper insights into the physiological, biochemical, and molecular mechanisms of nanoparticles in plants, shedding light on how they influence plant growth and resilience. Understanding these intricate mechanisms is essential to fully harness the benefits of nanotechnology, enabling plants to grow and thrive under nanoparticle exposure. This article explores the wide-ranging applications of nanotechnology in agriculture, including the development of nano-fertilizers, intelligent pesticides, nano-sensors, and soil remediation techniques. Nano-fertilizers enhance nutrient uptake and reduce nutrient loss, while intelligent pesticides increase pest control efficiency by targeting specific pest organisms, thus reducing unintended ecological damage. Nano-sensors

R. Ramesh
Department of Biotechnology, Al Ameen College, Aluva, Kerala, India

Post Graduate Department of Biosciences, M.E.S College, Marampally, Aluva, Kerala, India

T. R. Anju (✉)
Department of Biotechnology, Newman College, Thodupuzha, Kerala, India
e-mail: anju.tr@newmancollege.ac.in

J. M. Al-Khayri
Department of Agricultural Biotechnology, King Faisal University, Al-Ahsa, Saudi Arabia
e-mail: jkhayri@kfu.edu.sa

S. M. Jain
Department of Agricultural Sciences, University of Helsinki, Helsinki, Finland

J. M. Al-Khayri et al. (eds.), *Nanotechnology Applications in Modern Agriculture*, Nanotechnology in Plant Sciences 3, https://doi.org/10.1007/978-3-031-90506-3_1

provide real-time data on plant health and soil conditions, allowing for precise and timely interventions. Additionally, nanotechnology-based soil remediation techniques can remove contaminants, improving soil health and crop quality. The integration of nanotechnology into agriculture thus has immense potential for promoting sustainable farming practices. By enhancing resource efficiency, increasing crop yields, and fostering better environmental stewardship, nanotechnology is poised to play a crucial role in addressing global food security challenges and supporting environmentally responsible agricultural systems.

Keywords Agricultural nanomaterials · Crop productivity · Environmental sustainability · Intelligent pesticides · Nano-fertilizers · Nano-sensors for plants · Sustainable agriculture

1.1 Introduction

Agriculture serves as a fundamental pillar for numerous emerging nations, supporting more than 60% of their community livelihoods. Agricultural researchers face a multitude of challenges in addition to the farmers leaving their profession, such as inadequate crop yields, inefficient nutrient utilization, decreasing levels of soil organic matter, deficits in numerous essential nutrients, the impact of climate change, diminishing availability of arable land and water resources, and a shortage of labor. To satisfy the demands of food security, 4% sustainable agricultural development is needed, even though faced with the enormous challenges. The Food and Agriculture Organization (FAO 2017) projects that by 2050, there will be 10 billion people on Earth, and thereby increasing the food demand by 50%, particularly in the developing countries. Furthermore, it is estimated that 815 million people worldwide are suffering from the malnutrition, and by 2050, an additional 2 billion will fall into this group. This predicament necessitates significant modifications to the world's food production systems. Recent studies have shown that nanotechnology has the ability to significantly increase agricultural output and improve security (Khan et al. 2022). This is achieved by boosting the efficiency of agricultural inputs and offering solutions for environmental and agricultural challenges. Research on cutting-edge technologies like "nanotechnology," can accurately detect and deliver the optimum amount of nutrients and pesticides to enhance production while preserving higher efficient use and environmental safety, is necessary to address these problems (Jogaiah et al. 2020). This chapter provides a summary of the research results from several publications that cover different aspects of agricultural nanotechnology and how nanotechnology is transforming agriculture, from tailored nutrition delivery to advanced crop protection measures. Figure 1.1, illustrates the diverse applications of nanotechnology in modern agriculture, highlighting its role in enhancing crop yield, precision farming, pest and disease control, and soil health monitoring.

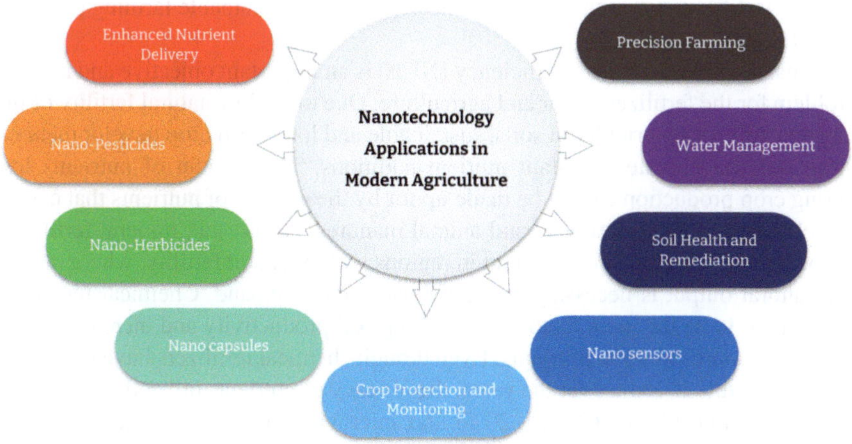

Fig. 1.1 Nanotechnology applications in modern agriculture (constructed by Rahimol Ramesh)

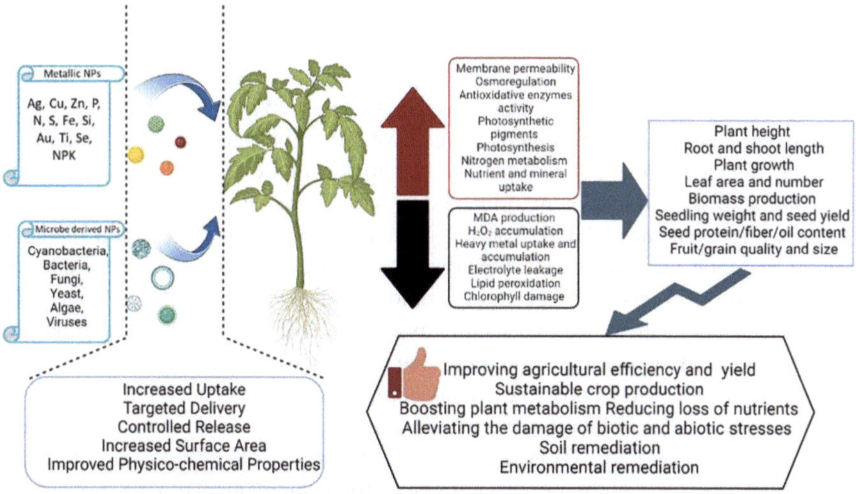

Fig. 1.2 Diagrammatic depiction of nanoparticle-based delivery system and its effects on plant growth. (Reproduced from Goyal et al. 2023)

1.2 Enhanced Nutrient Delivery

One of the key challenges in agriculture is ensuring plants receive optimal nutrition while minimizing wastage. Nano-sized fertilizers are engineered to address this challenge, releasing nutrients gradually to plants. This targeted delivery not only increases nutrient uptake but also moderates the environmental impact of excess fertilizers leaching into the soil and waterways (Yadav et al. 2023). The development of nano-fertilizers represents a significant step towards sustainable farming practices (Fig. 1.2).

Enhancing Nutrient usage efficiency (NUE) is an important objective and a major problem for the fertilizer sector and agriculture. Due to the low natural fertility of the majority of Indian agricultural soils, sustainable and long-term crop development on these soils necessitates constant nutrient additions. The amount of nutrients lost during crop production cannot be made up for by the amount of nutrients that can be recycled through crop leftovers and animal manures. As a result, mineral fertilizers are becoming increasingly important in regions with poor soil fertility, where raising agricultural output is necessary to satisfy rising food demand. Chemical fertilizers are thought to be the main factor in sustaining soil productivity and increasing crop yields as a source of plant nutrients. Even though chemical fertilizers have been used more frequently in India over the years, the efficacy of nutrients utilized as fertilizers particularly Phosphorus (P) (15–20%) and micronutrients (2–5%) like Zinc (Zn), Iron (Fe) and Copper (Cu) remains dreadfully poor. Inefficient utilization of fertilizer inputs raises the risk of biosphere pollution as well as cultivation costs. Thus, the ecology and economy draw attention to the compulsive need for crop production to use nutrients more effectively. Given the high cost and widespread usage of fertilizer nutrients nationwide, any improvement in use efficiency will result in a significant reduction in nutrient requirements as well as significant economic benefits. Yield per unit input is one way to define nutrient usage efficiency. While in scientific literature the NUE is frequently represented as fresh weight or product yield per content of nutrient, in agriculture this is typically tied to the input of fertilizer. Expanding agricultural productivity on marginal soils with inadequate nutrient availability requires improvement of NUE. The net nutrient exchange of agricultural plants is contingent upon various factors, including soil nutrient uptake efficiency, internal plant utilization, transit, storage, mobilization, and environmental conditions. To comprehend NUE, there are two main ways to go about it. First, to determine the systems impacted by this stress and those that might help to maintain development at low nutrient input, it is possible to investigate how plants respond to nutrient deficiency stress. The utilization of artificial or natural genetic variation is the second strategy (Mukhopadhyay 2014). These nanostructures could serve as an excellent medium for plant growth by providing micronutrient supplements to the roots of the plant, according to many researches like the regulation of Zn discharge in a solution by a hollow-core shell of nano-sized Mn-carbonate loaded with ZnSO4. By satisfying the root needs of plants through ion exchange processes, these nano-formulated structures also produced a slower release or discharge of zinc than its salt, ZnSO4, which clarifies the idea of nanoencapsulation (Yuvaraj and Subramanian 2015).

The properties of nano encapsulated micronutrients that facilitate higher nutrient use efficiency are:

- Their small size results in a significant surface area, which aids in the differentiation of metabolic activities within the plant system.
- They have a high reactivity with other compounds due to their small size.
- Water is a highly soluble solvent for them.

- Nano fertilizers' particle size is less than 100 nm, so they penetrate better into plants.
- The large surface area of nano fertilizers improves their uptake capacity and nutrient utilization efficiency.
- Fertilizer nanoparticles encapsulating nutrients within themselves will improve the plants' bioavailability (Tarafdar et al. 2012)

1.3 Smart Pesticides and Herbicides

Nanotechnology will enhance plant growth and fruiting, reduce the life cycle duration, improve soil quality, and increase resource efficiency. Nanotechnology seeks to boost the agriculture industry through novel nanotools that facilitate rapid disease diagnosis and augment plants' nutritional absorption capabilities, among other applications. Precision farming is introducing nanotechnology to the agricultural sector, since it may significantly enhance the framework of sustainable agriculture by providing efficient and eco-friendly technologies. Recent advancements in nanotechnology for drugs and pharmaceuticals have unveiled new opportunities to implement its principles in the agricultural sector. The fundamental objectives of nanotechnology are to maximize output (crop yields), minimize input (fertilizers, insecticides, and herbicides), evaluate environmental conditions using sensors, and implement targeted actions (Chaudhary and Mamta 2022).

1.4 Nano-Pesticides

Nanoparticles have the capability to significantly enhance the stability and effectiveness of entire cells, enzymes, and other natural products utilized in biopesticides. However, there are several obstacles to overcome in the field when applying Nanoparticles for the delivery of pesticides and biopesticides, including numerous environmental disturbances, a wide area that needs to be sprayed, and ultimately cost-effectiveness (Pan et al. 2022). For ease of application, a high-volume, low-value preparation is sprayed on the entire crop during the standard spraying routine. In contrast, low-volume, high-value applications are expected for preparations based on nanomaterials. Such regulated nanoparticulate delivery systems require targeted distribution tactics based on disease or pest life cycle and behavior (Manjunatha et al. 2016). Nano-biopesticides offer several advantages compared to biopesticides and traditional approaches. These advantages include their ecologically benign behavior, rapid application, easy distribution to plants, biodegradability, and gradual release from the vector (Sayed et al. 2017). Furthermore, when merged with pesticides that are easily assimilated by plants, their diminutive dimensions provide them a highly effective carrier. An additional benefit of nano-biopesticides is their lack of negative impact on soil microorganisms. In addition,

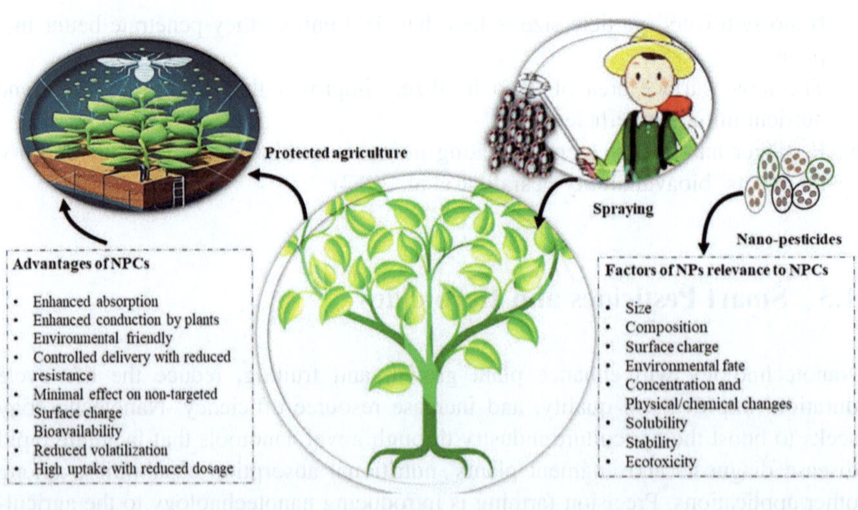

Fig. 1.3 Pictorial representation of factors affecting nano pesticides (NPCs) and its advantages. (Reproduced from Ali et al. 2024)

Ag-based nanoparticles' phototoxicity was reduced by nanocoating them with biocompatible polyvinyl pyrrole compounds (Lu et al. 2010). There are two methods that can be employed in the production of nano-biopesticides. The initial approach is isolating the biologically active pesticidal component (APC) from plants, merging it with nanoparticles, and then integrating it into a suitable polymer to serve as a supportive substance. The second method entails the secretion of a metallic salt by the APC, which then binds to and breaks down nanoparticles (NPs) to generate a suitable polymer. In conjunction with the nanoparticles, the antigen-presenting cells constituted a viable vector of micelles, liposomes, nanospheres, polymers, and nanofibers. To safeguard food, these substances were sprayed on insect pests (Ragaei and Sabry 2014). Figure 1.3 depicts the factors of nanoparticles (NPs) that affect nano pesticides and its advantages.

1.5 Nano-Herbicides

Nano-herbicides are produced by employing nanosized preparations or herbicide formulations composed of nanomaterials, which exploit the nanotechnological capabilities to achieve effective delivery of chemical or biological pesticides. Nanomaterial or nanostructure-based formulations have the potential to enhance the effectiveness, solubility, and safety profile of herbicides as compared to traditional herbicides. Utilizing nanoparticle-based systems to release herbicides early in the weed management process may reduce the likelihood of herbicide resistance, maintain the effectiveness of the active ingredient, and prolong the duration of herbicide release (Manjunatha et al. 2016). The synthesis of a particular herbicide compound enclosed within a nanoparticle is designed to selectively bind to certain receptors located at the root of the specific weed. The nanoparticle generated successfully infiltrates the root system of the weed and undergoes translocation, thereby executing its intended function of inhibiting glycolysis inside the plant root system. The deliberate action results in the plant succumbing to starvation. These nano-herbicides can even be used in rainy areas, as the absence of soil moisture leads to the evaporation and dispersion of herbicides through vaporization. The eradication of weeds can be achieved by employing herbicides that are distributed in a slow manner via encapsulation. Adjuvants commonly employed in combination with herbicides to enhance their herbicidal efficacy are now purportedly incorporating nanoparticles. The addition of a nanotechnology-based surfactant to a soybean micelle rendered a previously resistant crop susceptible to glyphosate (Abigail and Chidambaram 2017). Nanoparticles, when used in conjunction with herbicides, can serve as efficient carriers and generate nano-formulations that address the main drawback of the herbicide industry, namely the rise of herbicide-resistant plant species. A significant proportion of polymeric materials employed in nanoparticle systems for herbicide delivery exhibit biodegradability and generates metabolites that are non-toxic in nature.

1.6 Crop Protection and Monitoring

Nanotechnology is instrumental in enhancing crop protection through the use of nano capsules and nano sensors. Nano capsules deliver bioactive compounds directly to plants, protecting against diseases and pests. Nano sensors, embedded in the soil or on plants, offer real-time monitoring of environmental conditions, alerting farmers to potential threats such as pathogens or stress factors. This proactive approach to crop management helps prevent crop losses and ensures healthier yields.

1.7 Nano Capsules

Nano capsules consist of a polymeric protective shell and an interior solid or liquid core that contains active ingredients or insecticides (Fig. 1.4). According to Mora-Huertas et al. (2010), they could have an active ingredient in molecular dispersion, a polymeric matrix coated in a polymeric membrane, or a liquid core encircled by a polymeric membrane. Additional modified procedures include melt dispersion, emulsion polymerization, interfacial polymerization, interfacial deposition method, solvent displacement technique, and emulsion evaporation. The most often employed of these techniques is the interfacial deposition of premade polymer. This approach involves mixing the medication with a water-miscible solvent, regardless of the presence or absence of a surfactant, and then adding oil that is

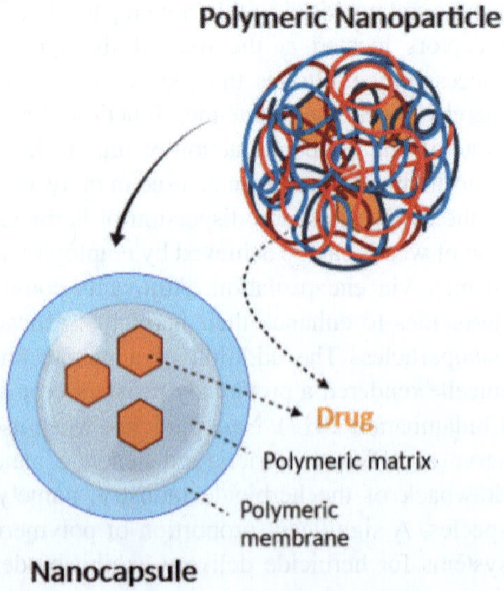

Fig. 1.4 Schematic representation showing nano-capsule structure. (Reproduced from Zielińska et al. 2020)

similarly solvent-miscible on top. Once the mixture is agitated, hydrophilic surfactant in an aqueous phase is added, forming nano capsules (Nuruzzaman et al. 2016). In one study, carboxymethyl chitosan was combined with amphiphilic bio copolymer and azido benzaldehyde. After photo crosslinking and polymer self-assembly in an aqueous environment, aqueous core nano capsules were created. After sonication, methomyl that had been disseminated in a pH 4.0 aqueous solution was added to the nano capsules. Methomyl was released at pH 6.0, with a half-release time ($t_{1/2}$) ranging from 36.3 to 69.5 hours. According to Sun et al. (2014), the methomyl-loaded insecticidal action showed enhanced efficacy against armyworm larvae (Sun et al. 2014).

1.8 Nano Sensors

Real bioanalytical applications could not previously be investigated, but nano biosensors open up new avenues for both fundamental research and bioanalytical applications. Nano biosensors can be included in lab-on-a-chips for molecular analysis; they support both molecular analysis and the monitoring of several metabolites. A variety of microorganisms and several analytes, such as pesticides, urea, glucose, etc., can be detected by them. Because they are lightweight and portable, they are ideal for use in the field and can also make laboratory settings easier. By improving crop output through better timing, fertilization, and environmental control, nano sensors can aid in precision farming (Ghaffar et al. 2020; Magnabosco et al. 2023).

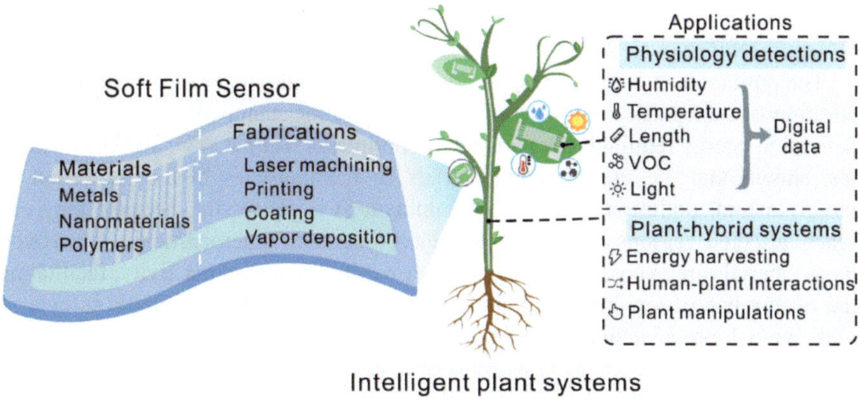

Fig. 1.5 Schematic representation of nano sensor fabrication to make intelligent plant systems. (Reproduced from Jiang et al. 2023)

The following lists the applications of nano sensors and devices in agriculture (Fig. 1.5):

1. Accurate measurements may be made of environmental parameters like temperature, humidity, air velocity, and air quality.
2. The field's water level and the amount of time needed to replenish it can be measured
3. Beneficial for monitoring the effective application of pesticides, herbicides, fertilizers, and other treatments
4. It is possible to research the soil's fertility and appropriateness for certain crops.
5. Crop nutrient status is analyzed
6. Nanodevices strewn around the field may detect the presence of any pathogen, including insects, pests, and microorganisms.
7. It is possible to gather evidence of stressors like drought or salt stress.
8. The best time to grow crops and harvest them can be ascertained

Nano sensors gather the necessary data from the farmed fields in real-time for precision farming monitoring. To gather data, such as geographic data, wireless nano sensors are connected to satellite systems, allowing remote sensing equipment to be distributed throughout the area (Soussi et al. 2024).

1.9 Soil Health and Remediation

Soil degradation and contamination pose significant challenges to agriculture. Nanotechnology offers solutions for soil health and remediation through nano remediation techniques. The pollutants in the soil undergo a process of adsorption, breakdown, or modification upon the addition of nanoparticles. By acting as a catalyst, the nanoparticles can lower the activation energy required to separate the pollutants (Vázquez-Núñez et al. 2020).

The pollutants in the soil undergo adsorption, breakdown, or modification by the addition of nanoparticles. By acting as a catalyst, the nanoparticles can lower the activation energy required to separate these pollutants (Rajput et al. 2022). Research has shown that the use of nanomaterials or nanoparticles in the resolution of environmental pollution (like soil contamination) is successful, because they can reach affected regions that are often inaccessible to microparticles. Moreover, nanomaterials are exposed to pollutants they exhibit increased reactivity, facilitating the occurrence of a redox reaction (Rajput et al. 2022). One drawback associated with nano bioremediation is the potential for nanoparticles or nanomaterials to permeate the plants parts such as roots and leaves, therefore altering the morphogenic capabilities of the plant species. This change could have a negative impact on the plant's health. Applying nanoparticles to soil for pollution remediation can make it more harmful since they produce reactive oxygen species. (ROS), which disrupt the electron transport within plant cells' mitochondria and chloroplasts. This disruption can potentially impact plant growth and vegetation, ultimately affecting human health indirectly (Rad et al. 2023). The investigation of the health impacts of nano bioremediation on plants has resulted in the implementation of safety protocols aimed at ensuring its overall safety. Safety protocols involve the creation of nanoparticles that are compatible with living organisms and the optimization of their size and concentration to improve their ability to be absorbed while minimizing any negative effects. Thorough surveillance and evaluations monitor the presence of nanoparticles in the environment and analyses their effects on ecological systems (Rad et al. 2023). These endeavors guarantee the continued safety and efficacy of nano bioremediation as a means of environmental remediation, hence minimizing any potential negative impacts on plant species and the surrounding ecological systems (Kumari et al. 2023). Figure 1.6 depicts the few nano remediation options for environment enrichment.

1.10 Water Management

Efficient water use is critical for sustainable agriculture, and nanotechnology contributes to water management solutions. Nano sensors accurately monitor soil moisture levels, enabling precise irrigation practices. Nanomaterials improve water filtration systems, removing contaminants from irrigation water and improving water quality. These advancements in water management not only conserve water resources but also contribute to healthier plants and increased yields.

Presently, the global task of ensuring a plentiful and uncontaminated supply of fresh water is of utmost significance for both human consumption and industrial sectors (Vörösmarty et al. 2010). Based on a study, the global population of those lacking access to clean water exceeds one billion, furthermore, the situation has deteriorated. It is projected that the average water supply per individual will decrease

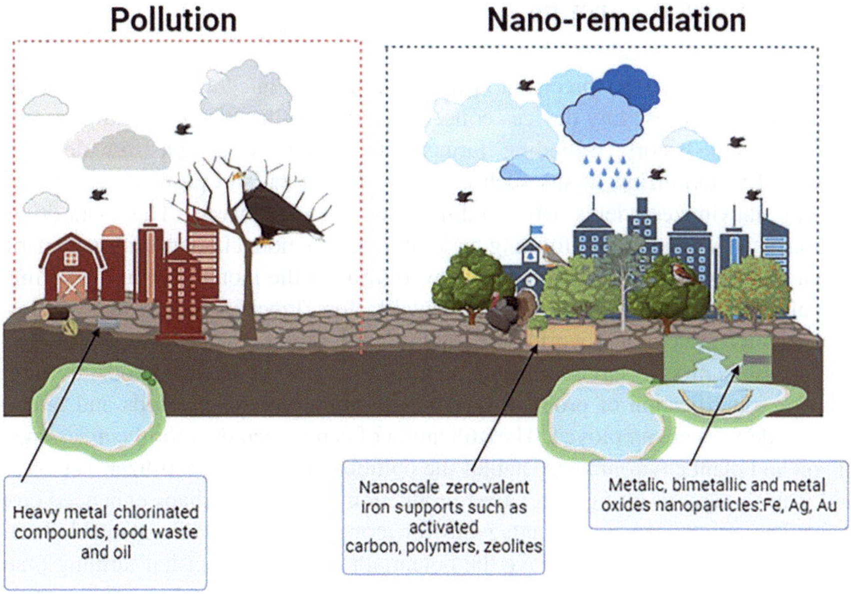

Fig. 1.6 Nano-remediation of pollution. (Reproduced from Dhanapal et al. 2024)

by a fraction of 1/3, leading to the preventable premature mortality of lots of individuals. (Cross et al. 2009). Agriculture necessitates a substantial quantity of potable water, which subsequently leads to groundwater contamination due to the application of fertilizers and pesticides. In order to address this issue, it will be necessary to develop innovative, environmentally friendly, and economically viable solutions for the management of the substantial volume of wastewater generated. During the wastewater treatment process, it is imperative to take into account various crucial factors such as water's quantity and quality, reuse and treatment methods, safety concerns arising from chemical and biological hazards, as well as the implementation of monitoring systems and sensors. The field of nanotechnology has facilitated the identification of innovative and economically viable approaches for the remediation and purification of wastewater. Water resources that can be accessed are often contaminated with harmful microorganisms such as cryptosporidium, coliform bacteria, and viruses. They also contain various salts and metals like Copper (Cu), Lead (Pb), and Arsenic (As). Additionally, large number of pharmaceuticals and personal care products (PPCP), runoff agricultural chemicals, endocrine disrupting compounds (EDC), and radioactive pollutants are also present. These contaminants may be present naturally or as a result of human activity, mining, and oil and gas production. Nanotechnology has developed as a promising possibility for enhancing water quality through the implementation of various innovative solutions. These include the utilization of nano-oligodynamic metallic particles, wireless nano-sensors, photocatalysis, heavy metal removal, desalination, and the detection of pollutants and impurities (Pérez et al. 2023).

1.11 Precision Farming

Nanotechnology enables precision farming through the integration of nano-sensors and nanochips. These tiny devices collect data on plant health, soil conditions, and environmental factors, providing farmers with valuable insights for decision-making. By optimizing inputs such as water, fertilizers, and pesticides, precision farming maximizes yields while minimizing resource waste. This data-driven approach to farming is transforming agriculture into a more efficient and sustainable endeavor (Elizabath 2019). This is achieved through the monitoring and coordination of interventions targeting environmental factors. Precision agriculture employs computer technology and worldwide satellite positioning systems (Fig. 1.7).

To assess precise environmental parameters, such as overall productivity or accurate identification of problem type and location, various methods and remote sensing devices are employed. The utilization of centralized data for the assessment of soil and plant environments enables the optimization of seed, fertilizer, chemical, and water usage, resulting in reduced production costs and perhaps enhanced crop yield. These outcomes are advantageous for farmers. Nanotechnology-enabled sensors and monitoring systems have the potential to enhance precision farming practices through the implementation of new agricultural approaches. This can be achieved by integrating a GPS, which facilitates real-time tracking (Hamad et al. 2020; Hoffmann et al. 2007; Kim et al. 2006; Lai et al. 2017; Salamanca-Buentello et al. 2005). To examine soil conditions and plant growth, it is possible to deploy Nano sensors across the field. Remote detectors are currently being utilized in certain regions of the US and Australia. Nanotechnology-enabled gadgets primarily boost

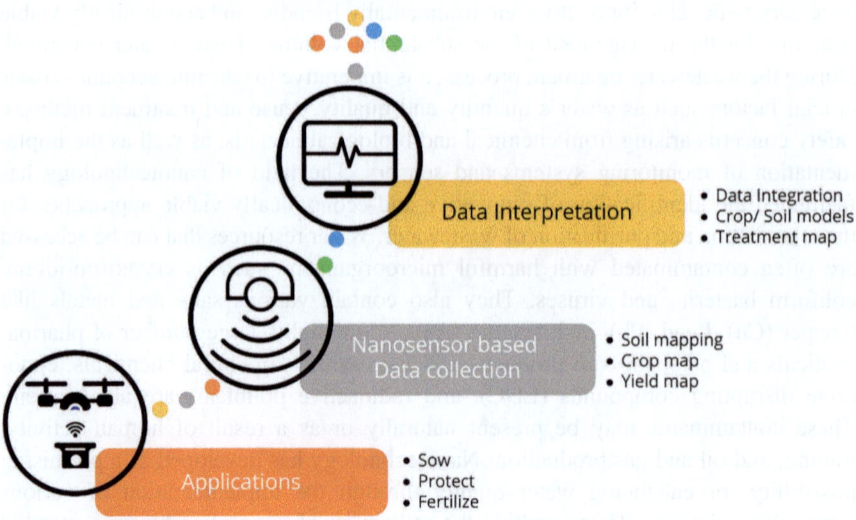

Fig. 1.7 Schematic representation of Nano sensor based precision agriculture. (Reproduced from Yadav et al. 2023)

the utilization of autonomous sensors that are connected to a GPS system, allowing for real-time tracking. By 2010, it is projected that almost seven billion USD will be expended. The convergence of Nanotechnology and Biotechnology has the potential to facilitate the expansion of highly sensitive equipment, hence enhancing the speed and efficacy of responses to external stimuli (Alatabe and Al-sharify 2019).

1.12 Conclusion and Prospects

Nanotechnology is revolutionizing agriculture, offering solutions to some of the most pressing challenges faced by farmers. From improved nutrient delivery and crop protection to enhanced soil health and water management, the applications of nanotechnology in agriculture are vast and promising. By harnessing the power of nanotechnology, we can move towards a more sustainable and efficient agricultural system, ensuring food security for future generations while preserving our environment. Furthermore, the majority of research conducted on nano-assisted agriculture predominantly relies on controlled experiments, with limited data accessible regarding their practical implementation in real-world settings. Enhanced expertise at the field level would be immensely beneficial for the widespread adoption of nano-based strategies.

As research in nanotechnology continues to advance, the future of agriculture holds exciting possibilities. Concepts such as nanorobotics for targeted delivery within plants, biodegradable nano pesticides, and nanomaterials for controlled release of nutrients offer promising avenues for sustainable farming. The integration of nanotechnology into agriculture represents a paradigm shift towards more efficient, environmentally friendly, and productive farming practices.

References

Abigail EA, Chidambaram R (2017) Nanotechnology in herbicide resistance. InTech eBooks. https://doi.org/10.5772/intechopen.68355

Alatabe MJA, Al-sharify ZT (2019) Utilization of low cost adsorbents for the adsorption process of lead ions: a review. Int J Modern Res Eng Technol 4(11):29–48

Ali S, Ahmad N, Dar MA, Manan S, Rani A, Alghanem SMS, Khan KA, Sethupathy S, Elboughdiri N, Mostafa YS, Alamri SA, Hashem M, Shahid M, Zhu D (2024) Nano-agrochemicals as substitutes for pesticides: prospects and risks. Plan Theory 13(1):1. https://doi.org/10.3390/plants13010109

Chaudhary A, Mamta. (2022) Nanopesticides: a promising nanoagrochemicals in crop protection [Review Article]. Chem Sci Rev Lett 11(44):455–464. https://doi.org/10.37273/chesci.cs205306500

Cross KM, Lu Y, Zheng T, Zhan J, McPherson G, John V (2009) Chapter 24—Water decontamination using iron and iron oxide nanoparticles. In: Savage N, Diallo M, Duncan J, Street A, Sustich R (eds) Nanotechnology applications for clean water. William Andrew Publishing, pp 347–364. https://doi.org/10.1016/B978-0-8155-1578-4.50033-0

Dhanapal AR, Thiruvengadam M, Vairavanathan J, Venkidasamy B, Easwaran M, Ghorbanpour M (2024) Nanotechnology approaches for the remediation of agricultural polluted soils. ACS Omega 9(12):13522–13533. https://doi.org/10.1021/acsomega.3c09776

Elizabath A (2019) Application of nanotechnology in agriculture. Int J Pure Appl Biosci 7(2): 131–139. https://doi.org/10.18782/2320-7051.6493

Ghaffar N, Farrukh MA, Naz S (2020) Applications of nanobiosensors in agriculture. Springer eBooks, pp 179–196. https://doi.org/10.1007/978-3-030-41275-3_10

Goyal V, Rani D, Ritika, Mehrotra S, Deng C, Wang Y (2023) Unlocking the potential of nano-enabled precision agriculture for efficient and sustainable farming. Plan Theory 12(21):21. https://doi.org/10.3390/plants12213744

Hamad HT, Al-Sharify ZT, Al-Najjar SZ, Gadooa ZA (2020) A review on nanotechnology and its applications on fluid flow in agriculture and water recourses. IOP Conf Ser Mater Sci Eng 870(1):012038. https://doi.org/10.1088/1757-899X/870/1/012038

Hoffmann WC, Walker TW, Smith VL, Martin DE, Fritz BK (2007) Droplet-size characterization of handheld atomization equipment typically used in vector control. J Am Mosq Control Assoc 23(3):315–320. https://doi.org/10.2987/8756-971X(2007)23[315:DCOHAE]2.0.CO;2

Jiang Q, Chai Z, Zong Z, Hu Z, Zhang S, Wu Z (2023) Micro/Nano soft film sensors for intelligent plant systems: materials, fabrications, and applications. Chemosensors 11(3):3. https://doi.org/10.3390/chemosensors11030197

Jogaiah S, Singh HB, Fernandes L, De Lima R (2020) Advances in nano-fertilizers and nano-pesticides in agriculture: a smart delivery system for crop improvement. Woodhead Publishing

Khan F, Pandey P, Upadhyay TK (2022) Applications of nanotechnology-based agrochemicals in food security and sustainable agriculture: an overview. Agriculture 12(10):1672. https://doi.org/10.3390/agriculture12101672

Kim J, Grate JW, Wang P (2006) Nanostructures for enzyme stabilization. Chem Eng Sci 61(3): 1017–1026. https://doi.org/10.1016/j.ces.2005.05.067

Kumari R, Suman K, Karmakar S, Mishra V, Lakra SG, Saurav GK, Mahto BK (2023) Regulation and safety measures for nanotechnology-based agri-products. Front Genome Ed 5. https://doi.org/10.3389/fgeed.2023.1200987

Lai F, Wissing SA, Müller RH, Fadda AM (2017) Artemisia arborescens L essential oil-loaded solid lipid nanoparticles for potential agricultural application: preparation and characterization. AAPS PharmSciTech 7(1):2. https://doi.org/10.1208/pt070102

Lu W, Senapati D, Wang S, Tovmachenko O, Singh AK, Yu H, Ray PC (2010) Effect of surface coating on the toxicity of silver nanomaterials on human skin keratinocytes. Chem Phys Lett 487(1–3). https://doi.org/10.1016/j.cplett.2010.01.027

Magnabosco P, Masi A, Shukla R, Bansal V, Carletti P (2023) Advancing the impact of plant biostimulants to sustainable agriculture through nanotechnologies. Chem Biol Technol Agric 10(1). https://doi.org/10.1186/s40538-023-00491-8

Manjunatha SB, Biradar D, Aladakatti Y (2016) Nanotechnology and its applications in agriculture: a review. J Farm Sci 29:1–13

Mukhopadhyay SS (2014) Nanotechnology in agriculture: prospects and constraints. Nanotechnol Sci Appl 63. https://doi.org/10.2147/NSA.S39409

Nuruzzaman M, Rahman MM, Liu Y, Naidu R (2016) Nanoencapsulation, nano-guard for pesticides: a new window for safe application. J Agric Food Chem 64. https://doi.org/10.1021/acs.jafc.5b05214

Pan X, Guo X, Zhai T, Zhang D, Rao W, Cao F, Guan X (2022) Nanobiopesticides in sustainable agriculture: developments, challenges, and perspectives. Environ Sci Nano 10(1):41–61. https://doi.org/10.1039/d2en00605g

Pérez H, García OJQ, Amezcua-Allieri MA, Vázquez RR (2023) Nanotechnology as an efficient and effective alternative for wastewater treatment: an overview. Water Sci Technol 87(12): 2971–3001. https://doi.org/10.2166/wst.2023.179

Rad SA, Nobaharan K, Pashapoor N, Pandey J, Dehghanian Z, Senapathi V, Minkina T, Ren W, Rajput VD, Lajayer BA (2023) Nano-microbial remediation of polluted soil: a brief insight. Sustainability 15(1):876. https://doi.org/10.3390/su15010876

Ragaei M, Sabry AH (2014) Nanotechnology for insect pest control. Int J Sci Environ Technol 3(2): 528–545

Rajput VD, Minkina T, Upadhyay SK, Kumari A, Ranjan A, Mandzhieva S, Sushkova S, Singh RK, Verma KK (2022) Nanotechnology in the restoration of polluted soil. Nanomaterials 12(5): 5. https://doi.org/10.3390/nano12050769

Salamanca-Buentello F, Persad DL, Court EB, Martin DK, Daar AS, Singer PA (2005) Nanotechnology and the developing world. PLoS Med 2(5):e97. https://doi.org/10.1371/journal.pmed.0020097

Sayed AMM, Kim S, Behle RW (2017) Characterisation of silver nanoparticles synthesised by Bacillus thuringiensis as a nanobiopesticide for insect pest control. Biocontrol Sci Tech 27(11): 1308–1326. https://doi.org/10.1080/09583157.2017.1397597

Soussi A, Zero E, Sacile R, Trinchero D, Fossa M (2024) Smart sensors and smart data for precision agriculture: a review. Sensors 24(8):2647. https://doi.org/10.3390/s24082647

Sun C, Shu K, Wang W, Ye Z, Liu T, Gao Y, Zheng H, He G, Yin Y (2014) Encapsulation and controlled release of hydrophilic pesticide in shell cross-linked nanocapsules containing aqueous core. Int J Pharm 463(1):108–114. https://doi.org/10.1016/j.ijpharm.2013.12.050

Tarafdar J, Xiong Y, Wang W-N, Quin D, Biswas P (2012) Standardization of size, shape and concentration of nanoparticle for plant application. Appl Biol Res 14:138–144

Vázquez-Núñez E, Molina-Guerrero CE, Peña-Castro JM, Fernández-Luqueño F, de la Rosa-Álvarez MG (2020) Use of nanotechnology for the bioremediation of contaminants: a review. Processes 8(7):7. https://doi.org/10.3390/pr8070826

Vörösmarty CJ, McIntyre PB, Gessner MO, Dudgeon D, Prusevich A, Green P, Glidden S, Bunn SE, Sullivan CA, Liermann CR, Davies PM (2010) Global threats to human water security and river biodiversity. Nature 467(7315):555–561. https://doi.org/10.1038/nature09440

Yadav A, Yadav K, Ahmad R, Abd-Elsalam KA (2023) Emerging frontiers in nanotechnology for precision agriculture: advancements, hurdles and prospects. Agrochemicals 2(2):2. https://doi.org/10.3390/agrochemicals2020016

Yuvaraj M, Subramanian KS (2015) Controlled-release fertilizer of zinc encapsulated by a manganese hollow core shell. Soil Sci Plant Nutr 61(2):319–326. https://doi.org/10.1080/00380768.2014.979327

Zielińska A, Carreiró F, Oliveira AM, Neves A, Pires B, Venkatesh DN, Durazzo A, Lucarini M, Eder P, Silva AM, Santini A, Souto EB (2020) Polymeric nanoparticles: production, characterization, toxicology and ecotoxicology. Molecules 25(16):16. https://doi.org/10.3390/molecules25163731

Chapter 2
Nanoparticle Synthesis and Modification for Plant Applications

Panangattu Sankarji Jayalakshmi, Sanu Mathew Simon, and Aji Abraham

Abstract Nanoparticles (NPs) have biological activity against microorganisms and plants, though they also have an aggregating tendency. Consequently, NPs serve as an important source of soluble metals, thereby exposing organisms to toxicities that are specific to both ions and particles. This chapter delves into the synthesis and modification processes of nanoparticles (NPs), shedding light on how various synthesis methods and surface alterations can influence NP properties and their interactions with plants and soil bacteria. It examines the substantial influence of NPs on crucial biological processes vital for soil fertility and plant yield, whether they are deliberately or accidentally introduced. Despite these challenges, the coordination among soil pore water, plant root exudates, and microbial cell surfaces serves to diminish NP bioactivity. Interestingly, the interactions of NPs with microbial cells and plants offer unexpected opportunities for extracting commercially valuable metabolites. Furthermore, the chapter delves into the potential utilization of engineered nanomaterials (ENMs) to enhance plant health and nutrient availability, thereby boosting plant productivity. Engineered nanomaterials (ENMs) find wide-ranging applications in many aspects of the plant system, such as quality improvement, growth promotion, and genetic preservation. However, it's crucial to note that their impact may also include adverse effects on plant physiology and growth suppression, emphasizing the importance of a comprehensive understanding of NP-plant interactions. Overall, this chapter clarifies the complex effects of NPs on soil ecosystems and plant health by summarizing the most recent research findings.

Keywords Carbon nanotubes · Engineered nanoparticles · Green synthesis · Modification · Nanofertilizers · Plant applications

---

P. S. Jayalakshmi · A. Abraham (✉)
Department of Botany, Mar Athanasius College (Autonomous), Kothamangalam, Kerala, India
e-mail: jayalakshmips@macollege.in; ajisijo@macollege.in

S. M. Simon
Department of Physics, Mar Athanasius College (Autonomous), Kothamangalam, Kerala, India
e-mail: sanumathewsimon@macollege.in

J. M. Al-Khayri et al. (eds.), *Nanotechnology Applications in Modern Agriculture*, Nanotechnology in Plant Sciences 3, https://doi.org/10.1007/978-3-031-90506-3_2

2.1 Introduction

Nanoparticles are minute particles of diverse shapes with size ranging from 1 to 100 nm. On account of their nano scale dimensions and large surface to mass ratio, they have distinctive physico-chemical characters. Physical, chemical as well as biological methods are employed for the synthesis of nanoparticles. Both physical and chemical methods are employed for the synthesis of monodispersed nanoparticles. Physical methods require significant time and expensive equipment, whereas chemical methods pose risk of contamination of nanoparticles with toxic chemicals and release of harmful byproducts. Hence green synthesis, being a rapid cost effective and environment friendly method is frequently employed for synthesis of nanoparticles.

Based on the form and size of metals within the solvent, nanoparticle synthesis occurs by either bottom-up pathway where synthesis of nanoparticles (NPs) happens through the aggregation of atoms, molecules, and/or clusters facilitated by reducing agents of chemical or biotic origin and top-down pathway where nanoparticles are formed by reduction in size of large molecules (Meyers et al. 2006). Nanoparticles formed by bottom-up pathway are homogenous and have few structural defects and hence preferred over top-down approach. This involves reduction of metal ions by phyto molecules, growth and amalgamation of small particles and their stabilization by an organic coating (Khan et al. 2019a). Various secondary metabolites present in plant extracts function as reducing and stabilizing agents which facilitate the reduction of metal ions to metal nanoparticles. Accordingly, nanoparticles of copper, cobalt, gold, silver, platinum, palladium and zinc oxide are synthesized by employing plant extracts (Kuppusamy et al. 2016).

Nanoparticles are modified using liposomes, quantum dots, and functionalized dendrimers. These nanoparticles can be metalloids, metallic oxides, nonmetals, or carbon nanomaterials. These modifications enable them to operate as nano fertilizers and bactericides/fungicides. They can also be designed to serve as carriers of genetic material, probes, and agrichemicals, as well as biosensors for the diagnosis of plant diseases. Through enhanced targeted delivery and absorption efficiency, nanoparticles that have been engineered, may be able to minimize environmental contamination associated with the number of agrichemicals required for crop production. Furthermore, it is possible to modify nanoparticles to have particular characteristics, such as increased solubility, stability, and bioavailability, all of which are essential for their success in applications involving plants (Ahmad et al. 2023; Sun et al. 2023).

Nanoparticles which gain entry mainly through roots and leaves of plants, can profoundly influence their growth and development. Interaction of these nanoparticles at the cellular and subcellular levels induces physiological and morphological changes, which may have suppressive or stimulatory effects on plants. Some of them induce phytotoxicity particularly at elevated concentrations, while others stimulate seed germination, root growth as well as photosynthetic efficiency. According to Khan, nanoparticle interactions with plants are complex and multifaceted, involving both beneficial and detrimental effects. Khan and co-workers intricate two-way traffic between nanoparticles and plants, exploring how nanoparticles influence plant processes and how plants, in turn, affect nanoparticle behavior (Khan et al. 2019a, b).

2.2 Green Synthesis of Nanoparticles

Green synthesis is an environment friendly and cost-effective approach for producing chemical compounds or nanoparticles which are considerably nontoxic and stable. Various plant parts like roots, leaves, flowers and seeds as well as microorganisms like bacteria and fungi can be used as reducing agents and stabilizers for the synthesis of silver and gold nanoparticles. Biologically synthesized nanoparticles often demonstrate better physical and biochemical properties compared to those created by physical and chemical methods (Kumar et al. 2023). Hence green synthesis is therefore often favoured over other methods for the large scale production of nanoparticles.

2.2.1 Synthesis of Nanoparticles Using Plants

Green synthesis is a simple, reliable, sustainable, economical and ecofriendly method for nanoparticle synthesis. Phytosynthesis is such a technique where extracts of different plant parts are employed for nanoparticle synthesis. Various parts of the plants like stem, leaves, flowers as well as fruits have been employed for nanoparticle synthesis. Alsammarraie (2018) reported the production of silver nanoparticles with an average dimension of 18 ± 0.5 nm using aqueous extract of turmeric powder. They also reported that these synthesized nanoparticles exhibited antimicrobial potential against food borne pathogens like *Escherichia coli* O157:H7 and *Listeria monocytogenes*. The methanolic stem bark extract of *Callicarpa maingayi* was used for the synthesis of silver nanoparticles by Shameli et al. (2012) who opined that the reduction of silver ions into metallic silver nanoparticles was brought about by aldehyde group contained in the plant extract. Similarly, Vanaja et al. (2013) documented the production of silver nanoparticles utilizing stem extracts of *Cissus quadrangularis* which exhibited significant antibacterial activity against *Klebsiella planticola* and *Bacillus subtilis.*

Leaves of plants also find use in the synthesis of nanoparticles. Vijayakumar et al. (2013) reported a simple ecofriendly method for the synthesis of silver nanoparticles from superior leaf part of *Artemisia nilagirica* wherein silver nitrate and hydrazine hydrate are used as metal precursor and reducing agent respectively. They also found that the synthesized silver nanoparticles showed antimicrobial activity against *Bacillus subtilis, Staphylococcus aureus, Escherichia coli* and *Proteus subtilis* invitro. *Platanus orientalis* leaf extracts has been employed by Devi et al. (2019) for the

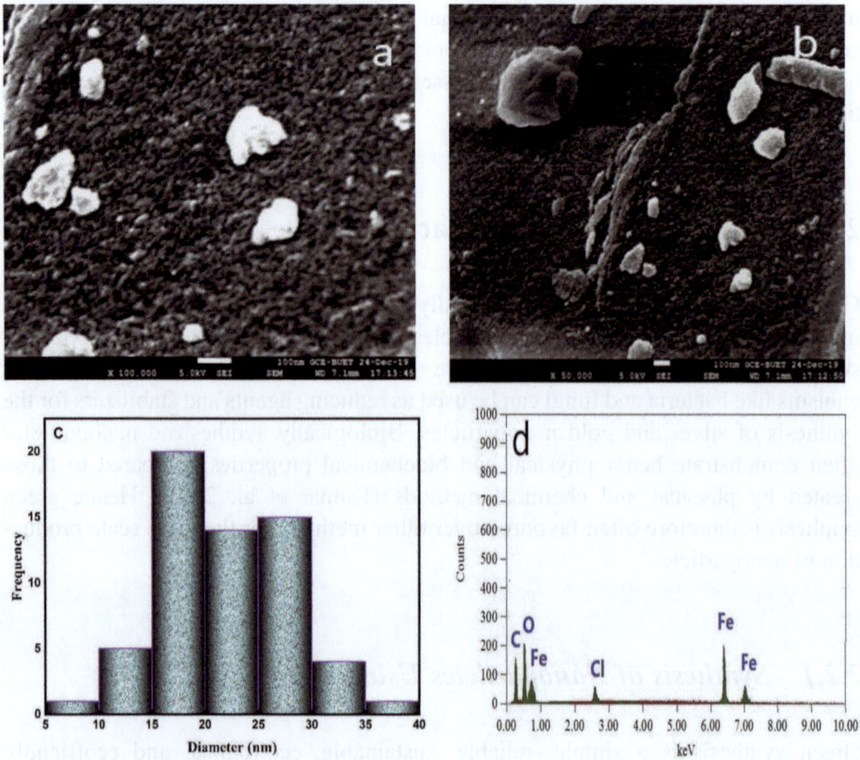

Fig. 2.1 (**a, b**) FESEM image (**c**) Size distribution histogram (**d**) EDX data of green synthesized α-Fe_2O_3 nanoparticles. (Reproduced from Bhuiyan et al. 2020)

synthesis of Iron oxide nanoparticles who also documented that nanoscale iron oxide nanoparticles of uniform size has significant antifungal properties and can be used as fungal antidote. The leaf extract of *Argemone mexicana* a weed plant was utilized for silver nanoparticle synthesis by Singh et al. (2010) who documented antimicrobial activity of these nanoparticles. The aqueous leaf extract of *Peganum harmala* was utilized both as a reducing and capping agent in the synthesis of silver nanoparticles. Alomar et al. (2020) reported that AgNPs synthesized using *Peganum harmala* leaf extract exhibited antibacterial potential against *E. coli*, *Bacillus cereus* and *Staphylococcus aureus*. Bhuiyan et al. (2020) documented the production of Iron oxide nanoparticles (Fig. 2.1) utilizing leaf extract of *Carica papaya*. Even though the nanoparticles were of diverse sizes and showed a tendency to agglomerate, they exhibited antibacterial properties. They were also found to act as a photocatalyst to degrade Remazol yellow RR dye, a reactive azo dye used for coloring fabrics and yarns.

Ukidave and Ingale (2022) synthesized zinc oxide nanoparticles from Zinc acetate solution using *Coriandrum sativum* leaf extracts and studied their impact on seed germination and growth of Bengal gram, green gram, and Turkish gram. The study results indicated that the deficiency of zinc led to a decrease in the rate of seed germination. Furthermore, the researchers observed that ZnO nanoparticles had a beneficial impact on the growth and development of green gram and Turkish gram compared to Bengal gram. The research suggests that zinc is essential for seed germination and reveals variability among plant species in their response to ZnO nanoparticles. Certain species demonstrate more pronounced improvements in growth and development when supplemented with zinc compared to others.

In another work by Asmat-Campos et al. (2022b), *Coriandrum sativum* leaf extract was employed for the production of zinc oxide nanoparticle from zinc acetate dihydrate. The effect of these nanoparticles on the seed germination of *Lycopersicon esculentum* Mill was also analyzed. They documented that concentrations nearing 100 ppm of the nanoparticles are optimal for treating *Lycopersicon esculentum* Mill seeds. At this concentration, the nanoparticles were shown to enhance enzymatic and metabolic activities within the seeds, resulting in cell elongation. Abbas et al. (2020) synthesized copper nanoparticles from copper acetate precursor *using Azadirachta indica* leaf extract and from aqueous copper sulphate solution using *Coriandrum sativum* leaf extracts. He suggested that copper nanoparticles synthesized using *Azadirachta indica* leaf extract possess remarkable potential as biosorbents due to their outstanding ability to remove naphthalene from spiked aqueous media. This emphasizes their suitability for environmental remediation purposes. Saka et al. (2022) utilized shell extracts of *Carica papaya* to synthesize Titanium dioxide nanoparticles from Titanium isopropoxide. They also documented that these nanoparticles exhibited antifungal properties and have the potential to influence antioxidant metabolisms in seed germinations. Another method for the biosynthesis of titanium dioxide nanoparticles was reported by Dobrucka (2017) where in the aqueous extract of *Echinacea purpurea* herb was used as a bio reductant. This method was reported to be ecofriendly and quick, but the pH of the solution affected the size of the nanoparticles. Suman et al. (2014) utilized the aqueous root extract of *Morinda citrifolia* as a reducing agent for the synthesis of gold nanoparticles from chloroauric acid. They documented that proteins in the extract functioned as both capping and reducing agents, facilitating the formation of crystalline gold nanoparticles with sizes ranging from 12.17 to 38.26 nm (Fig. 2.2).

Niraimathee and Subha (2016) observed the formation of iron oxide nanoparticles when the aqueous root extract of *Mimosa pudica* was exposed to ferrous sulphate solution. They reported that the mimosine present in the root extract act as reducing agent, facilitating the formation of roughly spherical nanoparticles of 60–80 nm size.

Devanesan and AlSalhi (2021) employed the aqueous flower extract of *Abelmoschus esculentus* for the production of silver nanoparticles. SEM-EDX analysis indicated that the nanoparticles had an average size of 18.24 nm and exhibited predominantly spherical shapes. These nanoparticles demonstrated remarkable antibacterial efficacy, particularly against gram-negative bacteria. Das et al. (2011) utilized the ethanolic flower extract of *Nyctanthes arbortristis* as a reducing agent to synthesize spherical gold nanoparticles of 19.8 ± 5.0 nm average size. The study reported that these gold nanoparticles demonstrated long term stability attributed to the adsorption of biomolecules from the flower extract on

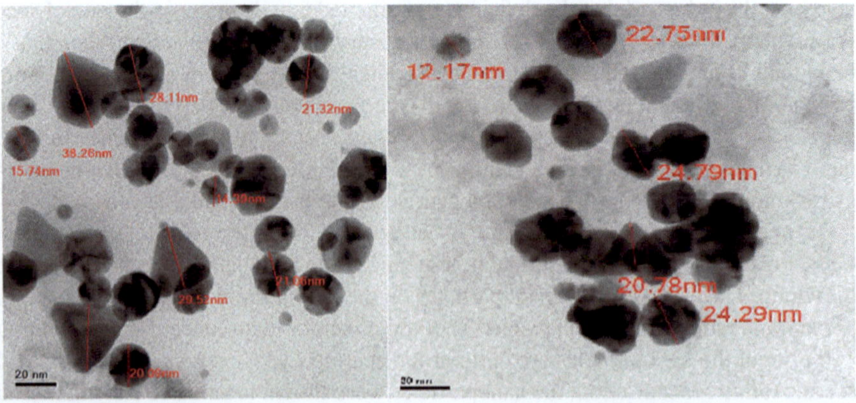

Fig. 2.2 High Resolution TEM images of gold nanoparticles synthesized using aqueous root extract of *Morinda citrifolia*. (Reproduced with permission from Suman et al. 2014; Elsevier)

their surface. Suriyakala et al. (2022) utilized $AuCl_3$ as a precursor and the flower extract *of Jatropha integerrima* Jacq. as both reducing and stabilizing agent to synthesize gold nanoparticles. The study indicated that these synthesized nanoparticles possess antibacterial properties.

Dhar et al. (2021) synthesized silver nanoparticles from silver nitrate precursors through an eco-friendly method wherein *Phyllanthus emblica* fruit extract was used as a reducing and stabilizing agent. FESEM analysis proved that the nanoparticles were of spherical shape with an average dimension of 60–80 nm.

Silver nanoparticles were synthesized from aqueous solution of silver nitrate using Grape pomace and orange extracts. Soto et al. (2019) reported that these nanoparticles exhibited antibacterial properties against *Escherichia coli* O157:H7 and *Listeria monocytogenes*. They also opined that secondary metabolites, especially phenolic compounds present in fruit extracts acted as capping agents promoting the reduction of silver ions.

Nava et al. (2017) reported the synthesis of zinc oxide nanoparticles from zinc nitrate using *Lycopersicon esculentum*, *Citrus aurantifolia*, *Citrus paradisi* and *Citrus sinensis* fruit peel extracts. Even though all the nanoparticles were polyhedral, their size and shape varied with the type of fruit peel extract used for synthesis.

The aqueous orange peel extract was used for synthesizing calcium carbonate nanoparticles from calcium chloride & sodium carbonate solution. Garg et al. (2021) opined that the synthesized nanoparticles were crystalline polymorphs.

Sreedevi et al. (2020) reported an ecofriendly and cost-effective method for the synthesis of silver nanoparticles from silver nitrate solution using the aqueous extract of seeds of *Sorghum bicolor*. The nanoparticles were of spherical shape and showed antibacterial potential against Gram-positive as well as Gram-negative bacteria even in microgram concentrations. It was recorded that the antimicrobial effect of silver nanoparticles was on account of them attaching to and penetrating the bacterial cell wall.

Rautela et al. (2019) employed *Tectona grandis* seed extract for producing silver nanoparticles from silver nitrate solution. They documented that the resulting silver nanoparticles were crystalline and spherical with size ranging from 10 to 30 nm. They proposed that the antibacterial effectiveness of these nanoparticles against *Staphylococcus aureus, E. coli,* and *Bacillus cereus* could be attributed to their ability to alter membrane permeability.

Ahmed et al. (2022) utilized aqueous extracts from flower buds of *Syzygium aromaticum* to synthesize zinc nanoparticles from zinc sulphate solution. Their investigation into the effects on *Pisum sativum* revealed that zinc nanoparticles promoted growth and productivity in *Pisum sativum* up to a specific concentration. However, beyond this threshold, they observed a decline in both plant growth and yield.

2.2.2 Synthesis of Nanoparticles Using Algae

Algae have the capacity to accumulate heavy metal ions at elevated levels and remodel them to flexible shapes on account of which they have been suggested as exemplary organisms for the production of bio-nanomaterials. The rapid growth rate and effective carbon dioxide sequestration of microalgae as well as their ability to produce biomolecules acting as both reducing and capping agents further solidify their potential as promising candidates for nanoparticle synthesis (Khan et al. 2022). Parial and Pal (2015) reported the intracellular production of spherical gold nanoparticles of 16 nm size when *Rhizoclonium fontinale*, a eukaryotic green alga was subjected to a solution containing 15 mg L^{-1} auric chloride at pH 9 for a duration of 72 hours. Silver nanoparticles synthesized by algae have been reported to have excellent antimicrobial efficacy even at low levels. *Nostoc ellipsosporum*, a cyanophycean alga has been utilized in the laboratory for the intracellular biosynthesis of gold nanorods from auric ions (Parial et al. 2012).

Algae can facilitate the rapid synthesis of nanoparticles and Rahman et al. (2020) reported that silver nanoparticles synthesized using algae have hydrophilic surface groups like sulphate, carboxyl and hydroxyl groups which confer them unique properties. Subramaniyam et al. (2015) synthesized highly reactive and stable iron nanoparticles utilizing the soil microalga *Chlorococcum* sp. MM11. They opined that these nanoparticles could be used in chromium remediation because of their increased effectiveness in reducing chromium ions compared to bulk iron.

Bifurcaria bifurcate, a brown alga, was employed in the biosynthesis of copper oxide nanoparticles ranging from 5 to 45 nm, from aqueous copper sulphate solution. Abboud et al. (2014) reported that these nanoparticles exhibited antimicrobial activity against both Gram-negative and Gram-positive bacteria, characteristically *Enterobacter aerogenes* and *Staphylococcus aureus*.

Stoechospermum marginatum, a brown alga, has been utilized in the production of crystalline gold nanoparticles, which displayed shapes ranging from spherical to hexagonal or triangular. Rajathi et al. (2012) reported that the hydroxyl groups in the diterpenoids of this alga facilitated the reduction of gold ions. Furthermore, they documented that these nanoparticles exhibited photoluminescence and demonstrated antibacterial properties.

2.2.3 *Synthesis of Nanoparticles Using Fungi*

Many fungal species are employed for the production of nanoparticles on account of their potential for biogenic synthesis. The most common nanoparticles synthesized by fungi are silver and gold. The extracellular and intracellular mechanisms of fungi are involved in reducing metal ions and stabilizing the resulting nanoparticles. Nanoparticles synthesized by fungi may have diverse shapes and sizes based on the fungal species employed and the conditions of the synthesis process.

Balaji et al. (2009) reported that *Cladosporium cladosporioides* (Fig. 2.3) could be utilized for the reduction of $AgNO_3$ solution to nanoparticles of silver. They also opined that, proteins, polysaccharides and organic acids released extracellularly by the fungus adds to the stability of nanoparticles. The synthesis of stable and biocompatible silver nanoparticles using *Guignardia mangiferae*, an endophytic fungus isolated from medicinal plant leaves, was reported by Balakumaran et al. (2015). They found that silver nanoparticles synthesized using *Guignardia mangiferae* demonstrated excellent antibacterial and antifungal activities. Silver nanoparticles were found to accumulate on the cell wall surface of *Aspergillus flavus* when the fungus was exposed to silver nitrate solution in an experiment by Vigneshwaran et al. (2007). They documented that the stability of these silver nanoparticles was attributed to the surface binding of stabilizing materials,

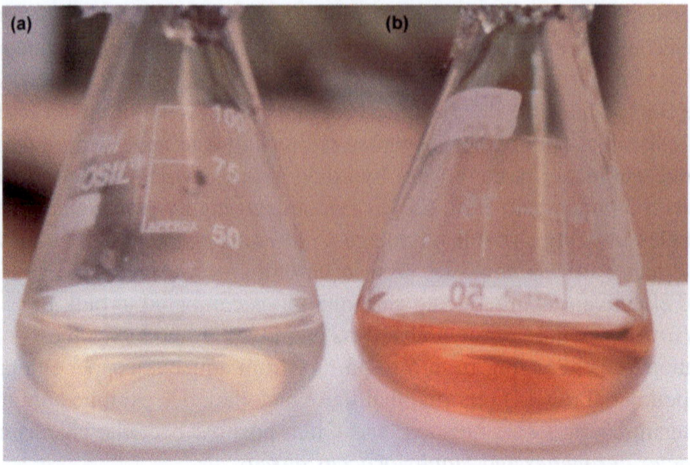

Fig. 2.3 Conical flasks containing the extracellular filtrate of the Cladosporium cladosporioides biomass in aqueous solution of silver nitrate at the start of the reaction (flask labeled as a) and after one day of reaction (flask labeled as b). (Reproduced with permission from Balaji et al. 2009; Elsevier)

predominantly proteins, secreted by the fungus. *Fusarium oxysporum* was documented to facilitate the reduction of aqueous silver ions, resulting in the production of stable silver nanoparticles. In an experiment by Ahmad et al. (2003) it was noticed that the reduction of silver ions took place extracellularly and was facilitated by the fungal secreted enzymes. The stability of silver nanoparticles was attributed to the fungal secreted proteins bound to the nanoparticle surfaces. Saravanan et al. (2018) reported the synthesis of silver nanoparticles ranging in size from 34 to 90 nm by subjecting cell-free extract of *Phenerochaete chrysosphorium* to 1 mM $AgNO_3$. The silver nanoparticles synthesized had pronounced antibacterial potential against *Pseudomonas aeruginosa* and *Klebsiella pneumoniae*. Hence, they opined that *Phenerochaete chrysosphorium* can be considered as potential bio factories for the synthesis of silver nanoparticles. Farzana Fathima et al. (2022) reported the formation of gold nanoparticles when the aqueous extract of *Laetiporus versisporus* was added to gold chloride solution and kept in dark.

The synthesis of Iron oxide nanoparticles from $FeCl_3$ using *Aspergillus oryzae* was reported by Tarafdar and Raliya (2013) who opined that this bio-based synthesis is reproducible, reliable and ecofriendly. Mohamed et al. (2015) reported an ecofriendly method for the production of iron nanoparticles from aqueous iron nitrate solution using *Alternaria alternata* as reducing agent. The iron nanoparticles were of cubic shape with an average dimension of 9 ± 3 nm and showed antibacterial effects against both Gram-positive and Gram-negative bacteria.

2.2.4 Synthesis of Nanoparticles Using Bacteria

Bacteria on account of their versatility, adaptability and biocompatibility are convenient targets for nanoparticle synthesis. Wang et al. (2016) reported that *Bacillus methylotrophicus* DC3, isolated from the soil of Korean ginseng, a well-known oriental medicinal plant, synthesized spherical silver nanoparticles extracellularly. These nanoparticles exhibited antimicrobial activity against *Candida albicans, Salmonella enterica, Escherichia coli* and *Vibrio parahaemolyticus*. *Bacillus megaterium*, a soil bacterium, was observed to produce silver, lead, and cadmium nanoparticles when exposed to silver nitrate, lead nitrate, and cadmium nitrate, respectively. Prakash et al. (2010) observed that these nanoparticles accumulate on the surface of the bacterial cell wall. *Lactobacillus plantarum,* a gram-positive facultative anaerobic bacterium, was employed as both a reducing and capping agent in the biosynthesis of nearly spherical ZnO nanoparticles ranging from 7 to 19 nm in size. Selvarajan and Mohanasrinivasan (2013) reported that the negative electrokinetic potential of the bacterium serves as a stimulus, attracting cations and initiating the biosynthesis of ZnO nanoparticles. Cell free culture supernatants of psychrophilic bacteria like *Pseudomonas antarctica* and *Arthrobacter kerguelensis* have been used for the synthesis of silver nanoparticles of dimensions ranging from 6 nm to 13 nm. Shivaji et al. (2011) reported that factors

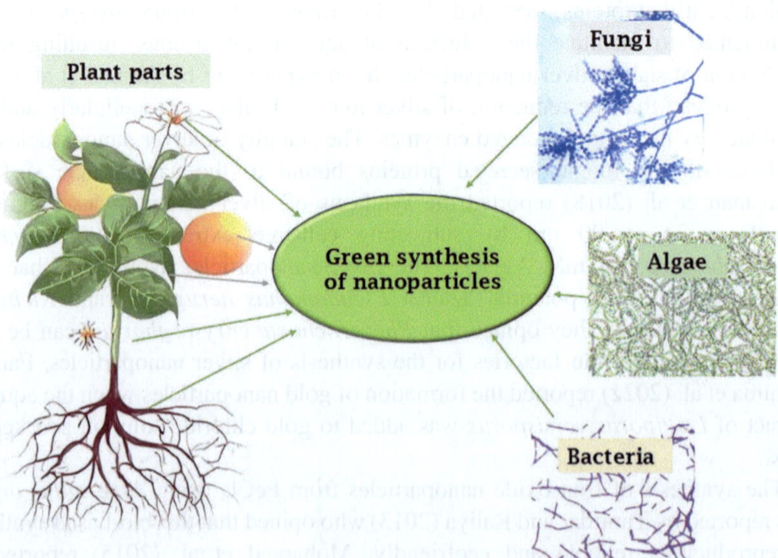

Fig. 2.4 Green Synthesis of Nanoparticles. (Figure prepared by Aji Abraham)

facilitating nanoparticle synthesis varied among species and stability of synthesized nanoparticles could be ascribed to proteinaceous capping agents. Extracellular synthesis of nanoparticles is preferred over intracellular synthesis due to ease of recovery of nanoparticles from solution (Singh et al. 2015). Singh and Kundu (2014) reported a simple environmentally friendly approach employing *Rhodopseudomonas capsulata* to synthesize stable gold nanoparticles (AuNPs). They also opined that pH is an important parameter which regulates the shape of AuNPs (Fig. 2.4).

2.3 Modification of Nanoparticles for Plant Applications

2.3.1 Need of the Engineered Nanoparticles

Increased use of nanoparticles in a variety of industrial sectors has prompted interest in creating biocompatible and environmentally safe production processes. The shapes of naturally occurring nanoparticles (NPs) are varied and irregular, while synthetic NPs can be shaped into desired forms like spherical balls, sheets, rods, and more complex designs like carbon multiwalled nanotubes and dendrimers (branching structures) (Husen and Iqbal 2019; Husen and Jawaid 2020). As a result, there is an increasing need to standardize procedures that have the least negative impact on the environment. And so, the preferred environmentally friendly way to synthesize nanoparticles is now green technology. One of the biggest challenges is the increasing global food consumption, which is putting varied disease ranges at risk and driving more food production. Another key concern is climate change (Chen et al. 2010). These days, managing plant diseases and keeping an eye on plant health depend heavily on these nanoparticles. Employing NPs reduces dependency on chemical inputs and reduces the possibility of pathogen over-detection, providing a sustainable solution to multiple risks to disease control and problems to food security (Chen and Mao 2007). Agrochemical synthesis presents a possible greener option with the use of nanoparticles (NPs). The compounds that bacteria release and their surfaces have a major influence in determining the charge of nanoparticles (NPs) when considering the interactions across NPs, plants, and bacteria. These substances function mainly as ionic surface coatings, acting as capping layers that influence the metal release from the NPs. The pH, microbial diversity, and soluble metal levels in the rhizosphere—the area of the soil where roots are found—are all determined by root exudates (Piccinno et al. 2012). Similarly, NPs and/or the metal ions they release are interacted with by bacterial extracellular polymeric materials (EPS) and secreted products such metal-chelating siderophores.

Metalloids like Boron (B) and Silicon (Si), non-metals like fullerenes, graphene oxide, quantum dots, and liposomes, as well as metallic oxide nanoparticles (NPs) like Zinc (Zn), Titanium (Ti), Nickel (Ni), Magnesium (Mg), Iron (Fe), Copper (Cu), Cerium (Ce), Silver (Ag), Gold (Au), and (Al) have all been studied for their possible applications in plant pathology (Elmer and White 2018; Zheng and Lu 2011; Alghuthaymi et al. 2021). By acting as nano fertilizers, nano fungicides, and nano bactericides in addition to acting as carriers of genetic materials and antimicrobial agents, these NPs show promise in the field of plant pathology. These NPs, which are engineered with certain designs and dimensions, can be incorporated into a range of composite arrays, increasing their usefulness and cross-disciplinary applicability.

2.3.2 Modification of Nanoparticles

The application of engineered/modified nanoparticles (ENPs) in agriculture has been heralded as the next big thing for maintaining and boosting crop output. There is a wealth of information available on the characteristics of several ENPs that promote growth. It has been highlighted in this scenario that ENPs can support seed setting, leaf development, and vegetative growth in addition to helping to lessen the effects of biotic and abiotic stressors. There has been an increase in discussion and concern lately over the phytotoxicity of ENPs. To address these issues, a more effective approach would be to incorporate modified or engineered nanoparticles into conventional laboratory sequencing technologies or molecular assays. This would greatly improve selectivity and sensitivity. There are five primary types of ENPs that are commonly found in the environmental landscape: metal oxides, carbonaceous nanoparticles, quantum dots, zero-valent metals, and nano polymers and this list is constantly being added to. According to Malandrakis et al. (2019), their distinct characteristics, like enhanced slow release of active ingredients, better drug delivery, decreased risk of disease resistance, cost-effectiveness, controlled release, multi-site mode of action, and increased effectiveness at lower doses, make them a feasible option. Target organisms can successfully use nanoparticles (NPs) due to the fact that they can construct complex systems with several desirable active elements. Through engineering, the targeted organisms' value is increased in a variety of functional domains (Rezaei et al. 2020; Fox et al. 2020; Singh et al. 2020; Dheyab et al. 2024).

ZnO nanoparticles can have their characteristics improved and altered by the widely utilized surface modification approach (Ong et al. 2018). Common environmental conditions regulate the stability of nanoparticle size and surface chemistry, which is essential for their function in any environment. Counter-ions neutralize the charge on the NP surface, creating an electrical double layer (Zabet-Khosousi 2008). The concentration and valency of the counter-ions, which can produce repulsive or attractive forces between charged NPs, are what govern the thickness of this double layer, also referred to as the Debye length. When the double layer's thickness is lowered, usually due to a rise in the valency of the electrolytes in the surrounding solution, the NPs begin to aggregate more readily (Doane et al. 2012). The modification of ZnO-NPs grants an interesting avenue for improving as well as expanding their functionality and applications respectively. By wisely modifying the surface chemistry of these nanoparticles, researchers can attain a notable transformation in their properties. This modification might include attaching organic molecules like carboxylic acids or silanes, integrating inorganic materials such as metal oxides, or even mixing polymers. Moreover, this strategic modification, not only improves the compatibility of ZnO-NPs with different environments but also significantly influences their performance features. The lifespan of these nanoparticles can be increased, their optical behaviour can be fine-tuned, and the overall efficiency can be enhanced. Additionally, modification benefits to minimize defects, paving the technique for greater stability and reduced energy consumption. By implementation of this approach, scientists are solving the full potential of ZnO-NPs, developing a new era of innovation in numerous fields, from medicine to electronics (Zhou et al. 2023).

Numerous gatekeepers identified with mesoporous silica nanoparticles (MSN) show great promise as frameworks for regulated pesticide release (Xu et al. 2018). By sealing the mesoporous surfaces or channels, these gatekeepers can control the release of cargo molecules and prohibit them from spilling too soon until the delivery goal is reached (Fig. 2.5). Furthermore, under particular internal or external stimuli, like temperature, pH, redox potential, light, magnetic field, and proteins, an ideal gatekeeper can be selectively eliminated, either completely or partially. This characteristic is very important for possible uses in precision farming. With the use of silica cross-linked with chitosan, Cao et al. (2017) developed a novel functionalized double-shelled avermectin microcapsule that showed improved controlled release capabilities. These microcapsules protect avermectin well against thermal and photodegradation. Using redox-responsive short-chain decanethiol as gatekeepers, Yi et al. (2015) developed functionalized mesoporous silica nanoparticles that were

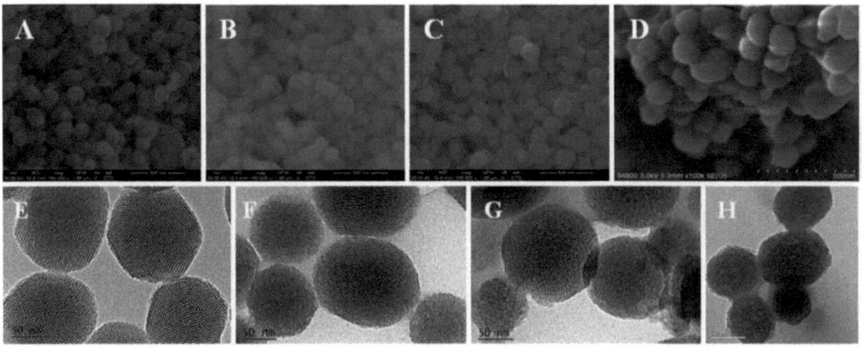

Fig. 2.5 SEM and TEM images of mesoporous silica nanoparticles for regulated pesticide release. (Reproduced with permission from Xu et al. 2018; Elsevier)

used to distribute the agrochemical salicylic acid (SA). When a particular concentration of glutathione (GSH) was present, the release of SA showed noticeably higher profiles than when GSH was absent. Recently, we used a post-grafting technique to add trimethylammonium groups to MSN, resulting in positive-charge-functionalized MSN loaded with 2,4-D sodium salt. Temperature, ionic strength, and pH also had an impact on the loading and release patterns of pesticides, which were mostly controlled by electrostatic interaction (Zhao et al. 2018; Kumar et al. 2014).

The increasing demand for consumer items based on nanotechnology is driving up the manufacturing of engineered nanoparticles, or ENPs (Baker et al. 2017). Unmodified nano-TiO2 has been shown in many research studies to exert a variety of effects on plant systems (Gardea-Torresdey et al. 2014; Ma et al. 2015a, b; Servin and White 2016). The tendency of nano-TiO2 and other ENPs to aggregate poses a serious obstacle to their use. Du et al. (2012) observed that pure nano-TiO_2 attached to the periderm cell wall of wheat in the form of aggregates, lowering plant biomass. This aggregation complicates even more toxicity assessments. Particles may be coated with a variety of organic and inorganic compounds to overcome this problem (Chen and Mao 2007). Furthermore, low interaction force molecules are coated on ENPs in order to stabilize surface energy (Kraynov and Müller 2011). Nano-TiO_2, featuring various surface modifications, holds applicability across multiple domains. Kurepa et al. (2010) observed the penetration of cell walls and subcellular accumulation of alizarin red S-stained nano-TiO_2 in *Arabidopsis thaliana*. Additionally, Larue et al. (2014) demonstrated that foliar application of paint containing aged nano-TiO_2 (hydrophobic) enabled penetration through leaf stomata and subsequent translocation to all plant tissues in lettuce (*Lactuca sativa*). Research involving other surface-coated engineered nanoparticles (ENPs) has revealed diverse physiological effects on plants which mainly observed that alginate-coated nano-ZnO increased the chlorophyll content of corn (*Zea mays*) seedlings grown in soil. Trujillo-Reyes et al. (2013) and Barrios et al. (2017) discovered that citric acid-coated nano-CeO_2 had a greater impact on the nutritional elements of radish (*Raphanus sativus*) and tomato (*Lycopersicon esculentum*) compared to uncoated NPs. Perreault et al.

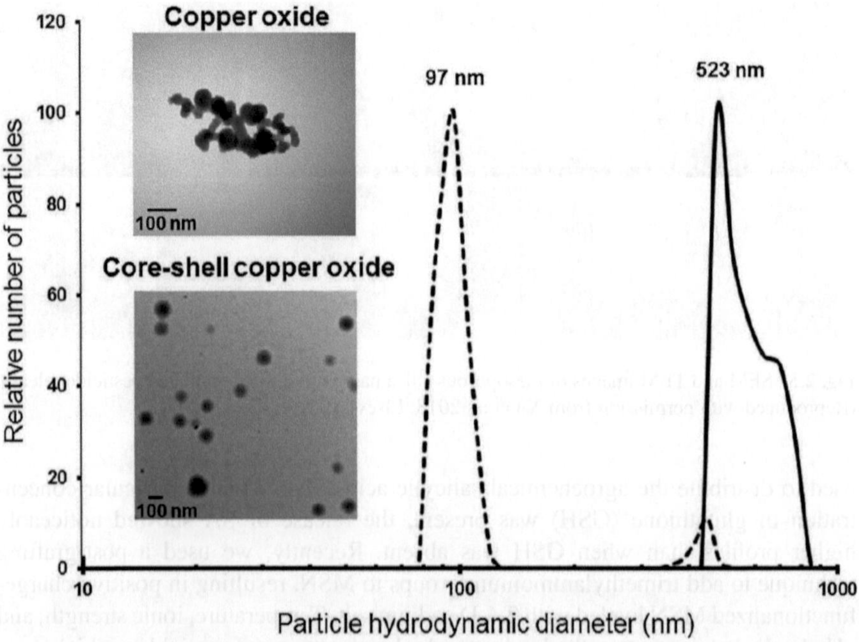

Fig. 2.6 TEM images of copper oxide nanoparticles and core-shell CuO NPs along with size distribution in the *L. gibba* culture medium, taken at the end of the 48-h exposure period. (Reproduced with permission from Perreault et al. 2014a, b; Elsevier)

(2014a, b) noted that nano-CuO with a poly(styrene-co-butyl acrylate) shell induced four times more oxidative stress in *Lemna gibba* than pristine NPs (Fig. 2.6).

2.4 Role of Nanoparticles on Plant Growth and Development

In recent years, there has been a surge in studies aimed at examining and elucidating the impact of nanoparticles (NPs) on the growth and development of plants. The exceptional reactivity and surface area of nanoparticles compared to their bulk counterparts, as noted by Schrand et al. (2010), stem from their exceptional physical and chemical properties. When nanoparticles are present in the environment and interact with biotic components, they inevitably undergo physicochemical changes such as dissolution, incidental coating by biomolecules, and redox reactions (Das et al. 2022).

Plants, due to their extensive interactions with the external environment, are significantly impacted by exposure to nanoparticles (Fig. 2.7) (Verma et al. 2018). These interactions induce various changes, including morphological and anatomical alterations, which are mainly influenced by the concentration and type of

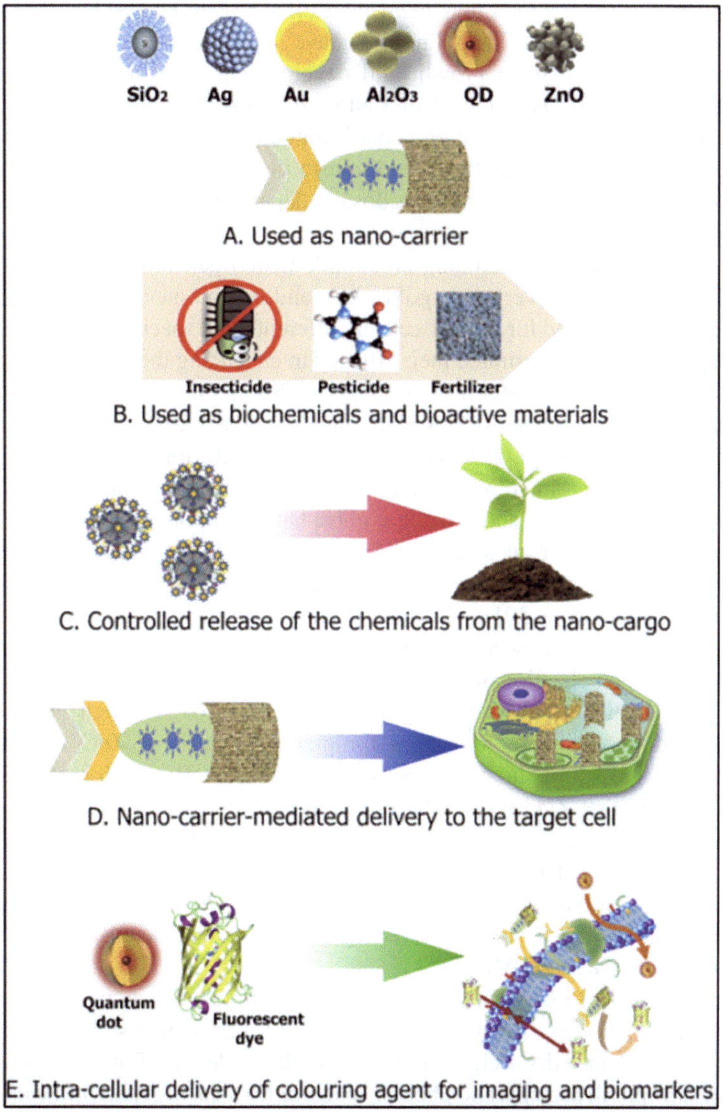

Fig. 2.7 Different Plant Applications of Synthesized Nanoparticles (**a**) Used as nano-carrier (**b**) Used as biochemicals and bioactive materials (**c**) Controlled release of the chemicals from the nano-cargo (**d**) Nano-carrier-mediated delivery to the target cell (**e**) Intra-cellular delivery of colouring agent for imaging and biomarkers. (Reproduced with permission from Verma et al. 2018; Elsevier)

nanoparticles involved (De la Rosa et al. 2021). Nanoparticles interact with plants, inducing various morphological and physiological changes that depend on the characteristics of the nanoparticles. The efficacy of nanoparticles is determined by factors such as their chemical composition, size, surface coating, reactivity, and notably, the dosage at which they exert their effects (Khodakovskaya et al. 2012). Researchers, drawing from their findings, have suggested both advantageous and detrimental effects on plant growth and development resulting from engineered nanoparticles (ENPs). The impact of ENPs on plants is contingent upon several factors, including the composition, concentration, size and physical as well as chemical attributes of the ENPs, in addition to the specific plant species involved (Ma et al. 2010). The efficacy of NPs is dosage-dependent and varies across different plant species.

Various nanoparticles, including zinc oxide (ZnO), silicon dioxide (SiO_2), titanium dioxide (TiO_2), silver (Ag), gold (Au), and carbon nanotubes (CNTs), have been extensively studied for their effects on different plant species. Zinc oxide (ZnO) nanoparticles have demonstrated their efficacy in enhancing the growth and yield of peanuts (*Arachis hypogaea* L.) (Prasad et al. 2012). Similarly, in cluster beans (*Cyamopsis tetragonoloba* L.), Raliya and Tarafdar (2013) observed notable improvements in chlorophyll levels, shoot-root growth, total soluble leaf protein content, rhizospheric microbial population, and phosphorus nutrient-mobilizing enzymes (phytase, acid and alkaline phosphatase) upon treatment with ZnO nanoparticles. Moreover, in wheat (*Triticum aestivum* L.), ZnO nanoparticles have been reported to enhance shoot-root growth, chlorophyll levels, and seed germination (Solanki and Laura 2018). Various studies have highlighted the biochemical changes induced by Zn seed treatment, including the breakdown of dormancy, hydrolysis or metabolization of inhibitors, imbibition, and enzyme activation (Harris et al. 2007; Samad et al. 2014; Solanki and Laura 2018). Singh et al. (2019) demonstrated that utilizing *Aloe vera* for green synthesis of ZnO nanoparticles facilitated the growth of shoot and root lengths in wheat. Research by Cyriac et al. (2020) revealed that ZnO nanoparticles, when modified with citric acid, not only increased the germination rate but also significantly enhanced root growth in pea (*Vigna unguiculata* L.) and rice (*Oryza sativa* L.) shown in Fig. 2.8. Asmat-Campos et al. (2022a, b) reported that employing *Coriandrum sativum* for green synthesis of ZnO nanoparticles led to improved germination characteristics in tomato (*Lycopersicon esculentum* Mill.).

Silicon dioxide (SiO_2) nanoparticles have exhibited beneficial effects across various plant species. Suriyaprabha et al. (2012) demonstrated that these nanoparticles enhance dry weight of plants and elevate levels of organic compounds such as chlorophyll, proteins and phenols in maize (*Zea mays* L.). SiO_2 nanoparticles have been observed to improve germination of seeds in tomatoes (*Lycopersicum esculentum* Mill.), as reported by Siddiqui and Al-Whaibi (2014). Alsaeedi et al. (2019b) found that nano-silica (NS) suspension treatments enhance seed germination and growth development in cucumbers (*Cucumis sativus* L.). Emamverdian et al. (2020) highlighted that SiO_2 NPs increase plant growth, particularly biomass, by augmenting antioxidant enzyme capacity in bamboo under Pb stress. Jiang et al. (2022) also noted that silicon dioxide and silicon carbide nanomaterials (NMs) promote shoot and root length in germinating seeds of rice (*Oryza sativa* L.).

Titanium dioxide (TiO_2) nanoparticles have been found to exert positive effects on the growth and development of spinach (*Spinacia oleracea* L.) and wheat

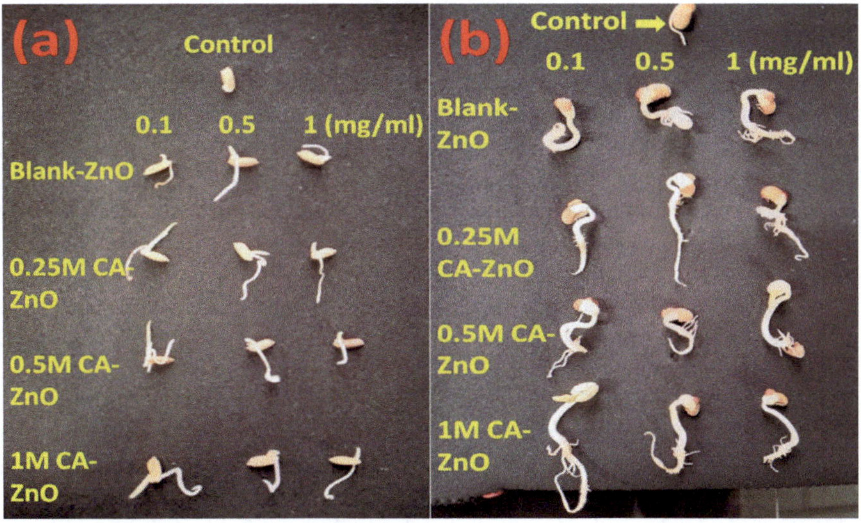

Fig. 2.8 Growth of *Oryza sativa* L. (**a**) and *Vigna unguiculata* L. (**b**) during germination in presence of control and various concentrations of bare and modified ZnO NPs. (Reproduced with permission from Cyriac et al. 2020; Elsevier)

(*Triticum aestivum* L.). Studies conducted by Zheng et al. (2005), Hong et al. (2005a, b) and Yang et al. (2006) have demonstrated that TiO_2 nanoparticles accelerate the germination of aged spinach seeds, promote nitrogen metabolism and photosynthesis, improve overall growth, and mitigate oxidative stress induced by UV-B radiation. Similarly, TiO_2 nanoparticles have been reported to expedite the germination of aged wheat seeds and enhance root elongation, as evidenced by studies conducted by Feizi et al. (2012) and Larue et al. (2012). Farahi et al. (2019) found that treatment with 200 $\mu g.ml^{-1}$ titanium dioxide nanoparticles had the most beneficial effect on the germination and seedling growth factors of *Vitex* plantlets. Basahi (2021) reported that the germinating seeds treated with titanium dioxide nanoparticles enhance water availability, facilitates mobilization of reserves, increases oxidative stress, and augments the antioxidant enzyme activity in pea (*Pisum sativum* L.).

Silver nanoparticles (AgNPs) and gold nanoparticles (AuNPs) have garnered attention for their potential applications in agriculture, chiefly in enhancing plant growth and stress tolerance. Sharma et al. (2012) highlighted the dose-dependent stimulatory effects of silver metal nanoparticles on the growth and antioxidant status of *Brassica juncea* (L.) Czern. ex Janch seedlings. Their findings demonstrate the positive influence of AgNPs on seedling growth parameters, vigor index, photosynthetic efficiency, and antioxidant enzyme activities, while concurrently reducing levels of reactive oxygen species and oxidative stress markers.

Green synthesized silver nanoparticles (AgNPs), derived from aqueous leaf extracts of *Cassia auriculata*, exhibited improve seed germination in *Pennisetum glaucum* (L.) R.Br. (Parveen and Rao 2015). Similarly, AgNPs synthesized using glucose enhanced shoot-root growth and promoted seed germination in wheat (Asanova et al. 2019). Investigating the impact of AgNPs on pea (*Pisum sativum* L.) growth and yield quality, Rahman et al. (2023) found significant enhancements in fruit and seed weight without adverse effects on Arbuscular Mycorrhizal Fungi (AMF) interaction. Moreover, the application of both silver nanoparticles (AgNPs) and gold nanoparticles (AuNPs) increased the activities of antioxidant enzymes, resulting in notable improvements in wheat growth parameters like fresh and dry biomass, leaf area, and chlorophyll contents (Manaf et al. 2021). Exposure to moderate concentrations of AuNPs enhanced growth in *Arabidopsis* seedlings, characterized by longer primary roots, increased lateral root numbers and length, and an expanded rosette diameter, while also reducing oxidative stress responses, suggesting a potential protective role of AuNPs in plant health (Ferrari et al. 2021).

Carbon nanotubes (CNTs) have demonstrated positive impacts on various plant species, showing potential to enhance water and nutrient uptake, thereby improving growth. Khodakovskaya et al. (2009) observed improvement in seed germination and root growth in tomatoes (*Lycopersicon esculentum* Mill.), while Wang et al. (2012) reported enhanced root growth in wheat (*Triticum aestivum* L.) treated with CNTs. Ghodake et al. (2010) and Mondal et al. (2011) found that carbon nanotubes improved germination and growth of seedling in mustard (*Brassica juncea* L.) and black gram (*Phaseolus mungo* L.), respectively. Nair et al. (2010) noted improved germination in rice (*Oryza sativa* L.) seeds treated with carbon nanotubes, while Tripathi et al. (2011) observed improved overall growth rates in common gram (*Cicer arietinum* L.) when treated with CNTs. Khodakovskaya et al. (2012) showed that activated carbon (AC) nanoparticles stimulated cell growth, resulting in a 16% increase, particularly at low concentrations (5 μg/mL) in tobacco (*Nicotiana tabacum* L.) cells. In a recent study, Hu et al. (2021) revealed that multiwalled carbon nanotubes (MWCNTs) stimulate plant growth by modulating crucial enzymes associated with carbon and nitrogen metabolism, leading to increased carbohydrate synthesis and nitrogen utilization, ultimately enhancing the growth of maize (*Zea mays* L.).

2.4.1 Role of Nanoparticles in Photosynthesis

Numerous studies have observed that the foliar application of metal nanoparticles (NPs) results in a notable rise in chlorophyll content in plants. This increase enables plants to synthesize more light-harvesting complexes (LHCII), which are pigment-protein complexes linked with photosystem I (PSI) and photosystem II (PSII). By synthesizing more LHCII, plants can capture a greater quantity of light energy, thereby enhancing photosynthesis (Kataria et al. 2019).

TiO_2 nanoparticles have been reported to have various effects on different plant species. In *Spinacia oleracea* L., they promoted chlorophyll content, increased Rubisco activity, and enhanced the rate of photosynthesis (Hong et al. 2005a, b). In *Linum usitatissimum* L., TiO_2 nanoparticles boosted chlorophyll and carotenoid content, enhanced growth, and positively affected yield attributes (Aghdam et al. 2016). In *Lycopersicon esculentum* L., TiO_2 improved photosynthesis by regulating energy dissipation and inducing stomatal opening, resulting in leaf cooling (Qi et al. 2013). TiO_2 nanoparticles enhanced the expression of Rubisco and chlorophyll-binding protein genes and increased the phosphoenolpyruvate carboxylase activity in *Cicer arietinum* L. (Hasanpour et al. 2015).

In recent studies, various types of nanoparticles, including silver (Ag), superparamagnetic iron oxide (SPIO), manganese (Mn), copper oxide (CuO), and gold (Au), have been investigated for their effects on plant photosynthesis and chlorophyll content. Sharma et al. (2012) found that silver (Ag) nanoparticles improved PSII quantum efficiency in *Brassica juncea* L. Hatami and Ghorbanpour (2014) reported an increase in leaf chlorophyll and carotenoid content, alongside a reduction in lipid peroxidation, in *Pelargonium zonale* (L.) L'Hér. ex Aiton treated with silver (Ag) nanoparticles. Falco et al. (2011) reported that gold (Au) nanoparticles enhanced absorbance and fluorescence quenching of chlorophyll a in *Glycine max* Merr. Pradhan et al. (2013) demonstrated that Mn nanoparticles improved the photophosphorylation activity of the electron transport chain in *Vigna radiata* (L.) R.Wilczek. Ghafariyan et al. (2013) observed that superparamagnetic iron oxide (SPIO) nanoparticles enhanced chlorophyll content in *Glycine max* Merr. Perreault et al. (2014a) noted a decrease in quantum yield, leading to the inhibition of the photosynthetic process in *Lemna gibba* L. upon exposure to CuO nanoparticles. Lalau et al. (2015) observed dilation of chloroplast membranes and destruction of stroma and grana in chloroplasts of *Landoltia punctata* treated with CuO nanoparticles.

The presence of ZnO-NPs in tomato (*Lycopersicon esculentum* L.) led to improvement in antioxidant systems and accelerated proline accumulation, potentially contributing to plant stability and improved photosynthetic efficiency (Faizan et al. 2018). Zinc oxide nanoparticles (ZnO NPs) with a silicon shell (ZnO-Si NPs) demonstrated beneficial effects on *Pisum sativum* L., enhancing the effective quantum yield of photosystem II (PSII) and the photochemistry of photosystem I (PS I) (Elshoky et al. 2021). The study conducted by Rai-Kalal and Jajoo (2021) demonstrated that priming wheat cultivar seeds with zinc oxide nanoparticles (ZnO NPs) resulted in improved overall primary photochemistry. This enhancement was achieved by boosting the performance of the water-splitting complex at the donor side of photosystem II (PSII), leading to increased activity of active reaction centers and more efficient electron transport. Furthermore, the nano priming approach reduced oxidative stress levels in plants, as indicated by decreased action of antioxidant enzymes such as peroxidase (POD), catalase (CAT), and superoxide dismutase (SOD), along with reduced lipid peroxidation.

The application of mesoporous silica nanoparticles (MSNs) in wheat and lupin plants resulted in the localization of MSNs to chloroplasts, resulting in a substantial increase in photosynthetic activity (Sun et al. 2016). Silicon dioxide and silicon carbide nanomaterials (NMs) elevated the levels of trace elements such as magnesium, copper, and zinc, thereby promoting the photosynthesis of rice (Jiang et al. 2022).

2.4.2 Role of Nanoparticles in Plant Protection

Nanoparticles (NPs) offer a promising avenue for delivering fertilizers, pesticides and other agrochemicals through the nano capsule production, which are very stable and biodegradable (Jha et al. 2009). The utilization of nanomaterials in crop protection presents promising and effective strategies for managing plant pathogens (Gopal et al. 2011). These particles adhere to the pathogen cell wall, inducing deformities through high-energy transfer, ultimately resulting in their demise (Sharma et al. 2016). Nanocomposites fulfill the two key requirements in disease management: maximum effectiveness with minimal ecological impact and reduced toxicity to humans (Sharma et al. 2016). Nanomaterials like zinc titanium, copper, silver, magnesium, gold and alginate have been developed proving exceptionally effective in combating a wide range of pathogens, including eukaryotic microorganisms, bacteria, and viruses (Guo et al. 2003).

Metallic nanoparticles are emerging as viable alternatives to traditional agrochemicals, showing promise in eliminating harmful microbes from plants, soils and hydroponic systems (Park et al. 2006; Ul Haq and Ijaz 2019). To assess the status and effect of metallic nanoparticles in plant disease management, two approaches can be considered: direct application to phytopathogens and formulation into nano fungicides (Khan and Rizvi 2014). These nano fungicides are commonly applied as foliar sprays to combat various plant diseases, with potential benefits for stimulating growth of plants (Agrawal and Rathore 2014). Lower concentrations of nanoparticles are typically suggested for effective disease control (Park et al. 2006). Moreover, nanoparticle application can be effective against microbes with reduced sensitivity to traditional antimicrobial agents due to their enhanced penetration through cell membranes (Vazquez-Muñoz et al. 2019).

2.4.3 Nano Fertilizers

Nano fertilizers are a recent agricultural innovation known for their nanoscale size and high surface area to volume ratio, making them more effective than conventional fertilizers (Chhipa 2019). Nanoparticles function as an effective "smart delivery system" due to their characteristics such as high surface area, sorption capacity, and controlled-release kinetics (Solanki et al. 2015). Nanostructured fertilizers have the capability to enhance nutrient use efficiency by employing mechanisms such as targeted delivery and gradual release, releasing active ingredients precisely in accordance with environmental and biological signals. Recent laboratory-scale studies have demonstrated that nano fertilizers can boost crop productivity by enhancing seed germination, seedling growth, photosynthetic activity, nitrogen metabolism, and protein and carbohydrate synthesis (Solanki et al. 2015; Azam et al. 2022; Channab et al. 2024; Semenova et al. 2024). A variety of nano fertilizers, including silver, zinc, iron, titanium, molybdenum, carbon nanotubes and silica, have been formulated and utilized across various crop systems.

Nanoparticles can enter plants through various pathways, including uptake through roots, absorption through leaf cuticles, entry through pores of stomata, or penetration through cuticle-free flowers (Ma et al. 2015a, b). The manner of application of nano fertilizers, such as soil amendment, foliar spray, or seed treatment, determines the specific uptake route. Once inside the plant, nanoparticles may travel within cells through symplastic pathways or diffuse through intercellular spaces, known as apoplastic routes (Kalia et al. 2019). They have the potential to revolutionize precision farming by enabling slow, controlled, or targeted delivery of plant nutrients, effectively acting as "smart fertilizers" (Raliya et al. 2017; Ghosh et al. 2023). These formulations not only enhance nutrient use competence but also minimize nutrient inputs while boosting crop productivity. Furthermore, nano-enabled sensors allow for active and real-time monitoring of soil nutrient concentrations, facilitating site-specific nutrient management through variable-rate application. This approach maximizes the efficiency of both conventional and nano "smart" fertilizers (Duhan et al. 2017; Chakraborty et al. 2023; Kaur et al. 2023).

2.4.4 Role of Nanoparticles in Mitigating Abiotic Response

The utilization of nanoparticles to ease abiotic stress in plants presents a promising avenue for enhancing crop resilience and productivity. Nanoparticles, such as titanium dioxide (TiO_2), silicon dioxide (SiO_2), zinc oxide (ZnO), silver nanoparticles (AgNPs) and carbon-based nanoparticles, have been investigated for their ability to mitigate various abiotic stresses, including drought, salinity, heat, and heavy metal toxicity (Thabet and Alqudah 2024).

Application of silica (SiO_2) in Soybean (*Glycine max* L.) has shown promising results in mitigating mercury toxicity by immobilizing and deactivating mercury (Li et al. 2020). Similarly, in Cucumber (*Cucumis sativus* L.), SiO_2 treatment has been found to improve growth and productivity, particularly under conditions of water deficit and salinity, by improving nutrient uptake, with notable increases in nitrogen and potassium contents, as reported by Alsaeedi et al. (2019a). In Common bean (*Phaseolus vulgaris* L.), supplementation with SiO_2 has been effective in alleviating Na+ stress, leading to improvements in various growth parameters like germination percentage, vigor index, and shoot and root lengths, as well as shoot and root dry masses, according to Alsaeedi et al. (2017). Studies on Wheat (*Triticum aestivum* L.) have demonstrated the efficacy of silicon (Si) nanoparticles in reducing oxidative stress and enhancing antioxidant enzyme activity. Si nanoparticles have also been shown to decrease cadmium (Cd) concentrations primarily in grains while increasing Si concentrations in plants under Cd toxicity conditions Hussain et al. (2019). Another study on Wheat emphasized that Si application increases superoxide dismutase (SOD) and peroxidase (POD) activity, particularly in response to Cd stress conditions Ali et al. (2019a).

In Licorice (*Glycyrrhiza glabra* L.) subjected to cold stress, the application of Titanium dioxide (TiO_2) resulted in a decrease in lipid peroxidation and hydrogen peroxide (H_2O_2) contents, while enhancing total protein, phenolics and osmolytes contents, as demonstrated by Ghabel and Karamian (2020). Similarly, in Chickpea (*Cicer arietinum* L.) under cold stress conditions, TiO_2 nanoparticles induced changes in metabolic pathways, as revealed by transcription profiling Amini et al. (2017). In Cotton (*Gossypium*) facing drought stress, the combined application of Titanium dioxide (TiO_2) and silicon dioxide (SiO_2) resulted in increased levels of total phenolics, total soluble proteins, total free amino acids, proline content, total antioxidant capacity, catalase (CAT) activity, peroxidase (POD), and superoxide dismutase (SOD) activity, as highlighted by Shallan et al. (2016).

Zinc oxide (ZnO) application demonstrated varied beneficial effects across different plant species under stress conditions. In Fenugreek (*Trigonella foenumgraecum* L.), ZnO application mitigated salinity stress by increasing proline levels and trigonelline content (Noohpisheh et al. 2021). For Wheat (*Triticum aestivum* L.) and Rice (*Oryza sativa* L.) facing cadmium toxicity, ZnO application led to increase in tissue dry weight, minimalized oxidative stress, and reduced cadmium accumulation (Khan et al. 2019a, b; Ali et al. 2019b). In Sunflower (*Helianthus annuus* L.) and Lupine (*Lupinus luteus* L.) under salinity stress, ZnO application enhanced biomass production, photosynthesis, and antioxidant enzyme activities while reducing stress-related parameters (Torabian et al. 2018; Abdel Latef et al. 2017).

Hashimoto et al. (2020) observed that Ag treatment in Soybean (*Glycine max* L.) under flooding stress significantly enhanced growth of plants and regulated protein quality. Likewise, Hojjat (2019) reported improvements in percentage of germination, shoot and root length, as well as seedling fresh and dry weight in Grass Pea (*Lathyrus sativus* L.) subjected to salinity stress following Ag treatment. Jaskulak et al. (2019) demonstrated that Ag application in Yellow-lupin (*Lupinus luteus* L.) grown in heavy metal contaminated soil enhanced glutathione peroxidase (GPX) activity and expression of metallothioneins, indicating a potential for metal stress mitigation. Iqbal et al. (2019) documented improved root and shoot length, root number, fresh and dry weight in Wheat (*Triticum aestivum* L.) under heat stress upon Ag treatment. Moreover, Hojjat and Ganjali (2016) highlighted the positive effects of Ag application on germination percentage, shoot length, and root length in Lentils (*Lens esculenta* (L.) Moench) experiencing drought stress. Therefore, the integration of nanoparticles in agricultural practices represents a promising avenue for addressing the challenges posed by abiotic stresses, ultimately contributing to food security and sustainable farming (Chaffai et al. 2024).

2.4.5 Role of Nanoparticles in In-vitro Culture

Nanoparticles contribute significantly to the advancement of *in-vitro* culture techniques by enhancing several aspects of plant tissue culture and micropropagation efficiency. They serve as effective carriers for nutrients and growth regulators, enhancing their bioavailability and enabling more efficient uptake by plant cells, thereby promoting growth rate and tissue development *in-vitro* (Ali et al. 2023; Balamurugan et al. 2024). A study by Ali et al. (2023) explored the use of phytomediated selenium nanoparticles and light regimes to enhance *in-vitro* callus cultures of *Caralluma tuberculata*, establishing a protocol that improves plant growth and antioxidant activity for sustainable bioactive compound production. Numerous studies have shown that incorporating nanoparticles in plant *in-vitro* culture diminishes bacterial contamination due to their antimicrobial properties, which effectively prevent contamination and enhance the overall success of tissue culture operations (Mandeh et al. 2012; Arab et al. 2014; Álvarez et al. 2019).

The use of nanoparticles in *in-vitro* culture enhances cell viability and proliferation by stabilizing cellular structures and protecting against oxidative stress, while also improving sterile explant percentages, survival rates, morphogenesis, and stimulating secondary metabolite synthesis to combat unfavorable environmental conditions in agriculture and forestry (Grodetskaya et al. 2024). Nanoparticles also facilitate the delivery of genetic material into plant cells, improving the efficiency of transformation techniques for developing genetically modified plants (Wang et al. 2019; Al-Qudah et al. 2022).

2.5 Conclusions and Prospects

The chapter explored various approaches to developing and modifying nanoparticles to meet the particular requirements of applications pertaining to plants. It looks at a number of strategies, such as "green synthesis," which makes the use of fungi, bacteria, plants, and algae to emphasize their environmentally beneficial qualities and potential for producing nanomaterials in a sustainable manner. In order to achieve targeted distribution, enhanced efficacy, and decreased environmental impact, it is imperative to improve the functionality of nanoparticles for use in plant applications. By optimizing the transport of nutrients, herbicides, and genetic material while minimizing negative environmental consequences, engineered nanoparticles have the potential to completely transform agriculture. The chapter also covers the critical role that nanoparticles play in the growth and development of plants, providing insight into the complex ways in which they interact with these organisms. It is noteworthy that studies have demonstrated the beneficial effects of nanoparticles on photosynthesis, increasing plant output and stress tolerance. This chapter provides a thorough overview of the synthesis, modification, and use of nanoparticles in agriculture. It also provides insights into the potential of these materials to transform crop productivity and address global food security issues in a way that is environmentally friendly. Nanoparticles are expected to play a pivotal role in enhancing various aspects of plant health and productivity. Future advancements may focus on improving nutrient uptake efficiency, fostering root development, and enhancing stress resistance to abiotic factors such as drought and salinity.

The integration of nanoparticles into pest management strategies could lead to more effective and targeted control of plant pathogens, reducing the need for chemical pesticides. Innovations in smart delivery systems may enable the precise application of genetic materials, paving the way for improved crop varieties that are resilient and high yielding. Overall, the prospects of nanoparticles in agriculture promise to significantly advance plant health and sustainability.

References

Abbas S, Nasreen S, Haroon A, Ashraf MA (2020) Synthesis of silver and copper nanoparticles from plants and application as adsorbents for naphthalene decontamination. Saudi J Biol Sci 27(4):1016–1023

Abboud Y, Saffaj T, Chagraoui A et al (2014) Biosynthesis, characterization and antimicrobial activity of copper oxide nanoparticles (CONPs) produced using brown alga extract (*Bifurcaria bifurcata*). Appl Nanosci 4:571–576. https://doi.org/10.1007/s13204-013-0233-x

Abdel Latef AA, Abu Alhmad MF, Abdelfattah KE (2017) The possible roles of priming with ZnO nanoparticles in mitigation of salinity stress in lupine (*Lupinus termis*) plants. J Plant Growth Regul 36:60–70

Aghdam MT, Mohammadi H, Ghorbanpour M (2016) Effects of nanoparticulate anatase titanium dioxide on physiological and biochemical performance of *Linum usitatissimum* (Linaceae) under well-watered and drought stress conditions. Braz J Bot 39:139–146

Agrawal S, Rathore P (2014) Nanotechnology pros and cons to agriculture: a review. Int J Curr Microbiol App Sci 3(3):43–55

Ahmad A, Mukherjee P, Senapati S et al (2003) Extracellular biosynthesis of silver nanoparticles using the fungus *Fusarium oxysporum*. Colloids Surf B Biointerfaces 28(4):313–318

Ahmad MA, Adeel M, Shakoor N, Ali I, Ishfaq M, Haider FU, Deng X (2023) Unraveling the roles of modified nanomaterials in nano enabled agriculture. Plant Physiol Biochem 202:107944

Ahmed S, Qasim S, Ansari M et al (2022) Green synthesis of zinc nanoparticles and their effects on growth and yield of *Pisum sativum*. J King Saud Univ Sci 34(6):102132. https://doi.org/10.1016/j.jksus.2022.102132

Alghuthaymi MA, Kalia A, Bhardwaj K et al (2021) Nanohybrid antifungals for control of plant diseases: current status and future perspectives. J Fungi 7(1):48

Ali S, Rizwan M, Hussain A et al (2019a) Silicon nanoparticles enhanced the growth and reduced the cadmium accumulation in grains of wheat (*Triticum aestivum* L.). Plant Physiol Biochem 140:1–8

Ali S, Rizwan M, Noureen S et al (2019b) Combined use of biochar and zinc oxide nanoparticle foliar spray improved the plant growth and decreased the cadmium accumulation in rice (*Oryza sativa* L.) plant. Environ Sci Pollut Res 26:11288–11299

Ali A, Mashwani ZUR, Raja NI et al (2023) Phytomediated selenium nanoparticles and light regimes elicited in vitro callus cultures for biomass accumulation and secondary metabolite production in *Caralluma tuberculata*. Front Plant Sci 14:1253193

Alomar TS, AlMasoud N, Awad MA et al (2020) An eco-friendly plant-mediated synthesis of silver nanoparticles: characterization, pharmaceutical and biomedical applications. Mater Chem Phys 249:123007. https://doi.org/10.1016/j.matchemphys.2020.123007

Al-Qudah T, Mahmood SH, Abu-Zurayk R et al (2022) Nanotechnology applications in plant tissue culture and molecular genetics: a holistic approach. Curr Nanosci 18(4):442–464

Alsaeedi AH, El-Ramady H, Alshaal T et al (2017) Engineered silica nanoparticles alleviate the detrimental effects of Na+ stress on germination and growth of common bean (*Phaseolus vulgaris*). Environ Sci Pollut Res 24:21917–21928

Alsaeedi A, El-Ramady H, Alshaal T et al (2019a) Silica nanoparticles boost growth and productivity of cucumber under water deficit and salinity stresses by balancing nutrients uptake. Plant Physiol Biochem 139:1

Alsaeedi AH, Elgarawany MM, El-Ramady H et al (2019b) Application of silica nanoparticles induces seed germination and growth of cucumber (*Cucumis sativus*). Met Environ Arid Land Agric Sci 28:57–68

Alsammarraie FK, Wang W, Zhou P, Mustapha A et al (2018) Green synthesis of silver nanoparticles using turmeric extracts and investigation of their antibacterial activities. Colloids Surf B 171:398–405. https://doi.org/10.1016/j.colsurfb.2018.07.059

Álvarez SP, Tapia MAM, Vega MEG et al (2019) Nanotechnology and plant tissue culture. In: Prasad R (ed) Plant nanobionics. Nanotechnology in the life sciences. Springer, Cham, pp 333–370

Amini S, Maali-Amiri R, Mohammadi R et al (2017) cDNA-AFLP analysis of transcripts induced in chickpea plants by TiO2 nanoparticles during cold stress. Plant Physiol Biochem 111:39–49

Arab MM, Yadollahi A, Hosseini-Mazinani M, Bagheri S (2014) Effects of antimicrobial activity of silver nanoparticles on in vitro establishment of G× N15 (hybrid of almond× peach) rootstock. J Genet Eng Biotechnol 12(2):103–110

Asanova AA, Yashin SE, Trofimova TV et al (2019) Application of silver nanoparticles to improve wheat seedlings growth. In: IOP Conference Series: Earth and Environmental Science, vol. 315(5). IOP Publishing, pp 052041

Asmat-Campos D, López-Medina E, De Oca-Vásquez GM et al (2022a) ZnO Nanoparticles obtained by green synthesis as an alternative to improve the germination characteristics of *L. esculentum*. Molecules 27(7):2343. https://doi.org/10.3390/molecules27072343

Asmat-Campos D, López-Medina E, Montes de Oca-Vásquez G et al (2022b) ZnO Nanoparticles obtained by green synthesis as an alternative to improve the germination characteristics of *L. esculentum*. Molecules 27(7):2343

Azam M, Bhatti HN, Khan A, Zafar L, Iqbal M (2022) Zinc oxide nano-fertilizer application (foliar and soil) effect on the growth, photosynthetic pigments and antioxidant system of maize cultivar. Biocatal Agric Biotechnol 42:102343

Baker S, Volova T, Prudnikova SV et al (2017) Nanoagroparticles emerging trends and future prospect in modern agriculture system. Environ Toxicol Pharmacol 53:10–17

Balaji DS, Basavaraja S, Deshpande R et al (2009) Extracellular biosynthesis of functionalized silver nanoparticles by strains of *Cladosporium cladosporioides* fungus. Colloids Surf B Biointerfaces 68:88–92. https://doi.org/10.1016/j.colsurfb.2008.09.022

Balakumaran MD, Ramachandran R, Kalaichelvan PT (2015) Exploitation of endophytic fungus, *Guignardia mangiferae* for extracellular synthesis of silver nanoparticles and their in vitro biological activities. Microbiol Res 178:9–17. https://doi.org/10.1016/j.micres.2015.05.009

Balamurugan V, Abdi G, Karthiksaran C et al (2024) A review: improvement of plant tissue culture applications by using nanoparticles. J Nanopart Res 26(8):188

Barrios AC, Medina-Velo IA, Zuverza-Mena N et al (2017) Nutritional quality assessment of tomato fruits after exposure to uncoated and citric acid coated cerium oxide nanoparticles, bulk cerium oxide, cerium acetate and citric acid. Plant Physiol Biochem 110:100–107

Basahi M (2021) Seed germination with titanium dioxide nanoparticles enhances water supply, reserve mobilization, oxidative stress and antioxidant enzyme activities in pea. Saudi J Biol Sci 28(11):6500–6507

Bhuiyan MSH, Miah MY, Paul SC et al (2020) Green synthesis of iron oxide nanoparticle using Carica papaya leaf extract: application for photocatalytic degradation of remazol yellow RR dye and antibacterial activity. Heliyon 6(8):e04603. https://doi.org/10.1016/j.heliyon.2020.e04603

Cao L, Zhou Z, Niu S et al (2017) Positive-charge functionalized mesoporous silica nanoparticles as nanocarriers for controlled 2, 4-dichlorophenoxy acetic acid sodium salt release. J Agric Food Chem 66(26):6594–6603

Chaffai R, Ganesan M, Cherif A (2024) Nanoparticles for sustainable agriculture and abiotic stress resistance. In: Plant adaptation to abiotic stress: from signaling pathways and microbiomes to molecular mechanisms. Springer Nature Singapore, pp 195–208

Chakraborty R, Mukhopadhyay A, Paul S et al (2023) Nanocomposite-based smart fertilizers: a boon to agricultural and environmental sustainability. Sci Total Environ 863:160859

Channab BE, Idrissi AE, Ammar A et al (2024) Recent advances in nano-fertilizers: synthesis, crop yield impact, and economic analysis. Nanoscale 9:1–30

Chen X, Mao SS (2007) Titanium dioxide nanomaterials: synthesis, properties, modifications, and applications. Chem Rev 107:2891–2959

Chen R, Ratnikova TA, Stone MB et al (2010) Differential uptake of carbon nanoparticles by plant and mammalian cells. Small 6:612–617

Chhipa H (2019) Applications of nanotechnology in agriculture. In: Methods in microbiology, vol 46. Academic Press, pp 115–142

Cyriac J, Melethil K, Thomas B et al (2020) Synthesis of biogenic ZnO nanoparticles and its impact on seed germination and root growth of *Oryza sativa* L. and *Vigna unguiculata* L. Mater Today Proc 25:224–229

Das RK, Gogoi N, Bora U (2011) Green synthesis of gold nanoparticles using *Nyctanthes arbortristis* flower extract. Bioproc Biosyst Eng 34:615–619. https://doi.org/10.1007/s00449-010-0510-y

Das S, Ghosh D, Kerkar K et al (2022) Effects of the transformation of metallic nanoparticles in the environment and its toxicity on aquatic and terrestrial life forms. In: Free radical biology and environmental toxicity. Springer International Publishing, Cham, pp 43–71

De la Rosa G, Vázquez-Núñez E, Molina-Guerrero C et al (2021) Interactions of nanomaterials and plants at the cellular level: current knowledge and relevant gaps. Nano-Environ Eng 6:1–9

Devanesan S, AlSalhi MS (2021) Green synthesis of silver nanoparticles using the flower extract of *Abelmoschus esculentus* for cytotoxicity and antimicrobial studies. Int J Nanomedicine:3343–3356. https://doi.org/10.2147/IJN.S307676

Devi HS, Boda MA, Shah MA et al (2019) Green synthesis of iron oxide nanoparticles using *Platanus orientalis* leaf extract for antifungal activity. Green Process Synth 8(1):38–45. https://doi.org/10.1515/gps-2017-0145

Dhar SA, Chowdhury RA, Das S, Nahian KM et al (2021) Plant-mediated green synthesis and characterization of silver nanoparticles using *Phyllanthus emblica* fruit extract. Mater Today Proc 42(5):1867–1871. https://doi.org/10.1016/j.matpr.2020.12.222

Dheyab MA, Oladzadabbasabadi N, Aziz AA, Khaniabadi PM, Al-ouqaili MT, Jameel MS et al (2024) Recent advances of plant-mediated metal nanoparticles: synthesis, properties, and emerging applications for wastewater treatment. J Environ Chem Eng 12:112345

Doane TL, Chuang CH, Hill RJ et al (2012) Nanoparticle ζ-potentials. Acc Chem Res 45(3): 317–326

Dobrucka R (2017) Synthesis of titanium dioxide nanoparticles using *Echinacea purpurea Herba*. Iran J Pharm Res 16(2):756–762

Du F, Zhang Q, Yu Q et al (2012) Soil contamination of *Toxoplasma gondii* oocysts in pig farms in central China. Vet Parasitol 187(1):53–56

Duhan JS, Kumar R, Kumar N et al (2017) Nanotechnology: the new perspective in precision agriculture. Biotechnol Rep 15:11–23

Elmer W, White JC (2018) The future of nanotechnology in plant pathology. Ann Rev Phytopathol 56:111–133

Elshoky HA, Yotsova E, Farghali MA et al (2021) Impact of foliar spray of zinc oxide nanoparticles on the photosynthesis of Pisum sativum L. under salt stress. Plant Physiol Biochem 167:607–618

Emamverdian A, Ding Y, Mokhberdoran F et al (2020) Silicon dioxide nanoparticles improve plant growth by enhancing antioxidant enzyme capacity in bamboo (*Pleioblastus pygmaeus*) under lead toxicity. Trees 34:469–481

Faizan M, Faraz A, Yusuf M et al (2018) Zinc oxide nanoparticle-mediated changes in photosynthetic efficiency and antioxidant system of tomato plants. Photosynthetica 56:678–686

Falco WF, Botero ER, Falcão EA et al (2011) In vivo observation of chlorophyll fluorescence quenching induced by gold nanoparticles. J Photochem Photobiol A Chem 225(1):65–71

Farahi SM, Iranbakhsh A, Mahmoodzadeh H et al (2019) Effects of titanium dioxide nanoparticles (TiO2) on germination and seedling growth of Vitex plants (*Vitex agnus-castus* L.). J Biosci Biotechnol 8(2):141–149

Farzana Fathima MR, Usha Raja Nanthini A, Al-Khattaf FS et al (2022) Mycosynthesis of noble metal nanoparticle using *Laetiporus versisporus* Mushroom and analysis of antioxidant activity. J Nanomater 2022:1–11

Feizi H, Rezvani Moghaddam P, Shahtahmassebi N et al (2012) Impact of bulk and nanosized titanium dioxide (TiO2) on wheat seed germination and seedling growth. Biol Trace Elem Res 146:101–106

Ferrari E, Barbero F, Busquets-Fité M et al (2021) Growth-promoting gold nanoparticles decrease stress responses in Arabidopsis seedlings. Nanomaterials 11(12):3161

Fox SE, Akmatbekov A, Harbert JL et al (2020) Pulmonary and cardiac pathology in African American patients with COVID-19: an autopsy series from New Orleans. Lancet Respir Med 8(7):681–686

Gardea-Torresdey JL, Rico CM, White JC (2014) Trophic transfer, transformation, and impact of engineered nanomaterials in terrestrial environments. Environ Sci Technol 48(5):2526–2540

Garg R, Kumari M, Kumar M et al (2021) Green synthesis of calcium carbonate nanoparticles using waste fruit peel extract. Mater Today Proc 46(15):6665–6668

Ghabel VK, Karamian R (2020) Effects of TiO2 nanoparticles and spermine on antioxidant responses of *Glycyrrhiza glabra* L. to cold stress. Acta Bot Croat 79(2):137–147

Ghafariyan MH, Malakouti MJ, Dadpour MR et al (2013) Effects of magnetite nanoparticles on soybean chlorophyll. Environ Sci Technol 47(18):10645–10652

Ghodake G, Seo YD, Park D et al (2010) Phytotoxicity of carbon nanotubes assessed by Brassica juncea and Phaseolus mungo. J Nanoelectron Optoelectron 5(2):157–160

Ghosh D, Sarkar MM, Roy S (2023) Smart fertilizers: the prospect of slow release nanofertilizers in modern agricultural practices. In: Nanofertilizers for sustainable agroecosystems: recent advances and future trends. Springer Nature Switzerland, Cham, pp 343–372

Gopal MA, Gogoi RO, Srivastava CH et al (2011) Nanotechnology and its application in plant protection. In: Plant pathology in India: vision 2030. Indian Phytopathological Society, New Delhi, pp 224–232

Grodetskaya TA, Fedorova OA, Zhuzhukin KV et al (2024) Nanoparticles in clonal micropropagation and plant stress control. In: Nanoparticles in plant biotic stress management. Springer Nature, Singapore, pp 315–353

Guo W, Li JJ, Wang YA et al (2003) Conjugation chemistry and bio applications of semiconductor box nanocrystals prepared via dendrimer bridging. Chem Mater 15(16):3125–3133

Harris D, Rashid A, Miraj G et al (2007) 'On-farm'seed priming with zinc sulphate solution—A cost-effective way to increase the maize yields of resource-poor farmers. Field Crop Res 102(2): 119–127

Hasanpour H, Maali-Amir R, Zeinali H (2015) Effect of TiO2 nanoparticles on metabolic limitations to photosynthesis under cold in chickpea. Russ J Plant Physiol 62:779–787

Hashimoto T, Mustafa G, Nishiuchi T et al (2020) Comparative analysis of the effect of inorganic and organic chemicals with silver nanoparticles on soybean under flooding stress. Int J Mol Sci 21(4):1300

Hatami M, Ghorbanpour M (2014) Defense enzyme activities and biochemical variations of Pelargonium zonale in response to nanosilver application and dark storage. Turk J Biol 38(1): 130–139

Hojjat SS (2019) Effect of interaction between Ag nanoparticles and salinity on germination stages of *Lathyrus sativus* L. Open Access J Environ Soil Sci 2:193–198

Hojjat SS, Ganjali A (2016) The effect of silver nanoparticle on lentil seed germination under drought stress. Int J Farm Allied Sci 5(3):208–212
Hong F, Yang F, Liu C et al (2005a) Influences of nano-TiO2 on the chloroplast aging of spinach under light. Biol Trace Elem Res 104:249–260
Hong F, Zhou J, Liu C (2005b) Effect of nano-TiO2 on photochemical reaction of chloroplasts of spinach. Biol Trace Elem Res 105:269–279
Hu Y, Zhang P, Zhang X et al (2021) Multi-wall carbon nanotubes promote the growth of maize (*Zea mays*) by regulating carbon and nitrogen metabolism in leaves. J Agric Food Chem 69(17): 4981–4991
Husen A, Iqbal M (2019) Nanomaterials and plant potential: an overview. Springer International Publishing, pp 3–29
Husen A, Jawaid M (eds) (2020) Nanomaterials for agriculture and forestry applications. Elsevier
Hussain A, Oves M, Alajmi MF et al (2019) Biogenesis of ZnO nanoparticles using Pandanus odorifer leaf extract: anticancer and antimicrobial activities. RSC Adv 9(27):15357–15369
Iqbal M, Raja NI, Mashwani ZU et al (2019) Effect of silver nanoparticles on growth of wheat under heat stress. Iran J Sci Technol Trans A Sci 43:387–395
Jaskulak M, Rorat A, Grobelak A et al (2019) Bioaccumulation, antioxidative response, and metallothionein expression in *Lupinus luteus* L. exposed to heavy metals and silver nanoparticles. Environ Sci Pollut Res 26:16040–16052
Jha AK, Prasad K, Prasad K (2009) A green low-cost biosynthesis of Sb2O3 nanoparticles. Biochem Eng J 43(3):303–306
Jiang Y, Yang J, Li M (2022) Effect of silica-based nanomaterials on seed germination and seedling growth of rice (*Oryza sativa* L.). Nanomaterials 12(23):4160
Kalia A, Sharma SP, Kaur H (2019) Nanoscale fertilizers: harnessing boons for enhanced nutrient use efficiency and crop productivity. In: Nanobiotechnology applications in plant protection, vol 2. Springer, pp 191–208
Kataria S, Jain M, Rastogi A et al (2019) Role of nanoparticles on photosynthesis: avenues and applications. In: Nanomaterials in plants, algae and microorganisms. Academic Press, pp 103–127
Kaur H, Hussain SJ, Mir RA et al (2023) Nanofertilizers–emerging smart fertilizers for modern and sustainable agriculture. Biocatal Agric Biotechnol 54:102921
Khan MR, Rizvi TF (2014) Nanotechnology: scope and application in plant disease management. Plant Pathol J 13(3):214–231
Khan MR, Adam V, Rizvi TF et al (2019a) Nanoparticle–plant interactions: two-way traffic. Small 15(37):1901794
Khan ZS, Rizwan M, Hafeez M et al (2019b) The accumulation of cadmium in wheat (*Triticum aestivum*) as influenced by zinc oxide nanoparticles and soil moisture conditions. Environ Sci Pollut Res 26:19859–19870
Khan F, Shahid A, Zhu H et al (2022) Prospects of algae-based green synthesis of nanoparticles for environmental applications. Chemosphere 293:133571
Khodakovskaya M, Dervishi E, Mahmood M (2009) Carbon nanotubes are able to penetrate plant seed coat and dramatically affect seed germination and plant growth. ACS Nano 3(10): 3221–3227
Khodakovskaya MV, De Silva K, Biris AS et al (2012) Carbon nanotubes induce growth enhancement of tobacco cells. ACS Nano 6(3):2128–2135
Kraynov A, Müller TE (2011) Concepts for the stabilization of metal nanoparticles in ionic liquids. In: Applications of ionic liquids in science and technology, vol 9. IntechOpen, pp 235–260
Kumar S, Bhanjana G, Sharma A et al (2014) Synthesis, characterization and on field evaluation of pesticide loaded sodium alginate nanoparticles. Carbohydr Polym 101:1061–1067
Kumar V, Kaushik NK, Tiwari SK, Singh D, Singh B (2023) Green synthesis of iron nanoparticles: sources and multifarious biotechnological applications. Int J Biol Macromol 253(4):127017

Kuppusamy P, Yusoff MM, Maniam GP, Govindan N (2016) Biosynthesis of metallic nanoparticles using plant derivatives and their new avenues in pharmacological applications– An updated report. Saudi Pharm J 24(4):473–484

Kurepa J, Paunesku T, Vogt S et al (2010) Uptake and distribution of ultrasmall anatase TiO2 Alizarin red S nanoconjugates in *Arabidopsis thaliana*. Nano Lett 10(7):2296–2302

Lalau CM, Mohedano RD, Schmidt ÉC et al (2015) Toxicological effects of copper oxide nanoparticles on the growth rate, photosynthetic pigment content, and cell morphology of the duckweed *Landoltia punctata*. Protoplasma 252:221–229

Larue C, Laurette J, Herlin-Boime N et al (2012) Accumulation, translocation and impact of TiO2 nanoparticles in wheat (*Triticum aestivum* spp.): influence of diameter and crystal phase. Sci Total Environ 431:197–208

Larue C, Castillo-Michel H, Sobanska S et al (2014) Foliar exposure of the crop Lactuca sativa to silver nanoparticles: evidence for internalization and changes in Ag speciation. J Hazard Mat 264:98–106

Li Y, Zhu N, Liang X et al (2020) Silica nanoparticles alleviate mercury toxicity via immobilization and inactivation of Hg (ii) in soybean (*Glycine max*). Environ Sci Nano 7(6):1807–1817

Ma X, Geiser-Lee J, Deng Y et al (2010) Interactions between engineered nanoparticles (ENPs) and plants: phytotoxicity, uptake and accumulation. Sci Total Environ 408(16):3053–3061

Ma C, White JC, Dhankher OP et al (2015a) Metal-based nanotoxicity and detoxification pathways in higher plants. Environ Sci Technol 49(12):7109–7122

Ma X, Zhang Q, Zhu Q et al (2015b) A robust CRISPR/Cas9 system for convenient, high-efficiency multiplex genome editing in monocot and dicot plants. Mol Plant 8(8):1274–1284

Malandrakis AA, Kavroulakis N, Chrysikopoulos CV (2019) Use of copper, silver and zinc nanoparticles against foliar and soil-borne plant pathogens. Sci Total Environ 670:292–299

Manaf A, Wang X, Tariq F et al (2021) Antioxidant enzyme activities correlated with growth parameters of wheat sprayed with silver and gold nanoparticle suspensions. Agronomy 11(8): 1494

Mandeh M, Omidi M, Rahaie M (2012) In vitro influences of TiO2 nanoparticles on barley (Hordeum vulgare L.) tissue culture. Biol Trace Elem Res 150:376–380

Meyers MA, Mishra A, Benson DJ (2006) Mechanical properties of nanocrystalline materials. Prog Mater Sci 51(4):427–556

Mohamed YM, Azzam AM, Amin BH, Safwat NA (2015) Mycosynthesis of iron nanoparticles by *Alternaria alternata* and its antibacterial activity. Afr J Biotechnol 14(14):1234–1241

Mondal A, Basu R, Das S et al (2011) Beneficial role of carbon nanotubes on mustard plant growth: an agricultural prospect. J Nanopart Res 13(10):4519–4528

Nair R, Varghese SH, Nair BG et al (2010) Nanoparticulate material delivery to plants. Plant Sci 179:154–163

Nava O, Soto-Robles CA, Gómez-Gutiérrez CM et al (2017) Fruit peel extract mediated green synthesis of zinc oxide nanoparticles. J Mol Struct 1147:1–6

Niraimathee VA, Subha V (2016) Green synthesis of iron oxide nanoparticles from *Mimosa pudica* root extract. Int J Environ Sustain Dev 15(3):227–240. https://doi.org/10.1504/IJESD.2016.077370

Noohpisheh Z, Amiri H, Mohammadi A, Farhadi S (2021) Effect of the foliar application of zinc oxide nanoparticles on some biochemical and physiological parameters of Tr*igonella foenum-graecum* under salinity stress. Plant Biosyst 155(2):267–280. https://doi.org/10.1080/11263504.2020.1739160

Ong CB, Ng LY, Mohammad AW (2018) A review of ZnO nanoparticles as solar photocatalysts: synthesis, mechanisms and applications. Renew Sust Energ Rev 81:536–551

Parial D, Pal R (2015) Biosynthesis of monodisperse gold nanoparticles by green alga Rhizoclonium and associated biochemical changes. J Appl Phycol 27:975–984

Parial D, Patra HK, Roychoudhury P et al (2012) Gold nanorod production by cyanobacteria—a green chemistry approach. J Appl Phycol 24:55–60

Park HJ, Kim SH, Kim HJ et al (2006) A new composition of nanosized silica-silver for control of various plant diseases. Plant Pathol J 22(3):295–302

Parveen A, Rao S (2015) Effect of nanosilver on seed germination and seedling growth in *Pennisetum glaucum*. J Clust Sci 26:693–701

Perreault F, Samadani M, Dewez D et al (2014a) Effect of soluble copper released from copper oxide nanoparticles solubilisation on growth and photosynthetic processes of *Lemna gibba* L. Nanotoxicology 8(4):374–382

Perreault F, Tousley ME, Elimelech M (2014b) Thin-film composite polyamide membranes functionalized with biocidal graphene oxide nanosheets. Environ SciTech Lett 1(1):71–76

Piccinno F, Gottschalk F, Seeger S et al (2012) Industrial production quantities and uses of ten engineered nanomaterials in Europe and the world. J Nanoparticle Res 14(9):1–11

Pradhan S, Patra P, Das S et al (2013) Photochemical modulation of biosafe manganese nanoparticles on Vigna radiata: a detailed molecular, biochemical, and biophysical study. Environ Sci Technol 47(22):13122–13131

Prakash A, Sharma S, Ahmad N et al (2010) Bacteria mediated extracellular synthesis of metallic nanoparticles. Int Res J Biotechnol 1(5):071–079

Prasad TN, Sudhakar P, Sreenivasulu Y et al (2012) Effect of nanoscale zinc oxide particles on the germination, growth and yield of peanut. J Plant Nutr 35(6):905–927

Qi M, Liu Y, Li T et al (2013) Nano-TiO2 improve the photosynthesis of tomato leaves under mild heat stress. Biol Trace Elem Res 156:323–328

Rahman A, Kumar S, Nawaz T (2020) Biosynthesis of nanomaterials using algae. In: Microalgae cultivation for biofuels production. Academic Press, Elsevier Inc, pp 265–279. https://doi.org/10.1016/B978-0-12-817536-1.00017-5

Rahman MS, Chakraborty A, Kibria A et al (2023) Effects of silver nanoparticles on seed germination and growth performance of pea (*Pisum sativum*). Plant Nano Biol 5:100042

Rai-Kalal P, Jajoo A (2021) Priming with zinc oxide nanoparticles improve germination and photosynthetic performance in wheat. Plant Physiol Biochem 160:341–351

Rajathi FA, Parthiban C, Kumar VG, Anantharaman P (2012) Biosynthesis of antibacterial gold nanoparticles using brown alga, *Stoechospermum marginatum* (kützing). Spectrochim Acta A Mol Biomol Spectrosc 99:166–173

Raliya R, Tarafdar JC (2013) ZnO nanoparticle biosynthesis and its effect on phosphorous-mobilizing enzyme secretion and gum contents in Clusterbean (*Cyamopsis tetragonoloba* L.). Agric Res 2:48–57

Raliya R, Saharan V, Dimkpa C et al (2017) Nanofertilizer for precision and sustainable agriculture: current state and future perspectives. J Agric Food Chem 66(26):6487–6503

Rautela A, Rani J, Debnath MC (2019) Green synthesis of silver nanoparticles from Tectona grandis seeds extract: characterization and mechanism of antimicrobial action on different microorganisms. J Anal Sci Technol 10:5. https://doi.org/10.1186/s40543-018-0163-z

Rezaei M, Azarmi M (2020) Deepsocial: social distancing monitoring and infection risk assessment in covid-19 pandemic. Appl Sci 10(21):7514

Saka A, Shifera Y, Jule LT et al (2022) Biosynthesis of TiO_2 nanoparticles by Caricaceae (Papaya) shell extracts for antifungal application. Sci Rep 12:15960. https://www.nature.com/articles/s41598-022-19440-w

Samad A, Khan MJ, Shah Z et al (2014) Determination of optimal duration and concentration of zinc and phosphorus for priming wheat seed. Sarhad J Agric 30(1):27–34

Saravanan M, Arokiyaraj S, Lakshmi T, Pugazhendhi A (2018) Synthesis of silver nanoparticles from *Phenerochaete chrysosporium* (MTCC-787) and their antibacterial activity against human pathogenic bacteria. Microb Pathog 117:68–72

Schrand AM, Rahman MF, Hussain SM et al (2010) Metal-based nanoparticles and their toxicity assessment. Nanomed Nanobiotech 2(5):544–568

Selvarajan E, Mohanasrinivasan VJ (2013) Biosynthesis and characterization of ZnO nanoparticles using *Lactobacillus plantarum* VITES07. Mater Lett 112:180–182

Semenova NA, Burmistrov DE, Shumeyko SA, Gudkov SV (2024) Fertilizers based on nanoparticles as sources of macro-and microelements for plant crop growth: a review. Agronomy 14(8):1646

Servin AD, White JC (2016) Nanotechnology in agriculture: next steps for understanding engineered nanoparticle exposure and risk. NanoImpact 1:9–12

Shallan MA, Hassan H, Namich AA et al (2016) Biochemical and physiological effects of TiO2 and SiO2 nanoparticles on cotton plant under drought stress. Res J Pharma Biol Chem Sci 7(4): 1540–1551

Shameli K, Ahmad MB, Al-Mulla EAJ et al (2012) Green biosynthesis of silver nanoparticles using *Callicarpa maingayi* stem bark extraction. Molecules 17:8506–8517. https://doi.org/10.3390/molecules17078506

Sharma P, Bhatt D, Zaidi MG et al (2012) Silver nanoparticle-mediated enhancement in growth and antioxidant status of *Brassica juncea*. Appl Biochem Biotechnol 167:2225–2233

Sharma R, Dewanjee S, Kole C et al (2016) Utilization of nanoparticles for plant protection. In: Plant nanotechnology: Princ Pract. Springer, pp 305–327

Shivaji S, Madhu S, Singh S (2011) Extracellular synthesis of antibacterial silver nanoparticles using psychrophilic bacteria. Process Biochem 46(9):1800–1807

Siddiqui MH, Al-Whaibi MH (2014) Role of nano-SiO2 in germination of tomato (*Lycopersicum esculentum* seeds Mill.). Saudi J Biol Sci 21(1):13–17

Singh PK, Kundu S (2014) Biosynthesis of gold nanoparticles using bacteria. Proc Natl Acad Sci India, Sect B Biol Sci 84:331–336. https://doi.org/10.1007/s40011-013-0230-6

Singh A, Jain D, Upadhyay MK et al (2010) Green synthesis of silver nanoparticles using Argemone mexicana leaf extract and evaluation of their antimicrobial activities. Dig J Nanomater Biosci 5:483–489

Singh R, Shedbalkar UU, Wadhwani SA, Chopade BA (2015) Bacteriagenic silver nanoparticles: synthesis, mechanism, and applications. Appl Microbiol Biotechnol 99:4579–4593. https://doi.org/10.1007/s00253-015-6622-1

Singh J, Kumar S, Alok A et al (2019) The potential of green synthesized zinc oxide nanoparticles as nutrient source for plant growth. J Clean Prod 214:1061–1070

Singh N, Tang Y, Zhang Z et al (2020) COVID-19 waste management: effective and successful measures in Wuhan, China. Resour Conserv Recycl 163:105071

Solanki P, Laura JS (2018) Effect of ZnO nanoparticles on seed germination and seedling growth in wheat (Triticum aestivum). J Pharmacogn Phytochem 7(5):2048–2052

Solanki P, Bhargava A, Chhipa H et al (2015) Nano-fertilizers and their smart delivery system. In: Nanotechnologies in food and agriculture. Springer, pp 81–101

Soto KM, Quezada-Cervantes CT, Hernández-Iturriaga M et al (2019) Fruit peels waste for the green synthesis of silver nanoparticles with antimicrobial activity against foodborne pathogens. LWT Food Sci Technol 103:293–300. https://doi.org/10.1016/j.lwt.2019.01.023

Sreedevi A, Manasa GL, Divya B et al (2020) Eco-friendly synthesis of silver nanoparticles using seeds of *Sorghum bicolor* research. J Pharm Tech 13(12):5935–5938. https://doi.org/10.5958/0974-360X.2020.01036.7

Subramaniyam V, Subashchandrabose SR, Thavamani P et al (2015) Chlorococcum sp. MM11—a novel phyco-nanofactory for the synthesis of iron nanoparticles. J Appl Phycol 27(5): 1861–1869

Suman TY, Rajasree SRR, Ramkumar R et al (2014) The Green synthesis of gold nanoparticles using an aqueous root extract of *Morinda citrifolia* L. Spectrochim Acta A Mol Biomol Spectrosc 118:11–16. https://doi.org/10.1016/j.saa.2013.08.066

Sun D, Hussain HI, Yi Z et al (2016) Mesoporous silica nanoparticles enhance seedling growth and photosynthesis in wheat and lupin. Chemosphere 152:81–91

Sun L, Liu H, Ye Y, Lei Y, Islam R, Tan S, Cai L (2023) Smart nanoparticles for cancer therapy. Signal Transduct Target Ther 8(1):418

Suriyakala G, Sathiyaraj S, Babujanarthanam R et al (2022) Green synthesis of gold nanoparticles using *Jatropha integerrima* Jacq. flower extract and their antibacterial activity. J King Saud Univ Sci 34(3):101830

Suriyaprabha R, Karunakaran G, Yuvakkumar R et al (2012) Silica nanoparticles for increased silica availability in maize (*Zea mays*. L) seeds under hydroponic conditions. Curr Nanosci 8(6): 902–908

Tarafdar JC, Raliya R (2013) Rapid, low-cost, and ecofriendly approach for iron nanoparticle synthesis using Aspergillus oryzae TFR9. J Nanoparticles 2013:1–4. https://doi.org/10.1155/2013/141274

Thabet SG, Alqudah AM (2024) Unraveling the role of nanoparticles in improving plant resilience under environmental stress condition. Plant Soil 503:1–8

Torabian S, Zahedi M, Khoshgoftarmanesh A (2018) Effect of foliar spray of zinc oxide on some antioxidant enzymes activity of sunflower under salt stress. J Agric Sci Technol 18:1013–1025

Tripathi S, Sonkar SK, Sarkar S (2011) Growth stimulation of gram (*Cicer arietinum*) plant by water soluble carbon nanotubes. Nanoscale 3:1176–1181

Trujillo-Reyes J, Vilchis-Nestor AR, Majumdar S et al (2013) Citric acid modifies surface properties commercial CeO2 nanoparticles reducing their toxicity and cerium uptake in radish (*Raphanus sativus*) seedlings. J Hazard Mater 263:677–684

Ukidave VV, Ingale L (2022) Green synthesis of zinc oxide nanoparticles from Coriandrum sativum and their use as fertilizer on bengal gram, turkish gram, and green gram plant growth. Int J Agron 2022:1–14. https://doi.org/10.1155/2022/8310038

Ul Haq I, Ijaz S (2019) Use of metallic nanoparticles and nanoformulations as nanofungicides for sustainable disease management in plants. In: Nanobiotechnology in bioformulations. Springer, pp 289–316

Vanaja M, Rajeshkumar S, Paulkumar K et al (2013) Phytosynthesis and characterization of silver nanoparticles using stem extract of *Coleus aromaticus*. Int J Mater Biomater Appl 3:1–4

Vazquez-Muñoz R, Meza-Villezcas A, Fournier PG et al (2019) Enhancement of antibiotics antimicrobial activity due to the silver nanoparticles impact on the cell membrane. PLoS One 14(11):e0224904

Verma SK, Das AK, Patel MK et al (2018) Engineered nanomaterials for plant growth and development: a perspective analysis. Sci Total Environ 630:1413–1435

Vigneshwaran N, Ashtaputre NM, Varadarajan PV et al (2007) Biological synthesis of silver nanoparticles using the fungus *Aspergillus flavus*. Mater Lett 61(6):1413–1418

Vijayakumar M, Priya K, Nancy FT et al (2013) Biosynthesis, characterisation and anti-bacterial effect of plant-mediated silver nanoparticles using *Artemisia nilagirica*. Ind Crop Prod 41:235–240. https://doi.org/10.1016/j.indcrop.2012.04.017

Wang X, Han H, Liu X et al (2012) Multi-walled carbon nanotubes can enhance root elongation of wheat (*Triticum aestivum*) plants. J Nanopart Res 14:1

Wang C, Kim YJ, Singh P et al (2016) Green synthesis of silver nanoparticles by *Bacillus methylotrophicus*, and their antimicrobial activity. Artif Cells Nanomed Biotechnol 44(4): 1127–1132

Wang JW, Grandio EG, Newkirk GM et al (2019) Nanoparticle-mediated genetic engineering of plants. Mol Plant 12(8):1037–1040

Xu C, Cao L, Zhao P et al (2018) Emulsion-based synchronous pesticide encapsulation and surface modification of mesoporous silica nanoparticles with carboxymethyl chitosan for controlled azoxystrobin release. Chem Eng J 348:244–254

Yang F, Hong F, You W et al (2006) Influence of nano-anatase TiO2 on the nitrogen metabolism of growing spinach. Biol Trace Elem Res 110:179–190

Yi Z, Hussain HI, Feng C et al (2015) Functionalized mesoporous silica nanoparticles with redox-responsive short-chain gatekeepers for agrochemical delivery. ACS Appl Mater Interfaces 7(18):9937–9946

Zabet-Khosousi A, Dhirani AA (2008) Charge transport in nanoparticle assemblies. Chem Rev 108(10):4072–4124

Zhao P, Cao L, Ma D et al (2018) Translocation, distribution and degradation of prochloraz-loaded mesoporous silica nanoparticles in cucumber plants. Nanoscale 10(4):1798–1806

Zheng H, Lu H (2011) Effect of microwave pretreatment on the kinetics of ascorbic acid degradation and peroxidase inactivation in different parts of green asparagus (*Asparagus officinalis* L.) during water blanching. Food Chem 128(4):1087–1093

Zheng L, Hong F, Lu S et al (2005) Effect of nano-TiO2 on strength of naturally aged seeds and growth of spinach. Biol Trace Elem Res 104:83–91

Zhou XQ, Hayat Z, Zhang DD, Li MY, Hu S, Wu Q et al (2023) Zinc oxide nanoparticles: synthesis, characterization, modification, and applications in food and agriculture. Processes 11(4):1193

Chapter 3
Nanoencapsulation of Agricultural Inputs

Gracesuganthi Jayaraj, Mohanapriya Balasubramaniam, and Kamaraj Raju

Abstract The integration of nanotechnology into agriculture has emerged as a transformative approach to enhance the efficiency and effectiveness of agricultural inputs. Nanoencapsulation of agricultural inputs such as nano fertilizers, nano pesticides, and nano herbicides offer a promising solution to several challenges faced in modern farming. Nanoencapsulation involves embedding these inputs within nanoscale carriers, which allows for controlled release, improved bioavailability, and reduced environmental impact. This method not only enhances the targeted delivery of agrochemicals, but also minimizes the quantities needed, thus lowering costs and mitigating potential adverse effects on non-target organisms and ecosystems. Recent advancements in nanomaterials, including nanoparticles, nano capsules, and nanocomposites, have shown remarkable potential in improving the stability and release profiles of agricultural products. By modifying the physical and chemical properties of these inputs at the nanoscale, it is possible to achieve more precise control over their release rates and interaction with plant systems. This leads to increased nutrient uptake, enhanced pest and disease control, and reduced leaching into groundwater. The development of nano encapsulated Agri-inputs is also aligned with the principles of sustainable agriculture, as it promotes the efficient use of resources and reduces the environmental footprint of conventional practices. This chapter addresses the current state of nanoencapsulation technologies, preparation techniques, their applications in agriculture, and the benefits they offer in the modern real-life practices. It also addresses the potential challenges and future directions in the field, emphasizing the need for continued research and innovation to fully realize the benefits of nanotechnology in agriculture.

Keywords Bioavailability · Nano capsules · Nanocomposites · Nano fertilizers · Sustainable agriculture · Targeted delivery

G. Jayaraj
Department of Biotechnology, Rathinam Technical Campus, Coimbatore, Tamil Nadu, India

M. Balasubramaniam · K. Raju (✉)
Department of Biotechnology, Rathinam College of Arts and Science, Coimbatore, Tamil Nadu, India

J. M. Al-Khayri et al. (eds.), *Nanotechnology Applications in Modern Agriculture*, Nanotechnology in Plant Sciences 3, https://doi.org/10.1007/978-3-031-90506-3_3

3.1 Introduction

The agricultural sector is at the critical juncture where it must adapt to the increasing demands for food production while simultaneously addressing environmental sustainability and resource efficiency. Traditional agricultural practices, reliant on conventional fertilizers, pesticides, and herbicides, have faced scrutiny due to their potential negative impacts on soil health, water resources, and ecosystems. In response to these challenges, the application of nanotechnology in agriculture has emerged as a revolutionary approach, promising to enhance the efficiency and sustainability of agricultural inputs through nanoencapsulation (Shaddel et al. 2024). Nanoencapsulation involves the incorporation of agricultural inputs such as nutrients, pest control agents, and herbicides into nanoscale carriers. These carriers can range from nanoparticles to nano capsules and nanocomposites, engineered to provide controlled release and targeted delivery of their contents. The primary advantage of nanoencapsulation lies in its ability to improve the stability and bioavailability of these inputs, allowing for precise application and reducing the quantities required. This results in minimized wastage, lower application rates, and reduced environmental impact (Aphibanthammakit and Kasemwong 2021).

One of the key benefits of nanoencapsulation is the enhancement of the targeted delivery of agrochemicals. Traditional methods often suffer from issues such as rapid degradation, leaching, and non-specific distribution, which can lead to inefficiencies and environmental harm. By contrast, nanoencapsulation ensures that the active ingredients are released gradually and precisely at the site of action, optimizing their efficacy and reducing off-target effects. This approach not only improves the performance of agricultural inputs but also supports sustainable farming practices by decreasing the overall load of chemicals applied to crops. Furthermore, the use of nano encapsulated inputs aligns with the principles of precision agriculture, which seeks to optimize resource use and minimize environmental impacts through advanced technologies. As research and development in nanotechnology continue to advance, the potential applications of nanoencapsulation in agriculture are expanding, offering solutions that promise to revolutionize crop management and agrochemical delivery (Ezhilarasi et al. 2013).

The main objective of this chapter is to investigate how nanotechnology can enhance the efficiency, delivery, and sustainability of agricultural products like fertilizers and pesticides. The chapter begins by defining nanoencapsulation and its significance in agriculture, followed by an explanation of the types of nanoparticles and their techniques, such as emulsification and spray-drying. It highlights the advantages of nanoencapsulation, including improved bioavailability, targeted delivery, and reduced environmental impact (Sheta et al. 2024). The chapter presents real-world applications, demonstrating the positive effects on crop yield and pest control. Additionally, it addresses challenges such as potential risks, regulatory issues, and economic considerations. Finally, it explores future directions for research and development, emphasizing the integration of nanoencapsulation with precision agriculture. Overall, the chapter aims to provide a comprehensive understanding of how this innovative technology can revolutionize agricultural practices.

3.2 Types of Nanoencapsulation Systems

Nanoencapsulation systems can be classified into several types based on their structure, composition, and method of preparation. Common systems include polymeric nanocarriers, liposomes, solid lipid nanoparticles, and inorganic nanoparticles. Polymeric nanocarriers, often made from biodegradable materials, provide controlled release and targeted delivery. Liposomes, formed by phospholipid bilayers, can encapsulate both hydrophilic and hydrophobic substances, enhancing bioavailability (Zhang et al. 2024). Solid lipid nanoparticles combine the benefits of solid matrices with lipid-based systems, offering stability and sustained release. Inorganic nanoparticles, such as silica or metal nanoparticles, are utilized for their unique properties and can serve as carriers for various agrochemicals. Each type of nanoencapsulation system offers distinct advantages and is tailored for specific agricultural applications, ultimately aiming to improve the efficiency and effectiveness of agricultural inputs (Albuquerque et al. 2022).

3.2.1 Nano Emulsions

Nano emulsions are stable dispersions composed of water and oil, which are held together by a surfactant layer and a cosurfactant. They have low viscosity, are transparent, and exhibit isotropic properties. The term “nano” is used to describe these particles because of their small size, which usually falls within the range of 5–100 nm. Nano emulsions possess numerous benefits compared to conventional emulsions as a result of their minute droplet size. One such advantage is their high optical clarity, wherein nano emulsions exhibit transparency or translucency due to the small droplet size, hence reducing light scattering (Huang et al. 2024). It provides excellent physical stability. The diminutive droplet size offers an expansive surface area for the absorption of surfactants, hence bolstering resistance against gravitational separation and the amalgamation of droplets. The reduced droplet size enhances the amount of surface area that can come into contact with biological membranes, resulting in enhanced absorption and bioavailability of encapsulated compounds (Anandharamakrishnan 2014).

In agriculture, nano emulsions have the potential to enhance the efficiency of pesticides and fertilizers by enabling controlled release, hence minimizing their environmental footprint. Nano emulsions have the capacity to encapsulate and transport hydrophobic medicines, hence enhancing their solubility, stability, and targeted delivery to specific locations. The food sector can benefit from the use of nano emulsions to increase the texture, look, and stability of food items (Delfanian et al. 2018). Additionally, nano emulsions can improve the bioavailability of nutrients and bioactive chemicals. Nano emulsions have the ability to improve the effectiveness and appearance of cosmetics and personal care products by enhancing the transport of active substances. Nano emulsions are effective in eliminating pollutants from water and soil because of their large surface area and capacity to dissolve hydrophobic substances. Nano emulsions can be prepared using various methods, including high-pressure homogenization method which forces the emulsion through a small orifice at high pressure, creating intense shear forces that break down large droplets into smaller ones (Shaddel et al. 2024). Ultrasonication is a method uses high-frequency sound waves to develop cavitation bubbles in the emulsion, which collapse and generate intense shear forces that break down droplets (Bamisaye et al. 2023).

3.2.2 Nanoparticles

Nanoparticles, due to their distinct physical and chemical characteristics, have significantly transformed multiple domains, including agriculture. When included into nanoencapsulation systems, these particles provide sophisticated solutions for the transportation of agrochemicals, improving effectiveness and environmental friendliness (Xie et al. 2024). Nanoencapsulation is the process of enclosing active chemicals within nanoparticles in order to regulate their release, shield them from environmental degradation, and enhance their bioavailability. Here, examines the function of nanoparticles in nanoencapsulation in agriculture, focusing on their many forms, advantages, uses, and potential future developments. Their small size allows nanoparticles to penetrate biological barriers, making them suitable for drug delivery and imaging (Mittal et al. 2020).

3.2.2.1 Types of Nanoparticles in Agriculture

(a) **Silver nanoparticles (AgNPs)** possess antibacterial properties and are utilized in nanoencapsulation for enhanced delivery of insecticides or fertilizers. Agrochemicals can be enhanced by stabilizers, which can also facilitate regulated release, hence enhancing the effectiveness of pest control and nutrient delivery (Dai et al. 2024).
(b) **Gold nanoparticles**, also known as AuNPs, are used in nanoencapsulation because they are biocompatible and easy to modify. They are utilized for administering plant growth regulators and fertilizers, providing improved absorption and minimized environmental impact (Xu et al. 2024).
(c) **Zinc Oxide Nanoparticles (ZnO NPs)** has antibacterial and UV-blocking characteristics, rendering them appropriate for the encapsulation of insecticides and fungicides. They aid in regulating the release of these compounds, amplifying their efficacy while diminishing the likelihood of resistance formation (Amin et al. 2023).
(d) **Titanium Dioxide Nanoparticles (TiO_2 NPs)** are utilized in nanoencapsulation due to their photocatalytic characteristics. They can be utilized to administer

agrochemicals in a regulated manner, especially when exposed to sunshine, which might trigger their capabilities and enhance their effectiveness (Mirara et al. 2024).

(e) **Polymeric nanoparticles** are formed by utilizing polymers like polylactic acid (PLA) and poly(lactic-co-glycolic acid) (PLGA) for the purpose of nanoencapsulation. These biodegradable organic nanoparticles can be customized to release agrochemicals in a regulated manner, thereby enhancing efficiency and minimizing the need for frequent applications (Vemula and Reddy 2023).

(f) **Lipid-based nanoparticles**, such as solid lipid nanoparticles (SLNs) and nanostructured lipid carriers (NLCs), are employed for the encapsulation of hydrophobic chemicals, such as insecticides and herbicides. These nanoparticles improve the capacity of the active substances to dissolve and remain stable, guaranteeing efficient delivery and regulated release (Alruwaili et al. 2024).

Nanoparticles play a crucial role in the nanoencapsulation of agrochemicals, offering significant benefits in terms of stability, bioavailability, controlled release, and reduced environmental impact. Their application in agriculture holds great promise for improving the efficiency and sustainability of farming practices. As research and development progress, the integration of nanoparticles in agricultural nanoencapsulation systems is likely to become increasingly prevalent, contributing to more effective and environmentally friendly agricultural solutions (Singh et al. 2021).

3.2.3 Nanogels

Nanogels represent a cutting-edge technology in nanomaterials, offering versatile and functional solutions across various industries. In agriculture, nanogels are emerging as a powerful tool for enhancing the delivery and efficiency of agrochemicals, improving soil health, and advancing sustainable practices. This explores the role of nanogels in agriculture, focusing on their properties, benefits, and applications, while also addressing potential challenges and future directions (Li et al. 2024). Nanogels are nanometer-sized hydrogel particles that can swell and contract in response to environmental stimuli such as pH, temperature, or light. They are composed of cross-linked polymer networks, which provide them with a high-water content and the ability to encapsulate and release various substances (Soni et al. 2016).

The unique properties of nanogels make them particularly suitable for applications in agriculture where controlled release and targeted delivery are critical. Nanogels have a substantial water holding capacity, which makes them ideal for encapsulating water-soluble substances such as nutrients, pesticides, and growth regulators. They can change their size and release rate in response to environmental stimuli, allowing for tailored release profiles that match the needs of crops or soil conditions. Many nanogels are made from biocompatible materials, which reduces the risk of toxicity to plants and soil organisms. The large surface area of nanogels enhances their ability to adsorb and release active substances effectively (Sindhu et al. 2022).

Nanogels offer innovative solutions for enhancing agricultural practices by improving the delivery, efficacy, and sustainability of agrochemicals. Their controlled release, enhanced bioavailability, and potential for soil conditioning make them valuable tools for modern agriculture. As research advances, addressing the challenges of cost, safety, and scalability will be essential for maximizing the benefits of nanogels. With continued innovation, nanogels are poised to play a significant role in the future of sustainable agriculture (Salvati et al. 2022; Li et al. 2024).

3.2.4 Nano Capsules

Nano capsules represent a significant advancement in agricultural technology, offering new possibilities for enhancing the delivery and effectiveness of agrochemicals. These nanoscale carriers, designed to encapsulate and release active substances, provide controlled and targeted delivery, improved stability, and enhanced bioavailability. Nano capsules are nanometer-sized carriers that encapsulate active substances within a protective shell. These carriers can be composed of various materials, including polymers, lipids, and inorganic substances. The core-shell structure of nano capsules allows for the controlled release of encapsulated agents, which can be tailored to meet specific agricultural needs (Sajjad Babaei et al. 2022; Masoumeh Ghorbani et al. 2024).

3.2.4.1 Types of Nano Capsules

(i) **Polymer-Based Nano capsules**

(a) **Polymeric Micelles:** These are self-assembled nano capsules formed from amphiphilic block copolymers. The hydrophobic core encapsulates the active substances, while the hydrophilic shell stabilizes the nano capsule in aqueous environments. They are used for encapsulating both hydrophobic and hydrophilic agents (Sun et al. 2024).

(b) **Nano capsules from Polymers:** Biodegradable polymers such as polylactic acid (PLA) and poly(lactic-co-glycolic acid) (PLGA) are commonly used. These nano capsules offer controlled and sustained release of agrochemicals, reducing the frequency of application (Xu et al. 2023).

(ii) **Lipid-Based Nano capsules**

(a) **Liposomes:** Liposomes are spherical vesicles with a phospholipid bilayer that can encapsulate hydrophilic substances in their aqueous core. They are effective for delivering water-soluble agrochemicals and protecting them from environmental degradation (Xing et al. 2016).

(b) **Solid Lipid Nanoparticles (SLNs) and Nanostructured Lipid Carriers (NLCs):** These lipid-based nano capsules have solid lipid cores and are used to encapsulate hydrophobic substances. They offer controlled release and stability for pesticides and fertilizers (Hamed et al. 2024).

(iii) **Inorganic Nano capsules**

(a) **Silica Nano capsules:** Silica-based nano capsules are known for their stability and ability to be functionalized. They are used for encapsulating a wide range of agrochemicals and can provide controlled release and targeted delivery (Fathi et al. 2024).

(b) **Metal Nano capsules:** Metal-based nano capsules, such as those made from gold or silver, can offer unique properties such as antimicrobial activity and enhanced stability. They are used in applications where additional functionality is required (Venezia et al. 2024).

Nano capsules offer a transformative approach to enhancing agricultural practices by improving the delivery, stability, and efficacy of agrochemicals. Their ability to provide controlled release, enhance bioavailability, and enable targeted delivery makes them valuable tools for modern agriculture. With continued innovation, nano capsules are poised to play a significant role in advancing sustainable and efficient agricultural practices (Staff et al. 2013).

3.3 Nanoencapsulation of Specific Agricultural Inputs

The techniques of nanoencapsulation of fertilizers is an advanced method that aims to enhance the delivery and usage of nutrients in agriculture. By encapsulating fertilizers in nanocarriers, it is feasible to optimize nutrient utilization and mitigate ecological consequences. This strategy is in line with the ideals of sustainable agriculture as it reduces waste and promotes more accurate nutrient management. Nanoencapsulation can greatly enhance the efficacy of nutrient absorption in plants (Aphibanthammakit and Kasemwong 2021). Conventional fertilizers frequently experience quick leaching and volatilization, resulting in the loss of nutrients and degradation of the environment. Nanoencapsulation resolves these concerns by offering precise and prolonged release of nutrients (Singh et al. 2024). Figure 3.1 depicts the application of various nano encapsulated products.

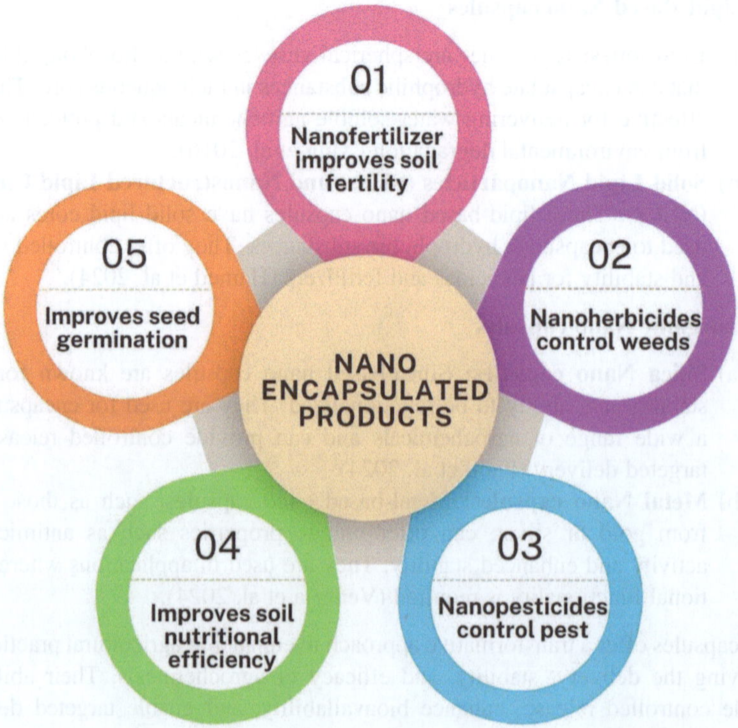

Fig. 3.1 Application of Nano encapsulated products. (Figure constructed by Gracesuganthi Jayaraj)

3.3.1 Nanoencapsulation of Fertilizers

Fertilizers with nanoencapsulation release nutrients progressively over time to correspond with the absorption rates of the plants and minimize the frequency of applications. This prolonged release lessens the impact on the ecosystem and minimizes nutrient leakage into groundwater. Nutrient solubility can be increased by nanoparticles, increasing the nutrients' accessibility to plant roots (Jeon et al. 2016). Because of its increased availability, nutrients are absorbed more efficiently, resulting in a lower need for fertilizer. Nanoencapsulation reduces runoff and nutrient leaching, which helps to ameliorate problems like soil degradation and eutrophication. This helps to promote more environmentally friendly farming methods (Mahato et al. 2021).

Due to the nano size of particles, they have a significant surface area, which facilitates various metabolic processes in the plant system, leading to increased synthesis of photosynthetic products. Due to their extensive surface area and extremely small size, these particles exhibit a high level of reactivity with other chemicals. They exhibit a high degree of solubility in solvents, particularly water.

Nano-fertilizers with particle sizes less than 100 nm have a greater ability to permeate into the plant. Nano fertilizers enhance the absorption capacity and efficiency of nutrient use by virtue of their substantial surface area. Nano-particles containing fertilizers will enhance the speed at which nutrients are made available and absorbed by plants (Tarafdar 2012).

Urea, a frequently utilized nitrogen fertilizer, is susceptible to rapid loss by volatilization. The process of encapsulating urea in either polymeric or inorganic nanoparticles allows for the controlled release of urea, thereby providing a prolonged availability of nitrogen to plants. Researchers had demonstrated that the application of urea-coated nanoparticles enhances the efficiency of nitrogen utilization and diminishes the release of nitrogen oxides. The application of urea coated with ZnO nanoparticles resulted in a rise in the nitrogen content in the shoot and a probable enhancement of the zinc content in wheat grain, thus potentially improving its nutritional value (Amin et al. 2023). Phosphorus and potassium are essential elements for the growth of plants. The process of nanoencapsulation improves the accessibility and absorption of these nutrients. Nano-formulated phosphorus fertilizers exhibit enhanced solubility and absorption by crops in comparison to conventional phosphorus fertilizers, resulting in superior growth and output (Kotp et al. 2018).

Nano-pores and stomatal apertures in leaves have been observed to facilitate the uptake and penetration of nanomaterials into leaves. The study determined that in broad/faba bean (*Vicia faba*), nano-sized particles (43 nm) were effective in deeply reaching the interior of the leaves in a greater quantity compared to larger particles measuring over 1.0 micrometers in size. Similarly, the size of the leaf stomata in Arabian coffee (*C. arabica*) was less than 2.5 nm, while the size of stomata in sour cherry (*P. cerasus*) was likewise less than 100 nm. Therefore, it was anticipated that nano-fertilizers would be successful in improving nutrient absorption (Naik et al. 2021).

There has been research done to develop an encapsulated fertilizer with a long-acting effect. It is demonstrated that the phosphate-containing coating's deep, tortuous nanopores permit soil moisture to enter the granule and subsequently dissolve the encapsulated granule's nitrogen core and organic portion of the shell (Yanovska et al. 2022). Because of their potential to increase crop productivity, decrease the use of chemical fertilizers, and lessen harmful effects on the soil, the development of nano-fertilizers is regarded as the most significant alternative to conventional fertilizers and pesticides.

3.3.2 Nanoencapsulation of Pesticides

Nanoencapsulation enables controlled and prolonged release of pesticides, so maintaining the efficacy of the active components for an extended duration. The implementation of this controlled release mechanism can result in a decrease in the frequency of pesticide application and a reduction in the overall amount of pesticide used. Nano encapsulated pesticides offer an extended-release mechanism, so prolonging their effectiveness and minimizing the necessity for frequent reapplications. As a result, there is improved pest management and a decrease in the overall utilization of pesticides. It also effectively decreases the highest levels of pesticides in the environment by regulating the rate at which they are released. This helps to minimize the harmful effects on non-target animals, such as beneficial insects and humans. Nanoencapsulation shields pesticides from environmental deterioration, such as hydrolysis and photodegradation, hence preserving their efficacy (Belagalla et al. 2024).

The researchers have shown that Copper oxide nanoparticles, which specifically target pests like *Fusarium oxysporum f. sp. Lycopersici* in tomatoes, have potent antifungal properties and are more effective than traditional copper-based fungicides (Bhattacharyya 2016). The efficacy of nano encapsulated deltamethrin against *Helicoverpaarmigera* and *Spodoptera litura* in cotton was demonstrated. Increased insecticidal efficacy and longer-lasting residual effect in comparison to traditional formulation (Chhipa 2016).

Ashfaq et al. investigated the potential of silver nanoparticles as a new type of insecticide to control the dengue vector mosquito, *Aedes aegypti*. The study revealed that silver nanoparticles displayed significant larvicidal and adulticidal properties, indicating their potential as a viable substitute for traditional insecticides in mosquito control (Ashfaq 2016). The effectiveness of nano-pesticides can be ascribed to several aspects, such as their improved ability to enter and move throughout plants, their ability to release chemicals in a regulated manner, and their ability to specifically target certain areas. The small size of the active substances at the nanoscale helps them to efficiently reach and interact with the specific pests or pathogens they are targeting. Additionally, being encapsulated within nanocarriers offers protection against early degradation and allows for a controlled and prolonged release over a period of time. Nevertheless, it is crucial to acknowledge that the effectiveness of nano-pesticides can differ based on the particular composition, intended insect or pathogen, crop type, and environmental circumstances (Kookana 2014). Hence, additional investigation is required to enhance the formulation and implementation strategies of nano-pesticides for various crop protection strategy.

Nano-pesticides have the potential to decrease the environmental consequences of pesticides by avoiding unintended effects and lowering the necessary application rates. Silver nanoparticles have exhibited potent antibacterial properties against a broad spectrum of plant diseases. Nano-fertilizers have the potential to enhance the efficiency of nutrient absorption and minimize nutrient wastage, resulting in increased agricultural productivity. Nano-sensors have the capability to identify and monitor pests and diseases in crops at an early stage, allowing for timely and focused interventions (Kah and Hofmann 2014; Mishra and Singh 2015). Nano-pesticides can be engineered to selectively target particular pests or pathogens, thereby minimizing their effects on beneficial organisms and biodiversity. Utilizing biodegradable polymers in nano-pesticide formulations can effectively reduce the long-lasting presence of pesticide residues in the environment. Nano-pesticides have the potential to enhance the efficiency of active ingredients by improving their penetration and translocation within plant tissues (de Oliveira et al. 2014; Wang et al. 2016).

The integration of nanotechnology into agriculture is a transformative advancement, particularly in the development of nano encapsulated insecticides, herbicides, and fungicides. This innovative method addresses critical challenges in contemporary crop management by improving the efficacy, safety, and environmental impact of agrochemicals. Nanoencapsulation technology provides insecticides with numerous benefits, such as enhanced controlled release and targeted delivery (Sun 2020). Nano encapsulated insecticides, which use nanoparticles of materials such as chitosan or silica, exhibit prolonged effectiveness by safeguarding the active ingredients from environmental degradation. For example, recent researchers had demonstrated that nano encapsulated insecticides can gradually release their active components, thereby reducing the frequency of applications and reducing the resistance development in pests. This method not only enhances pest control but also reduces the overall chemical burden in the environment (Graily-Moradi and Asgari Lajayer 2021).

Nanoparticles can encapsulate herbicides and release them in a regulated manner, increasing their efficacy and lowering the required dosage. The discovery of polymeric nanocarriers that transport herbicides specifically to target weeds has been a significant advancement in this field, reducing collateral damage to nearby crops and soil contamination. Precision application not only improves herbicide efficacy but also promotes sustainable agriculture practices. Nano-based herbicides can provide effective weed control while minimizing the impact on non-target plant species (Oliveira et al. 2015).

Nanoencapsulation also provides optimistic solutions for fungicides by enhancing their bioavailability, stability, and solubility. Encapsulated fungicides are capable of more effectively penetrating plant tissues and offer extended protection against fungal pathogens. Recent researchers have demonstrated that nano encapsulated fungicides can attain higher therapeutic indices than conventional formulations, resulting in improved disease control and reduced fungicide use. This method is particularly advantageous for the management of resistant fungal strains and the reduction of the environmental impact of fungicide applications. In general, the incorporation of nanoencapsulation technologies into the formulation of fungicides, herbicides, and insecticides is a substantial development in the field of agricultural chemistry. These technologies contribute to sustainable agricultural practices that are essential for meeting global food security demands while minimizing environmental impact by improving the precision and efficiency of agrochemical applications (Dias 2012; Bhattacharyya 2016).

Nano-pesticides are a promising method for improving the sustainability and efficacy of pest and disease management in agriculture. Compared to conventional pesticides, nano-pesticides have the potential to provide targeted delivery, controlled release, and a reduced environmental impact by utilizing the unique properties of nanomaterials.

3.3.3 Growth Regulators and Plant Hormones

The use of growth regulators and plant hormones in agriculture is being revolutionized by nanoencapsulation, which improves stress tolerance, enables targeted delivery, provides controlled and sustained release, and enhances their stability. These developments result in more effective and efficient crop management practices, which in turn contribute to increased yields and improved resilience to environmental stresses. Nanoencapsulation is on the brink of becoming an increasingly significant component of sustainable agriculture as technology continues to advance (Karabulut 2024).

(a) **Increased Stability and Durability**
The stability of plant hormones and growth regulators is significantly enhanced by nanoencapsulation. Numerous compounds are susceptible to degradation as a result of environmental factors, including heat, light, and oxygen. The hormones are protected from these adverse conditions by the encapsulation in nanomaterials, such as nanoparticles or nano capsules. For instance, nano encapsulated auxins demonstrate superior stability in comparison to free auxins, resulting in effective root development and protracted activity throughout a variety of environmental conditions (Edelmann 2022).

(b) **Sustained and Managed Release**
The capacity to regulate the release rate of encapsulated hormones is one of the primary benefits of nanoencapsulation. The need for frequent administrations can be reduced and efficiency can be improved by engineering nanoencapsulation systems to provide a sustained release of growth regulators over time. For example, encapsulated gibberellins provide a consistent release that encourages consistent fruit growth and yield, in contrast to conventional formulations that may necessitate multiple applications. This controlled release not only improves the hormones efficacy but also reduces their environmental impact (Ezhilarasi et al. 2013).

(c) **Targeted Delivery**
Nanoencapsulation facilitates the targeted delivery of growth regulators to specific plant tissues or developmental stages. The efficacy of the application can be enhanced by designing nanocarriers to discharge their contents precisely where and when they are required. For example, in order to improve germination and stress resistance, nanocarriers that encapsulate cytokinins can be directed to specific regions of the plant, such as stressed tissues or developing seeds. This targeted approach guarantees that the hormones are utilized more effectively and minimizes the probability of off-target effects (Noore et al. 2024).

(d) **Enhanced Stress Tolerance**
Nanoencapsulation improves the resistance of plants to environmental stresses. Encapsulated stress-responsive hormones, such as abscisic acid (ABA), offer long-term protection against abiotic stresses, including salinity and dehydration. Encapsulated ABA has been demonstrated to enhance overall stress tolerance by improving water use efficiency and reducing transpiration in

Table 3.1 Nano encapsulated products and their applications

Encapsulated Moieties	Nanocomponent	Applications	References
Nano fertilizer	Chitosan NPs; TiO2; Nanoparticles; Cu NPs	Plant growth and crop yield; Water retention and seed germination.	Chandrasekaran and Paramasivan (2024)
Nano pesticides	Porous hollow silica nanoparticles	Control pest at low dosage without damaging the crop.	Yadav et al. (2024)
Nano herbicides	Poly (e-caprolactone) nano capsule	Act as herbicides	Abd-Elsalam (2024)
Nanocarriers	Clay nanotubes	Delivery of fertilizers, pesticides, herbicides, plant growth regulators	Ullah et al. (2024)
Nano-bio stimulants	Ag-NPs; Nanosized Ag-silica hybrid complex (NSS);Carbon NP	Protection of crop plants from phytopathogens	Sadeghi et al. (2024)
Plant disease control	ZnO and MgO nanostructures	Anti-microbial/anti-bacterial	El-Sayed et al. (2024)

drought-stressed plants. Ultimately, this implementation of nanoencapsulation enhances crop productivity by enabling plants to more effectively withstand challenging conditions (Sampedro-Guerrero et al. 2023).

(e) **Novel Applications**

The development of innovative delivery systems that react to environmental stimuli is a recent development in nanoencapsulation technologies. For instance, pH-sensitive nanoparticles can release their contents in response to variations in soil pH, thereby guaranteeing that growth regulators are delivered at the optimal time for plant uptake. These advancements broaden the potential applications of nanoencapsulation in precision agriculture (Martins et al. 2022). Listed below the various application of encapsulated moieties with their application. Table 3.1 listed the nano encapsulated products and their applications.

3.4 Factors Influencing Nanoencapsulation Efficiency

Nanoencapsulation has become a critical technology for improving the efficacy of agrochemicals, such as plant hormones and growth regulators. Several critical factors, such as the stability and release kinetics of encapsulated compounds, the selection of appropriate nanocarrier materials, the choice of encapsulation techniques and parameters, and environmental factors that affect nano capsule performance, influence the efficiency of nanoencapsulation. It is essential to comprehend these factors in order to enhance the efficiency of nanoencapsulation processes and enhance their application in agriculture.

3.4.1 Selection of the Suitable Nanocarrier Materials

The efficacy of nanoencapsulation is contingent upon the selection of nanocarrier materials. Material that encapsulates active substances, nanocarriers serve to regulate their discharge and ensure them. Important factors to consider are as follows

(a) **Material Type**

Nanocarriers can be widely categorized as either inorganic (e.g., silica, metal nanoparticles) or organic (e.g., polymers, liposomes). Each type provides unique benefits to the active substances. Polymeric nanoparticles, which are organic nanocarriers, are frequently preferred due to their biodegradability and biocompatibility. For example, poly (lactic-co-glycolic acid) (PLGA) nanoparticles have been extensively employed for the encapsulation of plant hormones as a result of their minimal toxicity and controlled release property. Silica nanoparticles, for example, are inorganic nanocarriers that can be engineered to achieve specific release profiles and exhibit high stability (Hamed et al. 2024).

(b) **Surface Properties and Size**

The interaction between plant tissues and the efficacy of encapsulated substances is influenced by the surface characteristics and size of nanocarriers. Nanoparticles that are smaller in size (typically between 10 and 100 nm) are more effective in penetrating plant cell membranes. Surface modification, such as functionalization with ligands or polymers, can improve cellular uptake and targeting. In this regard, surface-functionalized silica nanoparticles have demonstrated enhanced targeting of plant roots, thereby improving nutrient uptake and growth regulation (Sinha et al. 2019).

(c) **Biodegradability and Toxicity**

The safety and environmental impact of nanocarriers are contingent upon their biocompatibility and biodegradability. Due to their environmental tolerance, biodegradable polymers are the preferred choice, as they decompose into non-toxic by-products. For instance, chitosan-based nanocarriers have been successfully employed to encapsulate growth regulators while posing minimal environmental risk and easily biodegradable (Mahato et al. 2021).

3.4.2 Encapsulation Techniques and Parameters

The efficiency of nanoencapsulation is impacted by the selection of encapsulation methodologies and process parameters. Multiple methodologies are frequently employed, each possessing distinct benefits based on the sample, which are mentioned below,

(a) **Solvent Evaporation**

This technique employed by dissolving the active substance and carrier material in a solvent, and subsequently eliminating the solvent, leading to the development of nano capsules. The selection of solvents, the rate of evaporation, and the temperature are crucial factors that influence the size and consistency of nanoparticles. An example of a successful method for developing uniform polymeric nanoparticles to encapsulate plant hormones such as auxins and gibberellins is solvent evaporation (Staff et al. 2013).

(b) **Coacervation**

Coacervation is the process of separating a solution that contains the active chemical and carrier material into different phases, resulting in the development of nano capsules. This method enables the achievement of a high level of encapsulation efficiency and may be fine-tuned by adjusting parameters such as temperature and pH. The technique of coacervation has proven effective in encapsulating several plant growth regulators, enhancing their stability and enabling regulated release (Suganya and Anuradha 2017).

(c) **High-Pressure Homogenization**

It is a method that utilizes high pressures to produce nano capsules by forcing the active component and carrier material through a small opening. It has a high level of effectiveness in generating nanoparticles that have consistent and homogeneous size distributions. High-pressure homogenization is a valuable technique for enclosing compounds that repel water and has been used to develop nanoparticles for the specific transport of growth regulators (Anandharamakrishnan 2014; Bashir et al. 2024).

(d) **Electrospinning**

It is a technique utilized to fabricate nanofibers capable of encapsulating active compounds. This method is beneficial for producing fibrous nanocarriers that have large surface areas. This, in turn, can increase the amount of encapsulated substances that can be loaded and improve the way they are released. Electrospinning has been investigated as a method for enclosing plant hormones, leading to enhanced regulation of their release and stimulation of plant development (Khajavi and Abbasipour 2017). Various preparation techniques are depicted in the Fig. 3.2.

3.4.3 Stability and Release Kinetics of Encapsulated Inputs

The stability and release kinetics of encapsulated compounds are critical factors influencing the performance of nanoencapsulation. The stability of encapsulated compounds relies on various factors, including the carrier material, encapsulation process, and storage conditions. Stable nanocarriers guarantee the preservation of the hormone activity over time. The shelf life and effectiveness of nano encapsulated plant hormones have been enhanced due to recent progress in stabilizing agents and coating materials. For instance, including antioxidant compounds into nanocarriers has been demonstrated to improve the stability of encapsulated auxins and gibberellins (Pereira et al. 2019). The release kinetics of encapsulated compounds dictate the efficiency with which they can gradually deliver their active constituents.

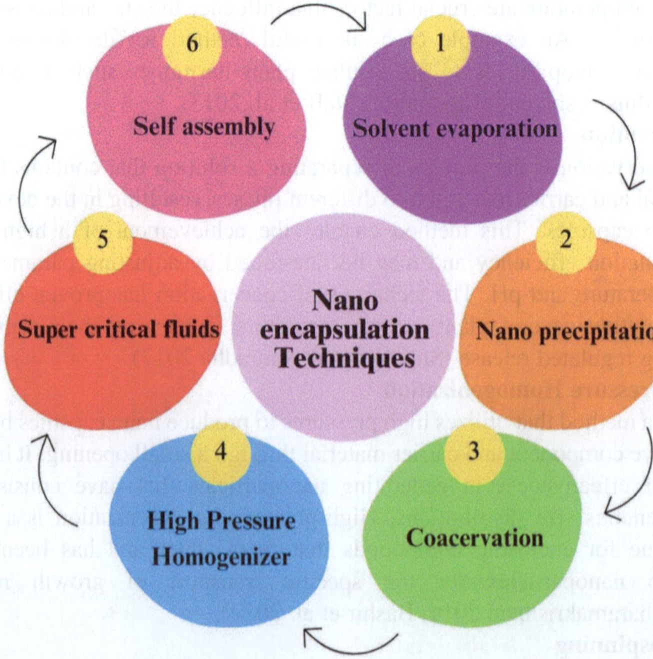

Fig. 3.2 Various nanoencapsulation preparation techniques. (Figure constructed by Gracesuganthi Jayaraj)

Controlled release is accomplished by carefully choosing suitable carrier materials and fine-tuning encapsulation characteristics. Modulating release kinetics can be achieved by altering elements such as the composition of the carrier matrix, the size of the particles, and the ambient conditions. For example, polymeric nanocarriers that have hydrophobic cores release growth regulators gradually, resulting in prolonged impacts on plant growth (Zahedi et al. 2024).

3.4.4 Environmental Factors Affecting Nanoencapsulation

Nanoencapsulation is significantly influenced by various environmental factors that can impact the stability, release profiles, and overall effectiveness of the encapsulated materials. Temperature plays a crucial role; higher temperatures can accelerate degradation processes and affect the solubility of both the core and shell materials (Singh et al. 2024). Humidity is another critical factor, as excessive moisture can lead to hydrolysis or agglomeration of nanoparticles, compromising their performance. pH levels in the surrounding environment can alter the charge and interaction of the nanocarriers, affecting their ability to encapsulate and release the active substances. Additionally, the presence of organic solvents or other chemicals can influence the encapsulation efficiency and stability of the nanoformulations (Sushil et al. 2021). Understanding and controlling these environmental parameters is essential for optimizing the nanoencapsulation process and ensuring the desired functionality of the final product. Several environmental factors can influence the activity of nano encapsulated compounds includes

(a) **pH and Temperature**

Alterations in pH and temperature can impact the stability and rate of release of nano capsules. Temperature variations can affect the pace at which encapsulated hormones are released, potentially changing their effectiveness. Similarly, alterations in pH levels within the soil or plant tissues might impact the breakdown and discharge of nanocarriers. Developing nano capsules with adaptive materials that can respond to changes in the environment can aid in reducing these impacts (Paula et al. 2016).

(b) **Soil and plant interactions**

The performance of nano capsules is influenced by their interaction with soil and plant tissues. Soil type, moisture content, and plant root properties can affect nano encapsulated chemicals are absorbed. Recent studies highlight the significance of nanocarriers to specific soil types and plant species in order to optimize their efficacy and reduce their ecological footprint (Masoumeh Ghorbani et al. 2024).

(c) **Biodegradation and environmental impact**

The assessment of the environmental impact of nanocarriers, which includes evaluating their possible toxicity and ability to persist in the environment, is a crucial factor to address. Biodegradable nanocarriers are favored due to their limited impact on the environment. Ongoing research is being conducted to investigate the environmental destiny and toxicity of nanocarriers, with the aim of ensuring that they do not present any dangers to ecosystems or human health (Sun et al. 2024).

3.5 Applications and Case Studies of Nano Encapsulated Inputs in Agriculture

Nanoencapsulation has greatly improved farming practices by boosting the delivery and effectiveness of many inputs, such as growth regulators, minerals, and insecticides. This technology has been utilized in the areas of foliar spray, seed treatment, and soil application, providing enhanced performance and sustainability (Masoumi et al. 2024). Described below are essential applications and practical cases of effective nanoencapsulation in the field of agriculture.

3.5.1 Foliar Application

Foliar application of nanoencapsulation involves the direct spraying of nano encapsulated inputs onto the leaves of plants. This approach delivers numerous benefits such as enhanced absorption and efficiency is one of the main applications of nanoencapsulation improves the absorption of active compounds by the leaf cuticle, hence enhancing the efficiency of delivering nutrients and hormones (Singh et al. 2024). For instance, nano encapsulated gibberellins have demonstrated enhanced leaf absorption and growth stimulation in comparison to traditional formulations. Nanoencapsulated controlled release was guaranteed a consistent delivery of active components, hence decreasing the need for frequent applications (Gouda et al. 2018).

Application of nano encapsulated growth regulators through foliar spraying has led to consistent enhancements in growth and production across different crop varieties. Applying nanoparticles to the leaves of plants enhances the efficiency of plant protection methods when compared to the conventional method of applying them to the soil and roots. Nanoparticles applied as a spray on the leaves primarily enter through the stomata and are then transported to various areas of the plant by both apoplastic and symplastic pathways. Foliar nanoparticles bolster the immune system and resilience against pests and diseases, while simultaneously enhancing the productivity and caliber of agricultural produce (Masoumi et al. 2024). Nevertheless, the precise processes by which foliar nanoparticles cause damage require additional clarification. Furthermore, it is crucial to explore key aspects that restrict the absorption of nanoparticles by foliage, such as the presence of wax deposits on leaf surfaces, ambient conditions (such as light, temperature, and humidity), and the physical and chemical characteristics of the nanoparticles. This research is necessary to enhance the effectiveness of this technology (Hong et al. 2021).

3.5.2 Seed Treatment

Nano encapsulated seed treatment comprises the application of nano encapsulated growth regulators or nutrients onto seeds prior to planting. This technique provides numerous advantages such as Nanoencapsulation enhances seed germination rates and early seedling growth through the controlled release of growth regulators. Nano encapsulated auxins and cytokinins have been employed to augment germination and seedling vigor in crops like maize and wheat (Yi et al. 2024). It also can encompass fungicides and insecticides as well, providing protective advantages throughout the initial stages of growth. Nano encapsulated fungicides have demonstrated efficacy in disease management while minimizing the impact on seedling vitality (Shelar et al. 2023).

Nanoencapsulation enables the effective transportation of active components straight to the seed, enhancing absorption and utilization. The nanoscale carriers can be customized to transport specific compounds to exact locations within the seed or soil (Hamed et al. 2024). This focused strategy minimizes inefficiency and guarantees that the active components are concentrated in the areas where they are most necessary, hence optimizing seed germination and early plant growth. Nanocarriers have the ability to safeguard seeds from diseases and pests, enhance germination rates, and promote robust root and shoot growth. Nanoencapsulation enhances the stability and prolongs the shelf life of seed treatments (Zhao et al. 2024b).

3.5.3 Soil Application

Nanoencapsulation enhances the effectiveness of inputs applied to soil by facilitating precise and gradual release, as well as focused delivery. This process guarantees the availability of nutrients, herbicides, and growth regulators to plants at the appropriate times, hence minimizing wastage and optimizing their effectiveness. Nanoencapsulation reduces the environmental impact of agricultural techniques by minimizing the amounts of pesticides used and enhancing their delivery. This method reduces the likelihood of runoff and leaching, which can lead to water contamination and soil damage. Nano encapsulated fertilizers and growth regulators have the potential to enhance soil health through the stimulation of beneficial microbial activity and the enhancement of soil structure (Tryfon et al. 2023). Increased nutrient accessibility and decreased chemical runoff leads to a more robust soil ecosystem. Optimal plant growth and development are facilitated by the proper administration of nutrients and growth regulators. Consequently, crops are more resilient to environmental stressors and able to achieve greater yields. Despite the higher initial cost, nanoencapsulation technology offers enhanced efficiency and less requirement for frequent applications, resulting in long-term cost benefits. Farmers can reap economic benefits from the prolonged efficacy of inputs and the minimized environmental footprint. For example, nano encapsulated growth hormones have been applied to the root zone to stimulate root development and improve overall plant health (Hafez and Khalil 2024).

3.5.4 Products of Nanoencapsulation

(a) **Nano encapsulated Gibberellins for Fruit Crops**
    In a study conducted on grapevines, nano encapsulated gibberellins were applied to enhance fruit size and yield. The results demonstrated that the nano encapsulated formulation provided a more consistent release of gibberellins compared to traditional methods, leading to improved fruit development and higher yields (Pereira et al. 2019).

Table 3.2 Listed the nano based Agri products producing companies

Commercial product	Components	Company
Nano Max NPK Fertilizer	Multiple major nutrients chelated with organic acids	JU Agri Sciences Pvt. Ltd., India
Nano Green	Extracts of corn, grain, soybeans and potatoes	Nano Green Sciences, Inc., India
Tag Nano	Micronutrients chelated with Protein-lactose-gluconate	Tropical AeroSystems India (P) Ltd., India
Biozar Nano-Fertilizer	Combination of micronutrients and macronutrients	Fanavar Nano-PazhooheshMarkazi Company, Iran
Master Nano Chitosan Organic Fertilizer	Water soluble liquid chitosan with macronutrients	Pannaraj Intertrade, Thailand

(b) **Nano encapsulated Seed Treatments in Maize**

A case study involving maize seed treatment with nano encapsulated auxins showed significant improvements in seed germination and seedling growth. The nanoencapsulation technology provided a controlled release of auxins, which enhanced root development and plant vigor, leading to better crop establishment and higher productivity (Mondal et al. 2023).

(c) **Nano encapsulated Fertilizers for Soil Application**

A field trial was conducted to assess the effects of nano encapsulated fertilizers on nitrogen absorption and crop productivity in wheat crops. The trial results demonstrated that the utilization of nano encapsulated fertilizers led to enhanced efficiency in nutrient utilization and a decrease in the negative effects on the environment when compared to traditional fertilizers. Consequently, this resulted in a higher yield of wheat and a reduction in leaching (Singh et al. 2024) (Table 3.2).

3.6 Safety and Regulatory Considerations for Nano-Enabled Agriculture

Nanoencapsulation technology integration in agriculture has several advantages, such as improved efficacy and sustainability. However, it also brings up significant safety and regulatory concerns. To ensure the appropriate utilization of nano-enabled agricultural products, it is crucial to address potential environmental hazards, human health issues, and effectively navigate the regulatory framework. Nano-enabled devices can present environmental hazards as a result of their diminutive size and distinctive characteristics, potentially causing unintentional interactions with ecosystems (Kumari et al. 2023). Possible hazards of nanoparticle as be listed below

(a) **Toxicity**

Nanoparticles have the potential to harm beneficial soil bacteria, insects, and aquatic life. Studies have demonstrated that metal nanoparticles can have a harmful effect on earthworms and soil microbes, which might potentially lead to disturbances in soil health and fertility. Conduct thorough environmental impact assessments and toxicity studies to understand the behavior and effects of nanoparticles in different ecosystems. This includes studying their degradation, bioavailability, and potential for accumulation (Gaviria et al. 2024).

(b) **Persistence and Mobility**

Nanoparticles have the ability to remain in the environment for a long time and move through soil and water systems. This can potentially result in the buildup of nanoparticles in the food chain. Design nanoencapsulation systems with controlled release mechanisms to limit the dispersion of nanoparticles in the environment. For example, using biodegradable carriers can help reduce long-term environmental impact (Youssef et al. 2024).

Nano-enabled agricultural products may pose health risks through exposure routes such as inhalation, ingestion, or skin contact. Concerns includes nanoparticles may have different toxicity profiles compared to their bulk counterparts, potentially causing adverse health effects. Studies have indicated that some nanoparticles can induce oxidative stress, inflammation, or cellular damage. Farmers and agricultural workers may be at higher risk of exposure during the application of nano-enabled products. There is also a concern about residue on food crops affecting consumers (Nur Fatin Nazurah et al. 2024).

To address these concerns, it is crucial to perform comprehensive studies to assess the potential health impacts of nano-enabled products, including acute and chronic exposure studies. Establish safety protocols for handling and applying nano-enabled products to minimize exposure risks. This includes appropriate protective equipment for workers and guidelines for safe application practices. Regulatory oversight of nano-enabled agricultural products is evolving (Mittal et al. 2020). Key aspects includes different countries have varying approaches to regulating nanomaterials. In the European Union, nano-enabled products are subject to the REACH (Registration, Evaluation, Authorization, and Restriction of Chemicals) regulation, which requires specific safety data for nanomaterials. In the U.S., the Environmental Protection Agency (EPA) oversees the regulation of nanomaterials under the Toxic Substances Control Act (TSCA), but there are calls for more specific guidelines.

As nano-technology in agriculture advances, future regulatory developments may include harmonized Standards for Development of international standards and guidelines to ensure consistency in the assessment and regulation of nano-enabled products across different jurisdictions. Increased transparency in the disclosure of safety data and risk assessments to build public trust and ensure informed decision-making. Safety and regulatory considerations are essential for the responsible use of nano-enabled agriculture. By addressing potential environmental risks, assessing human health impacts, and navigating evolving regulatory frameworks, stakeholders can ensure the safe and effective application of nanoencapsulation technology in agriculture (Wang et al. 2024; Song et al. 2024).

3.7 Future Trends and Challenges in Nanoencapsulation for Agriculture

Nanoencapsulation technology continues to evolve, offering promising advancements for agricultural applications. As the field progresses, several key trends and challenges are emerging, particularly in developing smart and responsive nanocarriers, improving cost-effectiveness and scalability, and addressing public perception and consumer acceptance. The future of nanoencapsulation in agriculture promises significant advancements, particularly through the development of smart nanocarriers, improved cost-effectiveness, and addressing public concerns. As these trends evolve, continued research and innovation will be essential in overcoming the challenges and harnessing the full potential of nanoencapsulation technology for sustainable agricultural practices (Hafez and Khalil 2024).

I. **Developing Smart and Responsive Nanocarriers:** Smart and responsive nanocarriers are at the forefront of innovation in nanoencapsulation. These advanced systems are designed to react to specific environmental stimuli, such as pH, temperature, or light, to release their payload in a controlled manner (Blinov et al. 2024). Key developments include such as

   (a) **Stimuli-Responsive Materials**

   Researchers are developing nanocarriers that can release their contents in response to environmental changes. For example, pH-responsive nanocarriers can release encapsulated nutrients or growth regulators in acidic or alkaline soil conditions, optimizing their effectiveness based on the soil's properties (Li et al. 2017).

   (b) **Targeted Delivery Systems**

   Smart nanocarriers can also be engineered to target specific plant tissues or pests. For instance, nanoparticles functionalized with plant-specific ligands can enhance the targeted delivery of growth regulators to specific plant organs, improving their efficacy and minimizing waste (Feng et al. 2024).

II. **Improving Cost-Effectiveness and Scalability:** Cost-effectiveness and scalability are critical for the widespread adoption of nanoencapsulation technologies. As the technology advances, addressing these aspects is crucial in developing more straightforward and cost-efficient production methods for nanocarriers is essential. Techniques such as one-pot synthesis and scalable electrospinning processes are being explored to reduce production costs while maintaining quality and performance (Zhao et al. 2024a).

(a) **Economies of Scale**

Achieving economies of scale is vital for making nanoencapsulation technologies commercially viable. Increased production capacity and the use of inexpensive raw materials can help lower costs and make these technologies accessible to a broader range of agricultural applications (Alsakhawy et al. 2022).

III. **Public Perception and Consumer Acceptance:** Public perception and consumer acceptance of nano-enabled agricultural products are critical factors influencing their adoption. Increasing public awareness about the benefits and safety of nanoencapsulation can help address concerns. Ensuring that nano encapsulated products meet stringent safety and regulatory standards can help alleviate public concerns. Certification from recognized bodies can provide assurance of the safety and efficacy of these products (Bhattacharyya 2016).

3.8 Conclusion and Prospects

Nano capsulation in agricultural products stands at the forefront of innovation, heralding a new era of precision and efficiency in farming practices. As this technology evolves, it promises to revolutionize how agricultural inputs are utilized, offering solutions to many of the challenges faced by modern agriculture. The future of nano capsulation in agriculture looks particularly promising. By enhancing the stability and controlled release of nutrients, pesticides, and herbicides, nano capsulation can significantly reduce input waste and environmental impact. This precision delivery system allows for more efficient use of resources, potentially leading to increased crop yields and improved quality. Furthermore, by minimizing the need for frequent applications and lowering overall chemical use, nano capsulation supports more sustainable agricultural practices, which is crucial in addressing global food security and environmental concerns.

Looking ahead, researchers play a pivotal role in unlocking the full potential of nano capsulation. Continued research is essential to refine the technology, ensuring that nano capsules are optimized for various agricultural applications. Investigations into the long-term effects of nanomaterials on ecosystems and human health are critical to ensuring that the benefits of nano capsulation are realized without unintended negative consequences. Researchers must also focus on developing cost-effective production methods and scalable solutions to facilitate widespread adoption in the agricultural sector. Moreover, interdisciplinary collaboration will be key to advancing this technology. By integrating insights from materials science, agriculture, environmental science, and toxicology, researchers can address complex challenges and drive innovation. Engaging with policymakers and industry stakeholders will also be important to develop regulatory frameworks that ensure the safe and effective use of nano capsules in agriculture.

In conclusion, the future of nano capsulation in agricultural products holds great promise for enhancing the efficiency and sustainability of farming practices. While significant progress has been made, ongoing research and development are crucial to addressing potential risks and maximizing the technology's benefits. As the field continues to evolve, it is essential for researchers to lead the way in ensuring that nano capsulation contributes positively to global agriculture and environmental stewardship. In brief, Nanoencapsulation technology offers significant benefits in agricultural applications, including foliar application, seed treatment, and soil application. By improving the delivery, stability, and controlled release of various inputs, nanoencapsulation enhances crop growth, yield, and environmental sustainability. Real-world case studies demonstrate the practical advantages of nanoencapsulation, underscoring its potential to transform modern agriculture.

References

Abd-Elsalam KA (2024) Special Issue: Agricultural nanotechnology. Plants 13:489. https://doi.org/10.3390/plants13040489

Albuquerque PM, Azevedo SG, de Andrade CP et al (2022) Biotechnological applications of nanoencapsulated essential oils: a review. Polymers (Basel) 14:5495. https://doi.org/10.3390/polym14245495

Alruwaili NK, Almalki WH, Almujri SS et al (2024) Hispolon-loaded lipid nanocapsules for the management of hepatocellular carcinoma: comparative study with solid lipid nanoparticles and suspension. Nanomedicine:1–22. https://doi.org/10.1080/17435889.2024.2406741

Alsakhawy SA, Baghdadi HH, El-Shenawy MA et al (2022) Encapsulation of thymus vulgaris essential oil in caseinate/gelatin nanocomposite hydrogel: in vitro antibacterial activity and in vivo wound healing potential. Int J Pharm 628:122280. https://doi.org/10.1016/j.ijpharm.2022.122280

Amin S, Aziz T, Zia-ur-Rehman M et al (2023) Zinc oxide nanoparticles coated urea enhances nitrogen efficiency and zinc bioavailability in wheat in alkaline calcareous soils. Environ Sci Pollut Res 30:70121–70130. https://doi.org/10.1007/s11356-023-27209-5

Anandharamakrishnan C (2014) Techniques for formation of nanoemulsions. In: Techniques for nanoencapsulation of food ingredients. Springer, pp 7–16

Aphibanthammakit C, Kasemwong K (2021) Nanoencapsulation in agricultural applications. In: Handbook of nanotechnology applications. Elsevier, pp 359–382

Ashfaq M (2016) Copper nanoparticles in agriculture: biological synthesis and antimicrobial activity. In: Nanoscience in food and agriculture. Springer, pp 129–146

Babaei S, Kahrizi D, Nosratti I et al (2022) Preparation and characterization of chloridazon-loaded alginate/chitosan nanocapsules. Cell Mol Biol 68:34–42. https://doi.org/10.14715/cmb/2022.68.3.5

Bamisaye A, Adegoke KA, Alli YA et al (2023) Recent advances in nanoemulsion for sustainable development of farm-to-fork systems. J Clean Prod 429:139226. https://doi.org/10.1016/j.jclepro.2023.139226

Bashir I, Wani SM, Jan N et al (2024) Optimizing ultrasonic parameters for development of vitamin D3-loaded gum arabicnanoemulsions—An approach for vitamin D3 fortification. Int J Biol Macromol 278:134894. https://doi.org/10.1016/j.ijbiomac.2024.134894

Belagalla N, Kumar A, Badekhan A et al (2024) Enhancing agricultural efficiency and biodiversity conservation through nano pesticides; a focus on food research. J Adv Biol Biotechnol 27:319–335. https://doi.org/10.9734/jabb/2024/v27i5791

Bhattacharyya A, Duraisamy P, Govindarajan M, Buhroo AA, Prasad R (2016) Nano-biofungicides: emerging trend in insect pest applications control. In: Advances and through fungal nanobiotechnology. Springer International Publishing, pp 307–319

Blinov A, Blinova A, Nagdalian A et al (2024) Sustainable detergent-disinfectant agent based on whey mineralizate and silver nanoparticles for cleaner production in dairy industry. Sci Rep 14: 23943. https://doi.org/10.1038/s41598-024-71542-9

Chandrasekaran M, Paramasivan M (2024) Chitosan derivatives act as a bio-stimulants in plants: a review. Int J Biol Macromol 271:132720. https://doi.org/10.1016/j.ijbiomac.2024.132720

Chhipa H (2016) Nanopesticide: a recent approach for pest control. Nanoscience in food and agriculture. Springer International Publishing, pp 69–98

Dai J-X, Yu Y, You L-X et al (2024) Integrated induction of silver resistance determinants and production of extracellular polymeric substances in Cupriavidusmetallidurans BS1 in response to silver ions and silver nanoparticles. Chemosphere 366:143503. https://doi.org/10.1016/j.chemosphere.2024.143503

Delfanian M, Razavi SMA, Haddad Khodaparast MH et al (2018) Influence of main emulsion components on the physicochemical and functional properties of W/O/W nano-emulsion: effect of polyphenols, Hi-Cap, basil seed gum, soy and whey protein isolates. Food Res Int 108:136–143. https://doi.org/10.1016/j.foodres.2018.03.043

Dias MC (2012) Phytotoxicity: an overview of the physiological responses of plants exposed to fungicides. J Bot:1–4. https://doi.org/10.1155/2012/135479

Edelmann HG (2022) Plant root development: is the classical theory for auxin-regulated root growth false? Protoplasma 259:823–832. https://doi.org/10.1007/s00709-021-01697-z

El-Sayed E-SR, El-Sayyad GS, Abdel-Fatah SS et al (2024) Novel nanoconjugates of metal oxides and natural red pigment from the endophyte Monascusruber using solid-state fermentation. Microb Cell Factories 23:259. https://doi.org/10.1186/s12934-024-02533-8

Ezhilarasi PN, Karthik P, Chhanwal N, Anandharamakrishnan C (2013) Nanoencapsulation techniques for food bioactive components: a review. Food Bioprocess Technol 6:628–647. https://doi.org/10.1007/s11947-012-0944-0

Fathi F, Machado TOX, de Kodel ACH et al (2024) Solid lipid nanoparticles (SLN) and nanostructured lipid carriers (NLC) for the delivery of bioactives sourced from plants: part I—composition and production methods. Expert Opin Drug Deliv 21:1479–1490. https://doi.org/10.1080/17425247.2024.2410951

Feng H, Yue Y, Zhang Y et al (2024) Plant-derived exosome-like nanoparticles: emerging nanosystems for enhanced tissue engineering. Int J Nanomedicine 19:1189–1204. https://doi.org/10.2147/IJN.S448905

Gaviria YAR, Chacon WDC, Cesca K et al (2024) Smart labels based on polyvinyl alcohol incorporated with chitosan nanoparticles loaded with grape extract: functionality, stability and food application. Int J Biol Macromol 263:130513. https://doi.org/10.1016/j.ijbiomac.2024.130513

Ghorbani M, Kahrizi D, Arkan E et al (2024) Enhancing rooting tobacco (Nicotiana tabacum) plant by loaded indole-3-butyric acid in alginate/chitosan nanocapsule. Cell Mol Biol 70:237–242. https://doi.org/10.14715/cmb/2024.70.7.34

Gouda S, Kerry RG, Das G et al (2018) Revitalization of plant growth promoting rhizobacteria for sustainable development in agriculture. Microbiol Res 206:131–140. https://doi.org/10.1016/j.micres.2017.08.016

Graily-Moradi F, Asgari Lajayer B (2021) Nanoinsecticides: preparation, application, and mode of action. In: Nanotechnology applications in health and environmental sciences. Springer, Cham, pp 385–404

Hafez M, Khalil HF (2024) Nanoparticles in sustainable agriculture: recent advances, challenges, and future prospects. Commun Soil Sci Plant Anal 55:2181–2196. https://doi.org/10.1080/00103624.2024.2339950

Hamed R, Jodeh S, Alkowni R (2024) Nano bio fertilizer capsules for sustainable agriculture. Sci Rep 14:13646. https://doi.org/10.1038/s41598-024-62973-5

Hong J, Wang C, Wagner DC et al (2021) Foliar application of nanoparticles: mechanisms of absorption, transfer, and multiple impacts. Environ Sci Nano 8:1196–1210. https://doi.org/10.1039/D0EN01129K

Huang Y, Huang H, Wu J et al (2024) Preparation and properties of PCL coaxial electrospinning films with shell loaded with CEO and core coated LEO nanoemulsions. Food Res Int 194: 114817. https://doi.org/10.1016/j.foodres.2024.114817

Jeon YO, Lee J-S, Lee HG (2016) Improving solubility, stability, and cellular uptake of resveratrol by nanoencapsulation with chitosan and γ-poly (glutamic acid). Colloids Surf B Biointerfaces 147:224–233. https://doi.org/10.1016/j.colsurfb.2016.07.062

Kah M, Hofmann T (2014) Nanopesticide research: current trends and future priorities. Environ Int 63:224–235. https://doi.org/10.1016/j.envint.2013.11.015

Karabulut F (2024) The impact of nanoparticles on plant growth, development, and stress tolerance through regulating phytohormones. In: Nanoparticles in plant biotic stress management. Springer Nature Singapore, Singapore, pp 87–105

Khajavi R, Abbasipour M (2017) Controlling nanofiber morphology by the electrospinning process. In: Electrospun nanofibers. Elsevier, pp 109–123

Kookana RS, Boxall AB, Reeves PT, Ashauer R, Beulke S, Chaudhry Q et al (2014) Nanopesticides: guiding principles for regulatory evaluation of environmental risks. J Agric Food Chem 62:4227–4240

Kotp YH, Eissa D, Hagab R (2018) Using nanotechnology for enhancing phosphorus fertilizer use efficiency of peanut bean grown in sandy soils. J Adv Pharm Educ Res 8(3):59–67

Kumari R, Suman K, Karmakar S et al (2023) Regulation and safety measures for nanotechnology-based agri-products. Front Genome Ed 5. https://doi.org/10.3389/fgeed.2023.1200987

Li Y, Gao J, Zhang C et al (2017) Stimuli-responsive polymeric nanocarriers for efficient gene delivery. Top Curr Chem 375:27. https://doi.org/10.1007/s41061-017-0119-6

Li X, Liu J, Dou J et al (2024) Enhanced cellular delivery of Tildipirosin by Xanthan Gum–Gelatin composite nanogels. Langmuir 40:4860–4870. https://doi.org/10.1021/acs.langmuir.3c03559

Mahato DK, Mishra AK, Kumar P (2021) Nanoencapsulation for agri-food applications and associated health and environmental concerns. Front Nutr 8. https://doi.org/10.3389/fnut.2021.663229

Martins VFR, Pintado ME, Morais RMSC, Morais AMMB (2022) Valorisation of micro/nanoencapsulated bioactive compounds from plant sources for food applications towards sustainability. Food Secur 12:32. https://doi.org/10.3390/foods12010032

Masoumi Z, Haghighi M, Mozafarian M (2024) Effects of foliar spraying with melatonin and chitosan Nano-encapsulated melatonin on tomato (Lycopersicon esculentum L. cv. Falcato) plants under salinity stress. BMC Plant Biol 24:961. https://doi.org/10.1186/s12870-024-05672-7

Mirara F, Dzidzienyo DK, Mwangi M (2024) Nano-enhanced defense: Titanium-enriched Alginate–Bentonite coating augments Bacillus amyloliquefaciens D203 efficacy against Magnaporthe oryzae in Kenyan rice cultivation. Heliyon 10:e36141. https://doi.org/10.1016/j.heliyon.2024.e36141

Mishra S, Singh HB (2015) Biosynthesized silver nanoparticles as a nanoweapon against phytopathogens: exploring their scope and potential in agriculture. Appl Microbiol Biotechnol 99: 1097–1107. https://doi.org/10.1007/s00253-014-6296-0

Mittal D, Kaur G, Singh P et al (2020) Nanoparticle-based sustainable agriculture and food science: recent advances and future outlook. Front Nanotechnol 2. https://doi.org/10.3389/fnano.2020.579954

Mondal S, Bose B, Panda D (2023) Effect of nano-priming on maize under normal and stressful environment. In: Molecular interventions for developing climate-smart crops: a forage perspective. Springer Nature Singapore, Singapore, pp 99–107

Naik BSS, Mahawar N, Rupesh T, Dhegavath S, Meena RS (2021) Nano technology based nano-fertilizer: a sustainable approach for enhancing crop productivity under climate changing situations. Curr Res Agric Farm 2:21–29

Noore S, Pathania S, Fuciños P et al (2024) Controlled release and target delivery of nanoencapsulated compounds. In: Nanocarriers for controlled release and target delivery of bioactive compounds. Springer Nature Switzerland, Cham, pp 63–65

Nur Fatin Nazurah R, Noranizan MA, Nor-Khaizura MAR, Nur Hanani ZA (2024) Chitosan nanoparticles incorporate with curry leaf essential oil: physicochemical characterization and in vitro release properties. Int J Biol Macromol 273:132972. https://doi.org/10.1016/j.ijbiomac.2024.132972

de Oliveira JL, Campos EVR, Bakshi M et al (2014) Application of nanotechnology for the encapsulation of botanical insecticides for sustainable agriculture: prospects and promises. Biotechnol Adv 32:1550–1561. https://doi.org/10.1016/j.biotechadv.2014.10.010

Oliveira HC, Stolf-Moreira R, Martinez CBR, Grillo R, de Jesus MB, Fraceto LF (2015) Nanoencapsulation enhances the post emergence herbicidal activity of atrazine against mustard plants. PLoS One 10(7):e0132971

Paula H, Oliveira E, Carneiro M, de Paula R (2016) Matrix effect on the spray drying nanoencapsulation of lippia sidoides essential oil in chitosan-native gum blends. Planta Med 83:392–397. https://doi.org/10.1055/s-0042-107470

do Pereira AES, Oliveira HC, Fraceto LF (2019) Polymeric nanoparticles as an alternative for application of gibberellic acid in sustainable agriculture: a field study. Sci Rep 9:7135. https://doi.org/10.1038/s41598-019-43494-y

Sadeghi E, Taghavi R, Hasanzadeh A, Rostamnia S (2024) Bactericidal behavior of silver nanoparticle decorated nano-sized magnetic hydroxyapatite. Nanoscale Adv. https://doi.org/10.1039/D4NA00183D

Salvati B, Santagapita P, Perullini M (2022) Exploring the conditions to generate alginate nanogels. J Solgel Sci Technol 102:142–150. https://doi.org/10.1007/s10971-021-05631-w

Sampedro-Guerrero J, Vives-Peris V, Gomez-Cadenas A, Clausell-Terol C (2023) Efficient strategies for controlled release of nano encapsulated phytohormones to improve plant stress tolerance. Plant Methods 19:47. https://doi.org/10.1186/s13007-023-01025-x

Shaddel R, Akbari-Alavijeh S, Cacciotti I et al (2024) Caffeine-loaded nano/micro-carriers: techniques, bioavailability, and applications. Crit Rev Food Sci Nutr 64:4940–4965. https://doi.org/10.1080/10408398.2022.2147143

Shelar A, Nile SH, Singh AV et al (2023) Recent advances in nano-enabled seed treatment strategies for sustainable agriculture: challenges, risk assessment, and future perspectives. Nanomicro Lett 15:54. https://doi.org/10.1007/s40820-023-01025-5

Sheta B, El-Zahed M, Nawareg M et al (2024) Nanoremediation of tilapia fish culture using iron oxide nanoparticles biosynthesized by Bacillus subtilis and immobilized in a free-floating macroporous cryogel. BMC Vet Res 20:455. https://doi.org/10.1186/s12917-024-04292-5

Sindhu RK, Gupta R, Wadhera G, Kumar P (2022) Modern herbal nanogels: formulation, delivery methods, and applications. Gels 8:97. https://doi.org/10.3390/gels8020097

Singh RP, Handa R, Manchanda G (2021) Nanoparticles in sustainable agriculture: an emerging opportunity. J Control Release 329:1234–1248. https://doi.org/10.1016/j.jconrel.2020.10.051

Singh S, Angurana R, Kaur H et al (2024) Nanoencapsulation of fertilizers. In: Nanofertilizer delivery, effects and application methods. Elsevier, pp 97–110

Sinha T, Bhagwatwar P, Krishnamoorthy C, Chidambaram R (2019) Polymer based micro- and nanoencapsulation of agrochemicals. In: Polymers for agri-food applications. Springer International Publishing, Cham, pp 5–28

Song W, Yuan Q, Xie Y et al (2024) Formulation and characterization of nanocapsules loaded with roselle anthocyanins extract and enhancement of anthocyanins bioaccessibility. Food Chem 459:140446. https://doi.org/10.1016/j.foodchem.2024.140446

Soni KS, Desale SS, Bronich TK (2016) Nanogels: an overview of properties, biomedical applications and obstacles to clinical translation. J Control Release 240:109–126. https://doi.org/10.1016/j.jconrel.2015.11.009

Staff RH, Landfester K, Crespy D (2013) Recent advances in the emulsion solvent evaporation technique for the preparation of nanoparticles and nanocapsules. In: Hierarchical macromolecular structures: 60 years after the Staudinger Nobel Prize II. Springer, pp 329–344

Suganya V, Anuradha V (2017) Microencapsulation and nanoencapsulation: a review. Int J Pharma Clin Res 9. https://doi.org/10.25258/ijpcr.v9i3.8324

Sun C (2020) Polymer-based nanoinsecticides: current developments, environmental risks and future challenges-a review. Biotechnol Agron Soc Environ 24:59–69

Sun H, Zhuang K, Du J et al (2024) Sustainable lignin-modified epoxy nanocarriers for enhanced foliar insecticide efficacy and food safety. Int J Biol Macromol 279:135262. https://doi.org/10.1016/j.ijbiomac.2024.135262

Sushil A, Kamla M, Nisha K et al (2021) Nano-enabled pesticides in agriculture: budding opportunities and challenges. J Nanosci Nanotechnol 21:3337–3350. https://doi.org/10.1166/jnn.2021.19018

Tarafdar JC, Xiang Y, Wang WN, Dong Q, Biswas P (2012) Standardization of size, shape and concentration of nanoparticle for plant application. Appl Biol Res 14:138–144

Tryfon P, Kamou NN, Pavlou A et al (2023) Nanocapsules of ZnO nanorods and geraniol as a novel mean for the effective control of botrytis cinerea in tomato and cucumber plants. Plants 12:1074. https://doi.org/10.3390/plants12051074

Ullah A, Yang H, Takemae K et al (2024) Sustainable bioactive food packaging based on electrospun zein-polycaprolactone nanofibers integrated with aster yomena extract loaded halloysite nanotubes. Int J Biol Macromol 267:131375. https://doi.org/10.1016/j.ijbiomac.2024.131375

Vemula M, Reddy AVB (2023) Polymeric nanoparticles as effective delivery systems in agriculture sustainability. Nanotechnol Environ Eng 8:805–814. https://doi.org/10.1007/s41204-023-00319-8

Venezia V, Pota G, Argenziano R et al (2024) Design of a hybrid nanoscaled skin photoprotector by boosting the antioxidant properties of food waste-derived lignin through molecular combination with TiO2 nanoparticles. Int J Biol Macromol 280:135946. https://doi.org/10.1016/j.ijbiomac.2024.135946

Wang P, Lombi E, Zhao F-J, Kopittke PM (2016) Nanotechnology: a new opportunity in plant sciences. Trends Plant Sci 21:699–712. https://doi.org/10.1016/j.tplants.2016.04.005

Wang R, Luan F, Wang Z et al (2024) Spontaneously formed ratio-tunable micro- and nano-capsule coexist system for precision and on-demand fungicide delivery in crop leave. Adv Mater. https://doi.org/10.1002/adma.202409839

Xie Y, Ding K, Xu S et al (2024) Citrus oil gland and cuticular wax inspired multifunctional gelatin film of OSA-starch nanoparticles-based nanoemulsions for preserving perishable fruit. Carbohydr Polym 342:122352. https://doi.org/10.1016/j.carbpol.2024.122352

Xing H, Hwang K, Lu Y (2016) Recent developments of liposomes as nanocarriers for theranostic applications. Theranostics 6:1336–1352. https://doi.org/10.7150/thno.15464

Xu Y, Chen H, Zhang L, Xu Y (2023) Clove essential oil loaded chitosan nanocapsules on quality and shelf-life of blueberries. Int J Biol Macromol 249:126091. https://doi.org/10.1016/j.ijbiomac.2023.126091

Xu T, Hao W, Du R et al (2024) Mercaptoimidazole-capped gold nanoparticles as a potent agent against plant pathogenic fungi. J Mater Chem B. https://doi.org/10.1039/D4TB01032A

Yadav A, Sohlot M, Sahu SR et al (2024) Determination of antifungal efficacy and phytotoxicity of a unique silica coated porous zinc oxide nanocomposite medium for slow-release agrochemicals. J Appl Microbiol 135. https://doi.org/10.1093/jambio/lxae153

Yanovska A, Artyukhov A, Vakal S et al (2022) Encapsulated organic–mineral fertilizers with nanoporous structure. Appl Nanosci 12:1275–1283. https://doi.org/10.1007/s13204-021-01893-6

Yi Y, Shan Y, Luan P et al (2024) Nanoencapsulation enhances the antimicrobial and antioxidant stability of cyclic lipopeptides for controlling Fusarium graminearum. Food Microbiol 124:104621. https://doi.org/10.1016/j.fm.2024.104621

Youssef JR, Boraie NA, Ismail FA et al (2024) Brain targeted lactoferrin coated lipid nanocapsules for the combined effects of apocynin and lavender essential oil in PTZ induced seizures. Drug Deliv Transl Res. https://doi.org/10.1007/s13346-024-01610-0

Zahedi Y, Shaddel R, Salamatian M, Szumny A (2024) Nanoliposomal encapsulation of capparis spinosa extract and its application in jelly formulation. Molecules 29:2804. https://doi.org/10.3390/molecules29122804

Zhang Z, Cheng D, Luo W et al (2024) Molecular dynamics simulation of lipid nanoparticles encapsulating mRNA. Molecules 29:4409. https://doi.org/10.3390/molecules29184409

Zhao L, Zhou L, Dansou DM et al (2024a) Development of an electrochemical sensor modified with gold nanoparticles for detecting fumonisin B1 in packaged foods. RSC Adv 14:29254–29259. https://doi.org/10.1039/D4RA04753B

Zhao L, Zhou X, Kang Z et al (2024b) Nano-enabled seed treatment: a new and sustainable approach to engineering climate-resilient crops. Sci Total Environ 910:168640. https://doi.org/10.1016/j.scitotenv.2023.168640

Chapter 4
Nanotechnology in Plant Disease Management

I. K. Nishitha, Tintu Jose Manicketh, and Arya P. Mohan

Abstract Agriculture faces significant challenges and limitations, including limited and depleting resources, the adverse impacts of climate change, and reduced productivity resulting from abiotic and biotic stresses. These factors collectively contribute to substantial losses in the agricultural sector, posing a serious threat to global food security. While conventional practices in managing agricultural challenges offer solutions, their effectiveness can be significantly improved through the integration of technology. Nanotechnology is one such emerging field that has profound influence in the agricultural field in managing plant diseases. By manipulating materials at the nanoscale, nanotechnology offers unprecedented opportunities for enhancing agricultural sustainability. Nanoscale particles are employed for the early disease detection, targeted pesticide delivery leading to higher efficacy, increased shelf life with low toxicity to the environment. This chapter discusses the key applications of nanotechnology in agriculture, with a particular focus on disease management. It describes the various tools and techniques employed in the application of nanotechnology in agriculture, giving details of the types of metal nanoparticles and its synthesis. It discusses green synthesis methods that are environmentally friendly touching upon the benefits and action of green synthesised nanoparticles. The chapter discusses the mechanisms by which the green-synthesized nanoparticles interact with microbes, highlighting their potential to enhance plant health. By integrating nanotechnology into agricultural practices, the sector can address pressing challenges more effectively, unlocking new avenues for sustainable and resilient food production systems ensuring abundant food supply for a growing global population.

Keywords Nanoparticles · Plant pathogens · Nano fertilizer · Agriculture · Green synthesis · Plant disease · Pest management

I. K. Nishitha · T. J. Manicketh · A. P. Mohan (✉)
Department of Botany, St. Teresa's College (Autonomous), Ernakulam, Kerala, India
e-mail: draryapmohan@teresas.ac.in

J. M. Al-Khayri et al. (eds.), *Nanotechnology Applications in Modern Agriculture*, Nanotechnology in Plant Sciences 3, https://doi.org/10.1007/978-3-031-90506-3_4

4.1 Introduction

Plant pathogens present a major threat to agricultural productivity, causing yearly economic losses exceeding $220 billion (FAO 2022). The reduced productivity in the agriculture sector, the depletion of natural resources, the significant post-farm losses, climate change and the rapid population expansion pose great challenges in ensuring food and nutritional security globally (Neme et al. 2021). Significant crop losses, or 20–40% annually, resulting from biotic factors are a big global issue (Worrall et al. 2018). Insects, nematodes, infections, weeds, and interactions between people and animals are the main biotic causes. Previous studies revealed that 20% and 26% of crop loss occurred due to disease and insect infestation (Bhattacharjee and Dey 2014). Conventional methods depend on chemical pesticides to control disease and pest management as they are readily available and produce fast results. However, the detrimental effects of these chemicals on the environment, hazards to beneficial insects and pest recurrence are major concerns (Balaure et al. 2017; Sinha et al. 2017). Keeping in mind all of the aforementioned reality-based issues, scientists are developing a new and promising concept and technique called nanotechnology, which is considered to have a positive effect on the agriculture and ecosystem (Fraceto et al. 2020).

The term "nanotechnology" was coined by Taniguchi in 1974. Nanotechnology is a science that describes the synthesis and application of nano-size particles 1–100 nm which are engineered to have unique physical, chemical, and biological properties (Sekhon 2014; Sai et al. 2018). Current applications of nanotechnology in pest disease management include targeted gene transfer, water management, seed germination, plant hormone delivery, nano barcoding, nano sensors, and controlled release of agrochemicals (Hayles 2017). Around 85–90% of applied pesticides are thought to be lost during or after application. Thus, nanotechnology would offer effective and environmentally friendly options for controlling plant diseases without endangering the environment (Sai et al. 2020).

The use of nanotechnology to enhance agricultural sustainability holds promise for increasing productivity by improving nutrient absorption, enhancing water quality, and enabling precise and timely application of crop-specific pesticides, ultimately supporting better soil health and quality (Gruere 2012; Mukhopadhyay 2014). Furthermore, nanotechnology in agriculture has a potential to use in genetic improvement of plants, gene delivery and drug molecules to particular locations within plants' and animals' cells, nano array-based technologies for regulating plant gene expression in response to stress, the development of sensors, and procedures for their use in precision farming, smart delivery systems for agrochemicals like pesticides and fertilizers, and early detection of diseases in plants (Mukhopadhyay 2014).

4.2 Tools of Nanotechnology in Agriculture

4.2.1 Nano Sensors

Nano-biosensors significantly contribute to the revolution in agriculture through the creation of diagnostic tools and techniques. It is a device designed to transmit data and information about the properties of nanoscale particles to the macroscopic scale (Parveen et al. 2022). Nano biosensors are accurate, quantitative, reliable, reproducible, robust and stable tools that enable the detection and monitoring of materials at extremely low concentrations. Integrated into precision farming, it plays a crucial role in diagnosing infections in asymptomatic plants and subsequently in administering and delivering targeted treatments. They can detect contaminants such as pests, microbes, plant stress and nutrient content due to insect or pathogen pressure. Biosensors are developed using various nanomaterials, including metal and metal oxide nanoparticles, quantum dots, carbon-based nanomaterials such as carbon nanotubes and graphene, along with other nanoscale materials (Sai et al. 2018). Other applications include the identification of heavy metal ions, pollutants, microbial load, and pathogens, in addition to rapid temperature, traceability, and humidity monitoring (Parveen et al. 2022). Therefore, they are important for protecting crops and advancing the idea of sustainable agriculture.

4.2.2 Nano Fertilizers

The term "nano fertilizer" refers to fertilizers that contain nutrients that have been coated or encapsulated in nanomaterials for controlled release and gradual soil diffusion. These synthetic nanoparticles are designed to regulate the delivery of macro and micronutrients to the plant rhizosphere (Nongbet et al. 2022). Furthermore, nano fertilizers have a high surface area-to-volume ratio and a high penetration ability, which makes them a viable substitute for chemical fertilizers. Nano fertilizers can increase crop output by promoting seed germination, stress tolerance, photosynthesis, nitrogen metabolism and protein synthesis. With their increased effectiveness and decreased environmental impact, nano fertilizers offer promise in addressing problems of overuse of traditional fertilisers and soil degradation. The development of nano fertilizers could become crucial in satisfying the increasing demand for food and minimizing environmental concerns since it is predicted that there will be more than 9 billion people on the planet by the year 2050 (Yadav et al. 2023).

4.2.3 Smart Delivery

The major challenge in drug delivery is releasing the pharmacological agents at the right time, usually to a predetermined target place, in a safe and repeatable manner. Nanoparticles are used in smart and targeted drug delivery. Paul Ehrlich developed the idea of using nanoparticles as "magic bullets" to target a pathogen specifically while sparing the host organism. Nanoparticles are useful in pest management, for instance, the delivery of pesticides to the target using encapsulated agrochemicals in nano-carriers provides targeted delivery of the required dose (i.e., reduced application dosages), time-controlled release, and less eco-toxicity (Slattery et al. 2019; Vega-Vásquez et al. 2020). A variety of materials, such as proteins, polysaccharides, synthetic polymers, and inorganic metallic salts, can be utilized to develop nano-carriers designed for drug delivery. Compared to conventional pesticides, nanomaterials as a pesticides or as carrier material have shown functional properties such as rigidity, permeability, crystallinity, thermal stability, and biodegradability (Bordes et al. 2009).

Nanoparticle enabled gene transfer into plant cells presents a viable avenue for genetic engineering for agriculture because of its small size (Abd-Elsalam and Alghuthaymi 2015). NPs are useful materials for delivering biomolecules into cells because they can traverse biological membranes, protect and release a variety of cargo, and enable multidimensional targeting through physical and chemical tunability. These characteristics have made it possible for NPs to completely transform controlled release and targeted delivery in mammalian systems. A range of nanomaterials can be used in plants for nanoparticle-mediated DNA transfer using carbon-based materials as well as gene-nanoparticle (NP) (Vega-Vasquez et al. 2020).

4.2.4 Quantum Dots (QD)

Semiconducting nanocrystals with a broad excitation spectrum, known as quantum dots, are three-dimensional nanoparticles that emit specific light wavelengths and convert the exposed light spectrum into a distinct frequency of energy (Wang et al. 2006; Poonam and Deo 2008). The usual range of QDs is 2–10 nm in diameter. QDs are made up of a semiconductor core that has been overcoated with a shell (such as ZnS) to enhance optical characteristics and a cap that allows for better solubility in aqueous buffers (Ghasemi et al. 2009). In comparison to conventional methods, quantum dots (QDs) have shown efficacy in pest management. QDs can be applied in intracellular and extracellular research to identify the molecular causes of disease, and novel therapeutic action mechanisms (Valizadeh et al. 2012).

4.3 Application of Nano Particles in Plant Disease Management

Nanoparticles can be applied to plants in two ways: either (a) they provide crop protection, or (b) they deliver insecticides or other active ingredients, such as double-stranded RNA (dsRNA), and can be sprayed on seeds, foliar tissue, or roots (Hayles 2017; Worrall et al. 2018) (Fig. 4.1).

Many benefits can be obtained by using nanoparticles as carriers, including (i) longer shelf life, (ii) better solubility of pesticides that are not very soluble in water, (iii) reduced toxicity, and (iv) enhanced site-specific absorption by the target pest. Another possible advantage of nanocarriers might be the enhanced activity and longevity of the nanopesticides under environmental conditions, which would reduce the number of applications, diminish harmful effects, and lower prices for them (Hayles 2017; Worrall et al. 2018).

Nanoparticles are employed as protectants to shield plants from harmful microbes. Since they have strong antimicrobial properties and exhibit antagonistic effects against all pathogenic bacteria, viruses, and fungi, a variety of metal nanoparticles, such as copper, silver, zinc oxide, and titanium dioxide, have been successfully tested as potential protectants (Gogos et al. 2012; Kah and Hoffmann

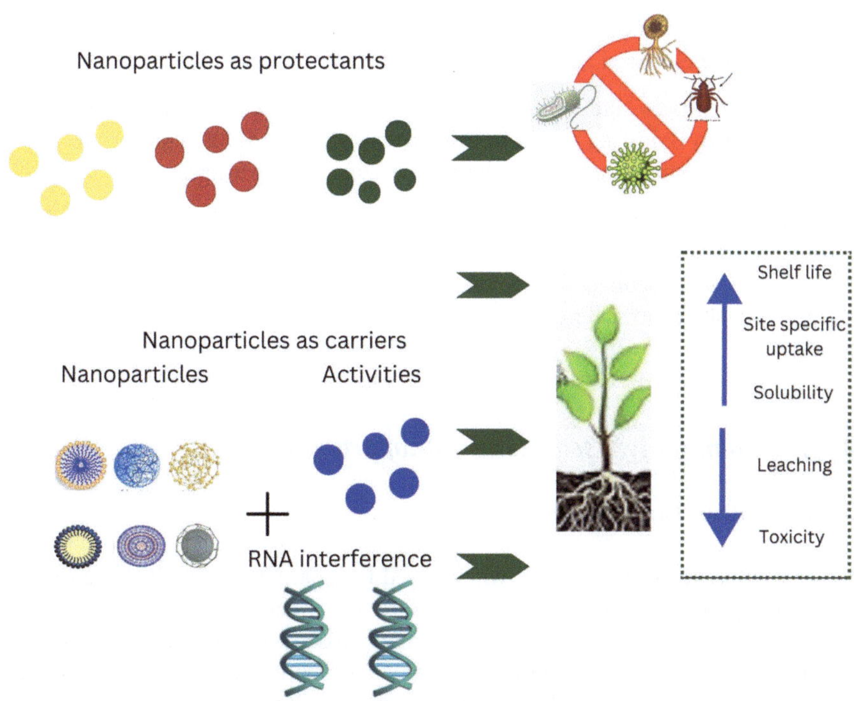

Fig. 4.1 Nanomaterials as protectants or carriers to provide crop protection

2014; Kim et al. 2018). These nanoparticles are easier to dissolve in water, have a longer shelf life, and display site-specific absorption when compared to conventional chemicals (Das and Pattanayak 2020).

4.4 Types of Nanoparticles

Among all types of NPs, metal and metal oxide NPs have been most thoroughly examined due to their excellent antimicrobial properties (Ingle et al. 2020). The most commonly used are silver (Ag), gold (Au), Zinc (Zn), zinc oxide (ZnO), titanium dioxide (TiO_2), cerium oxide (CeO_2), and copper oxide (CuO or Cu_2O) (Rico et al. 2015).

Metal nanoparticles have distinct physicochemical characteristics, including the following:

- A high ratio of surface area to volume.
- Facilitate chemical reactions very fast.
- Durable and resistant to deformation.
- Interesting Color.
- Display optical-like localised surface plasmon resonance in metal nanoparticles like gold and silver.
- Exhibit different melting points and thermal conductivity than bulk materials and have unique electrical and magnetic properties.

Metal nanoparticles are synthesized by reducing agents and metal oxide nanoparticles by oxidizing or precipitating agents (Sanchez-Dominguez et al. 2009).

There are two main approaches for synthesizing nano particles:

1. **Top-Down Approach:** In this subtractive process, a bulk starting material is broken down into smaller nanoparticles.
2. **Bottom-Up Approach:** In this additive process, precursor atoms or molecules mix to create nano scale structures (Fig. 4.2).

4.4.1 Commonly Used Metal Nanoparticles

4.4.1.1 Silver Nanoparticles

Silver nanoparticles are the most widely used nanoparticles (NPs) because of their distinctive property, namely their high surface to volume ratio which enhances the interaction with microbial pathogens (Thirunavoukkarasu et al. 2013). AgNPs are generally used in plant disease management against antibiotic resistant microorganisms (Panaek et al. 2006a, b). In agricultural practices, silver nanoparticles are

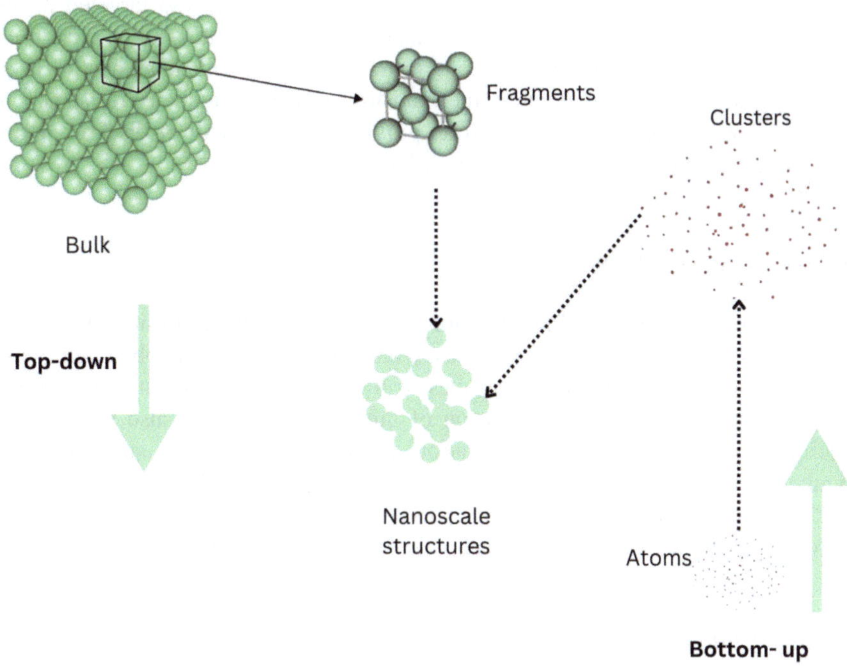

Fig. 4.2 Bottom-up and the top-down approaches in synthesis of carbon-based nanomaterials

employed as biopesticides, either through direct application to crops or by integrating them into seed coatings and soil treatments (Padmavathi and Chippada 2022).

4.4.1.2 Gold Nanoparticles

Gold nanoparticles (AuNPs) are significant in plant disease management because of their unique physicochemical properties and biocompatibility. A range of biomolecules can be functionalised with AuNPs to increase their efficacy and pathogen selectivity. By penetrating plant tissues and delivering antimicrobial chemicals directly to the site of infection, gold nanoparticles sprayed on crops can stop the spread of disease. Moreover, by generating systemic resistance, AuNPs might enhance plants' natural defence mechanisms (Prasad et al. 2017).

4.4.1.3 Zinc Nanoparticles

Zinc nanoparticles (ZnNPs) are useful in managing plant disease due to their wide spectrum antimicrobial abilities and essential role in plant nutrition. They are highly reactive and can interfere with the cellular functions of a variety of plant diseases, such as viruses, fungus, and bacteria. ZnNPs are applied to crops through foliar sprays, seed treatments, or soil amendments, ensuring direct and targeted action against pathogenic microorganisms. Zinc nanoparticles have antibacterial properties, but they also improve nutrient intake and growth, which benefits plant health. Their ability to stimulate plant defense mechanisms further strengthens resistance to infections. Zinc nanoparticles, therefore, provide a dual benefit of disease control and nutritional enhancement (Mittal et al. 2020).

4.4.1.4 Copper Nanoparticles

Copper nanoparticles (CuNPs) have an essential role in plant physiology and are useful in the treatment of plant diseases. CuNPs ruptures the microbial cell membranes, produce reactive oxygen species, and interferes with the enzymatic functions, effectively stops the progression of infections. Copper is also an essential micronutrient for plants, helping in various physiological functions such as photosynthesis, respiration, and the activation of defense enzymes. The dual functionality of CuNPs in providing both antimicrobial action and nutritional support makes them a promising alternative to traditional chemical pesticides (Antonio-Pérez et al. 2023).

4.4.1.5 Chitosan

Chitosan is a naturally occurring biopolymer that may have biological properties. Chitosan fragments are known to induce various defence responses in host plants in response to microbial infestations. Chitosan have gained more attention due to their biodegradability, compatibility and less toxicity. Due to its ease in preparation, it is a very helpful vehicle for gene transfer and medication delivery (El- Hadrami et al. 2010; Amanzadi et al. 2018).

4.4.2 Green Synthesis of Nano Particles

There are three different ways to create nanoparticles: chemically, physically, or biologically. However, creating nanoparticles by physical and chemical processes raises toxicity and environmental risks. The physical method requires a lot of space, a huge amount of energy, raises the temperature of the surrounding area and is time-consuming. The chemical method is beset with the use of toxic chemicals and solvents that might significantly harm the environment. Green nanotechnology, often referred to as green nanobiotechnology, has developed in response to the global demand for an alternative in the synthesis of nanoparticles by using biological systems. Green nanotechnology is simple, economical and eco-friendly and has since gained a lot of attention. Numerous nanoparticles produced using green nanotechnology have been successfully used in various fields (Pal et al. 2019) (Fig. 4.3).

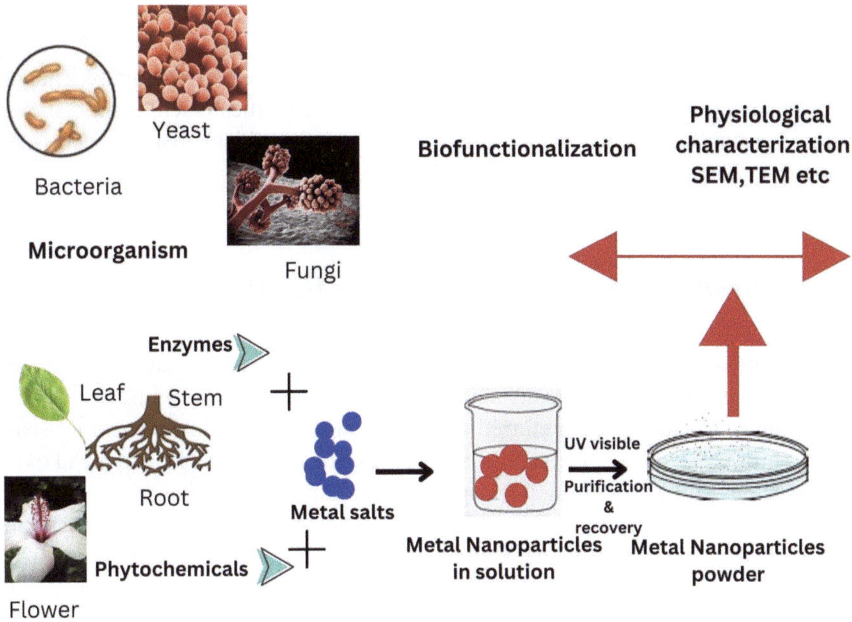

Fig. 4.3 Nanoparticle generation through green pathways

Green nanotechnology uses biological pathways such as bacteria, fungi, algae or plants for the synthesis of nanoparticles with the help of biotechnological techniques. The resulting nanoparticles are free of chemicals and are harmless to the environment (Pal et al. 2019).

4.4.3 Benefits of Green Synthesized NPs in Plant Disease Management

Nanomaterials are often synthesized using plant extracts, microorganisms, or other biological entities, ensuring that the production process is sustainable and minimizes ecological impact. Green NPs offer various advantages, which avoid the use of dangerous chemicals and are both economically and environmentally viable (Mehra et al. 2024). Green synthesized NPs help in targeted delivery and controlled release of antibiotics and deliver directly to the site of infection making it more effective at lower doses than traditional insecticides. By using this focused strategy, the overall chemical burden on the environment is decreased and collateral harm to helpful microbes is minimised (Weldick et al. 2022).

Green nanomaterials possess inherent antimicrobial properties. For example, it has been demonstrated that nanoparticles of zinc and silver oxide have potent fungicidal and bactericidal properties. These characteristics are further improved by functionalization, which makes them effective agents against a wide range of plant diseases.

Green nanomaterials which are synthesized using environmentally friendly methods and materials reported to possess unique properties (such as high surface area, adjustable size and shape, and specific functionalities) making them ideal candidates for targeted disease control (Wasule et al. 2024).

4.4.4 Synthesis of Nanoparticles Using Bacteria

Bacteria possess the unique ability to convert heavy metals into nanoparticles and have the vast capability to tolerate abiotic stress (Bridges et al. 1979). Bacteria due to their culture feasibility are preferred over other microbes like actinobacteria, yeast, fungi, virus etc. Different metal NPs, such as Ag, Au, Cu, Se and Fe and metal oxide NPs, such as silver oxide (Ag2O) copper oxide (CuO), zinc oxide (ZnO), titanium oxide (TiO2), manganese oxide (MnO2), magnesium oxide (MgO), iron oxide (Fe2O3), etc., have been manufactured using bacterial cells (Grasso et al. 2020). Bacteria can tolerate high metal concentrations and can reduce inorganic materials into NPs through extracellular and intracellular processes. Bacterial species like *Bacillus licheniformis*, *Pseudomonas deceptionensis*, *Pyrococcusfurio-sus*, *Pyrobaculum islandicum*, *Pseudomonas aeruginosa*, etc., have been recently explored synthesis of silver (Ag), gold (Au), iron (Fe), nickel (Ni), zinc (Zn) (Salem and Fouda 2021; Sanaeimehr et al. 2018; Saravanan et al. 2018).

4.4.5 Synthesis of NPs Using Fungi

Fungi are more efficient than bacteria due to their ability to handle more metal concentrations and accumulate metals. In “nano factories”, fungi may be readily cultivated on a massive scale, potentially producing NPs with exact dimensions and shapes (Azmath et al. 2016; Khan et al. 2017). The advantage of fungi over other microbes is that they are capable of producing massive proteins and enzymes that can be used for the sustainable production of NPs (Alghuthaymi et al. 2015). Metal NPs have been synthesized using *Trichoderma viride*, *Schizophyllum radiatuum*, *Fusarium oxysporum*, *F. solani*, *Cladosporium oxysporum*, *Rhizopus oryzae*, *Pleurotus ostreatus*, *Pencilium polonicum*, and others (Bhargava et al. 2016; Clarance et al. 2020; El Domany et al. 2018; Metuku et al. 2014; Mishra et al. 2014; Rajput et al. 2016).

4.4.6 *Synthesis of NPs Using Algae*

Algae can hyper-accumulate metals and convert them into NPs thus they are the ideal material for green synthesis (Dhillon et al. 2012). Some common kinds of algae, including blue-green algae (Cyanophyceae), brown algae (Phaeophyceae), green algae (Chlorophyceae), and red algae (Rhodophyceae), are used to create metallic and metal oxide NPs. Algae are inexpensive, simple to react, require little energy input, and don't require external reducing or capping makes them the best agents for green synthesis (Chaudhary et al. 2020). Micro algae such as *Lyngbya majuscule*, *Spirulina platensis*, and *Chlorella vulgaris* were used for the synthesis of silver NPs, which is a cost-effective method of nanoparticle synthesis (Chakraborty et al. 2009; Niu and Volesky 2000). Macro algae extract from *Ulva fasciata, Sargassum plagiophyllum*, *Caulerpa racemosa*, synthesis of silver NPs have done using as a reducing agent and these inhibited the growth of *Xanthomonas campestris* pv. *malvacearum* (Dhas et al. 2014; Edison et al. 2016). In addition to seaweeds, some diatoms (*Navicula atomus* and *Diadesmis gallica*) have the ability to synthesize gold NPs, gold, and silica–gold (Schrofel et al. 2011).

4.4.7 *Synthesis of NPs Using Plant Extracts*

Terpenoids, polyphenols, sugars, alkaloids, phenolic acids, and proteins are plant metabolites that can reduce metal ions to NPs (Makarov et al. 2014). NPs have been synthesized using a variety of plant parts, including leaves, roots, flowers, fruit, seeds, rhizomes, and more (Ahmad et al. 2017). Vanlalveni et al. (2021) have done extensive study on the production of NPs utilizing plant extracts from various plant components. Plant species to produce NPs has been termed a green technique and the most reliable environmentally sustainable approach (Khan et al. 2022). Biological synthesis includes the use of natural reducing, capping and stabilizing agents and using safe, sustainable and low consumption methods (Antunes et al. 2023; Khan et al. 2022).

4.5 Mechanism of Antimicrobial Activity by NPs

According to modern studies, NPs have shown hope in managing plant diseases through their antibacterial activity. According to Wang et al. (2017), the antimicrobial properties of NPs are due to the cooperative effect of numerous interrelated mechanisms, including metal ion release, non-oxidative processes, oxidative stress, and cell death. Blocking critical cellular processes like DNA replication and protein synthesis is one move to start antibacterial effects inside the cell (Aziz et al. 2015; Lee et al. 2019).

Silver NPs (Ag NPs) have been employed extensively in plant disease management due to their diverse antibacterial activity mechanisms (Cheng et al. 2016). According to experimental findings, silver NPs harm bacterial cells by producing too many Reactive Oxygen Species (ROS) (Qayyum et al. 2017). The bacterial plant pathogens *Ralstonia solanacearum* and *Xanthomonas axonopodis* pv. *punicae* have been the subject of research on this topic (Vanti et al. 2020). ROS are released by NPs and act with the lipids in the membrane, breaking off its integrity. This may result in a ruptured membrane and disturbance of the membrane's permeability, ultimately causing cell death (Paulkumar et al. 2014).

By creating strong connections with the thiol groups of bacterial enzymes, silver NPs also demonstrate their effectiveness. They actively interact with the plasma membrane's enzymes, changing the permeability of the cell membrane and impeding the generation of ATP. The enzymes succinate dehydrogenase and NADH dehydrogenase II are inhibited by silver NPs (Zhang et al. 2020). According to one such process, AgNPs elute Ag+ ions when they come into contact with moisture. The microbial development processes are inhibited by these silver ions (Tang et al. 2018). When used in agriculture, silver NPs are more successful than chemicals at controlling plant diseases at lower dosages, which has led to cheaper cultivation expenses and a more sustainable environment (Dutta et al. 2017).

With their exceptional antibacterial properties against a variety of microbes, including some that are resistant to several medications, copper NPs have become well-known antimicrobial agents. Additionally, the synthesis of copper NPs is a cost-effective procedure due to the abundance and affordability of copper. In the agricultural sector, copper NPs have been used to successfully inhibit the growth of fungus, such as *Phytophthora infestan*s, which causes tomato late blight disease. Copper metal's fungicidal qualities have been greatly improved by its conversion into NPs, increasing its effectiveness against several diseases like *Phytophthora* sp. and *Corticium salmonicolor* (Kanhed et al. 2014). By disrupting the transcription pathway, AuNPs cause harmful DNA damage (Rai et al. 2010). Likewise, research has shown that AuNPs' antibacterial action functions at different transcriptional and protein synthesis levels (Rim et al. 2013; Cui et al. 2012).

Viral illnesses can be efficiently managed with NPs. Studies have shown that NPs cause SAR (systemic acquired resistance) in plants and mostly disrupt the early stages of the virus replication cycle. The release of metal ions followed by oxidative and other non-oxidative stressors could be one of the inhibitory mechanisms that NPs employ. Nucleic acids (DNA or RNA), capsid protein, and other protein components are all inactivated and denaturated as part of the underlying mechanism. The NPs may also prevent virion fusion, binding, replication, and infectivity. The generation of reactive oxygen species (ROS) and subsequent protein oxidation result in inactivity and virucidal activity. Furthermore, the NPs may stop the virus from identifying and infiltrating the host plant. Glycoprotein receptors allow the NPs to bind with viral surface proteins.

Various biotic and abiotic stressors cause plants to release reactive oxygen species. When ROS levels surpass the threshold, the equilibrium between ROS and antioxidants is disrupted, and oxidative products are generated. The impacts of oxidants are offset by plants' antioxidant systems. The antioxidant system is composed of enzymes such as superoxide dismutase (SOD), catalase (CAT), ascorbate peroxidase (APX), and guaiacol peroxidase (GPX) (Tang et al. 2018). NPs affect cellular redox equilibrium by either raising or decreasing the incidence of oxidative stress (Soares et al. 2018). Previous studies have shown that NPs can either prevent the oxidative burst or promote the production of ROS, depending on the requirements of the host plant. NPs decrease the oxidative burst by inducing an antioxidant mechanism.

According to Rastogi et al. (2017), NPs are known to affect the hormonal balance of plants, causing the right defense reactions. Gibberellin (GA), strigolactone (SL), cytokinin (CK), auxin [indole-3-acetic acid (IAA)], abscisic acid (ABA), and brassinosteroids (BRs) are among the hormones that control plant growth and can control defense responses. Ethylene (ET), salicylic acid (SA), and jasmonic acid (JA) are the signaling chemicals involved in plant defense. Crosstalk between several plant hormones controls the balance between plant development and. According to studies, the many pathways are part of a complex signaling network that influences one another through both positive and negative regulatory networks rather than operating separately. When it comes to controlling plant diseases, NPs have the ability to upset this equilibrium. When exposed to CuONPs, Landa et al. (2017) observed increased levels of SA and JA as well as the transcription of genes involved in Arabidopsis thaliana. Cai et al. (2019) found that Nicotiana benthamiana had higher levels of SA and ABA but no change in JA. Therefore, the aforementioned findings imply that the expression of a specific plant hormone is dependent on the timing, dose, and unusual interaction between NPs and plants.

AgNP treatment given 24 hours after infection successfully avoided all of the harmful symptoms brought on by BYMV (Bean Yellow Mosaic), according to research by Elbeshehy et al. (2015). The virus concentration, infection rate, and disease severity were decreased in the treated plants. However, no effect was seen when the NPs were applied prior to the viral infection. These findings imply that because AgNP activity promotes the treated plants' systemic acquired resistance, it is more effective in the early stages of viral multiplication. After applying an aqueous solution of AgNPs at a concentration of 50 parts per million, Jain (2014) demonstrated that AgNPs are a potent antiviral agent and that the Sun hemp rosette virus (SHRV) illness was completely suppressed on cluster beans.

4.6 Limitations and Challenges of NPs in Agriculture

Nanotechnology has been described as multidisciplinary science of manipulating matters at the nanoscale with novel physical, chemical, and biological properties. It can be easily understood that designing a nanoparticle requires very specific structural and physicochemical properties. Any variation in size and structure can affect the function and performance of the designed nanoparticle conjugate (Mitter et al. 2017).

The biosafety and toxicity of nanopesticides to human health and the environment are significant concerns when they are applied for disease management. There is a chance that nanocomposites will land on leaves or flowers and affect animals, birds, honey bees, and other creatures. Additionally, NPs are thought to block stomatal pores, making it more difficult for pollen grains to reach the stigma. It is thought that the NPs enter the vascular tissues and block the movement of water and minerals. Several plant species were shown to be negatively impacted by nano-silver treatment, which also altered the biomass, extracellular enzyme activity, and composition of the microbial community (Abdelaal et al. 2020). Further research is required to understand how NPs enter the food chain, bioaccumulate there, and interact with other environmental contaminants by NPs (Marwal 2019). One crucial issue that needs investigation is how nano formulations interact with other environmental contaminants. The degree of toxicity of a given nanopesticide is closely correlated with its surface and surface characteristics, which might have negative impacts on humans, plants, and microorganisms at various trophic levels (Dutta et al. 2022). The fate, behavior, and impacts of nano formulations in soil with regard to their mobility and transformation under various environmental conditions are the subject of numerous research (Lewis et al. 2016; Queyrel et al. 2016).The plant system's absorption, translocation, and biotransformation pathways all affect how harmful a nanoformulation is in any given plant system. Adak et al. (2012) observed that compared to traditional pesticides, there was reduced buildup of nanopesticide on the soybean seeds and soil in an experiment using a PEG-coated imidacloprid nano formulation.

It has been demonstrated that when zinc NPs are introduced into the soil and absorbed by plants, they affect those plants. In the instance of *Vigna unguiculata*, Singh et al. (2023) discovered that the root membrane could degrade as a result of the cationic ions of ZnNPs moving across the cell membrane via endocytosis. Because NPs can damage biomolecules like proteins and nucleic acids and cause chemical reactions, their presence can be harmful to cells. Additionally, lipids and other membrane constituents may be oxidized by NPs, and reactive oxygen species may be produced in excess. For example, it has been discovered that zinc NPs are genotoxic, damaging DNA in plants such as onions, impairing mustard seed germination, and damaging soybean leaves at certain elevated doses (Ali et al. 2018).

4.7 Conclusion and Prospects

The degree of toxicity of many NPs is still unknown, even if there is a wealth of information about certain nanomaterials. As a result, the lack of knowledge on these materials' risk assessments restricts their application. The study emphasizes that scientific evaluations of the long-term advantages and hazards of nanotechnology should serve as the foundation for any future developments in this field.

References

Abdelaal KA, EL-Shawy EA, Hafez YM, Abdel Dayem SM, Chidya RCG, Saneoka H (2020) Nano-silver and non-traditional compounds mitigate the adverse effects of net blotch disease of barley in correlation with up-regulation of antioxidant enzymes. Pak J Bot 52(3):1065–1072

Abd-Elsalam KA, Alghuthaymi MA (2015) Nanobiofungicides: are they the next generation of fungicides? J Nanotechnol Mater Sci 2:1–2. https://doi.org/10.15436/2377-1372.15.013

Adak T, Kumar J, Dey D, Shakil NA, Walia S (2012) Residue and bioefficacy evaluation of controlled release formulations of imidacloprid against pests in soybean (Glycine max). J Environ Sci Health Part B 47:226–231. https://doi.org/10.1080/03601234.2012.634368

Ahmad A, Wei Y, Syed F, Tahir U, Rehman AU, Khan U, Ullah S, Yuan Q (2017) The effects of bacteria-nanoparticles interface on the antibacterial activity of green synthesized silver nanoparticles. Microb Pathog 102:133–142

Alghuthaymi MA, Almoammar H, Rai M, Said-Galiev E, Abd-Elsalam KA (2015) MycoNPs: synthesis and their role in phytopathogens management. Biotechnol Biotechnol Equip 29:221–236. https://doi.org/10.1080/13102818.2015.1008194

Ali S, Shafique O, Mahmood T, Hanif MA, Ahmed I, Khan BA (2018) A review about perspectives of nanotechnology in agriculture. Pak J Agric Res 30(2):116–121

Amanzadi B, Shariati S, Fanaei HR (2018) Nanoparticles and their potential application in plant disease management. J Plant Prot Res 58(2):93–108. https://doi.org/10.24425/119127

Antonio-Pérez A, Durán-Armenta LF, Pérez-Loredo MG, Torres-Huerta AL (2023) Biosynthesis of copper nanoparticles with medicinal plants extracts: from extraction methods to applications. Micromachines 14(10):1882. https://doi.org/10.3390/mi14101882

Antunes Filho S, Dos Santos MS, Dos Santos OAL, Backx BP, Soran ML, Opriş O, Lung I, Stegarescu A, Bououdina M (2023) Biosynthesis of nanoparticles using plant extracts and essential oils. Molecules 28(7):3060. https://doi.org/10.3390/molecules28073060

Aziz N, Faraz M, Pandey R, Shakir M, Fatma T, Varma A, Barman I, Prasad R (2015) Facile algae-derived route to biogenic silver NPs: synthesis, antibacterial, and photocatalytic properties. Langmuir 31(42):11605–11612. https://doi.org/10.1021/acs.langmuir.5b03081

Azmath P, Baker S, Rakshith D, Satish S (2016) Mycosynthesis of silver NPs bearing antibacterial activity. Saudi Pharm J 24:140–146. https://doi.org/10.1016/j.jsps.2015.01.008

Balaure PC, Gudovan D, Gudovan I (2017) Nanopesticides: a new paradigm in crop protection. In: New pesticides and soil sensors. Academic Press, pp 129–192

Bhargava A, Jain N, Khan MA, Pareek V, Dilip RV, Panwar J (2016) Utilizing metal tolerance potential of soil fungus for efficient synthesis of gold NPs with superior catalytic activity for degradation of rhodamine B. J Environ Manage 183:22–32. https://doi.org/10.1016/j.jenvman.2016.08.021

Bhattacharjee R, Dey U (2014) An overview of fungal and bacterial biopesticides to control plant pathogens/diseases. Afr J Microbiol Res 8(17):1749–1762

Bordes P, Pollet E, Avérous L (2009) Nano-biocomposites: biodegradable polyester/nanoclay systems. Prog Polym Sci 34:125–155. https://doi.org/10.1016/j.progpolymsci.2008.10.002

Bridges K, Kidson A, Lowbury E, Wilkins M (1979) Gentamicin-and silver-resistant pseudomonas in a burns unit. Br Med J 1:446–449

Cai L, Liu C, Fan G, Liu C, Sun X (2019) Preventing viral disease by ZnONPs through directly deactivating TMV and activating the plant immunity in Nicotiana benthamiana. Environ Sci Nano 6:3653–3669. https://doi.org/10.1039/C9EN00850K

Chakraborty N, Banerjee A, Lahiri S, Panda A, Ghosh AN, Pal R (2009) Biorecovery of gold using cyanobacteria and an eukaryotic alga with special reference to nanogold formation—a novel phenomenon. J Appl Phycol 21:145–152

Chaudhary R, Nawaz K, Khan AK, Hano C, Abbasi BH, Anjum S (2020) An overview of the algae-mediated biosynthesis of NPs and their biomedical applications. Biomol Ther 10(11):1498. https://doi.org/10.3390/biom10111498

Cheng G, Dai M, Ahmed S, Hao H, Wang X, Yuan Z (2016) Antimicrobial drugs in fighting against antimicrobial resistance. Front Microbiol 7:470. https://doi.org/10.3389/fmicb.2016.00470

Clarance P, Luvankar B, Sales J, Khusro A, Agastian P, Tack JC, Al Khulaifi MM, Al-Shwaiman HA, Elgorban AM, Syed A et al (2020) Green synthesis and characterization of gold NPs using endophytic fungi fusarium solani and its in-vitro anticancer and biomedical applications. Saudi J Biol Sci 27:706–712. https://doi.org/10.1016/j.sjbs.2019.12.026

Cui Y, Zhao Y, Tian Y, Zhang W, Lü X, Jiang X (2012) The molecular mechanism of action of bactericidal gold NPs on Escherichia coli. Biomaterials 33(7):2327–2333

Das S, Pattanayak S (2020) Nanotechnological approaches in sustainable agriculture and plant disease management. IntechOpen. https://doi.org/10.5772/intechopen.92463

Dhas TS, Kumar VG, Karthick V, Angel KJ, Govindaraju K (2014) Facile synthesis of silver chloride NPs using marine alga and its antibacterial efficacy. Spectrochim Acta Part A Mol Biomol Spectrosc 120:416–420. https://doi.org/10.1016/j.saa.2013.10.044

Dhillon GS, Brar SK, Kaur S, Verma M (2012) Green approach for nanoparticle biosynthesis by fungi: current trends and applications. Crit Rev Biotechnol 32:49–73. https://doi.org/10.3109/07388551.2010.550568

Dutta S, Das P, Dubey S, Maiti S, Kumar M, Varma S, Prasad R (2017) Nanotechnology applications in sustainable agriculture. Environ Chem Lett 15(4):591–605. https://doi.org/10.1007/s10311-017-0653-5

Dutta P, Arti K, Madhusmita M, Biswas KK, Agnieszka D, Thakuria ASA, Singh BS, Gomathy M, Sabarinathan KG, Mihir KM, Mazumdar N (2022) Advances in nanotechnology as a potential alternative. Front Microbiol 13. https://doi.org/10.3389/fmicb.2022.935193

Edison TNJI, Atchudan R, Kamal C, Lee YR (2016) Caulerpa racemosa: a marine green alga for eco-friendly synthesis of silver NPs and its catalytic degradation of methylene blue. Bioprocess Biosyst Eng 39:1401–1408. https://doi.org/10.1007/s00449-016-1616-7

El Domany EB, Essam TM, Ahmed AE, Farghali AA (2018) Biosynthesis physico-chemical optimization of gold NPs as anti-cancer and synergetic antimicrobial activity using Pleurotus ostreatus fungus. J Appl Pharma Sci 8:119–128

El Hadrami A, Adam LR, El Hadrami I, Daayf F (2010) Chitosan in plant protection. Mar Drugs 8(4):968–987. https://doi.org/10.3390/md8040968

Elbeshehy EK, Elazzazy AM, Aggelis G (2015) Silver NPs synthesis mediated by new isolates of Bacillus spp., nanoparticle characterization and their activity against Bean Yellow Mosaic Virus and human pathogens. Front Microbiol 6:453. https://doi.org/10.3389/fmicb.2015.00453

FAO (2022) FAOSTAT (Food and Agriculture Organization of the United Nations Statistical Database). Retrieved from https://www.fao.org/3/cc2211en/cc2211en.pdf

Fraceto LF, Grillo R, de Medeiros GA, Scognamiglio V, Rea G, Bartolucci C, Rai M (2020) Nanotechnology in agriculture: which innovation potential does it have? Front Environ Sci 8:4. https://doi.org/10.3389/fenvs.2020.00004

Ghasemi Y, Peymani P, Afifi S (2009) Quantum dot: magic nanoparticle for imaging, detection and targeting. Acta Biomed 80(2):156–165

Gogos A, Knauer K, Bucheli TD (2012) Nanomaterials in plant protection and fertilization: current state, foreseen applications, and research priorities. J Agric Food Chem 60:9781–9792

Grasso G, Zane D, Dragone R (2020) Microbial nanotechnology: challenges and prospects for green biocatalytic synthesis of nanoscale materials for sensoristic and biomedical applications. Nanomaterials 10(1):11. https://doi.org/10.3390/nano10010011

Gruere GP (2012) Implications of nanotechnology growth in food and agriculture in OECD countries. Food Pol 37:191–198. https://doi.org/10.1016/j.jhazmat.2014.05.079

Hayles K (2017) Nanotechnology applications in agriculture and food processing. In: Mukhopadhyay R (ed) Nanotechnology in agriculture. CRC Press, pp 67–89

Ingle AP, Biswas A, Vanlalveni C, Lalfakzuala R, Gupta I, Ingle P, Rokhum L, Rai M (2020) Biogenic synthesis of nanoparticles and their role in the management of plant pathogenic fungi. In: Microbial nanotechnology, pp 135–161

Jain D (2014) Green synthesis of silver NPs and their application in plant virus inhibition. J Mycol Plant Pathol 44:21–24

Kah M, Hofmann T (2014) Nanopesticide research: current trends and future priorities. Environ Int 63:224–235

Kanhed P, Birla S, Gaikwad S, Gade A, Seabra AB, Rubilar O (2014) In vitro antifungal efficacy of copper NPs against selected crop pathogenic fungi. Mater Lett 115:13–17

Khan NT, Khan MJ, Jameel J, Jameel N, Rheman SUA (2017) An overview: biological organisms that serve as nanofactories for metallic NPs synthesis and fungi being the most appropriate. Bioceramics Dev App 7:101

Khan F, Shariq M, Asif M, Siddiqui MA, Malan P, Ahmad F (2022) Green nanotechnology: plant-mediated nanoparticle synthesis and application. Nanomaterials 12:673. https://doi.org/10.3390/nano12040673

Kim DY, Kadam A, Shinde S, Saratale RG, Ghodake G et al (2018) Recent developments in nanotechnology transforming the agricultural sector: a transition replete with opportunities. J Sci Food Agr 98:849–864

Landa P, Dytrych P, Prerostova S, Petrova S, Vankova R (2017) Transcriptomic response of Arabidopsis thaliana exposed to CuO NPs, bulk material, and ionic copper. Environ Sci Tech 51:10814–10824. https://doi.org/10.1021/acs.est.7b02265

Lee NY, Ko WC, Hsueh PR (2019) NPs in the treatment of diseases caused by multi-drug resistant organisms. Front Pharm 10:1153

Lewis SE, Silburn DM, Kookana RS, Shaw M (2016) Pesticide behavior, fate, and effects in the tropics: an overview of the current state of knowledge. J Agric Food Chem 64:3917–3924. https://doi.org/10.1021/acs.jafc.6b01320

Makarov VV, Love AJ, Sinitsyna OV, Makarova SS, Yaminsky IV, Taliansky ME, Kalinina NO (2014) "Green" nanotechnologies: synthesis of metal NPs using plants. Act Naturae 6(1):35–44

Marwal A, Gaur R (2019) Nanophytovirology: an emerging field for disease management. In: Goyal A (ed) Plant diseases-current threats and management trends. IntechOpen. https://doi.org/10.5772/intechopen.86653

Mehra V, Kumar S, Tamang AM et al (2024) Green synthesis of gold NPs (AuNPs) by using plant extract wsand their biological application: a review. Bio NanoSci 15:18. https://doi.org/10.1007/s12668-024-01703-7

Metuku RP, Pabba S, Burra S, Gudikandula K, Charya MS (2014) Biosynthesis of silver NPs from Schizophyllum radiatum HE 863742.1: their characterization and antimicrobial activity. Biotech 4:227–234. https://doi.org/10.1007/s13205-013-0138-0

Mishra A, Kumari M, Pandey S, Chaudhry V, Gupta K, Nautiyal C (2014) Biocatalytic and antimicrobial activities of gold NPs synthesized by Trichoderma sp. Bioresour Technol 166: 235–242. https://doi.org/10.1016/j.biortech.2014.04.085

Mittal D, Kaur G, Singh P, Yadav K, Ali SA (2020) Nanoparticle-based sustainable agriculture and food science: recent advances and future outlook. Front Nanotechnol 2:579954. https://doi.org/10.3389/fnano.2020.579954

Mitter N, Worrall EA, Robinson KE, Jain LP, Taochy RGC (2017) Clay nanosheets for topical delivery of RNAi for sustained protection against plant viruses. Nat Plant 3:16207. https://doi.org/10.1038/nplants.2016.207

Mukhopadhyay R (ed) (2014) Nanotechnology in agriculture. CRC Press

Neme K, Nafady A, Uddin S, Tola TB (2021) Application of nanotechnology in agriculture, postharvest loss reduction and food processing: food security implication and challenges. Heliyon 7(12):1–12

Niu H, Volesky B (2000) Gold-cyanide biosorption with L-cysteine. J Chem Technol Biotechnol 75:436–442

Nongbet A, Mishra AK, Mohanta YK, Mahanta S, Ray MK, Khan M, Baek KH, Chakrabartty I (2022) Nanofertilizers: a smart and sustainable attribute to modern agriculture. Plants (Basel) 11(19):2587. https://doi.org/10.3390/plants11192587. PMID: 36235454; PMCID: PMC9573764

Padmavathi S, Chippada SA (2022) Nanotechnology in plant disease management - an overview. J Adv Sci Res 13(10):01–06

Pal G, Rai P, Pandey A (2019) Green synthesis of NPs: a greener approach for a cleaner future. In: Green synthesis, characterization and applications of NPs. Elsevier, pp 1–26. https://doi.org/10.1016/b978-0-08-102579-6.00001-0

Panaek A, Kvitek L, Prucek R, Kolar M, Veerova R, Pizurova N, Sharma VK, Nevena T, Zboril R (2006a) Silver colloid NPs: Synthesis, characterization and their antibacterial activity. J Phys Chem B 110:16248–16253

Panaek P, Sandbhy J, Kumar V (2006b) Application of nanotechnology in agriculture. J Nanotechnol Nanosci 6(3):245–257

Parveen S, Shaba A, Choudhary S, JaiswaL S (2022) Review: application of nanosensor in agriculture. Int Res J Modern Eng Technol Sci 4(9):2153–2157

Paulkumar K, Gnanajobitha G, Vanaja M, Rajeshkumar S, Malarkodi C, Pandian K, Annadurai G (2014) Piper nigrum leaf and stem assisted green synthesis of silver NPs and evaluation of its antibacterial activity against agricultural plant pathogens. Scientific World Journal 2014: 829894. https://doi.org/10.1155/2014/829894

Poonam P, Deo N (2008) Current correlation functions for chemical sensors based on DNA decorated carbon nanotube. N Sens Actuators B: Chem 135(1):327–335

Prasad R, Gupta N, Kumar M, Kumar V, Wang S, Abd-Elsalam KA (2017) Nanomaterials act as plant defense mechanism. In: Prasad R, Kumar V, Kumar M (eds) Nanotechnology. Springer, Singapore, pp 267–293. https://doi.org/10.1007/978-981-10-4678-0_14

Qayyum S, Oves M, Khan AU (2017) Obliteration of bacterial growth and biofilm through ROS generation by facilely synthesized green silver NPs. PLoS One 12(8):e0181363

Queyrel W, Habets F, Blanchoud H, Ripoche D, Launay M (2016) Pesticide fate modeling in soils with the crop model STICS: feasibility for assessment of agricultural practices. Sci Total Environ 542:787–802. https://doi.org/10.1016/j.scitotenv.2015.10.066

Rai A, Prabhune A, Perry CC (2010) Antibiotic mediated synthesis of gold NPs with potent antimicrobial activity and their application in antimicrobial coatings. J Mater Chem 20(32): 6789–6798

Rajput S, Werezuk R, Lange RM, McDermott MT (2016) Fungal isolate optimized for biogenesis of silver NPs with enhanced colloidal stability. Langmuir 32:8688–8697

Rastogi A, Zivcak M, Sytar O, Kalaji H, Xiaolan H, Mbarki S et al (2017) Impact of metal and metal oxide NPs on plant: a critical review. Front Chem 5:78. https://doi.org/10.3389/fchem.2017.00078

Rico CM, Peralta-Videa JR, Gardea-Torresdey JL (2015) Chemistry, biochemistry of NPs, and their role in antioxidant defense system in plants. In: Nanotechnology and plant sciences. Springer, pp 1–17

Rim KT, Song SW, Kim HY (2013) Oxidative DNA damage from nanoparticle exposure and its application to workers' health: a literature review. Saf Health Work 4:177–186. https://doi.org/10.1016/j.shaw.2013.07.006

Sai SK, Bharadwaj S, Das J, Dutta S (2018) Nano-bio interactions: an integrated approach. In: Hayles K (ed) Nanotechnology. Springer, Berlin, Heidelberg, pp 123–145. https://doi.org/10.1007/978-3-662-56789-9_6

Sai P, Kumar V, Garkot A (2020) Applications of nanotechnology in plant disease management. Int J Sci Res 9(4):1506–1508

Salem SS, Fouda A (2021) Green synthesis of metallic NPs and their prospective biotechnological applications: an overview. Biol Trace Elem Res 199:344–370. https://doi.org/10.1007/s12011-020-02138-3

Sanaeimehr Z, Javadi I, Namvar F (2018) Antiangiogenic and antiapoptotic effects of green-synthesized zinc oxide NPs using sargassum muticum algae extraction. Cancer Nanotechnol 9:1–16. https://doi.org/10.1186/s12645-018-0037-5

Sanchez-Dominguez M, Boutonnet M, Solans C (2009) A novel approach to metal and metal oxide nanoparticle synthesis: the oil-in-water microemulsion reaction method. J Nanopart Res 11(7): 1823–1829. https://doi.org/10.1007/s11051-009-9660-8

Saravanan M, Barik SK, MubarakAli D, Prakash P, Pugazhendhi A (2018) Synthesis of silver NPs from bacillus brevis (ncim 2533) and their antibacterial activity against pathogenic bacteria. Microb Pathog 116:221–226. https://doi.org/10.1016/j.micpath.2018.01.038

Schrofel A, Kratošová G, Krautová M, Dobročka E, Vávra I (2011) Biosynthesis of gold NPs using diatoms—silica-gold and EPS-gold bionanocomposite formation. J Nanopart Res 13:3207–3216

Sekhon BS (2014) Nanotechnology in agri-food production: an overview. Nanotechnol Sci Appl 7: 31–53

Singh LB, Zeshmarani S, Hijam C, Qutub MKM, Singh YH, Prabin S, Singh OS (2023) Nanotechnology and its role in plant pathology. Pharma Innov J 12(12):37–49

Sinha K, Ghosh J, Sil PC (2017) New pesticides: a cutting-edge view of contributions from nanotechnology for the development of sustainable agricultural pest control A2—Grumezescu, AlexandruMihai. In: New pesticides and soil sensors. Academic Press, Cambridge, p 5

Slattery M, Harper B, Harper S (2019) Pesticide encapsulation at the nanoscale drives changes to the hydrophobic partitioning and toxicity of an active ingredient. Nanomaterials 9:81. https://doi.org/10.3390/nano9010081

Soares S, Sousa J, Pais A, Vitorino C (2018) Nanomedicine: Principles, Properties, and Regulatory Issues. Frontiers in chemistry 6:360. https://doi.org/10.3389/fchem.2018.00360

Tang S, Vashisth T, Deshmukh R (2018) Nanotechnology in enhancing nutrient use efficiency in plants. Front Plant Sci 9:1127. https://doi.org/10.3389/fpls.2018.01127

Thirunavoukkarasu M, Balaji U, Behera S, Panda PK, Mishra BK (2013) Biosynthesis of silver nanoparticle from leaf extract of Desmodium gangeticum (L.) DC. and its biomedical potential. Spectrochim Acta Part A Mol Biomol Spectrosc 116:424–427

Valizadeh A, Mikaeili H, Samiei M, Farkhani SM, Zarghami N, Kouhi M, Akbarzadeh A (2012) Quantum dots: synthesis, bioapplications, and toxicity. Nanomed Nanotechnol Biol Med 8(2): 123–130. https://doi.org/10.1016/j.nano.2011.08.013

Vanlalveni C, Lallianrawna S, Biswas A, Selvaraj M, Changmai B, Rokhum SL (2021) Green synthesis of silver NPs using plant extracts and their antimicrobial activities: a review of recent literature. RSC Adv 11:2804–2834

Vanti GL, Kurjogi M, Basavesha KN, Teradal NL, Masaphy S, Nargund VB (2020) Synthesis and antibacterial activity of Solanum torvum mediated silver nanoparticle against Xanthomonas axonopodis pv. punicae and Ralstonia solanacearum. J Biotechnol 309:20–28

Vega-Vásquez P, Mosier NS, Irudayaraj J (2020) Nanoscale drug delivery systems: from medicine to agriculture. Front Bioeng Biotechnol 8:79. https://doi.org/10.3389/fbioe.2020.00079

Wang L, O'Donoghue MB, Tan W (2006) NPs for multiplex diagnostics and imaging. Nanomed 1(4):413–426

Wang L, Hu C, Shao L (2017) The antimicrobial activity of NPs: present situation and prospects for the future. Int J Nanomedicine 12:1227–1249

Wasule DL, Shingote PR, Saxena S (2024) Exploitation of functionalized green nanomaterials for plant disease management. Discover Nano 19:118. https://doi.org/10.1186/s11671-024-04063-z

Weldick PJ, Wang A, Halbus AF, Paunov VN (2022) Emerging nanotechnologies for targeting antimicrobial resistance. Nanoscale 14(11):4018–4041

Worrall EA, Hamid A, Mody KT, Mitter N, Pappu HR (2018) Nanotechnology for plant disease management. Agronomy 8(285):1–24

Yadav A, Kusum Y, Kamel Abd-Elsalam A (2023) Nanofertilizers: types, delivery and advantages in agricultural sustainability. Agrochemophysica 2(2):296–336. https://doi.org/10.3390/agrochemicals2020019

Zhang Y, Pan X, Liao S, Jiang C, Wang L, Tang Y, Wu G, Dai G, Chen L (2020) Quantitative proteomics reveals the mechanism of silver NPs against multidrug-resistant Pseudomonas aeruginosa biofilms. J Proteome Res 19(8):3109–3122. https://doi.org/10.1021/acs.jproteome.0c00114

Chapter 5
Nanosensors for Monitoring Plant Health

Anju Susan, K. R. Raghi, and T. M. Archana

Abstract Plant science maintains a firm position in major fields related to the environment and life on earth. The crucial role of plants is in food production and nurture environment balance and biodiversity. To fulfill the role it is important to have healthy plants. The growth and development are influenced by many components such as, environmental changes, like drought, cold, heat, salinity, presence of adequate amount of nutrients. In addition, the plants can be affected by various types of pathogens and may cause diseases which disturbs further growth. Large amounts of pesticides are used in order to overcome the pathogen attacks. Also, fertilizers are employed to deal with the nutrient deficit. However, inefficient use of these may result in contamination in plants. Conventional methods are available for the accurate determination of all these, though they are limited in applications due to the cost, time and man resources connected with these techniques. Progress in nanotechnology productively replaced these by nano sensors. Those are sensors in nano range and are capable of real-time monitoring with high-resolution and accuracy in less time. These provide micro-scale information, sometimes even at single molecule level, regarding the physiological processes and are able to interpret the same. This leap assists agricultural production and related areas. A brief overview of the application of nano sensors in detecting environmental stress, drought conditions, monitoring of nutrients, pathogens, plant metabolism is presented in this chapter.

Keywords Chemical contaminants · Environmental stress · Food security · Nutrient monitoring · Pathogens · Plant metabolism · Real-time analysis

---

A. Susan
Department of Chemistry, Mar Athanasius College (Autonomous), Ernakulam, Kerala, India

K. R. Raghi
Department of Integrated Chemistry, MES College, Marampally, Ernakulam, Kerala, India

T. M. Archana (✉)
Postgraduate and Research Department of Microbiology, Sree Sankara College, Kalady, Ernakulam, Kerala, India
e-mail: archana@ssc.edu.in

J. M. Al-Khayri et al. (eds.), *Nanotechnology Applications in Modern Agriculture*, Nanotechnology in Plant Sciences 3, https://doi.org/10.1007/978-3-031-90506-3_5

5.1 Introduction

The research in the field of plants is multidisciplinary in nature, since it serves important roles in the field of production of both food and materials, environmental management and biodiversity, and also in the field of energy production (Shaw and Honeychurch 2022). This implies that monitoring the plant health on a day-to-day basis is crucial in many aspects like assisting farmers to maximize the yield in crop production, reducing the risk of crop loss due to diseases, optimal dosage of pesticides, herbicides etc. The development of diseases in plants are caused by many factors, like environment (soil, climate), weather (temperature, rain, wind, and hail), age of the plant, genetics of the plants, single versus mixed infections, and genetics of the pathogen populations (Yusof and Isha 2020). The response of plants towards these interactions serves as a direct indicator of the plants' growth status and their ability to adapt to environmental changes. While traditional plant phenotypic technologies are available and those includes, remote sensing (RS), geographic information systems (GIS), global positioning systems (GPS), spectroscopy, machine vision, and digital image capture/analysis, excel at capturing macro-scale plant characteristics like chlorophyll and nitrogen levels, canopy details, leaf area index, and pest presence, they fall short in capturing crucial micro-nano scale information such as interplant communication, response to environmental stress, and internal physiological changes (Zhang et al. 2022). Most of these conventional methods are labor intensive, costly, and time consuming; hence unfeasible to monitor the plant health on a day-to-day basis (Khaled et al. 2018).

As noted earlier, plant growth/health can be adversely affected by external conditions which include environmental changes, nutrient deficiency, pathogen attacks etc. Early detection of the responses towards these is imperative to have a control on the growth, which may result in an improvement in crop production. These can be fulfilled by implementing advanced diagnostic tools in the field of plant science. The blending of nanotechnology and plant science efficiently replaced the classical detection tools by nano sensors. By definition these are sensing devices that detect the presence of certain substances or measure physical characteristics on a nanoscale with at least one of their sensing dimensions being not greater than 100 nm (Yusof and Isha 2020). Nano sensors, capable of real-time, high-resolution monitoring of individual plants at the micro-scale, interpret the signaling molecules emitted by plants using optical, wireless, or electrical signals. This advancement not only aids in intuitively understanding the physiological state of plants and the dynamic shifts in their environment but also addresses the pressing challenges in agricultural development. As a result, this breakthrough facilitates enhanced control over all aspects of agricultural production (Zhang et al. 2022).

Nano sensors in plants operate through a systematic process involving key components such as the analyte, probe, transducer, and detector. The main components of a typical nano sensor are represented in Fig. 5.1. The first phase begins with the binding of the bio element or analyte onto the sensor. This interaction triggers changes in the sensor, such as alterations in the frequency of vibration. In the second

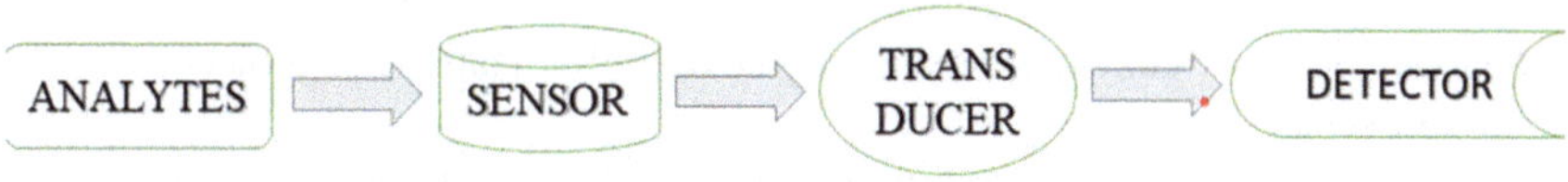

Fig. 5.1 Block Diagram to show the working principle of a nano sensor

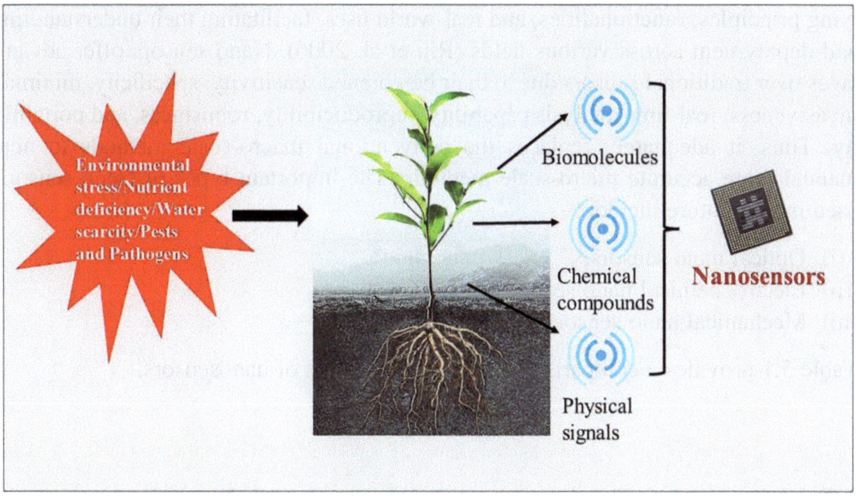

Fig. 5.2 Nanosensors for monitoring plant health, tracking their responses to environmental conditions, and detecting stress responses

phase, these changes are transduced into measurable signals through components like thermistors, piezoelectric crystals, optrodes, and electrodes, which comprise the transducer. These signals are then relayed to the farmers or monitoring systems, marking the final phase of the process. By providing real-time information on plant health, environmental conditions, and stress responses, nano sensors play a vital role in enabling informed decision-making for optimized agricultural practices (Adetunji et al. 2022). Figure 5.2 depicts nano sensors monitoring plant health, tracking their responses to environmental conditions, and detecting stress responses.

5.1.1 Classification of Nano Sensors

The most popular methods for classifying sensors center on the transducing mechanism, recognition principle, or application domain.

- Transducing mechanisms include electrical, optical, mass-based, thermal, piezoelectric, and others, which convert physical or chemical signals into measurable outputs.

- Recognition principles categorize sensors based on how they interact with target analytes, such as enzymatic reactions, DNA hybridization, or molecular recognition events.
- Sensors can be grouped based on their intended applications, spanning environmental monitoring, food quality assessment, medical diagnosis, and beyond.

This diverse classification framework helps categorize sensors based on their underlying principles, functionalities, and real-world uses, facilitating their understanding and deployment across various fields (Riu et al. 2006). Nano sensors offer advantages over traditional sensors due to their heightened sensitivity, specificity, minimal invasiveness, real-time analysis capability, reproducibility, robustness, and portability. Thus, it adequately replaces the conventional macro-scale methods to non manual more accurate micro-scale methods. The important types of Nano sensors used in agriculture include:

(i) Optical nano sensors,
(ii) Electrochemical nano sensors
(iii) Mechanical nano sensors

Table 5.1 provides a comparison of the various kinds of nanosensors.

5.2 Monitoring Environmental Stressors with Nano Sensors

Nanosensors offer a promising solution for monitoring environmental stressors, leveraging their exceptional sensitivity and specificity. These miniature devices accurately detect diverse stress-inducing factors, including pollutants, toxins, temperature fluctuations and humidity levels. By leveraging nanotechnology, researchers can develop nano sensors capable of real-time monitoring of environmental stressors in air, water, soil, and biological systems. These nano sensors enable early detection and continuous monitoring of stressors, providing valuable insights for environmental monitoring and management efforts. Additionally, nano sensors can be deployed in diverse settings, including industrial sites, agricultural fields, urban areas, and natural ecosystems, to assess the impact of environmental stressors and inform strategies for mitigation and adaptation (Rana et al. 2024; Vero et al. 2024).

5.2.1 Monitoring Water and Humidity

An optical-based sensor, featuring an aluminum oxide nano-porous ceramic plate, enables direct and continuous monitoring of soil water levels. This sensor integrates a nano-porous ceramic disc, silicon diaphragm and miniature optical displacement detection unit, complete with an integrated light source and photodetector. When

Table 5.1 A comparison of different types of nanosensors. The classification is based on the mode of action (Denizli et al. 2021; Kwak et al. 2017; Shaw and Honeychurch 2022; Cardoso et al. 2022; Baranwal et al. 2022)

Nanosensor	Working Principle	Classification based on the measuring property		Components	Analytes	Applications
Optical nanosensors	interaction between light and with the molecule. Changes in reflectance, absorbance etc.	Surface plasmon resonance	variation in wavelength, intensity, quenching of fluorescence	Plasmonic substrates	ATP, plant viruses, Calcium ions, metabolites Hormones, pesticides, pathogens	protein interactions, enzyme activity, and detection of biomolecules, pesticide detection, determination of biological active molecules like plant hormones, disease detection
		FRET	The analyte presence is quantified by the observed shift in the distance between the fluorophores (acceptor and donor)	genetically encoded fluorescent proteins or NPs like semiconducting quantum dots. Light sensitive molecules		
		SERS (Surface Enhanced Raman Scattering)	Raman scattering is enhanced when molecules are absorbed on rough metal surfaces. Variation in intensity	Raman active substrates		
Electrochemical nanosensors	Based on Nernst equilibrium, potential difference between electrodes	Amperometric	change in current at constant potential	three electrodes: reference, counter, and working.	Hormones, enzymes, ROS, ions like H^+, K^+, Na^+.	identification of redox active species and heavy metals in plants, soil content measurement, fructose content, pesticides
		Voltammetry	measure current by varying potential			
		Potentiometric	measurement of electrostatic potential			
Mechanical nanosensors	Converts mechanical stress or change in pressure, force, strain, temperature into electrical signals.	Piezoresistive	change in resistance under strain	Material with semiconducting and piezoelectric properties	Morphogenesis	Biomechanics of plants
		Piezoelectric	change in electric field under stress			

buried in unsaturated soil, the sensor generates negative pressure within the reservoir, inducing diaphragm bending. The resulting displacement measures soil saturation levels (Chen et al. 2019a). Additionally, a wearable, capacitive graphene-based plant humidity sensor, developed by Lan et al. (2020), employs laser direct writing technology on polyimide film. Featuring a graphene interdigital electrode (LIG-IDE), this sensor uses an aqueous graphene oxide solution as its sensing element. The flexible GO humidity nano-sensor attaches easily to plant leaves without harming growth.

5.2.2 Sensing Heavy Metals

Contamination of natural environments by heavy metals and ions presents substantial human health hazards. Electrochemical biosensors integrated with nanoparticles exhibit improved performance for heavy metal detection (Ejeian et al. 2018). Domínguez-Renedo et al. (2009) developed an amperometric urease biosensor utilizing screen-printed carbon (SPC) electrodes and gold nanoparticles (AuNPs) for mercury (Hg2+) detection. Building on advancements in DNA-based devices and nanotechnology, Ejeian et al. (2018) introduced high-precision nano-biosensors based on Förster Resonance Energy Transfer (FRET) for effective multiplexed detection of heavy metals. Wu et al. (2010) developed a novel Förster Resonance Energy Transfer (FRET) sensor utilizing quantum dots (QDs) and DNAzyme. This innovative design conjugated quencher-labeled DNAzymes onto carboxyl-silanized QD surfaces. In the presence of lead and copper ions, DNAzyme cleavage restored fluorescence emission, enabling rapid detection within 25 minutes using a single laser excitation source (Wu et al. 2010). The QD exhibited distinct fluorescence signals against various metal ions, allowing differentiation between them (Ejeian et al. 2018). Zhang et al. (2016) fabricated a glassy carbon electrode with self-doped polyaniline nanofibers, carbon nitride and bismuth for trace Cd^{2+} and Pb^{2+} detection using square wave anodic stripping voltammetry.

5.2.3 Detection of Plant Pathogens

Crop losses due to plant pathogens impose significant costs on farmers. Nano sensors offer a solution by enabling the detection of pathogens for timely containment. Currently, diagnosis relies on microbiological or PCR-based techniques. Researchers have employed fluorescent silica nanoparticles conjugated with specific antibody molecules to detect *Xanthomonas axonopodispv. vesicatoria*, the bacterial pathogen responsible for bacterial spot disease in Solanaceae plants. Similarly, a nano-sensor featuring fluorescently labeled DNA oligonucleotides coupled with 2-nm gold nanoparticles has been successfully employed for detecting phytoplasma associated with Flavescence dorée disease in Vitis vinifera (grapevines).

Furthermore, researchers have developed an electrochemical sensor utilizing gold nanoparticles, which effectively detects *Pseudomonas syringae infections in Arabidopsis thaliana (A. thaliana)* using differential pulse voltammetry. Nanosensors have been successfully utilized for mycotoxin detection, with the 4mycosensor demonstrating efficacy in identifying multiple mycotoxin residues in cereal crops, specifically corn, wheat, oats and barley. Furthermore, QD-based biosensors have been effective in detecting several plant viruses. Gold nanoparticle glycoconjugates, featuring functionalized sugars, have been developed for sensitive detection of *Penicillium italicum* spores and hyphae in various fruits. Julich et al. (2011) introduced a lab-on-a-chip approach for rapid nucleic acid-based diagnosis of *Phytophthora*, utilizing a single chip with integrated microchannels and DNA microarray components.

5.2.4 Detecting Toxins and Pollutants

Toxins like brevetoxins and microcystins stem from algal blooms fueled by cyanobacteria in eutrophicated aquatic systems. This underscores the necessity for reliable, cost-effective early detection systems (Justino et al. 2017). Utilizing molybdenum disulfide and Au nanorod nanocomposites, researchers devised an electrochemical immunosensor with high surface area and biocompatibility for detecting microcystin-LR in water. Similarly, transistor immune biosensors based on CNTs were effective in identifying domoic acid, associated with amnesic shellfish poisoning, in seawater samples (Marques et al. 2017). Bisphenol A (BPA), a common constituent in polycarbonate plastic and epoxy resins found in everyday items like water bottles and food cans, poses health risks as an endocrine disruptor, mimicking estradiol's functions (He et al. 2017). Researchers developed an aptamer-based "turn-off" fluorescent biosensor with AuNP for highly sensitive detection of BPA in water samples (Ragavan et al. 2013). Another approach involved a fluorescence-based aptasensor with molybdenum carbide nanotubes, developed by He et al. in 2017, for label-free detection of BPA in real water samples. Okadaic acid (OA), a potent diarrhetic shellfish poisoning toxin, poses significant health hazards and is often found in contaminated shellfish. A sensitive fluorescence immunosensor employing magnetic beads and quantum dots (QDs) was developed for OA detection (Pan et al. 2017).

5.3 Water Stress Detection

Plants continuously exchange gases and fluids with their environment through leaves and roots, making them susceptible to incessant environmental changes that directly impact growth, metabolism and development. The stress, due to the environmental changes, on the plant can be abiotic or biotic in nature. Abiotic stress mainly includes extreme conditions of temperature, water, heat, salinity etc. Whereas microorganisms, herbivores, weeds, other plants etc. contribute to biotic stress (Zain et al. 2023). Among these the stress due to water is the most highlighted one, because a large percentage of agricultural related products are grown in irrigated lands. In the life of a plant, water is the main source to maintain its physiological activities, since almost 50–90% of its weight comprises water. Similarly, water plays a key role in controlling the photosynthesis and transpiration process. The amount of water directly links with the process of photosynthesis, since the opening and closing of leaf stomata during the photosynthesis depends on the water content. However currently many of the lands are facing water scarcity and the increase in global temperature making the process faster and spreading to more areas. These point out the challenges facing out in the field of agriculture to maintain the growth of plants/crops under unfavorable environmental conditions, mainly under water deficient conditions. If the water intake is not able to fulfill the plant functioning on a daily basis then one can define the condition as water deficit. This is mainly attributed to the high evaporation, decreased ground water level and water retention by soil particles. If the dry weather continues for a long term, then the plant deficit is termed as drought and it will negatively affect the plant health (dos Santos et al. 2022).

In drought conditions both the root and leaves of the plant undergo many changes in order to survive the situation. The number of active roots will be increased to maximize the water absorption and further retention of it. Similarly, the root structure will be modified to increase the water absorption. In water logging the leaves become yellowing, falling off, changes in the thickness etc. to adapt in the condition. The stress in plants induces release of biochemical molecules which act as signaling molecules and has the potential to report the health of plants (Zain et al. 2023; Al-Khayri et al. 2021). The effects and the modifications adapted by the plants due to water stress are presented in Fig. 5.3.

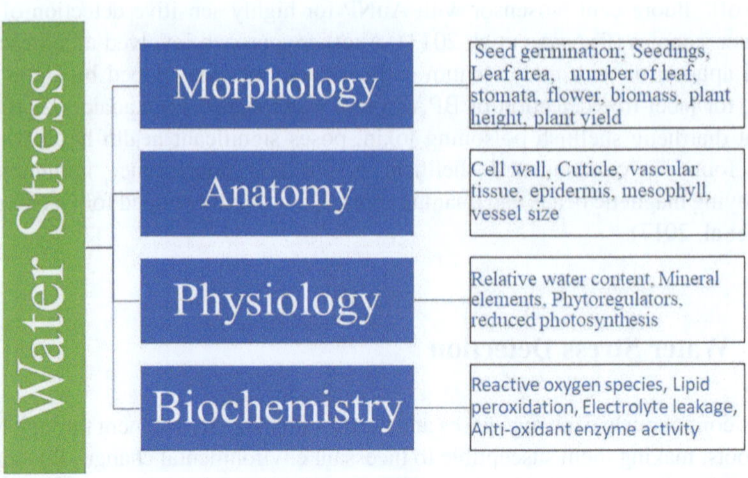

Fig. 5.3 The important changes in plants due to water stress

Morphological changes in the plants are not very prominent in the early stages of the water stress which makes on time precautions difficult. Hence regaining proper health and functioning is also difficult for plants. Though on time tracing of biochemical released signaling molecules is more efficient and successive in the onset of water stress. The first indication of water deficient condition in soil is by the synthesis of Abscisic acid (ABA). Abscisic acid triggers tissues and organelles to undergo morphological and anatomical modifications to cope up with water deficit conditions. ABA is considered as a short term response signaling molecule. If the water shortage continues for a long period, then few secondary messengers will get released, like Ca^{2+}, Volatile organic compounds (VOCs), reactive oxygen species (ROS), jasmonic acid etc. (Kim et al. 2010; Giraldo et al. 2019).

Proper tracking of these invisible chemical signals released in response to stress will be more effective. The incorporation of nanotechnology with plant science will open up a new scenario for the real-time monitoring of these molecules. In this sense nano sensors play a crucial role, they are able to convert these signals into optical, electrical or wireless signals which can be recorded and analyzed. This helps to take proper methods to increase the water intake by plants thus to diminish the effects of water stress.

Al_2O_3 is one of the functional materials which possess good water absorption capacity. In addition to this, it shows a variation in absorbance in accordance with the amount of absorbed water. Owing to these properties, Al_2O_3 based sensors for water are reported in research. An optical sensor which is loaded in between a nano-porous ceramic plate and silicon diaphragm for the direct continuous monitoring of soil water is reported (Chen et al. 2019a). The pressure variation induced by water content results in the bending of silicon diaphragm. This displacement is recorded and quantified using the optical detector. A low cost modified version of ceramic disc nano sensor is fabricated by another research group. In which the detection is based on near-infra-red and is connected through an optical fiber. There are two optional fibers, in which one is connected to a light source and the other is to the detector (Leone et al. 2022).

Considering the designing of effective nano sensors, the one with graphene oxide (GO) is found to be a constructive material. The two-dimensional structure and large surface area of GO offers good detection ability towards humidity, gas, toxicity etc. (Li et al. 2022). These sensor's exhibit great flexibility, high sensitivity, and long-term stability (Wei et al. 2024). The GO's nanolayers are composed of oxygen-containing functional groups, which interact well with water molecules in environmental objects. This property makes it a good choice for humidity sensing. Hence developing graphene and GO based sensors is a progressing field in research and technology. The electrical resistance of graphene alters with humidity and it is implemented for monitoring the evaporation of water from leaves by fabricating a graphene based plant wearable nano sensor. Graphene based nano sensor is connected to a resistance, inductance and capacitance meter. The resistance dynamics response of the graphene sensor in response to relative humidity is measured using this meter (Oren et al. 2017).

In addition to bare graphene oxide sensors, those supported on polymers are also reported in research. Long-term stability, low cost for manufacturing, water sensitivity makes polymers an attractive material. Besides, incorporation of polymers in an integrated circuit is effortless (Lee et al. 2008; Dai 2007). Among polymers, polyimide (PI) is the one which is implemented in the sensor fabrication. In accordance with the amount of absorbed water, PI exhibits variation in relative permittivity. The absorbed water can be monitored by measuring the variation in stoma size which in turn reflected as a change in PI capacitance. Laser direct writing technology was adapted for making graphene interdigital electrodes (LIG-IDE) on a PI film. Aqueous solution of graphene oxide placed on the surface of PI film acts as the sensing element for water. Here also the variation in capacitance gives an indication about water stress (Lan et al. 2020).

5.4 Nano Sensors for Detection of Plant Diseases and Pathogens

Due to insect invasion and pathogen infection, crop disease has become a major global concern and one of the primary challenges to agricultural production. The world economy loses $220 billion USD a year due to plant diseases brought on by several pests, such as bacteria, fungi, nematodes, and viruses (Singh et al. 2023; Savary et al. 2019). According to Savary et al. (2019), pathogen infections and pests cause up to 40% of the annual losses of major economic crops like wheat, potato, maize, rice, and soybean (Savary et al. 2019). To reduce such yield losses, two strategies are available: precision agriculture management and breeding for enhanced disease resistance, both of which depend on identifying and measuring plant disease signs and symptoms (Tanner et al. 2022; Zhang et al. 2024).

Nowadays, a lot of plant diseases can typically be recognized based on the visual condition of a plant and its leaf characteristics. This kind of plant disease identification often takes place at an advanced stage, complicating treatment and hindering effective recovery. Therefore, early detection is crucial for effective agrochemical treatment, tailored to specific plantations and infections. Several well-established methods exist for detecting pathogens through instrumentation, including: Enzyme-linked immunosorbent assay (ELISA), Polymerase chain reaction (PCR), Fluorescence in situ hybridization (FISH), Colony counting, and Flow cytometric detection (FCM), an immunology-based method (Cardoso et al. 2022). Although certain molecular biology reagents, like primers and enzymes, are costly and have very short shelf lives. Furthermore, the time-consuming nature of the aforementioned methods, along with their requirement for specialized equipment and laboratory setup, restrict their applicability in developing nations (Nizamani et al. 2024). Also, on-site individual plant monitoring is still a challenge to overcome.

Plant diseases can be detected and identified using both direct and indirect methods (Fang and Ramasamy 2015). Plant disease detection employs direct and indirect techniques. Direct methods involve examining biomolecular markers (nucleic acids, proteins, and carbohydrates) and plant pathogens (fungi, bacteria, and viruses) isolated from infected plant tissues. Indirect detection analyzes changes in physiological or histological factors, including: Leaf surface temperature and humidity, shape and growth rate, volatile organic compound (VOC) emissions, and spectroscopic features of plant tissues (Fang and Ramasamy 2015). Both approaches utilize various electrochemical, spectroscopic, and molecular technologies (Li et al. 2020). Meanwhile, to identify the characteristics of plant species, indirect methods of imaging techniques like thermography and hyperspectral have been applied in the fields under a variety of circumstances, including variations in temperature, morphology, transpiration rate, and the release of VOCs by infected plants (Li et al. 2020; Fang and Ramasamy 2015). But most of the plant disease detection methods have significant drawbacks, including a lack of specificity for disease strains or subtypes and being sensitive to environmental parameter changes (Li et al. 2020).

According to Calisgan et al. (2020), the conventional methods of detection of plant diseases and pathogens depend excessively on artificial intelligence for detection and decision-making, and as a result, the detection methods are only useful in cases of severe crop loss (Calisgan et al. 2020). At this time, the economic loss is irreversible. Thus, early detection of pests and disease is vital, as late-stage pest disease would result in much higher prevention and control costs. As a result, researchers are now using sensing technologies, which enable quick identification of pathogens in plants. This allows for the containment of these organisms. Among the various signal transduction mechanisms, the most commonly employed ones for the detection of plant disease sensors are electrical, or electrochemical, and optical techniques (Cardoso et al. 2022).

5.4.1 Electrical and Electrochemical Sensors

Nowadays, plant disease monitoring has become more and more dependent on electrical and electrochemical methods. Its simple operational methods, sensitivity, and selectivity against particular infections, and potential for making portable commercial devices that allow *in situ* measurements are only a few of its advantages over traditional approaches. Electrochemical techniques for plant disease detection predominantly employ biosensors to identify pathogens. Table 5.2 summarizes some works using nanomaterials in electrical and electrochemical methods to detect plant disease (Cardoso et al. 2022).

Table 5.2 Plant disease monitoring using nanomaterials in electrical and electrochemical methods

Target	Nanomaterial	Substrate	Detection method	Observation
VOC p-Ethyl phenol released by *Phytophthora cactorum*	Single-walled carbon nanotubes (SWCNTs)	Silicon Wafer cover with SiO_2	E-nose (Bioelectronic nose)	Identify of p-ethyl phenol using FET immobilized with SWCNTs and ssDNA (single strand DNAs) for the diagnoses of P. cactorum in strawberries (Wang et al. 2020).
Ganoderma boninense	Multiwalled carbon nanotube (MWCNT), gold nanoparticle, chitosan	Modified screen printed carbon electrode	Electrochemical detection, cyclic voltammetry and linear sweep voltammetry	A new sensitive, selective, cost-effective and environmentally friendly, multiwalled carbon nanotube electrode was developed for detection of Ganoderma boninense in oil palm (Isha et al. 2019).
A secreted protein biomarker —Sec-delivered effector 1 (SDE1)	Single walled carbon nanotubes (SWNTs)	Gold microelectrodes on to Si/SiO_2 wafer	Field effect transistor (FET)/ chemiresistor	Sensor detected SDE1 biomarker-confirmed the disease of citrus greening or Huanglongbing (HLB) (Tran et al. 2020)
Citrus Tristeza Virus (CTV) related nucleic acids	Electrodeposited gold nanoparticles (AuNPs)	Screen printed carbon electrode (SPCE)	Electrochemical impedance spectroscopy (EIS)	Designed biosensor of AuNPs label free modified SPCE with covalently bonded thiolated ssDNA selectively detected CTV related nucleic acids (Khater et al. 2019a)
Citrus Tristeza Virus (CTV)			Label free highly	Designed *in situ* RPA sensor

(continued)

Table 5.2 (continued)

Target	Nanomaterial	Substrate	Detection method	Observation
related nucleic acids	Gold nanoparticles (AuNPs)	Screen printed carbon electrode (SPCE)	integrated in situ RPA and EIS detection	exhibit better advantages in terms of sensitivity, portability, and simplicity (heat source and labelling is not required) together with allowing quantitative analysis of nucleic acids (Khater et al. 2019b).
PthA effector protein for citrus canker diagnosis	AuNPs	Glassy carbon electrode (GCE)	Fourier transformation square wave voltammetry method (FET-SWV)	Glassy carbon novel modification immunosensor was developed using GNP, BCNT-IL and PB for the detection of PthA protein for diagnosis of citrus canker disease (Haji-Hashemi et al. 2018).
VOCs exhaled by Aspergillus and Rhizopus fungal microorganisms	Multiwalled carbon nanotubes (MWCNTs), and doped multiwalled carbon nanotubes. Doping was done using nitrogen -and boron- N-MWCNTs) and B- MWCNTs)	Interdigitated ENIG (electroless nickel immersion gold) electrodes	Tristimulus analysis	By using a set of chemical sensors based on carbon nanostructures fungi infection in strawberries inoculated with *Rhizopus* sp. or with *Aspergillus*sp. section *Nigri* were detected (Greenshields et al. 2016).
Pantoea stewartii sbusp. *Stewartii* (PSS)	AuNPs	Glassy carbon electrode (GCE)	Electrochemical immunoassay	Large specific surface area of AuNPs and the catalytic ability of HRP along with the synergistic effect of the conductivity resulted in low LOD was (Zhao et al. 2014).

5.4.2 *Optical Sensors*

Optical sensors, which are being developed to detect plant diseases early on, are instruments that can measure and detect variations in the optical properties of a material and translate these changes into a measurable electrical signal. Some of the work using nanomaterials in optical methods to detect plant disease/pathogens has been tabulated in Table 5.3 (Cardoso et al. 2022).

RGB imaging (handheld instruments) is an optical technique for identifying plant diseases. Notably, photo sensor light sensitivity, optical focus and digital focus have shown significant annual improvements (Mahlein 2016). Researchers detect biotic stress by analyzing RGB color channel changes (red, green and blue). Advanced detection methods combine RGB, LAB (lightness and color opponent dimensions), YCBCR (luma, blue-difference and red-difference chroma) and HSV (hue, saturation, value) color information with spatial data, accurately identifying leaf spot, powdery mildew and fungal infections (Bock et al. 2010). This multimodal approach enhances crop monitoring, enabling timely interventions and optimized agricultural practices. Additional variables for detecting and identifying plant disease symptoms include color, gray levels, texture, dispersion, connectivity and shape parameters (Camargo and Smith 2009; Neumann et al. 2014; Mahlein 2016). Researchers have identified and detected plant diseases from RGB images using machine learning and pattern recognition techniques (Mahlein 2016).

Digital image analysis is a well- established technique for the assessment of plant diseases. For this, numerous software packages are available, including ASSESS 2.0, Leaf Doctor, Scion image software, and specially designed modules. In 2018, Arya and her co-worker used image processing and an Arduino genetic algorithm to identify fresh and diseased plant leaves, including those from pepper plants, potatoes, and tomatoes that had leaf spots and late blight (Arya et al. 2018). To determine the disease-affected area and color variations within, researchers utilized Matlab's image-processing toolkit and developed an algorithm to classify leaves. Observations indicated this approach effectively identifies early disease symptoms on plant leaves, streamlining agricultural crop monitoring.

5.5 Nutrient Monitoring

Nutrients are needed for the healthy growth of plants. Among the different nutrients 17 are considered as essential. The non-metal elements like C, H and O are taken up by plants either from atmosphere (air, CO_2) or as biomass or as water. Depending on the amount needed for the growth, these are classified as macro and micro. The macro nutrients are of two types, primary macronutrients which consist of nitrogen, phosphorous, and potassium. These are needed in larger quantities for better crop growth. However, those which are needed in lesser amounts and are usually present in soil are called secondary macronutrients (Calcium, magnesium, Sulphur).

Table 5.3 Optical sensors using nanomaterials to detect plant diseases

Target/culture	Nanomaterial	Bio-recognition	Detection method	Observations
Phytophthora infestans	Gold nanoparticles (AuNPs)	Cysteine (Cys) -capped	Colorimetric sensor array	Smartphone-based VOC- sensing system was able to detect Phytophthora infestans at very early stages (Li et al. 2019).
Xanthomonas campestris	Gold nanoparticles (AuNPs)	–	Colorimetric method	Based on the interaction between the thiolated M13KE phage and the target bacteria, a visible shift in SPR absorbance was observed (Peng and Chen 2018).
PhakopsoraPachyrhizi	Fluorescent nanoparticles	Polyclonal antibodies	fluorescence	Developed paper-based disposable immunosensor reaches the detection range of conventional-based test systems for the early diagnosis of Asian soybean rust (Miranda et al. 2013).
Tomato Yellow leaf curl *virus* (TYLCV)	AuNPs	DNA	LSPR -based colorimetric assay	Use of AuNPs as colorimetric probes allow sensitive and fast detection of TYLCV genome (Razmi et al. 2019)
Alternaria panax Whetz	AuNPs-SA	Mouse anti-Fam antibody and BSA Biotin	Single-tube nested PCR-lateral flow biosensor assay (STNPCR-LFBA)	Developed STNPCR-LFBA biosensor is more sensitive than traditional PCR (Wei et al. 2018)
Phytophthora infestans	AuNPs	streptavidin biotinylated T and C	Primer-mediated asymmetric PCR with AuNPs lateral flow assay (LFA)	Developed sensor has potential application in early prediction of late blight disease occur in potato (Zhan et al. 2018).

(continued)

Table 5.3 (continued)

Target/culture	Nanomaterial	Bio-recognition	Detection method	Observations
Potato Leafroll virus (PLRV)	AuNPs and silver	Anti-PLRV antibodies	Silver-enhanced lateral flow immunoassay	Detection sensitivity of Silver enhanced LFIA of PLRV was 15 times more than the conventional LFIA (Panferov et al. 2018)
Xanthomonas arboricolapv. Pruni	carbon nanoparticles	Polyclonal antibodies	Lateral flow immunoassay (LFIA)	Results obtained in less than 15 minutes. A comparison of the relative accuracy of LFIA with real time PCR and plate isolation demonstrates a strong correlation (Lopez-soriano et al. 2017).

Micronutrients are elements that play a role in the growth of plants, but required only in trace amounts, like zinc, nickel, chloride, etc. It is important to note that presence of higher concentration of these nutrients make the soil toxic for plants (Bharti et al. 2024).

The real-time monitoring of availability of nutrients in soils permits one to make conclusions in order to optimise and adapt appropriate fertilizing methods for different agriculture crops. Additionally, it offers guidance on how to prevent excessive nutrient runoff into surface and groundwater, which might negatively impact the environment (Nadporozhskaya et al. 2022; Doran and Zeiss 2000; Marios and Georgiou 2017; Bitutsky 2014; Troeh and Thompson 2005; Kashyap and Kumar 2021). Despite its shortcomings, which include lengthy analysis, high equipment costs, the need for skilled workers, and the requirement for sample pretreatment, standard laboratory procedures are currently used to analyze soil nutrients. Because of their sensitivity, selectivity, quick response time, and affordability, nanosensors are a beneficial addition to or replacement for these traditional instrumental techniques and wet chemical approaches. Indeed, the last 10 years have seen a notable increase in the amount of research interest in direct in-field soil analysis as well as sensory analysis of soil nutrients. Electrochemical sensors based on ion-selective electrodes (ISEs) measure primary ion activity in various sample matrices and have beneficial characteristics such as low cost, simplicity, ease of miniaturization, good reliability, portability, and timeliness. Compared to other methods, the sample pretreatment is not needed in ISE measurements. On account of these, they have been widely used for rapid ion detection in agricultural, environmental, biomedical, and industrial analyses. This kind of sensor consists of a working electrode and reference electrode and the underlying working principle is the same as that of electrochemical cells. The ion selective membrane (ISM), usually made with polymers, is then integrated with the working electrode. The potential difference between the working electrode and reference electrode gives the analyte concentration (Kim and Lee, 2022; Kashyap and Kumar 2021).

5.5.1 Nitrogen Detection

The building unit of proteins, amino acids, are made up primarily of the element nitrogen. There is a clear correlation between the growth rate and the protein content in plants. Thus, in order to encourage the growth of roots and vegetation, nitrogen is essential. In addition, nitrogen is an essential component of ATP, chlorophyll, and nucleic acids. All these characteristics make nitrogen a crucial role in plant development (Podar and Maathuis 2022). Nitrate (NO_3^-), nitrite (NO_2^-), and ammonia (NH_4^+) are the three forms of nitrogen found in soil; nitrate is the type that plants take nitrogen from the most. Table 5.4 contains a tabulation of the most recent nano sensors for nitrogen detection in detail.

5.5.2 Phosphorus Detection

Phosphorus in soil is taken as free phosphoric acid by the plants, though it's availability in soil is highly dependent on the pH of soil. Strong stems, robust root systems, and rapid vegetation development are all made possible by the phosphorus. It promotes the development of flowers and seeds and helps achieve the right ripening and maturity. Phosphorus serves as a pivotal component in ATP, phospholipids and in phosphoric acids (the backbone of nucleic acids). It is necessary for the Sun's energy to be properly absorbed and helps transform it into chemical energy (Podar and Maathuis 2022). Details of nano sensors for the phosphorus detection, mainly in the form of phosphate ion, are listed in Table 5.5.

5.5.3 Potassium Detection

Potassium is the main cation found in plants, and plants require the same amount of potassium from the soil as they do nitrogen. It acts as an osmotic and ionic regulator, along with, it controls water balance. It supports the cell's transportation system. Additional functions include stem strengthening and providing pest resistance, which promotes healthy plant growth (Podar and Maathuis 2022). Potassium ion detection based on nano sensors are tabulated in Table 5.6.

Table 5.4 Nano sensors for nitrate ion detection

Metal ion	Sensor description	Limit of Detection	Measuring Technique
Nitrate	The ion exchange for nitrate was tetradodecylammonium and 2-nitrophenyloctyl ether as a plasticizer. ISM is sheathed with Ni_2O_3/rGO nanocomposite (Sephra et al. 2024).	1.65 mg/L	Cyclic voltammetry
Nitrate	Stainless steel based microneedle electrochemical sensor which is coated with carbon nanotube/cellulose nanocrystal (CNT)/CNC) and further modified with silver nanoparticles (Mugo et al. 2022).	0.008 mM	Cyclic voltammetry
Nitrate	The nitrite-doped polypyrrole (PPy-NO_3) ISE is modified using a nanohybrid composite of electrochemically reduced GO and Au nanoparticles (Chen et al. 2020).	10^{-1}–10^{-5} M	Cyclic voltammetry
Nitrate	The working electrode comprises of nanocomposite with poly(3-octyl-thiophene) and molybdenum disulfide (POT–MoS_2). Further it is coated with Au electrode and ISM (Ali et al. 2019).	1–1500 ppm	Potentiometry
Nitrate	Reduced graphene oxide aerogel is used for modification of highly stable solid contact ion-selective electrode (SC-ISE) (Kim et al. 2021).	759 nM for NO_3^- and 186 nM for Ca^{2+}	Cyclic voltammetry and chronopotentiometry
Nitrate and nitrite	The glassy carbon electrode is designed with a Cu based nanocomposite. The components of nanocomposite are copper nanoparticles on MWNT with reduced graphene oxide (Bagheri et al. 2017).	0.1 to 75 μM	square wave voltammetry

5.6 Nano Sensors: Probing Plant Metabolism with Precision

Nano sensors offer a cutting-edge approach to monitoring plant metabolism, providing researchers and farmers with unprecedented insights into the biochemical processes that drive plant growth and development. Devices in the nanoscale are capable of detecting and quantifying key metabolites, such as sugars, amino acids, hormones, and secondary metabolites, in real-time and with high sensitivity. By continuously monitoring changes in metabolite levels, nano sensors enable the dynamic assessment of plant physiological status, responses to environmental stimuli, and stress tolerance (Konwarh and Sharma 2020; Tian et al. 2024). One of the significant advantages of nano sensors in monitoring plant metabolism is their ability to provide spatial and temporal resolution at the microscale level. This allows researchers to study metabolic activities at specific locations within plant tissues and organs, uncovering spatial heterogeneity and temporal dynamics in metabolic

Table 5.5 Nano sensors for phosphate ion detection

Metal ion	Sensor description	Limit of Detection	Measuring Technique
Phosphate	The working electrode is modified with Co nanoparticles to form phosphate ISE (polytetrafluoroethylene electrode Co_PTFE_ISE). Platinum and Ag/AgCl are used as counter and reference electrodes respectively. The response potential quantitatively measures phosphate ion (Xu et al. 2021).	–	Electrochemical Impedance
Phosphate	GQDs-Ce^{4+} or GQDs-Fe^{3+} aggregate formation leads to fluorescence quenching of GQDs. In presence of PO_4^{3-}, Ce^{4+} and Fe^{3+} free up GQDs thus regain fluorescence (Wang et al. 2022)	0.06 μM	Fluorescence
Phosphate	Stainless steel based microneedle electrochemical sensor which is coated with carbon nanotube/cellulose nanocrystal (CNT)/CNC) and further modified with silver nanoparticles. This on additional coating with polyaniline (pANI) and ammonium molybdenum tetrahydrate (AMT) detects phosphate ions (Mugo et al. 2022).	0.007 mM	Cyclic voltammetry

Table 5.6 Nano sensors for Potassium ion detection

Metal ion	Sensor description	Limit of Detection	Measuring Technique
K^+	A thin layer of valinomycin is coated on the surface of F-MWCNTs/ZnO/nanocomposite. This modified electrode is used for the detection of K^+ (Kumar et al. 2020).	5–25 mM	impedance
K^+	Valinomycin topped ZnO/rGO ISE (Tharini et al. 2023).	0.956 mM	cyclic voltammetry

processes. Additionally, nano sensors can be integrated with imaging techniques, such as fluorescence microscopy or scanning probe microscopy, to visualize metabolic activities in situ, further enhancing our understanding of plant metabolism in vivo. Furthermore, nano sensors offer the potential for non-destructive and minimally invasive monitoring of plant metabolism, allowing for continuous monitoring over extended periods without perturbing plant physiology. This capability is particularly valuable for studying metabolic changes in response to environmental factors, such as light, temperature, water availability, and nutrient status, as well as biotic and abiotic stressors (Konwarh and Sharma 2020; Tian et al. 2024). Some of the applications of nano sensors in plants, with a focus on understanding plant metabolism, include:

5.6.1 *Monitoring Key Biochemical Processes*

5.6.1.1 Photosynthesis

Photosynthesis is process occurs in plants in which the light energy is converted into chemical energy. This is considered as one of the fundamental processes in Earth. In contemporary times, scientists are actively seeking to enhance the relatively low efficiency of vascular plants through various techniques and genetic manipulations. In the realm of nanobiotechnology, researchers aspire to create bionic plants with improved photosynthetic efficiency and biochemical sensing capabilities. Giraldo et al. (2014) demonstrated that embedding semiconducting single-walled carbon nanotubes (SWCNTs) into isolated chloroplasts resulted in a threefold increase in photosynthetic activity compared to controls. Furthermore, this augmentation extended to enhanced maximum electron transport rates. Notably, the integration of SWCNTs also conferred the ability for plants to sense nitric oxide, a vital signaling molecule (Siddiqui et al. 2015).

5.6.1.2 Plant Morphogenesis

Unlike animal tissues, shape changes in plant tissues typically occur without cell migration or removal. Therefore, shaping a tissue involves coordinating three key processes at the cellular level: expanding the contact area between cells, altering cell shape, and creating new cell walls. To comprehend plant morphogenesis, scientists are adopting interdisciplinary approaches that combine biology and physics to explore the role of mechanical forces. Recent advancements in nanotechnology have revealed the potential of adapting piezoelectric materials to develop plant nano sensors. These sensors can monitor the mechanical environment within plants by detecting changes in electric signals (Kwak et al. 2017).

5.6.1.3 Plant Hormones and Signaling Mechanisms

Plant hormones play pivotal roles throughout a plant's life cycle, spanning from embryogenesis to reproductive development, as well as in biotic and abiotic stress responses. These are also referred to as phytohormones and are signaling molecules synthesized within the plants. Nano sensors offer a promising avenue for investigating plant hormones and their signaling mechanisms *in vivo*.

Strigolactones, a group of compounds produced by plant roots, regulate developmental processes and stress responses. An optical nano sensor called Yoshimulactone Green enables spatiotemporal monitoring of strigolactone levels, facilitating the identification of specific strigolactone receptors (Tsuchiya et al. 2015).

Ethylene, another vital plant hormone, influences various aspects of plant growth and stress responses. Nano sensors, such as those based on single-walled carbon nanotubes (SWNTs), demonstrate high sensitivity to ethylene and have been employed for detecting its presence in different fruit types (Esser et al. 2012).

Auxins, which regulate plant development, have been studied using sensors employing platinum black and carbon nanotubes or porous graphene bio nanocomposites for non-invasive monitoring and sensitive detection (McLamore et al. 2010).

Gibberellins, responsible for promoting organ growth, have been quantified using FRET-based nano sensors developed with plant hormone receptors as sensory domains (Rizza et al. 2017).

Additionally, salicylic acid, a key hormone in plant defense responses, has been detected in plant extracts using structure-switching aptamer-based nanopore thin film sensors (Chen et al. 2019b; Shaw and Honeychurch 2022).

5.6.2 Monitoring Plant Metabolites/Analytes

5.6.2.1 Adenosine Triphosphate (ATP)

ATP plays a crucial role as an energy source for various cellular processes. In metabolic activities, ATP is converted into adenosine diphosphate or adenosine monophosphate. Within plants, ATP synthesis primarily occurs in chloroplasts and mitochondria. Saito et al. (2012) engineered a chimera consisting of enhanced Renilla luciferase and Venus, a fluorescent protein known for its high bioluminescence resonance energy transfer efficiency. This research team developed a nano-lantern-based ATP indicator by incorporating the ε subunit of bacterial F0F1-ATP synthase into a modified nano-lantern (VRL10.0_S257G) at the 228–229 position. Referred to as ATP1, this nano-lantern, when expressed in live plant leaves, successfully detected reductions in cytoplasmic ATP levels upon the addition of oligomycin A, a mitochondrial ATP synthase inhibitor. This optical nano sensor, utilizing FRET, enabled the visualization of ATP production in chloroplasts during photosynthesis in transgenic Arabidopsis plants by targeting it to the chloroplast stroma through a transit peptide fusion.

5.6.2.2 Molecular Oxygen

Molecular oxygen serves as the terminal electron acceptor in aerobic respiration's electron transport chain. Two types of nano sensors, electrochemical and optical systems, are used to assess oxygen distribution in tissues. Optical nano sensors offer advantages over Clark-type polarographic electrode sensors, allowing nanoscale oxygen sensing and imaging over large areas. Probes encapsulated by biologically localized embedding (PEBBLEs) are a prominent type, with fluorescent dyes enclosed within an inert matrix to reduce leakage and maintain stability. However, PEBBLEs may emit red phosphorescence that interferes with plant chlorophyll autofluorescence. To address this, microbead-based probes with two-frequency phase modulation are utilized in plant and algae systems. However, their probe size, ranging from 20 to 600 nm in diameter, may cause cell damage, limiting their use in living plants (Buck et al. 2004; Schmälzlin et al. 2005).

5.6.2.3 Saccharide Fluxes

The detection of changes in subcellular compartmentalized steady-state sugar levels after cell or tissue perfusion has been made easier by a variety of FRET-based saccharide sensors. Stable glucose levels are influenced by metabolic rate, compartmentalization, and flux across the cell membrane. Nano sensors, with high signal-to-noise ratios expressed in *Arabidopsis* RNA-silencing mutants, were utilized to monitor glucose concentration modulations (Deuschle et al. 2006). Furthermore, papers describe how putative sugar-binding proteins from Agrobacterium are used as scaffolding to create genetically encoded FRET sensors that are unique to sucrose. Interestingly, FRET mutant nano sensors showed up to 50- and 11-fold increases in selectivity for sucrose, respectively, as compared to maltose and glucose (Lager et al. 2006).

5.6.2.4 Nitric Oxide

Throughout the plant life cycle, nitric oxide, a gaseous reactive nitrogen species, is an essential signaling molecule. It controls a variety of physiological processes in plants, such as senescence, programmed cell death, seed germination, and reaction to biotic and abiotic stressors. Semiconducting single-walled carbon nanotubes are used as signal transducers for nanosensors to assess nitric oxide levels. These nanotubes demonstrate effectiveness in monitoring A. thaliana's nitric oxide levels using near-infrared fluorescence. This technique uses the particular adsorption of a specifically created polymer at a nanoparticle interface to provide reliable detection, and it is based on corona phase molecular recognition (Giraldo et al. 2014).

5.6.2.5 Volatile Organic Compounds

The attack by insects in plants will result in the production of Volatile Organic Compounds (VOCs). Chemiresistive sensors pick up these signal molecules. Weerakoon et al. (2011) monitored the volatile chemical γ-terpinene using a poly (3-hexylthiophene)-coated chemiresistive sensor in order to detect bark beetle infestations early. A two-order-of-magnitude increase in resistance results from this molecule's interaction with the conducting polymer poly(3-hexylthiophene), which is released during infestation. The usefulness of the platform in identifying insect infestations was demonstrated by its detection limit, which was as low as 36 ppm.

5.7 Nano Sensors for the Detection of Pesticides

Pesticides are chemicals or biological substances, applied to agricultural crops to control various types of pests, including fungi, insects, nematodes, herbicides, and many others, to enhance the yield quality and quantity. These are necessary for the protection and production of crops, but their presence in soil beyond acceptable limits has raised concerns in recent decades as they have the potential to cause immunological and respiratory diseases, infertility, and bone marrow and nerve disorders, as well as being known to be carcinogenic and cytotoxic (Virk et al. 2024). Besides, it kills beneficial soil microbes, biomagnifies pesticides, cause pollinator decline, declines the fertility of soil, and obliterates the natural habitat of agricultural allies like birds (Duhan et al. 2017). The worldwide consumption of pesticides is approximately two million metric tons per year (Sharma et al. 2019). Conventional pesticide monitoring methods, like gas chromatography, high performance liquid chromatography, mass spectroscopy, and immunoassays are not impractical for on-site use. Moreover, most of these methods are single-signal assays that require expensive equipment, trained operators, and complex pretreatment of the samples; some are even prone to changes in the environmental conditions (Sharma et al. 2021).

By combining nanomaterials with various biosensing devices (nano-biosensors), it is possible to improve the sensitivity of pesticide detection and monitor even a single residue by using a variety of conjoining and multifunctional techniques (Dias et al. 2024). Recent advancements in nano sensors for pesticide monitoring includes the development of TiO_2 nanotubes for atrazine detection and pencil graphite electrodes modified with nanofilms herbicides of the phosphorus-containing amino acid type (Deuri et al. 2014). Fahimi Kashani and Hormozi-Nezhad (2016) have used hierarchical cluster analysis (HCA) and linear discriminant analysis (LDA) to identify organophosphate insecticides containing citrate-capped 13-mm gold nanoparticles (Fahimi Kashani and Hormozi-Nezhad 2016). At concentrations ranging from 120 to 400 ng/mL, AuNps-based biosensors have successfully detected five harmful organophosphate pesticides such as chlorpyrifos (CP), pirimiphos-methyl (PM), azinphos-methyl (AM), fenamiphos (FP), and phosalone (PS). Various nanomaterials, such as gold, silver, carbon nanotubes (CNTs), silica-coated carbon-dot assisted ratiometric fluorescent nano sensors, and apt sensors, can be used in the production of nano-biosensors as pesticides, herbicides, and insecticides (Virk et al. 2024). Herein, the most researched nano sensor systems- such as electrochemical and optical sensors, are discussed.

5.7.1 Electrochemical Sensors for Pesticide Detection in Plants

Electrochemical techniques evaluate the change in current or potential that occurs when an electrode comes into contact with the target analyte in order to identify pesticides. Differential pulse voltammetry, impedance spectroscopy, amperometry, and other efficient field-based transistor technologies are often employed electrochemical techniques (Zia et al. 2022). Platinum, gold, silver, graphite, carbon nanotubes, tungstate nanofiber graphene, nanoparticle composites, metals, and metal oxides can all be used to create solid electrodes for nanomaterial-based sensors (Zia et al. 2022). Aptamers, enzymes, or antibodies can be used to functionalize these electrodes as a biorecognition element. Direct spatial contact between the electrode and the recognition element results in a higher signal amplification and improved signal to noise ratio in contrast to traditional electrochemical methods. In addition to the characteristics of electrodes, the size and shape of nanomaterials can enhance sensor performance (Zia et al. 2022).

Compared to optical sensing methods, electrochemical techniques are simple to miniaturize and very accurate. Also, the magnetic nanomaterials -based electrochemical transducers have a minimal background signal, which makes them appropriate for detecting low analyte level in solutions. The simultaneous analysis of fenamiphos and carbaryl pesticides was carried out with the help of a modified electrode based on graphene oxide-Cu/CuO-Ag. The electrochemical efficacy of these two pesticides were examined by using differential pulse voltammetry and cyclic voltammetry on rGO/Cu/CuO-Ag/GCE. With a high level of sensitivity, the modified electrode showed two different oxidation signals when fenamiphos and carbaryl were detected. Using rGO/Cu/CuO-Ag/GCE under optimized conditions, carbaryl and fenamiphos were detected with a wide linear range of 0.05–20 and 0.01–30 μM, respectively, and a detection limit of 0.005 to 0.003 μM. Thus, as an affordable and user-friendly biosensor, an electrochemical platform has been effectively established for the detection of low levels of fenamiphos and carbaryl in fruit and vegetable samples (Hashemi et al. 2019; Zia et al. 2022).

5.7.2 Optical Sensors for Pesticide Detection in Plants

Quantum dots (QDs) are highly fluorescent materials owing to their optical properties as a result of the nanoconfinement. Among these, carbon dots (CDs) attain more attention due to their low-toxicity, high quantum yield, fast detection etc. The fluorescence of carbon dots gets quenched or endure changes in intensity with pesticide concentration. Hence these kinds of nano sensors are known as off-on sensors and are the focus of many research works. It is illustrated in a work that, presence of dialen super, duplosan, and confidor, turned off the fluorescence of CDs. However, it is not affected by bromoxynil, diazinon, glyphosate, deltamethrin and malathion. The limit of detection (LOD) reported is 10–8 ppm (Alvandi et al. 2021). A nanocomposite based on carbon quantum dots and gold nanoparticles (CQDs-GNPs) was able to detect the presence of malathion in cabbage. In the presence of malathion, the fluorescence intensity shows an increase and that of the absorbance spectrum exhibits a reduction. Noted changes in brightness and color of fluorescence aids in the visual detection of malathion. The acquired limit of detection is 10^{-9} M—10^{-2} M. The CDs detection capability towards diazinon, glyphosate, and amicarbazone is studied by Ashrafi Tafreshi et al. (2020). It is noted that, with increase in concentration of pesticides the fluorescence emissions of CDs decrease. The observed variation in intensity is more prominent for diazinon. Though at higher concentrations the other two pesticides were able to quench the fluorescence of CDs. The calculated limit of detection for diazinon, amicarbazone and glyphosate is 0.25, 0.5, and 2 ng ml^{-1} respectively (Ashrafi Tafreshi et al. 2020).

In addition to the bare QDs/CDs, the doped ones are also found to be good materials for pesticide detection. CdTe doped QDs found to be active towards the detection of carbaryl pesticide. A yellow emission is observed in these QDs, which changes to green in presence of the pesticide. At higher concentrations the fluorescence emission becomes intense blue. Thus, the visual detection of these color changes helps one to understand the contamination by carbaryl with a LOD 0.12 ng mL^{-1} (Shahdost-Fard et al. 2021). Similarly, boron and nitrogen doped CDs (B, N—CDs) are turned out to be a fluorescence probe for methyl-paraoxon. Upon adding the methyl-paraoxon to the CD suspension on CeO_2, it diminishes the blue emission of B, N—CDs. The recorded LOD is 24.7 ng mL^{-1} (Luo et al. 2020).

5.8 Conclusion and Prospects

Integration of Nano sensors to agricultural practices will be a transformative leap forward to maintain a sustainable agriculture and food security worldwide. These micro-scale devices are to revolutionize agriculture by real-time monitoring of nutrient status, plant metabolism, environmental stressors, pathogen attacks with great accuracy and precision. Nano sensors enable farmers and researchers to make informed decisions that maximize crop yield, reduce resource waste, and lessen the effects of various stressors on plant health by offering micro-scale insights into plant physiology and environmental dynamics. Additionally, the massive data as interpreted from the nano sensors may help the farmers and researcher to gain a deeper insight into the plant health, metabolism, signaling, responses to environmental conditions, optimize resource management, and to produce resilient crop varieties. Overall, integrating nano sensors to agriculture provides a route towards sustainable intensification, along with meeting the global food demand with less environmental impact and strengthening resilience to change in climatic conditions. The continued advancements in Nanotechnology bring about promising future prospects for nano sensors. With the development of innovative nano sensor technology, greater refinement and miniaturization of these sensors are launched, providing improved accuracy, versatility, and efficacy which benefits the various stages of the agricultural sector starting from seeding to marketing. Thus, Nano sensor technology will have a great influence on shaping the future of agricultural practices, guaranteeing food security for the generations to come.

References

Adetunji CO, Panpatte DG, Jhala YK (eds) (2022) Agricultural biotechnology: food security hot spots. CRC Press

Ali MA, Wang X, Chen Y, Jiao Y, Mahal NK, Moru S, Castellano MJ, Schnable JC, Schnable PS, Dong L (2019) Continuous monitoring of soil nitrate using a miniature sensor with poly (3-octyl-thiophene) and molybdenum disulfide nanocomposite. ACS Appl Mater Interfaces 11(32):29195–29206

Al-Khayri JM, Ansari MI, Singh AK (eds) (2021) Nanobiotechnology: mitigation of abiotic stress in plants. Springer Nature

Alvandi N, Assariha S, Esfandiari N, Jafari R (2021) Off–on sensor based on concentration-dependent multicolor fluorescent carbon dots for detecting pesticides. Nano-Struct Nano-Objects 26:100706

Arya MS, Anjali K, Unni D (2018) Detection of unhealthy plant leaves using image processing and genetic algorithm with Arduino. In: 2018 International Conference on Power, Signals, Control and Computation (EPSCICON). IEEE, pp 1–5

Ashrafi Tafreshi F, Fatahi Z, Ghasemi SF, Taherian A, Esfandiari N (2020) Ultrasensitive fluorescent detection of pesticides in real sample by using green carbon dots. Plos one 15(3):e0230646

Bagheri H, Hajian A, Rezaei M, Shirzadmehr A (2017) Composite of Cu metal nanoparticles-multiwall carbon nanotubes-reduced graphene oxide as a novel and high performance platform of the electrochemical sensor for simultaneous determination of nitrite and nitrate. J Hazard Mater 324:762–772

Baranwal J, Barse B, Gatto G, Broncova G, Kumar A (2022) Electrochemical sensors and their applications: a review. Chemosensors 10(9):363

Bharti A, Jain U, Chauhan N (2024) From lab to field: nano-biosensors for real-time plant nutrient tracking. Plant Nano Biology 9:100079

Bitutsky NP (2014) Mineral nutrition of plants. St. Petersburg University Press, St. Petersburg, p 540

Bock CH, Poole GH, Parker PE, Gottwald TR (2010) Plant disease severity estimated visually, by digital photography and image analysis, and by hyperspectral imaging. Crit Rev Plant Sci 29(2): 59–107

Buck SM, Xu H, Brasuel M, Philbert MA, Kopelman R (2004) Nanoscale probes encapsulated by biologically localized embedding (PEBBLEs) for ion sensing and imaging in live cells. Talanta 63(1):41–59

Calisgan SD, Rajaram V, Kang S, Risso A, Qian Z, Rinaldi M (2020) Micromechanical switch-based zero-power chemical detectors for plant health monitoring. J Microelectromech Syst 29(5):755–761

Camargo A, Smith JS (2009) Image pattern classification for the identification of disease causing agents in plants. Comput Electron Agric 66(2):121–125

Cardoso RM, Pereira TS, Facure MHM, dos Santos DM, Mercante LA, Mattoso LHC, Correa DS (2022) Current progress in plant pathogen detection enabled by nanomaterials-based (bio)-sensors. Sens Actuators Rep 4:100068

Chen Y, Tian Y, Wang X, Dong L (2019a) Miniaturized soil sensor for continuous, in-situ monitoring of soil water potential. In: 2019 20th International Conference on Solid-State

Sensors, Actuators and Microsystems &Eurosensors XXXIII (TRANSDUCERS & EUROSENSORS XXXIII). IEEE, pp 2025–2028

Chen C, Feng S, Zhou M, Ji C, Que L, Wang W (2019b) Development of a structure-switching aptamer-based nanosensor for salicylic acid detection. Biosens Bioelectron 140:111342

Chen M, Zhang M, Wang X, Yang Q, Wang M, Liu G, Yao L (2020) An all-solid-state nitrate ion-selective electrode with nanohybrids composite films for in-situ soil nutrient monitoring. Sensors 20:2270

Dai CL (2007) A capacitive humidity sensor integrated with micro heater and ring oscillator circuit fabricated by CMOS–MEMS technique. Sensors Actuators B Chem 122(2):375–380

Denizli A, Nguyen TA, Rajendran S, Yasin G, Nadda AK (eds) (2021) Nanosensors for smart agriculture. Elsevier

Deuri B, Sarma PP, Pathak J (2014) Beyond the microscope: nano sensors and its application in agriculture. researchgate.net 4(2):677–684

Deuschle K, Chaudhuri B, Okumoto S, Lager I, Lalonde S, Frommer WB (2006) Rapid metabolism of glucose detected with FRET glucose nanosensors in epidermal cells and intact roots of Arabidopsis RNA-silencing mutants. Plant Cell 18(9):2314–2325

Dias DA, Fernandes IA, Machado EP, Pedott L, Blassioli-Moraes MC, Borges M et al (2024) Nanosensors for detecting volatile compounds in pest management: a focus on agricultural sustainability. ACS Agric Sci Technol 4(11):1250–1259

Domínguez-Renedo O, Alonso-Lomillo MA, Ferreira-Gonçalves L, Arcos-Martínez MJ (2009) Development of urease based amperometric biosensors for the inhibitive determination of Hg (II). Talanta 79(5):1306–1310

Doran JW, Zeiss MR (2000) Soil health and sustainability: managing the biotic component of soil quality. Appl Soil Ecol 15:3

Duhan JS, Kumar R, Kumar N, Kaur P, Nehra K, Duhan S (2017) Nanotechnology: the new perspective in precision agriculture. Biotechnol Rep 15:11–23

Ejeian F, Etedali P, Mansouri-Tehrani HA, Soozanipour A, Low ZX, Asadnia M, Taheri-Kafrani A, Razmjou A (2018) Biosensors for wastewater monitoring: a review. Biosens Bioelectron 118: 66–79

Esser B, Schnorr JM, Swager TM (2012) Selective detection of ethylene gas using carbon nanotube-based devices: utility in determination of fruit ripeness. Chem Int Ed 51:5752–5756

Fahimi-Kashani N, Hormozi-Nezhad MR (2016) Gold-nanoparticle-based colorimetric sensor array for discrimination of organophosphate pesticides. Anal Chem 88(16):8099–8106

Fang Y, Ramasamy RP (2015) Current and prospective methods for plant disease detection. Biosensors 5(3):537–561

Giraldo JP, Landry MP, Faltermeier SM, McNicholas TP, Iverson NM, Boghossian AA, Reuel NF, Hilmer AJ, Sen F, Brew JA, Strano MS (2014) Plant nanobionics approach to augment photosynthesis and biochemical sensing. Nat Mater 13(4):400–408

Giraldo JP, Wu H, Newkirk GM, Kruss S (2019) Nanobiotechnology approaches for engineering smart plant sensors. Nat Nanotechnol 14:541–553

Greenshields MW, Cunha BB, Coville NJ, Pimentel IC, Zawadneak MA, Dobrovolski S, Souza MT, Hümmelgen IA (2016) Fungi active microbial metabolism detection of rhizopus sp. and aspergillus sp. section nigri on strawberry using a set of chemical sensors based on carbon nanostructures. Chemosensors 4(3):19

Haji-Hashemi H, Norouzi P, Safarnejad MR, Larijani B, Habibi MM, Raeisi H, Ganjali MR (2018) Sensitive electrochemical immunosensor for citrus bacterial canker disease detection using fast Fourier transformation square-wave voltammetry method. J Electroanal Chem 820:111–117

Hashemi P, Karimian N, Khoshsafar H, Arduini F, Mesri M, Afkhami A, Bagheri H (2019) Reduced graphene oxide decorated on Cu/CuO-Ag nanocomposite as a high-performance material for the construction of a non-enzymatic sensor: application to the determination of carbaryl and fenamiphos pesticides. Mater Sci Eng C 102:764–772

He MQ, Wang K, Wang J, Yu YL, He RH (2017) A sensitive aptasensor based on molybdenum carbide nanotubes and label-free aptamer for detection of bisphenol A. Anal Bioanal Chem 409: 1797–1803

Isha A, Akanbi FS, Yusof NA, Osman R, Mui-Yun W, Abdullah SNA (2019) An NMR metabolomics approach and detection of Ganoderma boninense-infected oil palm leaves using MWCNT-based electrochemical sensor. J Nanomater 2019:1–12

Julich S, Riedel M, Kielpinski M, Urban M, Kretschmer R, Wagner S, Fritzsche W, Henkel T, Möller R, Werres S (2011) Development of a lab-on-a-chip device for diagnosis of plant pathogens. Biosens Bioelectron 26(10):4070–4075

Justino CI, Duarte AC, Rocha-Santos TA (2017) Recent progress in biosensors for environmental monitoring: a review. Sensors 17(12):2918

Kashyap B, Kumar R (2021) Sensing methodologies in agriculture for soil moisture and nutrient monitoring. IEEE Access 9:14095–14121

Khaled AY, Abd Aziz S, Bejo SK, Nawi NM, Seman IA, Onwude DI (2018) Early detection of diseases in plant tissue using spectroscopy–applications and limitations. Appl Spectrosc Rev 53(1):36–64

Khater M, De La Escosura-Muñiz A, Quesada-González D, Merkoçi A (2019a) Electrochemical detection of plant virus using gold nanoparticle-modified electrodes. Anal Chim Acta 1046: 123–131

Khater M, Escosura-Muñiz ADL, Altet L, Merkoçi A (2019b) In situ plant virus nucleic acid isothermal amplification detection on gold nanoparticle-modified electrodes. Anal Chem 91(7): 4790–4796

Kim MY, Lee KH (2022) Electrochemical sensors for sustainable precision agriculture—a review. Front Chem 10:848320

Kim T-H, Böhmer M, Hu H, Nishimura N, Schroeder JI (2010) Guard cell signal transduction network: advances in understanding abscisic acid, CO_2, and Ca^{2+}signaling. Annu Rev Plant Biol 61:561–591

Kim M-Y, Lee J-W, Park DJ, Lee J-Y, Myung NV, Kwon SH, Lee KH (2021) Highly stable potentiometric sensor with reduced graphene oxide aerogel as a solid contact for detection of nitrate and calcium ions. J Electroanal Chem 897:115553

Konwarh R, Sharma PL (2020) Nanosensor platforms for surveillance of plant pathogens and phytometabolites/analytes vis-à-vis plant health status. In: Nanomaterials for agriculture and forestry applications. Elsevier, pp 357–385

Kumar AA, Kumar SN, Fernandez RE (2020) Real time sensing of soil potassium levels using zinc oxide-multiwall carbon nanotube-based sensors. IEEE Trans Nanobioscience 20(1):50–56

Kwak SY, Wong MH, Lew TTS, Bisker G, Lee MA, Kaplan A, Dong J, Liu AT, Koman VB, Sinclair R, Hamann C (2017) Nanosensor technology applied to living plant systems. Annu Rev Anal Chem 10:113–140

Lager I, Looger LL, Hilpert M, Lalonde S, Frommer WB (2006) Conversion of a putative Agrobacterium sugar-binding protein into a FRET sensor with high selectivity for sucrose. J Biol Chem 281(41):30875–30883

Lan L, Le X, Dong H, Xie J, Ying Y, Ping J (2020) One-step and large-scale fabrication of flexible and wearable humidity sensor based on laser-induced graphene for real-time tracking of plant transpiration at bio-interface. Biosens Bioelectron 165:112360

Lee C-Y, Wu G-W, Hsieh W-J (2008) Fabrication of micro sensors on a flexible substrate. Sens Actuators A Phys 147:173–176

Leone M, Consales M, Passeggio G, Buontempo S, Zaraket H, Youssef A, Persiano GV, Cutolo A, Cusano A (2022) Fiber optic soil water content sensor for precision farming. Opt Laser Technol 149:107816

Li Z, Paul R, Ba Tis T, Saville AC, Hansel JC, Yu T, Ristaino BJ, Wei Q (2019) Non-invasive plant disease diagnostics enabled by smartphone-based fingerprinting of leaf volatiles. Nat Plants 5(8):856–866

Li Z, Yu T, Paul R, Fan J, Yang Y, Wei Q (2020) Agricultural nanodiagnostics for plant diseases: recent advances and challenges. Nanoscale Adv 2(8):3083–3094
Li D, Li G, Li J, Xu S (2022) Wearable crop sensor based on nano-graphene oxide for noninvasive real-time monitoring of plant water. Membranes 12(4):358
Lopez-Soriano P, Noguera P, Gorris MT, Puchades R, Maquieira A, Marco-Noales E, López MM (2017) Lateral flow immunoassay for on-site detection of Xanthomonas arboricolapv. pruni in symptomatic field samples. PLoS One 12(4):e0176201
Luo M, Wei J, Zhao Y, Sun Y, Liang H, Wang S, Li P (2020) Fluorescent and visual detection of methyl-paraoxon by using boron-and nitrogen-doped carbon dots. Microchem J 154:104547
Mahlein AK (2016) Plant disease detection by imaging sensors–parallels and specific demands for precision agriculture and plant phenotyping. Plant Dis 100(2):241–251
Marios S, Georgiou J (2017) Precision agriculture: challenges in sensors and electronics for real-time soil and plant monitoring. In: 2017 IEEE Biomedical Circuits and Systems Conference, (BioCAS). IEEE, pp 1–4
Marques I, Pinto da Costa J, Justino C, Santos P, Duarte K, Freitas A, Cardoso S, Duarte A, Rocha-Santos T (2017) Carbon nanotube field effect transistor biosensor for the detection of toxins in seawater. Int J Environ Anal Chem 97(7):597–605
McLamore ES, Diggs A, Calvo Marzal P, Shi J, Blakeslee JJ, Peer WA, Murphy AS, Porterfield DM (2010) Non-invasive quantification of endogenous root auxin transport using an integrated flux microsensor technique. Plant J 63(6):1004–1016
Miranda BS, Linares EM, Thalhammer S, Kubota LT (2013) Development of a disposable and highly sensitive paper-based immunosensor for early diagnosis of Asian soybean rust. Biosens Bioelectron 45:123–128
Mugo SM, Lu W, Lemieux S (2022) Stainless steel electrochemical capacitive microneedle sensors for multiplexed simultaneous measurement of pH, nitrates, and phosphates. Microchim Acta 189:206
Nadporozhskaya M, Kovsh N, Paolesse R, Lvova L (2022) Recent advances in chemical sensors for soil analysis: a review. Chemosensors 10(1):35
Neumann M, Hallau L, Klatt B, Kersting K, Bauckhage C (2014, August) Erosion band features for cell phone image based plant disease classification. In: 2014 22nd International Conference on Pattern Recognition. IEEE, pp 3315–3320
Nizamani MM, Hughes AC, Zhang HL, Wang Y (2024) Revolutionizing agriculture with nano-technology: innovative approaches in fungal disease management and plant health monitoring. Sci Total Environ 928:172473
Oren S, Ceylan H, Schnable PS, Dong L (2017) High-resolution patterning and transferring of graphene-based nanomaterials onto tape toward Roll-to-Roll production of tape-based wearable sensors. Adv Mater Technol 2(12):1700223
Pan Y, Zhou J, Su K, Hu N, Wang P (2017) A novel quantum dot fluorescence immunosensor based on magnetic beads and portable flow cytometry for detection of okadaic acid. Proc Technol 27: 214–216
Panferov VG, Safenkova IV, Byzova NA, Varitsev YA, Zherdev AV, Dzantiev BB (2018) Silver-enhanced lateral flow immunoassay for highly-sensitive detection of potato leafroll virus. Food Agric Immunol 29(1):445–457
Peng H, Chen IA (2018) Rapid colorimetric detection of bacterial species through the capture of gold nanoparticles by chimeric phages. ACS Nano 13(2):1244–1252
Podar D, Maathuis FJM (2022) Primary nutrient sensors in plants. iScience 25(4):104029
Ragavan KV, Selvakumar LS, Thakur MS (2013) Functionalized aptamers as nano-bioprobes for ultrasensitive detection of bisphenol-A. Chem Commun 49(53):5960–5962
Rana K, Tripathi RM, Pudake RN (2024) Advances in the fabrication of nanosensors for management of abiotic stress in crop plants. In: Innovative methods in horticultural crop improvement: biosensors and nanosensors. Cham, Springer Nature Switzerland, pp 75–97

Razmi A, Golestanipour A, Nikkhah M, Bagheri A, Shamsbakhsh M, Malekzadeh-Shafaroudi S (2019) Localized surface plasmon resonance biosensing of tomato yellow leaf curl virus. J Virol Methods 267:1–7

Riu J, Maroto A, Rius FX (2006) Nanosensors in environmental analysis. Talanta 69(2):288–301

Rizza A, Walia A, Lanquar V, Frommer WB, Jones AM (2017) In vivo gibberellin gradients visualized in rapidly elongating tissues. Nat Plants 3(10):803–813

Saito K, Chang YF, Horikawa K, Hatsugai N, Higuchi Y, Hashida M, Yoshida Y, Matsuda T, Arai Y, Nagai T (2012) Luminescent proteins for high-speed single-cell and whole-body imaging. Nat Commun 3(1):1262

dos Santos TB, Ribas AF, de Souza SGS, Budzinski IGF, Domingues DS (2022) Physiological responses to drought, salinity, and heat stress in plants: a review. Stress 2:113–135

Savary S, Willocquet L, Pethybridge SJ, Esker P, McRoberts N, Nelson A (2019) The global burden of pathogens and pests on major food crops. Nat Ecol Evol 3(3):430–439

Schmälzlin E, Van Dongen JT, Klimant I, Marmodée B, Steup M, Fisahn J, Geigenberger P, Löhmannsröben HG (2005) An optical multifrequency phase-modulation method using microbeads for measuring intracellular oxygen concentrations in plants. Biophys J 89(2): 1339–1345

Sephra PJ, Chandrapagasam T, Sachdev A, Manikandan E (2024) An efficient sensing system using Ion-Selective Membrane on Ni2O3/rGO nanocomposite for electrochemical detection of nitrate ions. J Alloys Compd 980:173414

Shahdost-Fard F, Fahimi-Kashani N, Hormozi-Nezhad MR (2021) A ratiometric fluorescence nanoprobe using CdTe QDs for fast detection of carbaryl insecticide in apple. Talanta 221: 121467

Sharma A, Kumar V, Shahzad B, Tanveer M, Sidhu GPS, Handa N, Kohli SK, Yadav P, Bali AS, Parihar RD, Dar OI, Singh K, Jasrotia S, Bakshi P, Ramakrishnan M, Kumar S, Bhardwaj R, Thukral AK (2019) Worldwide pesticide usage and its impacts on ecosystem. SN Appl Sci 1:1–16

Sharma P, Pandey V, Sharma MMM, Patra A, Singh B, Mehta S, Husen A (2021) A review on biosensors and nanosensors application in agroecosystems. Nanoscale Res Lett 16:1–24

Shaw DS, Honeychurch KC (2022) Nanosensor applications in plant science. Biosensors 12(9):675

Siddiqui MH, Al-Whaibi MH, Firoz M, Al-Khaishany MY (2015) Role of nanoparticles in plants. In: Nanotechnology and plant sciences: nanoparticles and their impact on plants. Springer, pp 19–35

Singh BK, Delgado-Baquerizo M, Egidi E, Guirado E, Leach JE, Liu H, Trivedi P (2023) Climate change impacts on plant pathogens, food security and paths forward. Nat Rev Microbiol 21(10): 640–656

Tanner F, Tonn S, de Wit J, Van den Ackerveken G, Berger B, Plett D (2022) Sensor-based phenotyping of above-ground plant-pathogen interactions. Plant Methods 18(1):35

Tharini C, Iyappan G, Manikandan E, Sephra PJ (2023) Potentiometric sensing of potassium ion (K +) using valinomycin supported on ZnO/rGO nanocomposites. J Mater Sci Mater Electron 34(19):1474

Tian Y, Cong L, Sun H, Wei W, Wu F, Li F et al (2024) An in vivo surface-enhanced raman scattering nanosensor for detecting hydrogen peroxide in plants under abiotic stresses. ACS Appl Nano Mater 7(17):20308–20316

Tran TT, Clark K, Ma W, Mulchandani A (2020) Detection of a secreted protein biomarker for citrus Huanglongbing using a single-walled carbon nanotubes-based chemiresistive biosensor. Biosens Bioelectron 147:111766

Troeh FR, Thompson LM (2005) Soils and soil fertility, 6th edn. Wiley, Hoboken

Tsuchiya Y, Yoshimura M, Sato Y, Kuwata K, Toh S, Holbrook-Smith D, Zhang H, McCourt P, Itami K, Kinoshita T, Hagihara S (2015) Probing strigolactone receptors in Striga hermonthica with fluorescence. Science 349(6250):864–868

Vero V, Kukreja E, Tripathi RM, Pudake RN (2024) Recent advances in abiotic stress management with the help of plant wearable and in planta nanosensors. In: Nanotechnology for abiotic stress tolerance and management in crop plants. Academic Press, pp 127–139

Virk V, Deepak H, Taneja K, Srivastava R, Giri S (2024) Amelioration in nanobiosensors for the control of plant diseases: current status and future challenges. Front Nanotechnol 6:1310165

Wang H, Wang Y, Hou X, Xiong B (2020) Bioelectronic nose based on single-stranded DNA and single-walled carbon nanotube to identify a major plant volatile organic compound (p-ethylphenol) released by phytophthora cactorum infected strawberries. Nanomaterials 10(3):479

Wang Y, He Q, Zhao X, Yuan J, Zhao H, Wang G, Li M (2022) Synthesis of corn straw-based graphene quantum dots (GQDs) and their application in PO_4^{3-} detection. J Environ Chem Eng 10(2):107150

Weerakoon KA, Shu JH, Chin BA (2011) A chemiresistor sensor with a poly3-hexylthiophene active layer for the detection of insect infestation at early stages. IEEE Sensors J 11(7): 1617–1622

Wei S, Sun Y, Xi G, Zhang H, Xiao M, Yin R (2018) Development of a single-tube nested PCR-lateral flow biosensor assay for rapid and accurate detection of Alternaria panax Whetz. PLoS One 13(11):e0206462

Wei H, Liu K, Zhang H, Hou P, Pan D, Luo B, Li A, Zhao C (2024) Smart wearable flexible sensor based on laser-induced graphene/gold nanoparticles/black phosphorus nanosheets for in situ quercetin detection. Chem Eng J 497:154271

Wu CS, Khaing Oo MK, Fan X (2010) Highly sensitive multiplexed heavy metal detection using quantum-dot-labeledDNAzymes. ACS Nano 4(10):5897–5904

Xu F, Wang P, Bian S, Wei Y, Kong D, Wang H (2021) A Co-nanoparticles modified electrode for on-site and rapid phosphate detection in hydroponic solutions. Sensors 21(1):299

Yusof NA, Isha A (2020) Nanosensors for early detection of plant diseases. In: Nanomaterials for agriculture and forestry applications. Elsevier, pp 407–419

Zain M, Ma H, Nuruzzaman M, Chaudhary S, Nadeem M, Shakoor N, Azeem I, Duan A, Sun C, Ahamad T (2023) Nanotechnology based precision agriculture for alleviating biotic and abiotic stress in plants. Plant Stress 10:100239

Zhan F, Wang T, Iradukunda L, Zhan J (2018) A gold nanoparticle-based lateral flow biosensor for sensitive visual detection of the potato late blight pathogen, Phytophthora infestans. Anal Chim Acta 1036:153–161

Zhang C, Zhou Y, Tang L, Zeng G, Zhang J, Peng B, Xie X, Lai C, Long B, Zhu J (2016) Determination of Cd^{2+} and Pb^{2+} based on mesoporous carbon nitride/self-doped polyaniline nanofibers and square wave anodic stripping voltammetry. Nanomaterials 6(1):7

Zhang Q, Ying Y, Ping J (2022) Recent advances in plant nanoscience. Adv Sci 9(2):2103414

Zhang M, Ma W, Tao R, Fan Q, Zhang M, Qin D, Cao X, Li J, Xiong R, Huang C (2024) Nanomaterials: recent advances in plant disease diagnosis and treatment. NanoToday 57: 102326

Zhao Y, Liu L, Kong D, Kuang H, Wang L, Xu C (2014) Dual amplified electrochemical immunosensor for highly sensitive detection of Pantoea stewartii sbusp. stewartii. ACS Appl Mater Interfaces 6(23):21178–21183

Zia R, Taj A, Younis S, Bukhari SZ, Latif F, Feroz Y, Fatima K, Imran A, Bajwa SZ (2022) Application of nanosensors for pesticide detection. In: Nanosensors for smart agriculture. Elsevier, pp 259–302

Chapter 6
Nanotechnology for Soil Improvement and Plant Nutrition

Puthamohan Vinayaga Moorthi, Ravi Bavyataa, Kumar Vignesh, Divya Sri Kathiresan, and Rubadevi Balasubramani

Abstract The nanotechnology offered various applications in the field of health and industrial sectors. Its role in agricultural development such as crop improvement, crop protection, soil health improvisation is the new milestones reached by the nanomaterials. Such powerful nanomaterials are synthesized using various plants sources for the ecofriendly environment. This chapter brings the limelight on the plants that are used for the synthesis of nanoparticles, and the reducing agents used for it. The impactful usage of nanomaterials for the soil improvement and antioxidant enzyme inductions, attempted by the researchers, are highlighted in the present review.

Keywords Antioxidants · Nanoparticles · Silver nanoparticle · Soil health · Plant nutrition

6.1 Introduction

Soil is the base for the almost all the life forms. It encompasses as many as habitats and acting as head of all the habitats. There were different types of soil spread all over the world and according to the geographical existence they accommodate flora and fauna. This emphasis the importance of health of soil for rendering essential and basic nutritional nourishments to the plants and thereby harbour source of feed for herbivors and other fauna. The health of soil, due to pollutions of various types, is/was impaired a lot by the pollutants and this cause severe impact on the ecosystem as well as biodiversity. However, there were technologies, strategies and new components were implemented now and then for improving the health of the soil for nourishing essential nutrients to the plants. One such vital technology improving the sector of soil is the nanotechnology. As similar to that of the soil, nanotechnology is also accommodating all the fields of sciences and working on the improvement

P. V. Moorthi (✉) · R. Bavyataa · K. Vignesh · D. S. Kathiresan · R. Balasubramani
Department of Human Genetics and Molecular Biology, Bharathiar University, Coimbatore, Tamil Nadu, India
e-mail: pvmhgmb@buc.edu.in

J. M. Al-Khayri et al. (eds.), *Nanotechnology Applications in Modern Agriculture*, Nanotechnology in Plant Sciences 3, https://doi.org/10.1007/978-3-031-90506-3_6

and fine tuning. Nanotechnology revolutionized the industries, many health sectors, agriculture. The major concerns of the soils such as erosion, acidification, compaction and organic matter loss are encountered by the nanomaterials by nutrient cycling, microbial diversity and it enhances the soil fertility and strength (Bandala and Berli 2018). Fe3O4 capped with iron oxalate was known to enhance the basic essential physico-chemical qualities of the soil especially, alluvial soils (Das et al. 2016). Similarly, the hydrogen ion potential (pH) and porosity of the soil was improvised by a soil conditioner made of mineral based nano-submicron (Liu et al. 2017). On the other hand, a nanocomposite (Biochar) had demonstrated carbon sequestration, diminishing green house (soil-borne) and importantly enhanced water and nutrient retention by the soil (Song et al. 2019; El-Naggar et al. 2019; Hagemann et al. 2017). *Thuja occidentalis* based silver nanoparticle (from leaf extract) revolutionized the physicochemical nature of the soil (Das et al. 2018b). So, the nanoparticles and nanocomposites now play a major role in revolutionizing the soil health and nutrient. This chapter addressing the reforms done by the nanotechnology and bring limelight on it.

6.2 Synthesis of Nanoparticles

Nanoparticle synthesis occupied special attentions due to its size, charge, and other surface properties that decide its application in various field including agriculture with special drive in the area of soil and nutrient improvement. Silver nanoparticle (AgNps) was synthesized from various biological sources such as *Ocimum tenuiflorum* (Patil et al. 2012), *Syzygium cumini* (Prasad and Swamy 2013), *Thuja occidentalis* (Barua et al. 2013), *Elaeocarpus ganitrus* (Dwivedi et al. 2013), *Delphinium denudatum* (Suresh et al. 2014), *Tagetes erecta* (Padalia et al. 2014), *Catharanthus roseus* (Manisha et al. 2014), *Acacia leucophloea* (Murugan et al. 2014), *Melia azedarach* (Mehmood et al. 2014), *Quercus sp.* (Puišo et al. 2014), *Cochlospermum religiosum* (Sasikala et al. 2015), *Afzelia quanzensis* (Moyo et al. 2015), *Ficus benghalensis* (Nayak et al. 2015), *Syzygium alternifolium* (Yugandhar et al. 2015), *Pinus densiflora* (Velmurugan et al. 2015), *Cucurbita maxima*, *Moringa oleifera, Acorus Calamus* (Nayak et al. 2015), *Andrographis echioides* (Elangovan et al. 2015), *Areca catechu* (Rajan et al. 2015), *Azadirachta indica* (Velusamy et al. 2015), *Carica papaya* (Banala et al. 2015), *Hemidesmus indicus* (Latha et al. 2015), *Momordica charantia* (Ajitha et al. 2015), Potentilla fulgens (Mittal et al. 2015), Ocimum tenuiflorum, Solanum tricobatum, Syzygium cumini, Centella asiatica and Citrus sinensis (Logeswari et al. 2015), Ficus virens (Salar et al. 2015), Pinus thunbergia (Khullar et al. 2016), Taxus yunnanensis (Xia et al. 2016), *sugarcane* sap (Elemike et al. 2016), *Ocimum sanctum* (Chandra et al. 2016), *Cola nitida* (Lateef et al. 2016), essential oil (Vilas et al. 2016). *Caesalpinia pulcherrima* (Moteriya and Chanda 2017), *Allamanda cathartica* (Karunakaran et al. 2016), *Capparis spinosa* (Benakashani et al. 2016), *Azadirachta indica* (Verma et al. 2016), *Urtica dioica* (Jyoti et al. 2016), Hibiscus rosa-sinensis (Surya et al. 2016), *Tamarindus indica* (Jayaprakash et al. 2017), *Lantana camara* (Shriniwas and

Subhash 2017), *Butea monosperma* (Pattanayak et al. 2017), *Enteromorpha compressa* (Ramkumar et al. 2017), *Sapindus emarginatus* (Swarnavalli et al. 2017), *Ocimum gratissimum* (Das et al. 2017), *Melia azedarach L* (Mehmood et al. 2017), *Thuja occidentalis* (Barua et al. 2017; Das et al. 2018a, b), *Acacia rigidula* (Escárcega-González et al. 2018), rice grain; rice bran (RB), rice husk (RH), and rice germ (RG) (Suwan et al. 2018), turmeric powders (Alsammarraie et al. 2018), *Pleurotus citrinopileatus* (Bhardwaj et al. 2018), *Jurinea dolomiaea* (Riaz et al. 2018), *Rhodiola rosea* (Singh et al. 2018), *Diospyros montana* (Bharathi et al. 2018; Vasantharaj et al. 2018), *Ipomoea digitata* Linn (Varadavenkatesan et al. 2019), *Toxicodendron vernicifluum* (Saravanakumar et al. 2019), *Elaeagnus umbellate* (Ali et al. 2019), *Streptomyces capillispiralis Ca-1, Streptomyces zaomyceticus Oc-5, and Streptomyces pseudogriseolus Acv-11* (Fouda et al. 2019), *Pycnoporus sp* (Barbosa et al. 2019), *C. pentagyna* (Ebrahimzadeh et al. 2019), *Rheum Ribes* (Aygün et al. 2019), *Juniperus chinensis* (Al-Dhafri and Ching 2019), Cashew gum (Araújo et al. 2020), *Pinus wallichiana* (Khan et al. 2020), *Piper nigrum* (Kanniah et al. 2021), soybean (Ma et al. 2021), gum Arabic (Fadaka et al. 2022), *Benincasa hispida* (Baker et al. 2021), *Trichoderma harzianum* (Konappa et al. 2021), *Aspergillus japonicus* (Li et al. 2021), *Momordica charantia* (Moorthy et al. 2021), *Carissa carandas* (Singh et al. 2021), Ascorbic acid (Mahmood et al. 2021), *Ipomoea batatas* (Wibowo et al. 2021), *Ulva rigida, Cystoseira myrica*, *Gracilaria foliifera* (Algotiml et al. 2022), *Cuphea carthagenensis* (Rather et al. 2022), wild olive and *pistachio Olea* europaea subsp. and *Pistacia lentiscus* (Essghaier et al. 2022), *Zanthozylum armatum* (Riaz et al. 2022), cyanobacterium *Pseudanabaena/ Limnothrix* sp. (Karageorgou et al. 2022), *Gracilaria edulis* (Mohanta et al. 2022), *Tridax procumbens* (Pungle et al. 2022), bacterial cellulose (El-Sherbiny et al. 2022). Synthesis of Gold nanoparticles (AuNPs) from variety of biological sources such as *Plumeria alba* linn (Nagaraj et al. 2012), *Escherichia coli* (Gholami-Shabani et al. 2015), *Spirulina platensis* (Fatimah 2016), *Adiantum philippense* (Kalita et al. 2016), *Elettaria cardamomum* (Rajan et al. 2017), *Pinus kesiya* (Fernando et al. 2017), *Rhodiola rosea* (Singh et al. 2018), *Benincasa hispida* (Al Saqr et al. 2021), *Ricinus communis* (Rahman et al. 2022).

Copper Oxide NPs was synthesized from numerous biological source such as *Madhuca longifolia* (Das et al. 2018b), *Phyllanthus reticulatus*, *Conyza bonariensis* (Potbhare et al. 2019), *Mimusops elengi* (Sarah and Iyer 2019), Almond Gum (Nithiyavathi et al. 2021), *Cupressus macrocarpa* (Gad El-Rab et al. 2021), Iron nanoparticles was synthesized from biological source such as *Piliostigma thonningii* (Igwe and Nwamezie 2018), *Phoenix dactylifera* (Batool et al. 2021), polyherbal drug Liv52 (L52E) (Ansari and Asiri 2021). Magnesium Nanoparticles (Mg NPs) was synthesized from *Rosmarinus officinalis* (Abdallah et al. 2019). Nickel Oxide (NiO) NPs was synthesized from *Callistemon viminalis* (Sani et al. 2021), *Berberis pachyacantha* (Uddin et al. 2022). Platinum nanoparticles was synthesized from Chitosan (Sharma et al. 2021). Selenium NPs was synthesized from wide range of biological sources such as *Bacillus licheniformis* (Khiralla and El-Deeb 2015), ascorbic acid (Freire-Neto et al. 2022), *C. albicans* (Hosseini Bafghi et al. 2022), tung ting *oolong tea* (Malini et al. 2015), *Jacaranda mimosifolia* (Sharma et al. 2016), Oleaceae Nyctanthes arbor-tristis (Jamdagni et al. 2016), *Passiflora caerulea*

Table 6.1 Nanomaterials and its application in plant

Nanomaterial	Application	References
Silver nanoparticles (AgNPs)	Improve the pH and other properties of the soil	Das et al. (2018b)
Fe OCIO orthorhombic	Improve Carbon (C), Nitrogen (N) and Phosphorous (P) availability	Das et al. (2016)
Zinc Oxide	Retains the N	Aziz et al. (2019)
ZnO NP	Improve P uptake	Raliya et al. (2016)
Silica NP	Improve maize viability and NPK level	Karunakaran et al. (2013)
Nano zerovalent iron	rice seed vigor enhancer	Guha et al. (2018)
TiO2	Improve the dry weight, photosynthetic and chlorophyll a formation in spinach	Zheng et al. (2005)
Carbon nanotubes (MWCNT)	Tomato seeds to absorb more water	Khodakovskaya et al. (2009)
nano—TiO2	The protein content, chlorophyll, and carotenoids of Gycin max (L.) Merrill	Singh et al. (2018)
AgNp nanoparticle	Diminished the UV-B radiation and supporting the rate of photosynthesis	Gao et al. (2006a b), Hong et al. (2005a, b)
MWCNT	Make pore in tomato seeds for easy water transport	Khodakovskaya et al. (2009)

(Santhoshkumar et al. 2017), *Thymus vulgaris* (Zare et al. 2019), *Syzygium aromaticum* (Lakshmeesha et al. 2019), Elaeagnus angustifolia (Iqbal et al. 2021), pomegranate fruit (Khormali et al. 2021), *Moringa oleifera* (Gobinath et al. 2022), *Piper longum* (Asha et al. 2022), Pomegranate Fruit Peel (Abdelmigid et al. 2022). Cerium (IV) Oxide NPs was synthesized from *Gloriosa superba* (Arumugam et al. 2015), *Abelmoschus esculentus* (Ahmed et al. 2021). Silver and Gold (Ag and Au) NPs was synthesized from *Stereospermum suaveolens* (Francis et al. 2018), *Cinnamomum zeylanicum* (Soni and Prakash 2014). Ag & Zinc NPs was synthesized from source like *Acacia arabica* (Khan et al. 2021). Titanium dioxide (TiO) NPs was synthesized from source like *Coleus aromaticus* (Narayanan et al. 2021). Chromium NPs was synthesized from *Abutilon indicum* (Mohandoss et al. 2022). Zirconium NPs was synthesized from Pomegranate peel (Chau et al. 2022). An outline of few applications of nanomaterials in plants is given in Table 6.1.

6.3 Soil and Plant Nutrition Improvement

T. occidentalis leaf extract made AgNps greatly improved the pH and other properties of the soil (Das et al. 2018b). The essential components for the soil such as Carbon (C), Nitrogen (N) and Phosphorous (P) availability was improved by the Fe OCIO orthorhombic nanomaterial (Das et al. 2016). The Zinc Oxide (ZnO) nanoparticle efficiently retains the N which was comparatively higher than normal control soil (Aziz et al. 2019). The P uptake was fascinatingly improved by the ZnO NP by targeting and enhancing the performance of phytase and phosphatase enzyme that are solubilizing the P (Raliya et al. 2016). Similarly, the performance of urease was improved by the AgNps synthesized from *T. Occientalis* and exhibited mineralization of N and P and by the way supporting the productivity and growth of the plant (Das et al. 2018b). Similarly, gold nanoparticle decorated with citrate as well as polyvinylpyrrolidone had played an important role in soil enzyme induction (Asadishad et al. 2017). In saline-alkali soils, the impact of ZnO, TiO2, Fe3O4, CeO2 on the urease, invertase, phosphatase enzyme was interesting and are susceptible (You et al. 2018). There has been diversified behavior of graphen oxide (GO), and graphene (G) on the enzymes in different paddy soils such as red, yellow etc. (Rong et al. 2017).

In the tested soils the sorption of Ag+ ion was high compared to AgNps (Wang et al. 2018). It has been identified that the sorption was influenced by the precursor materials used for the synthesis such as trisodium citrate, PVP (Wang et al. 2018). On a calcareous soil, the sorption was also high when AgNps coating material was PVP (Rahmatpour et al. 2017). The soil physiology also represent that when pH increased it led to increased net negative charge thereby the soil outer and inner spheres exhibit higher adsorption (Robinson et al. 2009).

Nanocoating of the fertilizers such as sulfur, urea and phosphate enable the sustained release of the fertilizers and facilitate the soils to meet the demand such as soils with sulfur, urea and phosphate deficient (Brady and Weil 1999; Santoso et al. 1995). Chitosan nanoparticle supported the NPK fertilizer usage by adopting controlled release (Corradini et al. 2010). Similar to nanocoating, nano micronutrient formulations facilitate the nutrient uptake and plant and soil growth (Peteu et al. 2010). In Spinach, nano-anatase TiO2 induced the Rubisco activase that led to enhanced photosynthetic rate via Rubisco carboxulation (Gao et al. 2006a, b). The silica NP (50 nm) improved the maize viability and also improved the NPK level and helped seed germination efficiently.

6.4 Seed and Germination Improvement

A nano zerovalent iron was acting as rice seed vigor enhancer (Guha et al. 2018). TiO2 had improved the dry weight, photosynthetic and chlorophyll a formation in spinach seed that are naturally aged (Zheng et al. 2005). Similarly, carbon nanotubes (MWCNT) enable the tomato seeds to absorb more water and by the way helped the tomato seeds to be germinated in faster rate (Khodakovskaya et al. 2009). In Gycin max (L.) Merrill. the protein content, chlorophyll, and carotenoids were drastically altered by nano—TiO2 along with increasing the photosynthetic rate (Singh et al. 2018). TiO2 was also known for its seed and plant growth (Jun et al. 2016). Similarly, the cerium oxide (CeO) had notable impact on the morphology, and biochemical parameters (Rico et al. 2015). The AgNp nanoparticle diminished the

UV-B radiation and supporting the rate of photosynthesis (Gao et al. 2006a, b; Hong et al. 2005a, b). AgNps had influenced the enzymes and proteins of the Trachyspermum ammi i.e., nitrate reductase, carbonic anhydrase, leaf protein and nitrogen (Danish et al. 2021). The enhanced availability of photosynthetic pigments such as chl a, chl b, carotenoids, etc. made possible by the AgNps in *T. ammi* (Danish et al. 2021). In *Brassica juncea* was also witnessed the improved biochemical and growth parameters with nanoparticles (Salama 2012).

For boosting the yield of soybean iron oxide (nano) supported by the way of facilitating the nutrient reaching the apical area and the leaf and pods gain dry weight (Sheykhbaglou et al. 2010). It was also observed that the pore was made in seeds of tomato by the MWCNT for easy transport of water (Khodakovskaya et al. 2009). Similarly root entry of magnetic nanoparticle (carbon coated) was also observed by Cifuentes et al. (2010). Besides that, more than 200 nm longer MWCNT were able to reach the sub-cellular organelles while it was shorter length i.e., 30–100 nm MWCNT reaching the nucleus, vacuoles and plastids. On the other hand SWCNT get trapped on the surface of the roots of seedlings of the cucumber (Serag et al. 2013; Lou et al. 2011; Yuan et al. 2011), provided, in spinach, it was proved that the SWCNT appeared to be reaching the thylakoid via stromata (Giraldo et al. 2014). To withstand the extreme abiotic condition such as stress, nano-SiO2 facilitates the proline and other amino acids accumulation and support effective antioxidant nature (Siddiqui et al. 2015). Further, seed germination was enhanced by declining the oxidative damage via the reduced levels of malondialdehyde (MDA) and hydrogen peroxide (H2O2) (Siddiqui et al. 2015). Interestingly, epicuticular wax, that are prone to salinity, was found unaltered upon treatment with nano-SiO2 (Avestan et al. 2019). The CNT was found to regulate the essential oil content when under salt stress (Gohari et al. 2020). A well-coordinated magnification in electrons was pronounced due to plasmon resonance when treated with Ag and AuNP that are coupled with reaction center of the photosynthesis (Prasad et al. 2017). The state of abiotic stress protection was given by the AgNps by offering increased production of amino acids and sugars etc. (Karami Mehrian et al. 2015; Mohamed et al. 2017). The silica NP had increased 85% cytokine production as per the work of Kurdish et al. (2018). For the peanut growth the ZnO NP had promoted the seed germination and vigor (Prasad et al. 2012). The combination of *P. chlororaphis* and ZnO NP had promoted the higher siderophore production (Dimkpa et al. 2012).

6.5 Plant Protection

An induced immunity was exerted by the nanoparticle via inducing the vital antioxidant enzymes catalase, superoxide dismiutase, ascorbate peroxidase, peroxidase etc. and thereby suppressed the hallmarks of the nematode, Meloidohyne incognita, gall formation and no. of quality eggs (Kumar et al. 2018; Danish et al. 2021). This state of protection by the AgNps was noticed in egg plants (Abdellatif et al. 2016) and tomato (Hassan et al. 2016). In Bermudagrass, the root knot nematode was efficiently controlled by AgNps (Cromwell et al. 2014) and at low or sublethal doses it can efficiently reduce the rate of reproduction and also inhibit the growth (Roh et al. 2009; Lim et al. 2012; Meyer et al. 2010). AgNps induced ROS production was noticed in *Oryza sativa* (Mirzajani et al. 2013). Especially, production of superoxide and hydrogen peroxide was noticed during the application of AgNp in *Allium cepa* (Panda et al. 2011). Besides, in Castor, cell protecting antioxidants such as superoxide dismutase and peroxidase were witnessed apparently when treated with nanoparticles (Yasur and Rani 2013). The AgNP mediated oxidative stress tolerance by the phenol and phenolic acid was witnessed by Winkel-Shirley (2002), Krishnaraj et al. (2012), Yasur and Rani (2013). AgNp was used as a inducer for the metallathionein gene that are enforcing the detoxification in *Triticum aestivum* (Dimkpa et al. 2013). An electrostatic coupling was facilitated by the chitosan, when applied as nanomaterial, and by the way the cell membrane was made susceptible for leak that recognize the chitosan as potent antimicrobial nanomaterial (Kumaraswamy et al. 2018; Kashyap et al. 2015).

6.6 Plant Microbiome

Nanoparticle/nanomaterials widely employed for improving the soil inhabited agriculturally important bacteria. The nitrogen fixing bacterial population was strongly supported by the Ag-Nps (Das et al. 2018a, b). Besides, multiwalled carbon nanotubes behave benevolently with soil microbes (Xin et al. 2018; Das et al. 2016). Khati et al. (2017) highlighted the soil microbiota beneficial role of nanochitosan, nanoclay and nanozeolite. The Rhizobium and Azotobacter were not drastically affected for its nitrate reductase enzyme activity by the nano silver (Shahrokh et al. 2014). Similarly in activated sludge, upon nano silver treatment, no net change in the bacterial community essential for nitrifying (Zhang et al. 2014). In maize, the plant growth promoting bacteria was heavily supported by nanosilica and it also added higher biomass, soil nutrient as well (Rangaraj et al. 2014). The bacterial species (*Pseudomonas fluorescens* and *B. subtilis*) stuffed sodium alginate and CNT facilitated the micropropagation and root length of pistachio (Pour-Ali et al. 2019). The work of Boroumand et al. (2020) represented the improvement in root and shoot length, and as well as nitrogen and potassium as well when NP coupled with bacterial populations. In rice plant, the Zn coupled bacteria, increased the zinc, potassium, nitrogen and phosphorous and also the quality of grains (Gudadhe et al. 2018).

6.7 Nanopesticides

The nanoparticle dimensions and dynamics creates avenues for wide range of applications. Nanopesticide is one such a dimension in which they could control wide variety of plant pathogens such as *Colletotrichum* sp. *Fusarium* (Elmer and

White 2018) etc. Ag nanoparticles had been used for foliar application to curb the Colletotrichum sp. that infect the pepper plants (Lamsal et al. 2011). Similarly, Ag was also used against R. Solani infecting rice varieties (Nejad et al. 2016). Similarly, a watermelon infecting Fusarium oxysporum f. sp. niveum, was treated with CuO (Elmer and White 2018). In wheat plant, Fusarium poae, F. graminearum were efficiently treated with single, multiwalled, reduced graphene oxide and graphene oxide nano-groups (Wang et al. 2014). In tomato, 62% increase in fresh was made possible by application of CuO NP. Similarly, on Chrysanthemum, the infection by the *Fusarium* wilt was drastically reduced by the CuO NP application (Ma et al. 2019). The genes of (*PPO*, *PTI5*, *PR1A1*) defense mechanism was found upregulated by the Cu3(PO4)2 (Ma et al. 2019).

6.8 Pesticide Degradation

Nanoparticles are known to degrade the various types of pesticides (Zhang 2003; Hee Joo and Cheng 2006). The susceptibility of pesticides towards nanoparticle/nanomaterials were reported (Hee Joo and Cheng 2006). Incorporation of titania doped with Fe2O3 for the degrading the pesticide using photocatalytic process was witnessed (Sasson et al. 2007). A similar method was used for imidacloprid degradation with TiO2 and TiO2 modified with Ag and SDS (Guan et al. 2008).

6.9 Conclusion

The nanotechnology cultivated many powerful tiny molecules/particle possessing almost magical power to cure disease, kill cells, and penetrate very small spaces, highly reactive, slow release of drug, pesticides, water etc. and also plant defence through antioxidant enzyme production. This chapter brought in light exclusive role of synthesized nanoparticle in improving the plants structural and functional activities through various biochemical and physiological modifications. The nanomaterials have offered protection for the plants and make them free from disease and pest infestation as well. Besides that, the nanomaterials are involved in transforming the photosynthetic potential, water obtaining and retention potential, helping the essential minerals and other nutrients reaching the all the parts of the plants and thus promoting the nutrient absorption, growth and yield as well. Hence, the nanomaterials either metal, metal oxide, polymers etc. are supporting the plant growth and giving the enhanced protection, provided, the inevitable side effects are also witnessed in the current chapter. Thus, it is essential to judiciously supplemented or applied in the field for diminishing the non-target effects. Almost all the needs of the plants are offered by the nanomaterials, provided, the new challenges like climate change based structural and physiological changes and resistance to various factors like objectives also has to be addressed via rigorous in vitro and in vivo researches that may suggest emergence of new material and strategies that brings new guidelines for precisely using the nanotechnology in a safe manner to health and environment.

References

Abdallah Y, Ogunyemi SO, Abdelazez A, Zhang M, Hong X, Ibrahim E, Chen J (2019) The green synthesis of MgO nano-flowers using Rosmarinus officinalis L.(Rosemary) and the antibacterial activities against Xanthomonas oryzae pv. oryzae. Biomed Res Int 2019(1):5620989

Abdellatif KF, Hamouda RA, El-Ansary MSM (2016) Green nanoparticles engineering on root-knot nematode infecting eggplant plants and their effect on plant DNA modification. Iran J Biotechnol 14:250–259

Abdelmigid HM, Hussien NA, Alyamani AA, Morsi MM, AlSufyani NM, Kadi HA (2022) Green synthesis of zinc oxide nanoparticles using pomegranate fruit peel and solid coffee grounds vs. chemical method of synthesis, with their biocompatibility and antibacterial properties investigation. Molecules 27(4):1236

Ahmed HE, Iqbal Y, Aziz MH, Atif M, Batool Z, Hanif A et al (2021) Green synthesis of CeO2 nanoparticles from the Abelmoschus esculentus extract: evaluation of antioxidant, anticancer, antibacterial, and wound-healing activities. Molecules 26(15):4659

Ajitha B, Reddy YAK, Reddy PS (2015) Biosynthesis of silver nanoparticles using Momordica charantia leaf broth: evaluation of their innate antimicrobial and catalytic activities. J Photochem Photobiol B Biol 146:1–9

Al Saqr A, Khafagy ES, Alalaiwe A, Aldawsari MF, Alshahrani SM, Anwer MK et al (2021) Synthesis of gold nanoparticles by using green machinery: characterization and in vitro toxicity. Nanomaterials 11(3):808

Al-Dhafri K, Ching CL (2019) Phyto-synthesis of silver nanoparticles and its bioactivity response towards nosocomial bacterial pathogens. Biocatal Agric Biotechnol 18:101075

Algotiml R, Gab-Alla A, Seoudi R, Abulreesh HH, El-Readi MZ, Elbanna K (2022) Anticancer and antimicrobial activity of biosynthesized Red Sea marine algal silver nanoparticles. Sci Rep 12(1):2421

Ali H, Khan E, Ilahi I (2019) Environmental chemistry and ecotoxicology of hazardous heavy metals: environmental persistence, toxicity, and bioaccumulation. J Chem 2019(1):6730305

Alsammarraie FK, Wang W, Zhou P, Mustapha A, Lin M (2018) Green synthesis of silver nanoparticles using turmeric extracts and investigation of their antibacterial activities. Colloids Surf B: Biointerfaces 171:398–405

Ansari MA, Asiri SMM (2021) Green synthesis, antimicrobial, antibiofilm and antitumor activities of superparamagnetic γ-Fe2O3 NPs and their molecular docking study with cell wall mannoproteins and peptidoglycan. Int J Biol Macromol 171:44–58

Araújo CM, das Virgens Santana M, do Nascimento Cavalcante A, Nunes LCC, Bertolino LC, de Sousa Brito CAR et al (2020) Cashew-gum-based silver nanoparticles and palygorskite as green nanocomposites for antibacterial applications. Mater Sci Eng C 115:110927

Arumugam A, Karthikeyan C, Hameed ASH, Gopinath K, Gowri S, Karthika V (2015) Synthesis of cerium oxide nanoparticles using Gloriosa superba L. leaf extract and their structural, optical and antibacterial properties. Mater Sci Eng C 49:408–415

Asadishad B, Chahal S, Cianciarelli V, Zhou K, Tufenkji N (2017) Effect of gold nanoparticles on extracellular nutrient-cycling enzyme activity and bacterial community in soil slurries: role of nanoparticle size and surface coating. Environ Sci Nano 4(4):907–918

Asha S, Bessy TC, Sherin JJ, Kumar CV, Bindhu MR, Sureshkumar S et al (2022) Efficient photocatalytic degradation of industrial contaminants by Piper longum mediated ZnO nanoparticles. Environ Res 208:112686

Avestan S, Ghasemnezhad M, Esfahani M, Byrt CS (2019) Application of nano-silicon dioxide improves salt stress tolerance in strawberry plants. Agronomy 9(5):246

Aygün A, Gülbağça F, Nas MS, Alma MH, Çalımlı MH, Ustaoglu B, Altunoglu YC, Baloğlu MC, Cellat K, Şen F (2019) Biological synthesis of silver nanoparticles using Rheum ribes and evaluation of their anticarcinogenic and antimicrobial potential: a novel approach in phytonanotechnology. J Pharm Biomed Anal S0731-7085(19):32262–32269

Aziz A, Basheer F, Sengar A, Khan SU, Farooqi IH (2019) Biological wastewater treatment (anaerobic-aerobic) technologies for safe discharge of treated slaughterhouse and meat processing wastewater. Sci Total Environ 686:681–708

Baker A, Iram S, Syed A, Elgorban AM, Bahkali AH, Ahmad K et al (2021) Fruit derived potentially bioactive bioengineered silver nanoparticles. Int J Nanomedicine:7711–7726

Banala RR, Nagati VB, Karnati PR (2015) Green synthesis and characterization of Carica papaya leaf extract coated silver nanoparticles through X-ray diffraction, electron microscopy and evaluation of bactericidal properties. Saudi J Biol Sci 22(5):637–644

Bandala ER, Berli M (2018) Nanomaterials: new agrotechnology tools to improve soil quality? In: Agricultural nanobiotechnology: modern agriculture for a sustainable future. Springer, pp 127–140

Barbosa MC, Grosso RA, Fader CM (2019) Hallmarks of aging: an autophagic perspective. Front Endocrinol 9:416834

Barua S, Konwarh R, Bhattacharya SS, Das P, Devi KSP, Maiti TK et al (2013) Non-hazardous anticancerous and antibacterial colloidal 'green' silver nanoparticles. Colloids Surf B: Biointerfaces 105:37–42

Barua S, Banerjee PP, Sadhu A, Sengupta A, Chatterjee S, Sarkar S et al (2017) Silver nanoparticles as antibacterial and anticancer materials against human breast, cervical and oral cancer cells. J Nanosci Nanotechnol 17(2):968–976

Batool F, Iqbal MS, Khan SUD, Khan J, Ahmed B, Qadir MI (2021) Biologically synthesized iron nanoparticles (FeNPs) from Phoenix dactylifera have anti-bacterial activities. Sci Rep 11(1): 22132

Benakashani F, Allafchian AR, Jalali SAH (2016) Biosynthesis of silver nanoparticles using Capparis spinosa L. leaf extract and their antibacterial activity. Karbala Int J Mod Sci 2(4): 251–258

Bharathi D, Diviya Josebin M, Vasantharaj S, Bhuvaneshwari V (2018) Biosynthesis of silver nanoparticles using stem bark extracts of Diospyros montana and their antioxidant and antibacterial activities. J Nanostruct Chem 8(1):83–92

Bhardwaj AK, Shukla A, Maurya S, Singh SC, Uttam KN, Sundaram S et al (2018) Direct sunlight enabled photo-biochemical synthesis of silver nanoparticles and their Bactericidal Efficacy: photon energy as key for size and distribution control. J Photochem Photobiol B Biol 188:42–49

Boroumand N, Behbahani M, Dini G (2020) Combined effects of phosphate solubilizing bacteria and nanosilica on the growth of land cress plant. J Soil Sci Plant Nutr 20:232–243

Brady NC, Weil RR (1999) The nature and properties of soils. 12th Edition, Prentice Hall Publishers, London, 1-9, 453-536, 727, 739-740

Chandra S, Dwivedi P, Arti KM, Shinde LP (2016) An industrial cultivation of Tulsi (Ocimum sanctum) for medicinal use and rural empowerment. J Med Plants Stud 4(6):213–218

Chau TP, Veeraragavan GR, Narayanan M, Chinnathambi A, Alharbi SA, Subramani B et al (2022) Green synthesis of Zirconium nanoparticles using Punica granatum (pomegranate) peel extract and their antimicrobial and antioxidant potency. Environ Res 209:112771

Cifuentes Z, Custardoy L, de la Fuente JM, Marquina C, Ibarra MR, Rubiales D, Pérez-de-Luque A (2010) Absorption and translocation to the aerial part of magnetic carbon-coated nanoparticles through the root of different crop plants. J Nanobiotechnol 8:1–8

Corradini E, De Moura MR, Mattoso LHC (2010) A preliminary study of the incorparation of NPK fertilizer into chitosan nanoparticles. Express Polym Lett 4(8):509–515

Cromwell WA, Yang J, Starr JL, Jo YK (2014) Nematicidal effects of silver nanoparticles on root-knot nematode in bermudagrass. J Nematol 46(3):261–266. PMID: 25275999; PMCID: PMC4176408

Danish MSS, Estrella LL, Alemaida IMA, Lisin A, Moiseev N, Ahmadi M et al (2021) Photocatalytic applications of metal oxides for sustainable environmental remediation. Metals 11(1):80

Das P, Sarmah K, Hussain N, Pratihar S, Das S, Bhattacharyya P et al (2016) Novel synthesis of an iron oxalate capped iron oxide nanomaterial: a unique soil conditioner and slow release eco-friendly source of iron sustenance in plants. RSC Adv 6:103012–103025

Das S, Mondal K, Haque S (2017) A review on application of probiotic, prebiotic and synbiotic for sustainable development of aquaculture. Growth 14:15

Das S, Chakraborty J, Chatterjee S, Kumar H (2018a) Prospects of biosynthesized nanomaterials for the remediation of organic and inorganic environmental contaminants. Environ Sci Nano 5(12): 2784–2808

Das P, Ghosh S, Ghosh R, Dam S, Baskey M et al (2018b) Madhuca longifolia plant mediated green synthesis of cupric oxide nanoparticles: a promising environmentally sustainable material for waste water treatment and efficient antibacterial agent. J Photochem Photobiol B Biol 189:66–73

Dimkpa CO, McLean JE, Britt DW, Anderson AJ (2012) Bioactivity and biomodification of Ag, ZnO, and CuO nanoparticles with relevance to plant performance in agriculture. Ind Biotechnol 8(6):344–357

Dimkpa CO, McLean JE, Britt DW, Anderson AJ (2013) Antifungal activity of ZnO nanoparticles and their interactive effect with a biocontrol bacterium on growth antagonism of the plant pathogen Fusarium graminearum. Biometals 26:913–924

Dwivedi V, Manickam C, Binjawadagi B, Renukaradhya GJ (2013) PLGA nanoparticle entrapped killed porcine reproductive and respiratory syndrome virus vaccine helps in viral clearance in pigs. Veterinary Microbiology 166(1–2):47–58

Ebrahimzadeh MA, Mortazavi-Derazkola S, Zazouli MA (2019) Eco-friendly green synthesis and characterization of novel Fe3O4/SiO2/Cu2O–Ag nanocomposites using Crataegus pentagyna fruit extract for photocatalytic degradation of organic contaminants. J Mater Sci Mater Electron 30:10994–11004

Elangovan K, Elumalai D, Anupriya S, Shenbhagaraman R, Kaleena PK, Murugesan K (2015) Phyto mediated biogenic synthesis of silver nanoparticles using leaf extract of Andrographis echioides and its bio-efficacy on anticancer and antibacterial activities. J Photochem Photobiol B Biol 151:118–124

Elemike EE, Ibe KA, Onwuka JC (2016) Sugar cane cellulosic waste as a biofuel generating agricultural residue. Br J Renew Energy 1(02):0005–0007

Elmer W, White JC (2018) The future of nanotechnology in plant pathology. Annu Rev Phytopathol 56(1):111–133

El-Naggar A, El-Naggar AH, Shaheen SM, Sarkar B, Chang SX, Tsang DC et al (2019) Biochar composition-dependent impacts on soil nutrient release, carbon mineralization, and potential environmental risk: a review. J Environ Manag 241:458–467

El-Sherbiny GM, Abou El-Nour SA, Askar AA, Mohammad NH, Hammad AA (2022) Solar radiation-induced synthesis of bacterial cellulose/silver nanoparticles (BC/AgNPs) composite using BC as reducing and capping agent. Bioprocess Biosyst Eng 45:1–12

Escárcega-González CE, Garza-Cervantes JA, Vazquez-Rodríguez A, Montelongo-Peralta LZ, Treviño-Gonzalez MT, Díaz Barriga Castro E et al (2018) In vivo antimicrobial activity of silver nanoparticles produced via a green chemistry synthesis using Acacia rigidula as a reducing and capping agent. Int J Nanomed:2349–2363

Essghaier B, Ben Khedher G, Hannachi H, Dridi R, Zid MF, Chaffei C (2022) Green synthesis of silver nanoparticles using mixed leaves aqueous extract of wild olive and pistachio: characterization, antioxidant, antimicrobial and effect on virulence factors of Candida. Arch Microbiol 204(4):203

Fadaka AO, Meyer S, Ahmed O, Geerts G, Madiehe MA, Meyer M, Sibuyi NR (2022) Broad spectrum anti-bacterial activity and non-selective toxicity of gum Arabic silver nanoparticles. Int J Mol Sci 23(3):1799

Fatimah I (2016) Green synthesis of silver nanoparticles using extract of Parkia speciosa Hassk pods assisted by microwave irradiation. J Adv Res 7(6):961–969

Fernando, Somar Israel D, Khristina G. Judan-Cruz, and ACM de Guia (2017) Biologically synthesized gold nanoparticles (AuNP) using pine (Pinus kesiya) pollen extract show antifungal activity against Candida albicans. 13:2615–2622

Fouda A, Hassan SED, Eid AM, El-Din Ewais E (2019) The interaction between plants and bacterial endophytes under salinity stress. In: Endophytes and secondary metabolites. Springer, pp 1–18

Francis S, Koshy EP, Mathew B (2018) Green synthesis of Stereospermum suaveolens capped silver and gold nanoparticles and assessment of their innate antioxidant, antimicrobial and antiproliferative activities. Bioprocess Biosyst Eng 41:939–951

Freire-Neto FP, Teixeira DG, da Cunha DC, Morais IC, Tavares CP, Gurgel GP et al (2022) SARS-CoV-2 reinfections with BA. 1 (Omicron) variant among fully vaccinated individuals in northeastern Brazil. PLoS Negl Trop Dis 16(10):e0010337

Gad El-Rab SM, Ashour AA, Basha S, Alyamani AA, Felemban NH, Enan ET (2021) Well-orientation strategy biosynthesis of Cefuroxime-Silver nanoantibiotic for reinforced Biodentine™ and its dental application against Streptococcus mutans. Molecules 26(22):6832

Gao FQ, Hong FS, Liu C, Zheng L, Su MY, Wu X, Yang F, Wu C, Yang P (2006a) Mechanism of nanoanatase TiO2 on promoting photosynthetic carbon reaction of spinach: inducing complex of Rubisco-Rubiscoactivase. Biol Trace Elem Res 111:239–254

Gao F, Hong F, Liu C et al (2006b) Mechanism of nano-anatase TiO2 on promoting photosynthetic carbon reaction of spinach. Biol Trace Elem Res 111:239–253. https://doi.org/10.1385/BTER:111:1:239

Gholami-Shabani M, Shams-Ghahfarokhi M, Gholami-Shabani Z, Akbarzadeh A, Riazi G, Ajdari S et al (2015) Enzymatic synthesis of gold nanoparticles using sulfite reductase purified from Escherichia coli: a green eco-friendly approach. Process Biochem 50(7):1076–1085

Giraldo JP, Landry MP, Faltermeier SM, McNicholas TP, Iverson NM, Boghossian AA et al (2014) Plant nanobionics approach to augment photosynthesis and biochemical sensing. Nat Mater 13(4):400–408

Gobinath P, Packialakshmi P, Ali D, Alarifi S, Gurusamy R, Idhayadhulla A, SurendraKumar R (2022) [Retracted] Nanobased antibacterial drug discovery to treat skin infections of Staphylococcus aureus using Moringa oleifera-Assisted zinc oxide nanoparticle and molecular simulation study. Biomed Res Int 2022(1):7228259

Gohari G, Safai F, Panahirad S, Akbari A, Rasouli F, Dadpour MR, Fotopoulos V (2020) Modified multiwall carbon nanotubes display either phytotoxic or growth promoting and stress protecting activity in Ocimum basilicum L. in a concentration-dependent manner. Chemosphere 249:126171

Guan H, Chi D, Yu J, Li X (2008) A novel photodegradable insecticide: preparation, characterization and properties evaluation of nano-Imidacloprid Pesticide. Biochem Physiol 92:83–91

Gudadhe N, Raut A, Thanki JD, Imade PBS (2018) Agronomic technologies for enhancing pulse production ahead of climate change. Astral International Pvt. Ltd., New Delhi, pp 27–52

Guha T, Ravikumar KVG, Mukherjee A, Mukherjee A, Kundu R (2018) Nanopriming with zero valent iron (nZVI) enhances germination and growth in aromatic rice cultivar (Oryza sativa cv. Gobindabhog L.). Plant Physiol Biochem 127:403–413

Hagemann N, Joseph S, Schmidt HP, Kammann CI, Harter J, Borch T et al (2017) Organic coating on biochar explains its nutrient retention and stimulation of soil fertility. Nat Commun 8(1):1089

Hassan MEM, Zawam HS, El-Nahas SEM, Desoukey AF (2016) Comparative study between silver nanoparticles and two nematicides against Meloidogyne incognitaon tomato seedlings. Plant Pathol J 15:144–151

Hee Joo S, Cheng IF (eds) (2006) Nanotechnology for environmental remediation. Springer
Hong FS, Liu C, Zheng L, Wang XF, Wu K, Song WP, Lv SP, Tao Y, Zhao GW (2005a) Formation of complexes of rubiscoerubisco activase from La3+, Ce3+ treatment spinach. Sci China, Ser B: Chem 48:67–74
Hong FS, Yang P, Gao FQ, Liu C, Zheng L, Yan F, Zhou J (2005b) Effect of nano-TiO2on spectral characterization ofphotosystem II particles from spinach. Chem Res Chin Univ 21:196–200
Hosseini Bafghi M, Zarrinfar H, Darroudi M, Zargar M, Nazari R (2022) Green synthesis of selenium nanoparticles and evaluate their effect on the expression of ERG3, ERG11 and FKS1 antifungal resistance genes in Candida albicans and Candida glabrata. Lett Appl Microbiol 74(5):809–819
Igwe O, Nwamezie F (2018) Green synthesis of iron nanoparticles using flower extract of Piliostigma thonningii and antibacterial activity evaluation. Chem Int 4:60–66
Iqbal J, Abbasi BA, Yaseen T, Zahra SA, Shahbaz A, Shah SA et al (2021) Green synthesis of zinc oxide nanoparticles using Elaeagnus angustifolia L. leaf extracts and their multiple in vitro biological applications. Sci Rep 11(1):20988
Jamdagni P, Khatri P, Rana JS (2016) Nanoparticles based DNA conjugates for detection of pathogenic microorganisms. Int Nano Lett 6(3):139–146
Jayaprakash N, Vijaya JJ, Kaviyarasu K, Kombaiah K, Kennedy LJ, Ramalingam RJ et al (2017) Green synthesis of Ag nanoparticles using Tamarind fruit extract for the antibacterial studies. J Photochem Photobiol B Biol 169:178–185
Jun LIU, WANG LL, Fei ZY, Xian CHEN, Tang JH, Cui MF, Xu QIAO (2016) Structure and properties of amorphous CeO2@ TiO2 catalyst and its performance in the selective catalytic reduction of NO with NH3. J Fuel Chem Technol 44(8):954–960
Jyoti K, Baunthiyal M, Singh A (2016) Characterization of silver nanoparticles synthesized using Urtica dioica Linn. leaves and their synergistic effects with antibiotics. J Radiat Res Appl Sci 9(3):217–227
Kalita S, Kandimalla R, Sharma KK, Kataki AC, Deka M, Kotoky J (2016) Amoxicillin functionalized gold nanoparticles reverts MRSA resistance. Mater Sci Eng C 61:720–727
Kanniah P, Chelliah P, Thangapandi JR, Gnanadhas G, Mahendran V, Robert M (2021) Green synthesis of antibacterial and cytotoxic silver nanoparticles by Piper nigrum seed extract and development of antibacterial silver based chitosan nanocomposite. Int J Biol Macromol 189:18–33
Karageorgou D, Zygouri P, Tsakiridis T, Hammami MA, Chalmpes N, Subrati M et al (2022) Green synthesis and characterization of silver nanoparticles with high antibacterial activity using cell extracts of Cyanobacterium Pseudanabaena/Limnothrix sp. Nanomaterials 12(13):2296
Karami Mehrian S, Heidari R, Rahmani F (2015) Effect of silver nanoparticles on free amino acids content and antioxidant defense system of tomato plants. Indian J Plant Physiol 20:257–263
Karunakaran G, Suriyaprabha R, Manivasakan P, Yuvakkumar R, Rajendran V, Prabu P, Kannan N (2013) Effect of nanosilica and silicon sources on plant growth promoting rhizobacteria, soil nutrients and maize seed germination. IET Nanobiotechnol 7(3):70–77
Karunakaran G, Jagathambal M, Gusev A, Kolesnikov E, Mandal AR, Kuznetsov D (2016) Allamanda cathartica flower's aqueous extract-mediated green synthesis of silver nanoparticles with excellent antioxidant and antibacterial potential for biomedical application. MRS Commun 6(1):41–46
Kashyap PL, Xiang X, Heiden P (2015) Chitosan nanoparticle based delivery systems for sustainable agriculture. Int J Biol Macromol 77:36–51
Khan N, Khan I, Nadhman A, Azam S, Ullah I, Ahmad F, Khan HA (2020) Pinus wallichiana-synthesized silver nanoparticles as biomedical agents: in-vitro and in-vivo approach. Green Chem Lett Rev 13(2):69–82
Khan M, Yoo SJ, Clijsters M, Backaert W, Vanstapel A, Speleman K et al (2021) Visualizing in deceased COVID-19 patients how SARS-CoV-2 attacks the respiratory and olfactory mucosae but spares the olfactory bulb. Cell 184(24):5932–5949

Khati P, Chaudhary P, Gangola S, Bhatt P, Sharma A (2017) Nanochitosan supports growth of Zea mays and also maintains soil health following growth. 3 Biotech 7:1–9

Khiralla GM, El-Deeb BA (2015) Antimicrobial and antibiofilm effects of selenium nanoparticles on some foodborne pathogens. LWT-Food Sci Technol 63(2):1001–1007

Khodakovskaya M, Dervishi E, Mahmood M, Xu Y, Li Z, Watanabe F, Biris AS (2009) Carbon nanotubes are able to penetrate plant seed coat and dramatically affect seed germination and plant growth. ACS Nano 3(10):3221–3227

Khormali K, Mizwari ZM, Ghoreishi SM, Mortazavi-Derazkola S, Khezri B (2021) Novel Dy2O3/ZnO-Au ternary nanocomposites: green synthesis using pomegranate fruit extract, characterization and their photocatalytic and antibacterial properties. Bioorg Chem 115:105204

Khullar S, Dhillon H, Kaur G, Sharma R, Mehta K, Aggarwal R et al (2016) The prevalence and predictors of depression in type 2 diabetic population of Punjab. Community Ment Health J 52: 479–483

Konappa N, Udayashankar AC, Dhamodaran N, Krishnamurthy S, Jagannath S, Uzma F et al (2021) Ameliorated antibacterial and antioxidant properties by Trichoderma harzianum mediated green synthesis of silver nanoparticles. Biomol Ther 11(4):535

Krishnaraj C, Jagan G, Ramachandran R, Abirami SM, Mohan N, Kalaichelvan PT (2012) Effect of biologically synthesized silver nanoparticles on Bacopa monnieri L. Wettst plant growth metabolism. Process Biochem 47:651–658

Kumar V, Sharma M, Khare T, Wani SH (2018) Impact of nanoparticles on oxidative stress and responsive antioxidative defense in plants. In: Nanomaterials in plants, algae, and microorganisms. Academic Press, pp 393–406

Kumaraswamy RV, Kumari S, Choudhary RC, Pal A, Raliya R, Biswas P, Saharan V (2018) Engineered chitosan based nanomaterials: bioactivities, mechanisms and perspectives in plant protection and growth. Int J Biol Macromol 113:494–506

Kurdish I, Roy A, Chobotarov A et al (2018) Free-flowing complex bacterial preparation for crop and efficiency of its use in agro ecosystems. J Microbiol Biotechnol Food Sci 4:527–531. https://doi.org/10.15414/jmbfs.2015.4.6.527-531

Lakshmeesha TR, Kalagatur NK, Mudili V, Mohan CD, Rangappa S, Prasad BD et al (2019) Biofabrication of zinc oxide nanoparticles with Syzygium aromaticum flower buds extract and finding its novel application in controlling the growth and mycotoxins of Fusarium graminearum. Front Microbiol 10:1244

Lamsal K, Kim SW, Jung JH, Kim YS, Kim KS, Lee YS (2011) Application of silver nanoparticles for the control of Colletotrichum species in vitro and pepper anthracnose disease in field. Mycobiology 39(3):194–199

Lateef A, Azeez MA, Asafa TB, Yekeen TA, Akinboro A, Oladipo IC et al (2016) Biogenic synthesis of silver nanoparticles using a pod extract of Cola nitida: antibacterial and antioxidant activities and application as a paint additive. J Taibah Univ Sci 10(4):551–562

Latha M, Sumathi M, Manikandan R, Arumugam A, Prabhu NM (2015) Biocatalytic and antibacterial visualization of green synthesized silver nanoparticles using Hemidesmus indicus. Microb Pathog 82:43–49

Li PJ, Pan JJ, Tao LJ, Li X, Su DL, Shan Y, Li HY (2021) Green synthesis of silver nanoparticles by extracellular extracts from aspergillus japonicus PJ01. Molecules 26(15):4479

Lim D, Roh J-Y, Eom H-J, Hyun JW, Choi J (2012) Oxidative stress-related PMK-1 P38 MAPK activation as a mechanism for toxicity of silver nanoparticles to reproduction in the nematode Caenorhabditis elegans. Environ Toxicol Chem 31:585–592

Liu SK, Qi X, Han C, Liu JM, Sheng XB, Li H et al (2017) Novel nano submicron mineral-based soil conditioner for sustainable agricultural development. J Clean Prod 149:896–903

Logeswari P, Silambarasan S, Abraham J (2015) Synthesis of silver nanoparticles using plants extract and analysis of their antimicrobial property. J Saudi Chem Soc 19(3):311–317

Lou Y, Xu M, Wang W, Sun X, Zhao K (2011) Return rate of straw residue affects soil organic C sequestration by chemical fertilization. Soil Tillage Res 113(1):70–73

Ma P, Ma X, Suo Q, Chen F (2019) Cu NPs@ NiF electrode preparation by rapid one-step electrodeposition and its sensing performance for glucose. Sensors Actuators B Chem 292: 203–209

Ma L, Fang QX, Sima MW, Burkey KO, Harmel RD (2021) Simulated climate change effects on soybean production using two crop modules in RZWQM2. Agron J 113(2):1349–1365

Mahmood M, Rasheed A, Ayman I, Rasheed T, Munir S, Ajmal S et al (2021) Synthesis of ultrathin MnO2 nanowire-intercalated 2D-MXenes for high-performance hybrid supercapacitors. Energy Fuel 35(4):3469–3478

Malini M, Thirumavalavan M, Yang WY, Lee JF, Annadurai G (2015) A versatile chitosan/ZnO nanocomposite with enhanced antimicrobial properties. Int J Biol Macromol 80:121–129

Manisha DR, Alwala J, Kudle KR, Rudra MP (2014) Biosynthesis of silver nanoparticles using flower extracts of Catharanthus roseus and evaluation of its antibacterial efficacy. World J Pharm Pharm Sci 3:877–885

Mehmood A, Murtaza G, Bhatti TM, Raffi M, Kausar R (2014) Antibacterial efficacy of silver nanoparticles synthesized by a green method using bark extract of Melia azedarach L. J Pharm Innov 9:238–245

Mehmood A, Murtaza G, Bhatti TM, Kausar R (2017) Phyto-mediated synthesis of silver nanoparticles from Melia azedarach L. leaf extract: characterization and antibacterial activity. Arab J Chem 10:S3048–S3053

Meyer JN, Lord CA, Yang XY, Turner EA, Badireddy AR, Marinakos SM, Chilkoti A, Wiesner MR, Auffan M (2010) Intracellular uptake and associated toxicity of silver nanoparticles in Caenorhabditis elegans. Aquat Toxicol 100:140–150

Mirzajani F, Askari H, Hamzelou S, Farzaneh M, Ghassempour A (2013) Effect of silver nanoparticles on Oryza sativa L. and its rhizosphere bacteria. Ecotoxicol Environ Saf 88:48–54

Mittal AK, Tripathy D, Choudhary A, Aili PK, Chatterjee A, Singh IP, Banerjee UC (2015) Bio-synthesis of silver nanoparticles using Potentilla fulgens Wall. ex Hook. and its therapeutic evaluation as anticancer and antimicrobial agent. Mater Sci Eng C 53:120–127

Mohamed BA, Ellis N, Kim CS, Bi X (2017) The role of tailored biochar in increasing plant growth, and reducing bioavailability, phytotoxicity, and uptake of heavy metals in contaminated soil. Environ Pollut 230:329–338

Mohandoss S, Khanal HD, Palanisamy S, You S, Shim JJ, Lee YR (2022) Multiple heteroatom-doped photoluminescent carbon dots for ratiometric detection of Hg 2+ ions in cell imaging and environmental applications. Anal Methods 14(6):635–642

Mohanta YK, Mishra AK, Nayak D, Patra B, Bratovcic A, Avula SK et al (2022) Exploring dose-dependent cytotoxicity profile of Gracilaria edulis-mediated green synthesized silver nanoparticles against MDA-MB-231 breast carcinoma. Oxidative Med Cell Longev 2022(1): 3863138

Moorthy K, Chang KC, Wu WJ, Hsu JY, Yu PJ, Chiang CK (2021) Systematic evaluation of antioxidant efficiency and antibacterial mechanism of bitter gourd extract stabilized silver nanoparticles. Nanomaterials 11(9):2278

Moteriya P, Chanda S (2017) Synthesis and characterization of silver nanoparticles using Caesalpinia pulcherrima flower extract and assessment of their in vitro antimicrobial, antioxidant, cytotoxic, and genotoxic activities. Artif Cells Nanomed Biotechnol 45(8):1556–1567

Moyo M, Gomba M, Nharingo T (2015) Afzelia quanzensis bark extract for green synthesis of silver nanoparticles and study of their antibacterial activity. Int J Ind Chem 6(4):329–338

Murugan K, Senthilkumar B, Senbagam D, Al-Sohaibani S (2014) Biosynthesis of silver nanoparticles using Acacia leucophloea extract and their antibacterial activity. Int J Nanomedicine 9:2431–2438

Nagaraj B, Barasa Malakar, Divya TK, Krishnamurthy NB, Liny P, Dinesh R (2012) Environmental benign synthesis of gold nanoparticles from the flower extracts of *Plumeria Alba* Linn. (Frangipani) and evaluation of their biological activities. Int J Drug Dev & Res 4(1):144–150

Narayanan M, Vigneshwari P, Natarajan D, Kandasamy S, Alsehli M, Elfasakhany A, Pugazhendhi A (2021) Synthesis and characterization of TiO2 NPs by aqueous leaf extract of Coleus aromaticus and assess their antibacterial, larvicidal, and anticancer potential. Environ Res 200: 111335

Nayak AK, Raja R, Rao KS, Shukla AK, Mohanty S, Shahid M et al (2015) Effect of fly ash application on soil microbial response and heavy metal accumulation in soil and rice plant. Ecotoxicol Environ Saf 114:257–262

Nejad BN, Bamzadeh Z, Hejazi SH (2016) Antimicrobial effects of Ferula persica gum extract and gold nanoparticles on Pseudomonas aeruginosa. Avicenna J Clin Microbiol Infect 4(1):36646

Nithiyavathi R, Sundaram SJ, Anand GT, Kumar DR, Raj AD, Al Farraj DA et al (2021) Gum mediated synthesis and characterization of CuO nanoparticles towards infectious disease-causing antimicrobial resistance microbial pathogens. J Infect Public Health 14(12):1893–1902

Padalia H, Moteriya P, Chanda S (2014) Green synthesis of silver nanoparticles from marigold flower and its synergistic antimicrobial potential. Arab J Chem 8:732–741

Panda KK, Achary VMM, Krishnaveni R, Padhi BK, Sarangi SN, Sahu SN, Panda BB (2011) In vitro biosynthesis and genotoxicity bioassay of silver nanoparticles using plants. Toxicol Vitro 25:1097–1105

Patil RS, Kokate MR, Kolekar SS (2012) Bioinspired synthesis of highly stabilized silver nanoparticles using Ocimum tenuiflorum leaf extract and their antibacterial activity. Spectrochim Acta A Mol Biomol Spectrosc 91:234–238

Pattanayak S, Mollick MMR, Maity D, Chakraborty S, Dash SK, Chattopadhyay S et al (2017) Butea monosperma bark extract mediated green synthesis of silver nanoparticles: characterization and biomedical applications. J Saudi Chem Soc 21(6):673–684

Peteu SF, Oancea F, Sicuia OA, Constantinescu F, Dinu S (2010) Responsive polymers for crop protection. Polymers 2:229–251

Potbhare AK, Chaudhary RG, Chouke PB, Yerpude S, Mondal A, Sonkusare VN et al (2019) Phytosynthesis of nearly monodisperse CuO nanospheres using Phyllanthus reticulatus/Conyza bonariensis and its antioxidant/antibacterial assays. Mater Sci Eng C 99:783–793

Pour-Ali S, Kiani-Rashid AR, Babakhani A, Virtanen S, Nguyen NT, Hejazi S et al (2019) On the material characteristics of a high carbon cast austenitic stainless steel after solution annealing followed by quenching in a CNT nanofluid. Int J Mater Res 110(6):570–576

Prasad R, Swamy VS (2013) Antibacterial activity of silver nanoparticles synthesized by bark extract of Syzygium cumini. J Nanoparticles 2013(1):431218

Prasad TNVKV, Sudhakar P, Sreenivasulu Y, Latha P, Munaswamy V, Reddy KR et al (2012) Effect of nanoscale zinc oxide particles on the germination, growth and yield of peanut. J Plant Nutr 35(6):905–927

Prasad R, Bhattacharyya A, Nguyen QD (2017) Nanotechnology in sustainable agriculture: recent developments, challenges, and perspectives. Front Microbiol 8:1014

Puišo J, Jonkuvienė D, Mačionienė I, Šalomskienė J, Jasutienė I, Kondrotas R (2014) Biosynthesis of silver nanoparticles using lingonberry and cranberry juices and their antimicrobial activity. Colloids Surf B: Biointerfaces 121:214–221

Pungle R, Nile SH, Makwana N, Singh R, Singh RP, Kharat AS (2022) Green synthesis of silver nanoparticles using the Tridax Procumbens plant extract and screening of its antimicrobial and anticancer activities. Oxidative Med Cell Longev 2022(1):9671594

Rahman TU, Khan H, Liaqat W, Zeb MA (2022) Phytochemical screening, green synthesis of gold nanoparticles, and antibacterial activity using seeds extract of Ricinus communis L. Microsc Res Tech 85(1):202–208

Rahmatpour S, Shirvani M, Mosaddeghi MR, Bazarganipour M (2017) Retention of silver nanoparticles and silver ions in calcareous soils: influence of soil properties. J Environ Manag 193: 136–145

Rajan A, Vilas V, Philip D (2015) Catalytic and antioxidant properties of biogenic silver nanoparticles synthesized using Areca catechu nut. J Mol Liq 207:231–236

Rajan A, Rajan AR, Philip D (2017) Elettaria cardamomum seed mediated rapid synthesis of gold nanoparticles and its biological activities. OpenNano 2:1–8

Raliya R, Tarafdar JC, Biswas P (2016) Enhancing the mobilization of native phosphorus in the mung bean rhizosphere using ZnO nanoparticles synthesized by soil fungi. J Agric Food Chem 64(16):3111–3118

Ramkumar VS, Pugazhendhi A, Gopalakrishnan K, Sivagurunathan P, Saratale GD, Dung TNB, Kannapiran E (2017) Biofabrication and characterization of silver nanoparticles using aqueous extract of seaweed Enteromorpha compressa and its biomedical properties. Biotechnol Rep 14: 1–7

Rangaraj S, Gopalu K, Rathinam Y, Periasamy P, Venkatachalam R, Narayanasamy K (2014) Effect of silica nanoparticles on microbial biomass and silica availability in maize rhizosphere. Biotechnol Appl Biochem 61(6):668–675

Rather MA, Deori PJ, Gupta K, Daimary N, Deka D, Qureshi A et al (2022) Ecofriendly phytofabrication of silver nanoparticles using aqueous extract of Cuphea carthagenensis and their antioxidant potential and antibacterial activity against clinically important human pathogens. Chemosphere 300:134497

Riaz M, Altaf M, Khan MQ, Manzoor S, Shekheli MA, Shah MA et al (2018) Green synthesis of silver nanoparticles using Jurinea dolomiaea and biological activities. J Nanosci Nanotechnol 18(12):8386–8391

Riaz M, Altaf M, Ahmad P, Khandaker MU, Osman H, Eed EM, Shakir Y (2022) Biogenic synthesis of ag nanoparticles of 18.27 nm by Zanthozylum armatum and determination of biological potentials. Molecules 27(4):1166

Rico CM, Peralta-Videa JR, Gardea-Torresdey JL (2015) Chemistry, biochemistry of nanoparticles, and their role in antioxidant defense system in plants. In: Nanotechnology and plant sciences: nanoparticles and their impact on plants. Springer, pp 1–17

Robinson BH, Bañuelos G, Conesa HM, Evangelou MW, Schulin R (2009) The phytomanagement of trace elements in soil. Crit Rev Plant Sci 28(4):240–266

Roh JY, Sim SJ, Yi J, Park K, Chung KH, Ryu DY, Choi J (2009) Ecotoxicity of silver nanoparticles on the soil nematode Caenorhabditis elegans using functional ecotoxicogenomics. Environ Sci Technol 43(10):3933–3940

Rong Y, Wang Y, Guan Y, Ma J, Cai Z, Yang G, Zhao X (2017) Pyrosequencing reveals soil enzyme activities and bacterial communities impacted by graphene and its oxides. J Agric Food Chem 65(42):9191–9199

Salama HM (2012) Effects of silver nanoparticles in some crop plants, common bean (Phaseolus vulgaris L.) and corn (Zea mays L.). Int Res J Biotechnol 3(10):190–197

Salar RK, Sharma P, Kumar N (2015) Enhanced antibacterial activity of streptomycin against some human pathogens using green synthesized silver nanoparticles. Resour-Effic Technol 1(2): 106–115

Sani A, Hassan D, Khalil AT, Mughal A, El-Mallul A, Ayaz M et al (2021) Floral extracts-mediated green synthesis of NiO nanoparticles and their diverse pharmacological evaluations. J Biomol Struct Dyn 39(11):4133–4147

Santhoshkumar J, Kumar SV, Rajeshkumar S (2017) Synthesis of zinc oxide nanoparticles using plant leaf extract against urinary tract infection pathogen. Resour Effic Technol 3(4):459–465

Santoso D, Lefroy RD, Blair GJ (1995) Sulfur and phosphorus dynamics in an acid soil/crop system. Soil Res 33(1):113–124

Sarah SLR, Iyer PR (2019) Green synthesis of copper nanoparticles from the flowers of Mimusops elengi. Int J Recent Sci Res 10(2019):32956–32963

Saravanakumar K, Chelliah R, Hu X, Oh DH, Kathiresan K, Wang MH (2019) Antioxidant, anti-lung cancer, and anti-bacterial activities of Toxicodendron vernicifluum. Biomol Ther 9(4):127

Sasikala A, Linga Rao M, Savithramma N, Prasad TNVKV (2015) Synthesis of silver nanoparticles from stem bark of Cochlospermum religiosum (L.) Alston: an important medicinal plant and evaluation of their antimicrobial efficacy. Appl Nanosci 5:827–835

Sasson Y, Levy-Ruso G, Toledano O, Ishaaya I (2007) Nanosuspensions: emerging novel agrochemical formulations. In: Ishaaya I, Nauen R, Horowitz AR (eds) Insecticides design using advanced technologies. Springer-Verlag, Netherlands, pp 1–32

Serag MF, Kaji N, Habuchi S, Bianco A, Baba Y (2013) Nanobiotechnology meets plant cell biology: carbon nanotubes as organelle targeting nanocarriers. RSC Adv 3(15):4856–4862
Shahrokh S, Hosseinkhani B, Emtiazi G (2014) The impact of silver nanoparticles on bacterial aerobic nitrate reduction process. J Bioprocess Biotech 4(3):152
Sharma D, Sabela MI, Kanchi S, Mdluli PS, Singh G, Stenström TA, Bisetty K (2016) Biosynthesis of ZnO nanoparticles using Jacaranda mimosifolia flowers extract: synergistic antibacterial activity and molecular simulated facet specific adsorption studies. J Photochem Photobiol B Biol 162:199–207
Sharma N, Modak C, Singh PK, Kumar R, Khatri D, Singh SB (2021) Underscoring the immense potential of chitosan in fighting a wide spectrum of viruses: a plausible molecule against SARS-CoV-2? Int J Biol Macromol 179:33–44
Sheykhbaglou R, Sedghi M, Shishevan MT, Sharifi RS (2010) Effects of nano-iron oxide particles on agronomic traits of soybean. Notulae Scientia Biologicae 2(2):112–113
Shriniwas PP, Subhash TK (2017) Antioxidant, antibacterial and cytotoxic potential of silver nanoparticles synthesized using terpenes rich extract of Lantana camara L. leaves. Biochem Biophys Rep 10:76–81
Siddiqui MH, Al-Whaibi MH, Firoz M, Al-Khaishany MY (2015) Role of nanoparticles in plants. In: Nanotechnology and plant sciences: nanoparticles and their impact on plants. Springer, pp 19–35
Singh P, Pandit S, Beshay M, Mokkapati VRSS, Garnaes J, Olsson ME et al (2018) Anti-biofilm effects of gold and silver nanoparticles synthesized by the Rhodiola rosea rhizome extracts. Artif Cells Nanomed Biotechnol 46(sup3):886–899
Singh R, Hano C, Nath G, Sharma B (2021) Green biosynthesis of silver nanoparticles using leaf extract of Carissa carandas L. and their antioxidant and antimicrobial activity against human pathogenic bacteria. Biomol Ther 11(2):299
Song J, Price DJ, Guvenen F, Bloom N, Von Wachter T (2019) Firming up inequality. Q J Econ 134(1):1–50
Soni N, Prakash S (2014) Green nanoparticles for mosquito control. Sci World J 2014(1):496362
Suresh G, Gunasekar PH, Kokila D, Prabhu D, Dinesh D, Ravichandran N et al (2014) Green synthesis of silver nanoparticles using Delphinium denudatum root extract exhibits antibacterial and mosquito larvicidal activities. Spectrochim Acta A Mol Biomol Spectrosc 127:61–66
Surya S, Kumar GD, Rajakumar R (2016) Green synthesis of silver nanoparticles from flower extract of Hibiscus rosa-sinensis and its antibacterial activity. Int J Innov Res Sci Eng Technol 5(4):5242–5247
Suwan T, Khongkhunthian S, Okonogi S (2018) Green synthesis and inhibitory effects against oral pathogens of silver nanoparticles mediated by rice extracts. Drug Discov Ther 12(4):189–196
Swarnavalli GCJ, Dinakaran S, Raman N, Jegadeesh R, Pereira C (2017) Bio inspired synthesis of monodispersed silver nano particles using Sapindus emarginatus pericarp extract–Study of antibacterial efficacy. J Saudi Chem Soc 21(2):172–179
Uddin S, Iqbal J, Safdar LB, Ahmad S, Abbasi BA, Capasso R et al (2022) Green synthesis of BPL-NiONPs using leaf extract of Berberis pachyacantha: characterization and multiple in vitro biological applications. Molecules 27(7):2064
Varadavenkatesan T, Selvaraj R, Vinayagam R (2019) Dye degradation and antibacterial activity of green synthesized silver nanoparticles using Ipomoea digitata Linn. flower extract. Int J Environ Sci Technol 16:2395–2404
Vasantharaj S, Sripriya N, Shanmugavel M, Manikandan E, Gnanamani A, Senthilkumar P (2018) Surface active gold nanoparticles biosynthesis by new approach for bionanocatalytic activity. J Photochem Photobiol B Biol 179:119–125
Velmurugan P, Park JH, Lee SM, Jang JS, Lee KJ, Han SS et al (2015) Synthesis and characterization of nanosilver with antibacterial properties using Pinus densiflora young cone extract. J Photochem Photobiol B Biol 147:63–68

Velusamy P, Das J, Pachaiappan R, Vaseeharan B, Pandian K (2015) Greener approach for synthesis of antibacterial silver nanoparticles using aqueous solution of neem gum (Azadirachta indica L.). Ind Crop Prod 66:103–109
Verma V, Ravindran P, Kumar PP (2016) Plant hormone-mediated regulation of stress responses. BMC Plant Biol 16:1–10
Vilas V, Philip D, Mathew J (2016) Biosynthesis of Au and Au/Ag alloy nanoparticles using Coleus aromaticus essential oil and evaluation of their catalytic, antibacterial and antiradical activities. J Mol Liq 221:179–189
Wang YW, Cao A, Jiang Y, Zhang X, Liu JH, Liu Y, Wang H (2014) Superior antibacterial activity of zinc oxide/graphene oxide composites originating from high zinc concentration localized around bacteria. ACS Appl Mater Interfaces 6(4):2791–2798
Wang R, Du H, Wang Y, Wang D, Sun Q, Zhou D (2018) Retention of silver nanoparticles and silver ion to natural soils: effects of soil physi-cochemical properties. J Soils Sediments 18: 2491–2499. https://doi.org/10.1007/s11368-018-1918-2
Wibowo A, Tajalla GU, Marsudi MA, Cooper G, Asri LA, Liu F et al (2021) Green synthesis of silver nanoparticles using extract of Cilembu sweet potatoes (Ipomoea batatas L var. Rancing) as potential filler for 3D printed electroactive and anti-infection scaffolds. Molecules 26(7):2042
Winkel-Shirley B (2002) Biosynthesis of flavonoids and effects of stress. Curr Opin Plant Biol 5(3): 218–223
Xia QH, Ma YJ, Wang JW (2016) Biosynthesis of silver nanoparticles using Taxus yunnanensis callus and their antibacterial activity and cytotoxicity in human cancer cells. Nanomaterials 6(9): 160
Xin Y, Xu F, Wang M, Li T (2018) Synergistic effects of carbon nanotube/nano-MoS 2 hybrid on tribological performance of polyimide nanocomposite films. Tribol Lett 66:1–18
Yasur J, Rani P (2013) Environmental effects of nanosilver: impact on castor seed germination, seedling growth, and plant physiology. Environ Sci Pollut Res 20:8636–8648. https://doi.org/10.1007/s11356-013-1798-3
You D, Xu C, Qin F, Zhu Z, Manohari AG, Xu W et al (2018) Interface control for pure ultraviolet electroluminescence from nano-ZnO-based heterojunction devices. Sci Bull 63(1):38–45
Yuan ZY, Chen HY, Reich PB (2011) Global-scale latitudinal patterns of plant fine-root nitrogen and phosphorus. Nat Commun 2(1):344
Yugandhar P, Haribabu R, Savithramma N (2015) Synthesis, characterization and antimicrobial properties of green-synthesised silver nanoparticles from stem bark extract of Syzygium alternifolium (Wt.) Walp. 3 Biotech 5:1031–1039
Zare M, Namratha K, Thakur MS, Byrappa K (2019) Biocompatibility assessment and photocatalytic activity of bio-hydrothermal synthesis of ZnO nanoparticles by Thymus vulgaris leaf extract. Mater Res Bull 109:49–59
Zhang WX (2003) Nanoscale iron particles for environmental remediation: an overview. J Nanopart Res 5:323–332
Zhang C, Liang Z, Hu Z (2014) Bacterial response to a continuous long-term exposure of silver nanoparticles at sub-ppm silver concentrations in a membrane bioreactor activated sludge system. Water Res 50:350–358
Zheng L, Hong F, Lu S, Liu C (2005) Effect of nano-TiO2 on strength of naturally aged seeds and growth of spinach. Biol Trace Elem Res 104:83–91

Chapter 7
Nano Fertilizers for Precision Agriculture

Akhila Sen, P. Faseela, Julie Jacob, and T. Siju Thomas

Abstract Fertilizers play a crucial role in enhancing soil fertility and providing essential nutrients for plant growth and productivity, making them indispensable in agriculture. The excessive use of chemical fertilizers has led to significant environmental pollution and decreased nutrient efficiency. The increasing global population, predicted to reach approximately 10 billion by 2050, necessitates higher agricultural productivity, which is challenged by soil and water resource degradation and climate change. This has led to a call for more sustainable agricultural practices. Precision agriculture is a modern approach that uses data and technology to optimize resource use, improve crop management, and ensure environmental stability.

Nanotechnology has emerged as a transformative tool in precision agriculture, particularly through the development of nano fertilizers. These fertilizers, made of nanoparticles, offer enhanced nutrient delivery and efficiency compared to conventional fertilizers. Nano fertilizers are characterized by their high surface area and controlled release properties, ensuring sustained nutrient availability, and reducing environmental pollution. They can be classified based on their composition and application method, such as foliar application, seed nano priming and soil treatment. Research shows that nano fertilizers improve plant growth, yield, nutritional quality and stress resistance by enhancing nutrient uptake and utilization.

Various types of nano fertilizers, including nano encapsulated, nanostructured, and nanocoated fertilizers, are being developed to address specific agricultural requirements. These innovative fertilizers deliver nutrients precisely, minimize

A. Sen · T. Siju Thomas (✉)
Department of Botany, Mar Athanasius College (Autonomous), Kothamangalam, Kerala, India
e-mail: akhilasen@macollege.in; sijuthomas@macollege.in

P. Faseela
Department of Botany, Korambayil Ahamed Haji Memorial Unity Women's College, Manjeri, Kerala, India
e-mail: faseelap@unitywomenscollege.ac.in

J. Jacob
Department of Biochemistry, Mar Athanasius College (Autonomous), Kothamangalam, Kerala, India
e-mail: juliejacob@macollege.in

J. M. Al-Khayri et al. (eds.), *Nanotechnology Applications in Modern Agriculture*, Nanotechnology in Plant Sciences 3, https://doi.org/10.1007/978-3-031-90506-3_7

losses, and support sustainable agriculture. Studies have demonstrated the effectiveness of nano fertilizers in improving crop quality and productivity while reducing environmental impact. As agriculture faces growing challenges, the integration of nano fertilizers with precision agriculture offers promising solutions for achieving high yields and sustainability.

Keywords Abiotic stress · Climate change · Controlled release · Nano fertilizers · Nanoparticles · Nutrient enrichment · Precision agriculture · Stress tolerance

7.1 Introduction

Fertilizers enhance soil fertility and supply the essential nutrients to the plants they require for growth and high productivity. Hence fertilizers are imperative in any agricultural systems for high yield and quality produce. According to FAO and ITPS (2019), fertilizer is defined as "a chemical or natural substance or material that is used to provide nutrients to plants, usually via application to the soil, but also to foliage or through water in rice systems, fertigation or hydroponics or aquaculture operations". Chemical fertilizers are an essential component in modern agriculture to obtain maximum productivity from crop plants. Fertilizers are generally applied through soil, foliar spray, water (fertigation) or as seed treatment. Soil type, irrigation and water availability, type of crop and stage of plant growth are the factors considered while applying the fertilizer. In conventional application of fertilizers to soil, a considerable amount of nutrients is lost due to volatilization, leaching and fixation. The indiscriminate use of chemical fertilizers has resulted in grave environmental consequences leading to soil, water, and air pollution (Adnan et al. 2020).

According to the World Health Organization, the global population is projected to reach approximately 10 billion by 2050 (United Nations 2022). To feed this burgeoning population, agricultural productivity must be increased considerably. However, the degradation of soil and water resources presents a major hurdle in achieving increased agricultural productivity. The nutrient uptake and utilization by crop plants are negatively impacted by the excessive use of chemical fertilizers (Guo et al. 2018). Additionally, global warming and climate change pose serious threats to agricultural productivity and, consequently, to food security for the expanding population. All these call for a shift from chemical intensive agriculture to more sustainable practices in modern agriculture.

Precision agriculture, utilizing nano fertilizers, has emerged as a sustainable approach to enhance crop production while minimizing environmental impacts. This chapter examines the role of various nanoparticles as nano fertilizers, their modes of application and benefits over conventional fertilizers. It also explores the potential of nano fertilizers in precision agriculture to improve soil nutrient availability, increase crop yields, enhance nutritional quality, and strengthen plant resistance to both abiotic and biotic stresses.

7.2 Precision Agriculture

The production of agricultural crops requires a shift from conventional methods to more modern ones that might increase output through efficient resource use, strengthen farmer economies, and promote a clean, pollution-free environment. Precision agriculture (PA) is the analysis and use of data from various sources to improve crop management strategies by optimizing resource utilization. It is a new concept adopted throughout the world to increase production, reduce labor time, and ensure the effective management of fertilizers and irrigation processes. It uses a large amount of data and information to improve the use of agricultural resources, yields, and the quality of crops (Petrović et al. 2024). PA is an advanced innovation and optimized field level management strategy used in agriculture that aims to improve the productivity of resources on agriculture fields. In precision agriculture, optimized inputs such as water and fertilizer are provided to enhance productivity, quality, and yield (Sishodia et al. 2020). With its many potential benefits for crop quality, environmental protection, sustainability, food safety, economic development, profitability, and productivity, precision agriculture is transforming the agricultural industry globally (Zain et al. 2023a). In precision agriculture, technologies such as Geographical Information System, remote sensing, sensors, and robotics are integrated with agricultural management practices (Loures et al. 2020). With the information gathered from these sources, the crop productivity can be maximized by optimizing agricultural inputs like fertilizers, pesticides, and irrigation water. Thus, making the agricultural systems economically and environmentally more sustainable.

7.3 Nanotechnology in Precision Agriculture

Nanotechnology is an innovative scientific approach where matter is manipulated to the size of 1–100 nanometers. It has significantly impacted all fields of science, including agriculture. Application of nanotechnology in agriculture includes nano fertilizers, nano pesticides, nanobiotechnology, nanotoxicology and nano sensors. Nanotechnology has evolved as an integral part of precision agriculture with its capabilities in nutrition delivery, plant stress mitigation, pest management, nano sensor based soil health assessment and water management contributing to enhanced crop productivity (Duhan et al. 2017; Elizabath et al. 2019; Zain et al. 2023b).

7.3.1 Nanoparticles as Nano Fertilizers

Nano fertilizers (NF)are a cutting-edge category of fertilizers harnessing advanced nanotechnology to offer an effective and sustainable approach to crop manuring, replacing conventional bulk fertilizers. These developments specifically target plant mineral nutrition, soil health, and interactions with soil microorganisms. They exhibit enhanced properties that address existing challenges related to the bioavailability and uptake of minerals. This leads to accelerated plant growth, increased crop yield, decreased fertilizer wastage, biotic and abiotic stress management, and thus environmental protection. A nano-fertilizer refers to a product created through the direct manufacturing of nanoparticles or employed in a nanotechnology application with a size below 100 nm for efficient nutrient management, which is more ecofriendly and reduces environmental pollution (Ndaba et al. 2022). Due to their ultra-small size, nano fertilizers offer substantially greater surface area and surface area to volume ratio, allowing for higher nutrient content. The higher surface area and surface to volume ratio of nano fertilizers ensure slow and sustained release of nutrients into the soil facilitating enhanced nutrient availability to crop plants. This abundance of surface area enables nano fertilizers to release nutrients gradually for a longer time, in accordance with crop requirements, and plants quickly and thoroughly absorb nano-fertilizers, preventing fertilizer loss and reduced environmental pollution compared to conventional fertilizers (Emadjan 2017).

Nano fertilizers are formulated to deliver plant nutrients with precision, ensuring a controlled release that extends over time, thereby supplying essential elements to plants in a consistent and gradual manner. They are primarily manufactured by encapsulating using nano emulsions or coating nutrients using nanoparticles. These nanofertilizers act as carriers of nutrients and fall into various categories, including nano clays, hydroxyapatite nanoparticles, polymeric nanoparticles, carbon-based nanomaterials, mesoporous silica, and other miscellaneous materials. These are commonly coated with a specified chemical composition, enabling controlled release and targeted delivery of effective nanoscale components. Nanofertilizers exhibit efficient transport and delivery between cells through plasmodesmata channels, that are approximately 50–60 nm in size, which is a distinctive characteristic that contributes to enhanced plant productivity and improved nutritional quality (Iqbal et al. 2019; Verma et al. 2023).

The nano fertilizers offer targeted nutrient delivery to plants at the nanoscale level and proves their efficacy in terms of crop productivity and environmental sustainability over bulky chemical fertilizers. Nanofertilizers are classified variously considering different aspects such as the material used, the nutrient they contain, mode of action and method of application (Yadav et al. 2023a). Nano fertilizers can be categorized into three main groups such as nanoscale fertilizers composed of nanoparticles with a range of nutrients, including both macro and micro-nutrients; nanoscale additive fertilizers, which are conventional fertilizers enhanced with nanoscale additives; and nanoscale coating fertilizers, conventional fertilizers coated with nanoparticles, alternatively known as 'nutrient-loaded NFs' or 'nanomaterial-enhanced fertilizers' (Mejias et al. 2021; Mikkelsen 2018). Understanding the types and nature of nano fertilizer is crucial to find the best application method as it offers a wide range of benefits making them valuable for sustainable agriculture (Yadav et al. 2023b).

Nanotechnology encompasses various types of nano fertilizers such as nano encapsulated fertilizers, nanostructured fertilizers, nanocoated fertilizers, nano chelated fertilizers and smart nano fertilizers (Ndaba et al. 2022; Shaikh et al. 2021). Nano encapsulated fertilizers utilize nanoscale structures, such as polymer capsules or lipid bilayers, to shield nutrients from environmental degradation while enabling controlled release, thus ensuring efficient nutrient uptake by plants and minimizing losses through leaching or volatilization. Meanwhile, nanostructured fertilizers employ engineered nanomaterials like nanoparticles, nanotubes, or nanocomposites, leveraging their high surface area-to-volume ratios to expedite nutrient release and uptake by plant roots, ultimately enhancing nutrient utilization efficiency and crop yields (Yadav et al. 2023a, b). Nanocoated fertilizers, another variant, involve traditional fertilizers coated with nanomaterials such as nano clays or nano polymers, fortifying them against nutrient loss, improving soil adhesion, and facilitating sustained nutrient availability to plants. Nano chelated fertilizers, on the other hand, feature nutrients chelated with nanoscale chelating agents like nanoparticles or nano polymers, which enhance nutrient stability and bioavailability, thereby optimizing nutrient uptake by plants and mitigating risks of leaching or soil immobilization. Furthermore, the advent of smart nano fertilizers integrates nanotechnology with intelligent delivery systems, such as stimuli-responsive nanocarriers or nano sensors, allowing for precise nutrient release tailored to specific environmental cues or plant requirements. These smart nano fertilizers hold the promise of optimizing nutrient management in precision agriculture while minimizing environmental impact, thereby contributing to sustainable agricultural practices (Gade et al. 2023; Nongbet et al. 2022).

Based on nutrient and carrier selection which offer precise nutrient delivery systems, there are different types of nano fertilizers such as macronutrient-based, micronutrient-based, carbon-based, and polymer-based ones. Macronutrient-based nano fertilizers focus on essential elements like nitrogen, phosphorus, and potassium, encapsulating them within nano-sized carriers to enhance their uptake efficiency and minimize losses. These nano fertilizers hold immense potential in addressing nutrient deficiencies, optimizing crop growth, and reducing environmental impacts associated with traditional fertilizer application methods. Micronutrient-based nano fertilizers play a critical role in delivering trace elements necessary for plant health, such as iron, zinc, and copper. By utilizing nanotechnology, these fertilizers ensure precise and targeted delivery of micronutrients, mitigating deficiencies that can hamper crop productivity and quality. Their nano-sized carriers facilitate improved nutrient solubility and availability in the soil, leading to enhanced nutrient uptake by plants and better overall agricultural outcomes. Carbon-based nano fertilizers utilize materials like carbon nanotubes, graphene, and biochar as carriers, utilizing their unique properties such as high surface area and porosity to enhance nutrient adsorption and release. These nanomaterials not only provide a platform for nutrient delivery but also contribute to soil health and fertility by improving water retention and microbial activity. On the other hand, polymer-based nano fertilizers employ synthetic or natural polymers to encapsulate nutrients and control their release kinetics. These nano fertilizers offer versatility in addressing

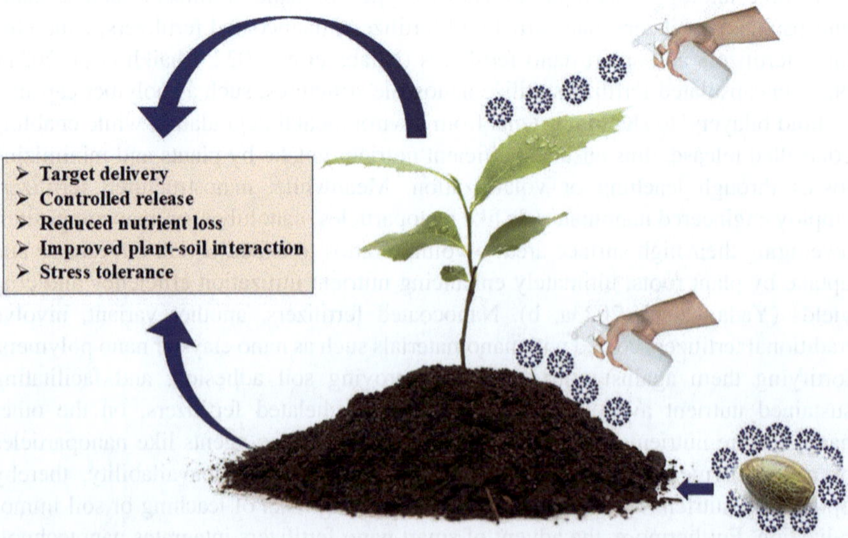

Fig. 7.1 Nanofertilizer application methods and their benefits. (Figure prepared by Akhila Sen)

specific agricultural challenges, such as improving nutrient availability in different soil types or under varying environmental conditions (Dixit et al. 2024; Farooq et al. 2023).

Three main approaches for nano fertilizer application include foliar application, seed nano priming and soil treatment. Figure 7.1 represents different methods of nano fertilizer application and associated benefits. Choosing the right method for nano fertilizer application is essential for achieving optimal plant growth, and the most suitable approach may vary based on the soil and climate conditions such as soil quality, nutrient availability, and climate factors which affect nutrient uptake and utilization (Babu et al. 2022; Yadav et al. 2023a, b).

Foliar application involves the direct spraying of liquid nano fertilizers onto plant leaves and it has emerged as a promising approach for delivering essential nutrients. This approach offers several advantages over traditional soil applications, such as quicker nutrient absorption, enhanced nutrient utilization, and reduced leaching and run-off. Notably, plant leaves, with their nanopores and stomatal openings, enable effective uptake of nanomaterials, allowing for deep penetration (Nongbet et al. 2022; Yadav et al. 2023a, b). Seed nano priming includes immersing seeds in nano fertilizers, a practice proven to cut fertilizer application in half while still yielding excellent results by boosting germination and development by permeating seed pores, dispersing internally, and activating plant hormones that stimulate growth. Incorporating nano fertilizers into the soil through methods like broadcasting or localized placing ensures gradual nutrient release while minimizing nutrient loss (Avila-Quezada et al. 2022; Mittal et al. 2020).

Research has explored the application of nano fertilizers in diverse aspects of plant production, involving the use of various nanoparticles which include metallic, inorganic, and organic nanoparticles. Nanoparticles, as potential enhancers of plant growth, possess distinctive attributes like a heightened sorption capacity, an increased surface-to-volume ratio, and controlled-release dynamics at specific locations. These distinctive characteristics have enabled the effective utilization of nano fertilizers as a proficient system for delivering plant nutrition in precision agriculture. Moreover, it elevated the activity of both macro and microorganisms within the nearby ecosystem in contrast to traditional fertilizers, contributing to the increased scale of agricultural output. Nano fertilizers based on nanoparticles have shown to have a substantial impact on several aspects of agriculture, such as plant growth and development, crop production, nutritional fortification, and plant stress management.

7.3.2 Nano Fertilizers in Soil Nutrient Management and Enhancing Crop Production

Normal growth and development of plants rely on continuous supply of essential nutrients. Soil nutrient deficiency is a major factor limiting crop productivity globally. Despite the application of large volume of fertilizers, the nutrient use efficiency remains constant for traditional fertilizers (Subramanian et al. 2015). Surplus usage of fertilizers results in leaching and eutrophication of water bodies. Nanoparticle based nano fertilizers ensure slow and sustained release of nutrients ensuring better nutrient use efficiency and crop yield.

The progress in nanotechnology has enhanced techniques for the mass production of nanoparticles containing crucial elements, which are now employed to enhance fertilizer compositions, resulting in enhanced absorption by plant cells and reduced nutrient wastage. Nanoparticles possess notable characteristics such as high surface area, sorption ability, and controlled release kinetics, rendering them as a "smart delivery system." Nanostructured fertilizers have the potential to amplify nutrient utilization efficiency through methods like targeted delivery and gradual or regulated release. They exhibit the capability to release their essential components precisely in response to environmental cues and biological requirements (El-Ghamry et al. 2018; ul Ain et al. 2023).

Nano fertilizers can bring many advantages to modern agriculture. Various studies reported that the application of nano fertilizers has the potential to enhance crop yield and quality through the stimulation of seed germination, shoot and root growth, nitrogen metabolism, photosynthesis, synthesis of proteins and carbohydrates, antioxidant defense, etc. (Bhaskar et al. 2023; Mejias et al. 2021). Abdel-Aziz et al. (2016) noted that, application of nano chitosan-NPK fertilizer through foliar application enhanced the yield attributes of wheat plants grown in sandy soils. The plants exhibited significant increases in harvest index, crop index and mobilization index. The foliar application of zinc oxide (ZnO) nano fertilizer improved growth and yield in wheat plants with minimum leaching of zinc into the soil compared to conventional zinc fertilizers (Sheoran et al. 2021). The application of iron oxide (FeO) nanoparticles to mulberry plants has improved the sprouting percentage, number of leaves, plant biomass and root length. These plants also had increased chlorophyll and sugar content compared to the control plants (Haydar et al. 2022). Green synthesized copper oxide (CuO) nanoparticles using macrofungal extracts at low concentration enhanced early seed growth and development in *Lepidium sativum*, *Raphanus sativus*, and *Zea mays* showing its potential as nano fertilizers (Sedefoglu et al. 2023). Soaking of fenugreek seeds in ZnO nanoparticles at a concentration of 30 mg/l improved the plant biomass, flavonoid and phenolic content and antioxidant activity of fenugreek seedlings (Elsherif et al. 2023). Chitosan based silicon nano fertilizer enhanced growth and yield of maize plants through slow and sustained release of silicon. The nano fertilizer promoted seedling vigor, induced plant antioxidant defense enzymes and enhanced photosynthesis of maize plants (Kumaraswamy et al. 2021).

Controlled release nano fertilizers can overcome the problems associated with traditional fertilizers such as nutrient loss and nutrient leaching. In controlled release nano fertilizers, nutrients are encapsulated in nanomaterials and their release is influenced by environmental cues such as temperature, soil pH and moisture or by biodegradation (Lawrencia et al. 2021; Vejan et al. 2021). The high solubility of nitrogen fertilizers makes them vulnerable to leaching and denitrification. Slow release and controlled release nano fertilizers developed using various synthetic or biopolymers proved to be effective in overcoming this problem (Fertahi et al. 2021; Firmanda et al. 2024).

Researchers have developed and experimented with a large number of nanoparticles based nano fertilizers over the years. Each nano fertilizer has its own properties and effects on plant growth and productivity (Table 7.1).

7.3.3 Nano Fertilizers in Nutritional Quality Improvement

Global population is expected to reach 8.5 billion in 2030, 9.7 billion in 2050 (United nations 2022). It is projected that 670 million people will be facing hunger in 2030 (UNEP 2023). The UN Food Systems Summit (2023) has advocated the need for transformative food systems to provide food security to the growing population. A multipronged approach has been envisioned to deliver healthy people and healthy planet. This objective can be achieved by use of a regenerative and resilient agricultural practices and circular food systems which incorporates the concept “reduce-reuse-recycle-remove”.

Commercial fertilizers, which have been in use for the past several years for improving crop yield has also been the cause for altered soil properties and undesirable changes in the environment. Nano fertilizer contains either natural or bioengineered nanoparticles with a diameter ranging from 1 to 100 nm. They have been found to increase soil fertility and crop productivity and found to cause low

Table 7.1 Effect of various nano fertilizer applications on crop plants

Sl. No.	Nano fertilizer/ Nanoparticles	Plant species	Mode of application	Plant responses	References
1	FeO	*Vigna radiata*	Soil addition	Enhancement of growth and yield	He et al. (2016)
2	$ZnSO_4$ and ZnO	*Coffea arabica*	Foliar spray	Fresh and dry weight of roots and leaves improved, enhanced fruit set and quality	Rossi et al. (2019)
3	Hydroxyapatite, i.e., calcium phosphate	*Rosmarinus officinalis*	Foliar spray	Enhanced the growth characters and essential oil production	Elsayed et al. (2022)
4	ZnO	*Cyamopsis tetragonoloba*	Foliar spray	Enhanced growth parameters, yield and resistance against various pathogens.	Rexlin et al. (2023)
5	Chiotsan-NPK	*Triticum aestivum*	Foliar spray	Increased plant growth and productivity	Abdel-Aziz et al. (2016)
6	TiO_2 and SiO_2	*Oryza sativa*	Foliar spray	Improved biomass, photosynthesis and diminishing Cd accumulation, reduced oxidative burst and improvement in antioxidant defense system.	Rizwan et al. (2019a)
7	ZnO	*Zea mays*	Soil addition and foliar spray	Enhanced plant growth, photosynthetic pigments, and antioxidant activity.	Azam et al. (2022)
8	Nano chitosan	*Phaseolus vulgaris*	Seed nano priming	Improved seed germination and seedling growth parameters under salt stress condition.	Zayed et al. (2017)
9	Ag	*Vigna unguiculata*	Foliar spray	Enhanced antimicrobial activity against plant pathogen *Xanthomonas axonopodis*pv. *malvacearum* and other harmful bacteria.	Vanti et al. (2019)
10	Cu-Fe bimetallic	*Oryza sativa*	Foliar spray	Increased vigor and photosynthetic efficiency.	Dey et al. (2023)
11	Nano chelate silicon fertilizes	*Solanum tuberosum*	Soil addition and	Increased total chlorophyll, dry matter, vegetative yield, fresh	Al-Juthery et al. (2019)

(continued)

Table 7.1 (continued)

Sl. No.	Nano fertilizer/ Nanoparticles	Plant species	Mode of application	Plant responses	References
			foliar spray	tubers yield, biological yield, starch, crude protein, ascorbic acid, agronomic efficiency and water use efficiency.	
12	FeO and ZnO	*Oryza sativa*	Soil addition and foliar spray	Physiological and biochemical parameters improved.	Afzal et al. (2022)
13	ZnO	*Arachis hypogea*	Soil addition	Improved plant growth and yield.	Oghenerume et al. (2020)
14	ZnO	*Triticum aestivum*	Seed nano priming	Enhanced seed water uptake and α-amylase activity, increased photosynthetic pigment content and overall primary photochemistry, decreased reactive oxygen species and antioxidant enzymes.	Rai-Kalal and Jajoo (2021)
15	$ZnSO_4$ and ZnO	*Oryza sativa*	Foliar spray	Increased grain per spike, total chlorophyll, soluble sugar, amylose, and soluble protein contents.	Parashar et al. (2023)
16	CuO	*Oryza sativa*	Foliar spray	Essential element accumulation, yield, sugar, starch, protein content, and the expression of auxin associated genes enhanced.	Deng et al. (2022)
17	ZnO	*Vigna mungo*	Seed nano priming	Decreased reactive oxygen species build-up and malondialdehyde content, enhanced accumulation of osmoregulatory and antioxidant enzymes, reduced arsenic induced toxicity by limiting its translocation from root to shoot.	Banerjee et al. (2023)

(continued)

Table 7.1 (continued)

Sl. No.	Nano fertilizer/ Nanoparticles	Plant species	Mode of application	Plant responses	References
18	CuO	*Solanum lycopersicum*	Soil addition	Increased growth and fruit yield. Improved activities of SOD, CAT and peroxidase. Increased photosynthesis and carbohydrate content.	Wang et al. (2021)
19	Molybdenum (Mo)	*Cicer arietinum*	Seed nano priming	Improved germination, peroxidize activity and reduced lipid peroxidation.	Shcherbakova et al. (2017)
20	Fe_3O_4 and FeO (OH)	*Raphanus sativus* L.	Soil addition	Enhanced nutrient uptake, increased essential amino acid content and Vitamin C content.	Shakoor et al. (2022)

environmental damage (Tarafder et al. 2020). Nano fertilizer can be designed to provide 14 essential elements required for plant nutrition which include nitrogen, phosphorus, potassium, calcium, magnesium, sulfur, iron, manganese, boron, zinc, copper, molybdenum, chlorine, and nickel. The micronutrients can be applied through three ways; soil application, foliar method, and seed treatment (Tondey et al. 2022).

In plants, the nano fertilizer may be distributed either by apoplastic or symplastic mode (Verma et al. 2022). A meta-analysis study by Wang et al. (2020) has shown that exposure to nanoparticles improved the nutritional quality of edible parts of crop by 2.4%. Study by Zahra et al. (2017) has shown that application of titanium dioxide (TiO_2) nanoparticles (750 mg/kg) in phosphorus deficient soil leads to a 1.3 fold increase of phosphorus content in rice grain. An increase in amino acid, palmitic acid and glycerol content was observed. Also, no TiO_2 was present in rice grain, indicating that the rice grain was safe for consumption. Zhang et al. (2022) have reviewed the role of nano particles in increasing the synthesis of secondary metabolites in plants. It was shown that the application of copper nanoparticles (800 mg/kg) in cucumber led to an increase in p-coumaric acid content by 225 times.

López-Vargas et al. (2018) has reported that foliar application of copper nanoparticles resulted in an increase in vitamin C content and lycopene in tomatoes. Combined action of copper nanoparticles (50 mg/L) and Selenium nanoparticle (1 mg/L) resulted in a 36% increase in vitamin C content in tomatoes (Hernández-Hernández et al. 2019). Urea modified hydroxyapatite containing copper, iron and zinc nanoparticles were found to result in increased amount of these minerals in the fruit of ladies' finger *(Abelmoschus esculentus)* (Tarafder et al. 2020).

Studies have shown that providing nano nutrients in wheat cultivation has resulted in an increase in protein content by 6% (Jayara et al. 2023). Application of zinc complexed chitosan nanoparticles in two durum wheat types have resulted in enrichment of grain with zinc, an essential micronutrient (Dapkekar et al. 2018). Similar results were obtained by Munir et al. (2018) by applying zinc oxide (ZnO) nanoparticles for seed priming in wheat. Foliar application of nano iron (2000 mg/L) led to an increase in carotenoid content as well as iron and calcium content in quince fruit (*Cydonia oblonga* Mill) as shown by Rahemi et al. (2020). Zinc oxide nanoparticles have been found to enhance the yield and the zinc content of wheat grain in wheat grown in zinc deficient soil when compared with the use of traditional zinc sulphate (Singh et al. 2021).

Use of nitrogen and zinc nano fertilizers along with organic farming methods resulted in improved yield of wheat, pearl millet, sesame, and mustard (Kumar et al. 2022). Treatment of maize seeds with zinc oxide nanoparticles (40 mg/L) resulted in increased starch, total soluble proteins, nitrogen, phosphorus, potassium and zinc in seed (Tondey et al. 2022). The crude protein content of maize grains were increased on application of nano zeo-urea to maize grown in two different types of soil, inceptisol-clay loam and alfisol-sandy loam. The increase in protein content was higher in sandy loam soil (Manikandan and Subramanian 2016).

Application of zinc oxide nanoparticles on leaves of turmeric (*Curcuma longa)* increased the yield as well as the protein, starch, and several bioactive compounds. Foliar application of 5 mg/L of ZnO nanoparticles lead to an increase in the concentration of coumaran, 4-Hydroxy-3-methylacetophenone, vanillin, alpha curcumene and beta-sesquiphellandrene, as compared with control. However, the concentration of alpha curcumene and beta-sesquiphellandrene showed a decrease when treated with increased concentration of ZnO nanoparticles (10, 20, and 40 mg/L),while Ar-tumerone showed a decrease when treated with ZnO nanoparticles. GC MS analysis of ethanolic rhizome extracts of turmeric showed that application of ZnO nanoparticles in plant leaves resulted in the production of several phytochemicals not found in untreated plants such as Zingiberene, P hydroxybenzaldehyde, cis oleic acid, cis, cis- Linoleic acid, Humulane 1,6 dien-3-ol,oleic acid ethyl ester, Pentadecanoic acid and thymol. Further, the expression of genes involved in biosynthesis of curcuminoid such as CURS1,-2,-3 and DCS were increased when the plants were sprayed with ZnO nanoparticles (Khattab et al. 2023).

Abd-Elrahman et al. (2023) highlighted the role of nano chitosan loaded with potassium on the yield and nutritional quality of strawberry grown in sandy soil. Application of 112.5 kg K_2SO_4 acre^{-1} (75% of recommended dosage) + 1000 mg $^{L-1}$ nano-chitosan-K and 112.5 kg K_2SO_4 acre^{-1} + 750 mg $^{L-1}$ nano-chitosan-K treatments were found to enhance the yield and firmness of strawberry fruits. Total soluble solids, vitamin C, total acidity and anthocyanin content of the fruits showed an increase by these two treatments.

An increase in total phenols, flavonoids and ascorbic acid were observed in pear fruits grown in a 20 year old orchard on foliar application of copper nanoparticles (250 ppm) and selenium nanoparticles (50 ppm) whereas, a decrease in the above compounds was seen on application of zinc nanoparticles (250 ppm) when compared to control (Hernández-Fuentes et al. 2023a). Similar results were observed in native kidney shape tomato fruits (Mexico) on application of copper nanoparticles

(Cu-NPs) in chitosan-polyvinyl alcohol (Cs-PVA) hydrogels under greenhouse conditions. The plants were treated with 0, 0.57, 1 and 2 mg of Cu-NPs g^{-1} of Cs-PVA per plant and it was observed that the concentrations of bioactive compounds such as lycopene, β-carotene, total phenols, and flavonoids increased than the untreated control. Also, the fruits treated with 0.57 and 1 mg of Cu-NPs g^{-1} of Cs-PVA showed increased total soluble solids (Hernández-Fuentes et al. 2023b).

Shebl et al. (2020) had reported the effect of manganese zinc ferrite nanoparticles (MnO.5ZnO.5Fe2O4 NPs) in squash (*Cucurbita pepo* L)plants prepared by a template-free microwave-assisted hydrothermal green synthesis technique at different temperatures (100, 120, 140, 160 and 180 °C). The prepared nanomaterials were applied by foliar method at different concentrations (0, 10, 20, and 30 ppm) 20 days after sowing. The yield of squash fruits (kg/plant and t/ha) showed an increase in plants treated with nano fertilizer prepared at 160 °C and at a concentration of 10 ppm whereas the fruit length and dimeter were higher in plants exposed to nano fertilizer prepared at 140 °C and at a concentration of 20 ppm. Similarly, the proximate analysis of fruits showed a variation according to the temperature at which the nano fertilizer was synthesized as well as their concentration. The amount of protein and fiber were increased in plants treated with T-180 at the concentrations of 30 ppm and 10 ppm respectively, while plants treated with nano fertilizer prepared at 160 °C and at a concentration of 20 ppm showed an increased fruit content of organic matter, lipids, and energy. Whereas the carbohydrate content of squash fruits improved on application of the nano fertilizer prepared at 120 °C at 30 ppm. The nitrogen and potassium content of fruits increased with treatment of fertilizer prepared at 180 °C at 30 ppm, while phosphorus and manganese content were enhanced by treatment with T-160 at 20 and 30 ppm respectively. The highest zinc content was seen in plants applied by fertilizer prepared at T-100 °C at 30 ppm, whereas the iron content of fruits increased with treatment with T-140 °C at 30 ppm.

The biological actions of nano materials are mediated by increased gene expression of transporters, enzymes and phytohormones. They are also found to regulate secondary metabolic pathway in plants (Wang et al. 2020). It was seen that Fe_2O_3 nano fertilizer upregulated the activities of ferric chelate reductase (FRO_2) gene and iron regulated ferrous iron transporter (IRT1) gene, thus enhancing the absorption and transport of iron in *Citrus maxima*. Similarly, selenium nano materials were reported to increase the expression of sulfate/selenate transporters (Sultr1;2) genes as well as ATP sulfurylase (ATPS) leading to increased selenium uptake into edible plant parts.

Although nanoparticles have shown to improve the nutritional quality of crops, adverse effects have also been reported. Nanoparticles were found to cause DNA damage and increased production of reactive oxygen species (Verma et al. 2022).

Further studies are required to gain understanding of interactions of nanoparticles with plants to design nano fertilizer which can improve the nutritional quality of food without any harmful effect on plants, humans or the environment thus promoting the concept of One Health.

7.3.4 *Nano Fertilizers in Abiotic Stress Management*

Global warming, human activity, and other inescapable factors have contributed to climate change, which causes various abiotic pressures that lower agricultural output and degrade land and water resources. Stress factors, mostly induced by drought, low temperature and salinity have been found to significantly reduce the yield of numerous crops in several studies (Tripathi et al. 2015). To cope with the detrimental physiological changes brought on by environmental stresses that they experienced at different growth stages, plants developed a variety of defense mechanisms. Numerous investigations have demonstrated that abiotic stressors elicit a wide range of biochemical, physiological, and molecular reactions that impact numerous plant cellular functions (Khan et al. 2023). Figure 7.2 depicts the impacts of nano fertilizer on plant growth and development.

7.3.4.1 Impact of Nanoparticles on Plants Under Heavy Metal Stress

Presently, the pollution of heavy metals (HMs) has emerged as a significant global concern (Adnan et al. 2024). Contamination of heavy metals can occur as a result of several activities such as atmospheric deposition, rapid industrialization, mining, accidental spills of petrochemicals, waste incineration and the disposal of high metal waste during industrial processes (Irshad et al. 2021). Their extended duration and possible harmfulness in polluted soils have been significant environmental concerns that detrimentally impact the viability of living forms such as plants, animals, and humans (Rizwan et al. 2019b). Of all the elements, cadmium (Cd) holds substantial environmental significance because it is very harmful to both humans and plants

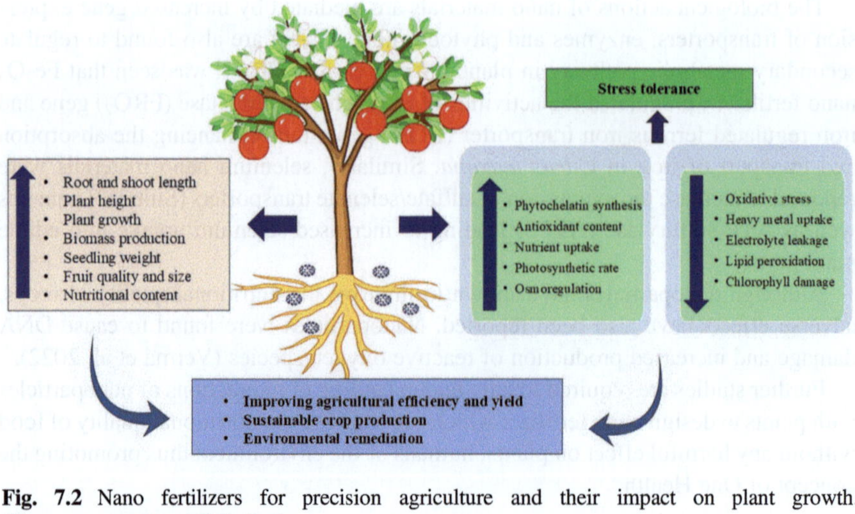

Fig. 7.2 Nano fertilizers for precision agriculture and their impact on plant growth. (Figure prepared by Akhila Sen)

(Hussain et al. 2021). The plant uptake of Cd leads to cellular oxidative damage caused by elevated concentrations of reactive oxygen species (ROS). This oxidative damage inhibits photosynthesis, affects the balance of ions and makes it harder for plants to absorb nutrients, which leads to slower plant growth (Li et al. 2020).

When plants are exposed to high levels of heavy metals and experience stress, applying nanoparticles (NPs) to the soil or leaves can help counteract the negative effects of stress. This can lead to improved plant growth and photosynthesis, as well as a reduction in toxicity caused by oxidative stress. Hence, the use of nanoparticles aids in the restoration of habitats contaminated with HMs. When plants are exposed to high levels of HMs, applying NPs can decrease the amount of HMs in the soil. This process also controls the expression of genes responsible for transferring HMs within plants, enhances the activity of different antioxidant enzymes in plants, enhances body processes and encourages the creation of protective substances such as root secretions, organic acids and phytochelatin (Zhou 2020). Applying silicon nanoparticles to maize plants experiencing arsenic (As) stress alleviated the negative impacts of As stress on photochemical quenching, non-photochemical quenching of photosystem II and maximum quantum efficiency (Tripathi et al. 2015). Manzoor et al. (2022) conducted a recent study showing that iron oxide (FeO) NPs improved the growth and development of wheat plants by reducing the harmful effects of Cd toxicity. This was achieved by enhancing biomass production, increased antioxidant enzyme activity and chlorophyll levels. These nanoparticles could reduce the buildup of harmful ions in plant cells and provide protection against stress. Efficacy of NPs in mitigating HM stress in plants is attributed to their diminutive size and expansive surface area. It has also been proved that silicon nanoparticles can alleviate the damage caused by heavy metals in rice (Gao et al. 2018), in wheat (Rehman et al. 2017) and in pea (Rehman et al. 2018) plants. Furthermore, Konate et al. (2017) discovered that FeO nanoparticles helped wheat plants grow by reducing the cellular oxidative stress that was caused by the harmful effects of lead (Pb) and cadmium (Cd). Similarly, the presence of magnetite NPs was found to improve the growth and establishment of wheat seedlings by reducing the negative effects of chromium (Cr) and cadmium (Cd) toxicity, as reported by López-Luna et al. (2015). Similarly, FeO NPs have been effectively employed to improve plant development and mitigate the effects of heavy metal stress (Adrees et al. 2020). Titanium dioxide (TiO_2) nanoparticles application to soil can significantly reduce the harmful effects of cadmium stress in soybean by enhancing plants' metabolic and photosynthetic rate. Therefore, TiO_2 nanoparticles are important for mitigating the oxidative stress caused by heavy metals (Singh and Lee 2016). Treatment with Silicon dioxide (SiO_2) NPs resulted in elevated antioxidant enzyme activities, specifically superoxide dismutase (SOD) and ascorbate peroxidase (APX) on exposure to Cr stress in pea (Tripathi et al. 2015). In addition, de Sousa et al. (2019) discovered that silicon nanoparticles could mitigate aluminum (Al) toxicity in maize plants by stimulating the antioxidant mechanism. Nevertheless, it is imperative to create innovative nano remediation methods to mitigate the adverse effects of heavy metals on plant growth and development.

7.3.4.2 Impact of Nanoparticles on Plants Under Drought Stress

Drought is a major issue in the environment that has the attention of experts in both agriculture and the environment. It is an important issue in agriculture around the world that makes plants grow slower and crops produce less. The economic sector is impacted by the effects of drought stress on plant growth factors (Kumar et al. 2022). Insufficient moisture content decreases cell size, disrupts the integrity of membranes, triggers oxidative stress, and leads to leaf senescence, ultimately resulting in a decrease in agricultural output (Qiao et al. 2024). Research from the past has shown that silicon nanoparticles assisted plants to tolerate drought stress better. For instance, the ability of hawthorn plants to withstand drought improved when they were provided with silicon (Si) nanoparticles. However, the specific physiological parameters associated with defense mechanisms exhibited differences based on how severe the drought is and the concentration of Si NPs applied, as reported by Ashkavand et al. (2018). Similarly, Si NPs showed promise in aiding the recovery of plants after a drought by influencing the physical and physiological characteristics of barley plants (Ghorbanpour et al. 2020). Alsaeedi et al. (2019) reported that, the application of Si NPs enhanced the growth and yield of cucumbers when they were exposed to conditions of limited water and high salinity.

In response to drought stress, Behboudi et al. (2019) reported that chitosan nanoparticles improved the water content as well as CAT and SOD activities there by increased photosynthetic rate and yield of wheat plants. It has been discovered that applying iron nanoparticles to safflower cultivars lessens the adverse effects of drought stress (Davar et al. 2014; Naseer et al. 2021). Similar to this, it has been demonstrated that adding cerium oxide nanoparticles to the soil at a concentration of 100 mg/kg increases the photosynthetic rate and promotes plant development in soybean plants via controlling water usage efficiency (Cao et al. 2017). Das and Das (2019) provided evidence that the application of silver nanoparticles to lentil plants reduced the negative consequences of drought stress. According to Sun et al. (2017), drought resistance in Arabidopsis can be effectively increased by using silicon nanoparticles to transport abscisic acid.

7.3.4.3 Impact of Nanoparticles on Plants Under Salt Stress

Due to ionic toxicity and ionic balance disruption, soil salinity considerably reduces crop productivity worldwide (Etesami et al. 2021). Saline conditions cause sodium (Na^+) and chlorine (Cl^-) ions to build up in plant cells, which causes an ionic imbalance and negative consequences (Khan et al. 2021). Plants under salt stress exhibit a large increase in the buildup of Na^+ in cytosols and a significant release of potassium ion (K^+) from leaf mesophyll cells. As a result, growth rate and production are reduced in saline environments due to the reduced process of carbon dioxide (CO_2) absorption (Shahid et al. 2020). Applying different kinds of nanoparticles is another way to deal with salt stress and lessen its negative consequences. When Si

NPs were applied to cherry tomato plants, it was discovered that these plants significantly improved in several characteristics such as, the antioxidant defense system, leaf turgor, carbon absorption, and seed germination (Haghighi and Pessarakli 2013).

Similarly, FeO NPs were found to help lessen the detrimental effects of salt stress on wheat plants (Manzoor et al. 2021). This was accomplished by boosting the plants' growth, raising their chlorophyll content, and boosting the activity of antioxidant enzymes. Sabaghnia and Janmohammadi (2015) conducted a study which revealed that lentil plants treated with Si NPs exhibited improved seed germination and seedling growth when exposed to salt stress. In a different study, Ye et al. (2020) shown that by controlling chemical reactions, manganese nanoparticles applied to seeds can successfully reduce the detrimental effects of salt stress on pepper plants. Zhao et al. (2019) discovered that rapeseed was better able to withstand high salt levels when multi-walled carbon nanotubes were added to the soil. This improvement was linked to a drop in the levels of thiobarbituric acid and the Na^+/K^+ ratio, as well as a reduction in the generation of reactive oxygen species (ROS). To comprehend how NPs improve plants' tolerance to salinity, more physiochemical and molecular research will be required.

7.3.4.4 Impact of Nanoparticles on Plants Under Heat Stress

Among all the different environmental stresses, one of the most detrimental kinds of stress is believed to be rising temperatures (Kan et al. 2023). Oxidative stress is caused by heat stress, which also raises the production of reactive oxygen species. As a result, various metabolic processes are hampered, cellular homeostasis is upset, and membrane lipids degrade. Crop plants ultimately die as a result of these effects (Chaudhry and Sidhu 2022; Savicka and Škute 2010). Furthermore, heat stress disrupts electron flow, carbon fixation, photosystem II, and causes chlorophyll degradation. Plant growth is eventually hampered by this disruption, which causes the photosynthetic process to slow down (Li et al. 2021). The agriculture industry has undergone a transformation due to advancements in nanotechnology, which have the ability to improve plant growth and development under adverse situations (Rana et al. 2021). Numerous research has shown how nanoparticles may be used to improve crop plants' resistance to heat stress (Thakur et al. 2021). Haghighi et al. (2014) found that using selenium (Se) nanoparticles could significantly lessen heat stress in tomato plants. By improving the plants' overall development, chlorophyll content, and hydration potential, this was accomplished. Furthermore, Qi et al. (2013) found that the addition of TiO_2 NPs significantly reduced heat stress in tomato plants by encouraging stomatal opening. El-Saadony et al. (2021) reported that the application of 100 μg/ml of naturally occurring selenium nanoparticles improved wheat development by improving the plants' resistance to heat stress. Iqbal et al. (2019) showed that using silver nanoparticles significantly improved the morphology of wheat plants under heat stress. In conclusion, using metallic nanoparticles as nano fertilizers may improve plants' ability to withstand heat stress, thereby advancing sustainable agriculture.

7.3.5 *Nanofertilizers in Biotic Stress Management*

Exposure to biotic stressors like insects, pests and pathogenic microorganisms limits plant productivity to a great extent. Application of chemical pesticides adversely affects soil microflora and contributes to environmental pollution. Nanoparticles with their controlled release and target specific activity offer a good alternative to conventional method of plant disease management.

The applications of nanoparticles as fungicides, bactericides or nano fertilizers for plant disease management have been extensively reviewed (Elmer and White 2018). By directly interacting with pathogens through mechanisms such as the production of reactive oxygen species (ROS), these materials can decrease crop diseases. By strengthening plant defense mechanisms and promoting crop nutrition, nanoparticles may also indirectly inhibit disease. Utilizing nanoparticles effectively can either supplement or replace traditional pesticides and fertilizers, lowering the environmental impact of agricultural activities (Adisa et al. 2019).

In young transplants of tomato and egg plants, foliar application of nano-Cu could suppress root diseases (Elmer and White 2016). Application of nano-Cu was found to be effective in managing soil borne pathogens such as *Fusarium oxysporum*. In a study conducted on chrysanthemum for 3 years, up to 50% decrease in disease severity was observed in transplants treated with nano copper oxide (nano-CuO) (Elmer and White 2016). Studies revealed that, application of CuO nanoparticles to seedlings have promoted the production and uploading of enzymes involved in plant defense systems and transcription of many genes contributing to stress tolerance (Ma et al. 2020; Shen et al. 2020). These results probably suggest the underlying mechanism of CuO nanoparticle induced disease resistance to crop plants.

Nano fertilizer applications of other metal ion nanoparticles were also found to be effective against plant pathogenic microorganisms. Treatment of watermelons plants with mesoporous nano-silicon decreased disease severity of soil fungus and increased biomass (Buchman et al. 2019). Zinc based nanomaterials have been experimented as plant protectants against many bacterial and fungal pathogens owing to their antimicrobial properties. ZnO nanoparticles were found to be effective in managing bacterial disease in citrus (Graham et al. 2016). The ZnO nanoparticles showed higher mobility compared to the bulk zinc oxide, hence these particles have higher chance of reaching the phytopathogens restricted to vascular tissue (Naranjo et al. 2020). The antibacterial activity of ZnO nanoparticles can be ascribed to the production of reactive oxygen species (ROS) such as superoxides and peroxides upon reaction with water. These ROS species contribute to lipid peroxidation of bacterial membrane leading to leakage of intracellular materials and cause cell death in pathogens (Dwivedi et al. 2014; Sirelkhatim et al. 2015). The application of nanoparticles as nano fertilizers shows good potential to reduce the infection of pathogenic microorganisms in crop plants.

7.4 Conclusions and Prospects

Agricultural systems conventionally rely on chemical inputs such as fertilizers and pesticides to attain maximum productivity for ensuring food security to the increasing global population. Intensive use of chemical fertilizers and pesticides over a long period of time has resulted in environmental issues like eutrophication, soil, air, and water pollution. Excessive use of these chemicals adversely affected soil microbiota, contributing to a decrease in soil fertility. Additionally, global warming and climate change are also posing threat to the productivity of agricultural systems. To overcome these challenges, new technological interventions need to be applied in agriculture which are ecologically and economically sustainable.

Precision agriculture is gaining attention in agriculture scenario as an alternative to conventional farming practices. Precision agriculture aims to increase crop productivity through optimizing the use of fertilizers, pesticides, and irrigation water. It ensures enhanced crop production, environmental protection, and profitability. Nanotechnology is utilized in precision agriculture for the development of nano sensor based smart delivery systems, nano fertilizers and nano pesticides to help the farmers to reduce agricultural inputs or resources. Nano fertilizers are an integral component of precision agriculture as they are able to release nutrients to the soil in a slow and sustained manner, ensuring soil nutrient availability for a longer duration. Different types of nano fertilizers have been synthesized and experimented for agricultural applications and the results are promising. Nano fertilizers are synthesized for both macro and micro nutrients, which can be applied to soil, through leaves or to the seeds. Studies show that application of nano fertilizers can increase productivity and nutrient enrichment in agricultural crops. They also demonstrate capacity in enhancing tolerance to biotic and abiotic stresses.

Incorporating nanomaterials as fertilizers in precision agriculture can transform the agriculture sector by enhancing fertilizer efficiency and effectiveness while minimizing environmental impact. However, extensive research is essential to entirely comprehend the long-lasting effects of nanomaterials on the environment and human health, as well as to develop sustainable methods for their production and application.

References

Abdel-Aziz HM, Hasaneen MN, Omer AM (2016) Nano chitosan-NPK fertilizer enhances the growth and productivity of wheat plants grown in sandy soil. Spanish J Agric Res 14(1):e0902

Abd-Elrahman SH, El-Gabry YA, Hashem FA et al (2023) Influence of nano-chitosan loaded with potassium on potassium fractionation in sandy soil and strawberry productivity and quality. Agronomy 13(4):1126

Adisa IO, Pullagurala VLR, Peralta-Videa JR et al (2019) Recent advances in nano-enabled fertilizers and pesticides: a critical review of mechanisms of action. Environ Sci-Nano 6(7): 2002–2030

Adnan N, Nordin SM, Anwar A (2020) Transition pathways for Malaysian paddy farmers to sustainable agricultural practices: an integrated exhibiting tactics to adopt green fertilizer. Land Use Pol 90:104255

Adnan M, Xiao B, Ali MU et al (2024) Heavy metals pollution from smelting activities: a threat to soil and ground water. Ecotoxicol Environ Saf 274:116189

Adrees M, Khan ZS, Ali S et al (2020) Simultaneous mitigation of cadmium and drought stress in wheat by soil application of iron nanoparticles. Chemosphere 238:124681

Afzal S, Aftab T, Singh NK (2022) Impact of zinc oxide and iron oxide nanoparticles on uptake, translocation, and physiological effects in *Oryza sativa* L. J Plant Growth Regul 41(4): 14451461

Al-Juthery HWA, Al-taee RAHG, Al-Obaidi ZHH et al (2019) Influence of foliar application of some nano-fertilizers in growth and yield of potato under drip irrigation. In: Journal of physics: conference series, vol 1294(9). IOP Publishing, Bristol, UK, p 092024

Alsaeedi A, El-Ramady H, Alshaal T et al (2019) Silica nanoparticles boost growth and productivity of cucumber under water deficit and salinity stresses by balancing nutrients uptake. Plant Physiol Biochem 139:1–10

Ashkavand P, Tabari M, Zarafshar M et al (2018) Effect of SiO_2 nanoparticles on drought resistance in hawthorn seedlings. For Res Pap 76:350–359

Avila-Quezada GD, Ingle AP, Golińska P, Rai M (2022) Strategic applications of nano-fertilizers for sustainable agriculture: benefits and bottlenecks. Nanotechnol Rev 11(1):2123–2140

Azam M, Bhatti HN, Khan A (2022) Zinc oxide nano-fertilizer application (foliar and soil) effect on the growth, photosynthetic pigments and antioxidant system of maize cultivar. Biocatal Agric Biotechnol 42:102343

Babu S, Singh R, Yadav D et al (2022) Nanofertilizers for agricultural and environmental sustainability. Chemosphere 292:133451

Banerjee S, Islam J, Mondal S (2023) Proactive attenuation of arsenic-stress by nano-priming: zinc oxide nanoparticles in *Vigna mungo* (L.) Hepper trigger antioxidant defense response and reduce root-shoot arsenic translocation. J Hazard Mater 446:130735

Behboudi F, Tahmasebi-Sarvestani Z, Kassaee MZ et al (2019) Evaluation of chitosan nanoparticles effects with two application methods on wheat under drought stress. J Plant Nutr 42:1439–1451

Bhaskar M, Kumar A, Rani R (2023) Application of nano formulations in agriculture. Biocatal Agric Biotechnol 54:102934

Buchman JT, Elmer WH, Ma C et al (2019) Chitosan-coated mesoporous silica nanoparticle treatment of *Citrullus lanatus* (Watermelon): enhanced fungal disease suppression and modulated expression of stress-related genes. ACS Sustain Chem Eng 7(24):19649–19659

Cao Z, Stowers C, Rossi L et al (2017) Physiological effects of cerium oxide nanoparticles on the photosynthesis and water use efficiency of soybean (*Glycine max* (L) Merr). Environ Sci Nano 4:1086–1094

Chaudhry S, Sidhu GP (2022) Climate change regulated abiotic stress mechanisms in plants: a comprehensive review. Plant Cell Rep 41(1):1–31

Dapkekar A, Deshpande P, Oak MD et al (2018) Zinc use efficiency is enhanced in wheat through nanofertilization. Sci Rep 8(1):6832

Das A, Das B (2019) Nanotechnology a potential tool to mitigate abiotic stress in crop plants. In: De Oliveira A (ed) Abiotic and biotic stress in plants. IntechOpen, London, UK

Davar ZF, Roozbahani A, Hosnamidi A (2014) Evaluation the effect of water stress and foliar application of Fe nanoparticles on yield, yield components and oil percentage of safflower (*Carthamus tinctorious* L). Int J Adv Biol Biomed 16:1150–1159

de Sousa A, Saleh AM, Habeeb TH et al (2019) Silicon dioxide nanoparticles ameliorate the phytotoxic hazards of aluminum in maize grown on acidic soil. Sci Total Environ 693:133636

Deng C, Wang Y, Navarro G et al (2022) Copper oxide (CuO) nanoparticles affect yield, nutritional quality, and auxin associated gene expression in weedy and cultivated rice (*Oryza sativa* L.) grains. Sci Total Environ 810:152260

Dey S, Nath S, Ansari TA, Biswas A (2023) Application of green synthesized bimetallic nZVI-cu nanoparticle as a sustainable alternative to chemical fertilizers to enhance growth and photosynthetic efficiency of rice seedlings. Plant Physiol Biochem 201:107837

Dixit M, Bhuyan M, Ghoshal D et al (2024) Mode of action of nanofertilizers and their role in nutrient budgeting: quality and attribute perspectives. In: Abd-Elsalam KA (ed) Nanofertilizer synthesis, methods and types, Nanobiotechnology for plant protection. Elsevier, Amsterdam, Netherlands, pp 375–398

Duhan JS, Kumar R, Kumar N et al (2017) Nanotechnology: the new perspective in precision agriculture. Biotechnol Rep 15:11–23

Dwivedi S, Wahab R, Khan F et al (2014) Reactive oxygen species mediated bacterial biofilm inhibition via zinc oxide nanoparticles and their statistical determination. PLoS One 9(11): e111289. https://doi.org/10.1371/journal.pone.0111289

El-Ghamry AM, Mosa AA, Alshaal TA, El-Ramady HR (2018) Nanofertilizers vs. biofertilizers: new insights. Environ Biodiv Soil Secur 2:51–72

Elizabath A, Babychan M, Mathew AM, Syriac GM (2019) Application of nanotechnology in agriculture. Int J Pure App Biosci 7(2):131–139

Elmer WH, White JC (2016) The use of metallic oxide nanoparticles to enhance growth of tomatoes and eggplants in disease infested soil or soilless medium. Environ Sci Nano 3(5):1072–1079

Elmer W, White JC (2018) The future of nanotechnology in plant pathology. In: Leach JE, Lindow SE (eds) Annual review of phytopathology, vol 56. Springer, pp 111–133

Elmer WH, White JC (2016). The use of metallic oxide nanoparticles to enhance growth of tomatoes and eggplants in disease infested soil or soilless medium. Environmental Science: Nano, 3(5):1072–1079. Royal Society of Chemistry, Cambridge, UK.

El-Saadony MT, Saad AM, Najjar AA et al (2021) The use of biological selenium nanoparticles to suppress *Triticum aestivum* L crown and root rot diseases induced by fusarium species and improve yield under drought and heat stress. Saudi J Biol Sci 28:4461–4471

Elsayed AA, Ahmed EG, Taha ZK et al (2022) Hydroxyapatite nanoparticles as novel nanofertilizer for production of rosemary plants. Sci Hortic 295:110851

Elsherif DE, Abd-ElShafy E, Khalifa AM (2023) Impacts of ZnO as a nanofertilizer on fenugreek: some biochemical parameters and SCoT analysis. J Genet Eng Biotechnol 21(1):52

Emadjan SE (2017) Physiological responses of Loblolly Pine (*Finustaeda* L.) to silicon and water stress. Ph.D. Thesis, Texas A &M University, pp 27–37

Etesami H, Fatemi H, Rizwan M (2021) Interactions of nanoparticles and salinity stress at physiological, biochemical and molecular levels in plants: a review. Ecotoxicol Environ Saf 225:112769

FAO, ITPS (2019) The international code of conduct for the sustainable use and management of fertilizers. FAO, Rome, Italy

Farooq T, Hameed A, Hameed A (2023) Emerging concept of nanofertilizers for sustainable crop plants growth and production. In: Husen A (ed) Engineered nanomaterials for sustainable agricultural production, soil improvement and stress management. Academic, Academic Press, Cambridge, MA, USA, pp 273–310

Fertahi S, Ilsouk M, Zeroual Y et al (2021) Recent trends in organic coating based on biopolymers and biomass for controlled and slow release fertilizers. J Control Release 330:341–361. https://doi.org/10.1016/j.jconrel.2020.12.026

Firmanda A, Fahma F, Syamsu K et al (2024) Biopolymer-based slow/controlled-release fertilizer (SRF/CRF): nutrient release mechanism and agricultural sustainability. J Environ Chem Eng 8: 112177. https://doi.org/10.1016/j.jece.2024.112177

Gade A, Ingle P, Nimbalkar U et al (2023) Nanofertilizers: the next generation of agrochemicals for long-term impact on sustainability in farming systems. Agrochemicals 2(2):257–278

Gao M, Zhou J, Liu H et al (2018) Foliar spraying with silicon and selenium reduces cadmium uptake and mitigates cadmium toxicity in rice. Sci Total Environ 631–632:1100–1108

Ghorbanpour M, Mohammadi H, Kariman K (2020) Nanosilicon-based recovery of barley (*Hordeum vulgare*) plants subjected to drought stress. Environ Sci Nano 7:443–461

Graham JH, Johnson EG, Myers ME et al (2016) Potential of nano-formulated zinc oxide for control of citrus canker on grapefruit trees. Plant Dis 100(12):2442–2447

Guo H, White JC, Wang Z, Xing B (2018) Nano-enabled fertilizers to control the release and use efficiency of nutrients. Curr Opin Environ Sci Health 6:77–83

Haghighi M, Pessarakli M (2013) Influence of silicon and nano-silicon on salinity tolerance of cherry tomatoes (*Solanum lycopersicum* L) at early growth stage. Sci Hortic 161:111–117

Haghighi M, Abolghasemi R, da Silva JA (2014) Low and high temperature stress affect the growth characteristics of tomato in hydroponic culture with Se and nano-Se amendment. Sci Hortic 178: 231–240

Haydar MS, Ghosh S, Mandal P (2022) Application of iron oxide nanoparticles as micronutrient fertilizer in mulberry propagation. J Plant Growth Regul 41:1726–1746

He S, Feng Y, Ni J et al (2016) Different responses of soil microbial metabolic activity to silver and iron oxide nanoparticles. Chemosphere 147:195–202. https://doi.org/10.1016/j.chemosphere.2015.12.055

Hernández-Fuentes AD, Arroyo-Aguilar JE, Gutiérrez-Tlahque J et al (2023a) Application of Cu nanoparticles in chitosan-PVA hydrogels in a native tomato genotype: evaluation of the postharvest behavior of the physicochemical and bioactive components of the fruits. Food Bioprocess Technol 16(9):1953–1962. https://doi.org/10.1007/s11947-023-03044-x

Hernández-Fuentes AD, Montaño-Herrera A, Pinedo-Espinoza JM et al (2023b) Changes of the antioxidant system in pear (*Pyrus communis* L.) fruits by foliar application of copper, selenium, iron, and zinc nanoparticles. J Agric Food Res 14:100885. https://doi.org/10.1016/j.jafr.2023.100885

Hernández-Hernández H, Quiterio-Gutiérrez T, Cadenas-Pliego G et al (2019) Impact of selenium and copper nanoparticles on yield, antioxidant system, and fruit quality of tomato plants. Plan Theory 8(10):355

Hussain A, Rizwan M, Ali S et al (2021) Combined use of different nanoparticles effectively decreased cadmium (Cd) concentration in grains of wheat grown in a field contaminated with Cd. Ecotoxicol Environ Saf 215:112139

Iqbal M, Raja NI, Mashwani ZUR et al (2019) Effect of silver nanoparticles on growth of wheat under heat stress. Iran J Sci Technol Trans A: Sci 43:387–395

Irshad MA, Nawaz R, Rehman MZ et al (2021) Synthesis, characterization and advanced sustainable applications of titanium dioxide nanoparticles: a review. Ecotoxicol Environ Saf 212: 111978

Jayara AS, Kumar R, Pandey P (2023) Nanofertilizers in wheat crop nutrition: a review. J Cereal Res 15(1):24–30

Kan Y, Mu XR, Gao J et al (2023) The molecular basis of heat stress responses in plants. Mol Plant 16(10):1612–1634

Khan I, Awan SA, Raza MA et al (2021) Silver nanoparticles improved the plant growth and reduced the sodium and chlorine accumulation in pearl millet: a life cycle study. Environ Sci Pollut Res 28:13712–13724

Khan F, Siddique AB, Shabala S et al (2023) Phosphorus plays key roles in regulating plants' physiological responses to abiotic stresses. Plan Theory 12(15):2861

Khattab S, Alkuwayti MA, Yap YK et al (2023) Foliar spraying of ZnO nanoparticals on *Curcuma longa* had increased growth, yield, expression of curcuminoid synthesis genes, and curcuminoid accumulation. Horticulturae 9(3):355

Konate A, He X, Zhang Z et al (2017) Magnetic (Fe_3O_4) nanoparticles reduce heavy metals uptake and mitigate their toxicity in wheat seedling. Sustain For 9:790

Kumar A, Singh K, Verma P et al (2022) Effect of nitrogen and zinc nanofertilizer with the organic farming practices on cereal and oil seed crops. Sci Rep 12(1):6938

Kumaraswamy RV, Saharan V, Kumari S et al (2021) Chitosan-silicon nanofertilizer to enhance plant growth and yield in maize (*Zea mays* L.). Plant Physiol Biochem 159:53–66

Lawrencia D, Wong SK, Low DY et al (2021) Controlled release fertilizers: a review on coating materials and mechanism of release. Plan Theory 10(2):238. https://doi.org/10.3390/plants10020238

Li Y, Guo P, Liu Y et al (2020) Effects of sulfur on the toxicity of cadmium to *Folsomia candida* in red earth and paddy soil in southern Fujian. J Hazard Mater 387:121683

Li D, Wang M, Zhang T et al (2021) Glycinebetaine mitigated the photoinhibition of photosystem II at high temperature in transgenic tomato plants. Photosynth Res 147:301–315

López-Luna J, Silva-Silva MJ, Martinez-Vargas S et al (2015) Magnetite nanoparticle (NP) uptake by wheat plants and its effect on cadmium and chromium toxicological behavior. Sci Total Environ 565:941–950

López-Vargas ER, Ortega-Ortíz H, Cadenas-Pliego G et al (2018) Foliar application of copper nanoparticles increases the fruit quality and the content of bioactive compounds in tomatoes. Appl Sci 8(7):1020

Loures L, Chamizo A, Ferreira P et al (2020) Assessing the effectiveness of precision agriculture management systems in mediterranean small farms. Sustain For 12(9):3765

Ma CX, Borgatta J, Hudson BG et al (2020) Advanced material modulation of nutritional and phytohormone status alleviates damage from soybean sudden death syndrome. Nat Nanotechnol 15(12):1033–1042

Manikandan A, Subramanian KS (2016) Evaluation of zeolite based nitrogen nano-fertilizers on maize growth, yield and quality on inceptisols and alfisols. Int J Plant Soil Sci 9(4):1–9. https://doi.org/10.9734/IJPSS/2016/22103

Manzoor N, Ahmed T, Noman M et al (2021) Iron oxide nanoparticles ameliorated the cadmium and salinity stresses in wheat plants, facilitating photosynthetic pigments and restricting cadmium uptake. Sci Total Environ 769:145221

Manzoor N, Ali L, Ahmed T et al (2022) Recent advancements and development in nano-enabled agriculture for improving abiotic stress tolerance in plants. Front Plant Sci 13:951752

Mejias JH, Salazar F, Perez Amaro L et al (2021) Nanofertilizers: a cutting-edge approach to increase nitrogen use efficiency in grasslands. Front Environ Sci 9:635114

Mikkelsen R (2018) Nanofertilizer and nanotechnology: a quick look. Better Crops 102(3):18–19

Mittal D, Kaur G, Singh P et al (2020) Nanoparticle-based sustainable agriculture and food science: recent advances and future outlook. Front Nanotechnol 2:579954

Munir T, Rizwan M, Kashif M et al (2018) Effect of zinc oxide nanoparticles on the growth and Zn uptake in wheat (*Triticum aestivum* L.) by seed priming method. Digest J Nanomater Biostruct 13(1):315–323

Naranjo E, Merfa MV, Santra S et al (2020) Zinkicide is a ZnO-based nanoformulation with bactericidal activity against *Liberibactercrescens* in batch cultures and in microfluidic chambers simulating plant vascular systems. Appl Environ Microbiol 86(16):e00788–e00720

Naseer I, Javad S, Singh A et al (2021) Alleviation mechanism of drought stress in plants using metal nanoparticles–a perspective analysis. In: Faizan M, Hayat S, Yu F (eds) Sustainable agriculture reviews, vol 53. Springer, Cham, pp 115–149. https://doi.org/10.1007/978-3-030-86876-5_5

Ndaba B, Roopnarain A, Haripriya RAMA, Maaza M (2022) Biosynthesized metallic nanoparticles as fertilizers: an emerging precision agriculture strategy. J Integr Agric 21(5):1225–1242

Nongbet A, Mishra AK, Mohanta YK et al (2022) Nanofertilizers: a smart and sustainable attribute to modern agriculture. Plan Theory 11(19):2587

Oghenerume P, Eduok S, Ita B et al (2020) Impact of zinc oxide nanoparticles amended organic manure on *Arachis hypogaea* growth response and rhizosphere bacterial community. Int J Plant Soil Sci 32(5):24–35. https://doi.org/10.9734/ijpss/2020/v32i530279

Parashar R, Afzal S, Mishra M (2023) Improving biofortification success rates and productivity through zinc nanocomposites in rice (*Oryza sativa* L.). Environ Sci Pollut Res Int 30(15): 44223–44233

Petrović B, Bumbálek R, Zoubek T et al (2024) Application of precision agriculture technologies in Central Europe-review. J Agric Food Res 15:101048. https://doi.org/10.1016/j.jafr.2024.101048

Qi M, Liu Y, Li T (2013) Nano-TiO_2 improve the photosynthesis of tomato leaves under mild heat stress. Biol Trace Elem Res 156:323–328

Qiao M, Hong C, Jiao Y et al (2024) Impacts of drought on photosynthesis in major food crops and the related mechanisms of plant responses to drought. Plan Theory 13(13):1808

Rahemi M, Gharechahi SR, Sedaghat S (2020) The application of nano-iron chelate and iron chelate to soil and as foliar application: treatments against chlorosis and fruit quality in quince. Int J Fruit Sci 20:300–313. https://doi.org/10.1080/15538362.2019.1628684

Rai-Kalal P, Jajoo A (2021) Priming with zinc oxide nanoparticles improve germination and photosynthetic performance in wheat. Plant Physiol Biochem 160:341–351

Rana RA, Siddiqui MN, Skalicky M et al (2021) Prospects of nanotechnology in improving the productivity and quality of horticultural crops. Horticulturae 7:1–19

Rehman MZ, Khalid H, Akmal F et al (2017) Effect of limestone, lignite and biochar applied alone and combined on cadmium uptake in wheat and rice under rotation in an effluent irrigated field. Environ Pollut 227:560–568

Rehman MZ, Rizwan M, Hussain A et al (2018) Alleviation of cadmium (cd) toxicity and minimizing its uptake in wheat (*Triticum aestivum*) by using organic carbon sources in cd-spiked soil. Environ Pollut 241:557–565

Rexlin J, Vijayakumar SM, Vidhya E et al (2023) Bioengineered ZnO nanoparticles as a nano priming agent in *Cyamopsis tetragonoloba* (L) Taub. to improve yield and disease resistance. Appl Nanosci 13(9):5993–6001

Rizwan M, Ali S, Rehman MZ et al (2019a) Effect of foliar applications of silicon and titanium dioxide nanoparticles on growth, oxidative stress, and cadmium accumulation by rice (*Oryza sativa*). Acta Physiol Plant 41:1–12

Rizwan M, Ali S, Rehman MZ et al (2019b) Alleviation of cadmium accumulation in maize (Zea mays L) by foliar spray of zinc oxide nanoparticles and biochar to contaminated soil. Environ Pollut 248:358–367

Rossi L, Fedenia LN, Sharifan H et al (2019) Effects of foliar application of zinc sulfate and zinc nanoparticles in coffee (*Coffea arabica* L.) plants. Plant Physiol Biochem 135:160–166

Zhou P, Adeel M, Shakoor N, Guo M, Hao Y, Azeem I, ... Rui Y (2020) Application of nanoparticles alleviates heavy metals stress and promotes plant growth: An overview. Nanomater 11(1):26. https://doi.org/10.3390/nano11010026

Sabaghnia N, Janmohammadi M (2015) Effect of nano-silicon particles application on salinity tolerance in early growth of some lentil genotypes. Ann Univ Mariae Curie-Sklodowska, Sect C—Biol 69:39

Savicka M, Škute N (2010) Effects of high temperature on malondialdehyde content, superoxide production and growth changes in wheat seedlings (*Triticum aestivum* L). Ekologija 56:26–33

Sedefoglu N, Er S, Veryer K et al (2023) Green synthesized CuO nanoparticles using macrofungi extracts: characterization, nanofertilizer and antibacterial effects. Mater Chem Phys 309:128393. https://doi.org/10.1016/j.matchemphys.2023.128393

Shahid MA, Sarkhosh A, Khan N et al (2020) Insights into the physiological and biochemical impacts of salt stress on plant growth and development. Agronomy 10(7):938

Shaikh A, Meroliya H, Dagade-Gadale S, Waghmode S (2021) Applications of nanotechnology in precision agriculture: a review. Res Rev Biotechnol Biosci 8:105–117

Shakoor N, Adeel M, Zain M et al (2022) Exposure of cherry radish (*Raphanus sativus* L. var. Radculus Pers) to iron-based nanoparticles enhances its nutritional quality by trigging the essential elements. NanoImpact 25:100388

Shcherbakova EN, Shcherbakov AV, Andronov EE et al (2017) Combined pre-seed treatment with microbial inoculants and Mo nanoparticles changes composition of root exudates and rhizosphere microbiome structure of chickpea (*Cicer arietinum* L.) plants. Symbiosis 73(1):57–69

Shebl A, Hassan AA, Salama DM et al (2020) Template-free microwave-assisted hydrothermal synthesis of manganese zinc ferrite as a nanofertilizer for squash plant (*Cucurbita pepo* L). Heliyon 6(3):e03596. https://doi.org/10.1016/j.heliyon.2020.e03596

Shen Y, Borgatta J, Ma CX et al (2020) Nanomaterial morphology and composition control foliar transfer through the cuticle and mediate resistance to root fungal disease in tomato (*Solanum lycopersicum*). J Agric Food Chem 68(41):11327–11338

Sheoran P, Grewal S, Kumari S, Goel S (2021) Enhancement of growth and yield, leaching reduction in *Triticum aestivum* using biogenic synthesized zinc oxide nanofertilizer. Biocatal Agric Biotechnol 32:101938

Singh J, Lee BK (2016) Influence of nano-TiO_2 particles on the bioaccumulation of Cd in soybean plants (*Glycine max*): a possible mechanism for the removal of Cd from the contaminated soil. J Environ Manag 170:88–96

Singh K, Madhusudanan M, Verma AK et al (2021) Engineered zinc oxide nanoparticles: an alternative to conventional zinc sulphate in neutral and alkaline soils for sustainable wheat production. 3 Biotech 11(7):322

Sirelkhatim A, Mahmud S, Seeni A et al (2015) Review on zinc oxide nanoparticles: antibacterial activity and toxicity mechanism. Nanomicro Lett 7(3):219–242

Sishodia RP, Ray RL, Singh SK (2020) Applications of remote sensing in precision agriculture: a review. Remote Sens 12(19):3136. https://doi.org/10.3390/rs12193136

Subramanian KS, Manikandan A, Thirunavukkarasu M, Rahale CS (2015) Nano-fertilizers for balanced crop nutrition. In: Rai M, Ribeiro C, Mattoso L, Duran N (eds) Nanotechnologies in food and agriculture. Springer, Cham, pp 69–80

Sun D, Hussain HI, Yi Z et al (2017) Delivery of abscisic acid to plants using glutathione responsive mesoporous silica nanoparticles. J Nanosci Nanotechnol 18:1615–1625

Tarafder C, Daizy M, Alam MM et al (2020) Formulation of a hybrid nanofertilizer for slow and sustainable release of micronutrients. ACS Omega 5(37):23960–23966

Thakur S, Asthir B, Kaur G et al (2021) Zinc oxide and titanium dioxide nanoparticles influence heat stress tolerance mediated by antioxidant defense system in wheat. Cereal Res Commun 49: 1–12

Tondey M, Kalia A, Singh A et al (2022) A comparative evaluation of the effects of seed invigoration treatments with precursor zinc salt and nano-sized zinc oxide (ZnO) particles on vegetative growth, grain yield, and quality characteristics of *Zea mays*. J Anal Sci Technol 13(1):40. https://doi.org/10.1186/s40543-022-00346-1

Tripathi DK, Singh VP, Prasad SM et al (2015) Silicon nanoparticles (SiNp) alleviate chromium (VI) phytotoxicity in *Pisum sativum* (L) seedlings. Plant Physiol Biochem 96:189–198

ul Ain Q, Hussain HA, Zhang Q (2023) Use of nano-fertilizers to improve the nutrient use efficiencies in plants. In: Aftab T, Hakeem KR (eds) Sustainable plant nutrition. Academic, Academic Press, Cambridge, MA, USA, pp 299–321

UNEP (2023) UN Food Systems Summit +2 Stocktaking Moment (UNFSS+2). https://www.unep.org/un-food-systems-summit. Accessed 30 Mar 2024

United Nations (2022) United Nations Department of Economic and Social Affairs, Population Division. World Population Prospects 2022: Summary of Results. UN DESA/POP/2022/TR/NO.3. https://www.un.org/development/desa/pd/sites/www.un.org.development.desa.pd/files/wpp2022_summary_of_results.pdf. Accessed 20 Mar 2024

Vanti GL, Nargund VB, Basavesha KN et al (2019) Synthesis of *Gossypium hirsutum*-derived silver nanoparticles and their antibacterial efficacy against plant pathogens. Appl Organomet Chem 33(1):e4630

Vejan P, Khadiran T, Abdullah R, Ahmad N (2021) Controlled release fertilizer: a review on developments, applications and potential in agriculture. J Control Release 339:321–334. https://doi.org/10.1016/j.jconrel.2021.10.003

Verma KK, Song XP, Joshi A et al (2022) Nanofertilizer possibilities for healthy soil, water, and food in future: an overview. Front. Plant Sci 13:865048. https://doi.org/10.3389/fpls.2022.865048

Verma KK, Song XP, Degu HD et al (2023) Recent advances in nitrogen and nano-nitrogen fertilizers for sustainable crop production: a mini-review. Chem Biol Technol Agric 10(1):111

Wang Z, Yue L, Dhankher OP, Xing B (2020) Nano-enabled improvements of growth and nutritional quality in food plants driven by rhizosphere processes. Environ Int 142:105831. https://doi.org/10.1016/j.envint.2020.105831

Wang C, Liu X, Li J et al (2021) Copper nanoclusters promote tomato (*Solanum lycopersicum* L.) yield and quality through improving photosynthesis and roots growth. Environ Pollut 289: 117912

Yadav A, Yadav K, Abd-Elsalam KA (2023a) Exploring the potential of nanofertilizers for a sustainable agriculture. Plant Nano Biol 5(1):100044

Yadav A, Yadav K, Abd-Elsalam KA (2023b) Nanofertilizers: types, delivery and advantages in agricultural sustainability. Agrochemicals 2:296–336. https://doi.org/10.3390/agrochemicals2020019

Ye Y, Cota-Ruiz K, Hernández-Viezcas JA et al (2020) Manganese nanoparticles control salinity-modulated molecular responses in *Capsicum annuum* L through priming: a sustainable approach for agriculture. ACS Sustain Chem Eng 8:1427–1436

Zahra Z, Waseem N, Zahra R et al (2017) Growth and metabolic responses of rice (*Oryza sativa* L.) cultivated in phosphorus-deficient soil amended with TiO_2 nanoparticles. J Agric Food Chem 65(28):5598–5606. https://doi.org/10.1021/acs.jafc.7b01843

Zain M, Ma H, Chaudhary S et al (2023a) Nanotechnology in precision agriculture: advancing towards sustainable crop production. Plant Physiol Biochem 206:108244. https://doi.org/10.1016/j.plaphy.2023.108244

Zain M, Ma H, Nuruzzaman M et al (2023b) Nanotechnology based precision agriculture for alleviating biotic and abiotic stress in plants. Plant Stress 10:100239. https://doi.org/10.1016/j.stress.2023.100239

Zayed MM, Elkafafi SH, Zedan AM, Dawoud SF (2017) Effect of nano chitosan on growth, physiological and biochemical parameters of *Phaseolus vulgaris* under salt stress. Int J Plant Prod 8(5):577–585

Zhang Y, Qi G, Yao L et al (2022) Effects of metal nanoparticles and other preparative materials in the environment on plants: from the perspective of improving secondary metabolites. J Agric Food Chem 70:916–933. https://doi.org/10.1021/acs.jafc.1c05152

Zhao G, Zhao Y, Lou W, Su J et al (2019) Nitrate reductase-dependent nitric oxide is crucial for multi-walled carbon nanotube-induced plant tolerance against salinity. Nanoscale 11:10511–10523

Chapter 8
Nanoparticle-Based Delivery Systems for Plant Nutrients and Stress Mitigation

Jaya Mary Jacob, G. K. Shamnamol, Shynu Mary John, and P. Fathima Rini

Abstract The rapid expansion of resource-intensive lifestyles, driven by population growth, has raised serious concerns about the depletion of natural resources and the escalating impacts of pollution, climate change, and land-use changes. In response, nanobionic systems have emerged as a promising technological intervention, aiming to engineer plant functions using nanoparticles to enhance stress tolerance, photosynthesis, and overall plant growth. This chapter explores the dual role of plants in nanotechnology: as bio-factories for nanoparticle synthesis and as recipients of nanoparticles for functional enhancement. Plant-mediated biosynthesis of nanoparticles offers a cost-effective and eco-friendly approach, leveraging plant biomolecules to facilitate nanoparticle synthesis with applications in antimicrobial activity, bioimaging, biosensing, and catalysis. Among various nanomaterials, carbon quantum dots (CQDs) have gained significant attention due to their excellent conductivity, low toxicity, and remarkable optoelectronic properties. The chapter delves into CQD synthesis through hydrothermal treatment of biomass, particularly citrus waste, and their characterization using advanced spectroscopic and microscopic techniques. Furthermore, the potential of nanoparticles to be integrated into plant systems for improved physiological functions and stress resilience is examined. While the field of plant nanobionics is still evolving, dedicated research efforts are necessary to harness the full potential of nanotechnology in agriculture and environmental sustainability. This chapter provides an overview of nanobionic systems, carbon-based nanomaterials, their synthesis, and their applications, offering insights into future research directions for developing sustainable plant-nanoparticle interactions.

J. M. Jacob (✉)
Kerala State Council for Science Technology and Environment (KSCSTE), Thiruvananthapuam, Kerala, India
e-mail: jaya.6100545@kerala.gov.in

G. K. Shamnamol · S. M. John · P. Fathima Rini
Department of Biotechnology and Biochemical Engineering, Sree Buddha College of Engineering, Pattoor, APJ Abdul Kalam Technological University, Thiruvananthapuram, Kerala, India

J. M. Al-Khayri et al. (eds.), *Nanotechnology Applications in Modern Agriculture*, Nanotechnology in Plant Sciences 3, https://doi.org/10.1007/978-3-031-90506-3_8

Keywords Agro-biotechnology · Climate change · Food security · Participatory breeding · Traditional breeding

8.1 Introduction

The globe is quickly transitioning to a resource-intensive lifestyle due to population growth, and the dynamics of using the natural resources that are now available are concerning, particularly in light of the combined negative consequences of pollution, climate change, and changes in land use patterns (Jacob et al. 2021). As per recent reports, the CO_2 emissions due to pollution is expected to double in the next 30 years and policy makers in climate change management have sought the adaptation of intensive trans-disciplinary holistic approaches to tackle the problem. A noteworthy intervention is the development of Nanobionic systems that aim at engineering plant functions using nanoparticles for improved stress tolerance, photosynthesis and enhanced growth characteristics. In response to these growing issues, an abundance of technological advances and advancements have been developed recently in the field of nanobionics, notably in agriculture (Avila-Quezada et al. 2022). Nano bionics is a promising intervention and hence, to effectively contribute to this upcoming tool, it is important that a researcher understands the nano and the bionic components of nanobionics.

In this context, it is important to understand that the concept of plants being used for nanoparticle synthesis and nanoparticles being used to improve plant functions are quite distinct. For years, plants have served as bio-sources for nanoparticle synthesis. The biomolecules within plant extracts serve as effective redox platforms that aid nanoparticle synthesis and render improved functional attributes to the plant derived nanoparticles. The plant biomolecules usually form a bio-compatible layer over the nanoparticle, that often aid enhanced target specific adhesion, improved photostability etc. Also, plant-based nanoparticle synthesis is cost effective and eco-friendly. The plant derived nanoparticles have reported to have excellent ani-microbial, bio-film inhibiting, anti-corrosive, dye degrading potential in addition to the size dependent photo-emissive properties. The biocompatible coating over the plant derived nanoparticles further broaden the application of these nanoparticles in bio-imaging, bio sensing and other biological applications. The schematic of the plant mediated biosynthesis of nanoparticles is depicted in Fig. 8.1. It is noteworthy that the plant derived molecules aid the reduction of the metal precursor to the associated metal ion and further facilitates nucleation and subsequent stabilization.

Carbon quantum dots (CQDs) are a novel class of "zero-dimensional" nanomaterials that have garnered significant attention recently due to their exceptional features, which include good conductivity, minimal toxicity, and unusual optical, photoelectronic, and fluorescent performances (Jing et al. 2019). With these benefits, they are a great substitute for hazardous metallic quantum dots in photocatalysis, electrocatalysis, chemical probe, bioimaging, drug delivery, and LED applications. There are two general approaches to the manufacturing of CQDs: "top-down" and "bottom-up." Graphite, active carbon, carbon nanotubes,

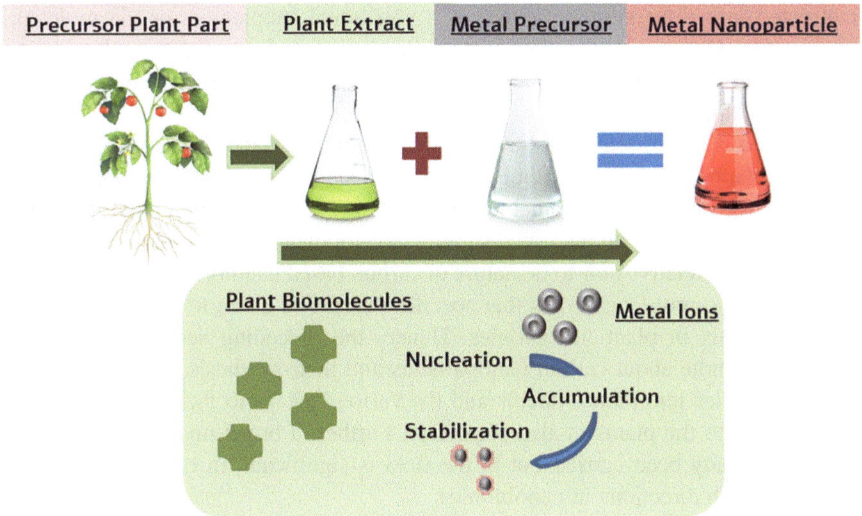

Fig. 8.1 Schematic representation of the plant mediated biosynthesis of nanoparticles

coal, soot, carbonized waste carbon paper, and biomass are examples of macroscopic carbon materials that are broken down using the top-down method. These materials are treated seriously using techniques like arc discharge, laser ablation, electronic oxidation, chemical oxidation, ultrasonic treatment, and solvothermal treatment. The process of creating CQDs from precursors such organic acid, saccharides, amino acids, and some naturally occurring items (like lemon peel and foxtail millet) using a microwave, thermal breakdown, hydrothermal/solvothermal treatment, template, and plasma is known as the "bottom-up method." Hydrothermal treatment is a potent, easy-to-use, and sustainable technology that may be used to synthesize CQDs from a variety of carbon sources in comparison to other approaches.

Among numerous precursors, biomass is the most promising one due to its low cost, renewability, abundance, environmental friendliness, and natural doping of multiheteroatom (including N, S, P, etc.). It is generally acknowledged that the production of CQDs by hydrothermal carbonization of biomass is an environmentally friendly process. Taking into consideration, the immense scope of CQDs in various applications, the present work focusses on the hydrothermal based biosynthesis of CQDs from citrus waste. Further, the biogenic CQDs would be characterized using advanced microscopic and spectroscopic techniques and the application of these technically relevant entities would be explored in photocatalysis, biosensing and in plant nano-bionics.

Until the last decade, researchers in biotechnology were largely focused on harnessing the potential of plants as sources for nanoparticle synthesis. However, with the advent of biocompatible nanoparticles, the question of using these counterparts to directly intervene and enhance plant functions arose as a new research motivation. It was hypothesized that relatively non-toxic nanoparticles could be integrated into plant systems; either as a nutritional supplement or via molecular engineering. Once inside the plant system, the nanoparticles can effectively aid plant functions and in fact provide unique sensing and stress management potential to the recipient plant.

The area is fast developing; however the potential of nano bionic systems is still largely untapped and would require more dedicated research efforts to extend the prospects for building a sustainable tomorrow. It is therefore imperative that plant scientists are aware of all the aspects in the development of a nanobionic system. In this regard, the relative non-toxic nature of carbon based nanomaterials and excellent opto-electronic properties urges that specific interest be given to the application of these materials in plant nanobionics. Hence, the preceding sections provide an introductory note about carbon nanoparticles and their synthesis. Further the entry of nanoparticles into plant systems and the various means to facilitate the entry of nanoparticle to the plants is also important. Further, a broad understanding of the research already been carried out in the field is significant, in order to streamline future research directions in nanobionics.

8.1.1 Carbon Quantum Dots

As a novel class of "zero-dimensional" nanomaterials, carbon quantum dots (CQDs) have attracted a lot of attention recently because of their exceptional properties, which include strong conductivity, low toxicity, and special optical, photoelectronic, and fluorescent capabilities. With these benefits, they are a great substitute for hazardous metallic quantum dots in applications such as drug administration, chemical probing, photocatalysis, electrocatalysis, bioimaging, and Light-emitting diode (LED) (Ashok Kumar et al. 2023; Dong et al. 2017; Fernando et al. 2015; He et al. 2020; Soumya et al. 2023; Tian et al. 2021). In recent years, luminescent carbon quantum dots (CQDs) have gained widespread attention as a new type of nanocarbon material, especially in chemical sensors, nanomedicine, and solar cells. The organic nature of these nano counterparts facilitates their entry and effective integration for biological applications, with lower degrees of rejection by the host system.

There are two main categories for the creation of CQDs: "top-down" and "bottom-up" methods (Fig. 8.2). Graphite, active carbon, carbon nanotubes, coal, soot, carbonized waste carbon paper, and biomass are examples of macroscopic carbon materials that are broken down using the top-down method. These materials are treated seriously using techniques like arc discharge, laser ablation, electronic oxidation, chemical oxidation, ultrasonic treatment, and solvothermal treatment. The process of synthesizing carbon quantum dots (CQDs) from precursors including organic acid, saccharides, amino acids, and some natural items (such lemon peel and foxtail millet) using a microwave, thermal breakdown, hydrothermal/solvothermal treatment, template, and plasma is known as the "bottom-up" approach. Hydrothermal treatment offers a simple, sustainable, and effective means to produce CQDs

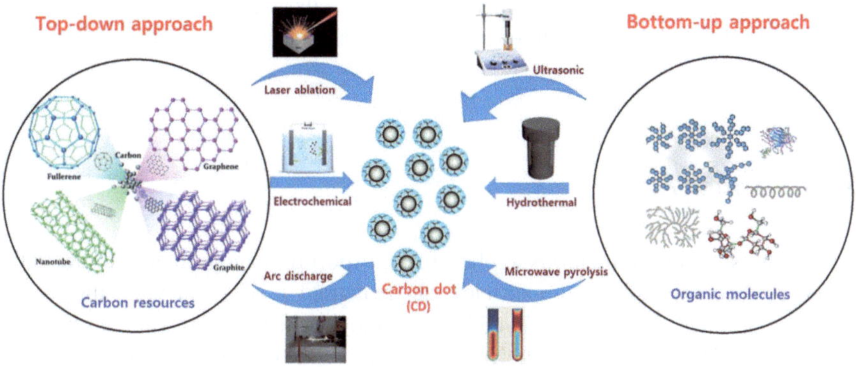

Fig. 8.2 Different methods for synthesis carbon quantum dots (Sciortino et al. 2018)

from a range of carbon sources when compared to other procedures. The most promising of the several precursors is biomass because of its abundance, low cost, renewability, environmental friendliness, and natural doping of multiple heteroatoms (such as N, S, P, etc.). In general, hydrothermal carbonization of biomass is regarded as an environmentally friendly way to produce CQDs. The engineering functions of plants can benefit greatly from the use of biogenic CQDs. The experimental results on nanoengineering plant functionalities with biogenic CQDs are elaborated in the proceedings sections. Nonetheless, it is essential to talk about issues pertaining to the entry and behavior of nanoparticles into the plant system before attempting to engineer plant functions with CQDs.

8.2 Entry of Nanoparticles Inside a Plant

After selecting appropriate nanomaterial, it is been introduced into the plant cell for altering of amplifying the suitable function inside the various cell organelles. The entry of nanoparticle and their function inside the plant cells will depends upon the concentration of nanoparticle and the type of nanoparticle which we are introducing into the plants. It is necessary to incorporate highly stable and pure nanoparticles into the physiological milieu. The introduction of nanoparticles into the cell can be accomplished in three primary ways. These three are chloroplast isolation, root entrance, and foliar entry. There are three distinct procedures in all. However, different types of nanoparticles enter different plant tissues and cells in different ways. These ways primarily involve binding to carrier proteins, binding to organic chemicals found in the environment, binding through aquaporins, ion channels, or endocytosis (Rai et al. 2018). In addition to this, the higher surface area-to-mass ratio makes the nanoparticles more reactive to their environment. As a result, they can form complexes with root exudates or membrane transporters, which make it easier for the particles to enter plant cells. Through plasmodesmata, nanoparticles can move apoplastically or symplastically across plant cells after they have entered them. From there they enter into different types cell then to organelles and interfere their functions.

8.2.1 *Foliar Entry*

It is the most common methods applied for transporting nanoparticle into the cells. When applied topically, nanoparticles can penetrate plant cells via cuticles, trichomes, stomata, stigma, and hydathodes, among other organs and tissues (Rai et al. 2018). The stomata or cuticles of the leaves are the primary entry points for the nanoparticles administered using this technique. As the main leaf barrier, the cuticle prevents particles smaller than five nanometers from entering the leaf. The plant's vascular system receives the nanoparticles that enter through the stomata and are transported throughout the cell via apoplastic and symplastic pathways. The cytoplasm of the next cell facilitates the transport of nanoparticles between 10 and 50 nm (the symplastic route). Larger nanoparticles, ranging in size from 50 to 200, are thereby translocated among the cells via the apoplastic route (Ali et al. 2021). The foliar entry is most commonly used because in the case of root entry the efficiency of nanoparticle transport decreases and the chloroplast isolation technique is quite expensive (Fig. 8.3).

8.2.2 *Root Entry*

The second most common method in the transfer of nanoparticle is root entry. When nanoparticles are absorbed by plants through their roots, they do so via the tips, lateral roots, hairs on the roots, wounds on the roots, and junctions between the roots. Small nanoparticles, with diameters ranging from 3 to 5 nanometers, can enter plant roots by means of capillary forces, osmotic pressure, or direct penetration of the root epidermal cells. Large nanoparticles are constrained by the semipermeable root cell wall epidermal cells, which have tiny pores (Avila-Quezada et al. 2022). Some nanoparticles induced new pores in the epidermal cell wall, which facilitated its entrance.

8.3 Nanoparticle in Plant Cell

In order to enter in to an organelle, nanoparticle have to overcome many barriers like some membrane molecules and some other factors. The main membrane the nanoparticle has to overcome are cell wall, plasma membrane, chloroplast membrane (Pérez-de-Luque 2017). In the cytoplasm there are lots of biomolecules which

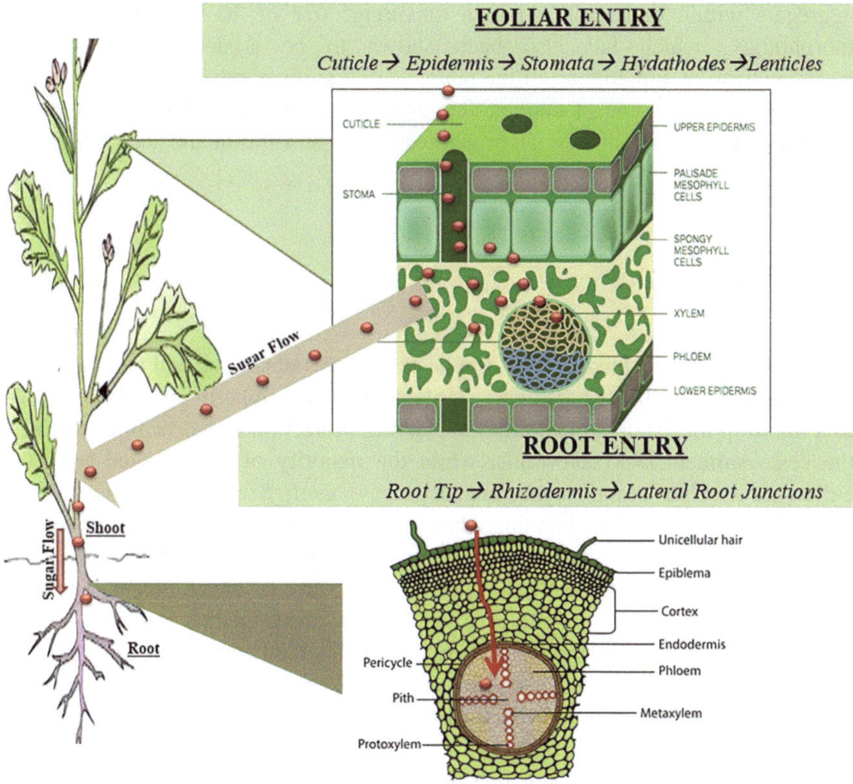

Fig. 8.3 Foliar and root entry of nanoparticles within a plant

present which will also inhibit the transfer of nanoparticles. Variables affecting a plant's ability to absorb is uptake rate, transport, and penetrate nanoparticles. That is, the characteristics of the nanoparticles influence both the application technique and how well they are absorbed and transported throughout the plant. Additionally, nanoparticles in the soil have the potential to interact with chemicals and microbes, either helping or hindering their absorption. Depending on the entry site (roots or leaves), a number of tissues (epidermis, endodermis) and barriers (Casparian strip, cuticle) must be passed before reaching the vascular tissues. Subsequently, nanomaterials have the ability to migrate along the apoplastic and/or symplastic pathways, as well as radially, to switch between the two paths.

The internalization of nanoparticles inside of cells has finally been linked to a number of processes, including pore formation, endocytosis, carrier protein-mediated internalization, and plasmodesmata. It's obvious that the characteristics of the nanoparticles will have a big impact on how it behaves and, consequently, whether the plant can absorb it. One of the primary obstacles to penetration into plant tissues appears to be size. Plants typically allow nanoparticles to migrate and aggregate within their cells up to a maximum size of 40–50 nm. Aside from morphology, which has also been shown to be a determinant in some functionalization and coating of the nanomaterial surface, other factors influencing uptake include the type of nanoparticle and its chemical composition, which can significantly alter and change the properties for the absorption and accumulation by the plant.

8.3.1 Cell Wall

Certain types of cells have a structural layer directly outside the cell membrane that is called the cell wall. It is resilient, adaptable, and occasionally inflexible. It serves as a filtration mechanism in addition to offering protection and structural support to the cell. Animals lack cell walls, while the majority of prokaryotes and other eukaryotes, such as plants, algae, and fungi, have them. Serving as pressure vessels to stop the cell from overexpanding when water enters is one of their main purposes. Different taxonomic groups and species have different cell wall compositions, which can also vary depending on the kind of cell and stage of development. The polysaccharides cellulose, hemicelluloses, and pectin make up the main cell wall of terrestrial plants. Plant cell walls are frequently embedded with or anchored by other polymers like lignin, suberin, or cutin. Land plants lack the glycoproteins and polysaccharides like agar and carrageenan that make up the cell walls of algae (Romaniuk and Cegelski 2015).

In order to enter the nanoparticle inside the plants, cell wall is the first most significant barrier that nanoparticle have to face. The average diameter of a cell wall is between 5 and 20 nanometers. There are some nanoparticle that have a diameter in between the diameter of the cell wall. Example of such nanoparticle is carbon nanotubes. The diameter of carbon nanotube is between 1 and 2 nanometer. Therefore the carbon nanotube can easily perceive directly through a spontaneous leakage into apoplast. This direct method is confronted only for small nanoparticle. If the diameter of nanoparticle is greater than that of the diameter of cell wall, another method is adopted for the transfer. That is cellulose is immobilized on the surface of the nanoparticle, these immobilization induces some local lesions on the cell wall which facilitate the transfer or uptake of nanoparticle.

8.3.2 Plasma Membrane

The biological membrane that isolates a cell's interior from the external world and shields it from it is called the cell membrane, sometimes referred to as the plasma membrane or cytoplasmic membrane. The lipid bilayer that makes up the cell membrane is composed of two layers of phospholipids with cholesterol strewn in between to maintain the proper fluidity of the membrane at different temperatures.

Additionally, membrane proteins are present in the membrane. These proteins include integral proteins, which cross the membrane and function as membrane transporters, and peripheral proteins, which cling slackly to the cell membrane's exterior and function as enzymes to promote interaction with the surrounding milieu. The function of glycolipids imbedded in the outer lipid layer is comparable. The selective permeability of the cell membrane to ions and organic molecules regulates the flow of substances into and out of cells and organelles. Living cells have cytoplasm that is surrounded by a membrane that physically divides the external environment from the intracellular components. The transfer of materials required for survival is made easier by the cell membrane's selective permeability and ability to control what enters and leaves the cell.

After passing the cell wall, the second barrier that is faced by nanoparticle is the plasma membrane. It is easier to transport across the plasma membrane than cell wall. There are three method involved in the transfer of nanoparticle through the plasma membrane. They are penetration, endocytosis, adhesion.

8.3.2.1 Penetration

It is the most common method used in the transfer of nanoparticles through the cell membrane. This only applicable if the particle is small enough. Because this process cannot induce by any agents. Penetration is a process in which the nanoparticle is simply slip through the cell membrane if the nanoparticle is small enough.

8.3.2.2 Endocytosis

It one of the most common method which employed for the nanoparticle transfer across the cell membrane. Here the membrane starts wrapping around the nanoparticle until it is fully inside. The wrapped particle is called vesicles. The endocytosis is of two types, endocytosis and semi endocytosis. In endocytosis the membrane first wrappes the nanoparticle and until it is fully inside and it goes to different organelle to alter their functions. But in the semi endocytosis, the membrane wraps the nanoparticle and it stay on the plasma membrane and it does not involve in any function or organelle other than the cell membrane.

8.3.2.3 Adhesion

It is process of used for the transfer nanoparticle. In the adhesion process the nanoparticle sticks to the membrane. It for an efficient transfer it is not used, because it does not moves further, it only stay on the membrane. The reason behind the adhesion of nanoparticle on the membrane is due to the opposite charges of nanoparticle and the membrane. Cationic nanoparticle is used here since the charge of the lipid bilayer is negative.

8.3.2.4 Cytosol

The cytoplasm of a eukaryotic cell includes the mitochondria, plastids, and other organelles (but not their internal fluids and structures); the cell nucleus is located outside of this cytoplasm. The cytosol is encircled by the cell membrane. Therefore, the cytosol surrounds the organelles like a liquid matrix. While many metabolic processes still take place in the cytosol in eukaryotes, some metabolic processes happen inside organelles. Large concentrations of macromolecules, known as macromolecular crowding, are also present in the cytosol and can change the behavior of molecules. The cytosol is the location of several different cellular processes rather than having a single purpose. These processes include the transmission of signals from the cell membrane to organelles or other internal locations like the cell nucleus. Following the nuclear membrane's disintegration during mitosis, this compartment is also the location of numerous cytokinesis processes. Transporting metabolites from the place of production to the site of usage is one of the cytosol's other primary roles.

The nanoparticle penetrates the cytosol and travels through the cell wall and membrane before entering various organelles and interfering with their normal functioning. Once inside various organelles, nanoparticles modify or enhance their activities. Here, increasing photosynthetic efficiency is our goal, hence nanoparticles must penetrate the chloroplast.

8.3.3 Chloroplast Membrane

A chloroplast is a kind of plastid, a membrane-bound organelle that is mostly found in plant and algal cells and is responsible for photosynthesis. While releasing oxygen from the water in the cells, the photosynthetic pigment chlorophyll absorbs solar energy, transforms it, and stores it in the energy-storage molecules ATP and NADPH. The Calvin cycle, which uses ATP and NADPH to convert carbon dioxide into organic molecules, is the next step. Other tasks performed by chloroplasts include the synthesis of fatty acids, many amino acids, and the plant immune system. Chloroplasts in land plants are typically lens-shaped, measuring 3–10 μm in diameter and 1–3 μm in thickness. There are three membrane systems present in all chloroplasts: the thylakoid system, the inner membrane, and the outer membrane. The thylakoid system floats in the chloroplast stroma, a semi-gel-like fluid that makes up a large portion of a chloroplast's volume and is located inside the outer and inner chloroplast membranes.

Chloroplast have double layered lipid bilayer composed of galactolipids and phospholipids in different composition. Highly positively and negatively charged nanoparticles interact with the exposed lipid in the membrane, allowing diffusion and eventual kinetic trapping into the chloroplast (Newkirk et al. 2021). Carbon nanotubes and carbon dots have high efficiency of penetrating to the chloroplast.

After entering into chloroplast, the nanoparticle interfere the photosynthetic machinery and alter their properties in order to improve photosynthetic efficiency. The improvement of efficiency of photosynthesis depends upon which nanoparticle we are introducing in the chloroplast.

8.4 Nanobionics in Photosynthesis

An essential process for plants on Earth is photosynthesis, which converts light energy into chemical energy. Plants converts only 2–4% of the available energy in radiation into new growth. For speeding up the photosynthesis, nanoparticle is used. Plants that contain nanoparticle is called nanobionic plants. Nanobionic plants have a better photosynthesis efficiency and biochemical sensing. Additionally, the photosynthetic activity of the nanobionic plant is three times higher than that of regular plants, and its maximal electron transfer rates are also increased. Photosynthetic organelles can now have new and improved functional features thanks to nanobionic plants. This study addresses the biosynthesis of CQDs and examines its effect on the growth of mung bean sprouts (Vigna radiata) by employing dried leaves as a precursor (Wang et al. 2018). The biogenic CQDs would be characterized using advanced microscopic and spectroscopic techniques.

The leaves of tamarind (Tamarindus indica) and nutmeg plants (Myristica fragrans) were collected from the surrounding areas. The obtained leaves were initially washed in running tap water and then with distilled water. Then, the leaves were then thoroughly dried in the sun until 85% of the moisture was removed. The resultant dried leaves were ground into powder using a blender and then used for further processing.

8.5 Fabrication of Carbon Quantum Dots by Hydrothermal Treatment

The leaves washed and dried in the sunlight initially for 2 days, and then it was kept in the hot air oven at 150 °C for 2 hours. The dehydrated leaves were blended into a powder. In short, 100 mL of double deionized water (DDW) was combined with 2.5 g of each sample (dry leaf powder from tamarind and nutmeg). The mixture was then put into a 150 mL autoclave walled with poly(tetrafluoroethylene) and heated at 200 °C for 6 hours in a hot air oven. After the procedure, the autoclave was cooled to room temperature. To prepare it for use later, the brown solution was diluted to 200 milliliters and filtered using a 0.45 μm membrane filter.

8.6 Characterization of Carbon Quantum Dots

The CQDS are characterized by their chemical composition, size, shape and structure; therefore numerous efforts/tools have been developed to obtain reliable and robust characterization for C-dots. With the use of spectroscopic methods like UV-Vis spectroscopy and fluorescence spectrophotometer, the optical characteristics of the CQDs were recorded. According to a survey of the literature, all varieties of carbon dots, are active in the UV-Vis portion of the electromagnetic spectrum, and the behavior of their fluorescence emission is excitation wavelength dependent. Given that nanoparticles have optical properties that are sensitive to form, concentration, refractive index, size, and agglomeration, UV-Vis spectroscopy is a crucial tool for characterizing, identifying, and researching these materials. The carbon core's π-π* transitions of C=C bonds are responsible for the UV absorption peak of C-Dots, which is centered about 230–270 nm. On the other hand, the n–π* transition of C=O surface groups is usually responsible for the absorption that occurs between 300 and 340 nm (Rooj and Mandal 2023).

The morphology and size characterization of CQDs have been studied using several nondestructive imaging and microscope techniques. The microscopic techniques that have been developed for measurements of C-dots include transmission electron microscopy (TEM), High resolution TEM (HRTEM) and X-ray photoelectron spectroscopy (XPS). TEM analysis can be used to evaluate the size and shape of the C-Dots which has an average size of <10 nm. TEM images can also be used to investigate particle agglomeration or dispersion. In recent years, high-resolution TEM (HRTEM) has been extensively and successfully used to an analyze C-dot structures and lattice imperfections using both transmitted and scattered beams. The surface properties of C-Dots are examined by X-ray photoelectron spectroscopy (XPS). Moreover, it can be used to evaluate the elements in C-Dots such as nitrogen and sulfur, as well as vibration modes of bonding such as OH, COOH, and NH_2 forming between elements (Lim et al. 2018).

8.7 Development of CQD Engineered *Vigna Radiata*

To further assess the potential of CQDs in the activation of plant growth of *Vigna Radiata*, we carried out an experiment by using two sets of mung bean seeds grown in different media. Set 1 plants are cultivated in aqueous medium, whereas Set 2 plants are grown in soil media. Mung bean seeds of identical size were used as the experiment model in both groups. After that, seeds from each set were soaked in equal parts pure water and CDs solution (0.02–0.12 mg/mL) in beakers, and they were grown for 3 days at 25 °C without light. Every day, we replaced the CDs solution in the experimental group and pure water in the control group. After 3 days, the Set 1 mung bean beakers were put in a constant light incubator at 25 °C for 3 days, and the Set 2 sprouts were put in plastic pots with five seeds each pot of

0.75 kg of dried soil. The plants were grown in controlled settings for 7 days. The effects of CQDs on plant growth, including variables like shoot length, root length, leaf area and leaf chlorophyll content, were also assessed.

Root and shoot lengths were measured manually with a ruler for growth assessment. Subsequently, each sample was taken after 6 days and the fresh weight (FW) of mung beans was determined. The leaf area was determined using ImageJ software. As for the pigments, chlorophylls a and b (Chl a and Chl b) were estimated in fresh leaves using a previously established technique (Lichtenthaler and Wellburn 1983).

8.8 Measurement of Chlorophyll Content

The youngest fully developed leaves were homogenized in ice-cold 100% (v/v) acetone (1.5 mL for a 250-mg sample) and extracted for 24 hours at 4 °C, samples were centrifuged at 5000 g for 15 min (Wodala et al. 2012). The pellet was extracted again with 80% (v/v) acetone (1.5 mL for a 250-mg sample) for 24 hours. The supernatants were collected after centrifugation (5000 g, 15 min, 4 °C).

$$\text{Chla (mg/L)} = 13.95 \times OD_{665} - 6.88 \times OD_{649}$$

$$\text{Chlb (mg/L)} = 24.96 \times OD_{649} - 7.32 \times OD_{665}$$

$$C_T\ \text{(mg/L)} = (Chla + Chlb) \times ilution\ ratio$$

8.9 Results and Discussion

The CQDs were fabricated by the carbonization of dried leaves residues through an easy as well as cost-effective hydrothermal fabrication approach for their application to the growth of *Vigna Radiata.*

8.9.1 *Fabrication of Carbon Quantum Dots by Hydrothermal Treatment*

Initially, in the present study, attempts were made to synthesis CQDs from dried leaves at temperature and pressure ranges below hydrothermal conditions. As depicted in Fig. 8.4, a visible colour change from pale green to brown was visible for the synthesis reactions initiated for precursors under mild conditions. This preliminary observation has been reported earlier by researchers as a first-line indication for the formation of CQDs.

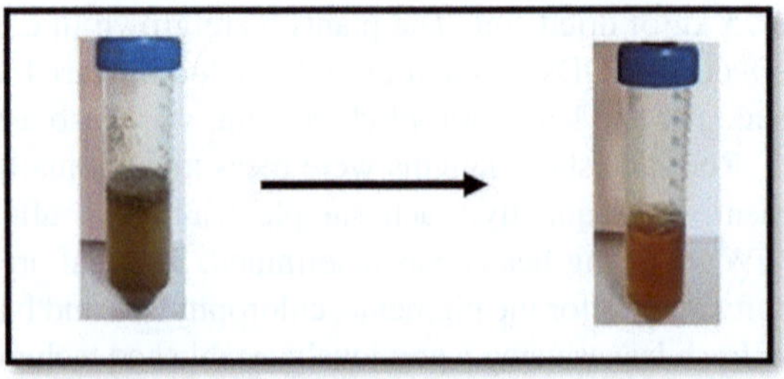

Fig. 8.4 Visible colour change from pale green to brown indicating synthesis of CQDs. (Data from the authors' work)

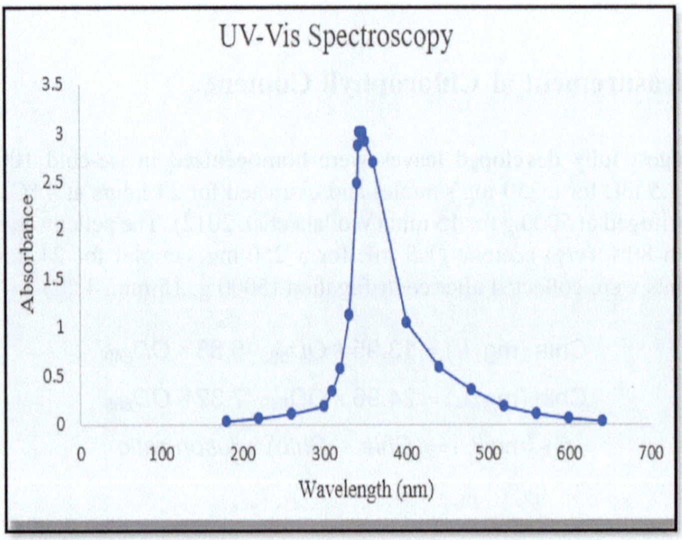

Fig. 8.5 Absorption spectra of the CQDs. (Data from the authors' work)

8.9.2 Characterization of Carbon Quantum Dots

8.9.2.1 UV-Vis Spectroscopy

However, to confirm the synthesis, the absorbance data of the as-synthesised CQDs were recorded using a UV-Visible spectrophotometer. Figure 8.5 shows the absorption spectra of the CQDs. It could be noted that dried leaves residues show an absorption peak at 349 nm, and according to L C Sim and his coworkers, CQDs exhibit absorption maxima at similar wavelength ranges (Sim et al. 2020). However, these results indicate the probable synthesis of CQDs, observations like the apparent aggregation of the resultant red-brown solution on storage and the broad absorption spectra appeals to optimise the operating conditions needed for synthesis. However, based on preliminary studies, it was concluded that lemon peel residues, when subjected to high temperature and pressure-mediated carbonisation using alternative approaches like hydrothermal treatment, could yield CQDs with uniform sizes and other optical properties. The figure presents the UV-Vis absorption spectra of the CQDs initiated by hydrothermal conditions. Also, it depicts the UV-Vis absorption spectra of CDs solution in the range of 200–800 nm. The CDs have only one absorption peak at 349 nm, which is attributable to n-π^* transition of C=O bonds in the CDs.

8.9.2.2 Fluorescence Spectra of CQDs Synthesis

The Excitation dependent fluorescence spectra of CQDs synthesis initiated at hydrothermal conditions. CQDs exhibited maximum fluorescent emission at 520 nm upon the excitation at 400 nm. The findings of this study are comparable to those of H. Wang, which showed that fluorescence spectra of CDs exhibited strongest emission located at 457 nm in the 330 nm excitation (Wang et al. 2017). Figure 8.6 shows the fluorescence spectra of CQDs synthesised.

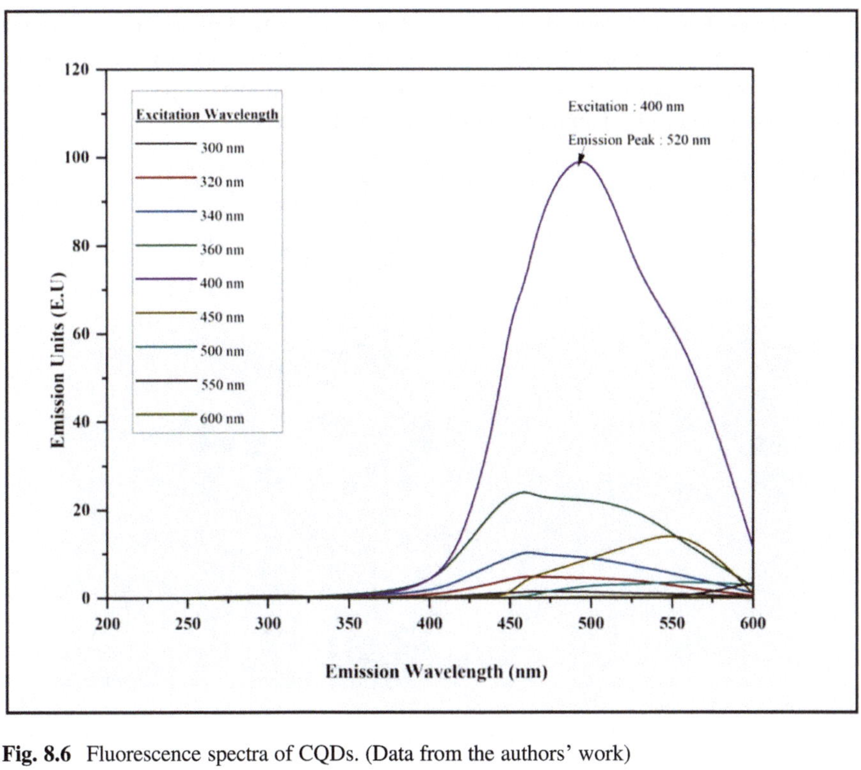

Fig. 8.6 Fluorescence spectra of CQDs. (Data from the authors' work)

8.9.2.3 High-Resolution Transmission Electron Microscopy (HRTEM)

After that, dried leaf residues were utilized to create CQDs using a one-step hydrothermal method. TEM was used to characterize the sizes and morphologies of the resulting CQDs, as shown in Fig. 8.7a. The synthesized CQDs had a spherical morphology with an average diameter of 4.8 nm, were uniformly distributed, and were isolated from one another. As seen in Fig. 8.7b, the high-resolution TEM picture revealed that the CQDs had an interplanar spacing of 0.21 nm, which is consistent with graphite (100) diffraction planes. The findings showed that the CQDs match nanocrystalline cores of sp2 graphitic carbon (Sharma et al. 2020).

8.9.2.4 X-Ray Photoelectron Spectroscopy (XPS)

XP spectra were recorded to characterise the surface functional groups and contents of atoms in the generated CQDs in Fig. 8.8a. The XP spectrum showed three typical peaks at 285, 400, and 531 eV, attributed to C 1s (eV), N 1s (eV), and O 1s (eV), respectively [14] in Fig. 8.8a. The relative atomic percentages of these significant elements were calculated (C (64.32%), O (25.11%), and N (5.69%)). The C 1s spectrum in Fig. 8.8b showed three peaks at 284.5, 285.8, and 287.6 eV, attributed to C–C/C=C, C–N/C–O, and C=O, respectively. The N 1s spectrum in Fig. 8.8c can be deconvoluted into multiple peaks in the range 404–396 eV that represent pyridinic N, amino N and pyrrolic N. The deconvolution of the O 1s spectrum. The pigment composition was determined with a double-beam spectrophotometer using the method of Lichtenthaler and Wellburn (1983).

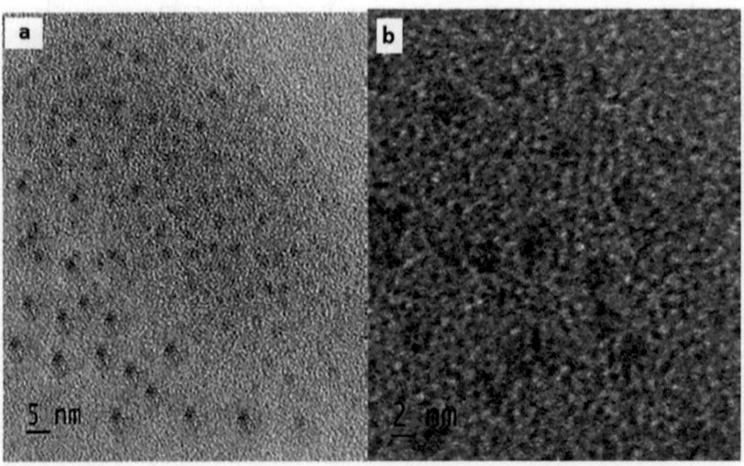

Fig. 8.7 TEM images to characterize the sizes and morphologies of CQDs. (Data from the authors' work)

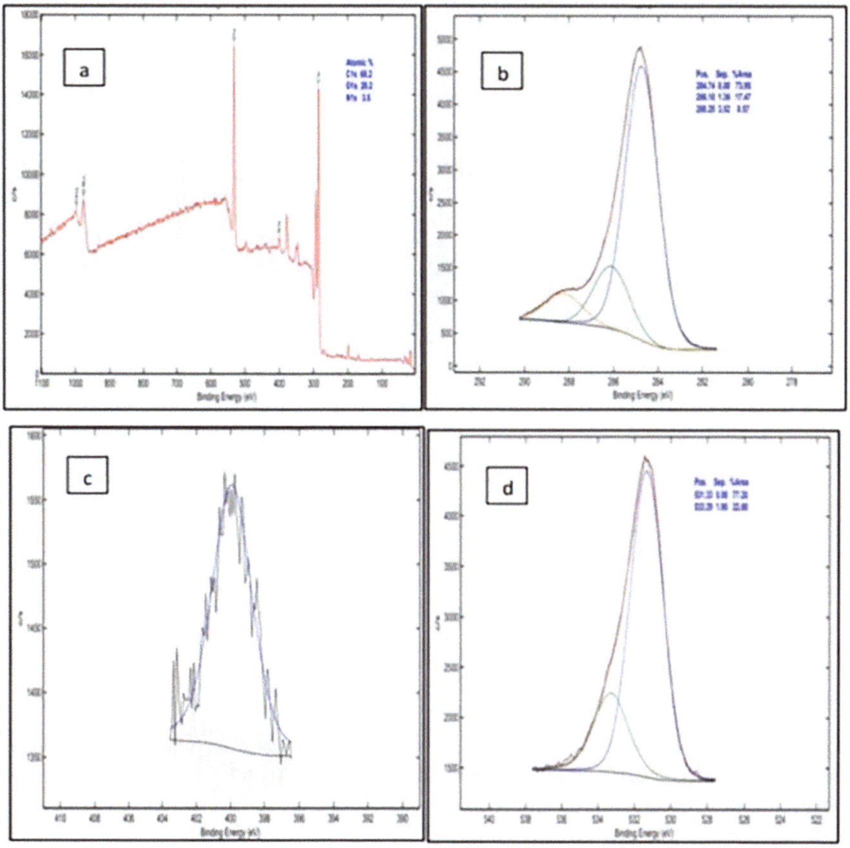

Fig. 8.8 X-ray photoelectron spectra to characterize the surface functional groups and contents of atoms in the generated CQDs

8.9.3 Development of CQD Engineered Vigna Radiata

A series of experiments were conducted to determine whether carbon dots affect plant growth and development. Mung beans were chosen as the model plant to study the physiological effects of CDs on plants (Wang et al. 2018). The mung bean was cultured with CDs. Moreover, seeds were incubated with different concentrations of CDs solution (experiment group) and pure water (control group), respectively. Mung bean sprouts were harvested after 6 days of incubation.

Three measurements were made of the mung bean sprouts' biomass, root length, and stem length in order to get precise results regarding the aqueous medium mung bean culture. The root elongation was found to be 27 dependent on the concentration of CDs, as indicated in Fig. 8.9. In comparison to the control, the roots of mung bean subjected CDs (0.02 mg/mL and 0.04 mg/mL) are longer by 67.1% and 58.3%,

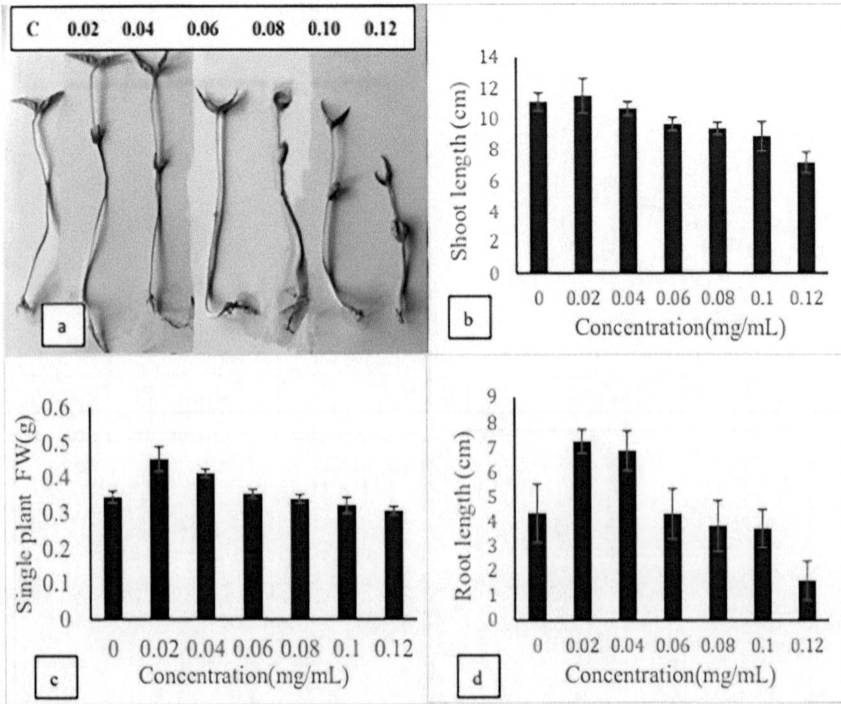

Fig. 8.9 Measurement of plant growth characteristics in aqueous medium (**a**) photograph of mung beans exposed to CDs after 6 days (**b**) Shoot length (**c**) Single plant fresh weight [FW] and (**d**) Root length

respectively. Figure 8.9d illustrates how CDs administration has a beneficial effect on growth when concentration is less than 0.02 mg/mL and increases with concentration. CDs can no longer stimulate the growth of mung bean sprouts at concentrations of 0.12 mg/mL. The ideal concentration of CDs for mung bean sprout growth is 0.02 mg/mL, which greatly enhances the root, stem, and biomass length of the sprouts.

Additionally, as seen in Fig. 8.9b, CDs have a comparable effect on stem elongation. Mung bean exposed CDs have stem lengths that are 3.96% and 3.6% longer than the control, respectively, at 0.02 mg/mL and 0.04 mg/mL. At high CD concentrations, stem elongation does not appear to be regulated by CDs. The buildup of biomass is a crucial plant development indicator. A notable increase in the total fresh biomass of mung bean sprouts treated with CDs (0.02 mg/mL and 0.04 mg/mL) is approximately 28.57% and 17.14% higher than control, respectively. The single plant fresh weight of mung bean sprouts is displayed in Fig. 8.9c, indicating that the presence of CDs can accelerate the biomass accumulation. Thus, our research indicates that CDs may enhance mung bean sprout growth.

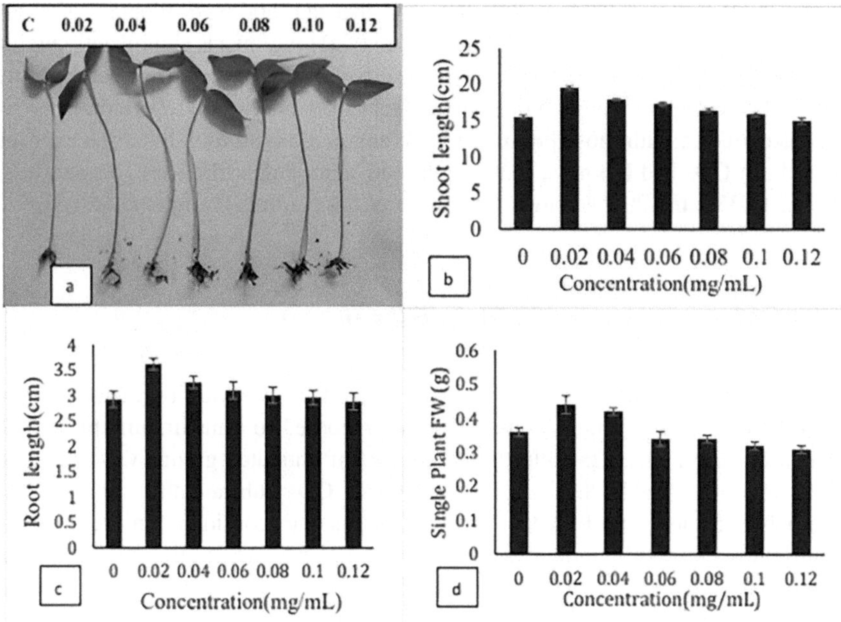

Fig. 8.10 Measurement of plant growth characteristics in soil media (**a**) photograph of mung beans exposed to CDs after 6 days (**b**) Shoot length (**c**) Single plant fresh weight [FW] and (**d**) Root length

Similarly, we grew mung beans in soil medium treated with CDs solution and investigated the impact of CQDs on plant growth characteristics such as individual plant fresh weight, shoot length, and root length. In addition, the amount of chlorophyll and leaf area were measured. A pigment called chlorophyll is utilized in photosynthesis to transform light energy into chemical energy. Plants with higher chlorophyll contents absorb light more efficiently and produce more carbohydrates during photosynthesis. Using a double-beam spectrophotometer and the Lichtenthaler and Wellburn technique, the chlorophyll content was ascertained. The concentration of CDs had an effect on root elongation, as seen by Fig. 8.10c. Furthermore, it indicates that CDs administration has a beneficial effect on growth when concentration is less than 0.02 mg/mL and increases with concentration. At 0.12 mg/mL, CDs are no longer able to promote the growth of mung bean sprouts. The ideal concentration of CDs for mung bean sprout growth is 0.02 mg/mL, which greatly enhances the root, stem, and biomass length of the sprouts.

Moreover, as Fig. 8.10b illustrates, CDs have a comparable effect on stem elongation. The stem lengths of the CDs subjected to mung beans (0.02 mg/mL and 0.04 mg/mL) are, respectively, 26.4% and 15.9% longer than the control. CDs do not seem to influence stem elongation, even at high quantities. The buildup of biomass is a crucial sign of plant development. Figure 8.10d depicts the fresh weight (FW) of a single plant of mung bean sprouts. It indicates that the presence of CDs can hasten the accumulation of biomass, and that the total fresh biomass of mung bean sprouts treated with CDs (0.02 mg/mL and 0.04 mg/mL) is significantly higher than the control, at 16.6% and 22.2%, respectively. Therefore, our research shows that CDs can improve the growth of mung bean sprouts. The study's findings are in good accordance with those published by Wang et al., who found that the exposed mung bean CDs had biomass, root length, and stem length that were, respectively, 14.9%, 18.3%, and 29.9% longer than those of the control (Wang et al. 2018).

8.9.4 Measurement of Chlorophyll Content

The chlorophyll content in untreated and treated groups was estimated. It was observed that 0.02 mg/mL CD treated group recorded to a maximum chlorophyll content of 10.6 mg/L. The chlorophyll content of untreated groups was comparatively very less (Table 8.1). Higher content of CDs substantially improved the chlorophyll content in all plants; however, taking into consideration the decrease in attributes like root length, shoot length etc. higher concentration of CDs is not preferred for the plant growth. Similar observations were reported by Zhang et al. on algal growth (Chlorella vulgaris) where the Chlorophyll content of CDs treated group was 12.6 mg/L (Zhang et al. 2018).

8.9.5 Measurement of Leaf Area in Different Concentrations of CDs

Chlorophyll is the major component present on the plant. To further validate our results, leaf area of all treated and untreated samples was measured and statistically analyzed samples using image J software. The results are shown in Table 8.2. It can be inferred that for 0.02 concentration, area is 5.57 cm^2. These findings further confirm that 0.02 mg/mL of CQDs results in enhanced plant growth, chlorophyll levels and leaf area. These results can also indicate an increased photosynthetic potential of the CQD treated plants.

Table 8.1 Measurement of Chlorophyll content in different concentrations of CDs

Concentration	At 649 nm	At 665 nm	CA	CB	CT
Control	0.008	0	0.1	0.2	0.3
0.02	0.337	0.178	3.5	7.1	10.6
0.04	0.194	0.054	2.3	4.4	6.8
0.06	0.109	0.057	1.1	2.3	3.4
0.08	0.28	0.229	2.3	5.3	7.6
0.10	0.103	0.02	1.3	2.4	3.7
0.12	0.139	0.06	1.5	3.0	4.6

Table 8.2 Area of leaf in different concentrations of CDs

Label	
Concentration (mg/mL)	Area of leaf (cm^2)
Control	4.89
0.02	5.57
0.04	4.52
0.06	3.78
0.08	4.29
0.10	4.28
0.12	3.12

8.10 Conclusion

Plant nano-bionics is an upcoming research arena that holds immense possibilities to uplift plant biotechnology for creating impactful interventions in agriculture. Preliminary research work in this area using bio-derived carbon quantum dots show exciting results that open wide applications in agricultural biotechnology. The as-synthesized CQDs demonstrated remarkable fluorescence and optical properties and were found to be biocompatible and hence extending their applications for biological and sensing purposes. The plant derived-CQDs exhibited uniform size distribution in nano ranges, and the sizes were well below 10 nm, i.e. in the quantum size regime. Further the treatment of mung bean sprouts with biosynthesized CDs was found to have a significant impact on the growth of the plant. In particular, the root elongation, stem elongation, and biomass of mung beans in different concentrations of different CDs demonstrate a dose-response relationship. Hence the study is of immense application as it opens up the development of new methods based on green synthesized CQDs to improve the photosynthetic carbon dioxide reduction potential of the selected plants. Future research may concentrate on exploring the photosynthetic responses of the nano engineered plants and the development of a promising strategy for promoting the development of novel systems in the field of plant nanobionics.

References

Ali S, Mehmood A, Khan N (2021) Uptake, translocation, and consequences of nanomaterials on plant growth and stress adaptation. J Nanomater 2021. https://doi.org/10.1155/2021/6677616

Ashok Kumar S, Dheeraj Kumar M, Saikia M, Renuga Devi N, Subramania A (2023) A review on plant derived carbon quantum dots for bio-imaging. Mater Adv 4(18):3951–3966. https://doi.org/10.1039/d3ma00254c

Avila-Quezada GD, Ingle AP, Golińska P, Rai M (2022) Strategic applications of nano-fertilizers for sustainable agriculture: benefits and bottlenecks. Nanotechnol Rev 11(1):2123–2140. https://doi.org/10.1515/ntrev-2022-0126

Dong S, Yuan Z, Zhang L, Lin Y, Lu C (2017) Rapid screening of oxygen states in carbon quantum dots by chemiluminescence probe. Anal Chem 89(22):12520–12526. https://doi.org/10.1021/acs.analchem.7b03711

Fernando KAS, Sahu S, Liu Y, Lewis WK, Guliants EA, Jafariyan A, Wang P, Bunker CE, Sun YP (2015) Carbon quantum dots and applications in photocatalytic energy conversion. ACS Appl Mater Interfaces 7(16):8363–8376. https://doi.org/10.1021/acsami.5b00448

He P, Shi Y, Meng T, Yuan T, Li Y, Li X, Zhang Y, Fan L, Yang S (2020) Recent advances in white light-emitting diodes of carbon quantum dots. Nanoscale 12(8):4826–4832. https://doi.org/10.1039/c9nr10958g

Jacob JM, Ravindran R, Narayanan M, Samuel SM, Pugazhendhi A, Kumar G (2021) Microalgae: a prospective low cost green alternative for nanoparticle synthesis. Curr Opin Environ Sci Health 20:1–5. https://doi.org/10.1016/j.coesh.2019.12.005

Jing S, Zhao Y, Sun R, Zhong L, Peng X (2019) Facile and high-yield synthesis of carbon quantum dots from biomass-derived carbons at mild condition [Research-article]. ACS Sustain Chem Eng 7:7833–7843. https://doi.org/10.1021/acssuschemeng.9b00027

Lichtenthaler HK, Wellburn AR (1983) Determinations of total carotenoids and chlorophylls a and b of leaf extracts in different solvents. Biochem Soc Trans 11(5):591–592. https://doi.org/10.1042/bst0110591

Lim H, Liu Y, Kim HY, Son DI (2018) Facile synthesis and characterization of carbon quantum dots and photovoltaic applications. Thin Solid Films 660(April):672–677. https://doi.org/10.1016/j.tsf.2018.04.019

Newkirk GM, de Allende P, Jinkerson RE, Giraldo JP (2021) Nanotechnology approaches for chloroplast biotechnology advancements. Front Plant Sci 12(July). https://doi.org/10.3389/fpls.2021.691295

Pérez-de-Luque A (2017) Interaction of nanomaterials with plants: what do we need for real applications in agriculture? Front Environ Sci 5(Apr):1–7. https://doi.org/10.3389/fenvs.2017.00012

Rai M, Ingle AP, Pandit R, Paralikar P, Shende S, Gupta I, Biswas JK, Da Silva SS (2018) Copper and copper nanoparticles: role in management of insect-pests and pathogenic microbes. Nanotechnol Rev 7(4):303–315. https://doi.org/10.1515/ntrev-2018-0031

Romaniuk JAH, Cegelski L (2015) Bacterial cell wall composition and the influence of antibiotics by cell-wall and whole-cell NMR. Philos Trans R Soc B: Biol Sci 370(1679). https://doi.org/10.1098/rstb.2015.0024

Rooj B, Mandal U (2023) A review on characterization of carbon quantum dots. 61(6):693–718. https://doi.org/10.1002/vjch.202300022

Sciortino A, Cannizzo A, Messina F (2018) Carbon nanodots: a review—from the current understanding of the fundamental photophysics to the full control of the optical response. C 4(4):67. https://doi.org/10.3390/c4040067

Sharma N, Das GS, Yun K (2020) Green synthesis of multipurpose carbon quantum dots from red cabbage and estimation of their antioxidant potential and bio-labeling activity. Appl Microbiol Biotechnol 104(16):7187–7200. https://doi.org/10.1007/s00253-020-10726-5

Sim LC, Khor JM, Leong KH, Saravanan P (2020) Synthesis and characterization of dried leaves derived carbon quantum dots and g-C3N4 composite. IOP Conf Ser Mater Sci Eng 894(1). https://doi.org/10.1088/1757-899X/894/1/012003

Soumya K, More N, Choppadandi M, Aishwarya DA, Singh G, Kapusetti G (2023) A comprehensive review on carbon quantum dots as an effective photosensitizer and drug delivery system for cancer treatment. Biomed Technol 4(November 2022):11–20. https://doi.org/10.1016/j.bmt.2023.01.005

Tian L, Li Z, Wang P, Zhai X, Wang X, Li T (2021) Carbon quantum dots for advanced electrocatalysis. J Energy Chem 55:279–294. https://doi.org/10.1016/j.jechem.2020.06.057

Wang R, Lu KQ, Tang ZR, Xu YJ (2017) Recent progress in carbon quantum dots: synthesis, properties and applications in photocatalysis. J Mater Chem A 5(8):3717–3734. https://doi.org/10.1039/c6ta08660h

Wang H, Zhang M, Song Y, Li H, Huang H, Shao M, Liu Y, Kang Z (2018) Carbon dots promote the growth and photosynthesis of mung bean sprouts. Carbon 136:94–102. https://doi.org/10.1016/j.carbon.2018.04.051

Wodala B, Eitel G, Gyula TN, Ördög A, Horváth F (2012) Monitoring moderate Cu and Cd toxicity by chlorophyll fluorescence and P 700 absorbance in pea leaves. Photosynthetica 50(3): 380–386. https://doi.org/10.1007/s11099-012-0045-3

Zhang M, Wang H, Song Y, Huang H, Shao M, Liu Y, Li H, Kang Z (2018) Pristine carbon dots boost the growth of chlorella vulgaris by enhancing photosynthesis. ACS Appl Bio Mater 1(3): 894–902. https://doi.org/10.1021/acsabm.8b00319

Chapter 9
Engineered Nanomaterials for Abiotic and Biotic Stress Tolerance in Plants

Mawuli K. Azameti, Muazu Issifu, Abdul-Wahab Imoro, Wadzani Palnam Dauda, and Kishor Prabhakar Panzade

Abstract With increasing plant biotic and abiotic stresses, there is a growing need for innovative strategies to increase plants' tolerance to these stresses. To address some of the biotic and abiotic challenges, nanotechnology has come to light as a fascinating field of study with possible applications in agriculture. Numerous studies indicate that nanomaterials play roles in the management of plants' stress. In recent times, large-scale production of nanoparticles and nanostructured materials has gained a lot of notice, as they are essential to several sectors, including the agricultural sector. Engineered nanomaterials have shown great potential in enhancing the growth and stress tolerance of plants under unfavorable conditions. Nanomaterials have unique properties that make them ideal tools for addressing plant stress tolerance. These properties include their small size, large surface area, high reactivity, and ability to penetrate plant tissues effectively. Furthermore, engineered nanomaterials interact with plants in specific ways, such as delivering nutrients or

M. K. Azameti (✉)
Department of Applied Biology, C. K. Tedam University of Technology and Applied Sciences, Navrongo, Ghana
e-mail: mkazameti@cktutas.edu.gh

M. Issifu
Department of Agriculture Biotechnology, University for Development Studies, Nyankpala Campus, Tamale, Ghana
e-mail: muazu.issifu@student.jkuat.ac.ke

A.-W. Imoro
Department of Forestry and Forest Resources Management, University for Development Studies, Tamale, Ghana
e-mail: iwahab@uds.edu.gh

W. P. Dauda
Crop Science Unit, Department of Agronomy, Federal University, Gashua, Nigeria
e-mail: wadzanidaudap@fugashua.edu.gh

K. P. Panzade
Department of Plant Biotechnology, SDMVM College of Agricultural Biotechnology, Georai Tanda, Maharashtra, India
e-mail: kppanzade@coabgt.org

J. M. Al-Khayri et al. (eds.), *Nanotechnology Applications in Modern Agriculture*, Nanotechnology in Plant Sciences 3, https://doi.org/10.1007/978-3-031-90506-3_9

enhancing antioxidant capacity. Incorporating nanomaterials into plant systems has been found to enhance plant resistance to various abiotic stresses. The use of engineered nanomaterials helps to augment non-native properties in plants, leading to enhanced stress tolerance. In effect, engineered nanomaterials hold great promise for enhancing abiotic stress tolerance in plants. This chapter highlights the synthesis of nanomaterials and the significant potential of engineered nanomaterials in enhancing biotic and abiotic stress tolerance in plants.

Keywords Abiotic · Biotic · Engineered · Nanomaterials · Plants · Stress · Tolerance · Nanomaterial · Nanotechnology · Heat · Drought · Heavy metals · Nanofertilizers · Salt stress · Stunted · Growth · Nutrient · Deficiency · Oxidative stress · Hormones · Imbalances · Growth · Photosynthesis · Bottom-up · Top-down · Hybrid · Mechanical · Physical · Chemical · Biological · Physicochemical · Nanofabrication · Bacteria · Fungi

9.1 Introduction

Biotic and abiotic stressors negatively affect global agricultural production in the present period. Based on the crop type, kind of stress, and duration of exposure, plants respond to stress through changes in a variety of morphological, physiological, and biochemical processes (Abd El-Mageed et al. 2022; Rady et al. 2021a; Shaaban et al. 2022). Researchers are working to improve agricultural yield and food production, reduce the adverse impacts of biotic and abiotic stresses, and attain sustainability and food security to fulfil the growing worldwide demand for food. Therefore, scientists need to keep creating novel technologies or solutions to deal with these pressing global issues.

Rapidly expanding, nanotechnology has produced many advancements (Abd El-Ghany et al. 2021; El-Saadony et al. 2020). Specifically, agriculture-related issues may be effectively resolved with nanotechnology, ensuring a safe and sustainable agricultural future.

Numerous everyday uses exist for nanotechnology. The agricultural sectors began using nanotechnology in various ways since 2003 (Xing et al. 2016). Studies involving nanotechnology have increased tremendously in the last decade. Among the many benefits of nanotechnology is enhanced food safety and quality (Nandini et al. 2017). By employing nanotechnology, scientists may create particles with special, adaptable properties for use in agriculture and other fields (Hochella et al. 2019). Engineered nanomaterials (ENMs) are materials with a maximum size of 100 nm. Their purpose is to possess specific physicochemical properties that can be utilized in several ways, including nano-fertilizers (Bandala and Berli 2019). Metallic oxides (such as Au, Ag, Fe_3O_4, and TiO_2), polymers, carbon ENMs (such as fullerenes, carbon nanotubes, and graphene), hybrid particles, and polymer nanocomposites are a few representative forms of ENMs (Dontsova et al. 2019). The most crucial aspects of ENMs are their size and shape, which control the kind and distribution of reactive sites in chemical reactions as well as their reactivity and transport mechanism. Changes in surface structure affect the extent to which nanomaterials react in the air, water, soil, or biota because of the high surface area-to-volume ratio (Coman et al. 2019). Dissolution, evaporation, and aggregation are three environmental conditions that affect ENMs (Hochella et al. 2019).

Over the past years, scientists have been investigating cutting-edge technologies, such as nanotechnology, to meet the population demands, and solve the world's agricultural problems (FAO 2017). Crop insecticides and fertilizers might be created using nanotechnology (Ammar 2018). By encasing ENMs in other materials, nutrient absorption is improved in plants, resulting in a reduction in agrochemical waste. For effective and regulated nutrient delivery, for instance, nano-encapsulated compounds offer improved stability and solubility (Shang et al. 2019). Encapsulated materials are kept from seeping into the ground by the barrier-like effect of nano capsules. Combined with other biological components, they may increase a plant's tolerance to abiotic challenges (Shang et al. 2019).

The development of ENMs is necessary to take advantage of their potential to boost plant growth, fend off biocides, and reduce stresses all of which are made worse by human activity and rising food demands (Ali et al. 2017). Due to inadequate nutrient availability and the drawbacks of conventional agricultural practices, crop output worldwide is becoming insufficient. However, these issues may be readily resolved by utilizing ENMs and new cultivation techniques (Hao et al. 2018a). In addition to being widely employed in contemporary plant technology, ENMs have significant potential in cutting-edge biotechnological applications. The mechanism of interaction between ENMs and organisms in diverse environmental conditions is influenced by a multitude of complicated events, including but not limited to particle dissolution, ion adsorption, cell contacts, and absorption by microbes (Schwabe et al. 2014).

To comprehend the function of ENMs in agriculture, this chapter expounds upon and accentuates the synthesis processes of ENMs. In the end, this chapter will assist in determining the main uses of ENMs in agriculture, involving the roles they play in plant stress tolerance, possible effects on sustainability, and the materials and pathways for ENM synthesis. This chapter also identifies future directions in ENM production and constraints of ENMs in the agriculture industry.

9.2 Engineered Nanomaterials

Engineered materials and nanoparticles with unique dimensionalities and features are produced for particular uses (Albalawi et al. 2021; Samyn and Barhoum 2018). For these nanomaterials to have as few unintended negative consequences as possible, they need to be well-characterized (Buzea et al. 2007). The aerosol (fumed silica) created in the 1940s was the source of the first commercial nanomaterials. Natural and incidental nanoparticles can have a variety of forms. Engineered nanoparticles often take the form of conventional objects like spheres, tubes, and rings.

9.2.1 Nanomaterials Engineering and Nanofabrication

Engineered nanoparticles are created and manufactured in a lab or industry, as opposed to natural nanomaterials. Their sizes, shapes and content can therefore be accurately regulated. This group also includes layers with complicated chemical compositions (e.g., a gold core coated by antibodies and encased in a silica shell). The branch of nanotechnology that involves creating ENMs for usage in agriculture is referred to as Nano-agriculture.

The methods for creating engineered nanomaterials are getting more and more complex. But basic chemical reactions that have been understood for many millennia are used in their synthesis. Their examination under a microscope showed that several distinct nanomaterials have previously been produced both intentionally and unintentionally (Bai and Ross 2016; Keller et al. 2013).

9.2.1.1 Classification of Synthesis Methods Based on the Raw Materials

The synthesis process employed determines the unique properties of nanomaterials. Generally, techniques are divided into "top-down" and "bottom-up" approaches without regard to the origin of the NP (Fig. 9.1).

(i) Bottom-up Strategies are techniques that produce nanoparticles through the assembling of individual atoms or molecules. These methods use physical and chemical interactions, such as hydrogen and ionic bonds, van der Waals forces, and water-mediated hydrogen bonds, and the self-assembly ability of atoms, ions, molecules, or nanoparticles to assemble basic building blocks into macroscopic structures (Iqbal et al. 2012; Nnaji et al. 2018). Examples of bottom-up techniques are wet chemical synthesis techniques, such as sol-gel, micro-emulsion, and co-precipitation. The primary benefit of these methods is the ability to create nanostructures with a high degree of chemical homogeneity and fewer flaws. On the other hand, their main drawbacks are the need for

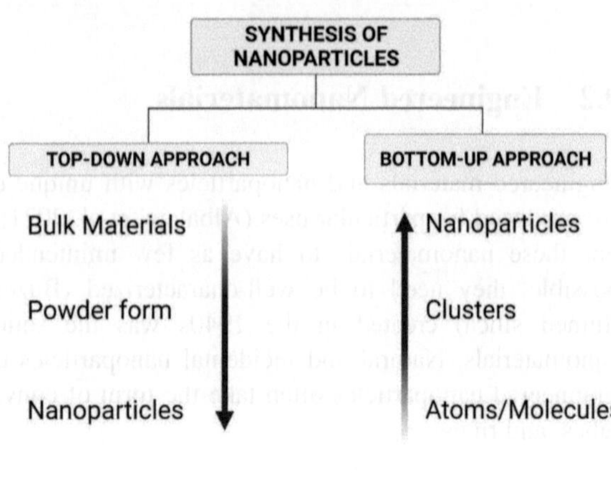

Fig. 9.1 The approaches used in the production of nanoparticles (NPs). Physical techniques like physical vapourdeposition are part of the top-down strategy, whilst chemical techniques like chemical precipitation and biological techniques like the utilization of plant extracts are part of the bottom-up strategy. (Source: Azameti and Imoro 2023)

suitable surfaces and molecules as well as the dearth of instruments for the efficient manipulation of atoms and molecules (Bayda et al. 2019).

(ii) Top-down Approaches: They involve breaking up big, homogeneous solid materials into smaller portions to create nanoparticles. Etching and lithography are appropriate techniques for creating nanostructures. Top-down methods may be used to create thin films and nanoparticles larger than 100 nm that have unique characteristics from their bulk counterparts. Because they facilitate integration and connectivity, top-down approaches are thought to be preferable for fabricating electronic circuits (Patel et al. 2021; Schmidt et al. 2002). The primary drawback of top-down techniques is the resolution limit of current technology (Ranjan et al. 2016).

(iii) Hybrid Approaches: These methods simultaneously use top-down and bottom-up techniques to generate nanomaterials. For instance, nanoparticles and nanofragments of various sizes are ejected by volcanoes. Smaller pieces are precursors that might subsequently take part in a "bottom-up" reaction to form another material type; these are "top-down" accidental nanomaterials. Lithography is the process of creating 1D and 2D nanostructures by employing techniques such as photolithography, electron beam lithography, focused ion beam, and others to replicate patterns by employing positive or negative masks over a substrate. It is primarily thought of as a hybrid technique, in which ion growth stacking is the bottom-up strategy and etching is the top-down method (Hoang 2011; Marcelo et al. 2020). The primary drawbacks of hybrid techniques are the expensive cost of equipment and the requirement for extremely hazardous chemicals (Jin et al. 2018).

9.2.1.2 Classification Based on the Nature of the Deriving Forces

Many research organisations have recommended using the originating force nature to classify the methods of nanomaterial synthesis (Harish et al. 2022). Based on this, the method of nanomaterial synthesis can be grouped as mechanical, physical, chemical, physicochemical, and biological.

(i) Mechanical Methods: Grinding and milling are part of this industrial process. Using high-efficiency and high-energy mills, the ball milling process grinds macro and micro-metric particles to produce particles of irregular dimensions. Its primary benefit is that waste emissions into the environment are eliminated, but its primary drawback is the inability to control the nanoparticle sizes. The following is a partial list of the most popular mechanical techniques: (1) Monitoring and verification, whereby the metal powder is first mechanically compacted, then cooled, and then melted to create nanoparticles. Atomic structures are created from the bulk material to generate nanoparticles (Kadhum Alghanimi and Hadi 2021). (2) Strong deformation procedures, which involve forcefully deforming crystalline materials (metals, porcelains, etc.) to increase the hardness and ductility of the resulting nanoparticles (Singh et al. 2020). (3) To create powdered nanoparticles, the initial material is treated to an exceptionally high energy and ground with steel balls. These nanoparticles range from 3 to 25 nm in size (Feng et al. 2019). (4) Scrubbing, which produces silicone particles on the surface of thin silicon strips by rubbing them against chemicals (such as hydrofluoric acid). After that, strips are submerged in a solution—such as isopropanol—at an ultrasonic apparatus to create nanoscale droplets (Iravani et al. 2014; Singh et al. 2020).

(ii) Physical Methods: These are "top-down" techniques that utilize energy exposure (such as mechanical, chemical, thermal, or laser irradiation) to break down bulk starting materials into smaller materials. These manufacturing techniques consist of spinning, physical vapourdeposition, milling, grinding, and laser ablation. The two major methods for combining two separate components are melt mixing and melt blending. In particular, polymers and nanoparticles help ensure the latter are properly dispersed within a given matrix. Physical approaches involve heating the initial material, subjecting it to an electron beam bombardment, or utilizing laser beams to dissolve it thermally. To prevent crystal formation, the obtained material is immediately placed on a cold surface after being cooled using a neutral gas to maximize saturation. Physical techniques enable the production of smaller, more pure nanoparticles in the required size. But the necessary dangerous chemicals are pricey, and the equipment is expensive. Two popular physical techniques are spinning and laser ablation (Barhoum et al. 2017; Jin et al. 2018). One example of a physical process is the use of radiofrequency heating coils to produce plasma. The metal is first ground in a pestle before being transferred into a vacuum chamber encircled by coils of radiofrequency heating. Here, it is heated with helium to a temperature over its evaporation point in order to create plasma (Sánchez-Ahijón et al. 2020). After nucleating on the atoms of helium gas, the metal vapourdiffuses to a cold collection rod, where the nanoparticles are gathered and oxygen gas is used to passivate them (Belaiche et al. 2021). During laser ablation, the beginning material is exposed to the intense energy that the pulsed laser emits. It results in the creation of plasma that deposits on the support to create thin films and the volatilization of material particles (Amendola and Meneghetti 2009). Using a variety of solvents, laser ablation synthesis in solution creates colloidal nanoparticle solutions (Janas and Koziol 2016). The bulk material is exposed to low pressure during the spinning process to deposit on a cool base. Afterwards, tiny particles that have accumulated on a support are eliminated by a magnetic field, creating a thin layer (Pesheck and Lorence 2009). Emulsion systems are commonly subjected to microwave irradiation, a form of electromagnetic irradiation that involves mobile electric charges (Panigrahi et al. 2004).

(iii) Chemical Methods: To create nanomaterials, chemical methods entail the nucleation and development of precursor species. The material vapourenters the Chemical vapourdeposition (CVD) reactor during vapourphase chemical reactions. The resultant particles then combine with other gases on a base at a certain temperature to produce a solid strip. The most common liquid media for these techniques is water and organic liquids. In order to create spherical nanoparticles with regulated dimensions, double chemical precipitation or hydrolysis are used. Alternatively, sol-gel methods and low-temperature colloidal solutions are employed. The ability to regulate the size and form of the nanomaterial as well as produce very stable nanomaterials are the main benefits of using chemical procedures. But to fabricate them, dangerous chemicals are needed (Barhoum et al. 2019; Nikam et al. 2018).

(iv) Physicochemical Methods: The technique combines chemical and physical processes to synthesize nanomaterials. The electrochemical processes used to create metal nanoparticles, in which a metallic anode dissolves in an aprotic solvent, are an illustration of a physicochemical technique. Physicochemical techniques include hydrothermal, solvothermal, templating CVD, microwave irradiation, combustion, thermal breakdown, and pulsed laser deposition (Campos et al. 2015; Tadic et al. 2019; Wani et al. 2021). A physicochemical technique might potentially be used to create nanoparticles by combining chemical and biological processes. Through the modulation of particular size, shape, crystallinity, and stability of nanoparticles, physicochemical processes shorten the time required for reactions. However, they require dangerous chemicals and expensive, high-tech equipment (Barhoum et al. 2014; Komal et al. 2019).

(v) Biological Methods: Using certain bacteria and plants' activities as a starting point, this approach advances the creation of nanomaterials. Because biological processes produce ecologically benign goods and byproducts, need less energy, and do not require costly chemicals, they are considered environmentally friendly ways. Because they need less upkeep, plant and plant extract-based methods are preferable to those using cells and microorganisms (Parsons et al. 2007).

Biomolecules such as telomers, actin filaments, and enzymes are commonly utilized as catalysts for the formation of nanoparticles (NPs), whereas biological entities including bacteria, fungi, and cells function as active units for producing NPs. It is yet unknown how NP grows in various biological systems. However, it might be explained by specific biosynthetic products such as reduced cofactors or a-hydroxy carboxylic acids, which decrease metal salt precursors (Jeevanandam et al. 2016). The greatest benefit of biological synthesis is its potential to lessen the toxicity of NPs while increasing their biological qualities, bioavailability, bioreactivity, and biocompatibility. The main disadvantage, though, is that the NPs produced might not be as stable as those produced by physical and chemical means.

9.2.1.3 Classification Based on the Reaction Phase

According to Harish et al. (2022), the synthesis techniques for nanomaterials may also be categorized according to the reaction phase (gas, liquid, or solid). These methods are; gasphase synthesis, plasmaphase synthesis, liquidphase synthesis, supercritical fluid synthesis, and solid phase synthesis.

(i) Gas Phase Synthesis: The production of non-volatile products by homogeneous nucleation, condensation, growth, and coagulation is the primary basis for the gas-phase synthesis of nanomaterials employing bottom-up approaches. When compared to liquid-phase synthesis, the primary benefits of gas-phase synthesis are as follows: (1) it can yield complex chemical structures; (2) there is continuous particle synthesis; and (3) it is more hygienic. When creating nanomaterials for electronic-grade semiconductors, it is crucial. Such contaminants can currently only be avoided by vacuum and gas phase synthesis techniques; (4) the ability to modify the characteristics of the nanomaterials. In contrast to the generated nanoparticulate material in the solid phase, the vapourcombination during gas phase production of nanomaterials is thermodynamically unstable (Barhoum et al. 2018). The process of creating plasma involves stripping at least one orbital electron from ions, atoms, or molecules, or adding an extra electron to them (Katouah and El-Metwaly 2021).

(ii) Plasma phase synthesis: Wastewater, carbon dioxide, nitrogen oxide, chlorine, and hydrogen chloride are among the pollutants produced by liquid-phase and gas-phase synthesis techniques (Prasad et al. 2018). Elevated temperatures and anhydrous conditions are employed in plasma phase synthesis to create particles with distinct surface characteristics and morphology that are unattainable by alternative methods. Utilizing such nanoparticles in materials applications necessitates these characteristics (Thimsen et al. 2014). In comparison to other gas phase techniques, thermoluminescence physical vapourdeposition (PLVD) may be a less expensive option for producing nanoparticles. However, there are a number of significant technical challenges (such as a short residence time, a small high-temperature processing zone, and an uneven high-temperature field) that must be addressed before PVPD can be considered a practical low-cost technology for producing nanoparticles commercially (Mantzaris 2005).

(iii) LiquidPhase Synthesis: Bottom-up techniques and regulated chemical processes can be used to create nanomaterials in the liquid phase. Furthermore, by adjusting the growth conditions, self-limiting self-assembly processes have been created (Bayda et al. 2019). In the natural world, erosion and chemical disintegration of biological or geological (clays) materials can produce nanostructures in the liquid phase. The most popular liquid phase synthesis techniques include sol-gel processing, micro-emulsions, hydrothermal, son chemical, chemical co-precipitation, and electrochemical techniques. To stop nanoparticle aggregation, all require a stabilizing agent (Canfarotta et al. 2016).

(iv) Supercritical Fluid Synthesis: A highly compressed fluid that combines the characteristics of a liquid and a gas is called a supercritical fluid. A phase in which a gas reaches a certain temperature and pressure at which it can no longer condense to a liquid is known as a supercritical fluid. Supercritical fluid synthesis makes use of the special qualities of liquefied gases, or solvents, to speed up processes and yield high-grade nanocrystals. Because supercritical fluids may function at mild temperatures, they are frequently referred to as "green" solvents. These techniques allow for the continuous synthesis of metal nanoparticles and have a great deal of promise for integration into high-volume commercial production processes. In addition to producing micro- or nanoparticles with a limited size range, supercritical procedures can also be utilized to co-crystallize with excipients or host molecules, microencapsulate active substance particles, or cover their surfaces with polymers. Phosphors, magnetic materials, and carbon nanotubes are just a few of the many nanomaterials that are created using supercritical CO_2 and H_2O. The most important prerequisite for using nanomaterials is controlling their size and shape, as these factors affect the nanoparticles' potential for use because their characteristics change dramatically with size.

(v) SolidPhase Synthesis: Top-down techniques are used in solid-phase synthesis. Mechanochemical reactions and mechanical attrition are mostly included (Nalluri et al. 2019). The formation of nanoparticles on a solid phase is dependent upon surface development in a vacuum or dry environment. Nanoparticles are formed when atoms or tiny clusters diffuse across appropriate surfaces to create islands.

9.2.2 Synthesis of Engineered Nanomaterials from Plants

Plant-mediated engineered nanomaterials (ENMs) are made using safe solvents and are a low-cost, fast, low-temperature, green technique (Fig. 9.2) (Chung et al. 2016; Munir et al. 2021).

To create silver (Ag) ENMs, Palithya et al. (2021) employed Croton Leaf Mysore Mallow (*Decaschistiacrotonifolia*) leaves. The synthesized ENMs' ($AgNO_3$) antibacterial activity was evaluated against *Escherichia coli*, *Bacillus subtilis*, gram-positive *Staphylococcus aureus*, and *Klebsiella pneumoniae*. When compared to the common antibiotic Amoxyclavhr, the biosynthesized ENMs showed modest effectiveness but higher antibacterial activity against all bacterial species. The ENMs exhibited optimal catalytic activity, facilitating the quick reduction of nitrophenol to aminophenol, hence increasing the colour degradation rate. As a result, these ENMs may find utility in a range of industrial and pharmacological applications. Wild cucumber (*Cucumis prophetarum*) leaf extract was utilized by Hemlata et al. (2020) to create silver (Ag) ENMs. The plant extract and 0.001 M $AgNO_3$ solution were combined in a 1:9 ratio. At 80 °C, the mixture was heated. After centrifuging and drying the precipitates, brown-colored silver (Ag) ENMs were produced.

Khaldari et al. (2021) used green tea and lavender leaves to create copper (II) oxide (CuO) ENMs. The biosynthesized CuO ENMs inhibited seed germination less than chemically synthesized ENMs did.

Ahmad et al. (2021) used the leaf of asthma weed (*E. hirta*) to create iron oxide ENMs. By measuring the size inhibitory zone on agar plates, the engineered ENMs' antibacterial properties were examined and shown to be directly correlated with their concentration. Using Mediterranean thyme (*T. spicata*), Erci and Cakir-Koc (2021) synthesized iron oxide ENMs (FeO).

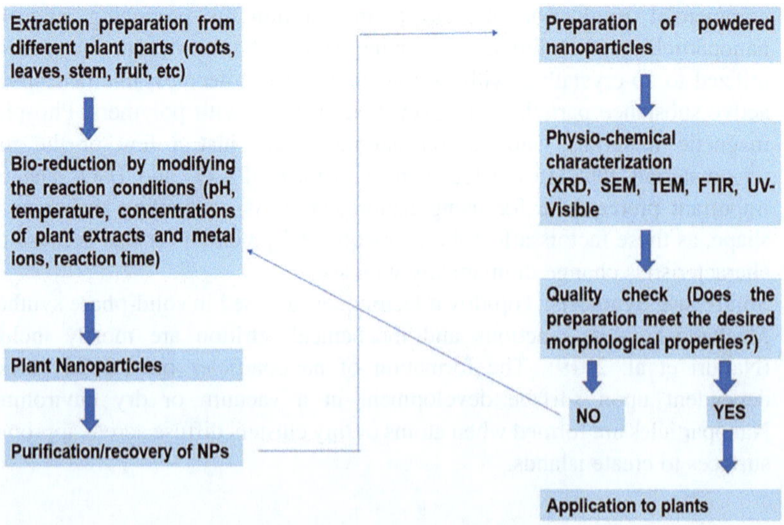

Fig. 9.2 Diagram showing the steps involved in creating green synthetic nanoparticles from extracts from numerous sections of plants. The bio-reduction-mediated synthesis of NPs depends on several factors, such as the quantity of metal ions and plant extracts, the pH of the solution, the length of the reaction, and the reaction temperature. To synthesize desired NPs, purifications and characterizations of NPs are essential. A quality assessment is performed to ascertain if the synthesized nanoparticles possess the intended morphological characteristics. (Source: Azameti and Imoro 2023)

9.2.3 Synthesis of Engineered Nanomaterials from Algae

When used as reducing agents for the manufacture of nanoparticles, algae are an abundant supply of metabolites, environmentally benign, and have the best capacity to absorb metals (Abideen et al. 2022; Hasnain et al. 2022). Furthermore, algae that use brackish water and saline soils can collect greater lipids and serve as a possible source of biodiesel. Aboelfetoh et al. (2017) used *Caulerpa serrulata* from the Egyptian shore of the Red Sea to synthesize silver ENMs. Initially, the algae were cleaned and allowed to dry naturally. Then, 1 g of the powdered algae was boiled for 10 minutes at 70 °C in 100 mL of water. Finally, 5–25 mL of *C. serrulata* extract was combined with 10^{-3} M $AgNO_3$ and the volume was adjusted to 100 mL. The colour of the resulting ENMs ranged from yellow to brown, according to the kind of algal extract employed, reaction duration, solution pH, and temperature. In the presence of sodium borohydride ($NaBH_4$) in a reducing aqueous solution of Congo Red dye, the catalytic activity of the Ag-NP was assessed at room temperature. It was shown that when biosynthesized Ag ENMs were introduced, dye degradation began within 5 minutes, and no dye degradation was seen even after 3 hours.

Abboud et al. (2014) synthesized copper ENMs using brown algae, *Bifurcariabifurcata*. Compared to the algal extract, which lacked antimicrobial action, the copper oxide ENMs showed improved antibacterial activity against both Gram-positive and Gram-negative bacterial species (*E. aerogenes* and *S. aureus*).

Green microalgae (*Chlorella vulgaris*) were used by Eroglu et al. (2013) to create spherical palladium nanocrystals by aqueous sodiumtetrachloro-palladate $Na_2(PdCl_4)$ photosynthetic processes. The nanocrystals catalyzed the Mizoroki-Heck cross-coupling process by being placed in electro-spun chitosan mats. Electro-spun cross-linked chitosan mats measuring 3 × 2 cm were immersed in a mature microalgal culture that had been growing for 1 month, containing 25 mgL^{-1} Na_2 $(PdCl_4)$. Chitosan mats treated with a 25 mgL^{-1} Na_2 $(PdCl_4)$ solution in the absence of microalgae made up the control sample.

Ishwarya et al. (2018) used marine seaweed (*Ulva Lactuca*) to synthesize zinc oxide (ZnO) ENMs. In both dark and visible light, the inhibition percentage of ZnO ENMs' antibacterial activity against Gram-positive and Gram-negative species ranged from 80% to 95%. Specifically, ZnO ENMs made from lactuca showed an 82% reduction rate for *B. licheniformis*, 84% for *B. pumilus,* 80% for *E. coli*, and 79% for *P. vulgaris* (79%). Furthermore, at 24 hours, *A. aegypti* in 50-mg mL^{-1}ZnO ENMs exhibited the maximum inhibitory percentage of 100%.

9.2.4 Synthesis of Engineered Nanomaterials from Fungi

Several unicellular and multicellular species of fungi have been employed in several processes since the dawn of civilization. They are crucial to ecosystems because they control the never-ending cycle of greenhouse gas emissions. Fungi affect our daily lives in both beneficial and harmful ways. They are employed in several industrial processes, including the synthesis of proteins and vitamins, organic acids (like citric acid), fermentation processes, hormones, steroids, and antibiotics (like cephalosporin and penicillin). Although fungi are responsible for a number of agricultural illnesses, certain species can enhance soil health and quality, which helps plants withstand drought and salinity stress (Ramasamy 2020). Numerous fungal species, such as certain mushrooms, morels, and truffles, are consumed for their high protein content. Plastics are the only materials that fungi cannot break down. Forest soils are home to a variety of fungal species that can endure a variety of conditions, including pH changes, the breakdown of organic wastes like cellulose and lignin, and the introduction of nutrients through hyphae that allow plants to absorb water, nitrogen, phosphorus, and micronutrients (Morris 2004).

A common single-celled fungus utilized in many different fermentation processes is yeast. Yeast synthesis of ENMs is a relatively low-cost, ecologically friendly method that streamlines the fermentation process at room temperature. Additionally, yeast synthesis has an edge over other microbial species since it is the world's most widely produced industrial microbe for various fermentation-based goods, including bread and wine. In their study, Bao et al. (2010) employed yeast cells to produce well-dispersed crystalline CdTe ENMs in a cubic zinc-blended form (3.6 nm).

Fernández et al. (2016) created Ag ENMs by combining *Rhodotorulaglutinis* and *Cryptococcus laurentii*, which were cultured for 24 hours at 100 rpm and 28 °C in

Muller-Hinton broth (MHB). The MHB contained 1.5 g of neutral pH corn starch, 17.5 g of acid casein peptone, and 2 g of beef infusion. The yeast cells were suspended in 2.0106 yeast cells and centrifuged before being used.

Eugenio et al. (2016) used yeasts isolated from the stomachs of 10 termites (*Cornitermescumulans*) to synthesize silver and silver chloride. After being put on Sabouraud agar plates, the digestive matter was incubated for 48 hours at 28 °C. The colonies were diluted and then re-incubated to incubate individual clones. To create AgCl ENMs, 15 of these distinct clones were employed; each isolated plate was infected for 24 hours at 30 °C with a liquid medium containing 4% glucose, 1% peptone, and 1% yeast extract at pH 6.5. After adding 3.5 M silver nitrate solution to the medium, it was incubated for 7 days at 30 °C, in the dark, and with continuous stirring at 150 rpm.

9.2.5 *Synthesis of Engineered Nanomaterials from Bacteria*

Using lactic acid bacteria as a reducing and stabilizing agent, Kanmani and Lim (2013) created silver ENMs. First, *Lactobacillus rhamnosus* GG was used to make exopolysaccharide (EPS). The bacteria were cultured in 1 L of MRS (De Man, Rogosa, and Sharpe) agar broth at 37 °C for 18 hours, and after that, it was heated to 100 °C for 15 minutes to deactivate the enzymes that break down EPS. Finally, probiotic cells and debris were removed by centrifuging the mixture. After combining the supernatant with 95% ethanol, it was chilled for 12 hours at 4 °C. After gathering the EPS precipitates, they were cleaned with distilled water and combined with 20 mL of EPS and 9 mM $AgNO_3$. After 60 days of dark storage, the combination produced a brownish solution after 10 hours of incubation, suggesting the development of silver ENMs. Biogenic silver ENMs exhibited dose-dependent antibacterial efficacy against L. *monocytogenes* K. *pneumonia*, P. *aeruginosa*, and *E. coli*. The growth of L. monocytogenes was only suppressed at high nanoparticle dosages (i.e., 2 mg mL^{-1}), but *P. aeruginosa* exhibited the strongest inhibition zones. Similar to this, the effectiveness of the synthesized ENMs against the fungal infections Aspergillus and Penicillium was dose-dependent, with Aspergillus being the most susceptible.

9.3 Role of Engineered Nanomaterials in Plants' Resistance to Abiotic Stress

Unfavorable environmental conditions limit plant productivity and development, which might cause the ecosystem as a whole to collapse. Plants have developed several defense mechanisms against abiotic challenges, and certain systemic signaling pathways, such as those involving reactive oxygen species (ROS), hydrodynamic waves, and phytohormones, have evolved to combat these stresses (Choudhury et al. 2018; Gilroy et al. 2014; Mittler and Blumwald 2015). These changes to the plant's systemic signaling might propagate fast across the whole thing (Kollist et al. 2018).

The presence of these system signals in plants' tissues triggers systemic acquired resistance (SAR), allowing the tissues to tolerate stress (Fichman and Mittler 2020). The impact of nanoparticles in enhancing plants' tolerance to abiotic stress has been documented (Table 9.1) by a few researchers (Azameti and Imoro 2023; Dilnawaz et al. 2023; El-Saadony et al. 2022; Hayat et al. 2023). NPs increase abiotic stress tolerance in plants by reducing ROS production (Abdel-Haliem et al. 2017; El-Shetehy et al. 2020; Namjoyan et al. 2020; Rangaraj et al. 2014; Suriyaprabha et al. 2014).

NPs are essential instruments that function as nano fertilizers, insecticides, herbicides, etc. for the enhanced growth and accelerated development of plants under a range of environmental circumstances (Siddiqui et al. 2015). NPs are used to mitigate stress-induced damage through several biochemical mechanisms, including enzymatic antioxidant activity and detoxification (Sarraf et al. 2022), as illustrated in Fig. 9.3. Major factors influencing NPs' reactivity include shape, size, composition, surface features, stability, chemical properties, purity, manufacturing method, and dosage delivered (Burman and Kumar 2018; Dykman and Shchyogolev 2017; Teske and Detweiler 2015). Supplementing with NP boosted the activity of many enzyme antioxidants, including SOD, CAT, APX, and POX; plants also produced more glutathione, proline, and phytochelatin (Rajput et al. 2021). Additionally, Mahato et al. (2021) revealed that NPs prevent the formation of oxidative stress by upregulating the antioxidant defense system in a variety of stressful circumstances, including temperature stress, UV exposure, drought stress, and salt stress. Accordingly, from this angle, plants become more tolerant to environmental obstacles due to the enhancement of the previously described characteristics brought about by NP supplementation.

9.3.1 Salt Stress

Nanoparticles have recently gained attention as a promising technique for increasing plant growth and performance under salinity stress (Ahmad 2019; Das 2019; Dut́an et al. 2013; Zulfiqar and Ashraf 2021). It has been demonstrated that the concentration at which Ag nanoparticles (Ag NPs) are applied affects the plants. When these NPs are used in excess of a specific concentration, they can have a major phytotoxic impact. Low concentrations of Ag NPs (30 μg/mL) in rice plants stimulated root growth without altering the morphology or structure of the cells. Conversely, a larger quantity of 60 μg/mL Ag NPs entered the cell wall, damaging its shape and structure and limiting root development (Mirzajani et al. 2014).

Application of Ag NP to the aquatic plant duckweed (*Lemna gibba*) causes the cells to produce free radicals, with the accompanied oxidative stress directly

Table 9.1 Role of some nanomaterials in enhancing abiotic stress tolerance in plants

Abiotic stress	Plant	Nanoparticle	Effect	References
Heavy metal	Barley	Calcium oxide NPs	Enhanced plant growth, photosynthesis efficiency, and antioxidant enzymes	Nazir et al. (2022)
Heavy metal	Wheat	Copper NPs	Improved plant growth and decrease in chromium availability	Noman et al. (2020)
Heavy metal and drought	Wheat	Iron NPs	Enhanced photosynthesis and reduced a biochemical nuisance like oxidative stress	Adrees et al. (2020)
Heavy metal and drought	Wheat	Zinc oxide NPs	Reduction in Cd and increase in tissue dry weight buildup	Khan et al. (2019)
Heavy metal	Wheat	Iron NPs	Increased plant growth and reduced oxidative stress	Rizwan et al. (2019)
Heavy metal and drought	Wheat	Silica NPs	Enhanced plant growth and development	Khan et al. (2019)
Drought and salinity	Cucumber	Silica NPs	Increased the growth and productivity of this plant by balancing nutrients uptake	Alsaeedi et al. (2019)
Drought	Wheat	Selenium NPs	Improved plant growth and development	Ikram et al. (2020)
Salinity and heavy metal	Wheat	Iron oxide NPs	Enhanced photosynthetic pigments and prevented cadmium absorption	Manzoor et al. (2021)
Salinity	Broad bean	Titanium oxide NPs	Improved growth and enhanced tolerance against salinity	Abdel Latef et al. (2018)
Salinity	Moldavian dragonhead	Titanium oxide NPs	Promoted plant growth and reduced salinity stress	Gohari et al. (2020)
Salinity	Cotton	Cerium oxide NPs	Improved the plant growth by maintaining cytosolic K^+/Na^+ ratio	Liu et al. (2021)
Salinity	Summer savory	Silver NPs	Increased germination and promote plant growth	Nejatzadeh (2021)
Salinity	Safflower	Zinc oxide NPs	Enhanced plant germination and ameliorate salinity stress	Yasmin et al. (2021)
Salinity	Sweet orange	Silicon NPs	Improved the oxidative stress tolerance	Mahmoud et al. (2022)
Salinity	Wheat	Gold NPs	Improved the plant defense systems	Wahid et al. (2021)
Salinity	Common bean	Selenium dioxide NPs	Enhanced the plant growth and yield	Rady et al. (2021b)
Salinity	Rice	Cerium oxide NPs	Improved crop yield by modulating the plant's physiological and biochemical mechanisms	Zhou et al. (2021)

Source: Azameti and Imoro (2023)

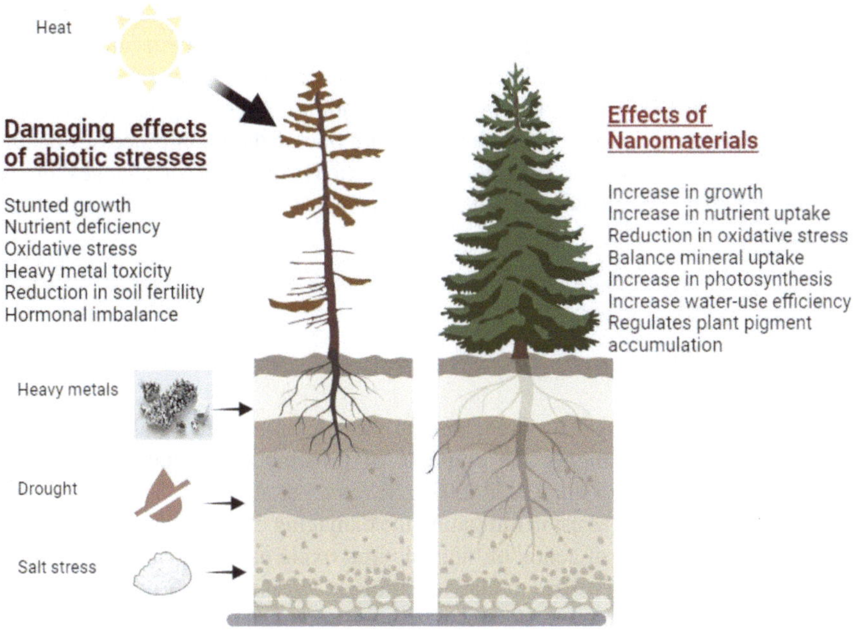

Fig. 9.3 The roles of nanomaterials on the plant exposed to abiotic stresses. Abiotic stresses in plants cause several effects, such as stunted growth, nutrient deficiency, oxidative stress, and hormonal imbalances, while the use of nanomaterials leads to positive effects such as an increase in the growth, in nutrient update and in photosynthesis of the plant. (Source: Azameti and Imoro 2023)

proportional to the concentration that was applied (Cvjetko et al. 2017; Oukarroum et al. 2013). When green synthesized silver nanoparticles (AgNPs) were exogenously applied towheat (*Triticum aestivum*),the average number of germinated seeds increased under salt stress (Wahid et al. 2020). When plants were co-treated with AgNP and NaCl, their H_2O_2 and thiobarbituric acid reactive substances (TBARS) contents were lower than those of the corresponding NaCl-treated plants. Furthermore, the use of AgNPs and NaCl dramatically increased the leakage of electrolytes in wheat plants under salt stress (Wahid et al. 2020). AgNPs effectively reduced the deleterious impacts of salt stress by, among other things, the modification of the K^+ ionic content, and the dynamic function of the antioxidant machinery. When all these characteristics were considered, the research on AgNPs demonstrated a viable method for promoting salt stress tolerance (Misra et al. 1996; Wahid et al. 2020). The impacts of NPs on the molecular characteristics of plants are greatly reliant on the biological activities of the plants (Ali et al. 2021).

During salt stress, NPs change gene expression, which impacts plant development. Kumar et al. (2013) reported that administering gold NPs (AuNPs) in *Arabidopsis thaliana* under salt stress led to improved total seed output, enhanced rate of seed germination, improved vegetative growth, and free radical scavenging activity. According to the authors, all of these alterations are correlated with the production of microRNAs (miRs), which are important plant regulatory molecules (Kumar et al. 2013). Heritable differences in gene expression patterns are studied in the field of epigenetics. In order to adapt to and endure abiotic stressors, plants undergo several DNA methylation-related alterations. According to Kumar et al. (2013), DNA methylation-related alterations are one of the main processes behind the epigenetic variation of stressed plants, which can result in changes in the phenotype, that are either transient, permanent, or heritable to the following generation.

9.3.2 Drought

Plants may endure drought by strengthening their defense systems, altering growth characteristics, and changing the amounts of proteins and metabolites through biochemical processes (Sanjari et al. 2021; Seleiman et al. 2021). The possible use of NP-based tactics can assist in mitigating the damaging effects of drought stress on plants. The mechanism of action of biogenic iron oxide nanoparticles (NPs), which are employed as nano fertilizer to alleviate drought stress, remains uncertain.

Iron oxide nanoparticles, or FeNP, are a novel approach to abiotic stress relief in plants that has not gotten much attention. It has been demonstrated that the marine algae *Chaetomorphaantennina* (green algae) may be used to create magnetite and iron oxide nanoparticles (Fe_3O_4-NP) using a green synthesis method, which can help reduce abiotic stress (Sreelakshmi et al. 2021). When these particles were applied to the drought-stressed Foxtail millet (*Setariaitalica*) plants on both bare and citrate-coated iron oxide NPs, it resulted in a simultaneous increase in chlorophyll. Moreover, the presence of soluble sugars suggests the successful absorption of the iron oxide NPs by the plants, which they then use to produce photo-assimilates. These iron oxide nanoparticles (NPs) were highly effective in promoting improved antioxidant activities, including CAT, SOD, POX, and higher osmolyte content. Moreover, there is an increase in biochemical content, including iron, H_2O_2, and proline, which indicates that the drought stress is being overcome. Consequently, in drought-stricken conditions, these environmentally benign synthetic nanoparticles may be utilized as nano fertilizers (Sreelakshmi et al. 2021). Selenium (Se) nanoparticles (NPs) were applied topically to wheat plants in different research to enhance their development during regulated watering and drought conditions. Se-NPs that were biosynthesized were produced utilizing garlic (*Allium sativum*) extract.

Exogenously varying amounts of Se-NPs were administered to the wheat cultivars that were prone to and tolerant to drought during the trifoliate stage. Different conditions with little water availability were used to irrigate the wheat plants. The use of Se-NPs significantly reduced lipid peroxidation and ionic leakage, which helped to lessen the cellular damage caused by drought stress (Ikram et al. 2020).

When silica nanoparticles (NPs) are applied to plants, they fortify their physical barrier by the deposition of the particles onto their cell walls. This enhances lodging tolerance and lowers oxidative stress by enhancing the plants' defense response against abiotic stressors (Ma and Yamaji 2008). Drought imposes restrictions on seed germination, reduces photosynthesis, and stunts plant development. Plants change their osmotic balance and accumulate suitable solutes in response to dryness (Blum 2017). The osmotic potential was able to be decreased by using silica nanoparticles.

9.3.3 Heavy Metals

The sensitivity and tolerance of plants to heavy metal contamination resulted in the identification of genes activated by heavy metal stress. Transcriptome research has revealed the presence of many transcription factors (TFs), including *bZIP*, under heavy metal stress in plants such as *Arabidopsis*, *Brassica*, and *Lycopersicum* (Hussain et al. 2021; Singh et al. 2016). Similarly, biomolecules, such as phenol, glutathione, and organic acids, enhance plants' tolerance to heavy metal stress (Collin et al. 2008; Singh and Shah 2014). The exogenous administration of calcium oxide nanoparticles (CaO-NPs) significantly mitigates the negative impact caused by As in two barley genotypes (Nazir et al. 2022).

To lessen the amount of copper transferred from the soil to wheat plants, biogenic copper nanoparticles (NPs) were created using the bacteria *Shigella flexneri* SNT22. When CuNPs were added to Cd-contaminated soil, wheat plants produced more biomass, absorbed nutrients more easily, and had lower Na^+ levels due to a reduction in acropetal Cd translocation. According to the study (Adrees et al. 2021; Nouman et al. 2014), it is possible to develop sustainable wheat growth in soil polluted with metals by using the biogenic-CuNPs technique. Adrees et al. (2021) investigated the effectiveness of applying zinc oxide nanoparticles (ZnO-NPs) topically to wheat plants grown in soil polluted with Cd under different moisture conditions and reported that the application of ZnO-NPs during normal moisture conditions resulted in enhanced leaf SOD and peroxidase activities with maximum Zn and minimum Cd concentrations, as well as better leaf chlorophyll contents and lowered oxidative stress. Si NPs, or silica nanoparticles, are crucial in lessening the harmful effects of heavy metals. Researchers looked at the connection between heavy metal stress and NP interaction, as well as its effects on physiological and biochemical processes in pea (*Pisum sativum* L.), to ascertain the effect of silica-NPs under Cr (VI) stress. Using chlorophyll fluorescence to measure photosynthesis, they assessed the growth by looking at oxidative stress, the antioxidant defense system, and the control of mineral elements.

According to Tripathi et al. (2015), Si-NPs reduce Cr(VI)-induced toxicity by reducing the extent of Cr accumulation in roots and shoots of plants. This was due to the reduced level of ROS, an enhanced antioxidant defense system that significantly regulates oxidative stress, and improved uptake of mineral nutrients that help protect the plants against oxidative stress. It was discovered that the best way to reduce the toxicity of lead and its buildup in rice grains was to apply silicon during the tillering stage of rice development. The rice exhibited the largest grain weight and lowest Cd content following foliar application of the Si sol, likely due to the rice's efficient absorption of Si during the tillering stage. Approximately two-thirds of the silicon in the rice plant as a whole and three-quarters of the silicon in the leaf blade were absorbed from tillering to booting.

It is possible that Cd sequestration in the shoot system is the mechanism by which Si foliar treatments reduce the toxicity of Cd buildup (Liu et al. 2009). In a further investigation, the impact of applying nano-Si topically on rice (*Oryza sativa* L.) seedlings cultivated in nutrient solution with or without $CdCl_2$ was assessed in order to mitigate Cd stress. By boosting the number of mineral elements, such as Mg, Fe, and Zn, and antioxidant capability, the post-application Cd accumulation was reduced, resulting in lessened Cd toxicity (Wang et al. 2015). Ali et al. (2019) conducted tests to learn more about the effects of Cd contamination on wheat (*Triticum aestivum* L.) growth. They evaluated the growth of wheat with soil amendment and foliar Si-NP addition following physiological maturity. The biomass of the plants stressed by Cd was greatly increased by the Si-NPs. Increased photosynthesis and decreased oxidative stress activity contributed to the wheat plants' improved development, which was attributed to the positive effects of abiotic stress (Ali et al. 2019). SiNPs accumulate on the cell walls, and Cd binds to the Si-NPs rather than entering the cells.

The level of Cd absorption and the expression level of the transport genes were likewise regulated by the Si-NP treatment. Reduced Cd absorption capacity and increased Cd tolerance were the results of rice's propensity to bind to Cd and obstruct its diffusion into cells. Due to the higher concentration of Cd on the cell surface, there was less Cd diffusion within the cells. Additionally, the absorption of Si may activate the Si transporter (OsLsi1), which may then increase the capacity to withstand Cd by causing the transporter (OsHMA3, tonoplast-localized transporter) to overexpress itself. By blocking the Cd transporters (OsLCT1, a cation transporter, and OsN- ramp5, a major Mn transporter), Si-NPs reduce the amount of Cd that is absorbed by cells (Cui et al. 2017). Si NPs can also lessen the toxicity of lead. Si-NPs' impact on coriander (*Coriandrum sativum* L.) cultivated in soil contaminated with lead (Pb) was assessed. The Si-NPs reduced the buildup of MDA in plant tissues by reducing the concentration of Pb, which was controlled by the activities of POD, CAT, and SOD, the foliar application of 1.5 mM (Fatemi et al. 2021).

According to reports, Si NPs can lessen the phytotoxicity that heavy metals cause in wheat (Gao et al. 2018), peas (Rehman et al. 2018), and rice (Gao et al. 2018). Furthermore, Konate et al. (2017) discovered that FeO NPs promoted wheat plant development by reducing cellular oxidative stress caused by Pb and Cd toxicity. FeO NPs have recently been effectively employed to reduce heavy metal and water stress in plants while also promoting plant development (Adrees et al. 2020; Hussain et al. 2019).

Pollution containing mercury (Hg) is a global problem. Both gaseous and particulate forms of mercury are found in the atmosphere (Sen and De 1987). According to Bishop et al. (1998), plants may take up mercury from soil and water via their roots as well as through the surfaces of their leaves. The impact of nano-SiO_2 (30 nm and 50 nm) on Hg accumulation in soybean plants was investigated by Li et al. (2020), and they reported that utilizing both sizes of nano-SiO_2 reduced the amount of mercury accumulation and growth inhibition in soybean seedlings. The amounts of mercury in the pericycle and epidermis of the roots and stems were significantly reduced by the nano-SiO_2 alterations. Additionally, it enhanced the quantity of chlorophyll and antioxidant enzyme activity that was brought on by mercury stress.

When compared to Hg-only exposed groups, the RS–Hg–SR bonding form in the roots was dramatically reduced in the nano-SiO_2 treated groups. The amounts of metals that are phytoavailable affect how much heavy metal builds up in plants. A rise in soil pH is associated with the immobilization of heavy metals in soil because it decreases root-shoot transfer and causes adsorption on the root surface. Utilizing materials high in silica can stop heavy metals from building up in plant cells, moving from roots to above-ground areas of the soil, and becoming bioavailable (Li et al. 2020).

Crop production is severely hampered by aluminium (Al) toxicity in acidic soils, which is primarily limited to non-phytotoxic forms (Kochian et al. 2015). When this element reaches micromolar quantities, it solubilizes into aluminium ions and becomes physiologically accessible, hindering the development of several plant species (Brunner and Sperisen 2013). Higher levels of lignification were discovered in the roots of sunflower plants. Several plants exhibit aluminium tolerance as a result of the exudation of phenolics and polypeptides as well as the secretion of mucilage (Brunner and Sperisen 2013; Kochian et al. 2015). It was studied how Si NPs reduced the toxicity of Al in maize (*Zea mays* L.) plants cultivated in acidic soil. The soil was combined with varying concentrations of $AlCl_3$ solution, either separately or in conjunction with Si-NPs suspension for the maize crops. Al concentrations in roots and shoots were not significantly impacted by Si NP treatment alone. When acute Al stress was applied to the leaves of maize plants, the rate of photosynthetic activity was reduced. When Al-nano-SiO_2 was applied in tandem, there was a greater build-up of intermediary metabolic components (de Sousa et al. 2019). Consequently, plants exposed to nano-SiO_2 had higher pigment levels and more Rubisco catalytic activity. Co-treatment of nano-SiO_2 could have promoted the buildup of organic acids (OAs) and been a primary factor in the Si-mediated reduction of Al toxicity. Thus, Si-NPs can lessen the detrimental effects of Al in maize plants by influencing the exudation of OAs and polyphenols in response to stress (de Sousa et al. 2019).

9.3.4 Heat Stress

The most dangerous abiotic stressor for crop growth has been predicted to be a gradual increase in air temperature. In order to impose heat stress during the reproductive stage, several planting dates were adjusted under field conditions in research carried out by Kareem et al. (2022). Mung beans (*Vigna radiata* L.) were exogenously treated with different concentrations of nano-ZnO by foliar spray, and the yield evaluation process was examined. Because the nano-ZnO increased the production of antioxidants and osmolytes, it greatly reduced the shocks caused by heat stress. Furthermore, by enhancing gas exchange, chlorophyll concentrations, and other agronomic characteristics, foliar application of nano-ZnO boosted grain yield under heat stress; nevertheless, in late-sown varieties that showed curtailment in lipid peroxidation, the formation of ROS was rebated (Kareem et al. 2022). Depending greatly on the species and exposure circumstances, when plants are exposed to certain concentrations, NPs react to stress by interacting with different enzymes and AOX metabolites through the antioxidant defense system. The primary mechanism of metal poisoning is still oxidative stress, and the warning signs are AOX stress responses.

9.4 Role of Engineered Nanomaterials in Plants' Resistance to Biotic Stress

The anti-phytopathogenic action of ENMs has been documented in several investigations (Table 9.2), both in vivo and in vitro (Chen et al. 2020; Hernández-Díaz et al. 2021; Jo et al. 2009). This is as a result of the metal ions that these ENMs release or by the nanoforms of the ENMs (Elmer and White 2018; Khan et al. 2018; Servin et al. 2015). In addition, bacteria that can inhibit these phytopathogens may be stimulated and increased in activity by ENMs. Because silver is an important antibacterial agent, several research has been done to examine its effects on various phytopathogens. For example, phytopathogenic fungi, known to infect ryegrass, were used to assess Ag NPs' inhibitory capabilities (Jo et al. 2009). In a similar vein, studies were conducted on the inhibitory characteristics of a number of ENMs, including MWCNTs, Fe_2O_3, etc., against the Tobacco and Turnip mosaic viruses (Hao et al. 2018b).

The viruses' coat proteins were considerably reduced (15–60%) by the ENMs, which lessened the viruses' pathogenicity. However, the author believes that the foliar spray's dosage (50–200 mg/L) was very high (Hao et al. 2018b). It was discovered that the ENMs enhanced the antifungal activity of *Pseudomonas protegens* CHA0 against *Candida albicans* when modest concentrations (500 ng/mL) of Ag, SiO_2, TiO_2, and ZnO were employed to evaluate their impact (Khan et al. 2018). ZnO NPs have been shown to suppress the plant disease *Fusarium graminearum* in vitro and in vivo (Dimkpa et al. 2013).

ZnO NP treatment considerably lowers the likelihood of *F. graminearum* infection in wheat plants. Comparing the *F. graminearum* colony forming unit to the control, a significant reduction was noted (Savi et al. 2015). The plant was not harmed by the Zn nanomaterial treatment, and the zinc content of the grains stayed within allowable bounds. In addition, a number of carbonaceous compounds were tested against two phytopathogenic fungi (*F. graminearum* and *F. poae*) (Wang et al. 2014). These materials included RGO, activated carbon, C60, MWCNTs, and others. With the exception of C60 and AC, it was discovered that every ENM

Table 9.2 Role of some nanomaterials in enhancing biotic stress tolerance in plants

Biotic stress	Crop	Nanomaterial	Effects	References
Magnporthe oryzae	Rice	ZnO	Inhibited the formation of appressorium, and upregulated the genes (*OsNAC4*, *OsPRO10*, *OsKSL4*,*OsPR1b*) involved in resistance	Qiu et al. (2023)
Bipolarissarokiniana	Wheat	Se	Increased chlorophyll content, membrane stability index, leaf surface area, and root length	Shahbaz et al. (2023)
Xanthomonas oryzae	Rice	Ag	Decreased ROS effect by enhancing cellular antioxidative system	Ahmed et al. (2020)
Rhyzoperthadominica and *Sitophilus granarius*	Wheat	CuO	Increased leaf pigment concentration and antioxidant enzymatic activities, leading to high insect mortality	Badawy et al. (2021)
Puccinia striiformis	Wheat	TiO_2	Suppressed the expression of proteins involved in ROS production	Satti et al. (2022)
Fusarium oxysporum and *Aspergillus niger*	Maize	Si	Increased phenolics	Suriyaprabha et al. (2014)
Fusarium fujikuroi	Rice	Si	Enhanced electrolyte leakage	Elamawi et al. (2020)
Fusarium oxysporum	Chickpea	ZnO	Enhanced antioxidant activity	Farhana Munis et al. (2022)
Phytophthora nicotianae and *Thielaviopsis basicola*	Tobacco	MgO	Induced ROS production	Chen et al. (2020)
Meloidogyne incognita, *Pectobacteriumbetavasculorum*, and *Rhizoctonia solani*	Beetroot	SiO_2	Increased chlorophyll content and improved activity of defense-related enzymes	Khan and Siddiqui (2020)
Rhizoctonia solani	Potato	$Ca_3(PO_4)_2$, SiO_2 and CuO	Boosted activities of POD, PPO, CAT and chitinase enzymes	El-Shewy (2019)
Fusarium andiyazi	Tomato	Chitosan	Caused the upregulation of PR genes, and also activated SOD and related antioxidant genes	Chun and Chandrasekaran (2019)
Alternaria solani	Tomato	AgNP	Enhanced the activity of antioxidant enzymes	Ansari et al. (2023)

suppressed the development of the two fungi under examination. The primary causes of this inhibition were (i) plasmolysis, (ii) spore blockage brought on by the buildup of ENMs on the spore surface, and (iii) restriction of water absorption. Ag NPs were shown to suppress *Colletotrichum* spp., a phytopathogen that causes anthracnose, even in field testing (Lamsal et al. 2011).

Blast caused by *Magnaporthe oryzae* is one of the most devastating diseases of rice (*Oryza sativa* L.) (Azameti et al. 2020). Bio-formulated moringa chitosan nanoparticles (M-CsNPs) suppressed rice blast disease and effectively inhibited the blast pathogen proliferation, integrity, and colonisation.

It was shown that using nanoparticles prior to the disease's beginning significantly reduced the amount of *Colletotrichum spp.* infection in pepper plants. Therefore, it is clear from the available research that ENMs have a great deal of potential to protect many plant species against a variety of pathogen infections; nevertheless, their toxicity at greater concentrations has to be thoroughly investigated, which is still a major cause for worry.

9.5 Application of Nano Fertilizers for Abiotic Stress Tolerance

Presently, nano-fertilizers are employed to improve the qualitative traits of plants under abiotic stress by adjusting the various physiological and biochemical processes (Bashir et al. 2020; Mozafari et al. 2018; Rossi et al. 2017). According to Ali et al. (2019), nano fertilizers have proven to be more successful than conventional fertilizers in both shielding plants from environmental stress and enhancing plant development characteristics through increased nutrition efficiency.

Nano-enabled methods for reducing nutrient deficiencies, boosting agricultural yields, and altering biological processes are being used to manage environmental challenges on plants (Kah et al. 2019).

Effective micronutrient delivery agents, nano fertilizers are a superior, environmentally benign substitute for chemical fertilizers (Irshad et al. 2021; Kumar et al. 2020; Shah et al. 2021). The growing worldwide population puts significant pressure on freshwater supplies, which makes irrigation using salt water an obsessive choice. Because of the intimate relationship between salt and the root anatomy in nature, tolerance occurs. According to Rossi et al. (2017), cerium oxide (CeO_2NPs) can alter the structure of plant roots in Rapeseed (*Brassica napus*), increasing the plant's ability to withstand salt stress and enhancing photosynthesis. The acquired data unequivocally demonstrate that the presence of CeO_2NPs in the soil combined with NaCl altered how the apoplastic barriers formed in Brassica roots. Salt-stressed plants exhibit a reduction in the root apoplastic barrier, which facilitates increased movement of Na^+ towards the shoot and reduced buildup of Na^+. Brassica showed improved physiological performance with changed Na^+ fluxes and transport, which enhanced photosynthesis (Rossi et al. 2017).

Bashir et al. (2020) assessed the effectiveness of ZnO-NPs as a foliar spray at different doses in the aged Cd-contaminated agricultural soil. Different areas of the wheat showed reduced Cd concentrations and electrolyte leakage, yet there was a noticeable increase in yield, biomass, quantities of chlorophyll, and the activities of superoxide dismutase and peroxidase. Using organic materials and NPs, it was demonstrated that the quantities of Cd and other hazardous trace elements in wheat could be reduced (Bashir et al. 2020). In a different research, Mozaffari et al. (2018) assessed the combined impact of SA and iron oxide NPs in an in vitro strawberry growing system during a drought.

The findings showed that the administration of SA has neutralized the detrimental impacts of drought stress on the growth parameters of strawberry plantlets. Strawberry plantlets were given an iron nanoparticle treatment to help them better withstand harsh environments. The iron served as a micronutrient for plant development, which had a significant impact on the plantlets' growth characteristics and other qualities that were assessed (Mozafari et al. 2018). Therefore, before replanting plants in the field, a combination application of SA and iron oxide NPs can help give enhanced drought-adapting properties.

9.6 Conclusions and Prospects

Overcoming the adverse impacts of environmental stressors on agricultural productivity has become a major research goal in recent years. The production of engineered nanomaterials and their potential to protect agricultural plants from biotic and abiotic stressors have been clarified in this chapter. Nanoparticles have been extensively investigated for their capacity to break down harmful substances in the environment, therefore their use can serve several functions (Khan et al. 2022).

Although significant progress has been made in the use of nanomaterials in agriculture, crucial areas still need to be understood. ENMs have been widely explored at the experimental or pilot size. However, several studies on hectares of agricultural fields and several years of application are still required to confirm their potential long-term effects on the soil and its microbiome, plants, and animals (El-Samad et al. 2022a, b). More information is also needed on their safe and lucrative commercial application, as well as the safety of ENM-produced foods (Kranjc and Drobne 2019).

The problems in using nanomaterials to treat biotic stress include the need to improve and further assess nanomaterial absorption, transport, and modification in plant tissue. Another significant obstacle is the public acceptability of this technology, as humans and/or animals are the ultimate consumers of plants treated with nanomaterials, and the consequences of these technologies on human and animal health must be assessed. In this regard, ecotoxicological studies of nanoparticles are crucial, as is the development of a regulatory framework for the use of nanomaterials in large-scale agricultural production. Future study into the safe use of nanomaterials on a wide scale in agricultural production is encouraged.

Industrial sectors must be established in order to govern the regulatory environment, develop application procedures, educate farmers about the use of engineered nanomaterials, and increase the production of nanoproducts.

References

Abboud Y, Saffaj T, Chagraoui A, El Bouari A, Brouzi K, Tanane O, Ihssane B (2014) Biosynthesis, characterization and antimicrobial activity of copper oxide nanoparticles (CONPs) produced using brown alga extract (*Bifurcariabifurcata*). Appl Nanosci 4(5):571–576

Abd El-Ghany WA, Shaalan M, Salem HM (2021) Nanoparticles applications in poultry production: an updated review. Worlds Poult Sci J 77(4):1001–1025

Abd El-mageed TA, Rady MOA, Abd El-Wahed MH, Abd El-mageed SA, Omran WM, Aljuaid BS, El-Shehawi AM, El-Tahan AM, El-Saadony MT et al (2022) Consecutive seasonal effect on yield and water productivity of drip deficit irrigated sorghum in saline soils. Saudi J Biol Sci 29(4):2683–2690

Abdel-Haliem MEF, Hegazy HS, Hassan NS, Naguib DM (2017) Effect of silica ions and nano silica on rice plants under salinity stress. Ecol Eng 99:282–289

Abdel Latef AA, Srivastava H, El-sadek A et al (2018) Titanium dioxide nanoparticles improve growth and enhance tolerance of broad bean plants under saline soil conditions. L Degrad Dev 29:1065–1073. https://doi.org/10.1002/ldr.2780

Abideen Z, Waqif H, Munir N, El-Keblawy A, Hasnain M, Radicetti E, Mancinelli R, Nielsen BL, Haider G (2022) Algal-mediated nanoparticles, phycochar, and biofertilizers for mitigating abiotic stresses in plants: a review. Agronomy 12(8):1788

Aboelfetoh EF, El-Shenody RA, Ghobara MM (2017) Eco-friendly synthesis of silver nanoparticles using green algae (*Caulerpa serrulata*): reaction optimization, catalytic and antibacterial activities. Environ Monit Assess 189(7):349

Adrees M, Khan ZS, Ali S, Hafeez M, Khalid S, Ur Rehman MZ, Hussain A, Hussain K, Shahid Chatha SA et al (2020) Simultaneous mitigation of cadmium and drought stress in wheat by soil application of iron nanoparticles. Chemosphere 238:124681

Adrees M, Khan ZS, Hafeez M, Rizwan M, Hussain K, Asrar M, Alyemeni MN, Wijaya L, Ali S (2021) Foliar exposure of zinc oxide nanoparticles improved the growth of wheat (*Triticum aestivum L.*) and decreased cadmium concentration in grains under simultaneous cd and water-deficient stress. Ecotoxicol Environ Saf 208:111627

Ahmad I (2019) Use of nanoparticles in alleviating salt stress. In: Salt stress, microbes, and plant interactions: causes and solution. Springer Nature Singapore Pte Ltd., Singapore. pp 199–215

Ahmad W, Kumar Jaiswal K, Amjad M (2021) *Euphorbia herita* leaf extract as a reducing agent in a facile green synthesis of iron oxide nanoparticles and antimicrobial activity evaluation. Inorg Nano Met Chem 51(9):1–8

Ahmed T, Shahid M, Noman M, Niazi MBK, Mahmood F, Manzoor I, Zhang Y, Li B, Yang Y, Yan C et al (2020) Silver nanoparticles synthesized by using Bacillus cereus SZT1 ameliorated the damage of bacterial leaf blight pathogen in rice. Pathogens 9:160

Albalawi F, Hussein MZ, Fakurazi S, Masarudin MJ (2021) Engineered nanomaterials: the challenges and opportunities for nanomedicines. Int J Nanomedicine 16:161–184

Ali S, Liu Y, Ishaq M, Shah T, Abdullah A, Ilyas A, Din IU (2017) Climate change and its impact on the yield of major food crops: evidence from Pakistan. Food Secur 6(6):39

Ali S, Rizwan M, Hussain A, Zia Ur Rehman MZ, Ali B, Yousaf B, Wijaya L, Alyemeni MN, Ahmad P (2019) Silicon nanoparticles enhanced the growth and reduced the cadmium accumulation in grains of wheat (*Triticum aestivum L.*). Plant Physiol Biochem 140:1–8

Ali S, Mehmood A, Khan N (2021) Uptake, translocation, and consequences of nanomaterials on plant growth and stress adaptation. J Nanomater 2021:1–17

Alsaeedi A, El-Ramady H, Alshaal T, El-Garawany M, Elhawat N, Al-Otaibi A (2019) Silica nanoparticles boost growth and productivity of cucumber under water deficit and salinity stresses by balancing nutrients uptake. Plant Physiol Biochem 139:1–10. https://doi.org/10.1016/j.plaphy.2019.03.008

Amendola V, Meneghetti M (2009) Laser ablation synthesis in solution and size manipulation of noble metal nanoparticles. Phys Chem Chem Phys 11(20):3805–3821

Ammar AS (2018) Nanotechnologies associated to floral resources in agri-food sector. Acta Agron 67:146–159

Ansari M, Ahmed S, Abbasi A, Hamad NA, Ali HM, Khan MT, Haq IU, Zaman QU (2023) Green synthesized silver nanoparticles: a novel approach for the enhanced growth and yield of tomato against early blight disease. Microorganisms 11:886

Azameti MK, Imoro AM (2023) Nanotechnology: a promising field in enhancing abiotic stress tolerance in plants. Crop Des 2(2):100037

Azameti MK, Vishalakshi B, Umakanth B et al (2020) Molecular characterization of popular rice (*Oryza sativa* L.) varieties of India and association analysis for blast resistance. Genet Resour Crop Evol 67:2225–2236. https://doi.org/10.1007/s10722-020-00976-7

Badawy AA, Abdelfattah NAH, Salem SS, Awad MF, Fouda A (2021) Efficacy assessment of biosynthesized copper oxide nanoparticles (CuO-NPs) on stored grain insects and their impacts on morphological and physiological traits of wheat (*Triticum aestivum* L.) plant. Biology 10:233

Bai W, Ross CA (2016) Functional nanostructured materials based on self-assembly of block copolymers. MRS Bull 41(2):100–107

Bandala ER, Berli M (2019) Engineered nanomaterials (ENMs) and their role at the nexus of food, energy, and water. Mater Sci Energy Technol 2(1):29–40

Bao H, Hao N, Yang Y, Zhao D (2010) Biosynthesis of biocompatible cadmium telluride quantum dots using yeast cells. Nano Res 3(7):481–489

Barhoum A, Ibrahim HM, Hassanein TF, Hill G, Reniers F, Dufour T, Delplancke MP, Van Assche G, Rahier H (2014) Preparation and characterization of ultra-hydrophobic calcium carbonate nanoparticles. IOP Conf Ser Mater Sci Eng 64:012037

Barhoum A, Samyn P, Öhlund T, Dufresne A (2017) Review of recent research on flexible multifunctional nanopapers. Nanoscale 9(40):15181–15205

Barhoum A, Rahier H, Benelmekki M, Assche G (2018) Van recent trends in nanostructured particles: synthesis, functionalization, and applications. In: Fundamentals of nanoparticles. Elsevier, pp 605–639

Barhoum A, Bechelany M, Makhlouf ASH (2019) Handbook of nanofibers, 1st edn. Springer International Publishing, Switzerland. isbn:978-3-319-53654-5

Bashir A, Rizwan M, Ali S, Adrees M, Rehman MZU, Qayyum MF (2020) Effect of composted organic amendments and zinc oxide nanoparticles on growth and cadmium accumulation by wheat; a life cycle study. Environ Sci Pollut Res Int 27(19):23926–23936

Bayda S, Adeel M, Tuccinardi T, Cordani M, Rizzolio F (2019) The history of nanoscience and nanotechnology: from chemical–physical applications to nanomedicine. Molecules 25(1):112

Belaiche Y, Khelef A, Laouini SE, Bouafia A, Tedjani ML, Barhoum A (2021) Green synthesis and characterization of silver/silver oxide nanoparticles using aqueous leaves extract of *Artemisia Herba-Alba* as reducing and capping agents. Rev Română Mater/Rom J Mater 51:342–352

Bishop KH, Lee YH, Munthe J, Dambrine E (1998) Xylem sap as a pathway for total mercury and methylmercury transport from soils to tree canopy in the boreal forest. Biogeochemistry 40(2/3): 101–113

Blum A (2017) Osmotic adjustment is a prime drought stress adaptive engine in support of plant production. Plant Cell Environ 40(1):4–10

Brunner I, Sperisen C (2013) Aluminum exclusion and aluminium tolerance in woody plants. Front Plant Sci 4(172):172

Burman U, Kumar P (2018) Plant response to engineered nanoparticles. In: Nanomaterials in plants, algae, and microorganisms. Academic, pp 103–118

Buzea C, Pacheco II, Robbie K (2007) Nanomaterials and nanoparticles: sources and toxicity. Biointerphases 2(4):MR17–MR71

Campos EA, Stockler Pinto DVBS, de Oliveira JIS, Mattos EDC, Dutra RDCL (2015) Synthesis, characterization and applications of iron oxide nanoparticles—a short review. J Aerosp Technol Manag 7(3):267–276

Canfarotta F, Poma A, Guerreiro A, Piletsky S (2016) Solid-phase synthesis of molecularly imprinted nanoparticles. Nat Protoc 11(3):443–455

Chen J, Wu L, Lu M, Lu S, Li Z, Ding W (2020) Comparative study on the fungicidal activity of metallic MgO nanoparticles and macroscale MgO against soilborne fungal phytopathogens. Front Microbiol 11:365

Choudhury FK, Devireddy AR, Azad RK, Shulaev V, Mittler R (2018) Local and systemic metabolic responses during light-induced rapid systemic signaling. Plant Physiol 178(4): 1461–1472

Chun S-C, Chandrasekaran M (2019) Chitosan and chitosan nanoparticles induced expression of pathogenesis-related proteins genes enhances biotic stress tolerance in tomato. Int J Biol Macromol 125:948–954

Chung IM, Park I, Seung-Hyun K, Thiruvengadam M, Rajakumar G (2016) Plant-mediated synthesis of silver nanoparticles: their characteristic properties and therapeutic applications. Nanoscale Res Lett 11(1):40

Collin VC, Eymery F, Genty B, Rey P, Havaux M (2008) Vitamin E is essential for the tolerance of *Arabidopsis thaliana* to metal-induced oxidative stress. Plant Cell Environ 31(2):244–257

Coman V, Oprea I, Leopold LF, Vodnar DC, Coman C (2019) Soybean interaction with engineered nanomaterials: a literature review of recent data. Nanomaterials (Basel) 9(9):1248

Cui J, Liu T, Li F, Yi J, Liu C, Yu H (2017) Silica nanoparticles alleviate cadmium toxicity in rice cells: mechanisms and side effects. Environ Pollut 228:363–369

Cvjetko P, Milošíc A, Domijan AMAM, VinkovícVřcek I, Tolíc S, PeharecŠtefaníc P, Letofsky-Papst I, Tkalec M, Balen B (2017) Toxicity of silver ions and differently coated silver nanoparticles in *Allium cepa* roots. Ecotoxicol Environ Saf 137:18–28

Das A (2019) Nanotechnology a potential tool to mitigate abiotic stress in crop plants. In: Abiotic and biotic stress in plants. I. IntechOpen

de Sousa A, Saleh AM, Habeeb TH, Hassan YM, Zrieq R, Wadaan MAM, Hozzein WN, Selim S, Matos M et al (2019) Silicon dioxide nanoparticles ameliorate the phytotoxic hazards of aluminum in maize grown on acidic soil. Sci Total Environ 693:133636

Dilnawaz F, Misra AN, Apostolova E (2023) Involvement of nanoparticles in mitigating plant's abiotic stress. Plant Stress 10:100280

Dimkpa CO, McLean JE, Britt DW, Anderson AJ (2013) Antifungal activity of ZnO nanoparticles and their interactive effect with a biocontrol bacterium on growth antagonism of the plant pathogen *Fusarium graminearum*. Biometals 26(6):913–924

Dontsova TA, Nahirniak SV, Astrelin IM (2019) Metaloxide nanomaterials and nanocomposites of ecological purpose. J Nanomater 2020:5942194

Dufan P, Acuña JJ, Jorquera MA, Azcoń R, Borie F, Cornejo P, Mora ML (2013) Enhanced selenium content in wheat grain by co-inoculation of selenobacteria and arbuscular mycorrhizal fungi: a preliminary study as a potential Se biofortification strategy. J Cereal Sci 57(3):275–280

Dykman LA, Shchyogolev SY (2017) Interactions of plants with noble metal nanoparticles (review). S-h biol 52(1):13–24

Elamawi RM, Tahoon AM, Elsharnoby DE, El-Shafey RA (2020) Bio-production of silica nanoparticles from rice husk and their impact on rice Bakanae disease and grain yield. Arch Phytopathol Plant Protect 53:459–478

Elmer W, White JC (2018) The future of nanotechnology in plant pathology. Annu Rev Phytopathol 56:111–133

El-Saadony MT, Abd El-Hack ME, Taha AE, Fouda MMG, Ajarem JS, Maodaa SN, Allam AA, Elshaer N (2020) Ecofriendly synthesis and insecticidal application of copper nanoparticles against the storage pest *Triboliumcastaneum*. Nanomaterials (Basel) 10(3):587

El-Saadony MT, Saad AM, Soliman SM, Salem HM, Desoky E-SM, Babalghith AO, El-Tahan AM, Ibrahim OM, Ebrahim AAM et al (2022) Role of nanoparticles in enhancing crop tolerance to abiotic stress: a comprehensive review. Front Plant Sci 13:946717

El-Samad LM, Hassan MA, Bakr NR, El-Ashram S, Radwan EH, Aziz KKA, Hussein HK, El Wakil A (2022a) Insights into Ag-NPs-mediated pathophysiology and ultrastructural aberrations in ovarian tissues of darkling beetles. Sci Rep 12:13899. https://doi.org/10.1038/s41598-022-17712-z

El-Samad LM, Bakr NR, El-Ashram S, Radwan EH, Aziz KKA, Hussein HK, El Wakil A, Hassan MA (2022b) Silver nanoparticles instigate physiological, genotoxicity, and ultrastructural anomalies in midgut tissues of beetles. Chem Biol Interact 367:110166. https://doi.org/10.1016/j.cbi.110166

El-Shetehy M, Moradi A, Maceroni M, Maceroni M, Reinhardt D, Petri-Fink A et al (2020) Silica nanoparticles enhance disease resistance in Arabidopsis plants. Nat Nanotechnol 16:344–353

El-Shewy E (2019) The efficacy of copper oxide, tri-calcium phosphate and silicon dioxide nanoparticles in controlling black scurf disease of potato. Ann Agric Sci Moshtohor 57:129–138

Erci F, Cakir-Koc R (2021) Rapid green synthesis of noncytotoxic iron oxide nanoparticles using aqueous leaf extract of *Thymbra spicata* and evaluation of their antibacterial, antibiofilm, and antioxidant activity. Inorg Nano Met Chem 51(5):683–692

Eroglu E, Chen X, Bradshaw M, Agarwal V, Zou J, Stewart SG, Duan X, Lamb RN, Smith SM et al (2013) Biogenic production of palladium nanocrystals using microalgae and their immobilization on chitosan nanofibers for catalytic applications. RSC Adv 3(4):1009–1012

Eugenio M, Müller N, Frasés S, Almeida-Paes R, Lima LMTR, Lemgruber L, Farina M, de Souza W, Sant'Anna C (2016) Yeast-derived biosynthesis of silver/silver chloride nanoparticles and their antiproliferative activity against bacteria. RSC Adv 6(12):9893–9904

Farhana Munis MFH, Alamer KH, Althobaiti AT, Kamal A, Liaquat F, Haroon U, Ahmed J, Chaudhary HJ, Attia H (2022) ZnO nanoparticle-mediated seed priming induces biochemical and antioxidant changes in chickpea to alleviate fusarium wilt. J Fungi 8:753

Fatemi H, Esmaiel Pour BE, Rizwan M (2021) Foliar application of silicon nanoparticles affected the growth, vitamin C, flavonoid, and antioxidant enzyme activities of coriander (*Coriandrum sativum* L.) plants grown in lead (Pb)-spiked soil. Environ Sci Pollut Res Int 28(2):1417–1425

Feng Y, Li Y, Cui L, Yan L, Zhao C, Dong Y (2019) Cold condensing scrubbing method for fine particle reduction from saturated flue gas. Energy 171:1193–1205

Fernández JG, Almeida AC, Fernández-Baldo MA, Felici E, Raba J, Sanz MI (2016) Development of nitrocellulose membrane filters impregnated with different biosynthesized silver nanoparticles applied to water purification. Talanta 146:237-243. https://doi.org/10.1016/j.talanta.2015.08.060

Fichman Y, Mittler R (2020) Rapid systemic signaling during abiotic and biotic stresses: is the ROS wave master of all trades? Plant J 102(5):887–896

Food and Agriculture Organization (2017) The future of food and agriculture: trends and challenges. In: 296: Gas: food and agriculture organization (Food and Agriculture Organization)-phase synthesis of nanoparticles. Wiley, pp 1–180

Gao M, Zhou J, Liu H, Zhang W, Hu Y, Liang J, Zhou J (2018) Foliar spraying with silicon and selenium reduces cadmium uptake and mitigates cadmium toxicity in rice. Sci Total Environ 631-632:1100–1108

Gilroy S, Suzuki N, Miller G, Choi WG, Toyota M, Devireddy AR, Mittler R (2014) A tidal wave of signals: calcium and ROS at the forefront of rapid systemic signaling. Trends Plant Sci 19(10):623–630

Gohari G, Mohammadi A, Akbari A, Panahirad S, Dadpour MR, Fotopoulos V et al (2020) Titanium dioxide nanoparticles (TiO2 NPs) promote growth and ameliorate salinity stress effects on essential oil profile and biochemical attributes of *Dracocephalummoldavica*. Sci Rep 10:912–914. https://doi.org/10.1038/s41598-020-57794-1

Hao Y, Ma C, Zhang Z, Song Y, Cao W, Guo J, Zhou G, Rui Y, Liu L et al (2018a) Carbon nanomaterials alter plant physiology and soil bacterial community composition in a rice-soil-bacterial ecosystem. Environ Pollut 232:123–136

Hao Y, Yuan W, Ma C, White JC, Zhang Z, Adeel M, Zhou T, Rui Y, Xing B (2018b) Engineered nanomaterials suppress turnip mosaic virus infection in tobacco (*Nicotiana benthamiana*). Environ Sci Nano 5(7):1685–1693

Harish V, Ansari MM, Tewari D, Gaur M, Yadav AB, García-Betancourt M-L, Abdel-Haleem FM, Bechelany M, Barhoum A (2022) Nanoparticle and nanostructure synthesis and controlled growth methods. Nano 12(18):3226. https://doi.org/10.3390/nano12183226

Hasnain M, Munir N, Siddiqui ZS, Ali F, El-Keblawy A, Abideen Z (2022) Integral approach for the evaluation of sugar cane bio-waste molasses and effects on algal lipids and biodiesel production. Waste Biomass Valor 1:20

Hayat F, Khanum F, Li J, Iqbal S, Khan U, Javed HU, Razzaq MK, Altaf MA, Peng Y et al (2023) Nanoparticles and their potential role in plant adaptation to abiotic stress in horticultural crops: a review. Sci Hortic 321:12285

Hemlata MPR, Singh AP, Tejavath KK (2020) Biosynthesis of silver nanoparticles using *Cucumis prophetarum* aqueous leaf extract and their antibacterial and antiproliferative activity against cancer cell lines. ACS Omega 5(10):5520–5528

Hernández-Díaz JA, Garza-García JJ, Zamudio-Ojeda A, León-Morales JM, López-Velázquez JC, García-Morales S (2021) Plant-mediated synthesis of nanoparticles and their antimicrobial activity against phytopathogens. J Sci Food Agric 101(4):1270–1287

Hoang VV (2011) Atomic mechanism of vitrification process in simple monatomic nanoparticles. Eur Phys J D 61(3):627–635

Hochella MF, Mogk DW, Ranville J, Allen IC, Luther GW, Marr LC, McGrail BP, Murayama M, Qafoku NP et al (2019) Natural, incidental, and engineered nanomaterials and their impacts on the earth system. Science 363(6434):eaau8299

Hussain A, Ali S, Rizwan M, Rehman MZU, Qayyum MF, Wang H, Rinklebe J (2019) Responses of wheat (Triticum aestivum) plants grown in a Cd contaminated soil to the application of iron oxide nanoparticles. Ecotoxicol Environ Saf. Qayyum HW, eds. 173:156–164

Hussain Q, Asim M, Zhang R, Khan R, Farooq S, Wu J (2021) Transcription factors interact with ABA through gene expression and signaling pathways to mitigate drought and salinity stress. Biomol Ther 11(8):1159

Ikram M, Raja NI, Javed B, Mashwani Z, Hussain M, Hussain M, Ehsan M, Rafique N, Malik K et al (2020) Foliar applications of bio-fabricated selenium nanoparticles to improve the growth of wheat plants under drought stress. Green Process Synth 9(1):706–714

Iqbal P, Preece JA, Mendes PM (2012) Nanotechnology: the "top-down" and "bottom-up" approaches. In: Supramolecular chemistry. Wiley

Iravani S, Korbekandi H, Mirmohammadi SV, Zolfaghari B (2014) Synthesis of silver nanoparticles: chemical, physical and biological methods. Res Pharm Sci 9(6):385–406

Irshad A, Rehman RNU, Abrar MM, Saeed Q, Sharif R, Hu T (2021) Contribution of rhizobium–legume symbiosis in salt stress tolerance in *Medicago truncatula* evaluated through photosynthesis, antioxidant enzymes, and compatible solutes accumulation. Sustain For 13(6):3369

Ishwarya R, Vaseeharan B, Kalyani S, Banumathi B, Govindarajan M, Alharbi NS, Kadaikunnan S, Al-Anbr MN, Khaled JM et al (2018) Facile green synthesis of zinc oxide nanoparticles using *Ulva Lactuca* seaweed extract and evaluation of their photocatalytic, antibiofilm and insecticidal activity. J Photochem Photobiol B 178:249–258

Janas D, Koziol KK (2016) Carbon nanotube fibers and films: synthesis, applications and perspectives of the direct-spinning method. Nanoscale 8(47):19475–19490

Jeevanandam J, Chan YS, Danquah MK (2016) Biosynthesis of metal and metal oxide nanoparticles. ChemBioEng Rev 3:55–67

Jin H, Guo C, Liu X, Liu J, Vasileff A, Jiao Y, Zheng Y, Qiao SZ (2018) Emerging two-dimensional nanomaterials for electrocatalysis. Chem Rev 118(13):6337–6408

Jo YK, Kim BH, Jung G (2009) Antifungal activity of silver ions and nanoparticles on phytopathogenic fungi. Plant Dis 93(10):1037–1043

Kadhum Alghanimi SM, Hadi SS (2021) The environmental effects of nanopowder against some microbes that isolated from oral cavity. IOP Conf Ser Earth Environ Sci 790(1):012067

Kah M, Tufenkji N, White JC (2019) Nano-enabled strategies to enhance crop nutrition and protection. Nat Nanotechnol 14(6):532–540

Kanmani P, Lim ST (2013) Synthesis and structural characterization of silver nanoparticles using bacterial exopolysaccharide and its antimicrobial activity against food and multidrug resistant pathogens. Process Biochem 48(7):1099–1106

Kareem HA, Saleem MF, Saleem S, Rather SA, Wani SH, Siddiqui MH, Alamri S, Kumar R, Gaikwad NB et al (2022) Zinc oxide nanoparticles interplay with physiological and biochemical attributes in terminal heat stress alleviation in mungbean (*Vigna radiata* L.). Front Plant Sci 13(842349):1–15

Katouah H, El-Metwaly NM (2021) Plasma treatment toward electrically conductive and superhydrophobic cotton fibers by in situ preparation of polypyrrole and silver nanoparticles. React Funct Polym 159:104810

Keller AA, McFerran S, Lazareva A, Suh S (2013) Global life cycle releases of engineered nanomaterials. J Nanopart Res 15(6):1–17

Khaldari I, Naghavi MR, Motamedi E (2021) Synthesis of green and pure copper oxide nanoparticles using two plant resources via solid-state route and their phytotoxicity assessment. RSC Adv 11(6):3346–3353

Khan MR, Siddiqui ZA (2020) Use of silicon dioxide nanoparticles for the management of *Meloidogyne incognita, Pectobacteriumbetavasculorum* and *Rhizoctonia solani* disease complex of beetroot (Beta Vulgaris L.). Sci Hortic 2020(265):109211

Khan ST, Ahmad J, Ahamed M, Jousset A (2018) Sub-lethal doses of widespread nanoparticles promote antifungal activity in Pseudomonas protegens CHA0. Sci Total Environ 627:658–662

Khan ZS, Rizwan M, Hafeez M, Ali S, Javed MR, Adrees M (2019) The accumulation of cadmium in wheat (Triticum aestivum) as influenced by zinc oxide nanoparticles and soil moisture conditions. Environ Sci Pollut Res 26:19859–19870. https://doi.org/10.1007/s11356-019-05333-5

Khan A, Roy A, Bhasin S, Bin Emran T, Khusro A, Eftekhari A, Moradi O, Rokni H, Karimi F (2022) Nanomaterials: an alternative source for biodegradation of toxic dyes. Food Chem Toxicol 164:112996. https://doi.org/10.1016/j.fct.2022.112996

Kochian LV, Piñeros MA, Liu J, Magalhaes JV (2015) Plant adaptation to acid soils: the molecular basis for crop aluminum resistance. Annu Rev Plant Biol 66:571–598

Kollist H, Zandalinas SI, SenGupta S, Nuhkat M, Kangasj J (2018) Rapid responses to abiotic stress: priming the landscape for the signal transduction network. Trends Plant Sci 24:25–37

Komal S, Sonia S, Kukreti S, Kaushik M (2019) Exploring the potential of environment friendly silver nanoparticles for DNA interaction: physicochemical approach. J Photochem Photobiol B 194:158–165

Konate A, He X, Zhang Z, Ma Y, Zhang P, Alugongo G, Rui Y (2017) Magnetic (Fe_3O_4) nanoparticles reduce heavy metals uptake and mitigate their toxicity in wheat seedling. Sustain For 9(5):790

Kranjc E, Drobne D (2019) Nanomaterials in plants: a review of hazard and applications in the agri-food sector. Nano 9:1094. https://doi.org/10.3390/nano9081094

Kumar V, Guleria P, Kumar V, Yadav SK (2013) Gold nanoparticle exposure induces growth and yield enhancement in Arabidopsis thaliana. Sci Total Environ 461–462:462–468

Kumar Y, Tiwari KN, Singh T, Sain NK, Laxmi S, Verma R, Sharma GC, Raliya R (2020) Nanofertilizers for enhancing nutrient use efficiency, crop productivity and economic returns in winter season crops of Rajasthan. Ann Plant Soil Res 22(4):324–335. Ann. Plant.

Lamsal K, Kim SW, Jung JH, Kim YS, Kim KS, Lee YS (2011) Application of silver nanoparticles for the control of Colletotrichum species in vitro and pepper anthracnose disease in field. Mycobiology 39(3):194–199

Li Y, Zhu N, Liang X, Bai X, Zheng L, Zhao J, Li Y-f, Zhang Z, Gao Y (2020) Silica nanoparticles alleviate mercury toxicity via immobilization and inactivation of Hg(ii) in soybean (Glycine max). Environ Sci Nano 7(6):1807–1817

Liu C, Li F, Luo C, Liu X, Wang S, Liu T, Li X (2009) Foliar application of two silica sols reduced cadmium accumulation in rice grains. J Hazard Mater 161(2–3):1466–1472

Liu J, Li G, Chen L, Gu J, Wu H, Li Z (2021) Cerium oxide nanoparticles improve cotton salt tolerance by enabling better ability to maintain cytosolic K+/NA+ ratio. J Nanobiotechnology 19:153. https://doi.org/10.1186/s12951-021-00892-7

Ma JF, Yamaji N (2008) Functions and transport of silicon in plants. Cell Mol Life Sci 65(19): 3049–3057

Mahato DK, Mishra AK, Kumar P (2021) Nanoencapsulation for Agri-food applications and associated health and environmental concerns. Front Nutr 8:663229

Mahmoud LM, Shalan AM, El-Boray MS, Vincent CI, El-Kady ME, Grosser JW et al (2022) Application of silicon nanoparticles enhances oxidative stress tolerance in salt stressed 'Valencia' sweet orange plants. Sci Hortic 295:110856. https://doi.org/10.1016/j.scienta.2021.110856

Mantzaris NV (2005) Liquid-phase synthesis of nanoparticles: particle size distribution dynamics and control. Chem Eng Sci 60(17):4749–4770

Manzoor N, Ahmed T, Noman M, Shahid M, Nazir MM, Ali L et al (2021) Iron oxide nanoparticles ameliorated the cadmium and salinity stresses in wheat plants, facilitating photosynthetic pigments and restricting cadmium uptake. Sci Total Environ 769:145221. https://doi.org/10.1016/j.scitotenv.2021.145221

Marcelo GA, Lodeiro C, Capelo JL, Lorenzo J, Oliveira E (2020) Magnetic, fluorescent and hybrid nanoparticles: from synthesis to application in biosystems. Mater Sci Eng C Mater Biol Appl 106:110104

Mirzajani F, Askari H, Hamzelou S, Schober Y, Römpp A, Ghassempour A, Spengler B (2014) Proteomics study of silver nanoparticles toxicity on *Oryza sativa* L. Ecotoxicol Environ Saf 108: 335–339

Misra AN, Murmu B, Singh P, Misra M (1996) Growth and proline accumulation in mungbean seedlings as affected by sodium chloride. Biol Plant 38(4):531–536

Mittler R, Blumwald E (2015) The roles of ROS and ABA in systemic acquired acclimation. Plant Cell 27(1):64–70

Morris LA (2004) Soil biology and tree growth | soil organic matter forms and functions. In: Burley J (ed) Encyclopedia of forest sciences. Elsevier, pp 1201–1207

Mozafari A, Havas F, Ghaderi N (2018) Application of iron nanoparticles and salicylic acid in in vitro culture of strawberries (Fragaria × ananassa Duch.) to cope with drought stress. Plant Cell Tissue Organ Cult 132(3):511–523

Munir N, Hanif M, Dias DA, Abideen Z (2021) The role of halophytic nanoparticles towards the remediation of degraded and saline agricultural lands. Environ Sci Pollut Res Int 28(43): 60383–60405

Nalluri SR, Nagarjuna R, Patra D, Ganesan R, Balaji G (2019) Large scale solid-state synthesis of catalytically active Fe3O4@M (M = Au, Ag and Au-Ag alloy) Core-shell nanostructures. Sci Rep 9(1):6603

Namjoyan S, Sorooshzadeh A, Rajabi A, Aghaalikhani M (2020) Nano-silicon protects sugar beet plants against water deficit stress by improving the antioxidant systems and compatible solutes. Acta Physiol Plant 42(10):157

Nandini B, Hariprasad P, Prakash HS, Shetty HS, Geetha N (2017) Trichogenic-selenium nanoparticles enhance disease suppressive ability of trichoderma against downy mildew disease caused by Sclerosporagraminicola in pearl millet. Sci Rep 7(1):2612–2612

Nazir MM, Li Q, Noman M, Ulhassan Z, Ali S, Ahmed T, Zeng F, Zhang G (2022) Calcium oxide nanoparticles have the role of alleviating arsenic toxicity of barley. Front Plant Sci 13:843795

Nejatzadeh F (2021) Effect of silver nanoparticles on salt tolerance of *Satureja hortensis* l. during in vitro and in vivo germination tests. Heliyon 7:e05981. https://doi.org/10.1016/j.heliyon.2021.e05981

Nikam AV, Prasad BLV, Kulkarni AA (2018) Wet chemical synthesis of metal oxide nanoparticles: a review. CrystEngComm 20(35):5091–5107

Nnaji CO, Jeevanandam J, Chan YS, Danquah MK, Pan S, Barhoum A (2018) Engineered nanomaterials for wastewater treatment: current and future trends. In: Fundamentals of nanoparticles. Elsevier, pp 129–168

Noman M, Ahmed T, Hussain S, Niazi MBK, Shahid M, Song F (2020) Biogenic copper nanoparticles synthesized by using a copper-resistant strain Shigella flexneri SNT22 reduced the translocation of cadmium from soil to wheat plants. J Hazard Mater 398:123175. https://doi.org/10.1016/j.jhazmat.2020.123175

Nouman W, Basra SMA, Yasmeen A, Gull T, Hussain SB, Zubair M, Gul R (2014) Seed priming improves the emergence potential, growth and antioxidant system of *Moringa oleifera* under saline conditions. Plant Growth Regul 73(3):267–278

Oukarroum A, Barhoumi L, Pirastru L, Dewez D (2013) Silver nanoparticle toxicity effect on growth and cellular viability of the aquatic plant *Lemna gibba*. Environ Toxicol Chem 32(4): 902–907

Palithya S, Gaddam SA, Kotakadi VS, Penchalaneni J, Challagundla VN (2021) Biosynthesis of silver nanoparticles using leaf extract of *Decaschistiacrotonifolia* and its antibacterial, antioxidant, and catalytic applications. Green Chem Lett Rev 14(1):137–152

Panigrahi S, Kundu S, Ghosh SK, Nath S, Pal T (2004) General method of synthesis for metal nanoparticles. J Nanopart Res 6(4):411–414

Parsons JG, Peralta-Videa JR, Gardea-Torresdey JL (2007) Use of plants in biotechnology: synthesis of metal nanoparticles by inactivated plant tissues, plant extracts, and living plants. Dev Environ Sci 5:463–485

Patel JK, Patel A, Bhatia D (2021) Introduction to nanomaterials and nanotechnology. In: Emerging technologies for nanoparticle manufacturing. Springer, pp 3–23

Pesheck P, Lorence M (eds) (2009) Development of packaging and products for use in microwave ovens. Elsevier

Prasad S, Kumar V, Kirubanandam S, Barhoum A (2018) Engineered nanomaterials: nanofabrication and surface functionalization. In: Emerging applications of nanoparticles and architectural nanostructures. Elsevier, Inc., pp 305–340. isbn:978-0-12-813516-7

Qiu J, Chen Y, Liu Z, Wen H, Jiang N, Shi H, Kou Y (2023) The application of zinc oxide nanoparticles: an effective strategy to protect rice from rice blast and abiotic stresses. Environ Pollut 331:121925

Rady MM, Boriek SHK, Abd El-mageed TA, Seif El-Yazal MA, Ali EF, Hassan FAS, Abdelkhalik A (2021a) Exogenous gibberellic acid or dilute bee honey boosts drought stress tolerance in *Vicia faba* by rebalancing osmoprotectants, antioxidants, nutrients, and phytohormones. Plants (Basel) 10(4):748

Rady MM, Desoky ESM, Ahmed SM, Majrashi A, Ali EF, Arnaout SMAI et al (2021b) Foliar nourishment with nano-selenium dioxide promotes physiology, biochemistry, antioxidant defenses, and salt tolerance in Phaseolus vulgaris. Plan Theory 10:1189. https://doi.org/10.3390/plants10061189

Rajput VD, Minkina T, Kumari A, Singh VK, Harish KK, Verma S, Mandzhieva S, Sushkova S, Srivastava C, Keswani (2021) Coping with the challenges of abiotic stress in plants: new dimensions in the field application of nanoparticles. Plan Theory 10(6):1221

Ramasamy MYAM (2020) Role of fungi in agriculture, biostimulants in plant science. Intech. Open, p 000

Rangaraj S, Gopalu K, Muthusamy P, Rathinam Y, Venkatachalam R, Narayanasamy K (2014) Augmented biocontrol action of silica nanoparticles and *Pseudomonas fluorescens* bioformulant in maize (Zea mays L.). RSC Adv 4(17):8461–8465

Ranjan S, Dasgupta N, Rajendran B, Avadhani GS, Ramalingam C, Kumar A (2016) Microwave-irradiation-assisted hybrid chemical approach for titanium dioxide nanoparticle synthesis: microbial and cytotoxicological evaluation. Environ Sci Pollut Res Int 23(12):12287–12302

Rehman MZU, Rizwan M, Hussain A, Saqib M, Ali S, Sohail MI, Shafiq M, Hafeez F (2018) Alleviation of cadmium (cd) toxicity and minimizing its uptake in wheat (Triticum aestivum) by using organic carbon sources in cd-spiked soil. Environ Pollut 241:557–565

Rizwan M, Ali S, Ali B, Adrees M, Arshad M, Hussain A et al (2019) Zinc and iron oxide nanoparticles improved the plant growth and reduced the oxidative stress and cadmium concentration in wheat. Chemosphere 214:269–277. https://doi.org/10.1016/j.chemosphere.2018.09.120

Rossi L, Zhang W, Ma X (2017) Cerium oxide nanoparticles alter the salt stress tolerance of Brassica napus L. by modifying the formation of root apoplastic barriers. Environ Pollut 229:132–138

Samyn P, Barhoum A (2018) Engineered nanomaterials for papermaking industry. In: Fundamentals of nanoparticles. Elsevier, Amsterdam, pp 245–277

Sánchez-Ahijón E, Marín-Gamero R, Molero-Sánchez B, Ávila-Brande D, Manjón-Sanz A, Fernández-Díaz MT, Morán E, Schmidt R, Prado-Gonjal J (2020) From theory to experiment: BaFe0.125 Co0.125 Zr0.75 O3 −δ, a highly promising cathode for intermediate temperature SOFCs. J Mater Chem A 8(6):3413–3420

Sanjari S, Shobbar ZS, Ghanati F, Afshari-Behbahanizadeh S, Farajpour M, Jokar M, Khazaei A, Shahbazi M (2021) Molecular, chemical, and physiological analyses of sorghum leaf wax under post-flowering drought stress. Plant Physiol Biochem 159:383–391

Sarraf M, Vishwakarma K, Kumar V, Arif N, Das S, Johnson R, Janeeshma E, Puthur JT, Aliniaeifard S et al (2022) Metal/metalloid-based nanomaterials for plant abiotic stress tolerance: an overview of the mechanisms. Plants (Basel) 11(3):316

Satti SH, Raja NI, Ikram M, Oraby HF, Mashwani Z-U-R, Mohamed AH, Singh A, Omar AA (2022) Plant-based titanium dioxide nanoparticles trigger biochemical and proteome modifications in Triticum Aestivum L. under biotic stress of *Puccinia striiformis*. Molecules 27:4274

Savi GD, Piacentini KC, de Souza SR, Costa ME, Santos CM, Scussel VM (2015) Efficacy of zinc compounds in controlling fusarium head blight and deoxynivalenol formation in wheat (Triticum aestivum L.). Int J Food Microbiol 205:98–104

Schmidt OG, Deneke C, Nakamura Y, Zapf-Gottwick R, Müller C, Jin-Phillipp NY (2002) Nanotechnology—bottom-up meets top-down. In: Advances in solid state physics. Springer, pp 231–240

Schwabe F, Schulin R, Rupper P, Rotzetter A, Stark W, Nowack B (2014) Dissolution and transformation of cerium oxide nanoparticles in plant growth media. J Nanopart Res 16(10):2668

Seleiman MF, Al-Suhaibani N, Ali N, Akmal M, Alotaibi M, Refay Y, Dindaroglu T, Abdul-Wajid HH, Battaglia ML (2021) Drought stress impacts on plants and different approaches to alleviate its adverse effects. Plants (Basel) 10(2):259

Sen AK, De AK (1987) Adsorption of mercury (II). Water Res 21(8):885–888

Servin A, Elmer W, Mukherjee A, De la Torre-Roche R, Hamdi H, White JC, Bindraban PS, Dimkpa C (2015) A review of the use of engineered nanomaterials to suppress plant disease and enhance crop yield. J Nanopart Res 17(2):92

Shaaban A, Al-Elwany OAAI, Abdou NM, Hemida KA, El-Sherif AMA, Abdel-Razek MA, Semida WM, Mohamed GF, Abd El-mageed TA (2022) Filter mud enhanced yield and soil properties of water-stressed *Lupinus termis* L. in Saline Calcareous Soil. J Soil Sci Plant Nutr 22(2):1572–1588

Shah AN, Tanveer M, Abbas A, Fahad S, Baloch MS, Ahmad MI, Saud S, Song Y (2021) Targeting salt stress coping mechanisms for stress tolerance in brassica: a research perspective. Plant Physiol Biochem 158:53–64

Shahbaz M, Akram A, Mehak A, Haq EU, Fatima N, Wareen G, Fitriatin BN, Sayyed RZ, Ilyas N, Sabullah MK (2023) Evaluation of selenium nanoparticles in inducing disease resistance against spot blotch disease and promoting growth in wheat under biotic stress. Plan Theory 12:761

Shang Y, Hasan MK, Ahammed GJ, Li M, Yin H, Zhou J (2019) Applications of nanotechnology in plant growth and crop protection: a review. Molecules 24(14):2558

Siddiqui MH, Al-Whaibi MH, Mohammad F (eds) (2015) Nanotechnology and plant science, first edn. Springer, Cham, pp 19–35
Singh I, Shah K (2014) Exogenous application of methyl jasmonate lowers the effect of cadmium-induced oxidative injury in rice seedlings. Phytochemistry 108:57–66
Singh S, Parihar P, Singh R, Singh VP, Prasad SM (2016) Heavy metal tolerance in plants: role of transcriptomics, proteomics, metabolomics, and ionomics. Front Plant Sci 6:1143
Singh AK, Kumar A, Sharma V, Kala P (2020) Sustainable techniques in grinding: state of the art review. J Clean Prod 269:121876
Sreelakshmi B, Induja S, Adarsh PP, Rahul HL, Arya SM, Aswana S, Haripriya R, Aswathy BR, Manoj PK et al (2021) Drought stress amelioration in plants using green synthesised iron oxide nanoparticles. Mater Today Proc 41(3):723–727
Suriyaprabha R, Karunakaran G, Kavitha K, Yuvakkumar R, Rajendran V, Kannan N (2014) Application of silica nanoparticle in maize to enhance fungal resistance. IET Nanobiotechnol 8(3):133–137
Tadic M, Trpkov D, Kopanja L, Vojnovic S, Panjan M (2019) Hydrothermal synthesis of hematite (α-Fe2O3) nanoparticle forms: synthesis conditions, structure, particle shape analysis, cytotoxicity and magnetic properties. J Alloys Compd 792:599–609
Teske SS, Detweiler CS (2015) The biomechanisms of metal and metal-oxide nanoparticles' interactions with cells. Int J Environ Res Public Health 12(2):1112–1134
Thimsen E, Kortshagen UR, Aydil ES (2014) Plasma synthesis of stoichiometric Cu_2S nanocrystals stabilized by oleylamine. Chem Commun (Camb) 50(61):8346–8349
Tripathi DK, Singh VP, Prasad SM, Chauhan DK, Dubey NK (2015) Silicon nanoparticles (SiNp) alleviate chromium (VI) phytotoxicity in *Pisum sativum* (L.) seedlings. Plant Physiol Biochem 96:189–198
Wahid I, Kumari S, Ahmad R, Hussain SJ, Alamri S, Siddiqui MH, Khan MIR (2020) Silver nanoparticle regulates salt tolerance in wheat through changes in ABA concentration, ion homeostasis, and defense systems. Biomol Ther 10(11):1506
Wahid I, Rani P, Kumari S, Ahmad R, Hussain SJ, Alamri S, Tripathy N, Khan MIR (2021) Biosynthesized gold nanoparticles maintained nitrogen metabolism, nitric oxide synthesis, ions balance, and stabilizes the defense systems to improve salt stress tolerance in wheat. Chemosphere 287:132142. https://doi.org/10.1016/j.chemosphere.2021.132142
Wang X, Liu X, Chen J, Han H, Yuan Z (2014) Evaluation and mechanism of antifungal effects of carbon nanomaterials in controlling plant fungal pathogen. Carbon 68:798–806
Wang S, Wang F, Gao S (2015) Foliar application with nano-silicon alleviates Cd toxicity in rice seedlings. Environ Sci Pollut Res Int 22(4):2837–2845
Wani TU, Pandith AH, Sheikh FA (2021) Polyelectrolytic nature of chitosan: influence on physicochemical properties and synthesis of nanoparticles. J Drug Deliv Sci Technol 65:102730
Xing K, Shen X, Zhu X, Ju X, Miao X, Tian J, Feng Z, Peng X, Jiang J et al (2016) Synthesis and in vitro antifungal efficacy of oleoyl-chitosan nanoparticles against plant pathogenic fungi. Int J Biol Macromol 82:830–836
Yasmin H, Mazher J, Azmat A, Nosheen A, Naz R, Hassan MN et al (2021) Combined application of zinc oxide nanoparticles and biofertilizer to induce salt resistance in safflower by regulating ion homeostasis and antioxidant defence responses. Ecotoxicol Environ Saf 218:112262. https://doi.org/10.1016/j.ecoenv.2021.112262
Zhou H, Wu H, Zhang F, Su Y, Guan W, Xie Y et al (2021) Molecular basis of cerium oxide nanoparticle enhancement of rice salt tolerance and yield. Environ Sci Nano 8:3294–3311. https://doi.org/10.1039/D1EN00390A
Zulfiqar F, Ashraf M (2021) Nanoparticles potentially mediate salt stress tolerance in plants. Plant Physiol Biochem 160:257–268

Chapter 10
Nanotechnology Application in Crop Improvement

Nisha Pallath, Tharakupeedikayil Abdul Majeed Sajeena, Aswathy Vijay, and Shalet Varghese

Abstract Nanotechnology has emerged as a transformative force in modern agriculture, offering innovative solutions to enhance crop productivity, sustainability, and resource efficiency. By manipulating materials at the nanoscale, researchers are developing advanced agrochemicals, precision delivery systems, and nanosensors that significantly improve soil health, pest management, and nutrient uptake. This paper explores the various applications of nanotechnology in agriculture, including the use and types of nanofertilizers that optimize nutrient release, nanopesticides that minimize chemical runoff and toxicity, and nanosensors that provide real-time monitoring of soil and crop conditions. Additionally, we discuss the potential benefits of these technologies in addressing global challenges such as food security, environmental sustainability, and climate change. Despite the promising advancements, the paper also highlights the ethical considerations and regulatory challenges associated with the adoption of nanotechnology in agriculture, emphasizing the need for responsible innovation. Overall, the integration of nanotechnology in agricultural practices presents a significant opportunity to revolutionize the sector, ensuring a more resilient and sustainable food system for the future.

Keywords Crop protection · Nanofertilizers · Nanopesticides · Nanosensors · Nanotechnology · Soil remediation

10.1 Introduction

Agriculture has been the bedrock of human civilization for millions of years. Agriculture has always been crucial to the provision of food and nourishment for people all over the world, from the earliest hunter-gatherer tribes to the intricate systems of contemporary industrial farming (Bowden 2022). The agricultural sector is confronted with a wide range of issues, including a lack of nutrients, a halt in crop

N. Pallath (✉) · T. A. M. Sajeena · A. Vijay · S. Varghese
Department of Biosciences, MES College Marampally, North Vazhakulam, Aluva, Ernakulam (Dist), Kerala, India
e-mail: nishap@mesmarampally.org

J. M. Al-Khayri et al. (eds.), *Nanotechnology Applications in Modern Agriculture*, Nanotechnology in Plant Sciences 3, https://doi.org/10.1007/978-3-031-90506-3_10

yield, a reduction soil organic matter, shifting climate patterns, a reduction in arable land and water availability, opposition to genetically modified organisms, and a labour shortage (Muzhinji and Ntuli 2021). Furthermore, the amount of land that can be used for farming is becoming increasingly limited due to political and economic issues such as population increase, climate change, and patterns of land use. Additionally, the increasing rate of urbanization is placing constant pressure on the amount of arable land that is available. Therefore, these elements would pose serious concerns to crop production since arable land is diminishing. Furthermore, over the past few decades, a lack of arable land has already impacted grain production (Hassan 2018).

Initiatives such as precision agriculture, agroecology, and organic farming are being undertaken in an attempt to address these issues and advance sustainable agricultural methods. Much of human history has seen the implementation of significant technological breakthroughs to boost agricultural production with limited resources (Misra et al. 2016). There is also a rising emphasis on strengthening farming systems and enhancing farmers' access to technology, knowledge, and resources, particularly in developing countries. Current studies have demonstrated the intriguing potential of nanotechnology to boost crop productivity and food security by boosting the efficacy of agricultural inputs and providing solutions to issues related to the environment and agriculture (Kah et al. 2018). Various roles of nanotechnology in agriculture are graphically represented in Fig. 10.1.

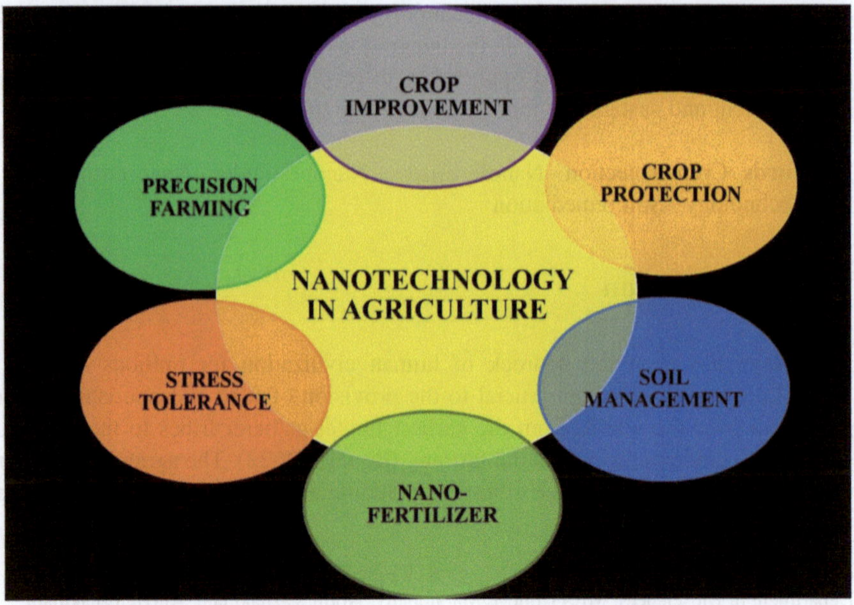

Fig. 10.1 Role of nanotechnology in agriculture. (Constructed by Nisha P)

Nanotechnology is considered the sixth breakthrough technology in the current period. Nanotechnology, the manipulation of matter at the atomic and molecular levels, has shown enormous promise in a variety of sectors, including medicine, electronics, and energy. The unique physicochemical characteristics of nanoparticles, such as their large surface area, high reactivity, and configurable pore size, offer a vast array of novel uses in the field of plant nutrition to meet the demands of the growing population (Abobatta 2018). Nanomaterials can interact either directly or indirectly with the majority of the biological system, based on their unique features. Plant-derived nanoparticles have significant advantages over other bio-systems due to their availability and cost-effectiveness of biogenic synthesis techniques (Sharma et al. 2019). Nanoparticles are classified depending on their qualities and materials. The nanoparticles of metallic oxides, metalloids, quantum dots, liposomes, dendrimers, non-metals, and carbon nanomaterials are commonly utilized for plant systems (Banerjee et al. 2021). There are currently few scientific studies regarding the application of nanoparticles in agriculture worldwide, even though it is thought to be a growing branch of science that is heavily influenced in many scientific domains and will likely play a major part in food science and agriculture in the future (Raliya et al. 2017). This chapter well explained about the various roles of nanotechnology in Modern agriculture especially in crop improvement and discussed the applications of nanotechnology not only boost crop productivity but also promote sustainable agricultural practices by reducing the reliance on chemical inputs and improving resource efficiency.

10.2 Nanotechnology for Crop Improvement

In recent years, researchers have begun to investigate the potential applications of nanotechnology in agriculture, notably crop development. Scientists can use nanomaterials and nanotechnology-based techniques to boost crop growth, increase yield, improve nutrient uptake, and protect plants from diseases and pests (Ghidan et al. 2017). Nanotechnology in agriculture is gaining global attention in the twenty-first century because of its widespread use in various agricultural systems. Researchers discovered that nanotechnology in agriculture could reduce the usage of water, fertilizers, and pesticides, while also increasing agricultural productivity (Shojaei et al. 2019). Nanotechnology can positively impact agriculture by improving food production (Fig. 10.2), security, and safety. The implementation of nanotechnology into agriculture is primarily intended to improve food quality through the use of techniques like disease identification, increased safeguarding of plants, waste reduction, and plant growth monitoring to promote agriculture sustainability (War et al. 2020). Nanomaterials in agriculture can improve fertilization, enhance yields through nutrient optimization, and reduce the need for plant protection substances (Thakur et al. 2018).

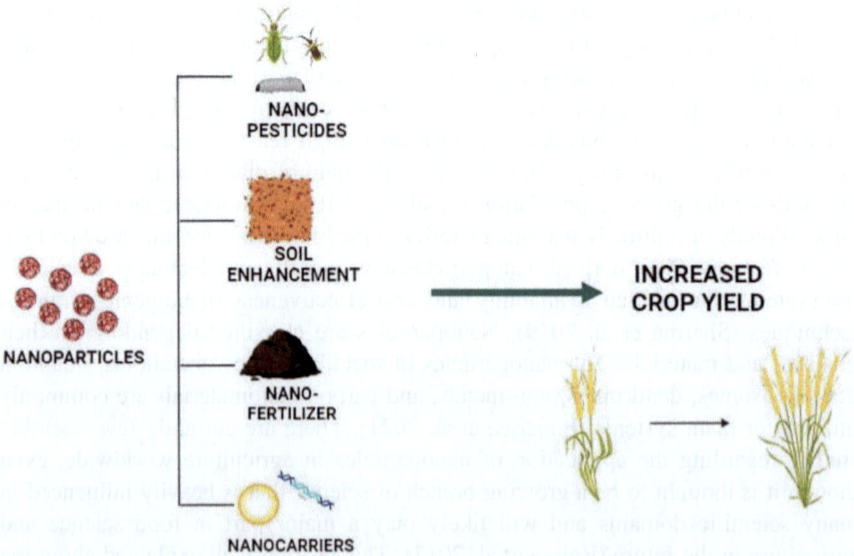

Fig. 10.2 Application of nanoparticles for crop production. (Constructed by Aswathy Vijay)

Nanotechnology has a wide range of potential uses in agricultural enhancement. Researchers can address some of the most pressing issues in agriculture today by leveraging the power of nanomaterials and methods, such as food security, environmental sustainability, and crop improvement (Sharma et al. 2021). The most crucial strategies for increasing crop productivity are applying fertilizers, insecticides, herbicides, and plant growth factors/regulators. The application of nanocarriers can accomplish the controlled release of herbicides, insecticides, and plant growth regulators (Oliveira et al. 2015). One of the most significant advantages of nanotechnology in crop enhancement is the invention of nano fertilizers. Traditional fertilizers frequently have low efficacy because they quickly seep into the soil and are not readily available for plant uptake (Chhipa 2019). Nano fertilizers, on the other hand, are intended to release nutrients gradually and in a controlled manner, ensuring that plants receive the nutrients they require at the appropriate moment. These nano fertilizers can also be customized to meet individual crop requirements, decreasing waste and environmental damage. It can be delivered and released gradually with the perfect use of nanoscale carriers. Such tactics, referred to as "precision farming," increase crop yields without compromising the health of the land or water (Duhan et al. 2017).

Nanotechnology also provides revolutionary crop protection technologies. Nanoparticles can be utilized to distribute pesticides more effectively, lowering the amount of chemicals required and reducing environmental impact (Singh et al. 2021a). Nanomaterials can also be utilized to form barriers that protect plants from pests and diseases, thereby enhancing crop health. Furthermore, nanotechnology-based sensors can detect early signals of plant stress, allowing for timely intervention and reducing crop loss (Shang et al. 2019).

In addition to nutrition delivery and pest management, nanotechnology can help crops become more resilient to environmental shocks (Mittal et al. 2020). Scientists are developing more resilient crops capable of thriving in changing climates by engineering nanomaterials that improve a plant's ability to endure drought, salinity, severe temperatures, and other difficulties (Kumari et al. 2022). The application of nanotechnology in agriculture has the potential to displace traditional farming methods by creating novel technologies that can be utilized to increase crop resilience to disease (Li and Yan 2020). Nanoparticles improve crop physiology by acting as a fertilizer and improving nitrogen uptake. Furthermore, by expressing genes linked to pathogenesis, proteins or genetic material that provides defence against plant pathogens can be transferred via nanoparticles (Ahmad et al. 2022). The use of nano sensors, which are also helpful in eliminating waste environmental pollutants, is another promising technology for pathogen detection (Mahmoud and Fawzy 2021).

Furthermore, nanotechnology can help improve crop quality and nutritional value. Nanoparticles can be utilized to increase the bioavailability of important nutrients in plants, making crops healthier for human consumption (Rana et al. 2021). Nanotechnology can also be utilized to make nano-sized carriers that transfer important substances like vitamins and antioxidants to plants, resulting in healthier and more profitable crops (Alfalahi and Abdulqahar 2021). Many aspects of agriculture have benefited from and been revolutionized by the novelty of nanotechnology, including crop improvement, genetically protected crop protection, crop management, target-specific transport of genes, and other biomolecules, controlled release of agricultural products, and the analysis of gene expressions in plants (War et al. 2020; Wahab et al. 2024).

10.3 Types of Nanoparticles and Their Interaction with Plants

10.3.1 Gold Nanoparticles (AuNPs)

Gold nanoparticles substantially improved seed germination and shoot size of tomato seedlings when compared to control plants. A 21.67% shoot size was elongated in Gold nanoparticle-treated plants, and nanoparticles with smaller sizes were shown to be more effective at provoking growth (Thakur et al. 2018). The AuNP-SCTA (Gold nanoparticles with sodium citrate and tannic acid) treatment increased the quantity and length of lateral roots by 1.7 and 1.5 times, respectively, compared to the controls. AuNP-SCTA exposure affects shoot formation in the same way that it affects the root system. Rosette diameters increased 1.3-fold as compared to control seedlings. These findings indicated that the AuNP-SCTA does not have adverse effects on plants, but rather has beneficial impacts on plant development (Ferrari et al. 2021).

10.3.2 Silica Nanoparticles (SiNPs)

The silica nanoparticles are commonly utilized as pesticide carriers to improve their efficacy, as they raise pesticide dissolution and plant absorption. The use of silica nanoparticle nanoparticles as carriers improved the dissolvability of insoluble pyraoxystrobin in cucumber plants (Xu et al. 2021). The study by Kang et al. (2021) examined the potency of the silica nanoparticles (Nanoparticles) to promote plant growth and prevent *Fusarium* wilt disease in Watermelon *Citrullus lanatus*). In a 2.5-month field investigation, the rapid dissolving NP treatment increased by 81.5% fruit yield to the untreated plants. These findings demonstrated that the colloidal nature of nanoparticles is crucial for nanoparticle-plant interactions resulting in disease control and plant health. Likewise, the foliar and soil feeding of silica nanoparticles have promoted plant growth, and flowering period, and also enhanced the traits of flowers in Marigold (*Tageteserecta*) (Attia and Elhawat 2021). The Silicon nanoparticles along with *Trichoderma* metabolites substantially controlled the cadmium stress in the Tomato (*Lycopersicumescultenum*) plants and the bacterial wilt too (Khan et al. 2023).

10.3.3 Silver Nanoparticles (AgNPs)

The silver nanoparticles can improve the seed germination, physiological indices, and yield of agricultural and horticulture plants. It suppressed the frequency of infections infecting germinating seeds, and its ability to boost physiological processes, plant production, and yields, particularly under environmental stress circumstances, thereby suggesting the sustainable role of silver nanoparticles in agriculture (Jaskulski et al. 2022). Silver nanoparticles along with *Bacillus cereus* LPR2 promoted plant development whereas LPR2 greatly prevented the growth of harmful fungal pathogens in *Zea mays* (Kumar et al. 2020). The findings of Tymoszuk (2021), revealed the beneficial role of silver nanoparticles in the seed germination and growth of plants such as *Brassica oleracea* var. *sabellica, Solanumlycopersicum,* and *Raphanus sativus* L. var. *sativus*.

10.3.4 Selenium Nanoparticles (SeNPs)

The foliar application of Selenium nanoparticles synthesized using *Allium sativum* improved the plant parameters such as root length, leaf length, shoot length, leaf number. at the concentration of 30 mg/L in the drought-stressed wheat plants (Ikram et al. 2020). Interestingly, the role of selenium nanoparticles in the germination of barley seeds was examined by Siddiqui et al. (2021). SeNPs synthesized using *Bacillus megaterium* ATCC 55000 have a potential against *R. solani* dampening and root rot diseases in *Vicia faba* and also serve as a significant plant growth promoter. Se-NPs were found to reduce *R. solani* damping-off both before and after emergence, as well as the extent of root rot disease (Hashem et al. 2021). Intriguingly Ahmed et al. (2020), designed a rapid and sustainable agricultural sensor from biosynthesized SeNPs for the identification of heavy metal toxicity. Heavy metal toxicity is confirmed by the colour of SeNPs turning from red to colourless.

10.3.5 Copper Nanoparticles (CuNPs)

The copper nanoparticles showed dual benefits in tomato plants as they minimized the *Fusarium* wilt severity by inhibiting the *Fusarium oxysporum f. sp. lycopersici* and also served as growth promoters. Moreover, the copper nanoparticles can be utilized as a nutrient source in the copper deficit soil (Lopez-Lima et al. 2020). Copper nanoparticles applied to the plant enhanced the overall number of seeds and grain production under drought-stress circumstances. Plants treated with copper nanoparticles had higher anthocyanin, chlorophyll, and carotenoid content than plants treated with water under drought conditions. Therefore, the Copper nanoparticles affect plant defensive systems related to drought tolerance (Van Nguyen et al. 2022). According to Jahagirdar et al. (2020), CuNPs produced from leaf extract of neem had a favourable influence on the development, nutrition, and seed germination of *Vigna radiata* L.

10.3.6 Zinc Oxide Nanoparticles (ZnONPs)

The foliar treatment of zinc oxide nanoparticles exhibited a significant effect on plant growth and crop yield of *Linumuitatissimum* L (Sadak and Bakry 2020). The zinc oxide nanoparticles in the concentration of 50 to 1600 mg/L^{-1} were applied for the germination of Allium cepa seeds without any adverse effects (Tymoszuk and Wojnarowicz 2020). A concentration of 100 ppm ZnO-NPs resulted in the most beneficial developmental variables, physiological features, yield characteristics, yield, and quality attributes for tomatoes. Foliar spraying with 100 ppm ZnO-NPs can be recommended to boost tomato production as well as quality under glasshouse soil conditions (Ahmed et al. 2023).

10.3.7 Manganese Nanoparticles (MnNPs)

Manganese oxide nanoparticles were tested as a nano priming agent for reducing salt stress in *Capsicum annuum* L. during sprouting at 100 mM NaCl. Mn NPs dramatically increased root development in salt-stressed as well as non-salt seedlings

(Ye et al. 2020). Foliar treatment of MnNPs revealed significant variations from soil treatment such as increased shoot and grain Mn concentrations, decreased soil nitrate content, and increased Phosphorous content in soil and shoot in *Triticum aestivum* L. Thus, delivering MnNPs via foliar may be an approach for modulating their modest effects in soil (Dimkpa et al. 2018). The study analyzed the seed-priming effects of manganese nanoparticles in the *Artemisia annua* plant (Salehi et al. 2023).

10.3.8 Titanium Oxide Nanoparticles (TiO_2NPs)

Plants treated with TiO_2NPs showed enhanced cell division, fewer chromosomal abnormalities, and less DNA damage in the *Vicia faba* plant. The increase of antioxidant genes lends credence to the anti-oxidative properties of TiO_2NPs. It has the potential to promote the overexpression of heat shock protein (HSP) genes and also to attenuate the osmotic and toxic consequences of salinity-induced stress in plants (Omar et al. 2023). TiO_2NPs along with *Bacillus mycoides* PM35 improved the morphological, physical, and biochemical parameters of Barley in cadmium stress conditions (Ma et al. 2023). Plants treated with TiO_2NPs demonstrated improved germination, morphology, and metabolic profile of wheat plants with and without salt stress. Wheat under salt stress showed the best response to 50 μg/mL TiO_2NPs (Badshah et al. 2023).

10.3.9 Iron Oxide Nanoparticles (Fe_2O_3NPs)

Ferric oxide nanoparticles derived from dried Guava (*Psidium guajava*) leaves at concentrations of 15, 10, and 5 mg/L demonstrated a concentration-related decrease in the extent of the disease and its occurrence in the Potato plant (*Solanum tuberosum*). This study also proved the effectiveness of iron oxide nanoparticles as biological control agents, without negative effects on systemic plant development (Anwaar et al. 2024). Likewise, iron oxide nanoparticles to avert the *Fusarium*-wilt disease caused by *Fusarium oxysporum* in eggplants. and also greatly increased the amounts of osmolytes such as proline, soluble sugar, and protein (Elbasuney et al. 2023). Ndou et al. (2023) reported the iron oxide nanoparticles synthesized from *Aspalathus linearis* enhanced the growth of *Sorghum bicolor* in drought conditions.

10.4 Applications of Nanotechnology in Crop Improvement

Nanotechnology has several applications in Crop Improvement, including:

- Nano-fertilizers: Nano-sized nutrient particles can enhance nutrient uptake efficiency, reduce nutrient leaching, and improve crop yields.

- Nano pesticides: Nano-formulations of pesticides can increase their efficacy, reduce environmental impact, and decrease the amount needed for effective pest control.
- Nano sensors: Nanotechnology enables the development of highly sensitive and selective sensors for detecting pathogens, toxins, and environmental conditions, aiding in early disease detection and optimal resource management.
- Nano-enabled delivery systems: Nanocarriers can protect and deliver agricultural inputs, such as nutrients, pesticides, and growth regulators, to target sites within plants, reducing waste and increasing efficiency.
- Nanostructured materials: Nanomaterials like nano clays and nanocomposites can improve soil properties, water retention, and nutrient availability, enhancing soil health and productivity.
- Precision agriculture: Nanotechnology facilitates the development of precision agriculture tools, such as nanoscale imaging and monitoring devices, enabling real-time assessment and management of crop health and environmental conditions.

Soil enhancement Nanomaterials can improve soil properties, such as water retention and nutrient availability, promoting healthier plant growth.

Genetic engineering Nanocarriers can facilitate the delivery of DNA, RNA, or other genetic materials into plant cells, enabling precise genetic modifications for desirable traits.

Disease detection Nanosensors can detect plant pathogens or nutrient deficiencies at early stages, allowing for timely intervention and minimizing crop losses.

Stress tolerance Nanotechnology can enhance plants' resilience to abiotic stresses (e.g., drought, salinity) by improving nutrient uptake or water retention in the soil.

Plant growth regulators Nanocarriers can improve the efficiency of growth regulators, promoting better growth and yield.

Improved seed treatments Nanotechnology can enhance seed coatings to protect against pathogens, improve germination rates, and provide nutrients directly to seedlings.

All these applications illustrate how nanotechnology can lead to more efficient, sustainable, and resilient agricultural practices.

In 1974 'Nanotechnology', was first coined by Norio Taniguchi which later on expanded tremendously. The matter in nano size exhibits some novel properties such as high reactivity, increased bioactivity as well as bioavailability, etc. mainly due to the diminished molecular size and alteration in interaction between the molecules. Nano technology can be employed in various fields such as medicine, electronics, food industry, energy production etc. Some important products are nano chips, green concrete, nano additives, solar cells, nano cells etc. In addition to this the nanotechnology is now widely employed in agriculture due to the novel properties mentioned above (Agrawal and Rathore 2014).

Even though the advancement in industry is playing a pivotal role in our economy, the beneficence of agriculture in the promotion of world's economy is to be considered. In addition to this agriculture play a pivotal role in generation of capital and employment, provision for food and also the supply of raw materials (Seth 2014). In recent years several studies revealed the intense application of nanotechnology utilizing nanoparticles of 1–100 nm in various fields such as pharmaceutical, medical as well as agricultural fields. Out of the various methods for the synthesis of nanoparticles, green synthesis from plants is easier, ecofriendly minimizing pollution problems by hazardous chemicals and less time consuming.

One of the most promising fields to enhance the research in agriculture and food manufacturing is nanotechnology. Recent studies found that by applying the nanomaterials in the field of agriculture we could diminish the dissemination of harmful chemicals, loss of nutrients, enhance the production etc. Food nanotechnology is an important field employing the nanocarriers for strengthening the bioactive components to provide an effective barrier to various alterations. It can also enhance the color, flavor, and texture of the food. Nanoparticles are widely employed in food manufacturing, processing, packaging, and food safety (Ghidan et al. 2019). Nanotechnology holds significant promise for revolutionizing the modern agriculture by promoting the crop productivity, reducing environmental impact as well as improving resource efficiency. Some key applications of nanotechnology in modern agriculture are the use of nano pesticides, nano fertilizers, and nano sensors (Fig. 10.3). Moreover, they are employed in soil remediation, crop improvement, water Management and for precision agriculture.

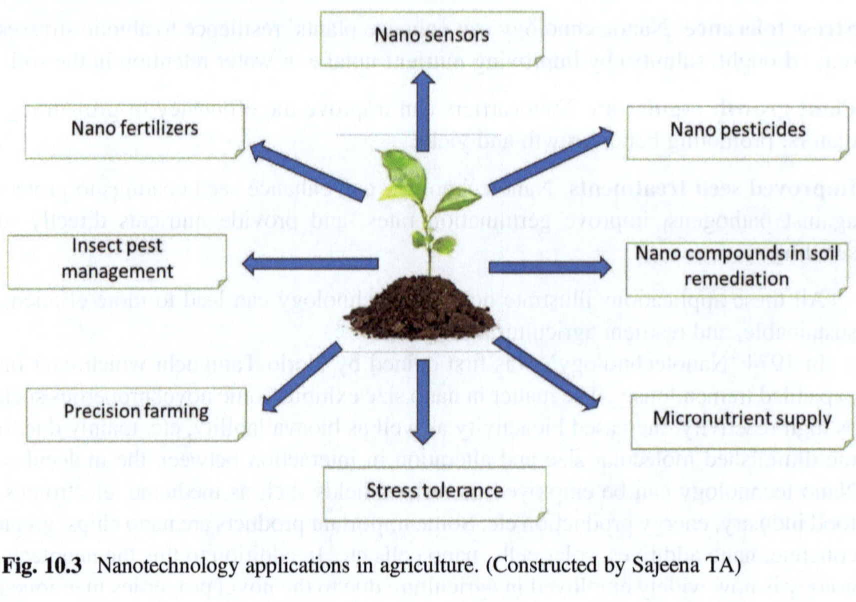

Fig. 10.3 Nanotechnology applications in agriculture. (Constructed by Sajeena TA)

The deficiency of Mult nutrients, stagnant crop yield, shortage of land and water, drastic climatic change etc. paved a way to the scientists in agricultural field to think about the introduction of nanotechnology in agriculture. The recent studies showed that majority of nano based products are from the countries such as France, China, Japan, Korea, USA, Switzerland etc. In India ICAR has taken initiatives for synthesis of nano compounds for improving farming, detection of plant pests and related diseases, introduced the concept of nano pheromones for controlling pests, etc. (RME, Sightsavers India, and Elizabath 2019).

10.4.1 Nano Fertilizers

Nano fertilizers are also called smart fertilizers which are designed to enhance the nutrient absorption, soil fertility, photosynthesis, plant health which automatically results in increased production. Other highlights of nano fertilizers are the reduction in soil toxicity and pollution hazards. They are applied less frequently when compared to the chemical fertilizers. The nanomaterial selection is to be done based on their size, concentration, properties and the nature of crop variety is also important for satisfactory result. Normally employed nanoparticles include silica, iron, oxides of zinc, titanium, cerium, aluminum etc. The nano fertilizers react with water to release nutrients in to soil. Coating with polymers helps to avoid the unnecessary nutrient loss (Singh et al. 2021a, b; Tyagi et al. 2022). In nano fertilizers production the desired nutrients are added either singly or in various combinations to an adsorbent of nano dimension which promotes the fast delivery of the nutrients to target sites (Badgar et al. 2022). The nonfertilizer can be made available as nanoscale-coating fertilizer, nanoscale-additive fertilizer or nanoporous substances (Al-Juthery et al. 2021). List of some Nano fertilizers are given below in Table 10.1,

Table 10.1 List of some Nano Bio fertilizers

Sl. No.	Nano fertilizers	Application	References
1	Hydroxyapatite-containing nano fertilizers	Supply both calcium and phosphorus to plants.	Ahmed et al. (2021)
2	Urea-loaded hydroxyapatite nanohybrids	Gradual release of nitrogen	Ahmed et al. (2021)
3	Mesoporous silica NPs	Enhance the crop quality	de Oliveira et al. (2014)
4	Carbon-based nanomaterials [carbon nanotubes fullerenes, fullerols etc.]	Increase the rate of germination, the chlorophyll content, and the protein content.	Ghormade et al. (2011)
5	Inorganic Nanoparticles such as ZnO	Increases the nutrient concentration and plant growth. Also reduces the diseases.	Raliya and Tarafdar (2013)

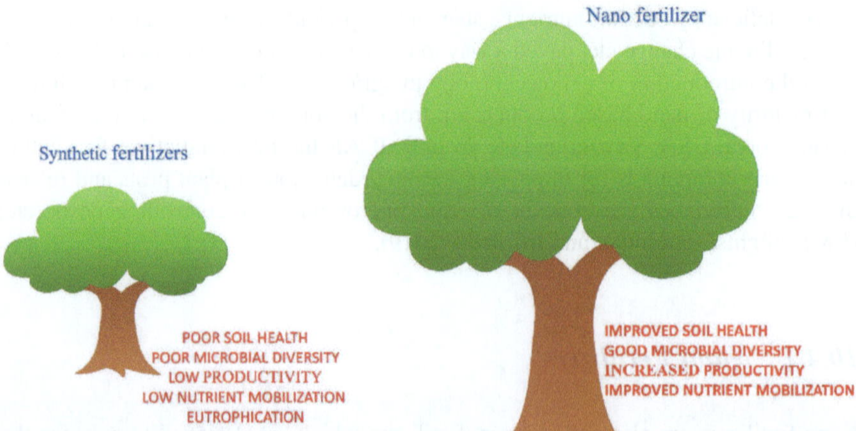

Fig. 10.4 Comparison of synthetic and nano fertilizers. (Constructed by Sajeena TA)

Also comparison between synthetic and Nano fertilizers are diagrammatically represented in Fig. 10.4.

The incorporation of nutrients in nano fertilizer is done by encapsulation, emulsion, nanocoating etc. The output of the nano fertilizer is determined by intrinsic factor (methods employed, size of particle), extrinsic factor (pH, temperature, texture, depth, organic matter and microbial content of soil) and route of administration. Three classes are available- nano formulation of micronutrients, nano formulations of macronutrients and nutrient loaded nanomaterials. Out of these nutrient loaded nanomaterials are more safe, ecofriendly and so more popular (Sindhu et al. 2020).

10.4.2 Nano Pesticides

Nano pesticides are a relatively new and innovative class of pesticides that utilize nanotechnology to enhance the effectiveness, targeted delivery, and efficiency of pest control agents. These nano pesticides typically consist of nano-sized particles that encapsulate or carry active ingredients such as insecticides, fungicides, or herbicides. Some of the types of nano pesticides are represented in Fig. 10.5.

The use of nanotechnology in pesticides has gained attention due to its potential to address some of the limitations associated with traditional pesticide formulations. Nanotechnology enables the development of more efficient and targeted pesticide delivery systems. Nanoencapsulation of pesticides improves their stability, solubility, and controlled release, thereby reducing the amount of chemical required and minimizing environmental pollution and health risks. The important advantages of nano pesticides are the following:

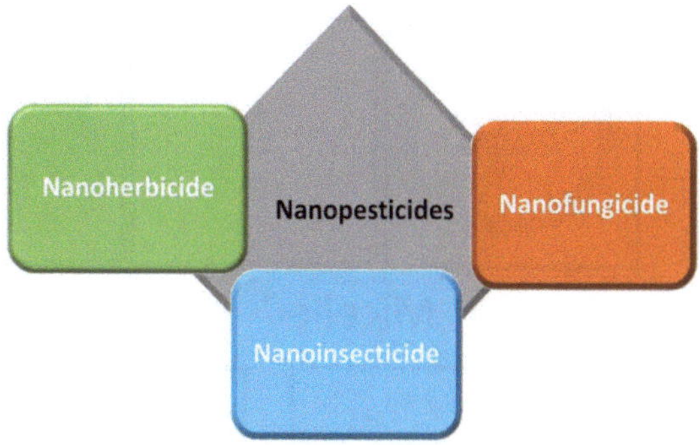

Fig. 10.5 Types of nano pesticides. (Constructed by Sajeena TA)

1. **Enhanced Efficacy:** Nanotechnology offers a precise control of particle size and surface properties, thereby enhancing the efficacy of active ingredients. This aids an improved pest control as well as reduced application rates (Mariyam et al. 2024).
2. **Targeted Delivery:** Another benefit of nano pesticide is the capability of targeted delivery. Nano pesticides can be designed to target specific pests or plant parts, so that we could minimize the off-target effects and reduce major environmental impacts. Development of controlled-release formulations enables the prolonged activity of pesticides, thereby reducing the frequency of applications (Wang et al. 2022).
3. **Reduced Environmental Impact:** By enhancing the efficiency of pesticide delivery and reducing the amount of active ingredients, nano pesticides have the potential to decrease environmental pollutions and significantly minimize risks to non-target organisms (Wang et al. 2022).
4. **Improved Stability:** When compared to the normal formulations the nano-formulations exhibit an improved stability and solubility of its active ingredients. This provides a longer shelf life (Preeti et al. 2023)

Some important nano formulations are listed below in Fig. 10.6.

10.4.3 Nano Emulsions

This includes an oil-in-water emulsion in which the main ingredient of the chemical is distributed as droplets of nano size in water, with surfactant molecules placed at the pesticide-water interface. Nano-emulsions get further classified based on the quantity and type of surfactants, as thermodynamically stable and kinetically stable. The partial solubility of the pesticide in aqueous phase enables the sudden and spontaneous formation of a stable emulsion whereas the insolubility makes the pesticide and surfactant to develop a system with two phases and only a continuous shearing process mix them together and the pesticide in nano emulsion can be seen dispersed for a long time (Mustafa and Hussein 2020).

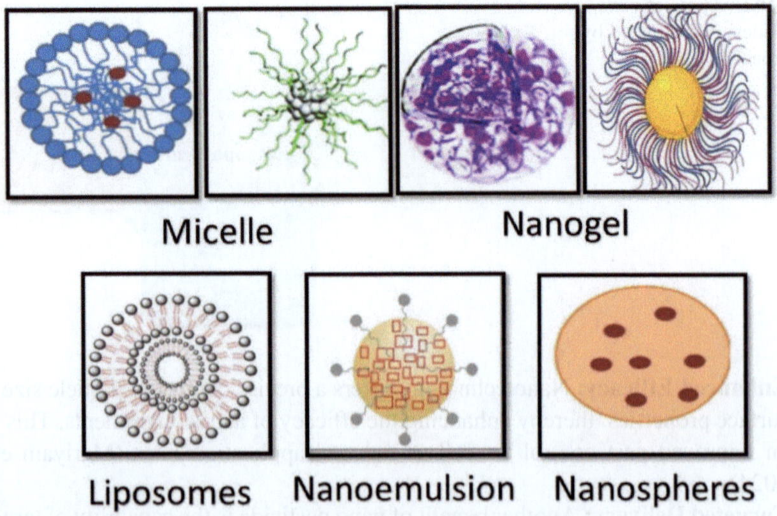

Fig. 10.6 Formulations of nano pesticides. (Constructed by Sajeena TA)

10.4.4 Nano-suspension

Otherwise known as 'nano-dispersions', can be synthesised by the dispersion of pesticide in aqueous medium as nanosized particles. The surfactant molecules can be seen boarded at the surface of the particle. The polar region extend in to the liquid phase into the aqueous solution and the non-polar region seen associated with the solid portion of pesticide.

10.4.5 Polymer Based Nano-particles

Helps in the slow and regulated release of pesticides to the target site. Nano-encapsulation, nano-spheres, nano-gels, nano-fibers, etc. are main examples. Nano-capsules have heterogeneous structure with central inner cavity. This cavity accommodates the hydrophobic or hydrophilic active ingredient with a polymer membrane surrounding (Wang et al. 2022). Nanospheres are vesicular forms with homogeneous nature. Here the active ingredient is distributed uniformly throughout a polymer matrix. in which the bioactive ingredient is uniformly dispersed throughout the polymer matrix. Nanogels or hydrogel are designed by the cross polymeric substances with hydrophilic groups thereby absorbing larger quantities of water molecules. Nano-fibers are another type which is designed by electrospinning and thermal induced phase separation (Li et al. 2021).

10.4.6 Nanoinsecticides

The nano insecticides are attractive when compared to chemical insecticides due to a variety of positive effects such as improved toxicity, high target specificity, efficiency, effectiveness, desirable half-life, and minimum requirement of active ingredient. Studies found that they can greatly enhance the toxicity effects on target insect pests (up to tenfold). Another development is the formulation of nano insecticides that are pH responsive, that is they are designed in such a way promoting the optimum release of active ingredients even in acidic/alkaline of pH conditions present in the gut of insect. The enzyme responsive preparations to enzymes of insects such as alkaline phosphatases, carboxylases etc. triggers the release of Active Ingredient of nano insecticides during feeding. Thermal responsive nanoformulation can also increase the release of Active Ingredients. Moreover researches states that nanostructured insecticides are causing less damage to human DNA and chromosome. Some nano insecticides available commercially are Asteroid, Subdue MAXX, Feenstra, Banner MAXX, Subdue MAXX, Bigender FC etc. (Manna et al. 2023)

Nano insecticides with nanoscale particles exhibits potential lethal reactions by altering the enzyme levels and hormone concentration. They can slow down the developmental cycle of insects. Another feature is their prolonged residual activity. Nano insecticides are available in various forms such as nanospheres, nano capsules, nanogels, micelles, liposomal preparations Nanoencapsulation can be achieved in carrier materials such as different polymers (Graily-Moradi and AsgariLajayer 2021).

Nanoencapsulation is the most common technique employed for production of nano insecticides. Here the Active Ingredient is encapsulated in a polymer/matrix of size range of nanoscale. The main advantage of this technique is the protection of the Active Ingredient from complications such as environmental degradation and rapid loss. A large number of solid nanoparticles have been recommended for formulation of nano insecticides like alumina, silica, clay etc. Their mode of action is by the disintegration of the waxy coating seen on the outer epidermis of the insects. This is achieved by scraping and surface assimilation. This destruction trigger insects to dehydrate, dry out and die. Out of these formulations the silica-based nanoparticles is highly recommended for the control of insect pests in agricultural land because of its capacity to increase the resistance power of the plants to both biological and non-biological stresses. Studies proved that this SNPs are highly efficient than silica. These silica nanoparticles destroy the cuticle of insect and destroy the insect completely. As they do not alter or modify the insect's gene expression and also found risk free for the human consumption, the World Health Organization announced unstructured silica as nano pesticide (Asif et al. 2021). The toxicity levels of some of the nano pesticides are given below in Table 10.2.

Table 10.2 Toxicity levels of some nano pesticides (Manna et al. 2023)

S. No.	Insecticides	Action
1	Nano-permethrins	Larvicidal effect on *Culexquinque fasciatus*
2	Nano-formulated Chlorantraniliprole and thiocyclam	Larvicidal effect on black cutworm, *Agrotisipsilon*
3	Pyridalyl nano suspension	Cytotoxicity on cotton bollworm, *Helicoverpaarmigera*
4	Indoxacarb	Neurotoxic effect lepidopteran larvae
5	Imidacloprid (neonicotinoid insecticide)	Exhibits neurotoxicity and arrests CNS of cotton leafworm, *S. littoralis.*
6	Spinetoram nano-suspension	Neurotoxic effect on diamondback moth, *Plutellaxylostella*

10.4.7 Nano Herbicides

The main problem associated with the usage of synthetic herbicide is that major portion is wasted through leaching, volatilization, chemical breakdown and biodegradation. Moreover, the uncontrolled usage of the same can induce the emergence of resistant weeds The maximum degree of resistance was seen in weeds of crops such as wheat, maize, rice, soyabean, spring barley, canola and cotton. Nanoencapsulation of herbicides is a very efficient delivery system where the Active ingredients embedded in nanoparticles are released slowly. Nanoencapsulation (Direct and Indirect methods), solvent evaporation and nano spray method are the methods employed for production (Kumar and Chinnamuthu 2020).

Nano herbicides can decrease the root and shoot growth more effectively. They are released very slowly through the entire season destroying the food reserves of weed seeds resulting in lesser regeneration of weeds. This provides a suitable atmosphere for better growth, development, and enhanced yield of crop through effective photosynthesis as well as increased nutrient uptake by crop. Three types of nano enabled herbicides are Organic (Table 10.3), inorganic and hybrid nano-enabled herbicides. Organic nanomaterials are based on proteins, polymers, lipids, dendrimers etc. Out of these polymers are outstanding due to their biodegradability and biocompatibility. Inorganic type is mainly based on metal, silica etc. Hybrid nano herbicide combines the advantages of both organic and inorganic. They are multifunctional available in a variety of properties, sizes, morphologies, chemical compositions etc. They have high traceability, targetability, and also stimuli-responsiveness properties (Sakarvadia 2022).

10.4.8 NanoFungicides

Fungal kingdom consists of a variety of phytopathogenic organisms that cannot be eliminated by synthetic fungicides. Another major concern associated with these fungicides is the negative impact imposed by them such as pollution, emergence of

Table 10.3 List of nano herbicides

Nano herbicides	Plants	Type
Nano emulsion containing glyphosate	*Eleusine Indica L*	Organic
Biochar_2,4-D	*Brassica sp.; Zea mays L.*	Organic
Nano-hydrogel_ glyphosate	*Weeds; Oryza sativa L.*	Organic
Carbon nanotubes_ Glyphosate	*Arabidopsis thaliana*	Organic
PCL_Metribuzin NPs	*Ipomoea grandifolia*	Organic
Layered double hydroxide Imazamox	*Brassica nigra L.*	Inorganic
Herbicide@ Hydrotalcites	*Convolvulus arvensis L*	Inorganic
Clay-imazaquin	*Brassica oleracea var. botrytis L.*	Inorganic
Nano formulation-sepiolite_Mesotrione	*Zea mays L.; Helianthus annuus L.*	Inorganic
AgNPs_CS(chitosan)_PQ(paraquat)	*Eichhorniacrassipes*	Hybrid
Magnetic-responsive controlled-release herbicide	*Cynodondactylon L*	Hybrid
GSNO (nitrosoglutathione)-containing alginate/ chitosan nanoparticles	*Zea mays L.; Glycine sp.*	Hybrid
Attapulgite-NH_4CO_3-Gly—Amino silicol oil—poly (vinyl alcohol)	*Zoysiamatrella L.*	Hybrid
Wheat gluten_organically modified montmorillonite_ethofumesate	*Lepidium sativum L*	Hybrid

pest resistance, reduction in biodiversity, poor delivery, and their presence in food commodities that we consume. Now a days metallic nanoparticles are widely employed in production of nano fungicides. Targeted delivery, increased bio accessibility, less toxicity etc. are some important features. The silver nanoparticles exhibit very effective antifungal activity. The Ag ions has high affinity for the thiol-group of the cell wall of the fungus leading to cell wall damage, mutation, inactivation of enzymes, and blockage of fungal respiration. (UlHaq and Ijaz 2019).

Silver nanoparticles exhibited a very high antimycotic potential against a variety of fungi (Fig. 10.7): such as *Botrytis cinerea, Bipolarissorokiniana, Magnaporthegrisea, Fusarium culmorum, Fusarium oxysporum.* This formulation can reduce the germination and length of sprouts. Similarly studies found that the Zinc oxide NPs formulations can inhibit the conidiophore and conidia development in *Penicillium expansum*, leading to the death of fungus (Abd-Elsalam 2012). A strong antifungal property against pathogens such as *Fusarium graminearum* and *Fusarium poae* is exhibited by carbon nano formulations. The nano fungicide enter into the fungus through three major steps -Direct internalization in to the fungal cell wall, Receptor mediated adsorption and Internalization through ion transporting proteins. The antifungal activity of nanomaterials include the following events (Alghuthaymi et al. 2021).

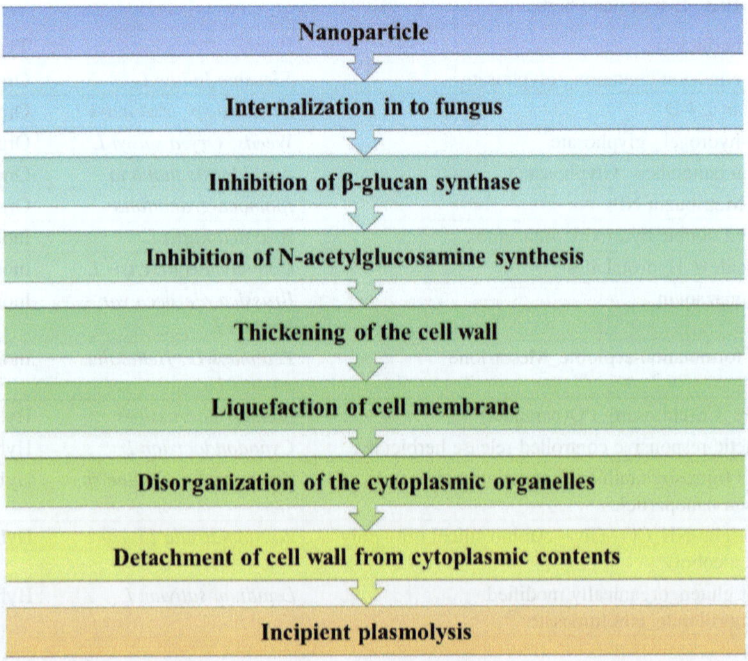

Fig. 10.7 Events accomplished with the antifungal activity of nanoparticles. (Constructed by Nisha. P)

10.5 Nano Sensors in Agriculture

Intensive farming and unsustainable agricultural practices have indeed led to a detrimental loss of soil and water quality, as well as environmental pollution. To address these challenges, several technological innovations have been developed, with nano sensors being one of them. Nano-sensors do offer advantages over conventional techniques, such as being inexpensive, simple to operate, and enabling visual detection, it is important to consider the specific application and requirements when choosing the appropriate sensing technology. Each technology has its own strengths and limitations that should be carefully evaluated to ensure optimal performance in a given scenario (Fan et al. 2021). Well-designed nano-sensors offer a self-contained setup where immobilized nanoparticles interact with the analyte, with a transduction approach converting signals into a detectable form. The use of inexpensive reagents, cost-effective manufacturing processes, enhanced sensitivity, and miniaturization make nano-sensors a compelling choice for in-field applications. These characteristics contribute to their suitability for various real-world scenarios where efficient and reliable detection is required (Rajakumaran et al. 2021). Nano sensors play a crucial role in modern agriculture by enabling real-time monitoring of soil and water quality, nutrient levels, and pest infestations. By leveraging nanotechnology, farmers can make more informed decisions to mitigate the negative impacts of intensive farming practices and move towards more sustainable agricultural methods (Garg et al. 2023).

Nano sensors are utilized to detect dynamic changes in the behaviour of nanosized molecules, serving as signals to indicate environmental changes. While synthetic nano sensors have been developed relatively recently, living organisms have been employing similar mechanisms since the inception of life (Shaw and Honeychurch 2022).

Nano sensors in agriculture are small devices that can detect and monitor various parameters such as soil moisture, nutrient levels, and pest infestations at a nanoscale level. These sensors can provide real-time data to farmers, allowing them to make more informed decisions about irrigation, fertilization, and pest control. Nano sensors have the potential to revolutionize agriculture by increasing efficiency, reducing waste, and improving crop yields.

Nanoscale sensors in agriculture play a crucial role in monitoring soil quality, nutrient levels, moisture content, and pest infestations in real-time. This data empowers farmers to make informed decisions regarding irrigation, fertilization, and pest management, ultimately leading to improved resource management and higher crop yields.

Nano sensors play a crucial role in precision agriculture by enabling the detection of various parameters such as soil humidity, pesticide residues, nutrient requirements, and plant pathogens in different environmental conditions. The advantages of nano sensors, including smaller detection limits, quick response time, high surface-to-volume ratio, and more reliable results, have demonstrated their superiority over conventional sensors. By utilizing nano sensors, farmers can make more precise and timely decisions to optimize resource management and enhance crop productivity in a sustainable manner.

10.5.1 Nano Sensors for the Detection of Pesticide Residue

Nano sensors are increasingly being utilized for the detection of pesticide residue in agricultural products. These small-scale devices can detect trace amounts of pesticides with high sensitivity and specificity, providing farmers and food safety regulators with valuable information about the presence of harmful chemicals. By using nano sensors for pesticide residue detection, stakeholders can ensure the safety and quality of agricultural products, protect consumer health, and promote sustainable farming practices. The repeated and unregulated use of pesticides in agriculture poses a significant threat to the environment, biodiversity, and ecosystem stability. Detecting pesticides in soil at lower concentrations remains a challenge, but nanotechnology has emerged as a promising solution in promoting sustainable agricultural practices. Nano sensors offer high sensitivity, the ability to detect low concentrations, and increased stability compared to traditional detection methods. Utilizing sensor technology for pesticide detection is a viable alternative to chromatographic techniques, contributing to the protection of soil quality, reduction of health risks, and preservation of the ecosystem. The development of nano sensors and nano biosensors play a crucial role in advancing the agricultural sector and addressing the urgent need for pesticide residue analysis in soil (Panoth et al. 2022). The proposed application of a simple disposable potentiometric biosensor for the determination of selected anticholinesterase pesticides and their degradation products is a promising approach. By assessing the relative degree of cholinesterase enzyme inhibition, this method enables the detection of pesticides such as methomyl, chlorpyrifos, prothiofos, and their degradation products. Variations in sensitivity towards these pesticides were observed, reflecting the inhibitory effects of each pesticide on the cholinesterase enzyme. This study highlights the potential of biosensors in providing valuable insights into pesticide detection and monitoring (Khaled et al. 2015). The integration of colorimetric and nanoparticle-based detection techniques in sensor technology is revolutionizing food analysis. These methods enhance the accuracy, sensitivity, and cost-effectiveness of detecting contaminants, contributing significantly to food safety and quality assurance (Tarannum et al. 2024).

10.5.2 Nano Sensors for the Detection of Pathogens

Common procedures for detection of plant pathogens have high cost for analyses, need a lot of time, are effortful and need new tools along with skilled staff. Therefore, to come cross such difficulties, it is essential to expand sensing methods including bio/nano sensors for detections and evaluations, which are quick, trustworthy, economical and beneficial. Nano sensors for detection of microorganisms and their toxins have been hugely probed with the emerging fields of nanotechnology and artificial intelligence. Nanomaterial-based sensors such as magnetic nanoparticles, gold nanoparticles, peptide nanotubes, quantum dots, etc. are the most common sensors with broad application for detection of pathogens and their toxins; Monitoring the quality/safety of food products and their contaminating factors can prevent and control foodborne diseases. Food transportation as well as storage conditions of products are the main factors which have a significant role in their physical, biological and chemical contamination. Among all the hazardous parameters which can affect food safety, pathogens and microbial toxins are the most important one (Sobhan et al. 2021).

In today's context, sensing techniques based on bio- or nanotechnology have emerged as a cost-effective substitute for conventional and pricey methods used to analyze a wide range of substances. The primary advantage lies in their simplicity of operation, wherein the focus is solely on converting biological responses into detectable signals (Chen et al. 2017). A typical sensing system consists of two fundamental parts: a receptor and a physicochemical transducer. Alternatively, these components can be integrated into a single sensor referred to as a receptor, a recognition-transduction component. Both the (bio)receptor or recognition element and the transducer work in harmony to convert the biochemical signal into different detection principles, such as electrochemical, optical, thermal, gravimetric, or others. Extensive investigations into different facets of bio/nano-sensors (bNSRs) have yielded numerous biotechnological advancements and innovations, particularly in the field of bacterial and viral detection and monitoring. The progress made in nanotechnology and nanoscience has paved the way for the development of highly promising bNSRs incorporating advanced nanomaterials (NMs). These cutting-edge biosensors hold significant potential in enhancing both sensitivity and selectivity, thereby revolutionizing this field (Dehghani et al. 2018). Applying nanoparticles in combination with high-affinity ligands and antibodies (Abds) as sensors (SRs), raise the developing simultaneous toxins detection. Several NMs like gold NPs (AuNPs), magnetic NPs (MNPs), silver NPs (AgNPs), silica NPs (SNPs), etc. have been widely used for detection of pathogen and microbial toxins because of their remarkable benefits in electrical conductivity, chemical or catalytic features, thermal stability, and mechanical strength (Majdinasab et al. 2021).

By applying NMs with special features, a broad spectrum of (bio)sensing strategies including electrochemical, SPR, surface-enhanced Raman scattering (SERS), fluorescence, calorimetric, chemiluminescence (CL), photoluminescence strategies, and image-based technologies are developed in bNSR systems to synthesize novel sensing platforms. Therefore, NMs-based BSRs have ignited a surge of unprecedented interest in the sensing field of pathogens and microbial toxins, The SR application in the food processing field has significantly improved the sensitivity/specificity of contaminants detection throughout the food chain. This review discusses several NM types and how they are exploited in various FbPs and their toxin identification systems. Researchers have been able to investigate and build new NM methods thanks to the substantial advances in nanotechnology.

Inorganic nanomaterials, including silver (Ag) and gold (Au) nanoparticles (NPs), as well as metal oxide NPs, have become pivotal in the development of nanobiosensors for detecting plant pathogens (Khaledian et al. 2017). demonstrated the use of Au NPs functionalized with single-stranded oligonucleotides to identify the genomic DNA of *Ralstonia solanacearum*, a major cause of potato bacterial wilt.

10.5.3 Nano Sensors for Nutrient Deficiency Detection Using

The use of nanomaterials as carriers for agrochemicals offers a novel approach to achieve site-targeted controlled delivery of nutrients, leading to enhanced crop protection. By enabling precise management and control of inputs such as fertilizers, pesticides, and herbicides, nanotools like nano biosensors play a crucial role in advancing high-tech agricultural practices. The integration of biology and nanotechnology in nano sensors has significantly expanded their capabilities to detect and identify environmental conditions or issues, thereby enhancing the overall efficiency and sustainability of modern farming techniques (Miguel-Rojas and Pérez-de-Luque 2023). The application of fertilizers is crucial for boosting agricultural production, but excessive usage can lead to irreversible changes in the soil's chemical ecology, limiting available land for crop cultivation. Sustainable agriculture emphasizes minimal agrochemical use to protect the environment and preserve biodiversity. Nanomaterials play a key role in enhancing crop productivity by improving the efficiency of agricultural inputs and enabling targeted nutrient delivery, reducing the need for excessive Agri-inputs. The integration of nanotechnology in plant protection products has shown promising results in increasing crop yields. Additionally, addressing challenges posed by climate change, such as extreme temperatures, water scarcity, salinity, alkalinity, and environmental pollution, while safeguarding sensitive ecosystems, remains a significant concern in agricultural production (Garg et al. 2023). The advancement and utilization of nano sensors in precision farming have significantly enhanced the monitoring and management of various aspects such as crop growth, soil conditions, diseases, application of agrochemicals, and environmental pollution. This technology has empowered humans to exercise greater control over soil and plant health, quality assurance, and safety measures, thereby playing a pivotal role in promoting sustainable agriculture and environmental systems (Garg et al. 2023). The integration of nano sensors in precision farming has proven to be instrumental in optimizing agricultural practices and ensuring environmental sustainability (Chen et al. 2016). In the realm of sustainable agriculture, safeguarding the environment against pollution is a paramount objective. Nanomaterials offer a promising avenue for improved management and conservation of inputs in plant production, potentially ushering in a new era of sustainable farming practices with reduced risks (Miguel-Rojas and Pérez-de-Luque 2023). Despite their potential benefits, there remain significant gaps in our understanding of factors such as uptake capacity, permissible limits, and ecotoxicity associated with various nanomaterials. Addressing these knowledge gaps is essential to ensure the safe and effective utilization of nanomaterials in agriculture while minimizing potential environmental impacts (He et al. 2018).

10.5.4 Nano Sensors for the Detection of Heavy Metals

The accumulation of hazardous elements in soil due to processes like fuel emissions, ore extractions, improper industrial waste disposal, and water and soil contamination poses a significant threat to human health through the consumption of crops grown in such soils. Therefore, there is a pressing need to scale up the production of nano sensors for on-field applications to detect and monitor heavy trace metals like, copper iron, arsenic, lead, mercury, palladium, platinum, rhodium, etc. in soil. Understanding the ideal parameters of fertile soil, recognizing the detrimental effects of soil contamination with heavy metals, and exploring new emerging techniques for developing advanced nano sensors are crucial steps in addressing this environmental and health concern (Miguel-Rojas and Pérez-de-Luque 2023). Researchers are continuously working towards the development of cutting-edge nano sensors to enable accurate and efficient detection of heavy metals in soil, contributing to environmental sustainability and human well-being (Potadar et al. 2022). Fabricated nano sensors exhibit high selectivity and sensitivity towards specific heavy metal pollutants. The optical properties of these nano sensors, influenced by surface plasmon resonance phenomena in metallic nanostructures like gold, platinum, and silver, or through the inner filter effect, can be significantly altered and enhanced. Additionally, novel nano sensors can serve as fluorescence quenchers to amplify fluorescence signals or as surface catalysers to induce autonomous motion. These unique mechanical, chemical, and electrical properties position many nano sensors as promising tools for environmental remediation. Various nano sensors, such as surface-enhanced Raman spectroscopy and nano jets, have been developed and utilized for heavy metal detection. This review provides an overview of recent advancements in the fabrication of innovative Nano sensors that offer potential solutions for heavy metal remediation (He et al. 2021).

10.6 Nanomaterials for Soil Remediation

Land pollution poses a significant threat to livelihoods, quality of life, and sustainable development. In the current era, with the pressures of a growing population and the reduction of arable lands due to human activities, conserving soil has become a critical necessity. Addressing land pollution and implementing measures to protect and preserve soil quality are essential for ensuring environmental sustainability and the well-being of present and future generations. The advancements in nanotechnology have significantly enhanced the potential for remediating and restoring polluted soil on a global scale. Nanotechnology offers an effective means to address soil pollution by mitigating the toxicities associated with various metals and metalloids. It is recognized as an environmentally cleaner technology with the ability to remediate pollutants in diverse environmental matrices, particularly in soils. Nanotechnological approaches have been extensively utilized in soil remediation, leveraging the unique properties of nanoparticles such as their large specific surface area. NPs can facilitate contaminant removal through mechanisms like adsorption, redox reactions, precipitation, and co-precipitation, showcasing the versatility and efficacy of nanotechnology in environmental remediation efforts (Rajput et al. 2020). Nano bioremediation combines nanotechnology and bioremediation to address environmental pollutants at the nanoscale. Through the unique physicochemical properties of nanoparticles (NPs), pollutants can be adsorbed, degraded, or modified, with NPs acting as catalysts to lower the activation energy required for compound breakdown. Carbon- and metal-based NPs are commonly utilized in nano bioremediation, along with polymeric NPs like nano capsules or nanospheres, which are effective in eliminating persistent pesticide compounds and long-chain hydrocarbons. However, the challenge shifts when dealing with heavy metals (HMs) due to their non-biodegradability and potential for entering biological systems and food chains, necessitating specialized approaches for their remediation (Briffa et al. 2020). The high reactivity and photo-catalytic activity of TiO_2 nanoparticles (NPs) make them effective in adsorbing contaminants on their surface area, thereby reducing the toxic behaviour of heavy metals (HMs) and maintaining their mobility. For instance, TiO_2 NPs were able to absorb 88.01% of Cu^{2+} and 70.67% of Cd^{2+} at a pH of 7.0 (Mahmoud et al. 2018) Additionally, other NPs, such as modified carbon, have shown to enhance plant growth and decrease the uptake of Cd and Ni in plants like *Suaeda salsa* (Khan and Bano 2016). Silver Nanoparticles have demonstrated the ability to reduce the uptake of Ni, Pb, Na, Zn, and Cu in maize plants, while also promoting plant growth through interactions with plant hormones like gibberellin and abscisic acid. When combined with *B. cereus* (LPR2), Ag NPs have been shown to significantly enhance the growth of maize plants. These findings underscore the importance of employing strategic approaches, such as utilizing hyperaccumulator plants in conjunction with compatible NPs, to effectively remediate contaminated soil and restore its health (Kumar et al. 2020). The utilization of polyvinylpyrrolidone (PVP)-coated FeO nanoparticles has shown success in enhancing the bioremediation of soil contaminated with Pb and Cd by the Gram-negative bacteria, *Halomonas* sp. This innovative approach has demonstrated remarkable efficiency, with nearly 100% removal of Pb within 24 hours and Cd within 48 hours, surpassing the removal capabilities of bacteria alone or nanoparticles alone. This study underscores the potential of nanoparticle-based strategies in improving the remediation of heavy metal-contaminated soil, showcasing promising results in the removal of toxic pollutants (Alabresm et al. 2018).

10.7 Nanotechnology in Gene Modification

Using nanoparticles as nanocarriers to transfer NPs into plant cells and form a binding complex with bio modifier molecules is a potent tool in the development of genetically modified organisms. Nanotechnology holds the genes for gene modification, because it genetic materials into the cells. The nanoparticles carry the gene editing tools, such as CRISPR (Clustered regularly interspaced short palindromic repeats) are genetic elements, they consist of short sequences that originate from viral genomes and have been incorporated into the bacterial genome. Nanotechnologies like CRISPR-Cas9, which allow for gene editing, are made possible in large part by nanotechnology. CRISPR components can be functionalized on nanoparticles to increase the specificity and efficiency of their delivery. Furthermore, the potential of nanomaterials like gold nanoparticles to improve CRISPR-Cas9-mediated gene editing by means of enhanced cellular absorption or localized heating has been investigated (Duan et al. 2021). Nanotechnology-based methods have potential for use in gene therapy applications to treat cancer, genetic abnormalities, and other illnesses. Therapeutic genes or gene-editing instruments can be delivered via nanoparticles to specific tissues or cells, providing hope for repairing genetic flaws or altering disease pathways.

Therapeutic genes or gene-editing instruments can be delivered via nanoparticles to specific tissues or cells, providing hope for repairing genetic flaws or altering disease pathways. Overall, nanotechnology provides versatile platforms for precise and efficient manipulation of genetic material, offering new opportunities for understanding biological processes, diagnosing diseases, and developing targeted therapies. However, challenges such as off-target effects, immunogenicity, and scalability need to be addressed to realize the full potential of nanotechnology in gene modification while ensuring safety and efficacy.

10.8 Advantages and Disadvantages of Nanotechnology in Agriculture

- **Precision Delivery**: Nanotechnology allows for the targeted delivery of nutrients, fertilizers, and pesticides, minimizing waste and reducing environmental impact. Nanoparticles are crucial for effective material delivery and enhancing plant growth, nanochips complement these technologies by enabling precise monitoring and data analysis, thus enhancing the overall capabilities of precision farming. Both play vital roles in advancing agricultural practices (Duhan et al. 2017; Garg et al. 2023).
- **Enhanced Nutrient Absorption**: Nanoparticles can improve the uptake of essential nutrients by plants, leading to better growth and yield. Nanofertilizers contains, nitrogen, phosphorus, potassium and micronutrients can increase the availability and enhance the nutrient uptake by the plants even in stress conditions too (Nongbet et al. 2022).
- **Disease Resistance**: Nanomaterials can be engineered to deliver genes or molecules that enhance plants' resistance to pathogens and pests, reducing the need for chemical treatments (Miguel-Rojas and Pérez-de-Luque 2023; Garg et al. 2023).
- **Stress Tolerance**: Nanotechnology can help develop crops that are more resilient to abiotic stresses such as drought, salinity, alkalinity, mineral toxicity. Metal toxicity and extreme temperatures (Mittal et al. 2020).
- **Improved Soil Quality**: Nanomaterials can be used to enhance soil properties, such as increasing microbial activity and improving soil structure, which supports healthier crop growth (Miguel-Rojas and Pérez-de-Luque 2023).
- **Smart Sensors**: Nanosensors can monitor plant health and soil conditions in real-time, allowing for precise management practices and early detection of problems (Duhan et al. 2017; Garg et al. 2023).
- **Reduced Chemical Use**: By improving the efficiency of inputs (like fertilizers and pesticides), nanotechnology can decrease the overall amount of chemicals needed, promoting more sustainable farming practices (Gupta et al. 2023).
- **Faster Breeding Programs**: Nanotechnology can speed up traditional breeding techniques by facilitating gene editing and marker-assisted selection (Ahmar et al. 2021).

- **Enhanced Crop Traits**: Nanotechnology enables the modification of crops for improved traits such as enhanced flavor, nutritional content, and shelf life (Gupta et al. 2023; Garg et al. 2023).
- **Sustainability**: By improving resource efficiency and reducing chemical use, nanotechnology contributes to more sustainable agricultural practices, benefiting the environment (Garg et al. 2023).

Nanotechnology offers significant potential benefits for Crop improvement, it also presents certain disadvantages:

- **Environmental Impact**: The long-term effects of nanoparticles on ecosystems are not fully understood, and there may be risks of soil and water contamination (Muzammil et al. 2023).
- **Health Risks**: Potential toxicity of nanoparticles to humans and non-target organisms, including beneficial insects and soil microbes, raises concerns (Ali et al. 2020).
- **Cost and Accessibility**: The technology may be expensive, limiting its accessibility for small-scale farmers and developing regions (Dhillon and Moncur 2023).
- **Regulatory Challenges**: Lack of clear regulations and guidelines for the use of nanotechnology in agriculture can hinder safe implementation (Dhillon and Moncur 2023).
- **Ethical Concerns**: Issues related to genetically modified organisms (GMOs) and the manipulation of natural systems can lead to public resistance (Dhillon and Moncur 2023).

Addressing these challenges requires careful assessment, regulation, and oversight to ensure the safe and responsible deployment of nanotechnology in agriculture while maximizing its benefits and minimizing potential risks.

10.9 Conclusion and Prospects

Nanotechnology has a significant role in the field of agriculture. It can increase crop production, and different nanoparticles can improve the yield. Nano fertilizers can increase the rate of germination, nutrient concentration, plant growth, chlorophyll content, and protein content and reduce diseases. Nano insecticides and Nano pesticides exhibit a wide range of potential in controlling insects and pests. Nano sensors are widely used for detecting various parameters such as humidity, pesticide residues, plant pathogens, nutrient requirements, etc. In conclusion, nanotechnology has a crucial role in the modern agricultural field. Nanotechnology has made significant strides in the field of agriculture, offering innovative solutions to enhance crop productivity, improve resource utilization, and address environmental challenges. Nanotechnology has indeed emerged as a promising tool in agriculture, offering a range of solutions to various challenges faced by the industry. Nanoparticles can be used to encapsulate pesticides, herbicides, and fertilizers, allowing for targeted delivery to plants. This reduces the amount of chemicals needed, minimizes environmental impact, and enhances effectiveness. Nano-sized nutrient delivery systems can improve the uptake of essential nutrients by plants, leading to better growth, development, and yield. Nanoparticles can be employed to remediate contaminated soils by absorbing, immobilizing, or degrading pollutants, thus improving soil quality and productivity. Nano sensors can provide real-time monitoring of soil and crop conditions, enabling farmers to make data-driven decisions regarding irrigation, fertilization, and pest management. Nanoparticles with antimicrobial properties can help control plant diseases, while nano-pesticides can target pests more effectively, reducing the need for broad-spectrum chemicals. Nanotechnology can contribute to water conservation efforts by developing materials for efficient irrigation systems, water purification, and moisture retention in soil. Nanotechnology techniques such as gene delivery systems can facilitate precise genetic modifications in plants, leading to traits like drought resistance, disease resistance, and increased nutritional value. Nano-enabled packaging materials can prolong the shelf life of agricultural products by providing barriers against moisture, gases, and microbes. These advancements have the potential to increase crop productivity and quality, and promote sustainable practices by reducing the environmental footprint of agriculture. However, it's important to ensure that the benefits of nanotechnology are balanced with consideration.

The prospects of nanotechnology in crop improvement are promising and could lead to significant advancements in agriculture. Nanotechnology is expected to make an impacts on some aspects including **Nutritional Enhancement, Nutrient Delivery, Genetic Modification and Breeding.** The integration of nanotechnology into crop improvement holds the potential to transform agriculture by increasing efficiency, sustainability, and resilience. Continued research and development, alongside careful consideration of environmental and health impacts, will be crucial in realizing these prospects.

References

Abd-Elsalam K (2012) Nanoplatforms for plant pathogenic fungi management. Fungal Genom Biol 2:2. https://doi.org/10.4172/2165-8056.1000e107

Abobatta WF (2018) Nanotechnology application in agriculture. Acta Sci Agric 2(6):99–102

Agrawal S, Rathore P (2014) Nanotechnology pros and cons to agriculture: a review. Int J Curr Microbiol Appl Sci 3:43–55

Ahmad Z, Tahseen S, Wasi A, Ganie IB, Shahzad A, Emamverdian A, Ramakrishnan M, Ding Y (2022) Nanotechnological interventions in agriculture. Nano 12(15):2667

Ahmar S, Mahmood T, Fiaz S, Mora-Poblete F, Shafique MS, Chattha MS, Jung K-H (2021) Advantage of nanotechnology-based genome editing system and its application in crop improvement. Front Plant Sci, Sec Plant Biotechnol 12

Ahmed F, Dwivedi S, Shaalan NM, Kumar S, Arshi N, Alshoaibi A, Husain FM (2020) Development of selenium nanoparticle based agriculture sensor for heavy metal toxicity detection. Agriculture 10(12):610

Ahmed DF, Isawi H, Badway NA, Elbayaa AA, Shawky H (2021) Graphene oxide incorporated cellulose triacetate/cellulose acetate nanocomposite membranes for forward osmosis desalination. Arab J Chem 14(3):102995

Ahmed R, Uddin MK, Quddus MA, Samad MY, Hossain MM, Haque AN (2023) Impact of foliar application of zinc and zinc oxide nanoparticles on growth, yield, nutrient uptake and quality of tomato. Horticulturae 9(2):162

Alabresm A, Chen YP, Decho AW, Lead J (2018) A novel method for the synergistic remediation of oil-water mixtures using nanoparticles and oil-degrading bacteria. Sci Total Environ 630: 1292–1297

Alfalahi AO, Abdulqahar FW (2021) Nanonutrients: plant nutritive and possible antioxidant regulators. In: Nanobiotechnology: mitigation of abiotic stress in plants, pp 471–498. https://doi.org/10.1007/978-3-030-73606-4_21

Alghuthaymi MACRPR, Kalia A, Bhardwaj K, Bhardwaj P, Abd-Elsalam KA, Valis M, Kuca K (2021) Nanohybrid antifungals for control of plant diseases: current status and future perspectives. J Fungi (Basel) 7(1):48

Ali M, Ahmed T, Wu W et al (2020) Advancements in plant and microbe-based synthesis of metallic nanoparticles and their antimicrobial activity against plant pathogens. Nanomaterials 10:1146

Al-Juthery HWA, Lahmod NR, Al-Taee RAHG (2021) Intelligent, nano-fertilizers: a new technology for improvement nutrient use efficiency (Article Review). IOP Conf Ser Earth Environ Sci 735(1):012086

Anwaar S, Ijaz DE, Anwar T, Qureshi H, Nazish M, Alrefaei AF, Almutairi MH, Alharbi SN (2024) Boosting *Solanum tuberosum* resistance to *Alternaria solani* through green synthesized ferric oxide (Fe2O3) nanoparticles. Sci Rep 14(1):2375

Asif M, Islam S, Malik MA, Dar ZM, Masood A, Shafi S, Rashid B, Sidique S, Asif M, Islam S, Malik MA, Dar ZM, Masood A, Shafi S, Rashid B, Sidique S (2021) Nano pesticides application in agriculture and their impact on environment. In: Agricultural development in Asia—potential use of nano-materials and nano-technology (pp. 1-11), IntechOpen https://doi.org/10.5772/intechopen.100690

Attia EA, Elhawat N (2021) Combined foliar and soil application of silica nanoparticles enhances the growth, flowering period and flower characteristics of marigold (Tageteserecta L.). Sci Hortic 282:110015

Badgar K, Abdalla N, El-Ramady H, Prokisch J (2022) Sustainable applications of nanofibers in agriculture and water treatment: a review. Sustain For 14(1):1

Badshah I, Mustafa N, Khan R, Mashwani ZU, Raja NI, Almutairi MH, Aleya L, Sayed AA, Zaman S, Sawati L (2023) Biogenic titanium dioxide nanoparticles ameliorate the effect of salinity stress in wheat crop. Agronomy 13(2):352

Banerjee A, Sarkar A, Acharya K, Chakraborty N (2021) Nanotechnology: an emerging hope in crop improvement. Lett Appl NanoBioScience 10(4):2784–2803

Bowden B (2022) Crops and the shaping of civilizations. In: Slavery, freedom and business endeavor: the reforging of Western civilization and the transformation of everyday life. Springer International Publishing, Cham, pp 55–71

Briffa J, Sinagra E, Blundell R (2020) Heavy metal pollution in the environment and their toxicological effects on humans. Heliyon 6:e04691

Chen YW, Lee HV, Juan JC, Phang SM (2016) Production of new cellulose nanomaterial from red algae marine biomass *Gelidium elegans*. Carbohydr Polym 151:1210–1219

Chen I-H et al (2017) Food control

Chhipa H (2019) Applications of nanotechnology in agriculture. In: Methods in microbiology, vol 46. Academic, pp 115–142

de Oliveira JL, Campos EVR, Bakshi M, Abhilash PC, Fraceto LF (2014) Application of nanotechnology for the encapsulation of botanical insecticides for sustainable agriculture: prospects and promises. Biotechnol Adv 32(8):1550–1561

Dehghani, S., Nosrati R., Yousefi M., Nezami A., Soltani F., Taghdisi S. M., Abnous K., Alibolandi M. and Ramezani M. (2018) Aptamer-based biosensors and nanosensors for the detection of vascular endothelial growth factor (VEGF): A review. Biosens Bioelectron 110:23–37

Dhillon R, Moncur Q (2023) Small-scale farming: a review of challenges and potential opportunities offered by technological advancements. Sustain For 15:15478

Dimkpa CO, Singh U, Adisa IO, Bindraban PS, Elmer WH, Gardea-Torresdey JL, White JC (2018) Effects of manganese nanoparticle exposure on nutrient acquisition in wheat (*Triticum aestivum* L.). Agronomy 8(9):158

Duan L, Ouyang K, Xu X, Xu L, Wen C, Zhou X, Qin Z, Xu Z, Sun W, Liang Y (2021) Nanoparticle delivery of CRISPR/Cas9 for genome editing. Front Genet 12:673286

Duhan JS, Kumar R, Kumar N, Kaur P, Nehra K, Duhan S (2017) Nanotechnology: the new perspective in precision agriculture. Biotechnol Rep (Amst) 1(15):11–23

Elbasuney S, El-Sayyad GS, Abdelaziz AM, Rizk SH, Tolba MM, Attia MS (2023) Stable colloidal iron oxide nanoparticles: a new green nanofertilizer and therapeutic nutrient for eggplant immune response against fusarium wilt disease. J Clust Sci:1–5

Fan Y et al (2021) Construction of pn heterojunctions by modifying MOF-derived α-Fe_2O_3 with partially covered cobalt tungstate for high-performance ethyl acetate detection. Sensor Actuator B Chem 344:130129

Ferrari E, Barbero F, Busquets-Fité M, Franz-Wachtel M, Köhler HR, Puntes V, Kemmerling B (2021) Growth-promoting gold nanoparticles decrease stress responses in Arabidopsis seedlings. Nano 11(12):3161

Garg S, Rumjit NP, Roy S (2023) Smart agriculture and nanotechnology: technology, challenges, and new perspective. Adv Agrochem 3(2):115–125

Ghidan AY, Al-Antary TM, Salem NM, Awwad AM (2017) Facile green synthetic route to the zinc oxide (ZnONPs) nanoparticles: effect on green peach aphid and antibacterial activity. J Agric Sci 9(2):131–138

Ghidan AY, Antary TMA, Ghidan AY, Antary TMA (2019) Applications of nanotechnology in agriculture. In: Applications of nanobiotechnology. IntechOpen SN - 978-1-78985-978-2

Ghormade V, Deshpande MV, Panikar KM (2011) Perspectives for nano-biotechnology enabled protection and nutrition of plants. Biotechnol Adv 29(6):792–803

Graily-Moradi F, AsgariLajayer B (2021) Nanoinsecticides: preparation, application, and mode of action. In: Saglam N, Korkusuz F, Prasad R (eds) Nanotechnology applications in health and environmental sciences. Nanotechnology in the Life Sciences. Springer, ChamSpringer International Publishing, pp 385–404

Gupta A, Rayeen F, Mishra R, Tripathi M, Pathak N (2023) Nanotechnology applications in sustainable agriculture: an emerging eco-friendly approach. Plant Nano Biol 4:100033

Hashem AH, Abdelaziz AM, Askar AA, Fouda HM, Khalil AM, Abd-Elsalam KA, Khaleil (2021) *Bacillus megaterium*-mediated synthesis of selenium nanoparticles and their antifungal activity against *rhizoctonia solani* in faba bean plants. J Fungi 7(3):195

Hassan QF (ed) (2018) Internet of things A to Z: technologies and applications. Wiley

He X, Deng H, Hwang H-M (2018) The current application of nanotechnology in food and agriculture. J Food Drug Anal 27:1–21

He X, Deng H, Hm H (2021) Nanosensors for heavy metal detection in environmental media: recent advances and future trends. In: Kumar V, Guleria P, Ranjan S, Dasgupta N, Lichtfouse E (eds) Nanosensors for environment, food and agriculture, vol. 1, Environmental chemistry for a sustainable world, vol. 60. Springer, Cham

Ikram M, Raja NI, Javed B, Mashwani ZU, Hussain M, Hussain M, Ehsan M, Rafique N, Malik K, Sultana T, Akram A (2020) Foliar applications of bio-fabricated selenium nanoparticles to improve the growth of wheat plants under drought stress. Green Process Synth 9(1):706–714

Jahagirdar AS, Shende S, Gade A, Rai M (2020) Bioinspired synthesis of copper nanoparticles and its efficacy on seed viability and seedling growth in mungbean (Vignaradiata L.). Curr Nanosci 16(2):246–252

Jaskulski D, Jaskulska I, Majewska J, Radziemska M, Bilgin A, Brtnicky M (2022) Silver nanoparticles (AgNPs) in urea solution in laboratory tests and field experiments with crops and vegeTables. Materials (Basel) 15(3):870

Kah M, Kookana RS, Gogos A, Bucheli TD (2018) A critical evaluation of nanopesticides and nanofertilizers against their conventional analogues. Nat Nanotechnol 13(8):677–684

Kang H, Elmer W, Shen Y, Zuverza-Mena N, Ma C, Botella P, White JC, Haynes CL (2021) Silica nanoparticle dissolution rate controls the suppression of fusarium wilt of watermelon (*Citrullus lanatus*). Environ Sci Technol 55(20):13513–13522

Khaled E, Kamel MS, Hassan HNA, Malhat FM, Abdel-Gawad H (2015) Rapid detection of methomyl and organophosphorous pesticides with portable potentiometric biosensor. Anal Chem Lett 5(3):117–126

Khan N, Bano A (2016) Modulation of phytoremediation and plant growth by the treatment with PGPR, Ag nanoparticle and untreated municipal wastewater. Int J Phytoremediation 18:1258–1269

Khan RA, Alam SS, Najeeb S, Ali A, Ahmad A, Shakoor A, Tong L (2023) Mitigating Cd and bacterial wilt stress in tomato plants through trico-synthesized silicon nanoparticles and Trichoderma metabolites. Environ Pollut 333:122041

Kumar P, Chinnamuthu. (2020) Synthesis and characterization of nanoencapsulated herbicide using direct and indirect method. J Pharmacogn Phytochem 9(5):1476–1479

Kumar P, Pahal V, Gupta A, Vadhan R, Chandra H, Dubey RC (2020) Effect of silver nanoparticles and Bacillus cereus LPR2 on the growth of Zea mays. Sci Rep 10(1):20409

Kumari A, Bhinda MS, Sharma B, Parihar M (2022) Climate change mitigation and nanotechnology: an overview. In: Sustainable agriculture reviews 53: nanoparticles: a new tool to enhance stress tolerance, pp 33–60

Li C, Yan B (2020) Opportunities and challenges of phyto-nanotechnology. Environ Sci Nano 7(10):2863–2874

Li C, Obireddy SR, Lai WF (2021) Preparation and use of nanogels as carriers of drugs. Drug Deliv 28(1):1594–1602

Lopez-Lima D, Mtz-Enriquez AI, Carrión G, Basurto-Cereceda S, Pariona N (2020) The bifunctional role of copper nanoparticles in tomato: Effective treatment for Fusarium wilt and plant growth promoter. Sci Hortic 277:109810

Ma J, Li Y, Chen F, Sun Y, Zhu Y, Wang L (2023) *Bacillus mycoides* PM35 in combination with titanium dioxide (TiO_2)– nanoparticles enhanced morpho-physio-biochemical attributes in Barley (*Hordeum vulgare* L.) under cadmium stress. Chemosphere 323:138224

Mahmoud, Alaa El Din AU - Fawzy, Manal ED - Makhlouf, Abdel Salam Hamdy ED - Ali, Gomaa A. M. PY - 2021 DA - 2021// TI - Nanosensors and Nanobiosensors for Monitoring the Environmental Pollutants BT - Waste Recycling Technologies for Nanomaterials Manufacturing SP - 229 EP - 246 PB - Springer International Publishing CY - Cham AB

Mahmoud ME, Abou Ali SAA, Elweshahy SMT (2018) Microwave functionalization of titanium oxide nanoparticles with chitosan nanolayer for instantaneous microwave sorption of Cu(II) and Cd(II) from water. Int J Biol Macromol 111:393–399

Majdinasab M, Daneshi M, Louis Marty J (2021) Recent developments in non-enzymatic (bio)-sensors for detection of pesticide residues: Focusing on antibody, aptamer and molecularly imprinted polymer. Talanta 232:122397, Elsevier

Manna S, Roy S, Dolai A, Ravula AR, Perumal V, Das A (2023) Current and future prospects of "all-organic" nanoinsecticides for agricultural insect pest management. Front Nanotechnol 4: 1082128

Mariyam S, Upadhyay SK, Chakraborty K, Verma KK, Duhan JS, Muneer S, Meena M, Sharma RK, Ghodake G, Seth CS (2024) Nanotechnology, a frontier in agricultural science, a novel approach in abiotic stress management and convergence with new age medicine-a review. Sci Total Environ 912:169097

Miguel-Rojas C, Pérez-de-Luque A (2023) Nanobiosensors and nanoformulations in agriculture: new advances and challenges for sustainable agriculture. Emerg Top Life Sci 7(2):229–238

Mittal D, Kaur G, Singh P, Yadav K, Ali SA (2020) Nanoparticle-based sustainable agriculture and food science: recent advances and future outlook. Front Nanotechnol 4(2):579954

Mustafa IF, Hussein MZ (2020) Synthesis and technology of nanoemulsion-based pesticide formulation. Nanomaterials (Basel) 10(8):1608

Muzammil S, Ashraf A, Siddique MH, Aslam B, Rasul I, Abbas R, Afzal M, Faisal M, Hayat S (2023) A review on toxicity of nanomaterials in agriculture: current scenario and future prospects. Sci Prog 106(4):368504231221672

Muzhinji N, Ntuli V (2021) Genetically modified organisms and food security in Southern Africa: conundrum and discourse. GM Crops Food 12(1):25–35

Ndou N, Rakgotho T, Nkuna M, Doumbia IZ, Mulaudzi T, Ajayi RF (2023) Green synthesis of iron oxide (Hematite) nanoparticles and their influence on sorghum bicolor growth under drought stress. Plan Theory 12(7):1425

Nongbet A, Mishra AK, Mohanta YK, Mahanta S, Ray MK, Khan M, Baek KH, Chakrabartty I (2022) Nanofertilizers: a smart and sustainable attribute to modern agriculture. Plan Theory 11: 2587

Oliveira HC, Stolf-Moreira R, Martinez CB, Grillo R, de Jesus MB, Fraceto LF (2015) Nanoencapsulation enhances the post-emergence herbicidal activity of atrazine against mustard plants. PLoS One 10(7):e0132971

Omar SA, Elsheery NI, Pashkovskiy P, Kuznetsov V, Allakhverdiev SI, Zedan AM (2023) Impact of titanium oxide nanoparticles on growth, pigment content, membrane stability, DNA damage, and stress-related gene expression in Viciafaba under saline conditions. Horticulturae 9(9):1030

Panoth D, Manikkoth ST, Jahan F, Thulasi KM, Paravannoor A, Vijayan BK (2022) Nanosensors for pesticide detection in soil. In: Nanosensors for smart agriculture micro and nano technologies, Nanosensors for Smart Agriculture, Elsevier. 237–258

Potadar RS, Disha N. Moholkar, Kiran D. Pawar, 14 - Nanosensors for the detection of heavy trace metals in soil, In Micro and Nano Technologies, Nanosensors for Smart Agriculture, Elsevier, 2022, Pages 329–353, ISBN 9780128245545

Preeti, Sambhakar S, Malik R, Bhatia S, Al Harrasi A, Rani C, Saharan R, Kumar S, Geeta, Sehrawat R (2023) Nanoemulsion: an emerging novel technology for improving the bioavailability of drugs. Scientifica (Cairo) 2023:6640103

Rajakumaran R et al (2021) Electrochemical investigation of zinc tungstate nanoparticles; a robust sensor platform for the selective detection of furazolidone in biological samples. Microchem J 160(Part B):105750

Rajput V, Minkina T, Semenkov I, Klink G, Tarigholizadeh S, Sushkova S (2020) Phylogenetic analysis of hyperaccumulator plant species for heavy metals and polycyclic aromatic hydrocarbons. Environ Geochem Health 16:68–75

Raliya R, Tarafdar JC (2013) ZnO nanoparticle biosynthesis and its effect on phosphorous-mobilizing enzyme secretion and gum contents in Clusterbean (*Cyamopsis tetragonoloba* L.). Agric Res 2(1):48–57

Raliya R, Saharan V, Dimkpa C, Biswas P (2017) Nanofertilizer for precision and sustainable agriculture: current state and future perspectives. J Agric Food Chem 66(26):6487–6503

Rana RA, Siddiqui MN, Skalicky M, Brestic M, Hossain A, Kayesh E, Popov M, Hejnak V, Gupta DR, Mahmud NU, Islam T (2021) Prospects of nanotechnology in improving the productivity and quality of horticultural crops. Horticulturae 7(10):332

RME, Sightsavers India, Elizabath A (2019) Application of nanotechnology in agriculture. Int J Pure Appl Biosci 7(2):131–139

Sadak MS, Bakry BA (2020) Zinc-oxide and nanoZnO oxide effects on growth, some biochemical aspects, yield quantity, and quality of flax (*Linumuitatissimum* L.) in absence and presence of compost under sandy soil. Bull Natl Res Cent 44:1–2

Sakarvadia, Hasmukh. (2022). Weed management with the use of nano-encapsulated herbicide formulations: A review RM Muchhadiya, PD Kumawat, HL Sakarvadia and PM Muchhadiya. 11:2068–2075

Salehi H, Cheheregani Rad A, Raza A, Djalovic I, Prasad PV (2023) The comparative effects of manganese nanoparticles and their counterparts (bulk and ionic) in Artemisia annua plants via seed priming and foliar application. Front Plant Sci 13:1098772

Seth T (2014) Importance of Agriculture in Indian Economy. Economics Discussion. https://www.economicsdiscussion.net/agricultural-economics/importance-of-agriculture-in-indian-economy/2088

Shang Y, Hasan MK, Ahammed GJ, Li M, Yin H, Zhou J (2019) Applications of nanotechnology in plant growth and crop protection: a review. Molecules 24(14):2558

Sharma S, Singh VK, Kumar A et al (2019) Effect of nanoparticles on oxidative damage and antioxidant defense system in plants. In: Roychoudhury A, Tripathi D (eds) Molecular plant abiotic stress: biology and biotechnology. Wiley, p 480

Sharma PK, Dorlikar S, Rawat P, Malik V, Vats N, Sharma M, Rhyee JS, Kaushik AK (2021) Nanotechnology and its application: a review. In: Nanotechnology in cancer management, 1–33

Shaw DS, Honeychurch KC (2022) Nanosensor applications in plant science. Biosensors (Basel) 12(9):675

Shojaei TR, Salleh MA, Tabatabaei M, Mobli H, Aghbashlo M, Rashid SA, Tan T (2019) Applications of nanotechnology and carbon nanoparticles in agriculture. In: Synthesis, technology and applications of carbon nanomaterials, 247–277

Siddiqui SA, Blinov AV, Serov AV, Gvozdenko AA, Kravtsov AA, Nagdalian AA, Raffa VV, Maglakelidze DG, Blinova AA, Kobina AV, Golik AB (2021) Effect of selenium nanoparticles on germination of HordéumVulgáre barley seeds. Coatings 11(7):862

Sindhu R, Chitkara M, Sandhu I, Bernela M, Rani R, Malik P, Mukherjee T (2020) Nanofertilizers: applications and future prospects. In: Nanotechnology: principles and applications edited, 978–979

Singh RP, Handa R, Manchanda G (2021a) Nanoparticles in sustainable agriculture: an emerging opportunity. J Control Release 329:1234–1248

Singh H, Sharma A, Bhardwaj SK, Arya SK, Bhardwaj N, Khatri M (2021b) Recent advances in the applications of nano-agrochemicals for sustainable agricultural development. Environ Sci: Processes Impacts 23(2):213–239

Sobhan A, Muthukumarappan K, Wei L (2021) Biosensors and biopolymer-based nanocomposites for smart food packaging: Challenges and opportunities. Food Packag Shelf Life 30:100745. ISSN 2214-2894. https://doi.org/10.1016/j.fpsl.2021.100745

Tarannum N, Gautam A, Chauhan T, Kumar D (2024) Nanomaterial based sensors for detection of food contaminants: a prospect. Sens Technol 2(1)

Thakur RK, Dhirta B, Shirkot P (2018) Studies on effect of gold nanoparticles on Meloidogyne incognita and tomato plants growth and development. BioRxiv. 428144

Tyagi J, Ahmad S, Malik M (2022) Nitrogenous fertilizers: impact on environment sustainability, mitigation strategies, and challenges. Int J Environ Sci Technol 19(11):11649–11672

Tymoszuk A (2021) Silver nanoparticles effects on in vitro germination, growth, and biochemical activity of tomato, radish, and kale seedlings. Materials (Basel) 14(18):5340

Tymoszuk A, Wojnarowicz J (2020) Zinc oxide and zinc oxide nanoparticles impact on in vitro germination and seedling growth in Allium cepa L. Materials (Basel) 13(12):2784. https://doi.org/10.3390/ma13122784

Ul Haq I, Siddra I, Ram P, Vivek K, Manoj K, Devendra C (2019) Use of metallic nanoparticles and nanoformulations as nanofungicides for sustainable disease management in plants. In: Nanobiotechnology in Bioformulations (pp. 289-316). Springer International Publishing

Van Nguyen D, Nguyen HM, Le NT, Nguyen KH, Nguyen HT, Le HM, Nguyen AT, Dinh NT, Hoang SA, Van Ha C (2022) Copper nanoparticle application enhances plant growth and grain yield in maize under drought stress conditions. J Plant Growth Regul 41(1):364–375

Wahab A, Muhammad M, Ullah S, Abdi G, Shah GM, Zaman W, Ayaz A (2024) Agriculture and environmental management through nanotechnology: eco-friendly nanomaterial synthesis for soil-plant systems, food safety, and sustainability. Sci Total Environ 926:171862

Wang D, Saleh NB, Byro A, Zepp R, Sahle-Demessie E, Luxton TP, Ho KT, Burgess RM, Flury M, White JC, Su C (2022) Nano-enabled pesticides for sustainable agriculture and global food security. Nat Nanotechnol 17(4):347–360

War JM, Fazili MA, Mushtaq W, Wani AH, Bhat MY (2020) Role of nanotechnology in crop improvement. In: Nanobiotechnology in agriculture: an approach towards sustainability, pp 63–97. https://doi.org/10.1007/978-3-030-39978-8_4

Xu Y, Xu C, Huang Q, Cao L, Teng F, Zhao P, Jia M (2021) Size effect of mesoporous silica nanoparticles on pesticide loading, release, and delivery in cucumber plants. Appl Sci 11(2): 575. https://doi.org/10.3390/app11020575

Ye Y, Cota-Ruiz K, Hernández-Viezcas JA, Valdes C, Medina-Velo IA, Turley RS, Peralta-Videa JR, Gardea-Torresdey JL (2020) Manganese nanoparticles control salinity-modulated molecular responses in Capsicum annuum L. through priming: a sustainable approach for agriculture. ACS Sustain Chem Eng 8(3):1427–1436. https://doi.org/10.1021/acssuschemeng.9b05615

Chapter 11
Nanoparticles for Crop Protection

K. R. Raghi, Anju Susan, and T. M. Archana

Abstract Research in nanotechnology within agriculture has led to the development of various nanoparticles, offering potential for pest control and serving as an environmentally safe strategy for crop protection. Nanotechnology introduces innovative formulations that seamlessly integrate from sowing to consumer acceptance, extending benefits to post-harvest storage and enhancing food shelf life while preserving nutritional quality. Recent advancements in nanotechnology envision diverse roles in plant protection, real-time monitoring of infections and pest threats, and boosting crop yields, thereby fostering sustainable farming practices. The development of nanoparticles shows promise for pathologists in detecting and treating plant diseases through nano based kits, pest detection via nano sensors, and improving plants' nutrient absorption. Monitoring environmental or physiological parameters crucial for agricultural productivity and crop protection is now feasible with nano based sensors. Compared to conventional methods, nano-agrochemical delivery systems such as nanospheres, nanocomposites, nano capsules, and micelles possess novel properties that increase efficiency, promote precise delivery of agrochemicals, and reduce residual pollution, crucial for crop protection. This delivery system enables targeted and controlled chemical delivery to essential sites, akin to nano drug delivery in humans. This article explores the array of nanomaterial applications in agriculture, particularly focusing on pest control,

K. R. Raghi
Department of Integrated Chemistry, MES College Marampally, Aluva, Ernakulam, Kerala, India
e-mail: raghi@mesmarampally.org

A. Susan
Department of Chemistry, Mar Athanasius College (Autonomous), Kothamangalam, Ernakulam, Kerala, India
e-mail: anjususan@macollege.in

T. M. Archana (✉)
Postgraduate and Research Department of Microbiology, Sree Sankara College, Kalady, Ernakulam, Kerala, India
e-mail: archana@ssc.edu.in

J. M. Al-Khayri et al. (eds.), *Nanotechnology Applications in Modern Agriculture*, Nanotechnology in Plant Sciences 3, https://doi.org/10.1007/978-3-031-90506-3_11

nanotechnology-based sensors, plant growth promotion, post-harvest management, and innovative formulations.

Keywords Nano pesticides · Nano sensors · Nano formulations · Post-harvest management · Plant growth promotion

11.1 Introduction

Agriculture serves as a crucial sustenance source for a substantial segment of the population in developing nations, with over 60% relying on it for their sustenance. Additionally, it serves as a primary source of income and employment for rural communities in many of these countries (Medina-Pérez et al. 2019). However, the agricultural sector faces numerous global challenges that threaten its sustainability and ability to meet rising food demands. Rapid population growth increases pressure on food production, while limited land and water resources constrain expansion. Climate change, with its extreme weather events and shifting growing seasons, further disrupts crop yields. Soil degradation due to unsustainable practices, along with increasing outbreaks of pests and diseases, exacerbates these difficulties. Additionally, the overuse of agrochemicals leads to environmental pollution and bioaccumulation, harming ecosystems (Pandey et al. 2018). In addressing these scenarios, encompassing innovative technologies such as high-yielding varieties, synthetic fertilizers, pesticides, genetically modified crops (GM crops), and hybrid seeds contributed a significant role in improving the sustainability of global agricultural. Regrettably, excessive reliance on these advancements has led to numerous adverse effects which includes

- Micro-nutrient imbalance, eutrophication and nitrate pollution as consequences of the excessive or improper application of fertilizers.
- The extensive use of pesticides poses significant risks to untargeted organisms, encompassing humans, beneficial microorganisms, earthworms, insects, aquatic life forms and birds, while also raising concerns such as bio-magnification, pesticide resistance development, and pest population resurgence.
- The vulnerability of high-yielding crops to diseases, insects, and environmental factors has amplified the need for fertilizers and pesticides.
- While GM crops represent a notable effort to reduce chemical treatments, pests may still develop resistance to them, as they have with many pesticides (Pandey et al. 2018).

In light of these challenges, researchers are pusuing alternative strategies to improve the agricultural sector comprehensively. Implementing various applications of nanotechnology offers promising solutions to address these pressing issues.

Nanoscience and nanotechnology represent cutting-edge tools in modern agriculture and are poised to emerge as significant economic drivers in the future. It offers a broad spectrum of applications in crop protection, reflecting its versatility and potential impact. By introducing novel agrochemicals and delivery methods for

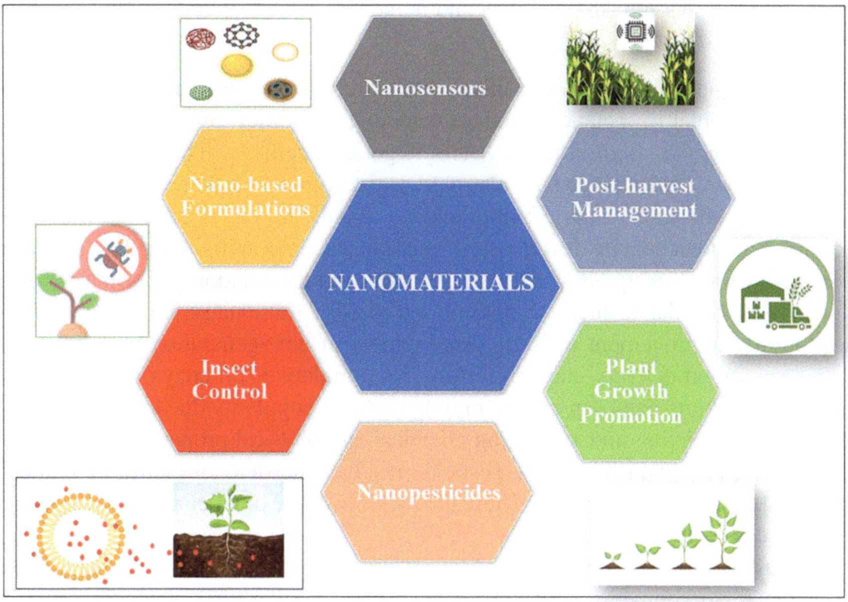

Fig. 11.1 A comprehensive overview of Nanotechnology applications in crop protection and pest control. (Figure constructed by T. M. Archana)

active compounds, nanotechnology has the potential to minimize and improve the efficiency of conventional products, including harmful pesticides. Its applications span various areas, including:

(a) developing pesticides and fertilizers tailored for enhanced crop performance,
(b) utilizing Nano sensors and nano biosensors for detecting pathogens or harmful residues,
(c) employing nanocarriers to enhance plant genome engineering and beneficial microorganisms,
(d) improving plant disease diagnostics,
(e) enhancing animal health and production, and
(f) optimizing postharvest handling practices (Medina-Pérez et al. 2019) (Fig. 11.1)

In the past decade, nanotechnology utilizing nanomaterials has made significant strides in the agricultural sector, aiming to enhance crop protection with the overarching goal of increasing overall yield. Field crops often face attacks from pathogenic microorganisms like bacteria, fungi, viruses, and insects, severely affecting productivity and sometimes resulting in complete crop loss. Conventional agricultural practices rely on various fungicides and insecticides to combat these pests, which indeed boost crop yield but come with adverse environmental strains resulting in soil and water contamination and reduced soil biodiversity (Tilman et al. 2002). Consequently, in agriculture, nanotechnology intervention aims to enhance yield by improving nutrient uptake efficiency and reducing pesticide quantities through nano formulations.

The development of diverse nano pesticide formulations has demonstrated superior antimicrobial, insecticidal, and nematocidal properties, along with controlled release capabilities, high efficiency, and targeted specificity. These characteristics make Nano pesticides promising alternatives to chemical pesticides and potentially more cost-effective than biopesticides (Paul et al. 2022). Moreover, nanoformulations require fewer active ingredients, thus mitigating their environmental impact. Consequently, the utilization of nano formulations in crop protection holds vast potential to increase agricultural yield. Additionally, crop productivity sees further improvement with the development of next-generation pesticides that enhance nutrient delivery and soil fertility by stimulating soil enzymes. In contemporary agriculture, nano sensors are rapidly rising in popularity due to the increasing demand for detecting and measuring chemical and physical properties at the nanoscale., including plant infections, pesticide residues, nutrient and moisture content, soil condition, and temperature. These characteristics are often challenging to assess through conventional means.

Insects constitute nearly two-thirds of all known animal species and can be found in various environments, consuming all types of vegetation, including weeds, medicinal plants, forest trees, and crop plants. They also contaminate stored food and other products, significantly reducing food stores and compromising quality. In their quest for sustenance, insects directly or indirectly damage plants and stored items. Arora (2017) reported that insect pests globally cause a loss of 18–20% of produced crops. Moreover, up to 30% of the global loss of major food crops can be attributed to weeds, insects, and plant pathogens, although abiotic factors like poor soils, excessive water use, and drought are not considered in this count (Rizzo et al. 2021).

The applications of nanotechnology in crop protection mark a significant leap forward in sustainable agriculture. Leveraging the distinct characteristics of nanomaterials, farmers can revolutionize pest management, enhance nutrient uptake, and minimize ecological footprints, fostering robust and high-yielding crop ecosystems. Nevertheless, research works currently underway to assess the safety and ecological ramifications of nanotechnology in agriculture remains imperative to guarantee its conscientious and sustainable integration. In this chapter, we will explore deeper into the multifaceted role of nanotechnology, particularly emphasizing its significance in crop protection and pest control.

11.2 Nano Scale Guardians: Safeguarding Crops and Combating Pests with Particle Power

Nanotechnology represents a revolutionary force driving sustainable agricultural practices and crop protection for the benefit of both human and animals. A paradigm shifts towards implementing nanotechnological principles, known as Nano-agrotechnology, is reshaping modern agriculture and food industries. Nano-agrotechnology introduces innovative formulations, seamlessly integrated from sowing to consumer acceptance, and extends its benefits to postharvest storage, enhancing food shelf life while preserving nutritional quality. Nanotechnology's recent strides envision diverse roles in plant protection, real-time monitoring of infections and pest threats, boosting crop yields, and fostering sustainable farming practices through nano-filtration for water management. Nano-agrotechnology holds promise in devising novel strategies, such as nano-formulations serving as effective pesticides. Utilizing nanoparticles in packaging, combined with spoilage detection systems, emerges as a pivotal factor in fostering consumer acceptance (Baker et al. 2019). Here, we delve into the array of nanomaterials applied in agricultural contexts, particularly focusing on pest control, sensors leveraging nanotechnology, promotion of plant growth using nanomaterials, post-harvest management, and innovative formulations driven by nanotechnology.

11.2.1 Nano Pesticides in Agriculture Applications

Nano pesticides represent any pesticide formulation intentionally including nanomaterials, whether they specially specifically attached to a hybrid substrate, immobilized in a functionalized nanocarriers or enzyme-mediated triggers. The category of nano-pesticides not only includes fungicides, herbicides, insecticides and other substances used to control pests, but also features active compounds like plant attractants and plant growth regulators, which can strengthen plant defenses against pests and diseases (Yin et al. 2023). The key characteristics of nano materials, such as small size, diverse configurations, high surface area-to-volume ration, and tailored chemical composition, make them suitable for pesticide applications, influencing their toxicity and effectiveness (Hussain et al. 2023; Ahmed et al. 2023; Hayles et al. 2017). They may have an impact on an array of plant growth characteristics, such as photosynthesis, root and shoot growth, fruit formation, seed germination, as well as crop yield, and also have the potential to improve crop protection from various bacterial and fungal infections (Mosquera et al. 2018; Kibbey and Strevett 2019). Some common benefits of nano formulations of pesticides are the increased solubility and dispersion of lipophilic pesticides in water, enhanced stability of formulation, avoiding the toxic organic solvents, increased mobility, and prevent their early degradation. Moreover, nanoencapsulation offers the ability to regulate the releasing capability of active ingredients, there by maintaining the effective pesticide concentration for an extended duration (Balaure et al. 2017). Due to this, only a small quantity of the active ingredients is needed to be applied for it to be effective for an extended period of time (Oliveira et al. 2019; Narayanan et al. 2017). Thus, it leads to a substantial decrease in the economic and environmental impacts associated with pesticide production and application. Further supporting their advantageous applications in agriculture are the observations of nanoparticle penetration into plants, uniform leaf coverage, and enhanced diffusion properties to soil (Hayles et al. 2017).

11.2.1.1 Classification of Nano Pesticides

Nano-pesticides can be categorized into two main types: Type I, which utilizes nanoparticles as direct active ingredients (AIs) in nano-dispersant emulsions, and Type II, which employs nanocarriers (clays, polymers, lipid droplets and zein nanoparticles) to encapsulate and deliver the active ingredients (Baliyarsingh and Pradhan 2023). Both types of pesticides generally incorporate with the possible advantages of conventional pesticides while also shielding, conveying, and controlling pesticide explosion (Shekhar et al. 2021). Most used Type I nano pesticides are metal-based, such as aluminum (Al), silver (Ag), copper (Cu), titanium (Ti), iron (Fe), and zinc (Zn). Of these, the most commonly used are Cu, Ag, and Ti-based nano pesticides (Ahmed et al. 2023; Wang et al. 2022; Mgadi et al. 2024). Generally, nanomaterials based on Ag, Ti, and Cu are commonly used as nano bactericides and nano fungicides; Ag-based nanomaterials are also applied as nano insecticides (Sherkhane et al. 2018; Elamawi et al. 2018; Akhtar et al. 2019; Ahmed et al. 2023). Moreover, through adhesion, dissolution (Ag^{+} and Cu^{2+} ions are examples), cytotoxicity, oxidative damage (reactive oxygen species, ROS), and genotoxicity-triggered cellular mortality, these nanomaterials exhibit antimicrobial activity. Some *in vitro* investigations have shown that few of these Ag and Cu-based nano pesticides can be effectively utilized against associated pathogens (Wang et al. 2022). Most of the Type 1 nano pesticides have efficient targeted delivery against the growing pathogens in the plant host. For example, *Citrus Huanglongbing* (a severe disease affecting citrus) caused by the bacterium *Candidatus liberibacter*, can be effectively combated using Ag-based Type 1 nano pesticides due to their precise or controlled targeted delivery within the plant host (Ahmed et al. 2023).

Type 2 nano pesticides possess innovative properties like encapsulation, enhanced loading, and controlled release of active ingredients. These features enable a synchronized release of active compounds, allowing for targeted pest control with precise timing and dosing adjustments. Thus, it places more attention on the nanoscale delivery platform, which could aid in achieving the objectives of sustainable agriculture (Vasseghian et al. 2022). In Type 2 nano pesticides, most active ingredients are traditional pesticides like glyphosate, avermectin, and atrazine. The main types of nanocarriers for these ingredients include polymer-based and clay-based materials, known for their biocompatibility, cost-effectiveness, and responsiveness to environmental triggers. Commonly used polymers, such as chitosan, polyethylene, and cellulose, are often utilized to create nano capsules, nanospheres, nanohydrogels, and nano micelles as carriers for active ingredients. Additionally, clay-based carriers like mesoporous silica nanoparticles (MSNs) and montmorillonite demonstrate a high capacity for encapsulating active ingredients, enhancing the effectiveness and control of pesticide release (Wang et al. 2022). Other emerging nanocarriers in clay-based include two-dimensional (2D) nanomaterials with high specific surface areas, carbon nanotubes, dendrimers, nanoliposomes, and nanocomposites, which make active ingredient delivery easier.

11.2.1.2 Preparation of Nano Pesticides

Over the past decade, nano-based pesticide formulations have become a prominent research focus in agriculture, aimed at improving pesticide efficiency while minimizing environmental impact. These formulations are designed to release precise amounts of active ingredients, allowing for controlled dosages that limit pesticide use and potential harm. (Yin et al. 2023).

11.2.1.3 Nano Pesticides Prepared as Inorganic and Organic Nanoparticles

Nanoparticles in these formulations are categorized as inorganic and organic nanomaterials (Fig. 11.2) Inorganic nanomaterials comprise substances like metal-organic frameworks, silica, graphene and clay minerals. Organic nanomaterials are further divided by source into natural polymers and synthetic organic compounds (Table 11.1). Natural polymers, which are environmentally compatible, cost-effective, and highly biodegradable, are abundant in nature, while synthetic organic materials offer the advantages of durability and adaptability.

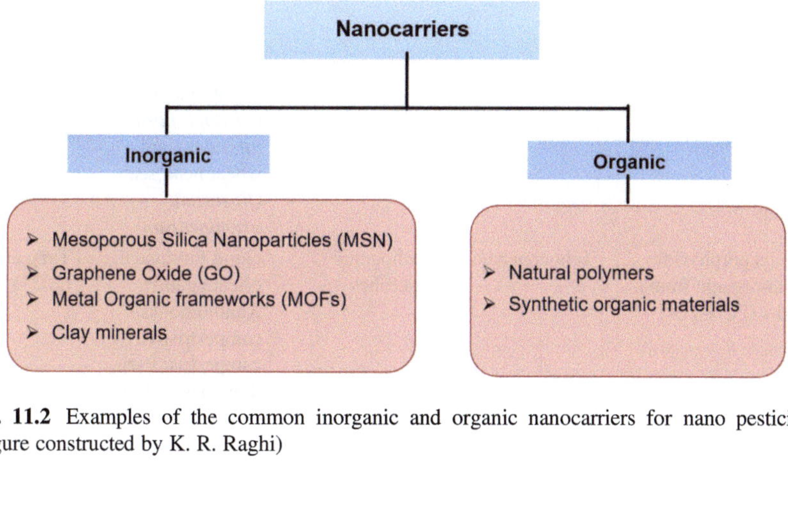

Fig. 11.2 Examples of the common inorganic and organic nanocarriers for nano pesticides. (Figure constructed by K. R. Raghi)

Table 11.1 Nano pesticides prepared as inorganic and organic nanoparticles (Yin et al. 2023)

Nanomaterials	Pesticide	Target	Performance	References
Hollow mesoporous silica (HMS) and TA-Cu metal-phenolic network as a caping reagent	Prochloraz (PRO)	Oil seed rape leaves	Demonstrates enhanced antifungal effectiveness against the pathogenic fungus *Sclerotinia sclerotiorum* while exhibiting reduced toxicity to zebrafish	Shi et al. (2022)
Disulphide-bond-bridged mesoporous organosilica (DMON)	Prochloraz (PRO)	Rice	Effectively reduced the photolysis of PRO by reflecting or absorbing UV light and delaying its premature leakage	Liang et al. (2022a)
Coordination polymers of copper and trimesic acid (CuBTC)	Avermectin	Pine wilt disease	Effectively enhanced the cytotoxicity and utilization of Avermectin against *Bursaphelenchus xylophilus*	Liu et al. (2022)
Mesoporous silica nanoparticles (MSN) (Pesticide carrier)	Pyraoxystrobin (pyr) (fungicide)	Cucumber plants	Improved the uptake and transmission of pyr to every part of the plant as well as the solubility of insoluble pyr	Xu et al. (2021)
Iron-based metal organic frameworks linked by 1,3,5-benzenetricarboxyltae (Fe-MIL-100)	Azoxystrobin	Wheat	Shows strong fungicidal activity against two major pathogenic fungi: tomato late blight (*Phytophthora infestans*) and wheat head scab (*Fusarium graminearum*)	Shan et al. (2020)
Porous porphyrinic metal-organic frameworks (PCN-224)	Tebuconazole	Pathogenic microbes	Good fungicidal effect against Xanthomonas campestrispv. campestris, Pseudomonas syringae	Tang et al. (2023)

(continued)

Table 11.1 (continued)

Nanomaterials	Pesticide	Target	Performance	References
			pv. Tomato and Alternaria alternate	
Zinc-based metal-organic frameworks (ZIF-90)	Kasugamycin (KSM)	Rice diseases	Demonstrates robust fungicidal activity and prolonged efficacy against *Magnaporthe oryzae*. Additionally, ZIF-90 nanocarriers provide effective protection for KSM, preventing its photodegradation	Liang et al. (2022b)
Sodium alginate (AL)/ polyethylene glycol (PEG)	Acetamiprid	Xanthogaleruca luteola	Enhanced the insecticidal effectiveness against *Xanthogaleruca luteola*	Ebadollahi et al. (2022)
Star polymer (SPc)	Dinotefuran	Oilseed rape	Improved distribution and increased absorption	Jiang et al. (2022)

11.2.1.4 Toxic Mechanism of Nano Pesticides

Effective integrated pest management (IPM) requires a comprehensive understanding of how nano-pesticides interact with plant species and move through the plant-soil system, including their transformation processes. Nano pesticides mainly act through two mechanisms when they come into contact with biological cells; (i) their small size and sharp surface enable explicit interaction with, or penetration through, cell walls and membranes; or (ii) they directly damage cell membranes, disrupting normal cellular processes, inhibiting photosynthesis, and ultimately inducing cell damage and death.

Nano pesticides may also obstruct the material transport channel, which inhibits nutrients from entering the cell and causes nutrient deficiencies and metabolic disorders in the cell. In pest control, certain of the nanomaterials have the ability to absorb water in addition to adhere to insect surfaces. For instance, SiO_2 nano pesticides have the same ability as silica to absorb water, which damages pest cuticles and cause surface damage, metabolic disorders, reducing water, and ultimately death (Granetto et al. 2022; Shoaib et al. 2018). SiO_2 nano pesticides also cause abrasion of mouthparts and intestines, lowering the efficiency of pest's digestion, and influencing their rate and speed of feeding. When used for plant control, nano pesticides can strongly adhere to the plant cell surfaces, impede gas exchange and nutrient uptake of the cells, as well as photosynthesis, which disrupts plant development and metabolism. Moreover, the ion exchange channels are blocked by the adsorption of nano pesticides on the plant cell surfaces, which results in cell death. In microbial control, new pores are formed as a result of interaction between cells and engineered nanoparticles (ENPs); these pores may be larger than normal, which increases the internalization of ENPs through the cell wall.

Some metal ions of metal or metal oxides containing nano pesticides, such as nano-copper, nano-zinc, and nano-silver, can be combined with the protein or fat in cells and release harmful ions of their own. The resultant consequences for organisms can be toxic (Yousef et al. 2023; Li et al. 2023). By interfering with antioxidants and detoxifying enzymes, silver nano pesticides reduce the function of acetylcholinesterase, which causes oxidative stress and cell death. Additionally, it can be shown that these inhibit gonadotropin release and protein synthesis, which harms development and impairs reproduction by either up- or down-regulating important insect genes (Jasrotia et al. 2022). Metal nano pesticides normally bind to the Sulphur from proteins or to phosphorous from DNA, which leads to reduced permeability of the cell membrane by denaturing organelles and enzymes and ultimately causing cell death (Benelli 2018). Nano pesticides made of silicon dioxide and aluminum oxide attach to the cuticle layers of insects and then cause the physisorption of lipids, and waxes, which dehydrates the insects (Jasrotia et al. 2022; Benelli 2018). According to Ibrahim and Salem (2019), the nano zeolite formulation adhered to the Tribolium confusum insect's body and produced scratching and cuticle splitting, which finally caused the insects to become dehydrated and die (Ibrahim and Salem 2019; Jasrotia et al. 2022).

A common toxicological mechanism behind cellular damage caused by nanoparticles is oxidative stress, which is triggered by free radicals. These radicals are primarily generated when the electron transport chain is disrupted. They can interact with intracellular proteins, leading to protein denaturation, and damage various cellular components such as mitochondria and nucleic acids. This process can also interfere with genetic information and disrupt biological signal transmission pathways. The primary source of these free radicals is the disruption of the electron transmission chain. These radicals can interact with intracellular proteins, denaturize proteins, damage mitochondria, nucleic acids, and other molecules, as well as obstruct the transfer of biological signals and gene information. Researchers reported that nano pesticides may greatly upregulate the expression of genes involved in stress response, antioxidant defense, and genome stability, all of which are associated with an increase in reactive oxygen species (ROS) in the cells (Li et al. 2023). A normal ROS level is necessary for several critical physiological activities, including gene expression, protein redox control, and cell communication; abnormalities and disruptions of normal functioning in these cellular processes are caused by elevated ROS levels, which ultimately lead to cell death.

11.2.1.5 Impact of Nano Pesticides on the Human

Besides offering benefits, the use of nano pesticides raises serious concerns due to possible environmental and human risks. They are more environmentally resilient and are being applied more often, thus posing a greater risk than conventional pesticides. When nano pesticides enter the cytoplasm or nucleus of non-target species in the environment or human body, they readily pass through the tissues and plasma membranes, with unpredictable consequences. Additionally, the active ingredients in nanomaterials and pesticides may exhibit synergistic harmful effects. The chemical and physical characteristics of the pesticides, the exposure pathway, the environmental fate of nanomaterials, and environmental factors can all have an impact on the harmful effects of nano pesticides on living things. Therefore, assuring the environmental safety of nano pesticides is far more difficult than ensuring the safety of particular nanomaterials or the development of conventional pesticides (Yin et al. 2023).

11.2.2 Nanoparticles for Insect Control

In recent years, the application of nanotechnology in insect pest management has surged in popularity at an impressive rate. It offers an alternative to conventional pesticide use and helps to mitigate some of the drawbacks associated with it. Many previous studies have clearly shown that nanoparticles have potent bactericidal, insecticidal, fungicidal, anti-feeding, and anti-reproductive effects on various insect species (Pittarate et al. 2023). Despite the vast amount of research on nanotoxicity to certain pests and vectors, there is a lack of trustworthy information regarding the potential mechanism of insecticidal action (Athanassiou et al. 2018).

The generally accepted theory for several kinds of nanoparticles is that they become toxic and induce oxidative stress in arthropod tissues. Furthermore, the penetration of nanoparticles through the exoskeleton may possibly be the cause of their toxicity. Then, organelles and enzymes are quickly denatured as a result of the nanoparticles binding to either phosphorous from DNA or Sulphur from proteins in the intracellular space (Amarasinghe et al. 2020; Benelli 2018). Consequently, the disruption in proton motive force and decrease in membrane permeability may eventually lead to cellular function loss and cell death (Benelli 2018). The mode of action of some of the nanoparticles against insects has been tabulated in Table 11.2.

Table 11.2 The mode of action of Nanoparticles against insects

Assessed nanomaterials	Target insects	Mode of action and/or morphological action	References
Silver nanoparticles synthesized with salicylic acid and 3,5-dinitrosalicylic acid	Aedes albopictus larvae and pupae	Instar larvae IVth showed a significant reduction in acetylcholine esterase, esterase, and phosphatase enzymes and total proteins	Ga'al et al. (2018)
Ag nanoparticles	Drosophila melanogaster	Accumulation of reactive oxygen species (ROS) in the fly tissues leading to ROS-mediated stress responses such as apoptosis, DNA damage, and autophagy; as well as the activation of the Nrf2-dependent antioxidant pathway	Mao et al. (2018)
Carbon-dot-silver nanohybrid	Anopheles stephensi and Culex quinquefasciatus mosquito species	Treated larvae shows severe cuticle membranes, and cellular organization damages-leads denaturation of organelles,—cause the inhibition of enzymatic activity. Finally, lost their intracellular functions and cell become dead	Sultana et al. (2018)
Ag nanoparticles synthesized using *Cassia fistula fruit pulp*	Aedes albopictus (Ae. Albopictus) and Culex Pipiens Pallens (Cx. pipiens)	Fourth instar larvae of *Ae. Albopictus and Cx. Pipiens* showed a decrease of total protein levels; acetylcholinesterase and α- and ß-carboxylesterase activity were also decreased by nano-Ag particles	Fouad et al. (2018)
Ag nanoparticles synthesized from pomegranate peel extract	Spodoptera litura (Fab)	Reduced the production of extracellular enzymes, weight, pH, and the overall population of heterotrophic bacteria; as well as a decrease in the activity of enzymes lipase, amylase, protease, and invertase	Bharani and Namasivayam (2017)
Nanostructured alumina (NSA)	Sitophilus oryzae (L)	Triboelectric forces caused NSA to adhere to the beetle's cuticle, sorbing its wax layer through surface area phenomena and causing the insect to become dehydrated	Stadler et al. (2017)
Gold nanoparticles	Cockroach Blattella germanica	Oral administration of gold nanoparticles interferes with	Small et al. (2016)

(continued)

Table 11.2 (continued)

Assessed nanomaterials	Target insects	Mode of action and/or morphological action	References
		reproductive and developmental processes	
Graphene oxide (GO) nanoparticles	Acheta domesticus	Following GO injection, the oxidative stress parameters (enzymatic activity of catalase and glutathione peroxidase) increased along with the heat shock protein and total antioxidant capacity levels	Dziewięcka et al. (2016)
Gold nanoparticles using latex of jatropha curcas	Aedes aegypti	Inhibit enzymatic activity of trypsin	Patil et al. (2016)
Polyethylene glycol coated nanoparticles with garlic essential oil	Tribolium castaneum	Shows significant difference in the controlling efficacy with the extension of the storage time of treated grain; speed of insecticidal loss became very slow	Yang et al. (2009)

11.2.3 Nano-based Pesticide Formulations

Recently, nanotechnology has been used to develop novel pesticide formulations that stay stable and active in the sun, heat, and rain. Nano formulated pesticides have increased the solubility of water-insoluble active ingredients, capability of controlled, targeted and prolonged release of the active ingredients, prevented premature degradation, increase bioavailability, diminish the likelihood of pest resistance, augment the uptake efficiency in target pests, and enhance both mobility and insecticidal effectiveness as a result of their reduced size, all while concurrently mitigating environmental contamination (Priyanka et al. 2019; Athanassiou et al. 2018). The most commonly used nano formulations are microemulsions, nanoencapsulation, nano emulsions, nanosuspension or nano dispersion, and polymer-based nano formulations (Fig. 11.3).

11.2.3.1 Microemulsions

Microemulsions are homogeneous, water-based, thermodynamically stable droplets smaller than 50 nm; here, the active ingredients are dissolved in a higher concentration of surfactants in the surfactant micelles (Knowles 2009). Microemulsions offer several advantages over the conventional formulations includes improved stability, decreased low flammability, decreased handler toxicity (because of the low solvent content), and above-all—increased efficacy because of better

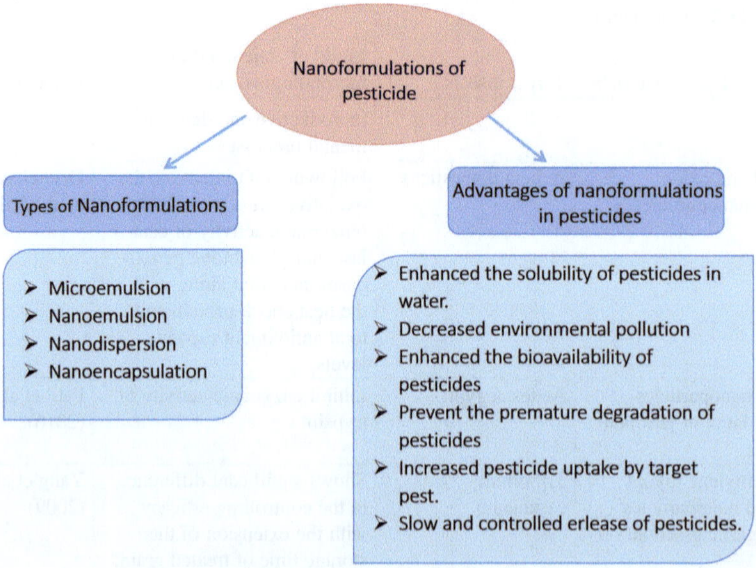

Fig. 11.3 Nano-based pesticide formulations (Priyanka et al. 2019). (Figure constructed by K. R. Raghi)

penetration or uptake brought about by surfactant's high solubilizing power. Nevertheless, these formulations have certain drawbacks, including a low quantity of active ingredient (<30%), high surfactant concentration (~20%) and a limited range of acceptable surfactant systems (Athanassiou et al. 2018).

11.2.3.2 Nano Emulsions

In pesticide nano emulsion formulations, the active ingredients that treat or protect crops that have an impact on crop yield are incorporated into the nano emulsion system. To attain pesticide delivery effectiveness maximum, the nano emulsion functions as a vector that transports and delivers the bioactive compounds to the targeted pests in the plants. In comparison to their counterparts, their attractive physicochemical characteristics, such as their larger surface area, enable the active ingredients to be released, accumulated, and absorbed more effectively. The active ingredient incorporation into the nano emulsion formulations has improved the solubility and kinetic stability, the dissolution of agrochemicals that are poorly soluble in water, improved wettability, and reduced surface tension. These factors all lead to significantly improved foliage adhesion, which allows the pesticide to remain on leaves or other vital parts of the plant for longer. Additionally, the pesticides might be coated with the nano emulsion to increase their resistance to photodegradation (Mustafa and Hussein 2020).

In nano emulsion pesticide formulations, an oil-in-water (O/W) emulsion is more frequently seen as a nano emulsion, in which the chemical's active component is distributed as nanoscale droplets in water and the surfactant molecules are trapped at the pesticide-water interface (Rajna and Paschapur 2019). Nano emulsions can be further categorized as thermodynamically and kinetically stable nano emulsions based on the type and quantity of surfactants utilized. When a non-polar pesticide has a strong repellent surfactant present at quantities greater than the critical micelle concentration (CMC) and is partially soluble in the aqueous phase, that is a thermodynamically stable nano emulsion. Because of their simple preparation techniques and chemical properties, these are highly recommended for use in commercial formulations. Despite these advantages, due to the requirement of a large quantity of surfactant and limited incorporation into the micelles, these types of nano emulsions are also very expensive and phytotoxic (Yadav et al. 2021). In contrast, the pesticide and surfactant in kinetically stable nano emulsions first form a two-phase system since the active components are insoluble. Because of this, they constantly mix through shearing, allowing the pesticide droplets in the nano emulsion to remain disseminated for an extended period of time. The key benefit of this approach is that a broader variety of surfactants can be used; however, the main disadvantage of these formulations is that the pesticide molecules must first be broken up into nanosized droplets in order for the surfactants to be coated for stabilizing purposes (Yadav et al. 2021; Rajna and Paschapur 2019).

11.2.3.3 Nano-suspension or Nano-dispersion

The dissolution of active ingredients, both crystalline and amorphous solid nanoparticles of 50 nm, is in a liquid medium. These are stabilized by remixing in water in the presence of the adsorbed surfactant molecules on the nanoparticles. The arrangement of the surfactant molecules is such that the nonpolar portion is linked to the solid pesticide nanoparticles and the polar portion extends into the aqueous solution. Unlike nano emulsions, in these formulations only kinetically stable nano dispersions with enhanced pesticide solubility and stability can be generated. These formulations enhance the solubility and dispersion of compounds that are either insoluble or fat-dispersible within a solution, a controlled release mechanism, enhanced chemical stability, improved pesticide bioavailability, and reduced costs due to their ease of large-scale manufacture (Yadav et al. 2021).

11.2.3.4 Nanoencapsulation

Nanoencapsulation is the process that generates pesticide-loaded or entrapped particles with a diameter in the nano range (1–100 nm), and it can be used to handle pesticides more safely and with less environmental exposure. Furthermore, this approach is safe because it reduces the need for pesticides, and enables a gradual and regulated release based on environmental factors like humidity, pH, temperature, reducing waste and pollution (Madhavi et al. 2020). Agrochemicals could be delivered in nano encapsulated materials by a variety of release mechanism, including dissolution, diffusion, or biodegradation (Sinha et al. 2019).

Encapsulated pesticides are expected to release more gradually over time, requiring less frequent application than very concentrated and potentially hazardous initial applications followed by repeated application (Athanassiou et al. 2018). Pesticides have previously been encapsulated using various nanomaterials, including solid lipid nanoparticles, polymer-based nanomaterials, inorganic porous nanomaterials, nanoclays, and layered double hydroxides (LDHs), etc. The pesticide is encapsulated by these substances, which are referred to as nanoencapsulation materials, to form various nanomaterials, such as liposomes, inorganic nanocages, micelles, nanospheres, nano capsules, gels, etc. (Nuruzzaman et al. 2016).

(a) Polymer-Based nanoencapsulation materials

A recently developed method for pesticide delivery is the use of polymeric nanoparticles. Since polymer nanocomposites (PNC) are made of a polymer with nanoparticles or nanofillers dispersed throughout the polymer matrix, the active ingredients are often enclosed with polymer. Natural source-derived polymers are relatively inexpensive, biodegradable and favorable to the environment, also do not produce any pollutants during decomposition. Because of these characteristics, they have shown to be effective encapsulating materials for active ingredients (Nuruzzaman et al. 2016). Based on the type of polymer used for nanoencapsulation, nanoparticles in the forms of nano capsules, nanospheres, micelles, and nanogels can function as nano pesticides (Table 11.3).

Table 11.3 Different structures of nanoencapsulation

Nano capsules	Nanospheres	Micelles	Nanogels
Composed of a polymeric membrane that encapsulates the active compounds within a liquid core at the nanoscale	Form the active nanocarrier system, in which the active compounds are uniformly dispersed and integrated within the polymeric matrix	Bioactive nanocarriers for encapsulating pesticides	Polymers undergo physical or chemical cross-linking to form a network of aqueous hydrogel particle dispersion
Synthesizing methods; emulsion-diffusion, layer by layer deposition, emulsification-coacervation, nanoprecipitation, double emulsification, solvent evaporation, etc.	Synthesizing methods; nanoprecipitation, interfacial or emulsion polymerization	Synthesizing methods; film casting method	Synthesizing methods; physical assembly of interactive polymers, template-assisted nanofabrication

Madhavi et al. (2020), Sinha et al. (2019), Nuruzzaman et al. (2016)

(b) Lipid-based nanomaterials encapsulation

Lipid-based nanomaterials have been reported to be used to deliver bioactive compounds with reduced toxicity and better encapsulation efficiency. Also, they have an extremely high potential for encapsulating active components that are hydrophilic, hydrophobic, and lipophilic. They frequently aid in the dispersion of hydrophobic active ingredients in aqueous solutions as well as the absorption of bioactive compounds through the cuticle of the insect's body. Solid-lipid and nanoliposome nanoparticles are among the various lipid—based nanoparticle types that have already been shown to be successful at encasing pesticidal active ingredients (Yadav et al. 2021; Kumar et al. 2019).

11.2.4 Empowering Plant Growth with Nanomaterials

The rapid advancement of nanotechnology has led to the extensive utilization of engineered nanomaterials (ENMs) across various aspects of plant systems, including improving quality, enhancing growth and nutritional value, and preserving genes. Nanotechnology offers significant improvements over conventional methods in addressing agricultural challenges. Research efforts worldwide are focusing on utilizing NMs to boost plant growth, increase crop production, and combat plant diseases. This surge of interest has spurred systematic investigations to confirm the beneficial impacts of ENMs on plant growth and development (Verma et al. 2018). Nanoparticles (NPs) are poised to revolutionize agriculture by enhancing germination, formulating nano fertilizers, leveraging nanoporous zeolites for gradual nutrient release, optimizing water and fertilizer application, delivering herbicides through nano capsules, and deploying Nano sensors for pest management (Ditta et al. 2015). In broad terms, engineered nanomaterials (ENMs) can be categorized into four primary groups (Srivastava et al. 2015):

(i) *Carbon-based nanomaterials*: This includes carbon nanotubes, including single-walled carbon nanotubes (SWCNTs) and multi-walled carbon nanotubes (MWCNTs), and fullerenes.
(ii) *Inorganic nanoparticles*: These include various metals such as gold, silver, an zero-valent iron, and aluminum; metal oxides like zinc oxide (ZnO), titanium dioxide (TiO_2), iron oxide (Fe_3O_4, Fe_2O_3), aluminum oxide (Al_2O_3), nickel oxide (NiO), cobalt oxide (CoO), cerium oxide (CeO_2), among others.
(iii) *Dendrimers*: These are branched-unit nanoscale polymer networks that can be tailored to perform specific chemical reactions.
(iv) *Composites*: This group includes hybrid structures, by combining nanomaterials with other larger or bulkier materials or with other nanoparticles itself.

The effects of four categories of ENMs is given in Table 11.4.

Table 11.4 Effects of engineered nanomaterials (ENMs) on plant systems

Nanomaterials	Types	Effects	References
Carbon-based nanomaterials	Multi-walled carbon nanotubes (MWCNTs)	In barley and soybean, MWCNTs facilitated seed germination without any adverse effects on subsequent plant development from the treated seeds. Also, an upregulation in the expression of genes responsible for water channel proteins was observed in the treated seeds compared to the control group (Lahiani et al. 2013).	Lahiani et al. (2013)
		Studies on maize plants showed MWCNTs to improve growth by facilitating enhanced water delivery. Comparable effects were also observed in mustard and gram plants.	Tiwari et al. (2014), Mondal et al. (2011), Tripathi et al. (2011)
		Khodakovskaya et al. (2012) illustrated that the growth of tobacco cell culture experienced enhancement across a broad spectrum of concentrations of multi-walled carbon nanotubes (MWCNTs). Cells exposed to MWCNTs exhibited an increase in the expression of the tobacco aquaporin gene 110, along with elevated production of the corresponding protein compared to the control group. Furthermore, treatment with MWCNTs led to increased expression of genes linked to cell division and cell wall expansion	Khodakovskaya et al. (2012)
	Single-walled carbon nanotubes (SWCNTs)	SWCNTs serve as nano transporters, facilitating the delivery of DNA and dye molecules into plant cells.	Srinivasan and Saraswathi (2010)
		According to Giraldo et al. (2014), chloroplasts isolated from plants showed a threefold increase in photosynthetic activity when single-walled carbon nanotubes (SWCNTs) were embedded within them compared to controls. Furthermore, the maximum electron transport rates were enhanced, and the presence of SWCNTs enabled plants to detect nitric oxide, a crucial signaling molecule.	Giraldo et al. (2014)
		SWCNTs were found to accelerate the growth of corn seminal roots by increasing the expression of genes associated with seminal root development. Additionally, SWCNTs selectively upregulated genes encoding epigenetic modification enzymes, leading to an overall deacetylation of histone H3, similar to plant responses to other stresses. It is hypothesized that the alteration in gene expression due to the interaction between nanoparticles and root cells influences the relative growth and development of roots.	Yan et al. (2013)

		Pourkhaloee et al. (2011) outlined the advantageous impacts of single-walled carbon nanotubes (SWCNTs) on seed germination and seedling growth in salvia, pepper, and tall fescue under both laboratory and greenhouse settings	Pourkhaloee et al. (2011)
	Fullerenes	De La Torre-Roche et al. (2013) investigated the influence of multi-walled carbon nanotubes (MWCNTs) or C60 fullerenes on the uptake of weathered pesticides by maize, zucchini, tomato, and soybean. Their findings revealed that pesticide accumulation varied depending on the plant species, type of nanomaterial, and its concentration. Research on the effects of fullerol on bitter melon (*Momordica charantia*) biomass, fruit yield, and phytomedicine content, as reported by Kole et al. (2013), demonstrated increased biomass, larger fruits, and enhanced content of anti-cancerous phytomedicines	De La Torre-Roche et al. (2013), Kole et al. (2013).
Inorganic nanoparticles	Metallic nanoparticles	*Gold nanoparticles*	
		Arora et al. (2012) studied the impact of foliar spray treatment with five different concentrations of gold nanoparticles (AuNPs) on Indian mustard (*Brassica juncea*) under field conditions. They found that 25 ppm AuNPs improved seed germination, possibly by enhancing seed capsule permeability. The number of leaves per plant increased, peaking at 10 ppm AuNPs, but average leaf area remained unchanged. Treated plants showed improved growth, including increased plant height, stem diameter, and number of branches, with the most significant enhancements seen at 10 ppm AuNPs. Additionally, an increased average yield, attributed to a higher number of pods per plant, was observed with NP treatment.	Arora et al. (2012)
		Kumar et al. (2013) highlighted the significant impact of gold nanoparticles (AuNPs) on seed germination and the antioxidant system in *Arabidopsis thaliana*. They also noted alterations in the expression levels of microRNAs, which regulate various morphological, physiological, and metabolic processes in plants.	Kumar et al. (2013)

(continued)

Table 11.4 (continued)

Nanomaterials	Types	Effects	References
		AuNPs enhance the number of plant height, leaves, leaf area, sugar content, and chlorophyll content, resulting in improved crop yield	Gopinath et al. (2014)
		Silver nanoparticles	
		Krishnaraj et al. (2012) investigated the impact of biologically synthesized silver nanoparticles (AgNPs) on the growth and metabolism of hydroponically grown *Bacopa monnieri*. Their study revealed that biosynthesized AgNPs significantly influenced seed germination, leading to increased protein and carbohydrate synthesis, while decreasing total phenol contents, catalase, and peroxidase activities.	Krishnaraj et al. (2012)
		Biologically synthesized silver nanoparticles (AgNPs) were found to enhance seed germination and seedling growth in *Boswellia ovaliofoliolata.*	Savithramma et al. (2012).
		AgNPs enhanced growth in *Brassica juncea*, common bean, and corn by increasing shoot and root length, leaf area, chlorophyll, carbohydrates, proteins, and antioxidant enzyme levels.	Salama (2012), Sharma et al. (2012)
		Gruyer et al. (2013) observed both positive and negative effects of AgNPs on root elongation, depending on the plant species. While root length increased in barley, it was inhibited in lettuce.	Gruyer et al. (2013)
		AgNPs were found to induce root growth in *Crocus sativus* by blocking ethylene signaling	Rezvani et al. (2012)
		Selenium nanoparticles	
		Haghighi et al. (2014) demonstrated that low concentrations of selenium nanoparticles (Se NPs) alleviated the adverse effects of heat stress in hydroponically grown tomato plants. The authors concluded that both bulk selenium and selenium nanoparticles could enhance certain growth parameters in tomato plants following short-term exposure to high and/or low-temperature stress	Haghighi et al. (2012)
		Iron nanoparticles	
			Kim et al. (2014)

		Kim et al. (2014) documented that nano zero-valent iron (nano-ZVI) promoted root elongation by triggering cell wall loosening through the induction of OH radicals.	
		It is evident from reports that iron nanoparticles (Fe NPs) possess the ability to induce antioxidant activities in plants and can be utilized to treat chlorosis. Significant findings indicate the translocation of Fe NPs from roots to endodermis via epidermis, effectively enhancing shoot and root elongation. However, the inductive effect of Fe NPs is largely dependent on their concentration and acts proportionally	Verma et al. (2018).
	Metal oxide nanoparticles	*Titanium dioxide (TiO_2) nanoparticles*	
		Larue et al. (2012) investigated the effects of TiO_2 nanoparticles (NPs) on wheat and rapeseed (*Brassica napus*) plantlets cultivated under hydroponic conditions. They observed increased root elongation during the early stages of development with TiO_2 NPs treatment; however, no significant effects on germination, evapotranspiration, or total plant biomass were noted.	Larue et al. (2012)
		Feizi et al. (2012) noted enhanced seed germination and vigor in wheat when appropriate concentrations of nano-TiO_2 were used. Conversely, inhibitory effects were observed with bulk TiO_2, while neutral effects were observed with higher concentrations of nano-TiO_2.	Feizi et al. (2012)
		Morteza et al. (2013) investigated the impact of nano-TiO_2 spraying at various stages of growth and development on photosynthetic pigments in maize plants. They observed significant effects on the total content of chlorophyll, carotenoids, and anthocyanin, with the most pronounced effect observed when nano-TiO_2 was sprayed during the reproductive stage of the plants	Morteza et al. (2013)
		Zinc oxide (ZnO) nanoparticles	
		Numerous research findings indicate that ZnO NPs promote plant growth and development across various plant species, such as soybean (Priester et al. 2012), cucumber (Zhao et al. 2013), peanut (Prasad et al. 2012), and green peas (Mukherjee et al. 2014).	Priester et al. (2012), Zhao et al. (2013), Prasad et al. (2012), Mukherjee et al. (2014)
		According to findings by Prasad et al. (2012), peanut seeds treated with a concentration of 1000 ppm of ZnO NPs with an average size of approximately	Prasad et al. (2012)

(continued)

Table 11.4 (continued)

Nanomaterials	Types	Effects	References
		25 nm showed improvements in seed germination rate and seedling vigor. Additionally, they exhibited early flowering and increased leaf chlorophyll content.	
		Research findings suggest that the application of zinc oxide nanoparticles (ZnO NPs) via foliar spray method at a concentration of 20 mg mL^{-1} can enhance the growth and biomass of tomato, cucumber, and alfalfa plants.	de la Rosa et al. (2013), Panwar et al. (2012)
		ZnO NPs application has been found to elevate the levels of photosynthetic pigments (chlorophyll and carotenoids) in pearl millet.	Tarafdar et al. (2014)
		Ramesh et al. (2014) noted that lower concentrations of ZnO NPs have a beneficial effect on seed germination in wheat.	Ramesh et al. (2014)
		Raskar and Laware (2014) reported a significant increase in root growth and dry weight in onions following ZnO NPs application	Raskar and Laware (2014)
		Silicon dioxide (SiO_2) nanoparticles	
		Lower concentrations of nano-SiO_2 improved seed germination in tomatoes.	Siddiqui and Al-Whaibi (2014)
		Suriyaprabha et al. (2014) observed that nano-SiO_2 increased seed germination by improving nutrient availability to maize seeds and adjusting the pH and conductivity of the growing medium.	Suriyaprabha et al. (2014)
		Bao-shan et al. (2004) applied exogenous nano-SiO_2 to Changbai larch (*Larix olgensis*) seedlings and noted improvements in seedling growth and quality, including increased height, root collar diameter, main root length, and the number of lateral roots, as well as enhanced chlorophyll synthesis. Nano-SiO2 was found to augment seed germination under abiotic stress conditions.	Bao-shan et al. (2004)
		Haghighi et al. (2012) observed enhanced seed germination and antioxidant system stimulation in tomatoes, while Siddiqui et al. (2014) reported similar effects in squash under NaCl stress	Haghighi et al. (2012), Siddiqui et al. (2014)
Dendrimers	PAMAM (Poly (amido amine)), PPI (Poly	Exposure of green microalgae to a PAMAM dendrimer resulted in elevated production of reactive oxygen species (ROS) and alterations in the expression of genes associated with antioxidant and photosynthetic processes.	Petit et al. (2012)

	(propylene imine)), POPAM (Poly(propylene amine))	Santiago-Morales et al. (2014) conducted a recent study demonstrating that amine-terminated G3 PAMAM dendrimers influenced the normal seed germination of both monocotyledonous and dicotyledonous species, including ryegrass, tomato, and lettuce	Santiago-Morales et al. (2014)
Composites		In a study by Pereira et al. (2015), nanocomposite nitrogen fertilizers were developed using various concentrations of polymers (hydrophilic or hydrophobic) combined with an exfoliated clay mineral to achieve slow-release properties. These nanocomposites, acting as a structural matrix, demonstrated high urea loading capacity (up to 75% w/w). Addition of PAM hydrogel and polycaprolactone (<4% w/w) influenced mechanical properties and urea release profiles. The incorporation of hydrophilic or hydrophobic polymers resulted in significantly reduced N_2O emissions in field applications.	Pereira et al. (2012)
		Another formulation developed by Berber et al. (2014) utilized a NO_3^--layered hydroxide material matrix for nitrogen delivery into soil. This formulation exhibited sustained and controlled release of NO_3^- into acidic soil for 16 days and into neutral soil for 20 days at 15 °C. Additionally, a slight increase in soil temperature led to enhanced NO_3^- release	Berber et al. (2014).

11.2.5 Nanoscale Sensors Revolutionizing Agriculture

Real-time monitoring of field conditions for prompt actions and efficient allocation of resources are the advantages offered by integrating Nano sensors into agricultural systems. The changes in nutrient concentrations, soil moisture levels, and pH, can be detected by nano sensors thus, enabling precise irrigation and fertilization strategies. Additionally, Nano sensors facilitate early pest detection and targeted pest management strategies, being equipped with specific receptors that can detect volatile organic compounds emitted by pests Moreover, nano sensors can be engineered to deliver pesticides or biocontrol agents precisely at the site of infestation, minimizing off-target effects and reducing environmental contamination. This integration of Nano sensors holds immense promise in enhancing crop protection and pest control practices, ultimately contributing to sustainable and resilient agricultural systems (Kaushal and Wani 2017).

The improved placement of autonomous nano sensors throughout farming fields, coupled with global positioning systems, is one of the main uses of nanotechnology devices, enabling thorough crop monitoring. According to El Beyrouthya and El Azzi (2014), this arrangement produces crucial information on the best times to sow and harvest crops, leading to improved agronomic practices. Large wireless sensor nodes, known as "motes," have been placed throughout an Oregon winery by Intel, which has produced nanoscale processors. A major step toward completely automated vineyard management, these sensors monitor temperature once per minute. Crossbow Technologies has also created motes for a variety of agricultural uses, including as bioremediation, harvest scheduling, pesticide application warnings, frost monitoring, irrigation control, and water quality evaluation. According to Li et al., effective agricultural production monitoring is made possible by nano barcodes and nano processing methods (Li et al. 2005).

11.2.6 Nanomaterials for Post-harvest Management

In recent decades, agriculture has experienced a surge of innovative advancements aimed at enhancing food production and security. The escalating global population necessitates a corresponding increase in crop production worldwide. Therefore, alongside crop production, the upkeep of post-harvest processing and storage is crucial. Approximately one-third of agricultural produce is lost annually due to post-harvest losses (Babu et al. 2022). Thus, the seamless integration of both processes is paramount, as neglecting either can exacerbate the situation significantly. Post-harvest processes in agriculture encompass a spectrum of activities aimed at preserving, handling, and enhancing the quality of harvested crops, thereby ensuring their longevity and market value. They ensure that the crops maintain their quality, freshness, and nutritional value from the time of harvest to consumption. Post-harvest processes typically include harvesting at the optimal stage of maturity, followed by cleaning to remove debris and foreign materials. Sorting and grading ensure uniformity and marketability, while proper packaging protects crops during transportation and storage. Storage facilities provide controlled environmental conditions to prolong shelf life and maintain quality. Transportation methods ensure timely delivery to markets or processing facilities while minimizing damage. Processing activities, such as washing and packaging, may be conducted to create value-added products. Continuous quality control monitoring detects and addresses any issues that may arise, ensuring that consumers receive fresh, high-quality produce (Maiti et al. 2018). Post-harvest losses primarily stem from microbial contamination, moisture content, degradation, and adverse effects of conventional storage methods (Babu et al. 2022).

Fumigation with aluminum phosphide (AlP) has traditionally been regarded as the most effective method for controlling storage insect pests. However, the indiscriminate use of AlP has led to the development of resistance among these pests, necessitating the search for alternative control options (Jasrotia et al. 2022). However, the current alternatives, such as fungicides like thiabendazole, pyrimethanil, fludioxonil, and imazalil, come with high costs and potential health risks to consumers, as well as environmental pollution. To address these challenges, alternative control measures for postharvest are being explored such as the use of natural compounds, biocontrol agents, irradiation, and applications of boiling water, air, or salt. Biological control utilizes yeast and bacterial antagonists, while physical treatments involve ozone, electrolyzed water, UV light, heating, and adjustments to atmospheric conditions. (Babu et al. 2022). Additionally, natural compounds such as chitosan and oligochitosan, as well as essential oils like thyme oil, tea tree oil, lemongrass oil, and oregano oil, are being investigated for their efficacy in managing fruits after harvesting (Babu et al. 2022). Given the limitations of traditional post-harvest techniques, such as their variable efficacy and the need for frequent applications to achieve consistent results, as well as their potential to compromise the quality and nutritional content of crops, there is an increasing need for innovative strategies to address these shortcomings.

Nanotechnology has emerged as a leading solution to address the limitations of traditional agricultural methods. From seed storage and priming to germination, fertigation, and post-harvest stages, nanotechnology holds the potential to enhance crop productivity and quality at every step of the agricultural process (Shukla et al. 2019). The different applications of nanomaterials in post-harvest management include:

11.2.6.1 Management of Insect-Pests

Nanotechnology offers innovative solutions in agriculture, especially for managing insect pests with precision and minimal environmental impact. Nano-formulations can enhance pesticide stability, improve solubility, and enable gradual release, making them highly effective against pests while also reducing the need for high chemical concentrations. Nano pesticides use nanoscale materials—such as silver, zinc oxide, and titanium dioxide nanoparticles.

A wide variety of nano pesticides is currently available, typically comprised of organic polymers, surfactants, and inorganic metal nanoparticles on the nanoscale. These nano-pesticides are often encapsulated in a thin polymer coating to improve dispersion in aqueous solutions and control the release of active ingredients. This encapsulation enables targeted, gradual release through mechanisms such as biodegradation, dissolution, osmotic pressure, and diffusion, often influenced by specific pH levels. Common encapsulating materials include poly(lactic) acid and polycaprolactone nanospheres. Ethiprole insecticide is commonly encapsulated in poly(lactic) acid and polycaprolactone nanospheres (Shukla et al. 2019). Similarly, nanoparticles from metals and metal oxides are also has been tested and found be effective in insect pest management. Silver nanoparticles (AgNPs) have shown considerable effectiveness in pest control, serving as carriers that precisely deliver agrochemicals to the target sites. With strong insecticidal, bactericidal, antifungal, and antiviral properties, AgNPs present a viable alternative to conventional pesticides by delivering concentrated doses directly to pest species. Research has demonstrated that AgNPs synthesized from plants like *Euphorbia prostrata (*Prostrate spurge) and *Avicennia marina* (Grey mangrove) significantly increase mortality rates in pests such as *S. oryzae* (Rice weevil), *T. castaneum* (Red flour beetle), and *R. dominica* (Lesser grain borer) (Jasrotia et al. 2022,Pathipati and Kanuparthi 2021). AgNPs derived from *Nerium oleander* (oleander) have also proven effective against larvae of *T. castaneum* (Red flour beetle) and *C. maculatus* (Cowpea seed beetle). Further studies confirm the high insecticidal impact of AgNPs produced from plants like rosemary, laurel, and cardamom, achieving notable pest mortality. When combined with traditional pesticides like Malathion, AgNPs enhance control against *T. castaneum* (Jasrotia et al. 2022). Biogenic nanoparticles, synthesized from *Daphne mucronata* (Spurge laurel) and *Monotheca buxifolia* (Afghan pistachio), also exhibit high efficacy in controlling pests, including *T. castaneum, R. dominica*, and *C. analis* (Pea beetle). Titanium dioxide nanoparticles (TiO_2NPs), known for their low toxicity and minimal impact on non-target organisms, achieved a 61.66% mortality rate in *S. oryzae* and a 60.66% rate in *S. zeamais* at a dosage of 1 g/kg over a 21-day period. Higher concentrations of TiO_2NPs further increased cumulative mortality rates in *T. castaneum* after 1, 3, and 5 days, reaching 15.30%, 23.57%, and 29.85%, respectively (Shukla et al. 2019; Jasrotia et al. 2022). Zinc oxide nanoparticles, valued for their antimicrobial properties, have been used in nanoformulations that yield high mortality rates against *S. oryzae* and *T. castaneum. Bacillus thuringiensis*–coated zinc oxide nanoparticles (Bt-ZnO NPs) have demonstrated complete mortality in pulse beetles (*C. maculatus*) at a concentration of 25 μg/ml, as well as a reduction in pest development and fecundity (Jasrotia et al. 2022). According to Hilal et al. (2021), Zinc oxide nanoparticles recorded an enhanced cumulative mortality against *T. castaneum* with values 20.42, 27.08, and 33.96%, upon exposure time of 1, 3, and 5 days respectively (Hilal et al. 2021). Dikbaş et al. (2021) found that ZnO nanoparticles integrated with chitinase from *Lactobacillus coryniformis* achieved complete lethality (100%) in the maize weevil,

Sitophilus zeamais, at a concentration of 6 mg/L, with an average lethal time frame of 2.4 days (Dikbaş et al. 2021). However, ZnO NPs are generally less effective against storage pests compared to AgNPs, aluminum oxide, and TiO_2.

Nano emulsions have been created to improve botanical insecticides for commercial applications, offering targeted action and cost-effectiveness through high-water solubility, which decreases the need for active and inert ingredients. Essential oil nano-formulations have demonstrated effectiveness against stored insect pests, with examples including nano emulsions of jojoba, purslane, oregano, and thyme oils. Polymer-based nano-formulations enhance the efficiency of active ingredients and lessen ecological impacts by reducing the use of organic solvents and surfactants (Jasrotia et al. 2022). A nanogel based on myristic acid and chitosan, loaded with *Carum copticum* (Ajwain seeds) essential oil, effectively targets *S. granarius* (granary weevil) and *T. confusum* (confused flour beetle), showing increased toxicity over time. Chitosan nanoparticles encapsulating peppermint oil have demonstrated high toxicity against *S. oryzae*, while nano-encapsulated *Melissa officinalis* (Lemon balm) essential oil in a chitosan matrix improved fumigant action and acted as an antifeedant against *T. castaneum.* Nano encapsulated essential oils from plants such as *C. cyminum,* (cumin), *R. officinalis* (rosemary), *Albizia procera* (Silk tree), eucalyptus, *Zataria multiflora* (Iranian wild mint), and *Artemisia haussknechtii* (worm wood) have also shown significant efficacy against various insect pests (Jasrotia et al. 2022).

11.2.6.2 Managing Food Security

Nanotechnology is increasingly recognized as a promising solution for addressing global food security challenges, particularly in light of the rising population and the impacts of climate change. Nanotechnology has emerged as a powerful tool for enhancing the solubility, stability, and bioavailability of phytochemicals in plant-based products. Techniques such as nanoliposomes, nanolipid particles, nano emulsions, and nanolipid carriers are now being used to protect sensitive phytochemicals from premature degradation, ensuring they reach their target within the body more effectively. In nutritional research, nanoparticles also offer insight into how bioactive components are distributed within cells or tissues, allowing for precise detection of nutrients and metabolites. In the field of food science, nanotechnology is employed to modify taste, color, and texture while boosting the nutritional content of foods. By enhancing the bioavailability and targeted delivery of dietary compounds, nanotechnology enables the creation of foods with optimized nutritional benefits. For instance, curcumin nano capsules have been developed to improve anti-inflammatory effects, while nano emulsion systems, including water-oil mixtures, solid-lipid, and biopolymer nano emulsions, have been shown to increase the bioactivity, solubility, and bioavailability of essential nutrients. Nanotubes derived from food proteins have specific applications in binding vitamins and enzymes, which is particularly useful in targeted nutritional therapies. Additionally, nanotechnology supports the structure-based design of nanocantilever tools for detecting biological and chemical contaminants, offering a significant step forward in food safety and quality control. These advancements allow for a highly precise approach to food safety monitoring, contributing to the overall assurance of food quality in the market (Panpatte and Jhala 2019; Sangeetha et al. 2019). Moreover, nano-coatings applied to food packaging can improve shelf life by preventing microbial growth and maintaining product quality. For example, silver nanoparticles embedded in packaging materials have shown antimicrobial properties that can extend the freshness of perishable goods (Prakash et al. 2024).

11.2.6.3 Role in Food Processing, Food Packaging and Storage

Nanoparticles play a significant role in food processing. They are utilized in various applications ranging from enhancing food quality and safety to improving nutritional content and shelf life. Nanotechnology has led to the development of nanostructured food products, such as nano mayonnaise used in ice cream production and nano-cochleates derived from soybean phosphatidylserine. These innovations serve as effective carriers that protect antioxidants and micronutrients from degradation during food processing and storage, ensuring that these beneficial compounds remain intact and bioavailable (Sangeetha et al. 2019). Nanotubes derived from milk proteins and various nanoencapsulation techniques are employed as nutrient carriers in beverages and food products. This technology allows for the effective delivery of vitamins and minerals, while also facilitating the encapsulation of live probiotic microorganisms, thereby enhancing gut health and overall wellness in consumers. Food packaging is essential for preserving the integrity of the food supply chain by protecting products from contaminants, microorganisms, and chemicals. An ideal packaging material should be safe, lightweight, affordable, easy to dispose of or reuse, and resilient against environmental stresses. Nanoparticles are emerging as a promising solution for extending the freshness and nutritional quality of food products over time (Sangeetha et al. 2019). Nanocomposites offer improved functionality for food packaging materials. These materials consist of multiphase solid structures with nanoscale dimensions that enhance barriers against water vapor, flavor compounds, oxygen, and carbon dioxide. Polymer nanocomposites (PNCs), formed by incorporating inert nanoscale fillers like carbon nanotubes, graphene, silica nanoparticles, and clay into a polymeric matrix, are at the forefront of this advancement The rise of biodegradable nanocomposites presents an exciting opportunity for environmentally sustainable packaging solutions. Materials such as polyhydroxybutyrate (PHB), polylactic acid (PLA), and polycaprolactone (PCL) can degrade naturally through microbial activity, leaving minimal environmental impact. Research indicates that incorporating nano clay into these biopolymers significantly enhances their oxygen barrier properties, making them suitable for food preservation (Prakash et al. 2024). Recent innovations have also led to the creation of nanocomposite films that incorporate materials like methyl cellulose, pediocin, and zinc oxide nanoparticles. These films exhibit antimicrobial properties, effectively combating pathogens such as Listeria monocytogenes and Staphylococcus aureus, thus improving food safety (Panpatte and Jhala 2019).

11.3 Conclusion and Prospects

To conclude, the wide applications of nanomaterials promise to tackle the key challenges in the agricultural sector, mainly crop production and management. From farm to market, the innovative solutions of nanotechnology form the backbone in contributing to sustainable agriculture. Nano sensors for real-time plant health and environmental condition monitoring, as well as herbicides and fertilizers based on nanotechnology that are optimized for improved crop performance, demonstrate how nanotechnology can be used to minimize ecological footprints and maximize resource efficiency. When nanocarriers are used for targeted delivery of active compounds, the applications of nano-formulations include efficient management of nutrients and provide crop protection. Thus, exploiting the advancements in nanotechnology, farmers can improve crop yields, reduce the dependence on conventional agrochemicals, thereby alleviating environmental impact. However, continued research is crucial to check the efficacy, toxicity, and environmental impact of nanomaterials to ensure responsible and sustainable agricultural practices upon their integration into the field. The multifaceted applications of nanotechnology in agriculture promise for tackling global food security issues and advancing the development of a more resilient and effective agricultural system as it continues to advance.

References

Ahmed A, He P, He P, Wu Y, He Y, Munir S (2023) Environmental effect of agriculture- related manufactured nano-objects on soil microbial communities. Environ Int 173:107819

Akhtar S, Shahzad K, Mushtaq S, Ali I, Rafe MH, Fazal-ul-Karim SM (2019) Antibacterial and antiviral potential of colloidal titanium dioxide (TiO2) nanoparticles suitable for biological applications. Mater Res Express 6(10):105409

Amarasinghe LD, Wickramarachchi PASR, Aberathna AAAU, Sithara WS, De Silva CR (2020) Comparative study on larvicidal activity of green synthesized silver nanoparticles and Annona glabra (Annonaceae) aqueous extract to control Aedes aegypti and Aedes albopictus (Diptera: Culicidae). Heliyon 6(6):e04322

Arora R (2017) Breeding insect resistant crops for sustainable agriculture. Springer Nature Singapore Pte Ltd.

Arora S, Sharma P, Kumar S, Nayan R, Khanna PK, Zaidi MGH (2012) Gold-nanoparticle induced enhancement in growth and seed yield of Brassica juncea. Plant Growth Regul 66:303–310

Athanassiou CG, Kavallieratos NG, Benelli G, Losic D, Usha Rani P, Desneux N (2018) Nanoparticles for pest control: current status and future perspectives. J Pest Sci 91:1–15

Babu PJ, Saranya S, Longchar B, Rajasekhar A (2022) Nanobiotechnology-mediated sustainable agriculture and post-harvest management. Curr Res Biotechnol 4:326–336

Baker S, Satish S, Prasad N, Chouhan RS (2019) Nano-agromaterials: influence on plant growth and crop protection. In: Industrial applications of nanomaterials. Elsevier, pp 341–363

Balaure PC, Gudovan D, Gudovan I (2017) Nanopesticides: a new paradigm in crop protection. In: New pesticides and soil sensors. Academic, pp 129–192

Baliyarsingh B, Pradhan CK (2023) Prospects of plant-derived metallic nanopesticides against storage pests-a review. J Agric Food Res 14:100687

Bao-Shan L, Shao-qi D, Chun-Hui L, Li-Jun F, Shu-Chun Q, Min Y (2004) Effect of TMS (nanostructured silicon dioxide) on growth of Changbai larch seedlings. J For Res 15:138–140

Benelli G (2018) Mode of action of nanoparticles against insects. Environ Sci Pollut Res 25(13): 12329–12341

Berber MR, Hafez IH, Minagawa K, Mori T (2014) A sustained controlled release formulation of soil nitrogen based on nitrate-layered double hydroxide nanoparticle material. J Soils Sediments 14:60–66

Bharani RA, Namasivayam SKR (2017) Biogenic silver nanoparticles mediated stress on developmental period and gut physiology of major lepidopteran pest Spodoptera litura (Fab.) (Lepidoptera: Noctuidae)—An eco-friendly approach of insect pest control. J Environ Chem Eng 5(1):453–467

de la Rosa G, López-Moreno ML, de Haro D, Botez CE, Peralta-Videa JR, Gardea- Torresdey JL (2013) Effects of ZnO nanoparticles in alfalfa, tomato, and cucumber at the germination stage: root development and X-ray absorption spectroscopy studies. Pure Appl Chem 85(12): 2161–2174

De La Torre-Roche R, Hawthorne J, Deng Y, Xing B, Cai W, Newman LA, Wang Q, Ma X, Hamdi H, White JC (2013) Multiwalled carbon nanotubes and C60 fullerenes differentially impact the accumulation of weathered pesticides in four agricultural plants. Environ Sci Technol 47(21):12539–12547

Dikbaş N, Uçar S, Tozlu G, Öznülüer Özer T, Kotan R (2021) Bacterial chitinase biochemical properties, immobilization on zinc oxide (ZnO) nanoparticle and its effect on Sitophilus zeamais as a potential insecticide. World J Microbiol Biotechnol 37:1–14

Ditta A, Arshad M, Ibrahim M (2015) Nanoparticles in sustainable agricultural crop production: applications and perspectives. In: Nanotechnology and plant sciences: nanoparticles and their impact on plants. Springer International Publishing, pp 55–75

Dziewięcka M, Karpeta-Kaczmarek J, Augustyniak M, Majchrzycki Ł, Augustyniak-Jabłokow MA (2016) Evaluation of in vivo graphene oxide toxicity for Acheta domesticus in relation to nanomaterial purity and time passed from the exposure. J Hazard Mater 305:30–40

Ebadollahi A, Valizadeh B, Panahandeh S, Mirhosseini H, Zolfaghari M, Changbunjong T (2022) Nanoencapsulation of acetamiprid by sodium alginate and polyethylene glycol enhanced its insecticidal efficiency. Nanomaterials 12(17):2971

El Beyrouthya M, El Azzi D (2014) Nanotechnologies: novel solutions for sustainable agriculture. Adv Crop Sci Technol 2(03):8863

Elamawi RM, Al-Harbi RE, Hendi AA (2018) Biosynthesis and characterization of silver nanoparticles using Trichoderma longibrachiatum and their effect on phytopathogenic fungi. Egypt J Biol Pest Control 28(1):1–11

Feizi H, Rezvani Moghaddam P, Shahtahmassebi N, Fotovat A (2012) Impact of bulk and nanosized titanium dioxide (TiO 2) on wheat seed germination and seedling growth. Biol Trace Elem Res 146:101–106

Fouad H, Hongjie L, Hosni D, Wei J, Abbas G, Ga'al H, Jianchu M (2018) Controlling Aedes albopictus and Culex pipiens pallens using silver nanoparticles synthesized from aqueous extract of Cassia fistula fruit pulp and its mode of action. Artif Cells Nanomed Biotechnol 46(3):558–567

Ga'al H, Fouad H, Tian J, Hu Y, Abbas G, Mo J (2018) Synthesis, characterization and efficacy of silver nanoparticles against Aedes albopictus larvae and pupae. Pestic Biochem Physiol 144:49–56

Giraldo JP, Landry MP, Faltermeier SM, McNicholas TP, Iverson NM, Boghossian AA, Reuel NF, Hilmer AJ, Sen F, Brew JA, Strano MS (2014) Plant nanobionics approach to augment photosynthesis and biochemical sensing. Nat Mater 13(4):400–408

Gopinath K, Gowri S, Karthika V, Arumugam A (2014) Green synthesis of gold nanoparticles from fruit extract of Terminalia arjuna, for the enhanced seed germination activity of Gloriosa superba. J Nanostructure Chem 4:1–11

Granetto M, Serpella L, Fogliatto S, Re L, Bianco C, Vidotto F, Tosco T (2022) Natural clay and biopolymer-based nanopesticides to control the environmental spread of a soluble herbicide. Sci Total Environ 806:151199

Gruyer N, Dorais M, Bastien C, Dassylva N, Triffault-Bouchet G (2013) Interaction between silver nanoparticles and plant growth. In: International symposium on new technologies for environment control, Energy-saving and crop production in greenhouse and plant 1037, pp 795–800

Haghighi M, Afifipour Z, Mozafarian M (2012) The effect of N-Si on tomato seed germination under salinity levels. J Biol Environ Sci 6(16)

Hayles J, Johnson L, Worthley C, Losic D (2017) Nanopesticides: a review of current research and perspectives. In: New pesticides and soil sensors. Elsevier, pp 193–225

Hilal SM, Mohmed AS, Barry NM, Ibrahim MH (2021) Entomotoxicity of TiO2 and ZnO nanoparticles against adults tribolium castaneum (Herbest) (Coleoptera: Tenebrionidae). In: IOP conference series: earth and environmental science, vol 910, No. 1. IOP Publishing, p 012088. https://doi.org/10.1088/1755-1315/910/1/012088

Hussain M, Zahra N, Lang T, Zain M, Raza M, Shakoor N, Adeel M, Zhou H (2023) Integrating nanotechnology with plant microbiome for next-generation crop health. Plant Physiol Biochem 196:703–711

Ibrahim SS, Salem NY (2019) Insecticidal efficacy of nano zeolite against Tribolium confusum (Col., Tenebrionidae) and Callosobruchus maculatus (Col., Bruchidae). Bull Natl Res Cent 43: 1–8

Jasrotia P, Nagpal M, Mishra CN, Sharma AK, Kumar S, Kamble U, Bhardwaj AK, Kashyap PL, Kumar S, Singh GP (2022) Nanomaterials for postharvest management of insect pests: current state and future perspectives. Front Nanotechnol 3:811056

Jiang Q, Xie Y, Peng M, Wang Z, Li T, Yin M et al (2022) A nanocarrier pesticide delivery system with promising benefits in the case of dinotefuran: strikingly enhanced bioactivity and reduced pesticide residue. Environ Sci Nano 9(3):988–999

Kaushal M, Wani SP (2017) Nanosensors: frontiers in precision agriculture. In: Nanotechnology: an agricultural paradigm, pp 279–291

Khodakovskaya MV, De Silva K, Biris AS, Dervishi E, Villagarcia H (2012) Carbon nanotubes induce growth enhancement of tobacco cells. ACS Nano 6(3):2128–2135

Kibbey TC, Strevett KA (2019) The effect of nanoparticles on soil and rhizosphere bacteria and plant growth in lettuce seedlings. Chemosphere 221:703–707

Kim JH, Lee Y, Kim EJ, Gu S, Sohn EJ, Seo YS, An HJ, Chang YS (2014) Exposure of iron nanoparticles to Arabidopsis thaliana enhances root elongation by triggering cell wall loosening. Environ Sci Technol 48(6):3477–3485

Knowles A (2009) Global trends in pesticide formulation technology: the development of safer formulations in China. Outlooks Pest Manag 20(4):165–170

Kole C, Kole P, Randunu KM, Choudhary P, Podila R, Ke PC, Rao AM, Marcus RK (2013) Nanobiotechnology can boost crop production and quality: first evidence from increased plant biomass, fruit yield and phytomedicine content in bitter melon (Momordica charantia). BMC Biotechnol 13:1–10

Krishnaraj C, Jagan EG, Ramachandran R, Abirami SM, Mohan N, Kalaichelvan PT (2012) Effect of biologically synthesized silver nanoparticles on Bacopa monnieri (Linn.) Wettst. plant growth metabolism. Process Biochem 47(4):651–658

Kumar V, Guleria P, Kumar V, Yadav SK (2013) Gold nanoparticle exposure induces growth and yield enhancement in Arabidopsis thaliana. Sci Total Environ 461:462–468

Kumar S, Nehra M, Dilbaghi N, Marrazza G, Hassan AA, Kim KH (2019) Nano-based smart pesticide formulations: emerging opportunities for agriculture. J Control Release 294:131–153

Lahiani MH, Dervishi E, Chen J, Nima Z, Gaume A, Biris AS, Khodakovskaya MV (2013) Impact of carbon nanotube exposure to seeds of valuable crops. ACS Appl Mater Interfaces 5(16): 7965–7973

Larue C, Veronesi G, Flank AM, Surble S, Herlin-Boime N, Carrière M (2012) Comparative uptake and impact of TiO2 nanoparticles in wheat and rapeseed. J Toxic Environ Health A 75(13–15): 722–734

Li Y, Cu YTH, Luo D (2005) Multiplexed detection of pathogen DNA with DNA-based fluorescence nanobarcodes. Nat Biotechnol 23(7):885–889

Li X, Chen Y, Xu J, Lynch I, Guo Z, Xie C, Zhang P (2023) Advanced nanopesticides: advantage and action mechanisms. Plant Physiol Biochem 108051:108051

Liang Y, Wang S, Yao Y, Yu S, Li A, Wang Y, Song J, Huo Z (2022a) Degradable self- destructive redox-responsive system based on mesoporous Organosilica Nano-vehicles for smart delivery of fungicide. Nanomaterials 12(23):4249

Liang Y, Wang S, Dong H, Yu S, Jia H, Wang J, Yao Y, Wang Y, Song J, Huo Z (2022b) Zeolitic imidazole framework-90-based pesticide smart-delivery system with enhanced antimicrobial performance. Nanomaterials 12(20):3622

Liu Y, Zhang Y, Xin X, Xu X, Wang G, Gao S, Qiao L, Yin S, Liu H, Jia C, Shen W, Xu L, Ji Y, Zhou C (2022) Design and preparation of avermectin nanopesticide for control and prevention of pine wilt disease. Nanomaterials 12(11):1863

Madhavi V, Reddy A, Madhavi G, Reddy NB (2020) Nanoencapsulation of pesticides: sustainable perspective in agriculture. In: AIP conference proceedings, vol 2280, No. 1. AIP Publishing

Maiti R, Thakur AK, Gupta A, Mandal D (2018) Postharvest management of agricultural produce. In: Research trends in bioresource management and technology, pp 137–166

Mao BH, Chen ZY, Wang YJ, Yan SJ (2018) Silver nanoparticles have lethal and sublethal adverse effects on development and longevity by inducing ROS-mediated stress responses. Sci Rep 8(1):2445

Medina-Pérez G, Fernández-Luqueño F, Campos-Montiel RG, Sánchez-López KB, Afanador-Barajas LN, Prince L (2019) Nanotechnology in crop protection: status and future trends. In: Nano-biopesticides today and future perspectives. Elsevier, pp 17–45

Mgadi K, Ndaba B, Roopnarain A, Rama H, Adeleke R (2024) Nanoparticle applications in agriculture: overview and response of plant-associated microorganisms. Front Microbiol 15: 1354440

Mondal A, Basu R, Das S, Nandy P (2011) Beneficial role of carbon nanotubes on mustard plant growth: an agricultural prospect. J Nanopart Res 13:4519–4528

Morteza E, Moaveni P, Farahani HA, Kiyani M (2013) Study of photosynthetic pigments changes of maize (Zea mays L.) under nano TiO2 spraying at various growth stages. Springerplus 2:1–5

Mosquera J, García I, Liz-Marzán LM (2018) Cellular uptake of nanoparticles versus small molecules: a matter of size. Acc Chem Res 51(9):2305–2313

Mukherjee A, Peralta-Videa JR, Bandyopadhyay S, Rico CM, Zhao L, Gardea-Torresdey JL (2014) Physiological effects of nanoparticulate ZnO in green peas (Pisum sativum L.) cultivated in soil. Metallomics 6(1):132–138

Mustafa IF, Hussein MZ (2020) Synthesis and technology of nanoemulsion-based pesticide formulation. Nanomaterials 10(8):1608

Narayanan N, Gupta S, Gajbhiye VT, Manjaiah KM (2017) Optimization of isotherm models for pesticide sorption on biopolymer-nanoclay composite by error analysis. Chemosphere 173:502–511

Nuruzzaman MD, Rahman MM, Liu Y, Naidu R (2016) Nanoencapsulation, nano-guard for pesticides: a new window for safe application. J Agric Food Chem 64(7):1447–1483

Oliveira CR, Domingues CE, de Melo NF, Roat TC, Malaspina O, Jones-Costa M, Silva-Zacarin EC, Fraceto LF (2019) Nanopesticide based on botanical insecticide pyrethrum and its potential effects on honeybees. Chemosphere 236:124282

Pandey S, Giri K, Kumar R, Mishra G, Raja Rishi R (2018) Nanopesticides: opportunities in crop protection and associated environmental risks. Proc Natl Acad Sci India Sect B Biol Sci 88: 1287–1308

Panpatte DG, Jhala YK (eds) (2019) Nanotechnology for agriculture: crop production & protection. Springer Nature

Panwar J, Jain N, Bhargaya A, Akhtar M, Yun Y (2012) Positive effect of zinc oxide nanoparticles on tomato plants: a step towards developing Nano-fertilizers. In: International conference on environmental research and technology (ICERT), Malaysia

Pathipati UR, Kanuparthi PL (2021) Silver nanoparticles for insect control: bioassays and mechanisms. In: Silver nanomaterials for Agri-food applications. Elsevier, pp 471–494

Patil CD, Borase HP, Suryawanshi RK, Patil SV (2016) Trypsin inactivation by latex fabricated gold nanoparticles: a new strategy towards insect control. Enzym Microb Technol 92:18–25

Paul SK, Sohrawardy H, Mahmud NU, Roy PC, Islam T (2022) Nanopesticides for crop protection. In: Agricultural nanobiotechnology. Elsevier, pp 389–438

Pereira EI, Minussi FB, da Cruz CC, Bernardi AC, Ribeiro C (2012) Urea–montmorillonite-extruded nanocomposites: a novel slow-release material. J Agric Food Chem 60(21):5267–5272

Petit AN, Debenest T, Eullaffroy P, Gagné F (2012) Effects of a cationic PAMAM dendrimer on photosynthesis and ROS production of Chlamydomonas reinhardtii. Nanotoxicology 6(3): 315–326

Pittarate S, Perumal V, Kannan S, Mekchay S, Thungrabeab M, Suttiprapan P, Sengottayan SN, Krutmuang P (2023) Insecticidal efficacy of nanoparticles against Spodoptera frugiperda (JE Smith) larvae and their impact in the soil. Heliyon 9(5):e16133

Pourkhaloee A, Haghighi M, Saharkhiz MJ, Jouzi H, Doroodmand MM (2011) Carbon nanotubes can promote seed germination via seed coat penetration. Seed Technol 32:155–169

Prakash S, Kumari M, Chauhan AK (2024) The intervention of nanotechnology in food packaging: a review. J Mater Sci 59:2585–2601. https://doi.org/10.1007/s10853-024-09360-7

Prasad TNVKV, Sudhakar P, Sreenivasulu Y, Latha P, Munaswamy V, Reddy KR, Sreeprasad TS, Sajanlal PR, Pradeep T (2012) Effect of nanoscale zinc oxide particles on the germination, growth and yield of peanut. J Plant Nutr 35(6):905–927

Priester JH, Ge Y, Mielke RE, Horst AM, Moritz SC, Espinosa K, Gelb J, Walker SL, Nisbet RM, An YJ, Schimel JP (2012) Soybean susceptibility to manufactured nanomaterials with evidence for food quality and soil fertility interruption. Proc Natl Acad Sci 109(37):E2451–E2456

Priyanka P, Kumar D, Yadav K, Yadav A (2019) Nanopesticides: synthesis, formulation and application in agriculture. In: nanobiotechnology applications in plant protection, vol 2. Springer International Publishing, pp 129–143

Rajna S, Paschapur AU (2019) Nanopesticides: its scope and utility in pest management. Indian Farmer 6:17

Ramesh M, Palanisamy K, Babu K, Sharma NK (2014) Effects of bulk & nano-titanium dioxide and zinc oxide on physio-morphological changes in Triticum aestivum Linn. J Glob Biosci 3(2): 415–422

Raskar SV, Laware SL (2014) Effect of zinc oxide nanoparticles on cytology and seed germination in onion. Int J Curr Microbiol App Sci 3(2):467–473

Rezvani N, Sorooshzadeh A, Farhadi N (2012) Effect of nano-silver on growth of saffron in flooding stress. Int J Agric Biosyst Eng 6(1):11–16

Rizzo DM, Lichtveld M, Mazet JA, Togami E, Miller SA (2021) Plant health and its effects on food safety and security in a one health framework: four case studies. One Health Outlook 3(1):6

Salama HM (2012) Effects of silver nanoparticles in some crop plants, common bean (Phaseolus vulgaris L.) and corn (Zea mays L.). Int Res J Biotechnol 3(10):190–197

Sangeetha J, Mundaragi A, Thangadurai D, Maxim SS, Pandhari RM, Alabhai JM (2019) Nanobiotechnology for agricultural productivity, food security and environmental sustainability. In: Panpatte D, Jhala Y (eds) Nanotechnology for agriculture: crop production & protection. Springer, Singapore. https://doi.org/10.1007/978-981-32-9374-8_1

Santiago-Morales J, Rosal R, Hernando MD, Ulaszewska MM, García-Calvo E, Fernández-Alba AR (2014) Fate and transformation products of amine-terminated PAMAM dendrimers under ozonation and irradiation. J Hazard Mater 266:102–113

Savithramma N, Ankanna S, Bhumi G (2012) Effect of nanoparticles on seed germination and seedling growth of Boswellia ovalifoliolata an endemic and endangered medicinal tree taxon. Nano Vision 2(1):2

Shan Y, Cao L, Muhammad B, Xu B, Zhao P, Cao C, Huang Q (2020) Iron-based porous metal–organic frameworks with crop nutritional function as carriers for controlled fungicide release. J Colloid Interface Sci 566:383–393

Sharma P, Bhatt D, Zaidi MGH, Saradhi PP, Khanna PK, Arora S (2012) Silver nanoparticle-mediated enhancement in growth and antioxidant status of Brassica juncea. Appl Biochem Biotechnol 167:2225–2233

Shekhar S, Sharma S, Kumar A, Taneja A, Sharma B (2021) The framework of nanopesticides: a paradigm in biodiversity. Mater Adv 2(20):6569–6588

Sherkhane AS, Suryawanshi HH, Mundada PS, Shinde BP (2018) Control of bacterial blight disease of pomegranate using silver nanoparticles. J Nanomed Nanotechnol 9(3):500

Shi L, Liang Q, Zang Q, Lv Z, Meng X, Feng J (2022) Construction of prochloraz-loaded hollow mesoporous silica nanoparticles coated with metal–phenolic networks for precise release and improved biosafety of pesticides. Nanomaterials 12(16):2885

Shoaib A, Elabasy A, Waqas M, Lin L, Cheng X, Zhang Q, Shi ZH (2018) Entomotoxic effect of silicon dioxide nanoparticles onPlutella xylostella(L.) (Lepidoptera: Plutellidae) under laboratory conditions. Toxicol Environ Chem 100(1):80–91

Shukla P, Chaurasia P, Younis K, Qadri OS, Faridi SA, Srivastava G (2019) Nanotechnology in sustainable agriculture: studies from seed priming to post-harvest management. Nanotechnol Environm Eng 4:1–15

Siddiqui MH, Al-Whaibi MH (2014) Role of nano-SiO2 in germination of tomato (Lycopersicum esculentum seeds Mill.). Saudi J Biol Sci 21(1):13–17

Siddiqui MH, Al-Whaibi MH, Faisal M, Al Sahli AA (2014) Nano-silicon dioxide mitigates the adverse effects of salt stress on Cucurbita pepo L. Environ Toxicol Chem 33(11):2429–2437

Sinha T, Bhagwatwar P, Krishnamoorthy C, Chidambaram R (2019) Polymer based micro-and nanoencapsulation of agrochemicals. In: Polymers for agri-food applications. Springer International Publishing, pp 5–28

Small T, Ochoa-Zapater MA, Gallello G, Ribera A, Romero FM, Torreblanca A, Garcerá MD (2016) Gold-nanoparticles ingestion disrupts reproduction and development in the German cockroach. Sci Total Environ 565:882–888

Srinivasan C, Saraswathi R (2010) Nano-agriculture-carbon nanotubes enhance tomato seed germination and plant growth. Curr Sci 99(3):274

Srivastava V, Gusain D, Sharma YC (2015) Critical review on the toxicity of some widely used engineered nanoparticles. Ind Eng Chem Res 54(24):6209–6233

Stadler T, López García GP, Gitto JG, Buteler M (2017) Nanostructured alumina: biocidal properties and mechanism of action of a novel insecticide powder. Bull Insectol 70(1):17–26

Sultana N, Raul PK, Goswami D, Das B, Gogoi HK, Raju PS (2018) Nanoweapon: control of mosquito breeding using carbon-dot-silver nanohybrid as a biolarvicide. Environ Chem Lett 16(3):1017–1023

Suriyaprabha R, Karunakaran G, Yuvakkumar R, Rajendran V, Kannan N (2012) Silica nanoparticles for increased silica availability in maize (Zea mays. L) Seeds under hydroponic conditions. Curr Nanosci 8(6):902–908

Tang Y, Zhao W, Zhu G, Tan Z, Huang L, Zhang P, Gao L, Rui Y (2023) Nano-pesticides and fertilizers: solutions for global food security. Nanomaterials 14(1):90

Tarafdar JC, Raliya R, Mahawar H, Rathore I (2014) Development of zinc nanofertilizer to enhance crop production in pearl millet (Pennisetum americanum). Agric Res 3:257–262

Tilman D, Cassman KG, Matson PA, Naylor R, Polasky S (2002) Agricultural sustainability and intensive production practices. Nature 418(6898):671–677

Tiwari DK, Dasgupta-Schubert N, Villaseñor Cendejas LM, Villegas J, Carreto Montoya L, Borjas García SE (2014) Interfacing carbon nanotubes (CNT) with plants: enhancement of growth, water and ionic nutrient uptake in maize (Zea mays) and implications for nanoagriculture. Appl Nanosci 4:577–591

Tripathi S, Sonkar SK, Sarkar S (2011) Growth stimulation of gram (Cicer arietinum) plant by water soluble carbon nanotubes. Nanoscale 3(3):1176–1181

Vasseghian Y, Arunkumar P, Joo SW, Gnanasekaran L, Kamyab H, Rajendran S et al (2022) Metal-organic framework-enabled pesticides are an emerging tool for sustainable cleaner production and environmental hazard reduction. J Clean Prod 373:133966

Verma SK, Das AK, Patel MK, Shah A, Kumar V, Gantait S (2018) Engineered nanomaterials for plant growth and development: a perspective analysis. Sci Total Environ 630:1413–1435

Wang D, Saleh NB, Byro A, Zepp R, Sahle-Demessie E, Luxton TP, Ho KT, Burgess RM, Flury M, White JC, Su C (2022) Nano-enabled pesticides for sustainable agriculture and global food security. Nat Nanotechnol 17(4):347–360

Xu Y, Xu C, Huang Q, Cao L, Teng F, Zhao P, Jia M (2021) Size effect of mesoporous silica nanoparticles on pesticide loading, release, and delivery in cucumber plants. Appl Sci 11(2):575

Yadav J, Jasrotia P, Kashyap PL, Bhardwaj AK, Kumar S, Singh M, Singh GP (2021) Nanopesticides: current status and scope for their application in agriculture. Plant Prot Sci 58(1):1–17

Yan S, Zhao L, Li H, Zhang Q, Tan J, Huang M, He S, Li L (2013) Single-walled carbon nanotubes selectively influence maize root tissue development accompanied by the change in the related gene expression. J Hazard Mater 246:110–118

Yang FL, Li XG, Zhu F, Lei CL (2009) Structural characterization of nanoparticles loaded with garlic essential oil and their insecticidal activity against Tribolium castaneum (Herbst)(Coleoptera: Tenebrionidae). J Agric Food Chem 57(21):10156–10162

Yin J, Su X, Yan S, Shen J (2023) Multifunctional nanoparticles and nanopesticides in agricultural application. Nano 13(7):1255

Yousef HA, Fahmy HM, Arafa FN, Abd Allah MY, Tawfik YM, El Halwany KK et al (2023) Nanotechnology in pest management: advantages, applications, and challenges. Int J Trop Insect Sci 43(5):1387–1399

Zhao L, Sun Y, Hernandez-Viezcas JA, Servin AD, Hong J, Niu G, Peralta-Videa JR, Duarte-Gardea M, Gardea-Torresdey JL (2013) Influence of CeO2 and ZnO nanoparticles on cucumber physiological markers and bioaccumulation of Ce and Zn: a life cycle study. J Agric Food Chem 61(49):11945–11951

Chapter 12
Nanotechnology in Crop Breeding and Genetics

Meril Sara Kurian, C. Akshaya Prakash, S. Sanjana, A. M. Shackira, Delse Parekkattil Sebastian, and Nair G. Sarath

Abstract The rapid growing human population has set increasing demands on food and agriculture. However, the present agricultural methods are insufficient to cater to the basic agricultural needs of humans. Enormous use of toxic chemicals like pesticides, insecticides and herbicides have deleterious health effects in living organisms including human beings. Moreover, pests or insects can become resistant to these chemicals rendering the usual pesticides/herbicides ineffective. Nanotechnology in agriculture has resulted in tremendous benefits in enhancing the agricultural productivity, nutritional value as well as shelf life of food substances. Nanotechnology has developed several nanomaterials such as nano-pesticides, nano emulsions and nano-insecticides for enhancing crop productivity and for replacing the conventional toxic chemicals used in agriculture. The use of nano-agricultural products has reduced the human dependency on toxic chemicals for agricultural purposes. In the current time, nanotechnology has also demonstrated its capability to alter the genetic makeup of plants through the introduction of new genes, leading to advancements in crop quality. This chapter provides an overview of the multifaceted role of nanotechnology in agriculture, highlighting its potential to revolutionize various aspects of crop production and precision breeding to advanced gene editing techniques and smart sensor technologies. Nanotechnology holds the key to unlocking sustainable and resilient agricultural practices for the future.

Keywords Agriculture · Crop productivity · Gene editing · Nanomaterials · Nanotechnology

M. S. Kurian · N. G. Sarath (✉)
Department of Botany, Mar Athanasius College (Autonomous), Kothamangalam, Kerala, India
e-mail: merilsara@macollege.in; sarathgnair@macollege.in

C. Akshaya Prakash · D. P. Sebastian
Department of Botany, St. Joseph's College (Autonomous), Devagiri, Kozhikode, Kerala, India

S. Sanjana · A. M. Shackira
Department of Botany, Sir Syed College, Kannur University, Kannur, Kerala, India

J. M. Al-Khayri et al. (eds.), *Nanotechnology Applications in Modern Agriculture*, Nanotechnology in Plant Sciences 3, https://doi.org/10.1007/978-3-031-90506-3_12

12.1 Introduction

Agriculture is the principal source of income for around 80% of impoverished individuals residing in rural regions, thus playing a critical role in the national economy. It not only ensures food security but also contributes to enhancing livelihoods and generating income for a sizeable portion of the population (Pinstrup-Andersen and Watson II 2011). However, the world will continue to face the formidable task of meeting the elevating requirement for food from an anticipated global population of 9.6 billion by the year 2050. Despite the interconnectedness of agriculture and food production with all United Nations Sustainable Development Goals, existing agricultural methods are insufficient to align with these objectives. Crop productivity is levelling off as demand for basic resources like electricity, water, and arable land rises.

The evolution of new technologies is imperative to elevate the sustainability, adaptability, and effectiveness of agricultural systems. Recent studies have pointed out the hopeful prospects of nanotechnology in enhancing the agricultural industry. This includes improving the effectiveness of agricultural inputs and delivering solutions to both agricultural and environmental challenges, aiming to boost food production and security. Nanomaterials have significant applications such as nano-pesticides, nano-fertilisers, nano-biosensors helps for the early detection of pollutants in soil and nanomaterials for soil remediation (Parisi et al. 2015).

Pesticides are commonly used in commercial agriculture, and the development of new, more effective pesticides is an ongoing process. Every year, many pesticides are evaluated, estimated to be around one million in 2009 (Resh and Cardé 2009). However, only a tiny fraction (0.1%) of these pesticides reaches their intended target pests, while the rest (99.9%) ends up contaminating the environment (Carriger et al. 2006), leading to serious repercussions on the food chain and human health. Additionally, widespread pesticide use has led to the emergence of pesticide-resistant weeds, insects, and pathogens (Rai and Ingle 2012). It is worth noting that without pesticides, the loss of life would be significantly higher (Lomborg 2002).

Biopesticides have emerged as a safer alternative to synthetic pesticides, but their effectiveness is hindered by slow action and dependency on environmental conditions. Nanopesticides show promise in overcoming these limitations by offering controlled release and prolonged efficacy (Chhipa 2017). They represent a valuable tool for sustainable pest management, potentially reducing reliance on synthetic chemicals and mitigating environmental threats.

Nano-pesticides operate dissimilarly from traditional pesticides, with nano-sized particles exhibiting unique transport mechanisms. This can lead to increased solubility and degradation of active ingredients by soil microorganisms (Kumpiene et al. 2008). Due to their enhanced solubility, nanoparticle-based pesticides are considered to have less environmental impact compared to traditional pesticides (Kah and Hofmann 2014). Nano pesticides offer advantages such as reduced energy and water usage, as they are administered in lesser quantities and less frequently compared to conventional pesticides. They also improve pesticide effectiveness and crop productivity by boosting yields and lowering input costs through reduced waste and labour. Nonetheless, nano-pesticides may pose health risks, as highlighted by the Environmental Protection Agency (EPA) in the USA (Ragaei and Sabry 2014). These risks include the potential for dermal absorption due to their small size, inhalation leading to deep lung penetration and potential translocation to the brain, environmental concerns related to the durability and reactivity of certain nanomaterials, and limited understanding of environmental exposure to engineered nanoparticles.

Over the past five decades, there has been a significant increment in crop yields, peculiarly in cereals, which has played a crucial role in meeting global nutritional needs. This increase is attributed to the widespread application of chemical fertilizers, which have boosted crop yields. However, the efficiency of chemical fertilizers is limited due to the loss of nutrients through volatilization and leaching, resulting in environmental contamination and elevated production costs (FAO 2017). For instance, a substantial portion (50–70%) of nitrogen from conventional fertilizers is lost to the environment (DeRosa 2010). Consequently, there is growing interest in developing alternative strategies to ensure sustainable nutrient use.

Nanotechnology is being employed to address these challenges by reducing nutrient losses, developing slow-release fertilizers, and enhancing the availability of nutrients. Nanofertilizers, either composed of nutrients themselves or acting as carriers/additives for nutrients, are being developed. These nano-fertilizers can encapsulate nutrients within nanomaterials (DeRosa 2010). Nano-fertilizers enhance crop yield and quality by improving nutrient use efficiency and decreasing production costs, thereby contributing to agricultural sustainability. Analysis by Kah et al. indicated that nano-fertilizers achieve a median efficacy gain of 18–29% with respect to conventional fertilizers. Phosphatic nano-fertilizers, for example, have been associated with augmented growth rates (32%) and seed yields (20%) in soybeans compared to conventional fertilizers (Liu and Lal 2014). Furthermore, nano-fertilizers enhance plant metabolism and nutrient absorption through nanoscale pores facilitated by molecular transporters or nanostructure cuticle pores (Rico et al. 2011; El-Saadony et al. 2021) (Fig. 12.1).

Nano-biosensors represent a significant advancement in sensor technology, capable of detecting various elements at extremely minute concentrations through a physiochemical transducer (Scognamiglio 2013). These sensors facilitate communication between plants and farmers (Giraldo et al. 2019; Wang et al. 2019), offering the potential for timely decision-making to enhance crop productivity through optimal management of water, land, fertilizers, and pesticides. Compared to older sensor models, nano-biosensors boast extended shelf-life due to their enhanced stability and sensitivity, rapid electron kinetics, and increased surface-to-volume ratio (Scognamiglio 2013). Several types of nano-biosensors, including plasmonic, fluorescence resonance energy transfer-based, carbon-based electrochemical, nanowire, and antibody nano-biosensors, have been utilized in plants. These sensors

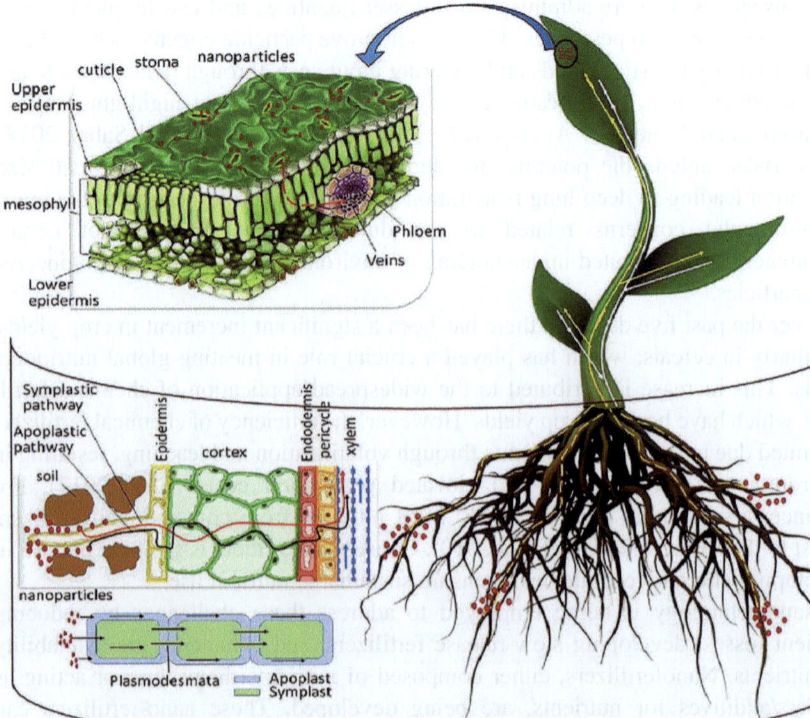

Fig. 12.1 The introduction and mechanism of action of nano-fertilizers in relation to plant development. (Reproduced with permission from El-Saadony et al. 2021; Elsevier)

can detect substances such as urea, glucose, and pesticides, monitor metabolites, and identify various microorganisms or pathogens (Rai et al. 2012).

During the last years, nanotechnology has opened avenues for enhancing food availability and developing new agricultural products advantageous to agriculture. This includes minimizing nutrient losses in fertilizers and enhancing food yield through improved pest and nutrient management, among other applications. Nanotechnology has demonstrated its capability to alter the genetic makeup of plants through the introduction of new genes, leading to advancements in crop quality. Jun et al. (2008) reported application involved the use of fluorescent-labelled starch nanoparticles as carriers for plant transgenes. These nanoparticles were developed to bind genes and transport them into plant cells by creating transient pore channels in the cell wall, cell membrane, and nuclear membrane using ultrasound assistance. A study revealed that carbon dots (nanoparticles) can serve as carriers for delivering plasmids into mature plants, enabling transient plant transformation. Application of plasmid-coated carbon dots via foliar spray on wheat plants was employed to deliver cas9, gRNA, and GFP, demonstrating the feasibility of straightforward, cost-effective, and rapid gene editing in plants through spray-on techniques (Doyle et al. 2019). El-Saadony et al. (2021) elucidated that the utilization of plant hormones facilitated the synthesis of Selenium and ZnO nanoparticles, aimed at examining their influence on seed germination amidst salt stress. These nanoparticles decreased abscisic acid levels and regulate gibberellic acid, consequently promoting early seedling growth and improving seed germination in salt-stressed rapeseed plants.

This chapter provides an overview of the multifaceted role of nanotechnology in agriculture, highlighting its potential to revolutionize various aspects of crop production and address critical challenges facing the global food system. From precision breeding to advanced gene editing techniques and smart sensor technologies, nanotechnology holds the key to unlocking sustainable and resilient agricultural practices for the future. However, it is important to consider the potential concerns and implications of real-world field applications of nanotechnology to ensure responsible and ethical deployment in agricultural systems.

12.2 Multifunctional Roles of Nanoparticles in Crop Growth and Productivity

Nanotechnology application in agriculture has resulted in tremendous benefits in enhancing the agricultural productivity, nutritional value as well as shelf life of food substances. Excessive use of toxic chemicals like pesticides, herbicides, insecticides etc. have deleterious health effects in living organisms including human beings. Moreover, pests or insects can become resistant to these chemicals rendering the usual pesticides/herbicides ineffective. Nanotechnology has developed several nanomaterials such as nano-pesticides, nano-emulsions, and nano-insecticides for enhancing crop productivity and for replacing the conventional toxic chemicals used in agriculture (Mali et al. 2020). There are reports on the reduced use of toxic chemicals in agriculture due to the application of nano-agricultural products (Echiegu 2017).

Nanoencapsulation of pesticides were an effective application of nanotechnology in pest control. Here, the pesticide of nano-meter size is encapsulated in a sac. This process leads to the slow release of the pesticide which makes the process of pest control efficient. Moreover, this technology reduces the quantity of pesticide required for pest control (Nuruzzaman et al. 2016). Another innovation of nanotechnology, nano-herbicides, can deliver nano-herbicides to weed plants without leaving remains of herbicides in the environment (Mali et al. 2020). Nano-insecticides are also known for their efficiency in removing insects and environment friendly nature due to their controlled release, enhanced photo degradation and lower requirement compared to toxic insecticides. Encapsulation of insecticides was done with polymer-based nano-formulations of chitosan, starch (Shah et al. 2016).

Nanoparticles as nano-bionics is also finding increasing application in crop growth and productivity. Nano-bionics deals with the use of nanomaterials for

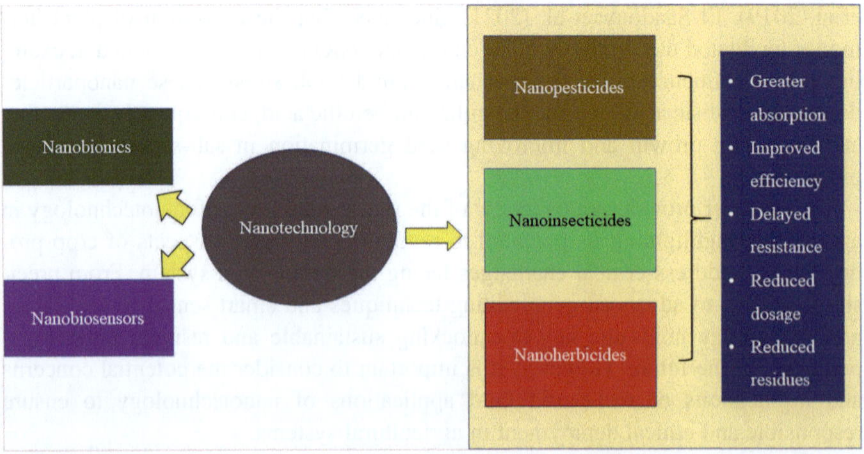

Fig. 12.2 Applications of nanotechnology in crop productivity. (Figure constructed by Akshaya Prakash C.)

improving the functions of plant cells or organs by evaluating the electronic interactions in living systems (Mali et al. 2020). The use of nanoparticles in this aspect has led to increased photosynthetic efficiency and nitrogen metabolism (Singh et al. 2015). TiO_2 nanoparticles and Cerium oxide nanoparticles enhanced the photosynthesis and (reactive oxygen species) ROS scavenging in Neem (*Azadirachtaindica*) and thale cress (*Arabidopsis thaliana*) respectively (Anjali et al. 2012; Wu et al. 2017). The use of nano-biosensors for detecting biotic or abiotic stresses in plants is also gaining momentum currently. Nano-biosensors can provide accurate information regarding the stress the crops are encountering and can help the farmers find solutions easily (Mali et al. 2020). Figure 12.2 represents the various applications of nanotechnology.

As mentioned earlier, nanotechnology has tremendous applications in the field of crop improvement and productivity. However, the negative effects of the technology must be analysed and considered before the application of this technology widely for sustainable development.

12.2.1 Nanotechnology in Plant Growth

The beneficial as well as detrimental effects of nanoparticles on plants were studied by various researchers. The impact of nanomaterials like graphene, fullerene, carbon nanotubes, metal, and metal oxide nanoparticles on the growth of plants have been studied (Aslani et al. 2014; Tripathi et al. 2015; Shweta et al. 2016). The growth of tobacco plants was reported to be increased with multiwalled carbon nanotubes (Khodakovskaya et al. 2012). Nanoparticles have been found to play a role in the upregulation of genes regulating cell division, water transport, seed germination and formation of cell walls thereby leading to increased growth in plants. The examples of some genes that are upregulated by nanoparticles include NtPIP1 (*Nicotiana tabacum* Plasma membrane Intrinsic Proteins 1, an aquaporin gene), CycB(B type cyclins involved in cell cycle regulation) and NtLRX1 (*Nicotiana tabacum* leucine-rich repeat extensin protein 1 involved in cell wall extension) (Khodakovskaya et al. 2012; Nile et al. 2022). Chowdappa and Gowda (2013) found that the germination rate of seeds of tomato increased by 20% upon the use of carbon nanotubes. Improved seed germination by using Fe/SiO_2 nanomaterials have been reported to augment seed germination in maize and barley (Disfani et al. 2017). In a study conducted by Raliya et al. (2016), the application of ZnO nanoparticles were found to increase the height of stem, root area, root length, biomass and yield of grains in mungbean. But a study conducted by Tripathi et al. (2017) reported detrimental effects of ZnO nanoparticles on wheat seedlings. The concentration of nanomaterial has a significant effect on growth of crops. At low concentrations, the nanoparticles resulted in positive effects on plant growth. However, at high concentrations, nanoparticles led to detrimental effects on crop growth (Disfani et al. 2017).

12.2.2 Nanotechnology in Crop Protection

Crop loss due to insects, pests, weeds, bacteria, fungi etc. has been an important issue of concern. The use of nanoparticles in crop protection is found to be a good measure to address some of these concerns. This is because use of nanomaterials in crop protection can reduce the quantity of chemicals used for pest and weed control, target specific delivery, slow release of active ingredients, reduced bioaccumulation of toxic chemicals and increasingly environment friendly (Ojha et al. 2018). Some of the nanobased crop protection systems include encapsulated nanosystems like biopolymers and synthetic polymer, nanocapsule, nanosphere, micelle, nanogels, lipid-based nanomaterials, inorganic porous nanomaterials, clay-based nanomaterials, nano encapsulated nanopesticides, nanocomposites etc. (Ojha et al. 2018). These are explained below:

1. Biopolymers: The sources of biopolymers are biological and include plants, animals, microbes etc. Cellulose, starch, chitin, and peptides are biopolymers. Biopolymers are eco-friendly, nontoxic and biodegradable as well as biocompatible (Sanchez-Vazquez et al. 2013; Perlatti et al. 2013).
2. Synthetic Polymer: Two or more polymer chains links together to form block polymers. Synthetic polymers include poly (d, l-lactide) and poly(-ε-caprolactone) (Kothamasu et al. 2012).
3. Nanocapsules: Nanocapsules are capsules that contain an inner active material like pesticide and an outer shell made of polymeric material (Mora-Huertas et al. 2010).
4. Nanosphere: A nanosphere consists of active ingredients that are uniformly distributed in the polymeric matrix (Letchford and Burt 2007).

5. Micelle: The agrochemicals that cannot be dissolved in water are delivered using micelles (Letchford and Burt 2007).
6. Nanogels: Crosslinking of hydrophilic polymers by van der Waal's forces results in the formation of nanogels (Sultana et al. 2013).
7. Lipid based nanomaterials: Lipid based nanomaterials are effective in targeting the cuticle of insects and release the active ingredients on to the insects resulting in destroying the insects. These include nanoliposomes and solid lipid nanoparticles (Nuruzzaman et al. 2016).
8. Inorganic porous nanomaterials: These are made by the absorption of active ingredients in nanoparticles without the formation of linkages. They are biocompatible, non-toxic with huge surface area (Lodha et al. 2012).
9. Clay based nanomaterials: Clay materials like aluminium silicates are intercalated with active ingredients (Suresh et al. 2010).
10. Nano-encapsulated nano-pesticides: Metallic or non-metallic nanoparticles are used as a pesticide delivery system in this case (Chowdappa and Gowda 2013).

Once these nanomaterials enter inside the cytoplasm of plant cells, they affect the cellular processes by binding to the cell organelles (Jia et al. 2005). The entry of nanomaterials into plants occurs via stomatal openings or by crossing the cell wall and cell membrane (Jia et al. 2005; Nair et al. 2010). The efficiency of these materials depends upon the controlled delivery of the active substance at the target area (Peteu et al. 2010).

12.2.3 Nanotechnology and Gene Editing in Plants

Nanotechnology has enlightened in delivery and editing of genes in plants. Nanoparticle mediated gene delivery and editing in plants is a promising, futuristic tool since it has been successfully overcome the hurdle that the inability of biomolecule transfer through inflexible and multi-layered plant cell wall (Gad et al. 2020; Kwak et al. 2019). In addition to this it is a cost effective, simple, easy method that protect the encapsuled genetic material without losing its integrity and it is also applicable towards a wide range of plant species; thereby helps to mitigate the limitations of conventional physical and biological transformation methods (Kwak et al. 2019; Rafsanjani et al. 2012). Moreover, the genetic material that are transported by nanotubules shows preserved functional activity as well as it gets easily dissociated from nanocarriers within the cell (Burlaka et al. 2015). Quantum dots, silica and mesoporous silica nanoparticles, carbon nanomaterial, metal and magnetic nanoparticles, peptide-based nanomaterial, and clay nanosheets serve as major gene delivery vectors. Among these carbon nanomaterials is a potential gene delivery tool due to its capability to traverse plant cell wall particularly organelle membranes (Gad et al. 2020; Law et al. 2023). The minimization of transgene silencing and improvement in efficiency and productivity can be achieved with nanotechnology-based biotransformation is combined with commonly applied approaches in transgenic plant development (Rafsanjani et al. 2012). Nanotechnology based gene delivery into chloroplast organelle is an efficient method to the overall performance of chloroplast organelle (Newkirk et al. 2021).

The chloroplast genome modifications or foreign gene introduction into chloroplast is conducted through the steps such as incorporation of gene of interest and selectable markers into the DNA construct & its introduction into the host, which is the chloroplast organelle. Biolistic, PEG-mediated, carbon nanotubes carriers, UV-laser microbeam, and Agrobacterium-mediated approaches are the typical methods of introducing gene into the chloroplast (Kumar and Ling 2021). In plants, nucleus-targeted gene delivery is more predominant than organelle such as mitochondria or chloroplast targeted gene delivery, it is not only because of that the nucleus is serving as the most abundant resource for genetic information but also because of the difficulty in obtaining homoplasmic transformants, small size & presence of double membrane in case of mitochondria and chloroplast. Eventhough the development of engineered nanomaterials can make changes in organelle-specific gene delivery in plants with improved delivery efficiencies and selectivity for the targeted organelle (Law et al. 2023).

Nanotechnology paved a new path for the advancements of chloroplast organelle focusing on the targeted delivery of genes & chemicals, nano-sensors for chloroplast biomolecule and chloroplast performance enhancement through nanotherapeutics (Newkirk et al. 2021). Therefore, the wider applications of chloroplast transformation establishing its significant role in various fields including agriculture, phytoremediation as well as biopharmaceuticals (Kumar and Ling 2021). The nanoparticle foliar delivery into plant cell and organelle is pivot on nanoparticle size and charge. The highest efficiency of foliar delivery into guard cell, extracellular space and chloroplast succeeded when a positive charge (>15Mv) in zeta potential with hydrophilic nanoparticle size less than 20 nm in case of cotton & 11 nm in case of maize (Hu et al. 2020). Chloroplast selective gene delivery are species independent, does not have the need of any particular equipment and are considered as in-situ gene delivery. However, some of the weakness of this method include it show only transient expression, potential limitation of plasmid DNA size per nanoparticles as well as it is only a proof of concept (Newkirk et al. 2021).

The emerging advancements in various techniques introduced the significant role of carbon nanotubes (CNTs) and carrier proteins as a gene delivery tools can overcome the limitations of previously practiced gene delivery approaches. CNTs serves as strong contender for plant gene delivery without any potential toxicity towards plant species whereas carrier protein-based methods are prominently practiced in animal cells (Nandy et al. 2020). Chloroplast solar energy harnessing and electron transport rates can be improved via semiconducting single-walled carbon nanotubes which eventually leads to the enhancement of chloroplast carbon capture. In addition, single-walled carbon nanotubes-nanoceria complexes suppress reactive oxygen species (ROS) thereby exhibit reduced level of ROS in extracted chloroplast (Giraldo et al. 2014).

Single-walled carbon nanotubes (SWCNTs) serve as the most suitable nanocarrier for genetic material delivery in plants than multi-walled carbon nanotubes (MWCNTs), because in most cases cellulose wall function as a barrier for MWCNTs penetration into the cell. Nonetheless, MWCNTs are applicable for transformation of protoplast whereas SWCNTs possess higher applicability with their immense ability for the transformation of protoplast as well as walled plant cells and these were successfully demonstrated in *Nicotiana tabacum* plant (Burlaka et al. 2015). Similarly, without the assistance of chemicals or biolistics, chitosan-complexed single-walled carbon nanotubes (CS-SWNTs) were efficiently used for the delivery of plasmid DNA (pDNA) to the chloroplast organelle of mature *Eruca sativa*, *Nasturtium officinale*, *Nicotiana tabacum* and *Spinacia oleracea* plants and in isolated *Arabidopsis thaliana* mesophyll protoplasts, implementing the mechanism of lipid exchange envelope penetration (LEEP) by SWNTs. The pDNA: SWNT ratio is a major factor in determining the trafficking efficiency of nanocomplexes into chloroplast (Kwak et al. 2019). The biorecognition-mediated delivery of chemical cargoes (TP-β-CDs) and plasmid DNA (TP-pATV1-SWCNTs) to chloroplast organelle is possible via targeted SWCNTs with higher delivery efficiency rate and was successfully demonstrated in Arabidopsis plant. This method of chloroplast-selective gene delivery created some impacts in Arabidopsis including reduction in chlorophyll contents, photosynthetic capacity, increased H_2O_2 levels that eventually leads to oxidative damages in whole plant cell DNA as well as transient effect on chloroplast DNA. At the same time, the percentage of dead cells showed no significant difference, and the integrity of cell membrane remain unaffected when exposed to targeted nanomaterials, but temporary disruption occurs in case of isolated chloroplast envelop by TP-pATV1-SWCNTs exposure (Santana et al. 2022).

12.2.4 Nanotechnology to Advance the CRISPR-Cas Technology

Clustered Regularly Interspaced Short Palindromic Repeats-Cas (CRISPR-Cas) technology is a fast, accurate and specific technology that can be exploited to induce new traits in the genome of desired organisms. It is a nuclease-based genome editing technology that is finding increased applications in agriculture and plant biology (Demirer et al. 2021). CRISPR-Cas system is basically a prokaryotic adaptive immune system (Hille and Charpentier 2016). The system comprises of CRISPR RNA (crRNA), trans-activating CRISPR RNA (tracrRNA), guide RNA, Cas proteins and Protospacer Adjacent Motif (PAM). The crRNAs are derived from the repetitive DNA sequences of bacterial genomes. They serve as guides for targeting specific DNA sequences and attach to those sequences (Nussenzweig and Marraffini 2020; Bhatia and Yadav 2023). Trans-activating CRISPR RNA helps in the creation of distinct crRNAs from precursor CRISPR transcripts (Liao and Beisel 2021). The guide RNA guides the Cas protein to the vicinity of the target DNA sequence (Liu et al. 2020). The Cas proteins along with CRISPR RNAs identify and cut target

DNA sequences leading to genetic modification of different organisms. Cas proteins mainly act as nucleases resulting in the cleavage of DNA at sites (Makarova et al. 2011; Sander and Joung 2014). Protospacer Adjacent Motif or PAM sequences help the Cas protein to recognise the target DNA. When a PAM sequence is close to the target DNA, the Cas protein can precisely recognise the targeted area for modification, avoiding accidental effects on non-target DNA (Yang et al. 2016; Gleditzsch et al. 2019).

The CRISPR-Cas system provides the prokaryote protection by providing sequence specific defence against foreign DNA or RNA (Klimek-Chodacka et al. 2018). It cleaves foreign DNA or RNA by identifying it and acting in a sequence-specific way. The mechanism underlying this involves three steps which are adaptation, crRNA biogenesis and target interference (Hille and Charpentier 2016). During the adaptation or spacer acquisition process, a sequence of the foreign DNA known as a protospacer is incorporated into the CRISPR array by certain Cas proteins. In some cases, the Protospacer Adjacent Motif, which is situated next to the protospacer is necessary for spacer acquisition and interference (Barrangou et al. 2007). In the second step, crRNA biogenesis, the precursor crRNA is produced from the CRISPR array. The precursor crRNA is then processed into mature crRNAs by tracrRNAs or other proteins like Cas5d (Carte et al. 2008; Deltcheva et al. 2011). In the final stage i.e. interference, crRNAs function as guides for interfering with the foreign nucleic acids. Cascade like complexes or single effector protein are employed for the degradation of the target sequence (Hale et al. 2009; Zetsche et al. 2015; Hille and Charpentier 2016). The CRISPR- Cas system can be exploited for genome editing for the introduction of desired features in organisms. The degradation of target sequence can result in double stranded break in the foreign DNA at the desired location. This break can be repaired either by nonhomologous end joining (NHEJ) which leads to knockout of gene by frameshift mechanism or by homology directed repair (HDR) which makes use of DNA templates having locations that suits the target sequence as a result of which error free restoration of DNA sequence or genome editing is possible (Salsman and Dellaire 2017; Yin et al. 2019; Jang et al. 2020).

The use of CRISPR-Cas technology has been found to improve the stress tolerance potential of crop plants as well as in the induction of desirable characteristics into crops (Demirer et al. 2021). The application of CRISPR Cas technology has led to the improvement of many crops like rice, sorghum, wheat, barley etc. (Dong et al. 2020; Char et al. 2020; Li et al. 2020; Zeng et al. 2020). But delivering CRISPR components in a timely, accurate, and safe manner is an impediment (Rui et al. 2020). Conventional methods of biomolecules like biolistic and Agrobacterium mediated transformation have various disadvantages like limited efficiency of gene transfer, tissue destruction etc. However, the use of nanotechnology has made it possible to get beyond the restrictions of conventional methods (Cunningham et al. 2018). Nanotechnology has increased the transformation efficiency for both temporary and permanent genetic alterations in different plant species (Serag et al. 2013; Vanhaeren et al. 2016; Novák et al. 2017). Using nanoparticles as nanocarriers of bio-modifier molecules like CRISPR-Cas system is a potent technique in the

Table 12.1 Application of nanotechnology and CRISPR- Cas system in crop improvement

Plant	Target gene	Nanoparticle and CRISPR-Cas	Effects	References
Cucumis sativus L.	*CsAKT1*	Nanoceria, CRISPR-Cas 9	Salt tolerance	Peng et al. (2022)
Oryza sativa	*OsPDS*	Carbon nanotubes, CRISPR-Cas 9	Downregulation of phytoene desaturase	Dunbar et al. (2022)

production of genetically modified organisms (Abd-Elsalam 2020; Demirer et al. 2021). Table 12.1 shows the application of Nanotechnology and CRISPR- Cas system in crop improvement.

Even though nanoparticles are incredibly effective in introducing CRISPR/Cas9, there are still plenty of issues that need to be worked out, such as limited transfection rates, weak encapsulation, scale-up issues, and bio-protection. In all areas of agricultural science, the use of nanomaterials in the CRISPR/Cas9 delivery is preferable than earlier delivery techniques. However, further research is required to guarantee that the CRISPR approach is delivered effectively and consistently and is used in the near future (Ahmar et al. 2021).

12.2.5 Nanotechnology in Fortification

Rapidly growing human population has induced an increasing demand for food production. But increasing human activities has led to reduced fertility of soil resulting in decreased crop productivity (Elemike et al. 2019). Nanotechnology, a recent advance in precision agriculture is used to create and direct approaches aimed at satisfying the rising need for food from the global population (Elemike et al. 2019). The use of nanotechnology for agricultural purposes is narrowing the gap between crop fortification and nutrient loss and can provide food security as well as good nutrition (Xiong et al. 2017). The nutrient content of crops can be improved while cultivating them and this process is known as biofortification (Bouis et al. 2011).

Nanotechnology can be employed for the biofortification of crops. Nano-fertilizers and nanoparticles can be used for this purpose. Nano-fertilizers have high surface area to volume ratio. This property makes nano-fertilizers advantageous over conventional fertilizers. Nutrient uptake of crops provided with nano-fertilizers is more efficient as nano-fertilizers aid in slow release of mineral nutrients. Encapsulated nanoparticles, nano-clays, and zeolites improve the effectiveness of fertiliser application, restore the fertility of soil and health of crops, and minimise contamination and agroecological destruction (Manjunatha et al. 2016). However, the species and size of the plant, concentration and chemical characteristics of nanoparticles affect the success of nano-fertilizers (Thakur et al. 2018). Nano-fertilizers may consist of zinc oxide nanoparticles, quantum dots, gold nanorods,

Table 12.2 Nanoparticles in biofortification

Crop	Nanoparticle	Effect	References
Tomato*(Solanum lycopersicum)*	Cu	Enhanced vitamin C and β- carotenoid content	Hernández-Fuentes et al. (2017)
Cumin *(Cuminumcyminum)*	Nano chelated Fe	Increased Fe in seeds	Sabet and Mortazaeinezhad (2018)
Sugar cane *(Saccharum officinarum)*	Cu	Increased Fe, Cu content	Tamez et al. (2019)
Wheat *(Triticum aestivum)*	ZnO	Enhanced Zn content in grains	Du et al. (2019)
Common bean *(Phaseolus vulgaris)*	ZnO	Enhanced Zn content in grains	Salama et al. (2019)
Jowar *(Sorghum bicolor)*	ZnO	Enhanced Zn content in grains	Dimkpa et al. 2019
Wheat *(Triticum aestivum)*	Fe	Enhanced Fe content in grains	Rizwan et al. (2019)
Maize *(Zea mays)*	ZnO	Enhanced Zn content in grains	Umar et al. (2020)
Spring onion *(Allium fistulosum)*	CuO	Increased Cu, Ca, Mg content in bulb	Wang et al. (2020)
Wheat *(Triticum aestivum)*	FeO	Enhanced Fe uptake	Manzoor et al. (2021)
Rice *(Oryza sativa)*	ZnO	Increased Zn in seeds	Sharma et al. (2021)
Finger millet (Eleusinecoracana)	ZnO	Enhanced Zn content in grains	Kumar et al. (2021)

TiO_2, Al_2O_3, CeO_2, FeO etc. (Prasad et al. 2017). The application of nano-fertilizers and nanoparticles in biofortification is summarized in Table 12.2.

Even though nanotechnology offers beneficial effects on crop fortification, and negative effects on environment has already been reported. As a result, a thorough and all-encompassing technique, life cycle analysis is required for improving comprehension, assessment, evaluation, and usage management of nanoparticles (Salieri et al. 2018). The effect of nanoparticle on the environment and the life cycle of organisms can be analysed with life cycle analysis. This can help in enhancing the ability and efficiency of nanoparticle in a sustainable manner (Nizam et al. 2021).

12.3 Concerns of Real-World Field Applications of Nanotechnology

During the past few decades nanotechnological inventions has gained much attention for their great application in various fields like agriculture, medicine, industry, etc. However, the application of nanoparticle in agricultural/environmental field is limited due to the limited field level trials and the social concern regarding the entry of nanoparticle to the food crops. Hence, the success rate of nano-enabled agricultural practices depends upon the social acceptance and to the fact that how the consumers perceive the nano-products in their daily life (Frewer et al. 2014; Hofmann et al. 2020). Potential challenges of nanotechnology in the agricultural field includes the strict scrutinization before the field trials, nature and the complexity of the target plant, unintended exposure to human and other live beings, limited reproducibility due to the application of complex technology, high production cost, discrepancies with the laboratory versus filed trials, etc. (Raemdonck and De Smedt 2015; Hofmann et al. 2020).

One of the main challenges in the employment of nanoparticle is the methods of delivery to the agricultural crops. As with the case of conventional application of fertilizers, foliar spray and soil inputs to the crops offers a great means to enhance the rate of uptake by plants. Still only half of the fertilizer may be reaching inside the plant tissue. But nanofertilizers when delivered by coating with a guiding biological molecule, the rate of uptake by plant tissue is much higher as compared to the direct exposure of the conventional agrochemicals (Cunningham et al. 2018). This in result may enhance the level of nutrient acquisition (like nitrogen, phosphorous, etc.) without inducing much toxicity as in the case of other agrochemicals leading the soil and water pollution. Moreover, direct soil application of nanomaterial is still challenging as the concentration of nanoparticle to make a positive response is very crucial and at the same time the cost of resource material for the viable application of nanomaterials is too high (Demirer et al. 2019; Wang et al. 2019).

Adding up to the practical difficulty of nanoparticle application at field level, the agricultural sector is highly regulated with safety protocols to ensure the safety of food supplements. The widespread nature of nanoparticle and the possible threat that may cause to the ecosystem and the human being is pointing to the fact that an apparent regulatory framework is mandatory for the field trials of the nanoparticles. This protocol must be amicable in all means as the inconsistencies in regulatory frameworks may result in trade conflicts among countries. Application of any innovative technology like nanotechnology in the food sector is really a challenge as people are always dynamic in nature (Adisa et al. 2019; Giraldo et al. 2019). In some cases, people may reject the nanomodified food as with the case of GM foods if there is a lack of proper knowledge dissemination regarding the nano-enabled agriculture. Hence, as a first step to the field trial, it must be ensured that the product will be consumed and for this the regulatory bodies should convince and create trust in consumers with high scrutiny of the product (Hofmann et al. 2020).

12.4 Conclusion

Nanotechnology has the potential to revolutionize agriculture by enabling resilient and sustainable techniques for the future. It has the potential to transform the crop yield, improve gene editing techniques, and enable precision breeding to highlight its vital role in determining the trajectory of agriculture in the coming years. Nanoparticle mediated gene delivery and editing in plants is a promising, futuristic tool since it has been successfully overcome the hurdle that the inability of biomolecule transfer through inflexible and multi-layered plant cell wall. In addition to this it is a cost effective, simple, easy method that protect the encapsuled genetic material without losing its integrity and it is also applicable towards a wide range of plant species; thereby helps to mitigate the limitations of conventional physical and biological transformation methods. The minimization of transgene silencing, improvement in efficiency and productivity can be achieved when nanotechnology-based biotransformation is combined with commonly applied approaches in transgenic plant development. Using nanoparticles as nanocarriers of bio-modifier molecules like CRISPR-Cas system is a potent technique in the production of genetically modified organisms. Even though nanotechnology offers beneficial effects on agriculture, the effect of nanoparticle on the environment and the life cycle of organisms should be analysed by life cycle analysis. This can help in enhancing the ability and efficiency of nanoparticle in a sustainable manner.

References

Abd-Elsalam KA (2020) Multifunctional hybrid nanomaterials for sustainable Agri-food and ecosystems: a note from the editor. Elsevier, Amsterdam

Adisa IO et al (2019) Recent advances in nano-enabled fertilizers and pesticides: a critical review of mechanisms of action. Environ Sci Nano 6:2002–2030

Ahmar S, Mahmood T, Mora-Poblete F, Shafique MS, Jung KH (2021) Advantage of nanotechnology-based genome editing system and its application in crop improvement. Front Plant Sci 12:663849

Anjali CH, Sharma Y, Mukherjee A, Chandrasekaran N (2012) Neem oil (Azadirachtaindica) nanoemulsion: a potent larvicidal agent against Culex quinquefasciatus. Pest Manag Sci 68: 158–163

Aslani F, Bagheri S, Muhd Julkapli N, Juraimi AS, Hashemi FSG, Baghdadi A (2014) Effects of engineered nanomaterials on plants growth: an overview. Sci World J 2014:1–28

Barrangou R, Fremaux C, Deveau H, Richards M, Boyaval P, Moineau S, Romero DA, Horvath P (2007) CRISPR provides acquired resistance against viruses in prokaryotes. Science 315:1709–1712. https://doi.org/10.1126/science.1138140

Bhatia S, Yadav SK (2023) CRISPR-Cas for genome editing: classification, mechanism, designing and applications. Int J Biol Macromol 238:124054

Bouis HE, Hotz C, McClafferty B, Meenakshi JV, Pfeiffer WH (2011) Biofortification: A new tool to reduce micronutrient malnutrition. Food Nutr Bull 32:31–40

Burlaka OM, Pirko YV, Yemets AI, Blume YB (2015) Plant genetic transformation using carbon nanotubes for DNA delivery. Tsitol Genet 49(6):3–12

Carriger JF, Rand GM, Gardinali PR, Perry WB, Tompkins MS, Fernandez AM (2006) Pesticides of potential ecological concern in sediment from South Florida canals: an ecological risk prioritization for aquatic arthropods. Soil Sediment Contam 15:21–45

Carte J, Wang R, Li H, Terns RM, Terns MP (2008) Cas6 is an endoribonuclease that generates guide RNAs for invader defense in prokaryotes. Genes Dev 22:3489–3496. https://doi.org/10.1101/gad.1742908

Char SN, Lee H, Yang B (2020) Use of CRISPR/Cas9 for targeted mutagenesis in Sorghum. Curr Protoc Plant Biol 5:e20112

Chhipa H (2017) Nanofertilizers and nanopesticides for agriculture. Environ Chem Lett 15:15–22

Chowdappa P, Gowda S (2013) Nanotechnology in crop protection: status and scope. Pest Manag Hortic Ecosyst 19(2):131–151

Cunningham FJ, Goh NS, Demirer GS, Matos JL (2018) Nanoparticle-mediated delivery towards advancing plant genetic engineering. Trends Biotechnol 36:882–897. https://doi.org/10.1016/j.tibtech.2018.03.009

Deltcheva E, Chylinski K, Sharma CM, Gonzales K, Chao Y, Pirzada ZA, Eckert MR, Vogel J, Charpentier E (2011) CRISPR RNA maturation by trans-encoded small RNA and host factor RNase III. Nature 471:602–607. https://doi.org/10.1038/nature09886

Demirer GS et al (2019) High aspect ratio nanomaterials enable biomolecule delivery and transgene expression or silencing in mature plants. Nat Nanotechnol 14:456–464

Demirer GS, Silva TN, Jackson CT, Thomas JB, Ehrhardt W, Rhee SY et al (2021) Nanotechnology to advance CRISPR–Cas genetic engineering of plants. Nat Nanotechnol 16(3):243–250

DeRosa MC (2010) Nanotechnology in fertilizers. Nat Nanotechnol 5:91

Dimkpa CO, Singh U, Bindraban PS, Elmer WH, Gardea-Torresdey JL, White JC (2019) Zinc oxide nanoparticles alleviate drought-induced alterations in sorghum performance, nutrient acquisition, and grain fortification. Sci Total Environ 688:926–934

Disfani MN, Mikhak A, Kassaee MZ, Maghari A (2017) Effects of nano Fe/SiO2 fertilizers on germination and growth of barley and maize. Arch Agron Soil Sci 63:817–826

Dong OX, Yu S, Jain R, Zhang N, Duong PQ, Butler C et al (2020) Marker-free carotenoid-enriched rice generated through targeted gene insertion using CRISPR-Cas9. Nat Commun 11:1178. https://doi.org/10.1038/s41467-020-14981-y

Doyle C, Higginbottom K, Swift TA, Winfield M, Bellas C, Benito-Alifonso D, Fletcher T, Galan MC, Edwards K, Whitney HM (2019) A simple method for spray-on gene editing in planta. bioRxiv 2019:805036

Du W, Yang J, Peng Q, Liang X, Mao H (2019) Comparison study of zinc nanoparticles and zinc sulphate on wheat growth: from toxicity and zinc biofortification. Chemosphere 227:109–116. https://doi.org/10.1016/j.chemosphere.2019.03.168

Dunbar T, Tsakirpaloglou N, Septiningsih EM, Thomson MJ (2022) Carbon nanotube-mediated plasmid DNA delivery in rice leaves and seeds. Int J Mol Sci 23(8):4081

Echiegu EA (2017) Nanotechnology applications in the food industry nanotechnology. Springer, Singapore, pp 153–171

Elemike EE, Uzoh IM, Onwudiw DC, Babalola OO (2019) The role of nanotechnology in the fortification of plant nutrients and improvement of crop production. Appl Sci 9(3):499

El-Saadony MT, Al-Moshadak AS, Shafi ME, Albaqami NM, Saad AM, El-Tahan AM et al (2021) Vital roles of sustainable nano-fertilizers in improving plant quality and quantity-an updated review. Saudi J Biol Sci 28(12):7349–7359

FAO (2017) The future of food and agriculture "Trends and challenges". FAO

Frewer LJ et al (2014) Consumer attitudes towards nanotechnologies applied to food production. Trends Food Sci Technol 40:211–225

Gad MA, Li M, Ahmed FK, Almoammar H (2020) In: Abd-Elsalam KA (ed) Multifunctional hybrid nanomaterials for sustainable Agri-food and ecosystems. Elsevier, pp 135–153

Giraldo JP, Landry MP, Faltermeier SM, McNicholas TP, Iverson NM, Boghossian AA, Reuel NF, Hilmer AJ, Sen F, Brew JA (2014) Plant nanobionics approach to augment photosynthesis and biochemical sensing. Nat Mater 13:400–408

Giraldo JP, Wu H, Newkirk GM, Kruss S (2019) Nanobiotechnology approaches for engineering smart plant sensors. Nat Nanotechnol 14:541–553

Gleditzsch D, Pausch P, Müller-Esparza H, Özcan A, Guo X, Bange G, Randau L (2019) PAM identification by CRISPR-Cas effector complexes: diversified mechanisms and structures. RNA Biol 16(4):504–517

Hale CR, Zhao P, Olson S, Duff MO, Graveley BR, Wells L, Terns RM, Terns MP (2009) RNA-guided RNA cleavage by a CRISPR RNA-Cas protein complex. Cell 139:945–956. https://doi.org/10.1016/j.cell.2009.07.040

Hernández-Fuentes AD, López-Vargas ER, Pinedo-Espinoza JM, Campos-Montiel RG, Valdés-Reyna J, Juárez-Maldonado A (2017) Postharvest behavior of bioactive compounds in tomato fruits treated with Cu nanoparticles and NaCl stress. Appl Sci 7:980

Hille F, Charpentier E (2016) CRISPR-Cas: biology, mechanisms and relevance. Philos Trans R Soc B Biol Sci 371(1707):20150496

Hofmann T, Victor Lowry G, Ghoshal S, Tufenkji N, Brambilla D, Dutcher J, Gilbertson LM, Giraldo JP, Kinsella JM, Landry MP, Lovell W, Naccache R, Paret M, Pedersen JA, Unrine JM, White JC, Wilkinson KJ (2020) Technology readiness and overcoming barriers to sustainably implement nanotechnology-enabled plant agriculture. Nat Food 1:416–425

Hu P, An J, Faulkner MM, Wu H, Li Z, Tian X, Giraldo JP (2020) Nanoparticle charge and size control foliar delivery efficiency to plant cells and organelles. ACS Nano 14(7):7970–7986

Jang HK, Song B, Hwang GH, Bae S (2020) Current trends in gene recovery mediated by the CRISPR-Cas system. Exp Mol Med 52(7):1016–1027

Jia G, Wang H, Yan L, Wang X, Pei R, Yan T, Zhao Y, Guo X (2005) Cytotoxicity of carbon nanomaterials: single-wall nanotube, multi-wall nanotube, and fullerene. Environ Sci Technol 39(5):1378–1383

Jun L, Feng-Hua W, Ling-ling W, Su-yao X, Chun-yi T, Dong-ying T, Xuan-ming L (2008) Preparation of fluorescence starch-nanoparticle and its application as plant transgenic vehicle. J Cent S Univ Technol 15:768–773

Kah M, Hofmann T (2014) Nanopesticide research: current trends and future priorities. Environ Int 63:224–235

Khodakovskaya MV, de Silva K, Biris AS, Dervishi E, Villagarcia H (2012) Carbon nanotubes induce growth enhancement of tobacco cells. ACS Nano 6(3):2128–2135

Klimek-Chodacka M, Oleszkiewicz T, Lowder LG, Qi Y, Baranski R (2018) Efficient CRISPR/Cas9-based genome editing in carrot cells. Plant Cell Rep 37:575–586

Kothamasu P, Kanumur H, Ravur N, Maddu C, Parasuramrajam R, Thangavel S (2012) Nanocapsules: the weapons for novel drug delivery systems. Bioimpacts 2(2):71–81

Kumar AU, Ling APK (2021) Gene introduction approaches in chloroplast transformation and its applications. J Genet Eng Biotechnol 19:148

Kumar A, Pandey D, Tiwari A, Gururani K, Agarwal A, Dhasmana A, Kumar A (2021) Metal based nanoparticles trigger the differential expression of key regulatory genes which regulate iron and zinc homeostasis mechanism in finger millet. J Cereal Sci 100:103235. https://doi.org/10.1016/j.jcs.2021.103235

Kumpiene J, Lagerkvist A, Maurice C (2008) Stabilization of As, Cr, Cu, Pb and Zn in soil using amendments–a review. Waste Manag 28:215–225

Kwak SY, Lew TTS, Sweeney CJ et al (2019) Chloroplast-selective gene delivery and expression in planta using chitosan-complexed single-walled carbon nanotube carriers. Nat Nanotechnol 14:447–455

Law SSY, Miyamoto T, Numata K (2023) Organelle-targeted gene delivery in plants by nanomaterials. Chem Commun 59:7166–7181

Letchford K, Burt H (2007) A review of the formation and classification of amphiphilic block copolymer nanoparticulate structures: micelles, nanospheres, nanocapsules and polymersomes. Eur J Pharm Biopharm 65(3):259–269

Li J, Wang Z, He G, Ma L, Deng XW (2020) CRISPR/Cas9-mediated disruption of TaNP1 genes results in complete male sterility in bread wheat. J Genet Genomics 47:263–272. https://doi.org/10.1016/j.jgg.2020.05.004

Liao C, Beisel CL (2021) The tracrRNA in CRISPR biology and technologies. Annu Rev Genet 55:161–181

Liu R, Lal R (2014) Synthetic apatite nanoparticles as a phosphorus fertilizer for soybean (Glycine max). Sci Rep 4:5686

Liu G, Zhang Y, Zhang T (2020) Computational approaches for effective CRISPR guide RNA design and evaluation. Comput Struct Biotechnol J 18:35–44

Lodha A, Lodha M, Patel A, Chaudhuri J, Dalal J, Edwards M, Douroumis D (2012) Synthesis of mesoporous silica nanoparticles and drug loading of poorly water-soluble drug cyclosporin A. J Pharm Bioallied Sci 4(Suppl. 1):S92–S94

Lomborg B (2002) The skeptical environmentalist. 2001. Cambridge University Press, Cambridge, p 515

Makarova KS, Aravind L, Wolf YI, Koonin EV (2011) Unification of Cas protein families and a simple scenario for the origin and evolution of CRISPR-Cas systems. Biol Direct 6:1–27

Mali SC, Raj S, Trivedi R (2020) Nanotechnology a novel approach to enhance crop productivity. Biochem Biophys Rep 24:100821

Manjunatha SB, Biradar DP, Aladakatti YR (2016) Nanotechnology and its applications in agriculture: a review. J Farm Sci 29:1–3

Manzoor N, Ahmed T, Noman M, Shahid M, Nazir MM, Ali L, Alnusaire TS, Li B, Schulin R, Wang G (2021) Iron oxide nanoparticles ameliorated the cadmium and salinity stresses in wheat plants, facilitating photosynthetic pigments and restricting cadmium uptake. Sci Total Environ 769:145221. https://doi.org/10.1016/j.scitotenv.2021.145221

Mora-Huertas CE, Fessi H, Elaissari A (2010) Polymer-based nanocapsules for drug delivery. Int J Pharm 385(1):113–142

Nair R, Varghese SH, Nair BG, Maekawa T, Yoshida Y, Kumar DS (2010) Nanoparticulate material delivery to plants. Plant Sci 179(3):154–163

Nandy D, Maity A, Mitra AK (2020) Target-specific gene delivery in plant systems and their expression: insights into recent developments. J Biosci 45:30. https://doi.org/10.1007/s12038-020-0008-y

Newkirk GM, de Allende P, Jinkerson RE, Giraldo JP (2021) Nanotechnology approaches for chloroplast biotechnology advancements. Front Plant Sci 12:691295. https://doi.org/10.3389/fpls.2021.691295

Nile SH, Thiruvengadam M, Wang Y et al (2022) Nano-priming as emerging seed priming technology for sustainable agriculture—recent developments and future perspectives. J Nanobiotechnol 20:254

Nizam NUM, Hanafiah MM, Woon KS (2021) A content review of life cycle assessment of nanomaterials: current practices, challenges, and future prospects. Nano 11:1–27. https://doi.org/10.3390/nano11123324

Novák O, Napier R, Ljung K (2017) Zooming in on plant hormone analysis: tissue- and cell-specific approaches. Annu Rev Plant Biol 68:323–348. https://doi.org/10.1146/annurev-arplant-042916-040812

Nuruzzaman M, Rahman MM, Liu Y, Naidu R (2016) Nanoencapsulation, nano-guard for pesticides: a new window for safe application. J Agric Food Chem 64(7):1447–1483

Nussenzweig PM, Marraffini LA (2020) Molecular mechanisms of CRISPR-Cas immunity in bacteria. Annu Rev Genet 54:93–120

Ojha S, Singh D, Sett A, Chetia H, Kabiraj D, Bora U (2018) Nanotechnology in crop protection. In: Nanomaterials in plants, algae, and microorganisms. Academic, pp 345–391

Parisi C, Vigani M, Rodríguez-Cerezo E (2015) Agricultural nanotechnologies: what are the current possibilities? Nano Today 10:124–127

Peng Y, Chen L, Zhu L, Cui L, Yang L, Wu H, Bie Z (2022) CsAKT1 is a key gene for the CeO2 nanoparticle's improved cucumber salt tolerance: a validation from CRISPR-Cas9 lines. Environ Sci Nano 9(12):4367–4381

Perlatti B, Fernandes JB, das Graças Fernandes da Silva MF, Forim MR, de Souza Bergo PL (2013) Polymeric nanoparticle-based insecticides: a controlled release purpose for agrochemicals. INTECH Open Access Publisher

Peteu SF, Oancea F, Sicuia OA, Constantinescu F, Dinu S (2010) Responsive polymers for crop protection. Polymers 2(3):229–251

Pinstrup-Andersen P, Watson DD II (2011) Food policy for developing countries: the role of government in global, national, and local food systems. Cornell University Press

Prasad R, Bhattacharyya A, Nguyen QD (2017) Nanotechnology in sustainable agriculture: recent developments, challenges, and perspectives. Front Microbiol 8:1–13

Raemdonck K, De Smedt SC (2015) Lessons in simplicity that should shape the future of drug delivery. Nat Biotechnol 33:1026–1027

Rafsanjani MSO, Alvari A, Samim M, Hejazi MA, Abdin MZ (2012) Application of novel nanotechnology strategies in plant biotransformation: A contemporary overview. Recent Pat Biotechnol 6:69–79

Ragaei M, Sabry A-K (2014) Nanotechnology for insect pest control. Int J Sci Environ Technol 3: 528–545

Rai M, Ingle A (2012) Role of nanotechnology in agriculture with special reference to management of insect pests. Appl Microbiol Biotechnol 94:287–293

Rai V, Acharya S, Dey N (2012) Implications of nanobiosensors in agriculture. J Biomater Nanobiotechnol 3:315–324

Raliya R, Tarafdar JC, Biswas P (2016) Enhancing the mobilization of native phosphorus in the mung bean rhizosphere using ZnO nanoparticles synthesized by soil fungi. J Agric Food Chem 64(16):3111–3118

Resh VH, Cardé RT (2009) Encyclopedia of insects. Academic

Rico CM, Majumdar S, Duarte-Gardea M, Peralta-Videa JR, Gardea-Torresdey JL (2011) Interaction of nanoparticles with edible plants and their possible implications in the food chain. J Agric Food Chem 59:3485–3498

Rizwan M, Ali S, Ali B, Adrees M, Arshad M, Hussain A, Zia ur Rehman M, Waris AA (2019) Zinc and iron oxide nanoparticles improved the plant growth and reduced the oxidative stress and cadmium concentration in wheat. Chemosphere 214:269–277. https://doi.org/10.1016/j.chemosphere.2018.09.120

Rui Y, Varanasi M, Mendes S, Yamagata HM, Wilson DR, Green JJ (2020) Poly(Beta-Amino Ester) nanoparticles enable nonviral delivery of CRISPR-Cas9 plasmids for gene knockout and gene deletion. Mol Ther Nucleic Acids 20:661–672. https://doi.org/10.1016/j.omtn.2020.04.005

Sabet H, Mortazaeinezhad F (2018) Yield, growth and Fe uptake of cumin (Cuminum cyminum L.) affected by Fe-nano, Fe-chelated and Fe-siderophore fertilization in the calcareous soils. J Trace Elem Med Biol 50:154–160. https://doi.org/10.1016/j.jtemb.2018.06.020

Salama DM, Osman SA, Abd El-Aziz ME, Abd Elwahed MSA, Shaaban EA (2019) Effect of zinc oxide nanoparticles on the growth, genomic DNA, production and the quality of common dry bean (Phaseolus vulgaris). Biocatal Agric Biotechnol 18:101083. https://doi.org/10.1016/j.bcab.2019.101083

Salieri B, Turner DA, Nowack B, Hischier R (2018) Life cycle assessment of manufactured nanomaterials: where are we? NanoImpact 10:108–120. https://doi.org/10.1016/j.impact.2017.12.003

Salsman J, Dellaire G (2017) Precision genome editing in the CRISPR era. Biochem Cell Biol 95(2):187–201

Sanchez-Vazquez SA, Hailes HC, Evans JRG (2013) Hydrophobic polymers from food waste: resources and synthesis. Polym Rev 53(4):627–694

Sander JD, Joung JK (2014) CRISPR-Cas systems for genome editing, regulation and targeting. Nat Biotechnol 32(4):347–355

Santana I, Jeon S-J, Kim HI, Islam MR, Castillo C, Garcia GFH, Newkirk GM, Giraldo JP (2022) Targeted carbon nanostructures for chemical and gene delivery to plant chloroplasts. ACS Nano 16:12156–12173

Scognamiglio V (2013) Nanotechnology in glucose monitoring: advances and challenges in the last 10 years. Biosens Bioelectron 47:12–25

Serag MF, Kaji N, Habuchi S, Bianco A, Baba Y (2013) Nanobiotechnology meets plant cell biology: carbon nanotubes as organelle targeting nanocarriers. RSC Adv 3:4856. https://doi.org/10.1039/c2ra22766e

Shah MA, Wani SH, Khan AA (2016) Nanotechnology and insecticidal formulations. J Food Bioeng Nanopross 1:285–310

Sharma D, Afzal S, Singh NK (2021) Nanopriming with phytosynthesized zinc oxide nanoparticles for promoting germination and starch metabolism in rice seeds. J Biotechnol 336:64–75. https://doi.org/10.1016/j.jbiotec.2021.06.014

Shweta TDK, Singh S, Singh S, Dubey NK, Chauhan DK (2016) Impact of nanoparticles on photosynthesis: challenges and opportunities. Mater Focus 5(5):405–411

Singh A, Singh NB, Hussain I, Singh H, Singh SC (2015) Plant-nanoparticle interaction: an approach to improve agricultural practices and plant productivity. Int J Pharm Sci Invent 4: 25–40

Sultana F, Imran-Ul-Haque M, Arafat M, Sharmin S (2013) An overview of nanogel drug delivery system. J Appl Pharm Sci 3:95–105

Suresh R, Borkar SN, Sawant VA, Shende VS, Dimble SK (2010) Nanoclay drug delivery system. Int J Pharm Sci Nanotechnol 3:901–905

Tamez C, Morelius EW, Hernandez-Viezcas JA, Peralta-Videa JR, Gardea-Torresdey J (2019) Biochemical and physiological effects of copper compounds/nanoparticles on sugarcane (Saccharum officinarum). Sci Total Environ 649:554–562

Thakur S, Thakur T, Kumar R (2018) Bio-nanotechnology and its role in agriculture and food industry. J Mol Genet Med 12:1–5

Tripathi DK, Singh VP, Prasad SM, Chauhan DK, Dubey NK (2015) Silicon nanoparticles (SiNp) alleviate chromium (VI) phytotoxicity in Pisum sativum (L.) seedlings. Plant Physiol Biochem 96:189–198

Tripathi DK, Mishra RK, Singh S, Singh S, Vishwakarma K, Sharma S, Singh VP, Singh PK, Prasad SM, Dubey NK, Pandey AC (2017) Nitric oxide ameliorates zinc oxide nanoparticles phytotoxicity in wheat seedlings: implication of the ascorbate–glutathione cycle. Front Plant Sci 8:1

Umar W, Hameed MK, Aziz T, Maqsood MA, Bilal HM, Rasheed N (2020) Synthesis, characterization and application of ZnO nanoparticles for improved growth and Zn biofortification in maize. Arch Agron Soil Sci 67:1164–1176. https://doi.org/10.1080/03650340.2020.1782893

Vanhaeren H, Inzé D, Gonzalez N (2016) Plant growth beyond limits. Trends Plant Sci 21:102–109. https://doi.org/10.1016/j.tplants.2015.11.012

Wang JW, Grandio EG, Newkirk GM, Demirer GS, Butrus S, Giraldo JP et al (2019) Nanoparticle-mediated genetic engineering of plants. Mol Plant 12:1037–1040

Wang K, Wang Y, Li K, Wan Y, Wang Q, Zhuang Z, Guo Y, Li H (2020) Uptake, translocation and biotransformation of selenium nanoparticles in rice seedlings (Oryza sativa L.). J Nanobiotechnol 18:1–15. https://doi.org/10.1186/s12951-020-00659-6

Wu H, Tito N, Giraldo JP (2017) Anionic cerium oxide nanoparticles protect plant photosynthesis from abiotic stress by scavenging reactive oxygen species. ACS Nano 11:11283–11297

Xiong T, Dumat C, Dappe V, Vezin H, Schreck E, Shahid M, Pierart A, Sobanska S (2017) Copper oxide nanoparticle foliar uptake, phytotoxicity, and consequences for sustainable urban agriculture. Environ Sci Technol 51:5242–5251

Yang H, Gao P, Rajashankar KR, Patel DJ (2016) PAM-dependent target DNA recognition and cleavage by C2c1 CRISPR-Cas endonuclease. Cell 167(7):1814–1828

Yin H, Xue W, Anderson DG (2019) CRISPR–Cas: a tool for cancer research and therapeutics. Nat Rev Clin Oncol 16(5):281–295

Zeng Z, Han N, Liu C, Buerte B, Zhou C, Chen J et al (2020) Functional dissection of HGGT and HPT in barley vitamin E biosynthesis via CRISPR/Cas9-enabled genome editing. Ann Bot 126: 929–942. https://doi.org/10.1093/aob/mcaa115

Zetsche B et al (2015) Cpf1 is a single RNA-guided endonuclease of a class 2 CRISPR-Cas system. Cell 163:759–771. https://doi.org/10.1016/j.cell.2015.09.038

Chapter 13
Nanoscale Nutrient Delivery for Hydroponics and Aeroponics

Rashmi Singh and Kamaraj Raju

Abstract The rapid growth of the global population over the last decade has prompted the agricultural industry to enhance crop yield to meet demands, particularly in emerging nations. Soil nutrient deficiency leads to economic losses and diminishes the nutritional quality and overall grain production. Extensive use of chemical fertilizers disrupts soil mineral balance and reduces fertility, causing irreversible harm to soil structure, mineral cycles, microbial flora, plants, and food chains and potentially inducing genetic alterations in subsequent generations of consumers. Nanotechnology has become more relevant in the fields of plant biology and the agricultural sector, enhancing methods for producing nanoparticles composed of metals crucial for physiological processes. These nanoparticles enhance fertilizer compositions, improving absorption by plant cells and reducing nutrient wastage. Nanostructured fertilizers possess significant surface area, sorption capacity, and controlled-release kinetics, functioning as intelligent delivery systems for targeted areas. They enhance nutrient utilization efficiency through precise delivery and regulated release, responding to environmental stimuli and biological needs. Integration of nanotechnology with soilless hydroponics and aeroponics offers a sustainable and productive alternative to traditional methods, accelerating seed germination, promoting seedling growth, and improving plant metabolism, plant resistance to diseases, modifying plant tissue function through nano bionics, and prolonging microgreen shelf life using nanoparticles. However, ethical and safety concerns necessitate careful evaluation before implementing nano-fertilizers in agriculture. This chapter discusses hydroponics and aeroponics nano-fertilizer delivery, absorption, movement, and plant effects, addressing ethical and safety problems as the global population grows and food demand increases, demanding creative and sustainable farming methods like nanotechnology.

R. Singh
Department of Zoology, Dayanand Vedic College, Orai, Uttar Pradesh, India

K. Raju (✉)
Department of Biotechnology, Rathinam College of Arts and Science, Coimbatore, Tamil Nadu, India

J. M. Al-Khayri et al. (eds.), *Nanotechnology Applications in Modern Agriculture*, Nanotechnology in Plant Sciences 3, https://doi.org/10.1007/978-3-031-90506-3_13

Keywords Aeroponics · Environmentally friendly · Hydroponics · Nonfertilizer · Nanotechnology

13.1 Introduction

In the fast-developing field of nanotechnology, nanomaterials play a pivotal role, a burgeoning technology that has made groundbreaking advancements in diverse industries like electronics, energy, cleanup, transportation, space technology, and life sciences. Nanotechnology research and development aim to comprehend and produce enhanced materials, technologies, and systems utilizing nanoscale features (Borm et al. 2006; Oberdorster et al. 2005). The name "nano" originates from the Greek word "nanos," denoting "dwarf," and it refers to a unit of measurement, which is equal to 10^{-9} meters. Nanoparticles are defined as particles that are smaller than 100 nm in at least one dimension (Thakkar et al. 2010). Nanoparticles possess a significant ratio in surface area and volume as well as display unique characteristics, making them highly versatile. Nanotechnology integrates science and engineering to create an interdisciplinary field in agricultural and food sciences with substantial potential for advancing agricultural research. Nanoparticles possess promising applications in agriculture, such as detecting contaminants, plant illnesses, pests, and pathogens, as well as controlling the distribution of herbicides, fertilizer products, minerals, and biological material; and serving as nano-architects in soil structure creation and bonding (Ghormade et al. 2011). Nanoparticles can cause changes in plant expression of genes and associated metabolic paths, which can eventually affect the progress and developmental stages of plants (Nair et al. 2010). Nanoparticles can be composed of a wide range of materials, including oxides of metals, ceramics, silicon dioxide, magnetic substances, quantum dots, fatty acids, polymer compounds, dendrimers, and fluid emulsion. The composition of nanoparticles is crucial in determining their applicability. Polymer-coated nanoparticles are utilized as carriers for agrochemicals because they can release chemicals in a controlled manner. Metal nanoparticles display size-dependent features, including magnetic, fluorescence detection, and photocatalytic oxidation. These properties make them valuable in the development of sensors, the breakdown of agrochemicals, and the restoration of soil (Ghormade et al. 2011).

The rapid increase in global population over the past decade has necessitated greater agricultural output to meet demands, particularly in emerging nations. Insufficient supply of nutrients in soils results in financial losses for producers and decreases the nutritive value and quantity of grain for both people and cattle. The fertilizers can enhance agricultural productivity but plants struggle to utilize nutrients in bulk chemical forms, with less than 50% typically utilized (Loomis and Connor 1992). Overuse of chemical fertilizers negatively impacts the soil's makeup, mineral cycles, microbiological flora, vegetation, and ecological systems, leading to irreversible damage. To address this, intelligent materials capable of controlled chemical release to precise plant locations are needed. "Smart delivery systems" combine precise targeting, strict control, remote regulation, and multifunctionality to overcome biological obstacles and accomplish accurate placement (Nair et al. 2010). Nanoparticles made from physiologically significant metals are utilized as "smart delivery systems" to enhance fertilizer formulation, reducing nutrient loss and promoting greater uptake in plant cells (Naderi and Danesh-Shahraki 2013). This chapter seeks to examine the evolution, dissemination, and effects of nano-fertilizers on plant growth, emphasizing their absorption, translocation, and consequences. It emphasizes the incorporation of nanotechnology in sustainable agriculture, namely in soilless and microgreen cultivation, providing creative methods to augment nutrient absorption, enhance plant disease resistance, and extend shelf life. Ethical and safety issues are considered, highlighting the capacity of nanoparticles to transform agricultural methods while also facilitating progress in many other domains due to their distinctive characteristics.

13.2 Nutrient Availability to Plants

Nutrient availability to plants is crucial for their growth, influenced by soil composition, root structure, pH, soil moisture, and microbial activity. To acquire minerals from soil, plants must be bioavailable (Barber 1995). The solid phase's quantity, content, and interaction with other nutrients determine plant availability. The soil-plant system's nutrient uptake is another definition. This includes nutrient release from the solid phase into the solution, transport, and plant absorption (Comerford 2005). The nutrient concentration in the soil solution is subject to the influence of multiple elements, including soil humidity, soil dimensions, pH levels, cation exchange capacity, redox capacity, biological substance, and microbial development. An excess of minerals in the soil may hamper growth of plants by limiting the accessibility of water and obtaining toxic heavy metals, which can be poisonous. Soil composition and root structure determine plant nutrient availability. Even in well-organized soils, many factors affect root-soil interaction. Soil aeration and fertility impact root respiration, soil mass density, and nutrient absorption (Marschner 1995). The availability of nutrients and plant root development is influenced by the pH of the soil. Acidic pH accelerates rock weathering by releasing potassium ions, magnesium ions, calcium ions and manganese ions. The enhanced solubility facilitates the uptake of calcium carbonates, sulphates, and phosphates by the root system of plants. Precipitation and organic breakdown lower the pH of soil (Taiz and Zeiger 2010). Plant roots' optimal development and maturation affect nutrient uptake. The root surface absorbs nutrients from the soil by diffusion or mass flow. Nutrients diffuse from high to low concentrations due to molecular mobility. Diffusion allows plant fluids to flow small distances between cells or from soil to roots. Root nutrient uptake, replenishment, and ion diffusion shape depletion zones surrounding roots throughout time. Ion mobility is measured via diffusion coefficient (Marschner 1995). Nano-fertilizers play important role in enhancing soil quality, nutrient uptake, and plant defense mechanisms for sustainable agriculture (Arora et al. 2022). Nanotechnology has potential to boost crop yield and nutrient use efficiency under various conditions. Pressure-induced Mass flow (molecular movement) is used for long-distance transfer by xylem and phloem. This mechanism depends on soil nutrients, transpiration (Mengel and Kirkby 2001) and variables like plant variety, age, and time of day (Marschner 1995). The quantity of nutrients in soil varies across sites, as shown by the importance of nutrients in plant systems. This emphasizes agricultural fertilizers' importance.

13.3 Conventional Fertilizers Versus Nano-fertilizers

Conventional fertilizers are typically administered to agricultural produce via spraying or disseminating techniques, but the amount of fertilizers reaching plant is often below the desired level due to processes like leaching, drift, runoff, and evaporation. A substantial proportion of nitrogen, phosphorous, and potassium present in fertilizers gets released into the environment, resulting in both ecological and financial consequences (Trenkel 1997). Global fertilizer consumption has shown significant recovery and growth, but excessive use has led to environmental contamination and negative impacts on natural ecosystems. Optimizing chemical fertilizer use is crucial to meet crop nutritional needs while reducing environmental contamination (Tilman et al. 2002). Exploring alternative fertilization techniques can help maximize crop production while preserving soil integrity and environmental cleanliness. Nanotechnology allows for the development of smart fertilizers, enhancing nutrient use efficiency and reducing pollution at a lower cost. Nano-fertilizers operate at the nanoscale and can be designed as carriers for nutrients, providing regulated release and targeted distribution to specific sites. They exhibit characteristics such as elevated surface area to volume ratio, exceptional dissolution, and precise placement skills because of their small size, low toxicity and fast motility (Sasson et al. 2007).

13.4 Formulations of Nano-fertilizers and Their Sophisticated Delivery Systems

Nano-fertilizers must be developed with certain attributes such as excellent solubility, equilibrium, efficiency, controlled release over time, improved targeted action at optimum levels, and reduced eco-toxicity. Additionally, they should have safe and convenient methods of administration and discharge. Nanoparticles have the capability to transport nutrients to precise locations within biological structures. Nutrient loading upon nanoparticles is commonly accomplished by the process of absorption, facilitated by ligand-mediated attachment, entrapment in polymeric nanoparticles, or synthesis of nanoparticles composed of the nutrient itself. Nanostructured formulations have the potential to decrease the pace at which fertilizer nutrients are lost into the soil by weathering and/or leakage.

The work conducted by Corradini et al. (2010) examined the relationship between resilience of nanoparticles of chitosan suspensions that included chemical fertilizers Nitrogen (N), Phosphorus (P), and Potassium (K). These suspensions have prospective applications in agriculture. In a similar manner, Kottegoda et al. (2011) synthesized hydroxyapatite (HA) nanoparticles treated with urea to provide a slow and continuous release of nitrogen during the growth of crops. The nano-fertilizers demonstrated an initial quick release of nitrogen, followed by a progressive release over a period of 60 days. This is in contrast to commercial fertilizers, which only release nitrogen for a maximum of 30 days. The large surface area of HA allows for substantial binding of urea, resulting in controlled release. Moreover, polymer-based mesoporous nanoparticles can function as efficient carrier systems for agrochemical chemicals, hence improving their effectiveness and cost efficiency. The mesoporous silica nanoparticles with a size of 150 nm exhibited efficient urea retention and regulated release in both soil and water environments. This release was prolonged by a considerable fivefold increase in duration, as reported by Wanyika et al. in 2012. In their study, Milani et al. (2012) investigated the dissolution and solubility dynamics of zinc oxide (ZnO) nanoparticles and bulk ZnO particles when applied as a coating on macronutrient fertilizers such as urea and monoammonium phosphate. The dissolving rates of the granules coated with monoammonium phosphate were increased. The efficacy of fertilizer application is influenced by the method used and the plant system. Several methods can be employed to provide nano-fertilizers to plants:

13.5 Nano Fertilizer Delivery Method

13.5.1 In Vitro Methods

13.5.1.1 Aeroponics

The method was initially documented by Weathers and Zobel in 1992. It involves suspending plant roots in the air and continuously spraying them with a nutritional solution, regulating the gaseous atmosphere around the roots (Fig. 13.1). Aeroponics is a sophisticated type of hydroponics, cultivating plants in a tank or vessel where plant roots are suspended and misted with nutrition. This method requires abundant nutrients to support accelerated plant growth. It leads to healthier plants, increased development, and higher production while utilizing fewer resources and providing oxygen around the roots. In the aeroponic system, Plants are cultivated in a suspended manner within a closed or partially closed setting, where their roots and lower stems are regularly exposed and sprayed to a fine atomized mist of liquid medium, which is rich in nutrient. This method is highly promising for the application of metal nanoparticles as nutrient substitutes or fertilizers. The plant's roots are effectively separated from the surrounding environment by the utilization of a support structure. This is commonly accomplished by compressing sealed-cell

Fig. 13.1 Setup of an aeroponic system. (Source: Created with AI)

foam around the base of the stem and subsequently introducing it into an aperture within the aeroponic container. Larger plants with heavy vegetation and fruits are typically suspended using trellising. Aeroponics offers several benefits over conventional tissue culture, including excellent aeration of the nutrient medium, enhanced stimulation of plant growth, and prevention of pathogen colonization. Disease transmission is limited as plant-to-plant contact is reduced, and the sterile nutrient medium used to spray the roots helps prevent infections. Commercial-scale aeroponic systems were first introduced by Stoner's company, GTi, and have since been utilized in various applications, including growing pharmaceutical medicine inside plants and genetically modified crops (Stoner and Clawson 1999–2000).

Advantages

1. Unrestricted and natural growth.
2. Disease-free environment.
3. Higher plant density.
4. Propagation can be achieved from a solitary stem.

Disadvantages

1. Demands continuous surveillance.
2. Prone to frequent power failures.
3. Demands expertise in technological matters.

13.5.1.2 Hydroponics

Initially suggested by this technique involves cultivating plants by submerging the plant roots in a nutrient-rich liquid medium, in absence of soil while the plant is supported by an inert medium (Fig. 13.2). Additional materials like sand and gravel are used in some commercial applications, with the nutritional solution flushed out and replaced periodically. Hydroponics offers several advantages over conventional farming, including efficient resource management, reduced pesticide use, and faster plant growth. It is particularly helpful in areas with environmental stress.

Fig. 13.2 Setup of a hydroponic system. (Source: Created with AI)

Commercial hydroponic systems are automated, reducing labor and eliminating many traditional agricultural practices. Water and nutrient preservation, as well as the ability to customize the environment, are significant benefits. However, hydroponics also has drawbacks such as high construction costs, the need for trained personnel, and increased electricity consumption. Overall, hydroponics enables maximum potential yield and cultivation even in areas with unsuitable soil conditions.

Advantages

1. Even in areas where the soil is unsuitable for plant growth, crops can be cultivated.
2. The potential yield is maximum.
3. Labor costs have been largely eliminated.
4. The preservation of nutrients and water.
5. The environment can be changed to suit our preferences.

Disadvantages

1. Construction cost is high.
2. Trained person must be required.
3. Consumption of electricity is high.

The approach has resemblance to the aeroponic system. Materials like as the mineral perlite or pebbles can provide physical support to the roots. The nutrient medium consists of both organic and inorganic components. Hydroponic systems can be fully automated to regulate water quantity, light intensity, and other parameters based on the needs of the plant species being grown. Figure 13.2 depicts an ideal setup of a hydroponic system, which has been widely used commercially for cultivating significant crops and leafy vegetables on a large scale. There are numerous advantages to growing plants in hydroponic systems compared to soil-based culture. Hydroponically grown crops are often reported to be healthier and exhibit better yields than soil-grown crops. Plants grow faster in hydroponic systems because nutrients are readily available in the water, which can be easily absorbed by the roots. Additionally, plants can be grown in close proximity, requiring only a fraction of the space and water compared to traditional soil-based methods. Considering the current water scarcity, hydroponics offers the significant advantage of water reuse. Furthermore, cultivation in a closed environment prevents weed growth and soil-borne pest infestations. Thus, hydroponic systems enable precise nutrient regulation and higher-density planting, resulting in increased yield per unit area and improved product quality.

Given the constraints of natural resources such as land and water, hydroponics is considered an ideal technique for meeting the growing demand for food in the future. Similar to the aeroponic system, nano-fertilizers can be added to the nutrient medium in hydroponics. Metal nanoparticles can replace their respective metal salts, as they have been shown to have better absorption efficiency by plants, leading to enhanced growth. Both Aeroponic and Hydroponic systems are limited only for terrestrial plants.

13.5.2 *In Vivo* Methods

13.5.2.1 Soil Application

To enhance nutrient level, soil application is a predominant method, which entails the utilization of chemical and organic fertilizers. The primary factors to consider when choosing a method for delivering this fertilizer are its longevity in the soil, soil texture, the soil salinity, the susceptibility of the plants to salts, the concentration of salts, and the pH level of the amendment. The influence of adverse soil particles on the absorption of mineral nutrients is generally acknowledged. Most agricultural soils have a relatively low anion exchange capacity when compared to their cation exchange capacity. Nitrate ion (NO_3^-), a negatively charged ion, has significant solubility in the liquid component of soil and it is prone to being leached away by the water that permeates via soil. Phosphate ions (PO_4^{-2}) stick to soil particles that have Aluminum (Al) or Iron (Fe) because of the existence of positively charged Ferric ion (Fe^{3+}), Ferrous ion (Fe^{2+}), and Aluminum ion (Al^{3+}) ions th at have a hydroxyl ion (OH^-) group capable of swapping with Phosphate ions (PO_4^{-2}). Hence, Phosphate ions (PO_4^{-2}) have the ability to form strong chemical bonds, which can hinder their movement and accessibility in soil, thereby restricting plant growth (Taiz and Zeiger 2010).

13.5.2.2 Foliar Application

The technique entails a direct administration of liquid fertilizers onto the foliage. It is frequently used to supply trace elements. Administering a substance directly onto a plant's leaves can accelerate the plant's uptake of the material during its rapid growth stage. Furthermore, it can effectively tackle the issue of limited absorption of a nutrient from the soil. This approach exhibits the capacity to enhance the efficiency of iron, manganese, and copper absorption in comparison to their application on the soil. This is because the nutrients are less likely to bind to soil particles and hence become more easily accessible to the roots (Taiz and Zeiger 2010). The Stomata and the epidermal cells of leaf have a crucial function in the absorption of nutrients. Hence, employing the foliar spraying technique for nano-fertilizers can yield agricultural advantages. However, it is essential to minimize leaf damage in these circumstances by implementing a standardized application protocol. One primary limitation of this approach is the requirement for precise timing of the sprinkling procedure, namely in the early hours of the day and evening, as the stomata are only open during these designated time windows. An additional drawback is to the potential for detrimental effects on plants in the absence of appropriate application of chemical fertilizers.

13.6 Effect of Nano-fertilizers on Plant Growth

The majority of current research has been concentrated on the adverse impacts of nanoparticles on plants. However, a limited number of studies have demonstrated that low dosages of nanoparticles can enhance plant growth and yield. According to Khodakovskaya et al. (2012), the utilization of multi-walled carbon nanotubes (MWCNTs) has been found to enhance the germination of tomato seeds and promote the growth of tobacco cells. According to Mondal et al. (2011), the utilization of MWCNTs resulted in enhanced development and growth of mustard plant seeds. Lower quantities of oxidized MWCNTs demonstrated greater efficacy in terms of germination index and relative root elongation. In their study, Sahandi et al. (2011) found that nano-silver exhibited superior efficacy compared to silver nitrate in terms of enhancing borage seed output and suppressing leaf abscission. Silver ions may prevent leaf abscission by displace copper ions from receptors, inhibiting ethylene action. Using foliar spray, nano-silver and silver nitrate were applied to different plant groupings. Nano silver was more effective than silver nitrate at lower doses. Namasivayam and Chitrakala (2011) explored how biosynthesized silver nanoparticles affect seedling emergence and growth indicators of economically important plant species.

The application of Zinc Oxide (ZnO) nanoparticles to seedlings resulted in increased growth of biomass, potentially attributed to decreased levels of malondialdehyde (MDA), indicating a reduction in reactive oxygen species (ROS). Plants were aided by nano zinc at a concentration of 1000 ppm. (Prasad et al. 2012). At 2000 ppm, it showed dangerous symptoms, emphasizing the necessity for careful use. In the conducted field trials, a significant decrease of 15-fold in ZnO nanoparticles was seen in comparison to the recommended dosage of Zinc Sulphate (ZnSO4), resulting in a notable increase of 29.5% in pod yield. In a study conducted by López-Moreno et al. in 2010, it was observed that ZnO nanoparticles promoted root elongation in soybean at a concentration of 500 ppm. However, their size decreased as the concentration increased. In their study, Zhao et al. (2014) examined the impact of 400 ppm ZnO and Selenium dioxide (CeO_2) nanoparticles on the quality of cucumber (*Cucumis sativus*) fruit. Both categories of nanoparticles exhibited an augmentation in starch content and demonstrated the potential to alter the carbohydrate pattern. The benefits of Silicon dioxide (SiO_2) and Titanium dioxide (TiO_2) nanoparticles for soybean (*Glycine max*) were demonstrated by Lu et al. (2002). This increased absorption of water, fertilizer, nitrate reductase, and antioxidant activity. Research has examined how nano-TiO_2 affects spinach photosynthesis and growth. These findings suggest that TiO_2 boosts visible and UV photosynthetic activity (Lei et al. 2007). Spinach seeds, when treated with TiO_2 nanoparticle, dry weight increased by 73%, photosynthetic rate by thrice, and chlorophyll a by 45%. Higher absorption of inorganic nutrients improves organic substance use and suppresses oxygen-free radicals, increasing photosynthetic rate Zheng et al. (2005). Mahmoodzadeh et al. (2013) demonstrated that TiO_2 nanoparticles at 2000 ppm improve Rapeseed (*Brassica napus*) seed germination and seedling vigor, contrary to most studies. Shah and Belozerova (2009) explored how silicon (Si), palladium (Pd), gold (Au), and copper (Cu) nanoparticles affect seed germination of lettuce. The researchers found nanoparticles useful at varying concentrations. Pb and Au were beneficial at low quantities, while Si and Cu were beneficial at larger quantities. A mix of Au and Cu also helped. Quoc Buu et al. (2014) found that soybean seed germination increased, when subjected to a relatively modest concentration of nanocrystalline powder containing iron (Fe), cobalt (Co), and copper (Cu) in comparison to the control group. Crop output, chlorophyll index, and nodule count increased significantly. In field studies, it was found that gold foliar spraying on mustard (*Brassica juncea*) plants is beneficial (Arora et al. 2012). This treatment boosted plant height, stem diameter, branches, pods, and seeds. Gold nanoparticles improved plant redox status and it was also found that SiO_2 nanoparticles greatly boosted maize plant dry weight (Suriyaprabha et al. 2012). Protein, chlorophyll, and phenol levels rose by following treatments.

13.7 Enhancement of Metal Nanoparticles via Uptake, Translocation, and Transformation in Plant Systems

The investigation into the assimilation and impacts of nano-fertilizers in plants is an increasingly popular subject of study. The processes of absorption, translocation, and accumulation of nanoparticles are influenced by several factors, such as the specific plant species, its developmental stage, environmental conditions, as well as the physical and chemical characteristics, functionalization, stability, and administration methods employed for the nanoparticles. Rico et al. (2011) depicted a graphic that illustrates the process by which various nanoparticles are taken in, moved, and altered within plants, as well as prospective techniques for cellular absorption.

According to Fleischer et al. (1999), nanoparticles are able to traverse the cell membrane by means of membrane pores that vary in size from 5 to 20 nanometers. Consequently, nanoparticles or clusters with dimensions smaller than the pore size of the plant cell wall have the ability to traverse the cell wall and access the plasma membrane (Moore 2006; Navarro et al. 2008). Functionalized nanoparticles increase cell wall hole size and uptake. Numerous scholarly studies have examined nanoparticle uptake into plant cells. Nanoparticles can enter plant cells via aquaporin, ion channels, or endocytosis (Nair et al. 2010). Create chemicals with membrane transporters or root exudates to deliver nanoparticles into plants (Kurepa et al. 2010). Recent studies indicate nanoparticles can penetrate leaf tissues via stomata or trichome bases (Eichert et al. 2008; Fernández and Eichert 2009; Uzu et al. 2010).

Kurepa et al. (2010) observed and recorded the process of absorption and mobility of the TiO_2-alizarin red S complex in *Arabidopsis thaliana* seedlings. It has been observed that the roots excrete mucilage, which combines with pectin hydrogel to form a complex around the root. This compound was found to facilitate the transit of the nanoparticle-dye complex. A recent study examined the mechanisms by which nanoparticles are absorbed and distributed into plants. The scientists employed silica nanoparticles that were uniformly sized and possessed fluorescent characteristics, which were specifically labelled for the experiment. It was shown that these nanoparticles penetrated the plant roots through two distinct pathways: the symplastic pathway, which entails movement through the linked plant cells, and the apoplastic pathway, which entails movement across the intercellular spaces then, the nanoparticles infiltrated the roots and were then conveyed to the plant's aerial sections, such as the stems and leaves, via the xylem tissue. This study was conducted in 2014 by Sun et al. Even, the exact method via which plants assimilate nanoparticles, it is still comprehensively investigated. Nanoparticles, when entered into the cell, possess the capacity to undergo either apoplastic or symplastic transport (Moore 2006). Rico et al. (2011) described that they can be transmitted by plasmodesmata intercellularly. The presence of nanoparticles in the cytoplasm leads to interactions with many cytoplasmic organelles, causing disruptions in a wide range of cellular metabolic activities. Larue et al. (2011) investigated in wheat crops the absorption of TiO_2 nanoparticles and identified their presence in the root's cells. In their study, it was studied that how ZnO nanoparticles are absorbed and transported within the cells of ryegrass (Lin and Xing (2008). The offered evidence showcased the capacity of ZnO nanoparticles to infiltrate the root cells of ryegrass and go towards the vascular tissues.

The study conducted by Lopez-Moreno et al. (2010) investigated the mechanism by which ZnO nanoparticles are absorbed and accumulated in *G. max* seedlings. The seeds underwent treated with ZnO nanoparticles at doses ranging from 500 to 4000 ppm. At a concentration of 500 ppm, the researchers noted a higher absorption of Zn. The authors put up the hypothesis that an augmentation in the number of nanoparticles results in the creation of agglomerates, hence impeding the penetration of nanoparticles into the seed via the pores present in the cell wall. Furthermore, the utilization of the X-ray absorption spectroscopy technique on seedlings treated with ZnO demonstrated the detection of Zn^{2+} ions in lieu of ZnO. This discovery suggests that the roots contribute to the ionization of ZnO on their surface. The utilization of high-magnification transmission electron microscopy allowed for the identification of ZnO nanoparticles in various plant structures, including the apoplast, cytoplasm, and nuclei found in endodermal cell and vascular cylinder.

According to a study, presence of magnetite nanoparticles was discovered in many parts of pumpkin (*Cucurbita maxima*), including the root, stem, and leaves (Zhu et al. 2008). The researchers found that the amount of nanoparticle absorption is influenced by the precise composition of the culture media. Plants grown in hydroponic media had a greater absorption rate in comparison to those grown in sand. Conversely, plants grown in soil showed minimal absorption, possibly attributed to the attachment of magnetite nanoparticles to both soil and sand particles. On the other hand, the lima bean (*Phaseolus lunatus*) plants subjected to treatment exhibited no signs of absorption, indicating that the uptake of nanoparticles is contingent upon the specific plant species. Wang et al. (2011) found that pumpkin plants were unable to uptake magnetite nanoparticles because of their substantial size. Corredor et al. (2009) examined the influence of functionalization on the absorption of nanoparticles. To accomplish this, Carbon-coated iron nanoparticles were administered onto the foliage of a pumpkin plant. The presence of nanoparticles was identified within the epidermal cells, but no nanoparticles were found in close proximity to the xylem, as reported by the researchers. A study to examine the absorption and movement of copper nanoparticles in mung bean and wheat plants grown on agar growth media.

Copper nanoparticles possess the capability to infiltrate the cell membrane and aggregate within the cell. The existing body of research on the effects of TiO_2 and ZnO nanoparticles is considerable. However, the predominant focus of these studies is on the absorption, translocation, and accumulation of these particles in plants, particularly during the initial stages of growth leading up to germination. Hence, a significant gap exists in comprehending the patterns of nanoparticle behavior inside the plant system (Rico et al. 2011; Wang et al. 2023).

13.8 Foliar Uptake and Translocation of Metal Nanoparticles

The cuticle is the outermost layer of waxy substance on the leaves that serves to prevent excessive water loss through transpiration. Therefore, this layer serves as the main obstacle for the absorption of metal nanoparticles.

The uptake of metal nanoparticles in the cuticular pathway occurs through two distinct routes:

1. The lipophilic pathway is used for non-polar solutes, where solutes are absorbed through diffusion and/or permeation.
2. The hydrophilic pathway is used for polar solutes, where solutes are absorbed through pores. The pores have a high affinity for nanoparticles with a size smaller than 4.8 nm. There are literature publications documenting the successful use and absorption of foliar treatments containing nanoparticles larger than 5 nm in plants. Therefore, it is believed that metal nanoparticles larger than 4.8 nm can enter through the stomatal aperture, which has a pore size of roughly 25 μm in length and 3–10 μm in breadth, as depicted in Fig. 3a (Eichert et al. 2008). Since the phloem system does not circulate, metal nanoparticles can enter the plant tissue and potentially be transferred through the vascular system, specifically the phloem vessels, from the leaves to the roots (Eichert et al. 2008). Many researchers have reported the existence of metal nanoparticles in roots following the application of a spray to the leaves. This suggests that the nanoparticles are transported through the stomatal pathway and phloem sieve tubes, ultimately reaching the roots (Wang et al. 2011, 2023).

13.9 Ethical and Safety Issues in Using Nano-fertilizers

While nanotechnology holds significant promise for transforming several facets of human existence, it is crucial to recognize that there may also be certain disadvantages associated with these advantages. Prior to adopting nanotechnology, a crucial issue it is imperative to address the fundamental question of whether the potential advantages of nanoparticles are greater than the unacknowledged hazards they present to the environment and human well-being. Before fully implementing this technology, it is imperative to evaluate the potential risks associated with the use of nanoparticles. The emergence of the subject of "nano-toxicology" has been prompted by this notion, which entails the assessment of the potential toxicity of nanoparticles and the promotion of their safe design and utilization (Oberdorster et al. 2005; Verma et al. 2022). The advancement of future nanotechnology applications can be facilitated by conducting a comprehensive and methodical quantitative investigation of the potential health effects, environmental compatibility, and suitable disposal techniques for nanoparticles. Currently, there is a lack of definitive data linking nanoparticles to human diseases. Recent work focuses on the potential human health risks from nanoparticles used in fertilizers. It highlights that nanoparticles can be absorbed through the skin, inhalation, or ingestion, potentially leading to toxic effects. Therefore, interdisciplinary research is critical to fully understand how nanoparticles interact with biological systems and ensure the safety of nano-fertilizers in food production. However, first experimental investigations suggest that nanoparticles possess the ability to elicit unfavorable biological reactions, which may result in toxicological consequences. Nanoparticles, including minuscule particulate matter, have the ability to infiltrate the human or animal body via the oral cavity, respiratory tract, or dermal layer. There exists a commonly accepted belief that nanoparticles possess the ability to readily penetrate tissues, cells, and organelles due to their diminutive dimensions, hence facilitating interactions with functional bio-molecular entities such as DNA and ribosomes. Artificial nanostructures exhibit physical dimensions that closely resemble those of biological molecules, including antibodies, proteins, and viruses. Therefore, the integration of nanoparticles into essential biological systems has the potential to result in adverse effects, which could have a significant influence on human health (Verma et al. 2022).

However, a particularly unpleasant situation occurs when there is a lack of precise scientific information regarding the toxicological effects of nanoparticles. This provides an opportunity for proponents and detractors of nanotechnology to make contradictory, non-scientific, and excessively broad assessments regarding the safety of nanoparticles. The presence of ambiguity in nanotechnology is the main cause of concern for skeptics. It is imperative to conduct precise research and comprehend the precise physical and chemical characteristics of nanoparticles. Additionally, it is crucial to build dependable methodologies for assessing their behavior in the environment, their absorption by living organisms, and their distribution within these organisms. By addressing these problems, it would be feasible to provide the optimal experimental conditions for assessing the potential harm of a specific nanoparticle on human health (Thomas and Sayre 2005). In order to ascertain the authentic hazards posed by nanotechnology, it is imperative for experts in materials science, environmental science, and biological science to engage in interdisciplinary research and surmount the current limitations. Unfortunately, the assessments of the hazards associated with nanotechnology are partly reliant on subjective viewpoints and are susceptible to significant political biases. Regrettably, because nanotechnology is varied and always changing, there is no universally applicable method for assessing risks and implementing safety measures that can be uniformly used to reach a definitive result. The absence of standardized protocols and benchmarks is a barrier when seeking to compare safety and toxicity evaluations carried out by several research and technological institutions. The ethical considerations will only apply to the current state of information and to a specific product and its associated circumstances of use. Furthermore, it is imperative to guarantee a meticulous level of accuracy when formulating experiments. In addition, Schulte and Salamanca-Buentello 2007 suggested alternative assessments that accurately represent ethical, social, and political values when creating legislation pertaining to nanotechnology. Prior to conducting an analysis of toxicological data, it is imperative to perform calculations and ascertain the anticipated quantities of nanoparticles that could potentially interact with the biological system or exist within the ecological system.

Nanotechnology plays a crucial role in agriculture, which holds immense importance owing to its direct influence on human beings (Bouwmeester et al. 2009). Nano-fertilizers enhance the incorporation of nanoparticles into the food chain, hence promoting their dispersion throughout all organisms within the food chain. Given some chemicals, including arsenic and table salt, could harm plants, animals, or men, when given specific quantities, this does not exclude their utilization in many applications that are carefully designed to ensure safe levels of exposure. Based on extensive study on the utilization of nanoparticles for enhancing plant development, especially in smaller amounts, it may be argued that the possible dangers to health and the environment are insignificant (Colvin 2003). Numerous nations have recognized the prospective application of nanotechnology in the fields of food and agriculture and are making significant expenditures to harness its use in food production. However, because our information about the possible effects of these technologies on human health is limited, these countries recognize the need to rapidly evaluate the food safety implications of nanotechnology. The EFSA Scientific Committee (2011) has advised that nanoparticles should be subjected to the risk assessment framework, that encompasses the identification of hazards, characterization of hazards, evaluation of exposure, and characterization of risks. However, while assessing the possible hazards of these nanoparticles in relation to food and feed, it is important to consider the unique attributes of the nanoparticles in issue, as well as the similarities they share with their non-nano counterparts. Various categories of nanoparticles are expected to display differences in their toxicological characteristics. The available evidence on the toxicity arising from oral exposure to individual nanoparticles is extremely limited. Most of the evidence about the toxicity of nanoparticles is either from laboratory-based research or studies that entail exposure through routes other than oral ingestion. The assessment of the potential risks linked to nanoparticles must be carried out on a case-by-case basis, taking into account the unique characteristics of each circumstance. The risk linked with the utilization of a particular nanoparticle in food and feed can be assessed by taking into account numerous features. The research areas encompass the investigation of the physicochemical properties of nanoparticles, their ability to withstand degradation in food and feed, the probability of nanoparticle exposure through food and feed, and the toxicokinetic (absorption, distribution, metabolism/biotransformation, excretion/elimination) within human and animal systems. The US Food and Drug Administration (FDA) has developed the Nanotechnology Regulatory Science Research Plan, which provides a detailed framework and strategy for effectively addressing regulatory science activities and issues concerning FDA-regulated products that contain nanoparticles or utilize nanotechnology. The primary aim of the plan is to rectify notable inadequacies in scientific knowledge, methodologies, and instruments necessary for conducting regulatory assessments of these entities (Chaudhry and Castle 2011).

13.10 Conclusion and Prospects

The prevalence of nutrient shortage in agricultural soils has led to substantial reductions in crop productivity and substantial economic repercussions within the agricultural industry. While the application of chemical fertilizers has the potential to increase agricultural productivity, it is not a viable and sustainable long-term solution on a significant scale. Furthermore, the nutrients present in conventional fertilizers are not completely accessible to plants as a result of their large chemical compositions. Furthermore, it is worth noting that the majority of macronutrients exhibit a rather poor utilization rate due to their conversion into insoluble forms within the soil. The application of nanotechnology in agriculture relies heavily on the transportation of agrochemicals, particularly fertilizers that are responsible for delivering crucial macro- and micronutrients to plants. The employment of nanoscale or nanostructured materials as carriers for fertilizers or as controlled-release vectors in the development of intelligent fertilizers has the potential to improve nutrient utilization efficiency and mitigate the costs associated with environmental contamination. Nano-fertilizers exhibit the ability to emit their active components with precision and selectivity in response to specific environmental stimuli and biological requirements. Nano-fertilizers can be delivered to plants using both in vitro and in vivo approaches. Nevertheless, the current understanding of the absorption, dispersion, and ultimate destiny of nanoparticles in plants remains limited. The insufficient comprehension in this domain has resulted in numerous ethical and safety concerns regarding the utilization of nano-fertilizers for enhancing plant productivity. Conducting a comprehensive and methodical quantitative assessment of the possible health effects, environmental compatibility, and proper disposal of nano-materials can contribute to the development of future uses for nano-fertilizers.

References

Arora S, Sharma P, Kumar S, Nayan R, Khanna PK, Zaidi MGH (2012) Gold-nanoparticle induced enhancement in growth and seed yield of Brassica juncea. Plant Growth Regul 66:303–310

Arora J, Minkina T, Li Y-R (2022) Recent trends in Nano-fertilizers for sustainable agriculture under climate change for global food security. Nanomaterials 12(1):173

Barber SA (1995) Soil nutrient bioavailability: a mechanistic approach, 2nd edn. Wiley-Blackwell, New Jersey, 384 pp

Borm PJA, Robbins D, Haubold S, Kuhlbusch T, Fissan H, Donaldson K, Schins R, Stone V, Kreyling W, Lademann J, Krutmann J, Warheit D, Oberdorster E (2006) The potential risks of nanomaterials: a review carried out for ECETOC. Part Fibre Toxicol 3:11

Bouwmeester H, Dekkers S, Noordam MY, Hagens WI, Bulder AS, de Heer C, ten Voorde SE, Wijnhoven SW, Marvin HJ, Sips AJ (2009) Review of health safety aspects of nanotechnologies in food production. Regul Toxicol Pharmacol 53:52–62

Chaudhry Q, Castle L (2011) Food applications of nanotechnologies: an overview of opportunities and challenges for developing countries. Trends Food Sci Technol 22:595–603

Comerford NB (2005) Soil factors affecting nutrient bioavailability ecological studies. In: Bassiri Rad H (ed) Nutrient acquisition by plants an ecological perspective. Springer, Berlin, pp 1–14

Corradini E, De Moura MR, Mattoso LHC (2010) A preliminary study of the incorporation of NPK fertilizer into chitosan nanoparticles. Express Polym Lett 4:509–515

Corredor E, Testillano PS, Coronado MJ, González-Melendi P, Fernández-Pacheco R, Marquina C, Ibarra MR, de la Fuente JM, Rubiales D, Pérez-de-Luque A, Risueno MC (2009) Nanoparticle penetration and transport in living pumpkin plants: in situ subcellular identification. BMC Plant Biol 9:45

EFSA Scientific Committee (2011) Guidance on the risk assessment of the application of nanoscience and nanotechnologies in the food and feed chain. EFSA J 9:2140–2176

Eichert T, Kurtz A, Steiner U, Goldbach HE (2008) Size exclusion limits and lateral heterogeneity of the stomatal foliar uptake pathway for aqueous solutes and water-suspended nanoparticles. Physiol Plant 134:151–160

Fernández V, Eichert T (2009) Uptake of hydrophilic solutes through plant leaves: current state of knowledge and perspectives of foliar fertilization. Crit Rev Plant Sci 28:36–68

Fleischer A, O'Neill MA, Ehwald R (1999) The pore size of non-graminaceous plant cell walls is rapidly decreased by borate ester cross-linking of the pectic polysaccharide rhamnogalacturonan II. Plant Physiol 121:829–838

Ghormade V, Deshpande MV, Paknikar KM (2011) Perspectives for nano-biotechnology enabled protection and nutrition of plants. Biotechnol Adv 29:792–803

Khodakovskaya M, Silva KD, Biris AS, Dervishi E, Villagarica H (2012) Carbon nanotubes induce growth enhancement in tobacco cells. ACS Nano 6:2128–2135

Kottegoda N, Munaweera I, Madusanka N, Karunaratne V (2011) A green slow-release fertilizer composition based on urea-modified hydroxyapatite nanoparticles encapsulated wood. Curr Sci 101:73–78

Kurepa J, Paunesku T, Vogt S, Arora H, Rabatic BM, Lu J, Wanzer MB, Woloschak GE, Smalle JA (2010) Uptake and distribution of ultrasmallanatase TiO_2 alizarin red S nanoconjugates in Arabidopsis thaliana. Nano Lett 10:2296–2302

Larue C, Khodja H, Herlin-Boime N, Brisset F, Flank AM, Fayard B, Chaillou S, Carriere M (2011) Investigation of titanium dioxide nanoparticles toxicity and uptake by plants. J Phys Conf Ser 304:012057

Lei Z, Mingyu S, Chao L, Liang C, Hao H, Xiao W, Xiaoqing L, Fan Y, Fengqing G, Fashui H (2007) Effects of nanoanatase TiO_2 on photosynthesis of spinach chloroplasts under different light illumination. Biol Trace Elem Res 119:68–76

Lin D, Xing B (2008) Root uptake and phytotoxicity of ZnO nanoparticles. Environ Sci Technol 42: 5580–5585

Loomis RS, Connor DJ (1992) Crop ecology: productivity and management in agricultural systems. Cambridge University Press, Cambridge, p 538

Lopez-Moreno ML, de la Rosa G, Hernandez-Viezcas JA, Castillo-Michel H, Botez CE, Peralta-Videa JR, Gardea-Torresdey JL (2010) Evidence of the differential biotransformation and genotoxicity of ZnO and CeO2 nanoparticles on soybean (Glycine max) plants. Environ Sci Technol 44:7315–7320

Lu CM, Zhang CY, Wu JQ, Tao MX (2002) Research of the effect of nanometer on germination and growth enhancement of Glycine max and its mechanism. Soybean Sci 21:168–172

Mahmoodzadeh H, Nabavi M, Kashefi H (2013) Effect of nanoscale titanium dioxide particles on the germination and growth of canola (Brassica napus). J Ornam Hortic Plants 3:25–32

Marschner H (1995) Mineral nutrition of higher plants, 2nd edn. Academic, New York, p 889

Mengel K, Kirkby EA (2001) Principles of plant nutrition, 5th edn. Kluwer Academic Publishers, Dordrecht, p 849

Milani N, McLaughlin MJ, Stacey SP, Kirkby JK, Hettiarachchi GM, Beak DG, Cornelis G (2012) Dissolution kinetics of macronutrient fertilizers coated with manufactured zinc oxide nanoparticles. J Agric Food Chem 60:3991–3998

Mondal A, Basu R, Das S, Nandy P (2011) Beneficial role of carbon nanotubes on mustard plant growth: an agricultural prospect. J Nanopart Res 13:4519–4528

Moore M (2006) Do nanoparticles present ecotoxicological risks for the health of the aquatic environment? Environ Int 32:967–976

Naderi MR, Danesh-Shahraki A (2013) Nanofertilizers and their role in sustainable agriculture. Int J Agric Crop Sci 5:2229–2232

Nair R, Varghese SH, Nair BG, Maekawa T, Yoshida Y, Kumar DS (2010) Nanoparticulate material delivery to plants. Plant Sci 179:154–163

Namasivayam SKR, Chitrakala K (2011) Ecotoxicological effect of Lecanicilliumlecanii (Ascomycota: Hypocreales) based silver nanoparticles on growth parameters of economically important plants. J Biopest 4:97–101

Navarro E, Baun A, Behra R, Hartmann NB, Filser J, Miao AJ, Quigg A, Santschi PH, Sigg L (2008) Environmental behavior and ecotoxicity of engineered nanoparticles to algae, plants, and fungi. Ecotoxicology 17:372–386

Oberdorster G, Oberdorster E, Oberdorster J (2005) Nanotoxicology: an emerging discipline evolving from studies of ultrafine particles. Environ Health Perspect 113:823–839

Prasad TNVKV, Sudhakar P, Sreenivasulu Y, Latha P, Munaswamy V, Reddy KR, Sreeprasad TS, Sajanlal PR, Pradeep T (2012) Effect of nanoscale zinc oxide particles on the germination, growth and yield of peanut. J Plant Nutr 35:905–927

Quoc Buu N, Trong Hien D, Hoai Chau N, Xuan Tin T, Tuong Van N, Thuy Duong K, Thi Ha H (2014) Effects of nanocrystalline powders (Fe, Co and Cu) on the germination, growth, crop yield and product quality of soybean (Vietnamese species DT-51). Adv Nat Sci: Nanotechnol 5: 015016

Rico CM, Majumdar S, Duarte-Gardea M, Peralta-Videa JR, Gardea-Torresdey JL (2011) Interaction of nanoparticles with edible plants and their possible implications in the food chain. J Agric Food Chem 59:3485–3498

Sahandi SM, Sorooshzadeh A, Rezazadeh HS, Naghdibadi HA (2011) Effect of nanosilver and silver nitrate on seed yield of borage. J Med Plants Res 5:706–710

Sasson Y, Levy-Ruso G, Toledano O, Ishaaya I (2007) Nanosuspensions: emerging novel agrochemical formulations. In: Ishaaya I, Horowitz AR, Nauen R (eds) Insecticides design using advanced technologies. Springer, Berlin, pp 1–39

Schulte PA, Salamanca-Buentello F (2007) Ethical and scientific issues of nanotechnology in the workplace. Environ Health Perspect 115:5–12

Shah V, Belozerova I (2009) Influence of metal nanoparticles on the soil microbial community and germination of lettuce seeds. Water Air Soil Pollut 197:143–148

Stoner RJ, Clawson JM (1999–2000) Low-mass, inflatable Aeroponic system for high performance food production. Principal Investigator, NASA SBIR NAS10-00017

Sun D, Hussain H, Yi Z, Siegele R, Cresswell T, Kong L, Cahill D (2014) Uptake and cellular distribution, in four plant species, of fluorescently labeled mesoporous silica nanoparticles. Plant Cell Rep 33:1389–1402

Suriyaprabha R, Karunakaran G, Yuvakkumar R, Prabu P, Rajendran V, Kannan N (2012) Growth and physiological responses of maize (Zea mays L.) to porous silica nanoparticles in soil. J Nanopart Res 14:1294–1308

Taiz L, Zeiger E (2010) Plant physiology, 5th edn. Sinauer Associates Inc., Massachusetts, 781 pp

Thakkar MN, Mhatre S, Parikh RY (2010) Biological synthesis of metallic nanoparticles. Nanotechol Biol Med 6:257–262

Thomas K, Sayre P (2005) Research strategies for safety evaluation of nanomaterials, Part I: evaluating the human health implications of exposure to nanoscale materials. Toxicol Sci 87: 316–321

Tilman D, Knops J, Wedin D, Reich P (2002) Plant diversity and composition: effects on productivity and nutrient dynamics of experimental grasslands. In: Loreau M, Naeem S, Inchausti P (eds) Biodiversity and ecosystem functioning. Oxford University Press, Oxford, pp 21–35

Trenkel ME (1997) Controlled-release and stabilized fertilizers in agriculture. International Fertilizer Industry Association, Paris

Uzu G, Sobanska S, Sarret G, Munoz M, Dumat C (2010) Foliar lead uptake by lettuce exposed to atmospheric fallouts. Environ Sci Technol 44:1036–1042

Verma KK, Song X-P, Joshi A, Tian D-D, Rajput VD, Singh M, Arora J, Minkina T, Li Y-R (2022) Recent trends in Nano-fertilizers for sustainable agriculture under climate change for global food security. Nano 12(1):173

Wang H, Kou X, Pei Z, Xiao JQ, Shan X, Xing B (2011) Physiological effects of magnetite (Fe3O4) nanoparticles on perennial ryegrass (Loliumperenne L.) and pumpkin (Cucurbitamixta) plants. Nanotoxicology 5:30–42

Wang Z, Wang S, Ma T, Liang Y, Huo Z, Yang F (2023) Enhancing plant stress resistance: a review. Agronomy 13(12):3060

Wanyika H, Gatebe E, Kioni P, Tang Z, Gao Y (2012) Mesoporous silica nanoparticles carrier for urea: potential applications in agrochemical delivery systems. J Nano Sci Nanotechnol 12: 2221–2228

Weathers PJ, Zobel RW (1992) Aeroponics for the culture of organisms, tissues and cells. Biotechnol Adv 10:93–115

Zhao L, Peralta-Videa JR, Rico CM, Hernandez-Viezcas JA, Sun Y, Niu G, Servin A, Nunez JE, Duarte-Gardea M, Gardea-Torresdey JL (2014) CeO2 and ZnO nanoparticles change the nutritional qualities of cucumber (Cucumis sativus). J Agric Food Chem 62:2752–2759

Zheng L, Hong F, Lu S, Liu C (2005) Effect of nano-TiO_2 on strength of naturally aged seeds and growth of spinach. Biol Trace Elem Res 104:83–92

Zhu H, Han J, Xiao JQ, Jin Y (2008) Uptake, translocation, and accumulation of manufactured iron oxide nanoparticles by pumpkin plants. J Environ Monit 10:713–717

Chapter 14
Nanotechnology Applications in Modern Agriculture

Mariat George, Amritha Krishnakumar, and T. R. Anju

Abstract Nanotechnology applications in modern agriculture are transforming traditional farming practices by enhancing efficiency, sustainability, and productivity. This abstract explores the multifaceted roles of nanotechnology in agriculture, focusing on its potential to address critical challenges such as nutrient management, disease detection, and pest control. Nanofertilizers, designed with nano-sized carriers, offer precise and controlled nutrient delivery, improving nutrient use efficiency and reducing environmental impact. Nanotechnology-based disease detection systems, including nanoparticle-based biosensors and nanopore sensors, enable rapid, sensitive, and specific identification of plant pathogens, facilitating early intervention and disease management. Nanomaterials are also used in pesticide formulations to increase their effectiveness and minimize adverse environmental effects. The integration of nanotechnology in precision agriculture provides farmers with advanced tools for monitoring soil health, optimizing water usage, and enhancing crop resilience to biotic and abiotic stresses. Despite the promising benefits, challenges such as regulatory hurdles, cost, and potential environmental risks need to be addressed to ensure the safe and widespread adoption of nanotechnology in agriculture. By fostering innovation and interdisciplinary collaboration, nanotechnology can significantly contribute to sustainable agricultural practices, ensuring food security and environmental conservation for future generations.

Keywords Nanotechnology · Agriculture · Nanofertilizers · Environmental effects

M. George (✉) · A. Krishnakumar
Department of Botany, Newman College, Thodupuzha, Kerala, India

T. R. Anju
Department of Biotechnology, Newman College, Thodupuzha, Kerala, India
e-mail: anju.tr@newmancollege.ac.in

J. M. Al-Khayri et al. (eds.), *Nanotechnology Applications in Modern Agriculture*, Nanotechnology in Plant Sciences 3, https://doi.org/10.1007/978-3-031-90506-3_14

14.1 Introduction

Nanotechnology, the science of manipulating matter at the atomic and molecular scale, has emerged as a groundbreaking field with transformative potential across various sectors. In agriculture, where the demand for sustainable practices, increased productivity, and resource efficiency is ever-growing, nanotechnology offers a suite of innovative solutions poised to revolutionize farming practices and address critical challenges facing the industry.

The application of nanotechnology in agriculture, often termed "nano-agriculture," encompasses a diverse array of technologies and approaches aimed at enhancing crop yield, mitigating environmental impacts, improving resource management, and bolstering food security. At the heart of these advancements lies the unique properties exhibited by nanomaterials and nanostructures, which enable precise control over interactions at the molecular level and the development of novel tools and systems tailored to the needs of modern agriculture (Yadav et al. 2019).

This chapter serves as a comprehensive exploration of the multifaceted role of nanotechnology in modern agriculture. It delves into the various applications of nanotechnology across different aspects of agricultural production, from enhancing soil health and nutrient delivery to monitoring plant health and combating pests and diseases (Prasad et al. 2017). By leveraging nanoscale innovations, farmers and researchers can unlock new opportunities for sustainable and efficient farming practices while minimizing environmental impacts and ensuring food security in the face of global challenges such as climate change and population growth. From nanomaterials designed to boost crop resilience and yield to precision agriculture tools empowered by nanosensors, each section will shed light on the transformative potential of nanotechnology in reshaping the agricultural landscape (Masurkar et al. 2024). Moreover, the chapter discusses the ethical, environmental, and regulatory considerations inherent in the adoption of nanotechnology in agriculture, emphasizing the importance of responsible innovation and sustainable practices.

14.2 Importance of Nanotechnology in Agriculture

The importance of nanotechnology in agriculture lies in its potential to address pressing challenges facing the global food system while promoting sustainability and efficiency. Nanotechnology offers unique opportunities to revolutionize traditional agricultural practices and enhance various aspects of crop production, soil management, and pest control (Kim et al. 2018). One key area where nanotechnology demonstrates its significance is in crop enhancement (Elemike et al. 2019). Nanomaterials, such as nanoparticles and nanocomposites, can be tailored to improve crop productivity, enhance nutrient uptake, and increase resistance to environmental stresses (Javed et al. 2022). By delivering nutrients and pesticides more effectively to plants, nanotechnology enables farmers to optimize resource utilization and minimize the environmental impact of agricultural inputs.

Precision agriculture, another critical aspect of modern farming, benefits greatly from nanotechnology applications. Nanosensors and nanodevices allow for real-time monitoring of soil conditions, crop health, and environmental parameters (John et al. 2022). This precise data collection enables farmers to make informed decisions about irrigation, fertilization, and pest management, leading to more efficient use of resources and higher yields. Furthermore, nanotechnology offers innovative solutions for pest and disease management in agriculture. Nanoparticles can be engineered to deliver targeted doses of pesticides or antimicrobial agents, reducing chemical usage and minimizing harm to beneficial organisms and the environment (Kashyap et al. 2015). Nanobiotechnology approaches, such as using nanomaterials for genetic modification or RNA interference, hold promise for developing sustainable and environmentally friendly methods of crop protection (Banerjee et al. 2021). In addition to improving crop production, nanotechnology plays a crucial role in soil health and environmental sustainability. Nanoremediation techniques help remediate contaminated soils and mitigate environmental pollution, restoring soil fertility and supporting long-term agricultural productivity (Dhanapal et al. 2024).

In short, the importance of nanotechnology in agriculture cannot be overstated. By leveraging the unique properties of nanomaterials and nanodevices, farmers can overcome existing challenges and build resilient agricultural systems capable of meeting the demands of a growing global population while minimizing environmental impact. As research in this field continues to advance, nanotechnology holds the potential to transform the future of agriculture, ensuring food security, environmental sustainability, and economic prosperity for generations to come.

14.3 Types of Nanomaterials Used in Agriculture

Nanomaterials used in agriculture encompass a wide range of substances engineered at the nanoscale for various applications. Here are some common types (Fig. 14.1).

- Nanoparticles: These are tiny particles with dimensions typically ranging from 1 to 100 nanometers. They can be made from various materials such as metals (e.g., silver, gold), metal oxides (e.g., zinc oxide, titanium dioxide), carbon-based materials (e.g., carbon nanotubes, graphene), and polymers. Nanoparticles are versatile and can be designed for applications such as nutrient delivery, pest management, and stress tolerance enhancement (Dolez 2015).
- Nanoemulsions: Nanoemulsions are stable dispersions of nanoscale droplets of one liquid within another immiscible liquid. In agriculture, nanoemulsions are often used as carriers for pesticides, herbicides, and fungicides, enabling better coverage and penetration on plant surfaces while reducing environmental impact and improving efficacy (Maali and Hamed 2013).

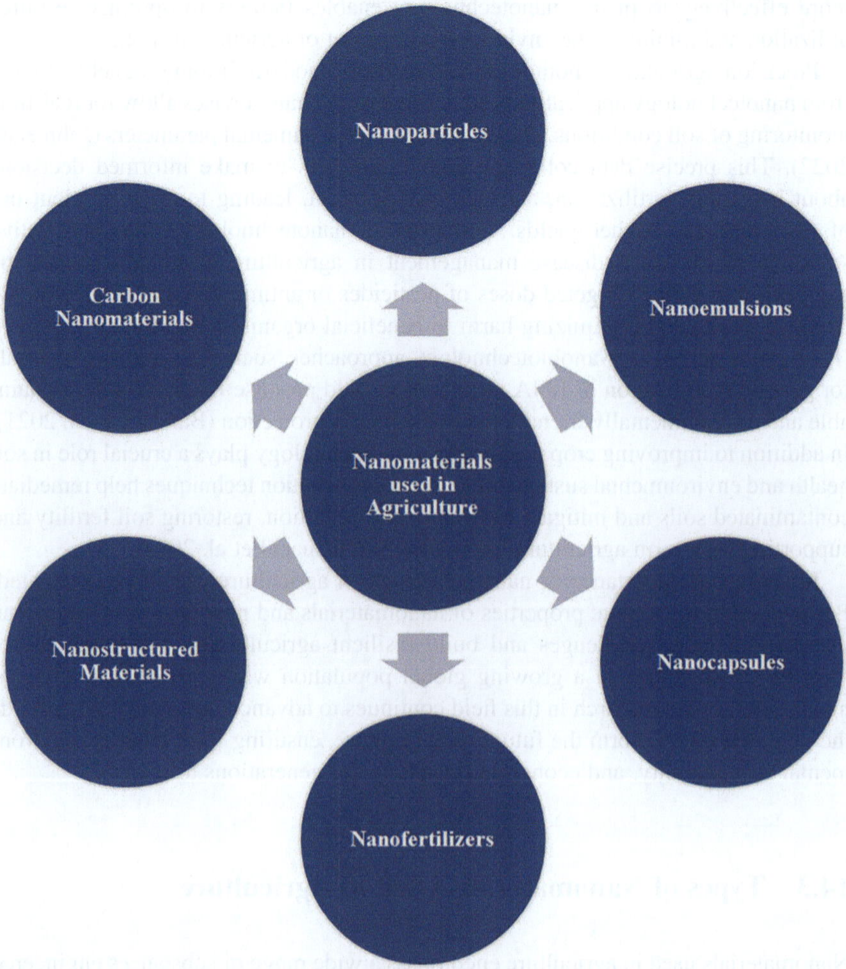

Fig. 14.1 Nanomaterials used in agricultural fields

- Nanocapsules: Nanocapsules are nano-sized capsules with a core-shell structure, where the active ingredient is encapsulated within a shell made of polymeric or lipid materials (Erdoğar et al. 2018). They are utilized for controlled release and targeted delivery of agrochemicals, nutrients, or bioactive compounds to plants, minimizing waste and enhancing efficiency.
- Nanofertilizers: Nanofertilizers are formulated using nanoparticles to encapsulate or deliver nutrients such as nitrogen, phosphorus, and potassium to plants (Solanki et al. 2015). They can improve nutrient uptake efficiency, reduce nutrient leaching, and enhance crop productivity while minimizing environmental pollution and fertilizer runoff.

- Nanostructured materials: These are materials with nanostructured surfaces or architectures, such as nanostructured films, membranes, and coatings (Sotiropoulou et al. 2008). Nanostructured materials can be applied in agriculture for purposes such as soil remediation, water purification, and controlled release of agrochemicals.
- Carbon nanomaterials: Carbon-based nanomaterials, including carbon nanotubes, graphene, and carbon dots, have unique properties such as high surface area, mechanical strength, and electrical conductivity (Díez-Pascual 2021). They find applications in soil improvement, plant growth promotion, and sensing of environmental parameters.
- Nanoencapsulated plant extracts: Natural plant extracts with pesticidal, fungicidal, or growth-promoting properties can be encapsulated within nanomaterials for enhanced stability, controlled release, and targeted delivery to plants. Nanoencapsulation preserves the bioactivity of plant extracts and improves their efficacy in agricultural applications (Guía-García et al. 2022).

These are few examples of nanomaterials used in agriculture, and ongoing research continues to explore new formulations and applications to address challenges in sustainable crop production and environmental management.

14.4 Nanomaterials for Crop Enhancement

Nanotechnology is being used in agriculture in a variety of ways, such as fertilizers to boost plant growth and yield, insecticides to control pests and diseases, and sensors to keep an eye on plant health and soil quality. Nanomaterials hold immense potential for revolutionizing agriculture and enhancing crop production. Nanomaterials play a significant role in improving crop yield and quality through various mechanisms and applications:

14.4.1 Enhanced Nutrient Management

Nanofertilizers can deliver nutrients more efficiently to plants, ensuring optimal uptake and utilization. Controlled-release nanofertilizers can provide nutrients gradually over time, reducing nutrient leaching and maximizing plant growth. This improves yield and enhances the nutritional quality of crops (Mahanta et al. 2019).

14.4.2 Improved Water Management

Nanomaterials can enhance water retention in soil, reducing water loss through evaporation and improving water availability to plants (Zhao et al. 2022). Nanoparticles can also modify soil properties to enhance water infiltration and root penetration, particularly in arid and drought-prone regions. Better water management leads to increased crop yield and quality.

14.4.3 Pest and Disease Control

Nanoparticles can be used to deliver pesticides, fungicides, and antimicrobial agents more effectively to plants. Nanoformulations provide better adhesion to plant surfaces, increased penetration into plant tissues, and controlled release of active ingredients, resulting in improved pest and disease management. This reduces crop damage and ensures higher-quality harvests (Guía-García et al. 2022).

14.4.4 Stress Tolerance

Abiotic stressors that affect plants include heat, cold, drought, salt, flooding, and heavy metals. Abiotic stressors typically have negative effects on plant growth and development, which lowers agricultural output, creates issues with food security, and results in financial losses. Novel technologies, such nanotechnology, have evolved to lessen the detrimental impacts of environmental stress on crop plants. Modern agriculture can benefit from the application of nanotechnology in a number of ways, including increased sustainability, reduced environmental pollution, plant disease prevention, improved water efficiency, and guaranteed food security (El-Saadony et al. 2022). Nanomaterials can enhance crop tolerance to various environmental stresses, including drought, salinity, heat, and heavy metals. Nanoparticles can scavenge reactive oxygen species, regulate plant hormone levels, and protect cellular structures, mitigating stress-induced damage and improving plant resilience. This leads to higher yields and better-quality produce under adverse growing conditions.

14.4.5 Enhanced Photosynthesis

Nanoparticles can modulate the photosynthetic process by improving light absorption, chlorophyll synthesis, and electron transport in plant cells. To be very specific, NPs either boost up the photosynthesis processes by improving LHC (Light

Harvesting Complex) in plants or hinder their pathways by blocking ETC (Electron Transport Chain) and they affect photosynthetic rate by change in several genes and enzymes like Carbonic anhydrase, RUBISCO and PEP caboxylase (Kataria et al. 2019). Nanostructured materials such as quantum dots and carbon nanotubes can enhance light-harvesting efficiency and electron transfer rates, boosting photosynthetic activity and biomass production. This results in increased crop yield and improved nutritional content (Shiyani and Bagchi 2020).

14.4.6 Genetic and Molecular Manipulation

Nanomaterials enable precise delivery of nucleic acids, including genes and RNA molecules, into plant cells for genetic engineering and gene regulation (Squire et al. 2023). This facilitates the development of transgenic crops with desirable traits such as improved yield, disease resistance, and nutritional quality. Nanotechnology-based gene editing tools such as CRISPR-Cas9 offer precise control over crop genetics, leading to enhanced yield and quality (Ahmar et al. 2021).

14.4.7 Post-harvest Preservation

Nanomaterials can extend the shelf life of harvested crops by inhibiting microbial growth, delaying ripening, and reducing post-harvest losses (Babu et al. 2022). Nanostructured films and coatings can provide barrier properties, moisture regulation, and controlled release of antimicrobial agents, preserving freshness and quality during storage and transportation (Zhang et al. 2021).

14.4.8 Nano-fertilizers

Nano-fertilizers have been demonstrated to improve nutrient uptake efficiency and enhance crop yield (Fig. 14.2). For example, a study published in the journal "Scientific Reports" showed that zinc oxide nanoparticles, when applied as a foliar spray on maize plants, increased the plant's zinc content and grain yield by 17–29% compared to conventional zinc fertilizers (Dapkekar et al. 2018). The nanoparticles were able to penetrate the leaf surface more effectively, leading to improved nutrient absorption (Jalal et al. 2023).

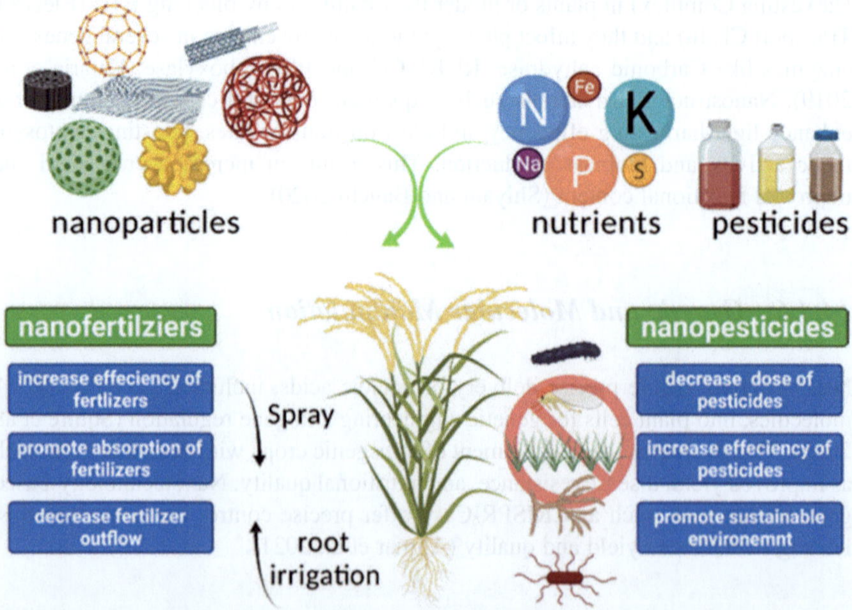

Fig. 14.2 Schematic representation of effect of nanofertilizers and nanopesticides. (Reproduced from Liu et al. 2021a, b)

14.4.9 Nano-pesticides

Nano-formulations of pesticides have shown promise in improving pest control while reducing environmental impact (Fig. 14.2). In one study published in "Journal of Agricultural and Food Chemistry," researchers developed nanoemulsion formulations of a commonly used insecticide, lambda-cyhalothrin (Qin et al. 2019). These nanoemulsions exhibited higher insecticidal activity against cotton aphids compared to the conventional formulation, with reduced application rates and environmental toxicity (Abdelmaksoud et al. 2024).

14.4.10 Stress Tolerance Enhancement

Nanomaterials have been utilized to enhance crop tolerance to abiotic stresses such as drought and salinity. In a study published in "Plant Physiology and Biochemistry," researchers investigated the role of cerium oxide nanoparticles (nanoceria) in alleviating drought stress in tomato plant (Thiruvengadam et al. 2024). They found that foliar application of nanoceria increased antioxidant enzyme activity and water retention capacity in plant tissues, resulting in improved drought tolerance and higher yield under water-deficit conditions.

14.5 Nanomaterials for Other Delivery of Bioactive Compounds or Genes

Nanomaterials hold immense potential for improving crop yield and quality by addressing various challenges in agriculture, including nutrient management, pest and disease control, environmental stress tolerance, and post-harvest preservation. Continued research and development in nanotechnology offer innovative solutions to enhance agricultural productivity and sustainability. Other than direct crop improvement strategies, nanomaterials can be used for many other applications in modern agriculture. The most significant ones are discussed below:

14.5.1 Nanoencapsulation of Bioactive Compounds

Nanoencapsulation allows for the controlled release and targeted delivery of bioactive compounds to plants, enhancing their growth and stress tolerance. In a study published in " *International Journal of Food Microbiology*," researchers developed nanoencapsulated essential oils derived from medicinal plants for controlling fungal diseases in strawberry plants. The nanoencapsulated essential oils exhibited superior antifungal activity compared to free essential oils, leading to better disease control and higher fruit yield (Javanmardi et al. 2023).

14.5.2 Gene Delivery and Plant Transformation

Nanoparticles have been employed for efficient gene delivery and plant transformation, enabling the development of transgenic crops with improved traits. In a study published in ". *Nature Protocols*," researchers demonstrated the use of carbon nanotubes for delivering DNA into plant cells. The carbon nanotube-based delivery system enabled targeted gene transfer and stable integration into the plant genome, resulting in the generation of transgenic tobacco plants with enhanced resistance to viral infections (Demirer et al. 2019).

14.6 Precision Agriculture: Monitoring and Management Through Nanosensors

Over the past 10 years, nanotechnology has gained traction and emerged as a significant tool in modern agriculture. Through the use of nanoformulations of fertilizers to improve nutrient use efficiency, agrochemicals for crop enhancement, disease detection and treatment, molecular host-parasite interactions through the use

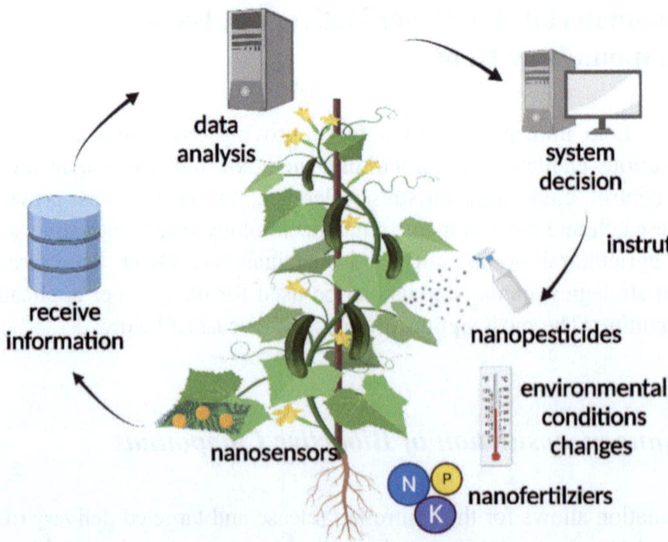

Fig. 14.3 Schematic representation of application of nanosensors. (Reproduced from Liu et al. 2021a, b)

of nanosensors, plant disease diagnostics, removal of contaminants from soil and water, postharvest management of vegetables and flowers, and reclamation of salt-affected soils, among other methods, nanotechnology can increase agricultural production. Additionally, a wide range of pathogens, fertilizers, moisture, and soil pH can be sensed using nanobiosensors with the goal of eliminating the need for plant protection products, minimizing nutrient loss, and improving crop yields through prudent nutrient management. Here, we examine the uses of nanotechnology in agriculture, namely the usage of metal oxide-based nanosensors to shield crops against bacterial infections (Kaushal and Wani 2017).

Precision agriculture, enabled by nanosensors, is revolutionizing the way we monitor and manage crops. Nanosensors are tiny devices capable of detecting and transmitting data about various environmental factors at a micro-level. Nanosensors provide real-time data on soil moisture, nutrient levels, temperature, and other crucial parameters (Fig. 14.3). This data is transmitted wirelessly to farmers' smartphones or computers, allowing them to make informed decisions promptly. With precise information on soil conditions and crop health, farmers can optimize resource use such as water, fertilizers, and pesticides. This not only reduces waste but also minimizes environmental impact (Yin et al. 2021; Shah and Wei 2019). Nanosensors can detect subtle changes in plant health indicative of pest infestations or diseases. Early detection allows farmers to take timely action, preventing widespread damage and reducing the need for chemical interventions. By analyzing data collected by nanosensors, farmers can tailor their management practices to specific crop requirements. This personalized approach maximizes yield and quality while minimizing input costs. By maintaining optimal growing conditions and promptly addressing issues, precision agriculture with nanosensors helps improve crop quality and yield. This is crucial for meeting the demands of a growing population (Yadav et al. 2023a, b; Zain et al. 2023).

14.6.1 Nanosensors for Monitoring Soil Conditions

The dynamic balance of the soil's chemical, biological, and physical characteristics is known as soil health. Effective soil management necessitates a real-time assessment of soil health. Precision agriculture relies heavily on sensors for precise regulation of moisture and nutrients. Because of their higher specific surface and greater chemical reactivity, nanomaterials appear to be crucial to the development of biosensors. Catalytic activity increases the use of nanobiosensors because it produces results with high specificity, precision, stability, fast reactivity, and unique physico-chemical properties. Mercury removal and detection are now possible with the development of multimodal nanosensors. A sensor device based on carbon nanodots and quantum dots has been created to detect heavy metals. Utilizing nanobiosensors to detect urea and track soil enzymatic activity has been produced in a lab (Mandal et al. 2020a, b).

Nanosensors are incredibly small devices, typically on the scale of nanometers (billionths of a meter). This small size allows them to be embedded directly into the soil without disturbing the natural ecosystem. Despite their size, nanosensors are highly sensitive and can detect subtle changes in soil properties. Nanosensors can monitor a wide range of soil conditions including moisture content, nutrient levels (such as nitrogen, phosphorus, and potassium), pH levels, salinity, temperature, and even the presence of pollutants or harmful chemicals (Khanna 2016; Yin et al. 2021).

These sensors provide real-time data on soil conditions, allowing farmers to make timely decisions about irrigation, fertilization, and other management practices. This real-time feedback loop is crucial for optimizing resource use and maximizing crop yield. Many nanosensors are equipped with wireless connectivity, enabling data transmission to a central database or farmers' smartphones or computers. This allows for remote monitoring and management, reducing the need for manual labor and on-site inspections. By accurately measuring soil moisture levels, nanosensors enable precision irrigation, where water is applied only when and where it is needed. This not only conserves water but also prevents overwatering, which can leach nutrients and lead to waterlogging (Yin et al. 2021; Yeshe et al. 2022).

Nanosensors can assess nutrient levels in the soil, allowing farmers to tailor fertilizer applications to the specific needs of their crops. This prevents both under-fertilization, which can limit crop growth, and over-fertilization, which can contribute to pollution and environmental degradation. Nanosensors can detect early signs of soil degradation, such as nutrient depletion or salinity buildup, allowing farmers to take corrective action before these problems impact crop health and productivity (Khanna 2016; Manjunatha et al. 2016). By enabling precise management of soil conditions, nanosensors contribute to environmental sustainability by reducing the use of water, fertilizers, and other inputs. This helps minimize the environmental impact of agriculture and promotes long-term soil health.

Nanosensors can be integrated into larger precision agriculture systems, where data from multiple sensors are combined and analyzed to provide comprehensive insights into crop and soil health. This integration enhances the effectiveness of precision agriculture practices. s nanotechnology continues to advance, the capabilities of nanosensors for soil monitoring will likely expand. Future developments may include sensors capable of detecting additional soil parameters, improved durability and longevity, and integration with autonomous agricultural systems (Khanna 2016).

14.6.2 Nanosensors for Assessing Plant Health and Nutrient Status

Nanosensors play a crucial role in assessing plant health and nutrient status by providing real-time, accurate data on various physiological and biochemical parameters. Nanosensors can measure nutrient levels within plants, including essential elements like nitrogen, phosphorus, potassium, and micronutrients. By monitoring these levels, farmers can adjust fertilizer applications to ensure optimal nutrient uptake by the plants, preventing deficiencies or excesses that can impact growth and yield. Nanosensors can assess the pH level and ionic balance within plant tissues. Maintaining the proper pH and ion balance is critical for various physiological processes, and deviations from the optimal range can indicate nutrient imbalances or environmental stressors (Yadav et al. 2023a, b; Cronan and Grigal 1995).

Some nanosensors are designed to measure photosynthetic activity within plants, including parameters such as chlorophyll content, photosynthetic efficiency, and stomatal conductance. Monitoring photosynthesis provides insights into overall plant health and productivity. Nanosensors can detect changes in plant water status, including water uptake, transpiration rates, and water stress. By monitoring these parameters, farmers can adjust irrigation schedules to prevent water stress while avoiding overwatering (Giraldo et al. 2019; Kwak et al. 2017).

Nanosensors can detect stress-related compounds within plants, such as reactive oxygen species (ROS) and stress hormones like abscisic acid (ABA). Elevated levels of these compounds can indicate various stressors, including drought, salinity, pathogens, or pests. Nanosensors can detect biomarkers associated with pest infestations or diseases before visible symptoms appear. Early detection allows farmers to implement timely pest management strategies, such as targeted pesticide applications or biological control methods, minimizing crop damage. Nanosensors can also assess environmental factors that impact plant health, such as temperature, humidity, light intensity, and air quality. By monitoring these parameters, farmers can optimize growing conditions within greenhouses or controlled environments (Kordrostami et al. 2021; Giraldo et al. 2019).

Data collected by nanosensors can be integrated with other agricultural data sources, such as weather data, soil data, and historical crop performance records. Advanced analytics and machine learning algorithms can then analyze this integrated data to provide actionable insights and recommendations for optimizing plant health and nutrient management. Nanosensors equipped with wireless connectivity allow for remote monitoring of plant health and nutrient status, enabling farmers to access real-time data from anywhere via smartphones or computers. This remote monitoring capability is particularly valuable for large-scale or geographically dispersed farming operations. By providing precise, localized data on plant health and nutrient status, nanosensors enable farmers to implement targeted management practices, such as site-specific fertilization, irrigation, and pest control. This precision approach maximizes resource efficiency while minimizing environmental impacts (Yeshe et al. 2022; Yadav et al. 2023a, b).

14.6.3 Real-Time Data Analysis and Decision-Making in Precision Agriculture

Real-time data analysis and decision-making are integral components of precision agriculture, enabling farmers to optimize crop management practices based on up-to-date information. Here's how real-time data analysis drives decision-making in precision agriculture:

- Various sensors, including nanosensors, drones, satellites, weather stations, and soil monitors, continuously collect data on factors such as soil moisture, temperature, nutrient levels, pest infestations, crop growth, and weather conditions (Yadav et al. 2023a, b).
- Data from multiple sources are integrated into a centralized system, often utilizing Internet of Things (IoT) technology. This integration allows for a comprehensive view of the farm, including spatial and temporal variability in environmental conditions and crop performance ((Boursianis et al. 2022)).
- Advanced analytics techniques, including machine learning algorithms, statistical models, and data visualization tools, are applied to the integrated data to identify patterns, trends, and anomalies. These analyses help extract meaningful insights from the vast amount of data collected.
- The analyzed data are continuously monitored in real-time, providing farmers with immediate feedback on the current state of their crops and the surrounding environment. This real-time monitoring allows for early detection of issues and rapid response to changing conditions (Monteiro et al. 2021).
- In addition to monitoring current conditions, predictive models are developed to forecast future outcomes based on historical data and current trends. These models can predict crop yields, pest outbreaks, irrigation requirements, and optimal planting times, allowing farmers to plan ahead and mitigate risks (Matthews et al. 2013).

- Real-time data analysis feeds into decision support systems (DSS) that provide actionable recommendations to farmers. These recommendations may include precise irrigation schedules, targeted fertilizer applications, optimal planting densities, and pest management strategies tailored to specific field conditions (Baldin et al. 2021).
- Real-time data analysis enables the implementation of VRT, where inputs such as water, fertilizers, and pesticides are applied at variable rates across the field according to spatial variability in soil and crop conditions. This maximizes resource efficiency and crop yields while minimizing environmental impacts (Pawase et al. 2023).
- Farmers can access real-time data and receive alerts remotely via smartphones, tablets, or computers. This remote monitoring capability allows farmers to stay connected to their farms from anywhere, enabling timely decision-making even when off-site (Visconti et al. 2020).
- Real-time data analysis facilitates a feedback loop where decisions made based on current data are evaluated for their effectiveness. By comparing predicted outcomes with actual results, farmers can refine their management practices over time, continuously improving farm performance.

Ultimately, real-time data analysis enables farmers to optimize their agricultural practices for maximum efficiency, productivity, and sustainability. By leveraging timely insights from data analytics, farmers can make informed decisions that lead to better resource management, higher yields, and improved profitability (Cheema and Khan 2019).

14.6.4 Efficient Delivery Systems: Nanocarriers for Agrochemicals

In the pursuit of sustainable agricultural practices, the effective delivery of agrochemicals is paramount. Agrochemicals, including pesticides, fertilizers, and plant growth regulators, play a pivotal role in enhancing crop yield and quality. However, conventional delivery methods often suffer from drawbacks such as uneven distribution, rapid degradation, and environmental contamination. In response to these challenges, the emergence of nanotechnology has ushered in a new era of precision and efficiency in agrochemical delivery.

Nanocarriers, colloidal structures with dimensions ranging from 1 to 100 nanometers, have emerged as promising platforms for the controlled delivery of agrochemicals. These nanocarriers can encapsulate active ingredients, protecting them from degradation and facilitating their targeted delivery to specific plant tissues. By encapsulating agrochemicals within nanocarriers, their solubility, stability, and bioavailability can be enhanced, resulting in improved efficacy and reduced environmental impact (Rehman et al. 2021; Martínez-Ballesta et al. 2018).

One of the key advantages of nanocarriers is their ability to overcome barriers to agrochemical delivery, such as limited penetration into plant tissues and poor adhesion to foliage. Nanocarriers can be engineered to exhibit desirable properties such as surface modification for enhanced adhesion, stimuli-responsive release triggered by environmental cues, and sustained release kinetics for prolonged efficacy. These tailored characteristics enable precise control over the release of agrochemicals, ensuring optimal dosage and minimizing off-target effects (Grillo et al. 2021). Furthermore, nanocarriers offer versatility in terms of the types of agrochemicals they can deliver. Whether hydrophilic or hydrophobic, small molecules or biomacromolecules, nanocarriers can accommodate a wide range of active ingredients, opening up possibilities for novel formulations and delivery strategies.

14.6.5 Nanocarriers for Targeted Delivery of Fertilizers and Pesticides

Nanocarriers represent a cutting-edge approach to the targeted delivery of fertilizers and pesticides in agriculture, offering numerous advantages over traditional application methods (Jogaiah et al. 2020). These nanoscale delivery systems, typically ranging from 1 to 100 nanometers in size, hold great promise for enhancing the efficiency, efficacy, and sustainability of agrochemical use.

Nanocarriers can encapsulate fertilizers and pesticides, protecting them from degradation and improving their solubility in water. This enhances the bioavailability of active ingredients and reduces losses due to leaching or volatilization, resulting in more effective and longer-lasting treatments (Jogaiah et al. 2020). Nanocarriers can be engineered to target specific plant tissues or pests, enabling precise delivery of fertilizers and pesticides where they are needed most. Functionalization of nanocarrier surfaces with ligands or antibodies allows for selective binding to receptors on plant surfaces or pest cells, ensuring efficient uptake and localized action (Grillo et al. 2021).

By delivering fertilizers and pesticides directly to target sites, nanocarriers minimize off-target effects and environmental contamination. This targeted approach reduces the amount of chemicals required for effective treatment, mitigating risks to non-target organisms, soil, and water quality (Wang et al. 2022). Nanocarriers can be designed to release fertilizers and pesticides in a controlled manner, prolonging their activity and reducing the need for frequent reapplication (Kalia et al. 2020). Stimuli-responsive nanocarriers can be triggered to release their cargo in response to specific environmental cues, such as pH, temperature, or enzymatic activity, further enhancing precision and efficacy. Nanocarriers allow for the encapsulation of multiple active ingredients, enabling the development of synergistic formulations that combine fertilizers with pesticides or other crop protection agents (Kalia et al. 2020). Co-delivery of complementary compounds enhances the overall efficacy of treatments while reducing the number of separate applications required.

By encapsulating fertilizers and pesticides within nanocarriers, their solubility, stability, and bioavailability are enhanced, leading to more efficient uptake by plants or pests (Wani et al. 2019). This targeted delivery approach maximizes the efficacy of treatments while minimizing wastage and off-target effects. Agrochemicals encapsulated in nanocarriers are shielded from degradation by environmental factors such as sunlight, moisture, and microbial activity. This enhanced stability prolongs the persistence of active ingredients in the soil or on plant surfaces, reducing the need for frequent reapplication and minimizing the risk of environmental contamination (Vurro et al. 2019).

Nanocarriers minimize environmental exposure to agrochemicals by delivering them directly to target sites, such as plant roots or pest habitats. This targeted approach reduces the amount of chemicals required for effective treatment, lowering the risk of runoff, leaching, and volatilization into surrounding ecosystems. By reducing environmental exposure, nanocarriers help protect non-target organisms, soil health, and water quality (Singh et al. 2022; Verma and Saroop 2024).

Nanocarriers can be engineered to release agrochemicals in a controlled manner, providing sustained release over time. This controlled release kinetics ensure that active ingredients remain available to plants or pests for an extended period, optimizing their efficacy while minimizing potential negative impacts on the environment. Stimuli-responsive nanocarriers can further enhance control by releasing agrochemicals in response to specific environmental cues, such as soil moisture or pH levels (Machado et al. 2022; Singh et al. 2020).

Nanocarriers allow for the encapsulation of multiple active ingredients, enabling the development of synergistic formulations that combine fertilizers with pesticides or other crop protection agents. Co-delivery of complementary compounds enhances the overall efficacy of treatments while reducing the number of separate applications required, further minimizing environmental impact (Fig. 14.4).

14.7 Combatting Plant Diseases: Nanobiotechnology Approaches

Plant diseases pose significant threats to global food security, causing yield losses, reducing crop quality, and compromising agricultural sustainability. Conventional methods of disease control, including the use of chemical pesticides and breeding for disease resistance, have limitations such as environmental pollution, development of resistance in pathogens, and lengthy breeding cycles. In response to these challenges, nanobiotechnology has emerged as a promising approach for combatting plant diseases. By harnessing the unique properties of nanomaterials and leveraging biological mechanisms, nanobiotechnology offers innovative solutions for disease management in agriculture (Yadav and Yadav 2018).

Nanobiotechnology involves the application of nanotechnology principles to biological systems, enabling precise manipulation and control at the nanoscale.

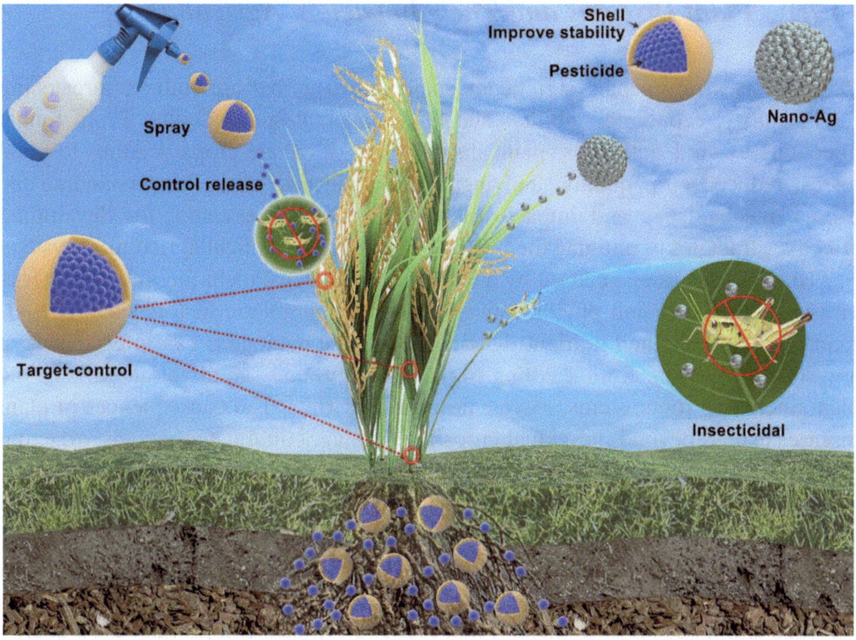

Fig. 14.4 Schematic representation of action of nanocarriers. (Reproduced from Jiang et al. 2021)

This interdisciplinary field brings together expertise in nanoscience, biology, chemistry, and engineering to develop novel strategies for disease diagnosis, monitoring, prevention, and treatment. In the context of plant diseases, nanobiotechnology offers a range of approaches that capitalize on nanomaterials' ability to interact with biological systems at the molecular level (Shah et al. 2015).

One of the key advantages of nanobiotechnology approaches is their potential for targeted delivery of bioactive compounds to plants or pathogens. Nanocarriers, such as nanoparticles and nanofibers, can encapsulate and deliver antimicrobial agents, genetic materials, or signaling molecules directly to the site of infection, minimizing off-target effects and enhancing efficacy. Additionally, nanomaterials can be functionalized with ligands or antibodies to specifically target pathogens or host plant cells, further increasing precision and efficiency (Martínez-Ballesta et al. 2018).

Nanobiotechnology also enables the development of sensitive and rapid diagnostic tools for detecting plant pathogens. Nanoscale biosensors and imaging techniques can detect pathogen-specific biomarkers or nucleic acids with high sensitivity and specificity, facilitating early disease detection and monitoring. Furthermore, nanomaterial-based formulations can enhance the stability and bioavailability of diagnostic reagents, improving their performance in field conditions (Khiyami et al. 2014; Nesakumar et al. 2021).

14.7.1 Plant Diseases and Their Impact on Agriculture

Plant diseases represent a significant challenge to global agriculture, threatening food security, economic stability, and environmental sustainability. From fungal infections to viral outbreaks, plant diseases can devastate crops, leading to yield losses, reduced crop quality, and increased production costs. Understanding the causes, mechanisms, and impacts of plant diseases is essential for developing effective strategies to mitigate their effects and ensure a resilient agricultural system.

Plant diseases can be caused by a variety of pathogens, including fungi, bacteria, viruses, nematodes, and phytoplasmas, as well as by abiotic factors such as environmental stressors and nutrient deficiencies. These pathogens can infect plants at any stage of growth, from seedling to mature plant, and can affect all parts of the plant, including roots, stems, leaves, flowers, and fruits. The consequences of plant diseases extend beyond individual crops to entire agricultural ecosystems, affecting soil health, biodiversity, and ecosystem services.

The impact of plant diseases on agriculture is multifaceted and far-reaching. Yield losses due to plant diseases can range from moderate to severe, depending on factors such as pathogen virulence, host susceptibility, and environmental conditions. In some cases, entire crops may be destroyed, leading to economic hardship for farmers and food shortages for consumers. Plant diseases can also compromise crop quality by affecting factors such as taste, appearance, and nutritional value, reducing marketability and profitability.

In addition to direct economic losses, plant diseases can have indirect impacts on agricultural productivity and sustainability. Farmers may be forced to increase their use of chemical pesticides and fungicides to control disease outbreaks, leading to environmental pollution, ecosystem disruption, and human health risks. Furthermore, the development of resistance in pathogens to chemical treatments poses ongoing challenges for disease management and control.

14.7.2 Nanotechnology-Based Strategies for Disease Detection

Nanotechnology-based strategies offer innovative solutions for detecting plant diseases with high sensitivity, specificity, and efficiency. These strategies leverage the unique properties of nanomaterials to develop sensitive biosensors, imaging techniques, and diagnostic assays for rapid and accurate detection of plant pathogens and disease biomarkers. Here are some key nanotechnology-based approaches for disease detection in plants:

14.7.2.1 Nanoparticle-Based Biosensors

Nanoparticle-based biosensors are versatile platforms that utilize the optical, electrical, or magnetic properties of nanoparticles to detect specific biomolecules associated with plant diseases. Functionalized nanoparticles can bind to pathogen-specific biomarkers, such as nucleic acids, proteins, or metabolites, generating detectable signals that indicate the presence of the pathogen. Nanoparticle-based biosensors offer high sensitivity, real-time detection, and multiplexing capabilities, making them suitable for on-site disease diagnosis and monitoring (Koedrith et al. 2015).

14.7.2.2 Nanoparticle-Enabled Imaging

Nanoparticles can serve as contrast agents for various imaging modalities, including fluorescence microscopy, magnetic resonance imaging (MRI), and surface-enhanced Raman spectroscopy (SERS). Functionalized nanoparticles can selectively target and label pathogen-infected tissues or cells, enabling non-invasive visualization and localization of plant diseases in vivo. Nanoparticle-enabled imaging techniques provide high spatial resolution, deep tissue penetration, and real-time monitoring capabilities, facilitating early disease detection and characterization (Balabanova et al. 2020).

14.7.2.3 Nanopore-Based Sensors

Nanopore sensors utilize nanoscale pores to detect changes in electrical conductivity caused by the passage of biomolecules through the pore. Functionalized nanopores can selectively capture and identify pathogen-specific nucleic acids or proteins, enabling rapid and label-free detection of plant pathogens. Nanopore-based sensors offer advantages such as high sensitivity, single-molecule detection, and compatibility with portable devices, making them suitable for point-of-care diagnostics and field-based disease surveillance (Shi et al. 2017; Baldi and La Porta 2020).

14.7.2.4 Nanomaterial-Enhanced Assays

Nanomaterials can be integrated into existing diagnostic assays, such as polymerase chain reaction (PCR), enzyme-linked immunosorbent assay (ELISA), and lateral flow immunoassay (LFIA), to enhance their performance (Sadeghi et al. 2021). Nanoparticles can improve the sensitivity, specificity, and speed of assays by increasing the signal-to-noise ratio, amplifying the detection signal, or facilitating signal readout. Nanomaterial-enhanced assays offer potential applications in high-throughput screening, multiplexed detection, and miniaturized diagnostic platforms for plant disease detection (Ranjan Srivastava et al. 2023).

14.7.2.5 Nanofiber-Based Devices

Nanofiber-based devices, such as electrospun nanofiber mats or paper-based sensors, offer simple and cost-effective platforms for plant disease detection (Iftikhar et al. 2023). Functionalized nanofibers can selectively capture and concentrate pathogen-specific biomolecules from plant samples, enabling sensitive and rapid detection using colorimetric, electrochemical, or fluorescence-based assays. Nanofiber-based devices are portable, disposable, and easy to use, making them suitable for on-site diagnostics in resource-limited settings.

14.8 Revolutionizing Soil Health: Nanoremediation Techniques

Contaminated soil and groundwater, especially in industrialized and urban areas, is a widespread problem that presents extreme risks to the environment and humans (Cheng et al. 2016; Thomé et al. 2015). Numerous studies have focused on the remediation of soil, groundwater, wastewater, and landfill leachate polluted by various contaminants (Mateas et al. 2017; Rajan 2011). Soil and groundwater remediation can be broadly classified according to the place of remediation, which can be ex situ or in situ. For ex situ remediation, the polluted soil or groundwater is recovered from the subsurface and treated on the same site or transferred to another site for treatment (Tsai et al. 2008). In contrast, in situ remediation is when the contaminated soil or groundwater is remediated directly in the subsurface. The in situ remediation process is often preferred because it is cheaper than the ex situ remediation process (Thomé et al. 2015; Honetschlägerová et al. 2019).

Reducing pollution to a desirable and safe level is the main target of soil and groundwater remediation processes. Physical, chemical, and biological technologies have been used to achieve this goal for soil and groundwater remediation. In general, several factors play a significant role in the selection of the optimal soil and groundwater remediation, including soil properties and contaminants and the nature of selected and designed remediation technology (Liu et al. 2021a, b).

14.8.1 Relationship Between Soil and Groundwater: Contaminants and Remediation

Soil and groundwater are susceptible to pollution by a wide array of pollutants such as petroleum hydrocarbon, chlorinated solvents, and heavy metals (Galdames et al. 2017) Selecting a proper remediation technology for a contaminated environment usually depends on contaminant characteristics and contaminated site characteristics such as physical, chemical, and biological properties. All these factors should be considered during the remediation process, design, and implication. Moreover, the time/cost constraints, the regulatory requirements, and the remediation mechanisms should be considered in the selection process.

Nevertheless, adopting risk-based management approaches is increasingly a focus of environmental researchers due to the high demand for sustainable responses to environmental pollutions (Karn et al. 2009; Thavamani et al. 2015). The polluted environments are usually surface water, sediments, soil, and groundwater, which are mainly contaminated with low and high molecular weight petroleum hydrocarbon compounds, semi-volatile organic compounds, volatile organic compounds, polycyclic aromatic hydrocarbons (PAHs), polychlorinated biphenyls, persistent organic pollutants, organochlorinated pesticides, NAPL, hydrophobic organic compounds, heavy metals, and xenobiotics (Anyika et al. 2015; Shayegan et al. 2020; Moreno-Sader et al. 2019). These pollutants may migrate or spread far from the source and seriously affect flora, fauna, and the ecosystem (Thavamani et al. 2015). Managing the polluted environments requires the selection of the proper remediation technology for the pollutants, destruction, and separation methods according to many ex situ and in situ remediation methods for surface water, sediments, soil, and groundwater comprising physicochemical, biological, chemical, thermal, electromagnetic, electric, and ultrasonic remediation technologies (Cameselle et al. 2013; Rodrigo et al. 2014; Chang et al. 2016). Remediation in an aqueous environment includes remediation of groundwater and surface water polluted by contaminants, whereas soil remediation includes remediation of sediment subsoil and topsoil polluted by contaminates (Camenzuli et al. 2013). Soil and groundwater remediation could be conducted separately or together, depending on the concentration of contaminants and the extent of pollution. The efficiency of remediation technology depends on the design and implication based on the characteristics of polluted soil and the remediation technique. Combining one remediation technology with others sequentially or simultaneously may enhance the overall remediation process through combined or synergistic effects (Ossai et al. 2020).

14.8.2 Nanoremediation Approaches for Soil Improvement

Nanoparticles offer distinct advantages over traditional soil remediation methods, primarily because of their size and surface area (Sathish et al. 2024).This small particle size enables remarkable efficacy in nanotechnology applications. This explores the utilization of nanotechnology in phytoremediation and explores its applications and future prospects. Various types of nanoparticles are effective in cleansing and detoxifying diverse pollutants, as discussed comprehensively herein. Ongoing research and interdisciplinary collaboration promise further research in this intriguing field. Although planetary reclamation remains challenging, nanophytoremediation has emerged as a promising solution to address these environmental issues. Soil pollution and degradation pose urgent global environmental challenges that impact agricultural productivity, food security, and human well-being (Gomiero 2016). The depletion of soil resources owing to escalating food production demands for the growing human population has contributed to widespread soil exploitation and deterioration (Gomiero 2016). Additionally, soil contamination with heavy metals, pesticides, and persistent organic pollutants (POPs) has intensified this crisis. Polluted soil containing these substances increases the risk of contaminating the food chain through the bioaccumulation of pollutants (Alengebawy et al. 2021).

Nanotechnology applications are cost-effective and user-friendly and offer efficient treatment and remediation approaches to significantly mitigate soil pollution (Ibrahim et al. 2016). This explores the potential of nanotechnology-based soil remediation, specifically addressing heavy metals, pesticides, their residues, and POPs while also examining their role in enhancing phytoremediation and bioremediation (Tufail et al. 2022) (Fig. 14.5).

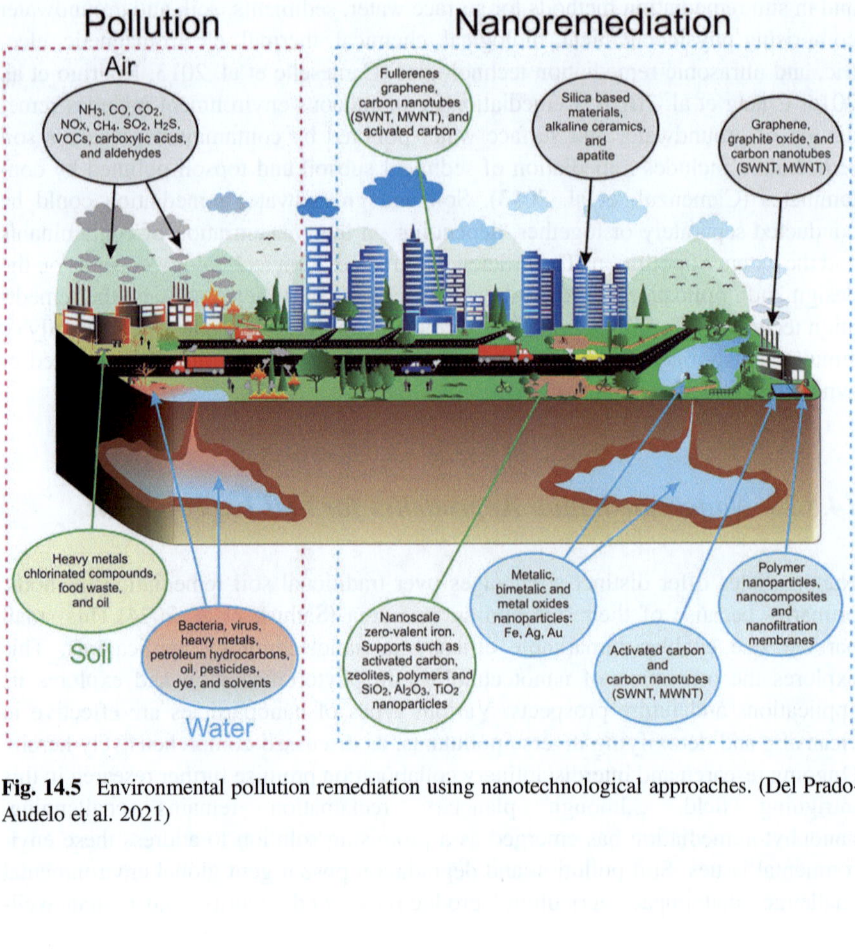

Fig. 14.5 Environmental pollution remediation using nanotechnological approaches. (Del Prado-Audelo et al. 2021)

14.9 Sustainability and Safety Concerns: Ethical and Environmental Implications of Nanoagriculture

Nano-agriculture involves the application of nanotechnology in agricultural practices to enhance crop production, improve soil health, and reduce the use of chemical inputs (Manjunatha et al. 2016). While it holds significant promise for advancing agricultural productivity and sustainability, there are several ethical and environmental implications that warrant careful consideration. Below are key sustainability and safety concerns associated with nano-agriculture:

14.9.1 Environmental Implications

- **Persistence in Soil**: Nanoparticles (NPs) can persist in the soil environment, potentially disrupting microbial communities and soil health. This persistence could affect nutrient cycling and soil fertility over time (Kalwani et al. 2022).
- **Bioaccumulation**: NPs can accumulate in plants and, subsequently, in the food chain, leading to potential toxicity in animals and humans. Long-term effects of bioaccumulation are not fully understood, raising concerns about ecological and human health (Murali et al. 2022).
- **Runoff and Leaching**: NPs used in fertilizers, pesticides, and soil conditioners can leach into groundwater or be carried by runoff into water bodies. This can lead to contamination of water resources, affecting aquatic ecosystems and potentially entering drinking water supplies (Rashmi et al. 2020).
- **Biodiversity**: NPs might adversely affect non-target organisms such as beneficial insects, soil fauna, and microorganisms (Zhang and Goss 2022). These impacts can disrupt ecosystems and reduce biodiversity, which is essential for ecosystem resilience and function.

14.9.2 Safety Concerns

- **Exposure Routes**: Humans can be exposed to NPs through inhalation, ingestion, or dermal contact. The health effects of chronic exposure to low levels of NPs are not well understood, necessitating comprehensive risk assessments (Sufian et al. 2017).
- **Toxicity**: Some NPs may exhibit toxic properties, leading to potential health risks such as respiratory issues, inflammatory responses, and even cellular damage (Ajdary et al. 2018).
- **Lack of Comprehensive Regulation**: The rapid development of nano-agriculture technologies has outpaced regulatory frameworks (Wernette and Nilsen 2008). There is a need for robust regulations to ensure the safe use of

NPs in agriculture, including standardized testing for toxicity and environmental impact.
- **Labeling and Transparency**: Lack of clear labeling of nano-enabled agricultural products can hinder informed decision-making by consumers and farmers, raising ethical concerns about transparency and consent (Cummings et al. 2021).

14.9.3 Ethical Implications

The major ethical implications are related to economic disparities and global inequities. The high cost of developing and implementing nanotechnology in agriculture might widen the gap between large agribusinesses and smallholder farmers, exacerbating economic disparities and access to technology (Sabourin 2015). Developing countries may face challenges in adopting and regulating nano-agriculture, leading to potential exploitation and uneven distribution of benefits and risks (Pandey 2018).

14.9.4 Informed Consent and Public Perception

There is often a lack of public awareness and understanding of nano-agriculture, leading to mistrust and resistance (Yadav et al. 2023a, b). Ensuring informed consent and engaging with the public about the benefits and risks is crucial for ethical deployment. Companies and researchers must maintain transparency regarding the development, testing, and use of NPs in agriculture to build public trust and address ethical concerns (Srivastava 2023). Communities located near agricultural areas using nanotechnology may face disproportionate exposure to potential risks, raising issues of environmental justice and the need for protective measures (Sandler 2007).

14.10 Future Directions: Nanotechnology's Role in the Evolution of Agriculture

Nanotechnology's role in agriculture is poised to expand significantly, offering transformative solutions to some of the most pressing challenges in the sector. An overview of nanotechnology application in agricultural systems is given in Fig. 14.6. Utilizing nanomaterials to develop fertilizers, pesticides, and herbicides that release their active ingredients in response to specific environmental conditions or plant needs (Ditta et al. 2015). This can optimize resource use and minimize environmental impact. Notably, applying nanocarriers as controlled release systems (CRSs) for agrochemicals can overcome the limitations of conventional products. A CRS for agrochemicals is an eco-friendly strategy for the ecosystem and human health

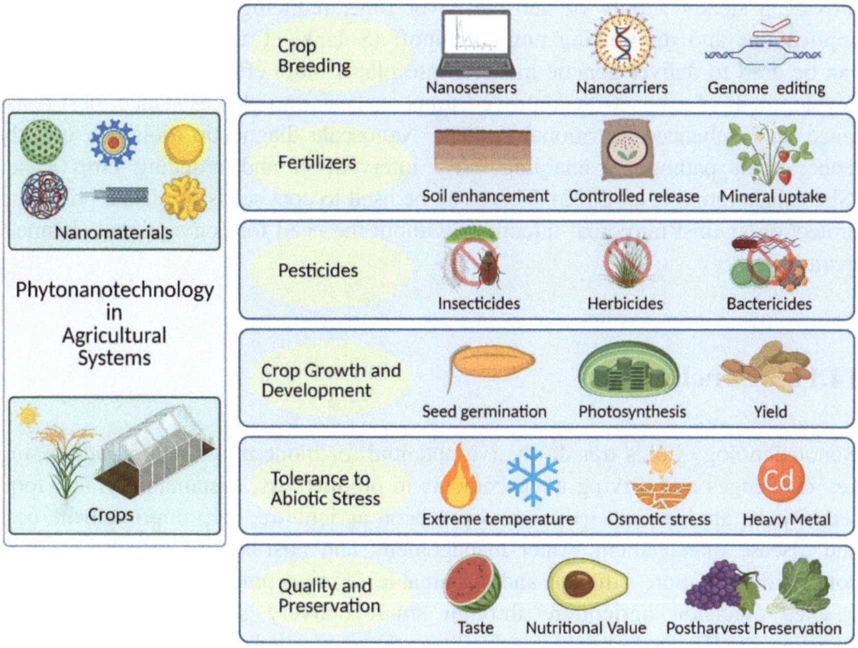

Fig. 14.6 Overview of nanotechnology applications in agriculture. (Reproduced from Jiang et al. 2021)

(Fincheira et al. 2023). Nanocarriers can deliver nutrients or pesticides directly to specific parts of the plant or pest, increasing effectiveness and reducing the quantity needed.

These tiny sensors can be embedded in soil or attached to plants to monitor various parameters such as moisture levels, nutrient status, and the presence of pathogens in real-time. This data can help farmers make informed decisions and apply inputs more precisely (Yin et al. 2021). Integrating nanosensors into wireless networks can provide comprehensive and real-time data on crop health and environmental conditions, enhancing precision farming practices.

Sustainable Practices can be encouraged by the application of nanomaterials in agriculture. Using nanoparticles to improve soil structure, enhance nutrient retention, and increase water-holding capacity. For example, nano-clays can help retain water and nutrients in sandy soils (Chen et al. 2022). Nanomaterials can be used to detoxify soils contaminated with heavy metals or organic pollutants, thus improving soil health and making it suitable for agriculture again. Nanosensors can monitor soil moisture levels with high precision, enabling more efficient irrigation practices and reducing water waste (Awasthi et al. 2022). Nanotechnology-based filtration systems can remove contaminants from irrigation water, ensuring crops receive clean water and reducing the risk of disease.

Nanofertilizers can provide nutrients in a form that is more easily absorbed by plants, leading to better growth and higher yields with less fertilizer input. These can provide a steady supply of nutrients over time, reducing the need for frequent applications and minimizing nutrient runoff (Solanki et al. 2015). Nanoparticles can be used to deliver genetic material into plants more efficiently, facilitating the development of crops with improved traits such as drought resistance, pest resistance, and enhanced nutritional content. Nanoscale diagnostic tools can quickly detect plant pathogens, enabling early intervention and reducing crop losses (Shivashakarappa et al. 2022). These can be used to coat seeds or plants, providing protection against microbial infections without the need for conventional chemical treatments.

14.11 Conclusion

Nanotechnology holds transformative potential for modern agriculture, addressing key challenges and driving advancements in productivity, sustainability, and food security. Its applications span across precision agriculture, crop improvement, pest and disease management, water management, and post-harvest processing, each contributing to more efficient and sustainable farming practices. Nanotechnology enables precision agriculture through smart delivery systems and advanced nanosensors. These technologies facilitate the controlled and targeted release of agricultural inputs, optimizing their use and minimizing environmental impact. Real-time monitoring of soil and crop health through nanosensors allows for precise management decisions, leading to better resource utilization and increased crop yields. Nanotechnology contributes to sustainable agriculture by improving soil health and water management. Nano-enhanced soil amendments and pollutant remediation techniques restore and maintain soil fertility. Efficient irrigation systems and water purification technologies ensure the judicious use of water resources, crucial in water-scarce regions. These advancements support long-term agricultural productivity while safeguarding natural ecosystems.

The development of nanofertilizers and advanced gene delivery systems can revolutionize crop improvement. Nanofertilizers enhance nutrient uptake and provide a steady nutrient supply, reducing the need for frequent applications and minimizing runoff. Nanotechnology-facilitated genetic engineering accelerates the development of crops with desirable traits such as drought resistance and improved nutritional content, contributing to food security and resilience against climate change. Nanopesticides and rapid diagnostic tools offer more effective and environmentally friendly solutions for pest and disease management. Nanopesticides improve pest control efficacy at lower doses, reducing chemical use and environmental contamination. Early disease detection and antimicrobial nanomaterials help protect crops from pathogens, minimizing crop losses and the reliance on conventional chemical treatments.

The integration of nanotechnology into modern agriculture presents a pathway to more sustainable, efficient, and resilient farming systems. However, it is crucial to address potential environmental and health risks through comprehensive research and robust regulatory frameworks. Ensuring equitable access to nanotechnology innovations will also be key to maximizing their benefits globally.

Collaboration among scientists, policymakers, industry stakeholders, and farmers will be essential to realize the full potential of nanotechnology in agriculture. By fostering innovation and ensuring responsible use, nanotechnology can play a pivotal role in addressing the global challenges of food security, environmental sustainability, and climate resilience, paving the way for the future of agriculture.

References

Abdelmaksoud NM, El-Bakry AM, Abdel-Aziz NF, Sammour EA, Salem HA-N (2024) Potency of emulsifiable concentrate and nanoemulsion formulations as green insecticides against two insects, Aphis craccivora and Liriomyza trifolii. Ind Crop Prod 208:117854

Ahmar S, Mahmood T, Fiaz S, Mora-Poblete F, Shafique MS, Chattha MS, Jung K-H (2021) Advantage of nanotechnology-based genome editing system and its application in crop improvement. Front Plant Sci 12:663849

Ajdary M, Moosavi MA, Rahmati M, Falahati M, Mahboubi M, Mandegary A, Jangjoo S, Mohammadinejad R, Varma RS (2018) Health concerns of various nanoparticles: a review of their in vitro and in vivo toxicity. Nanomaterials 8(9):634

Alengebawy A, Abdelkhalek ST, Qureshi SR, Wang MQ (2021) Heavy metals and pesticides toxicity in agricultural soil and plants: Ecological risks and human health implications. Toxics 9(3):42

Anyika C, Majid ZA, Ibrahim Z, Zakaria MP, Yahya A (2015) The impact of biochars on sorption and biodegradation of polycyclic aromatic hydrocarbons in soils—a review. Environ Sci Pollut Res 22:3314–3341

Awasthi G, Nagar V, Mandzhieva S, Minkina T, Sankhla MS, Pandit PP, Aseri V et al (2022) Sustainable amelioration of heavy metals in soil ecosystem: existing developments to emerging trends. Fortschr Mineral 12(1):85

Babu PJ, Saranya S, Longchar B, Rajasekhar A (2022) Nanobiotechnology-mediated sustainable agriculture and post-harvest management. Curr Res Biotechnol 4:326–336

Balabanova, Biljana, Trajče Stafilov, and Robert Sajn. "Tracking silent geochemical markers for polymetallic pollution in area with Cu-Pb-Zn minerals exploitation." (2020)

Baldi P, La Porta N (2020) Molecular approaches for low-cost point-of-care pathogen detection in agriculture and forestry. Front Plant Sci 11:570862

Baldin M, Breunig T, Cue R, De Vries A, Doornink M, Drevenak J, Fourdraine R et al (2021) Integrated decision support systems (IDSS) for dairy farming: a discussion on how to improve their sustained adoption. Animals 11(7):2025

Banerjee A, Sarkar A, Acharya K, Chakraborty N (2021) Nanotechnology: an emerging hope in crop improvement. Lett Appl Nano BioSci 10(4):2784–2803

Boursianis AD, Papadopoulou MS, Diamantoulakis P, Liopa-Tsakalidi A, Barouchas P, Salahas G, Karagiannidis G, Wan S, Goudos SK (2022) Internet of things (IoT) and agricultural unmanned aerial vehicles (UAVs) in smart farming: a comprehensive review. Internet of Things 18:100187

Camenzuli D, Freidman BL, Statham TM, Mumford KA, Gore DB (2013) On-site and in situ remediation technologies applicable to metal-contaminated sites in Antarctica and the Arctic: a review. Polar Res 32(1):21522

Cameselle C, Chirakkara RA, Reddy KR (2013) Electrokinetic-enhanced phytoremediation of soils: status and opportunities. Chemosphere 93(4):626–636

Chang K-S, Lo W-H, Lin W-M, Wen J-X, Yang S-C, Huang C-J, Hsieh H-Y (2016) Microwave-assisted thermal remediation of diesel contaminated soil. Eng J 20(4):93–100

Cheema MJM, Khan MA (2019) Information technology for sustainable agriculture. In: Innovations in sustainable agriculture. Springer International Publishing, pp 585–597

Chen X, Duan M, Zhou B, Cui L (2022) Effects of biochar nanoparticles as a soil amendment on the structure and hydraulic characteristics of a sandy loam soil. Soil Use Manag 38(1):836–849

Cheng M, Zeng G, Huang D, Lai C, Piao X, Zhang C, Liu Y (2016) Hydroxyl radicals based advanced oxidation processes (AOPs) for remediation of soils contaminated with organic compounds: a review. Chem Eng J 284:582–598

Cronan CS, Grigal DF (1995) Use of calcium/aluminum ratios as indicators of stress in forest ecosystems. J Environ Qual 24(2):209–226

Cummings CL, Kuzma J, Kokotovich A, Glas D, Grieger K (2021) Barriers to responsible innovation of nanotechnology applications in food and agriculture: a study of US experts and developers. NanoImpact 23:100326

Dapkekar A, Deshpande P, Oak MD, Paknikar KM, Rajwade JM (2018) Zinc use efficiency is enhanced in wheat through nanofertilization. Sci Rep 8(1):6832

Del Prado-Audelo ML, García Kerdan I, Escutia-Guadarrama L, Reyna-González JM, Magaña JJ, Leyva-Gómez G (2021) Nanoremediation: nanomaterials and nanotechnologies for environmental cleanup. Front Environ Sci 9:793765

Demirer GS, Zhang H, Goh NS, González-Grandío E, Landry MP (2019) Carbon nanotube–mediated DNA delivery without transgene integration in intact plants. Nat Protoc 14(10): 2954–2971

Dhanapal AR, Thiruvengadam M, Vairavanathan J, Venkidasamy B, Easwaran M, Ghorbanpour M (2024) Nanotechnology approaches for the remediation of agricultural polluted soils. ACS Omega 9(12):13522–13533

Díez-Pascual AM (2021) Carbon-based nanomaterials. Int J Mol Sci 22(14):7726

Ditta A, Arshad M, Ibrahim M (2015) Nanoparticles in sustainable agricultural crop production: applications and perspectives. In: Nanotechnology and plant sciences: nanoparticles and their impact on plants. Springer International Publishing, pp 55–75

Dolez PI (2015) Nanomaterials definitions, classifications, and applications. In: Nanoengineering. Elsevier, pp 3–40

Elemike EE, Uzoh IM, Onwudiwe DC, Babalola OO (2019) The role of nanotechnology in the fortification of plant nutrients and improvement of crop production. Appl Sci 9(3):499

El-Saadony MT, Saad AM, Soliman SM, Salem HM, Desoky E-SM, Babalghith AO, El-Tahan AM et al (2022) Role of nanoparticles in enhancing crop tolerance to abiotic stress: a comprehensive review. Front Plant Sci 13:946717

Erdoğar N, Akkın S, Bilensoy E (2018) Nanocapsules for drug delivery: an updated review of the last decade. Recent Pat Drug Deliv Formul 12(4):252–266

Fincheira P, Hoffmann N, Tortella G, Ruiz A, Cornejo P, Diez MC, Seabra AB, Benavides-Mendoza A, Rubilar O (2023) Eco-efficient systems based on nanocarriers for the controlled release of fertilizers and pesticides: toward smart agriculture. Nanomaterials 13(13):1978

Galdames A, Mendoza A, Orueta M, de Soto García IS, Sánchez M, Virto I, Vilas JL (2017) Development of new remediation technologies for contaminated soils based on the application of zero-valent iron nanoparticles and bioremediation with compost. Resour Effic Technol 3(2): 166–176

Giraldo JP, Honghong W, Newkirk GM, Kruss S (2019) Nanobiotechnology approaches for engineering smart plant sensors. Nat Nanotechnol 14(6):541–553

Gomiero T (2016) Soil degradation, land scarcity and food security: reviewing a complex challenge. Sustain For 8(3):281

Grillo R, Mattos BD, Antunes DR, Forini MML, Monikh FA, Rojas OJ (2021) Foliage adhesion and interactions with particulate delivery systems for plant nanobionics and intelligent agriculture. Nano Today 37:101078

Guía-García JL, Charles-Rodríguez AV, Humberto Reyes-Valdés M, Ramírez-Godina F, Robledo-Olivo A, García-Osuna HT, Cerqueira MA, Flores-López ML (2022) Micro and nanoencapsulation of bioactive compounds for agri-food applications: a review. Ind Crop Prod 186:115198

Honetschlägerová L, Martinec M, Škarohlíd R (2019) Coupling in situ chemical oxidation with bioremediation of chloroethenes: a review. Rev Environ Sci Biotechnol 18:699–714

Ibrahim RK, Hayyan M, AlSaadi MA, Hayyan A, Ibrahim S (2016) Environmental application of nanotechnology: air, soil, and water. Environ Sci Pollut Res 23:13754–13788

Iftikhar FJ, Shah A, Wali Q, Kokab T (2023) Advancements in nanofiber-based electrochemical biosensors for diagnostic applications. Biosensors 13(4):416

Jalal A, da Silva Oliveira CE, Fernandes GC, da Silva EC, da Costa KN, de Souza JS, da Silva Leite G, Biagini ALC, Galindo FS, Filho MCMT (2023) Integrated use of plant growth-promoting bacteria and nano-zinc foliar spray is a sustainable approach for wheat biofortification, yield, and zinc use efficiency. Front Plant Sci 14:1146808

Javanmardi Z, Saba MK, Nourbakhsh H, Amini J (2023) Efficiency of nanoemulsion of essential oils to control Botrytis cinerea on strawberry surface and prolong fruit shelf life. Int J Food Microbiol 384:109979

Javed T, Shabbir R, Hussain S, Naseer MA, Ejaz I, Ali MM, Ahmar S, Yousef AF (2022) Nanotechnology for endorsing abiotic stresses: a review on the role of nanoparticles and nanocompositions. Funct Plant Biol 50:831–849

Jiang M, Song Y, Kanwar MK et al (2021) Phytonanotechnology applications in modern agriculture. J Nanobiotechnol 19:430. https://doi.org/10.1186/s12951-021-01176-w

Jogaiah S, Singh HB, Fraceto LF, De Lima R (eds) (2020) Advances in nano-fertilizers and nano-pesticides in agriculture: a smart delivery system for crop improvement. Woodhead Publishing

John SA, Chattree A, Ramteke PW, Shanthy P, Nguyen TA, Rajendran S (2022) Nanosensors for plant health monitoring. In: Nanosensors for smart agriculture. Elsevier, pp 449–461

Kalia A, Sharma SP, Kaur H, Kaur H (2020) Novel nanocomposite-based controlled-release fertilizer and pesticide formulations: prospects and challenges. In: Multifunctional hybrid nanomaterials for sustainable agri-food and ecosystems, pp 99–134

Kalwani M, Chakdar H, Srivastava A, Pabbi S, Shukla P (2022) Effects of nanofertilizers on soil and plant-associated microbial communities: emerging trends and perspectives. Chemosphere 287:132107

Karn B, Kuiken T, Otto M (2009) Nanotechnology and in situ remediation: a review of the benefits and potential risks. Environ Health Perspect 117(12):1813–1831

Kashyap PL, Xiang X, Heiden P (2015) Chitosan nanoparticle based delivery systems for sustainable agriculture. Int J Biol Macromol 77:36–51

Kataria S, Jain M, Rastogi A, Živčák M, Brestic M, Liu S, Tripathi DK (2019) Role of nanoparticles on photosynthesis: avenues and applications. In: Nanomaterials in plants, algae and microorganisms. Academic, pp 103–127

Kaushal M, Wani SP (2017) Nanosensors: frontiers in precision agriculture. In: Nanotechnology: an agricultural paradigm. Springer Singapore, pp 279–291

Khanna V (2016) Nanosensors. CRC Press

Khiyami MA, Almoammar H, Awad YM, Alghuthaymi MA, Abd-Elsalam KA (2014) Plant pathogen nanodiagnostic techniques: forthcoming changes? Biotechnol Biotechnol Equip 28(5):775–785

Kim D-Y, Kadam A, Shinde S, Saratale RG, Patra J, Ghodake G (2018) Recent developments in nanotechnology transforming the agricultural sector: a transition replete with opportunities. J Sci Food Agric 98(3):849–864

Koedrith P, Thasiphu T, Weon J-I, Boonprasert R, Tuitemwong K, Tuitemwong P (2015) Recent trends in rapid environmental monitoring of pathogens and toxicants: potential of nanoparticle-based biosensor and applications. Sci World J 2015:510982

Kordrostami M, Mafakheri M, Al-Khayri JM (2021) Contributions of nano biosensors in managing environmental plant stress under climatic changing era. In: Nanobiotechnology: mitigation of abiotic stress in plants, pp 117–137

Kwak S-Y, Wong MH, Lew TTS, Bisker G, Lee MA, Kaplan A, Dong J et al (2017) Nanosensor technology applied to living plant systems. Annu Rev Anal Chem 10:113–140

Liu S, Wang X, Guo G, Yan Z (2021a) Status and environmental management of soil mercury pollution in China: a review. J Environ Manag 277:111442

Liu C, Zhou H, Zhou J (2021b) The applications of nanotechnology in crop production. Molecules 26:7070. https://doi.org/10.3390/molecules26237070

Maali A, Hamed Mosavian MT (2013) Preparation and application of nanoemulsions in the last decade (2000–2010). J Dispers Sci Technol 34(1):92–105

Machado TO, Grabow J, Sayer C, de Araújo PHH, Ehrenhard ML, Wurm FR (2022) Biopolymer-based nanocarriers for sustained release of agrochemicals: a review on materials and social science perspectives for a sustainable future of agri-and horticulture. Adv Colloid Interf Sci 303: 102645

Mahanta N, Dambale A, Rajkhowa M, Mahanta C, Mahanta N (2019) Nutrient use efficiency through nano fertilizers. Int J Chem Stud 7(3):2839–2842

Mandal A, Sarkar B, Mandal S, Vithanage M, Patra AK, Manna MC (2020a) Impact of agrochemicals on soil health. In: Agrochemicals detection, treatment and remediation. Butterworth-Heinemann, pp 161–187

Mandal N, Adhikary S, Rakshit R (2020b) Nanobiosensors: recent developments in soil health assessment. In: Soil analysis: recent trends and applications, pp 285–304

Manjunatha SB, Biradar DP, Aladakatti YR (2016) Nanotechnology and its applications in agriculture: a review. J Farm Sci 29(1):1–13

Martínez-Ballesta MC, Gil-Izquierdo Á, García-Viguera C, Domínguez-Perles R (2018) Nanoparticles and controlled delivery for bioactive compounds: outlining challenges for new "smart-foods" for health. Food Secur 7(5):72

Masurkar SA, Ghewari PG, Pathade AG (2024) Innovations in food technology: enhancing flavor, nutrition, and safety. Naturalista Campano 28(1):212–219

Mateas DJ, Tick GR, Carroll KC (2017) In situ stabilization of NAPL contaminant source-zones as a remediation technique to reduce mass discharge and flux to groundwater. J Contam Hydrol 204:40–56

Matthews RB, Rivington M, Muhammed S, Newton AC, Hallett PD (2013) Adapting crops and cropping systems to future climates to ensure food security: the role of crop modelling. Glob Food Sec 2(1):24–28

Monteiro A, Santos S, Gonçalves P (2021) Precision agriculture for crop and livestock farming—brief review. Animals 11(8):2345

Moreno-Sader K, García-Padilla A, Realpe A, Acevedo-Morantes M, Soares JBP (2019) Removal of heavy metal water pollutants (Co2+ and Ni2+) using polyacrylamide/sodium montmorillonite (PAM/Na-MMT) nanocomposites. ACS Omega 4(6):10834–10844

Murali M, Gowtham HG, Brijesh Singh S, Shilpa N, Aiyaz M, Alomary MN, Alshamrani M et al (2022) Fate, bioaccumulation and toxicity of engineered nanomaterials in plants: current challenges and future prospects. Sci Total Environ 811:152249

Nesakumar N, Lakshmanakumar M, Srinivasan S, Jbb AJ, Rayappan JBB (2021) Principles and recent advances in biosensors for pathogens detection. ChemistrySelect 6(37):10063–10091

Ossai IC, Ahmed A, Hassan A, Hamid FS (2020) Remediation of soil and water contaminated with petroleum hydrocarbon: a review. Environ Technol Innov 17:100526

Pandey G (2018) Challenges and future prospects of agri-nanotechnology for sustainable agriculture in India. Environm Technol Innov 11:299–307

Pawase PP, Nalawade SM, Bhanage GB, Walunj AA, Kadam PB, Durgude AG, Patil MG (2023) Variable rate fertilizer application technology for nutrient management: a review. Int J Agric Biol Eng 16(4):11–19

Prasad R, Bhattacharyya A, Nguyen QD (2017) Nanotechnology in sustainable agriculture: recent developments, challenges, and perspectives. Frontiers in microbiology 8:1014

Qin H, Zhou X, Danfei G, Li L, Kan C (2019) Preparation and characterization of a novel waterborne lambda-cyhalothrin/alkyd nanoemulsion. J Agric Food Chem 67(38):10587–10594

Rajan CS (2011) Nanotechnology in groundwater remediation. Int J Environ Sci Dev 2(3):182

Ranjan Srivastava V, Kumari R, Chandra P (2023) Miniaturized surface engineered technologies for multiplex biosensing devices. Electroanalysis 35(8):e202200355

Rashmi I, Roy T, Kartika KS, Pal R, Coumar V, Kala S, Shinoji KC (2020) Organic and inorganic fertilizer contaminants in agriculture: impact on soil and water resources. In: Contaminants in agriculture: sources, impacts and management. Springer International Publishing, pp 3–41

Rehman A, Feng J, Qunyi T, Korma SA, Assadpour E, Usman M, Han W, Jafari SM (2021) Pesticide-loaded colloidal nanodelivery systems; preparation, characterization, and applications. Adv Colloid Interf Sci 298:102552

Rodrigo MA, Oturan N, Oturan MA (2014) Electrochemically assisted remediation of pesticides in soils and water: a review. Chem Rev 114(17):8720–8745

Sabourin V (2015) Commercial opportunities and market demand for nanotechnologies in agribusiness sector. J Technol Manag Innov 10(1):40–51

Sadeghi P, Sohrabi H, Hejazi M, Jahanban-Esfahlan A, Baradaran B, Tohidast M, Majidi MR, Mokhtarzadeh A, Tavangar SM, de la Guardia M (2021) Lateral flow assays (LFA) as an alternative medical diagnosis method for detection of virus species: the intertwine of nanotechnology with sensing strategies. TrAC Trends Anal Chem 145:116460

Sandler R (2007) Nanotechnology and social context. Bull Sci Technol Soc 27(6):446–454

Sathish T, Ahalya N, Thirunavukkarasu M, Senthil TS, Zakir Hussain M, Siddiqui IH, Panchal H, Sadasivuni KK (2024) A comprehensive review on the novel approaches using nanomaterials for the remediation of soil and water pollution. Alex Eng J 86:373–385

Shah F, Wei W (2019) Soil and crop management strategies to ensure higher crop productivity within sustainable environments. Sustain For 11(5):1485

Shah SP, Hou P, Cheng X (2015) Durability of cement-based materials and nano-particles: a review. In: Nanotechnology in Construction: Proceedings of NICOM5, pp 15–24

Shayegan H, Ali GAM, Safarifard V (2020) Recent progress in the removal of heavy metal ions from water using metal-organic frameworks. ChemistrySelect 5(1):124–146

Shi W, Friedman AK, Baker LA (2017) Nanopore sensing. Anal Chem 89(1):157–188

Shivashakarappa K, Reddy V, Tupakula VK, Farnian A, Vuppula A, Gunnaiah R (2022) Nanotechnology for the detection of plant pathogens. Plant Nano Biol 2:100018

Shiyani T, Bagchi T (2020) Hybrid nanostructures for solar-energy-conversion applications. Nanomater Energy 9(1):39–46

Singh A, Dhiman N, Kar AK, Singh D, Purohit MP, Ghosh D, Patnaik S (2020) Advances in controlled release pesticide formulations: prospects to safer integrated pest management and sustainable agriculture. J Hazard Mater 385:121525

Singh G, Ramadass K, Sooriyakumar P, Hettithanthri O, Vithange M, Bolan N, Tavakkoli E, Van Zwieten L, Vinu A (2022) Nanoporous materials for pesticide formulation and delivery in the agricultural sector. J Control Release 343:187–206

Solanki P, Bhargava A, Chhipa H, Jain N, Panwar J (2015) Nano-fertilizers and their smart delivery system. In: Nanotechnologies in food and agriculture. Springer International Publishing, pp 81–101

Sotiropoulou S, Sierra-Sastre Y, Mark SS, Batt CA (2008) Biotemplated nanostructured materials. Chem Mater 20(3):821–834

Squire HJ, Tomatz S, Voke E, González-Grandío E, Landry M (2023) The emerging role of nanotechnology in plant genetic engineering. Nat Rev Bioeng 1(5):314–328

Srivastava S (2023) Ethical issues regarding the use of nanobiotechnology-based products. In: Nanobiotechnology for the livestock industry. Elsevier, pp 435–473

Sufian MM, Khattak JZK, Yousaf S, Rana MS (2017) Safety issues associated with the use of nanoparticles in human body. Photodiagn Photodyn Ther 19:67–72

Thavamani P, Smith E, Kavitha R, Mathieson G, Megharaj M, Srivastava P, Naidu R (2015) Risk based land management requires focus beyond the target contaminants—a case study involving weathered hydrocarbon contaminated soils. Environ Technol Innov 4:98–109

Thiruvengadam R, Easwaran M, Rethinam S, Madasamy S, Siddiqui SA, Kandhaswamy A, Venkidasamy B (2024) Boosting plant resilience: the promise of rare earth nanomaterials in growth, physiology, and stress mitigation. Plant Physiol Biochem 208:108519

Thomé A, Reddy KR, Reginatto C, Cecchin I (2015) Review of nanotechnology for soil and groundwater remediation: Brazilian perspectives. Water Air Soil Pollut 226:1–20

Tsai TT, Kao CM, Yeh TY, Lee MS (2008) Chemical oxidation of chlorinated solvents in contaminated groundwater. Pract Period Hazard Toxic Radioact Waste Manag 12(2):116–126

Tufail MA, Iltaf J, Zaheer T, Tariq L, Amir MB, Fatima R, Asbat A et al (2022) Recent advances in bioremediation of heavy metals and persistent organic pollutants: a review. Sci Total Environ 850:157961

Verma R, Saroop S (2024) Impact of nanopesticides in the environment: solutions, threats, and opportunities. In: Pesticides in a changing environment. Elsevier, pp 251–292

Visconti P, Giannoccaro NI, de Fazio R, Strazzella S, Cafagna D (2020) IoT-oriented software platform applied to sensors-based farming facility with smartphone farmer app. Bull Electr Eng Inform 9(3):1095–1105

Vurro M, Miguel-Rojas C, Pérez-de-Luque A (2019) Safe nanotechnologies for increasing the effectiveness of environmentally friendly natural agrochemicals. Pest Manag Sci 75(9): 2403–2412

Wang D, Saleh NB, Byro A, Zepp R, Sahle-Demessie E, Luxton TP, Ho KT et al (2022) Nano-enabled pesticides for sustainable agriculture and global food security. Nat Nanotechnol 17(4): 347–360

Wani TA, Masoodi FA, Baba WN, Ahmad M, Rahmanian N, Jafari SM (2019) Nanoencapsulation of agrochemicals, fertilizers, and pesticides for improved plant production. In: Advances in phytonanotechnology. Academic, pp 279–298

Wernette RC, Nilsen RL (2008) The legal risks of nanotechnologies. For The Defense 50(11): 73–81

Yadav A, Yadav K (2018) Nanoparticle-based plant disease management: tools for sustainable agriculture. In: Nanobiotechnology applications in plant protection. Cham: Springer International Publishing, pp 29–61

Yadav SHIVK, Lal S, Yadav S, Laxman J, Verma B, Sushma M, Choudhary R, Singh P, Singh S, Sharma V (2019) Use of nanotechnology in Agri-food sectors and apprehensions: an overview. Seed Res 47(2):99–149

Yadav A, Yadav K, Ahmad R, Abd-Elsalam KA (2023a) Emerging frontiers in nanotechnology for precision agriculture: advancements, hurdles and prospects. Agrochemicals 2(2):220–256

Yadav S, Sawarni N, Twinkle Dahiya JS, Rana MS, Batra B (2023b) Nanoagriculture: advantages and drawbacks. In: Agricultural and environmental nanotechnology: novel technologies and their ecological impact. Springer Nature Singapore, Singapore, pp 3–42

Yeshe AS, Vaidya PH, Shinde GU, Gourkhede PH (2022) Application of wireless Nano sensors network and nanotechnology in precision agriculture. Int J Adv Agric Sci Technol 9:36–65

Yin H, Cao Y, Marelli B, Zeng X, Mason AJ, Cao C (2021) Soil sensors and plant wearables for smart and precision agriculture. Adv Mater 33(20):2007764

Zain M, Ma H, Sadaf Chaudhary M, Nuruzaman IA, Mehmood F, Rahman SU, Aiwang D, Sun C (2023) Nanotechnology in precision agriculture: advancing towards sustainable crop production. Plant Physiol Biochem 206:108244

Zhang Y, Goss GG (2022) Nanotechnology in agriculture: comparison of the toxicity between conventional and nano-based agrochemicals on non-target aquatic species. J Hazard Mater 439: 129559

Zhang L, Dawei Y, Regenstein JM, Xia W, Dong J (2021) A comprehensive review on natural bioactive films with controlled release characteristics and their applications in foods and pharmaceuticals. Trends Food Sci Technol 112:690–707

Zhao L, Wang W, Xiaohong F, Liu A, Cao J, Liu J (2022) Graphene oxide, a novel nanomaterial as soil water retention agent, dramatically enhances drought stress tolerance in soybean plants. Front Plant Sci 13:810905

Chapter 15
Biosafety and Environmental Impact Assessment of Nanopesticides

Rashmi Singh and Kamaraj Raju

Abstract Effective delivery of agricultural chemicals is accomplished by the utilization of nanopesticides, which are distinguished by their nanostructures that range from 1 to 200 nanometers. In comparison to conventional insecticides, nanoparticles that contain active ingredients have a number of beneficial characteristics. There is, however, a new category of environmental contaminants that have emerged as a result of the rapid development of nanoparticles that are specialized for the purpose of pest management. A discussion of the negative effects that pesticide exposure has on both the environment and human health is included in this present article, which also examines the history of nanopesticides. In addition to this, it investigates the possible advantages of nanoparticles in terms of increasing agricultural output and issues regarding the environment. According to existing literature, there exist many approaches for achieving controlled dispersal and stimuli-responsive systems that can distribute agricultural chemicals and genetic material in a slow and targeted manner on a specific target. The study places an emphasis on the numerous sources of nanoparticles, the environmental consequences that they have in crop habitats, and the recent advancements in nanopesticides research that are based on the findings of experiments. Furthermore, this chapter covers concerns relating to the development, formulation, and toxicity of pesticide products within the framework of environmentally conscious and sustainable farming practices.

Keywords Agricultural chemicals · Controlled release · Ecotoxicological framework · Genotoxic · Nanopesticides · Sustainable farming practices

R. Singh
Department of Zoology, Dayanand Vedic College, Orai, Uttar Pradesh, India

K. Raju (✉)
Department of Biotechnology, Rathinam College of Arts and Science, Coimbatore, Tamil Nadu, India

J. M. Al-Khayri et al. (eds.), *Nanotechnology Applications in Modern Agriculture*, Nanotechnology in Plant Sciences 3, https://doi.org/10.1007/978-3-031-90506-3_15

15.1 Introduction

Due to rise in the world's population, it is imperative that there should be a sufficient supply of high-quality food that can satisfactorily meet the varied requirements of people living on every continent throughout the globe. It has been noticed that the increased demand for agricultural commodities in terms of economic investment has coincided with the increased consumption of agricultural chemicals such as fertilizers, insecticides, herbicides, and other agricultural chemicals in order to improve agricultural output results. Nevertheless, it is of the utmost importance for cultivators, producers of food, finance companies, and authorities to make the careful control of the adverse effects that are a consequence of the extensive use of agrochemical technology as a priority. It is necessary to consider the impact on the environment, ecosystems, customers, and all other sectors involved in the farming network, whether they are impacted directly or indirectly (Kah and Hofmann 2014). In order to increase food production on a worldwide scale, nanotechnology has developed nano agrochemicals, which have become a bio-stimulant for smart agriculture (Mgadi et al. 2024). Nanotechnology possesses a diverse array of possible uses that are poised to bring about a transformative impact on the realms of science and technology in the twenty-first century. These possible uses span across numerous sectors like as materials science, healthcare, the agricultural sector, and industry, and are anticipated to assume significant roles in various dynamic domains. Nanotechnology's utilization in plant science and agriculture is seeing rapid growth in the field of nano research. Nanoparticles are primarily employed in agriculture to efficiently transport external insecticidal agents, therefore enhancing plant well-being, broadening the range of insecticides, minimizing pesticide application, extending the duration of effectiveness, and mitigating environmental contamination. Nanoparticles have the potential to serve as adjuvants in the administration of chemical or botanical pesticides, hence enhancing their effectiveness and minimizing pesticide usage in addition to serving as adjuvants for chemical or botanical pesticides, nanoparticles possess the capability to transport biological pesticides, including ds-RNA (double-stranded RNAs) and deadly *Bacillus thuringiensis* (Bt) proteins. This versatile use has significant promise in the field of pest management.

A link between agricultural technology and agriculture is established all over the world through a number of international patents and licensed products. The primary focus of these connections is to ensure that various commodities, including food, fuel, and wood, are made available to the human population (Yadav et al. 2020). The primary objective of agrochemical technology is to protect agricultural regions from a variety of pests, including illnesses, harmful insects, and parasitic weeds, all of which have the potential to adversely affect production and productivity. Among them are the evaluation of the state of the soil and the plantation, as well as the management of livestock and goods from the fishing industry. However, the spraying of pesticides against hazardous insects and pests in an unregulated manner has also had a detrimental influence on the outcome of the production process. As a consequence of this, illnesses and insects have developed resistance to chemicals, there has been a surge in the need for new agrochemicals, and environmental imbalances have become even more severe (Bombo et al. 2019). It is possible for agrochemical nanoparticulate to have a greater surface area, which enables them to come into more contact with agricultural pests. This renders them both effective and unpredictable in their behavior. It is the illogical and disorderly use of agrichemicals that contributes to the escalation of disease resistance, the reduction of nitrogen fixation and biodiversity, and the amplification of the accumulation of pesticides in agricultural and animal products, as well as in aquatic species. Consequently, this not only constitutes a substantial and growing threat to the health of humans and the environment, but it also impedes the development of environmentally responsible agricultural practices (Osorio-Echavarría et al. 2021; Vigneshwaran et al. 2006). This paper discusses the primary approaches to presenting and assembling innovative nanopesticides, as well as the benefits and drawbacks of using nanomaterials for the management of agricultural pests, environmental risks, and the influence on the health of both humans and animals (Chaud et al. 2021). In the agrochemical industry, a significant amount of focus is placed on the major nanomaterials that are utilized, with a particular emphasis on their employment in crop cultivation. This is the case regardless of the safety qualities with which they are associated.

15.2 Nanotechnology for Pest Control

It has become clear that the implementation of delivery systems that are based on nanotechnology has given substantial opportunities for the rejuvenation of agricultural operations. It is possible that this is because of the intelligent controlled release qualities of fertilizers and agrochemicals, both of which are necessary for increasing crop output (Prasad et al. 2017; Zainab et al. 2024). Due to the fact that they improve the efficiency of fertilizers and agrochemicals, these systems are extremely important in the agricultural industry (Fraceto et al. 2016). In spite of the fact that nanopesticides have the potential to bring about advancements in agriculture, there is a significant cause for concern regarding the possibility of being exposed to hazardous agricultural chemicals that are capable to pass through biological barriers. These barriers include the blood-brain barrier, the blood-placental barrier, and the blood-retinal barrier. This is because there is a risk of irreparable damage to vital organs. In the present moment, there is a substantial amount of interest in doing research into the possible dangers that are connected to the exposure to dangerous nanopesticides. These nanopesticides also have the potential to have genotoxic and toxic effects on living organisms. The influence of nanopesticides on the chemical composition of the bulk material as well as their physicochemical features, such as size, electrical charge, and surface qualities, is currently being investigated by researchers (Zielinska et al. 2020; Grillo et al. 2012). Nanomaterials that are especially developed to protect plants, reduce the quantity of pesticide that is sprayed, increase leaf coverage, enhance stability, and reduce the number of components that are utilized in the formulation are referred to as innovative nanopesticides. Liposomes, dendrimers, metallic and bimetallic nanoparticles, and nanotubes are some examples of the self-organized systems that may be used to classify nanopesticide formulations. Active encapsulating components, such as nanoemulsions, polymeric nanoparticles, lipid nanoparticles, and nanotubes, are also included in this category.

15.3 Impact of Nanopesticides

In the context of agriculture, the word "nanopesticides" refers to pesticides that have been manufactured using nanoparticles in order to find uses in the field. These insecticides have the potential to be turned into nanocarriers that are functionalized for enzyme-mediated triggers, particularly when they are fixed on a hybrid substrate or encased in a matrix. It is believed that nanosized particles, in conjunction with their shape and distinctive characteristics, can be utilized to investigate pesticide activities in nanocarrier novel formulations that are generated from a numerous substance, like silica, lipids, polymers, copolymers, ceramic, metal, carbon, and other substances (Agostini et al. 2012). The formulations of nanopesticides have the ability to enhance water solubility and bioavailability, in addition to protecting agrochemicals against degradation caused by the environment. It is possible that this will completely transform the way that diseases, weeds, and insects are managed in agricultural settings (Fig. 15.1) (Yadav et al. 2020).

Nanomaterials, on the other hand, possess properties that present them with the potential to be both cytotoxic and genotoxic at the same time.

When pesticides are used in a manner that is not only irrational but also indiscriminate, it disrupts the existing environmental equilibrium, hence increasing the

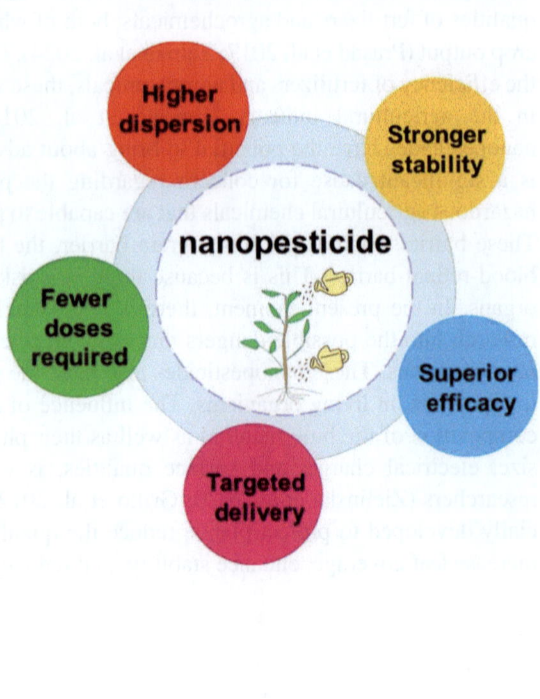

Fig. 15.1 Nanopesticides efficiency. (Figure constructed by Rashmi Singh-author)

risk to the health of all individuals. The consumption of residues of pesticides from food/water, whether due to work or inadvertent reasons, can lead to significant outcomes, including mortality or disability for long term (also called as disability-adjusted life years or DALY). These are the unfavorable effects that can arise as a consequence of pesticide residues. These effects might be of two types i.e. short-term also known as acute or long-term also known as chronic. In addition to being more prone to experiencing irreparable harm to their organs and tissues, youngsters are also more likely to be exposed to certain types of pesticides. Both the neurotoxicity of the central nervous system and the peripheral nervous system, as well as the effect on the loss of the blood's capacity to coagulate, is all regarded to be substantial grounds for worry (Kuhlbusch et al. 2018). It is a matter of great concern to acknowledge the impact of activity and toxicity of nanopesticides, which requires studying the benefits and drawbacks of nanopesticides in order to ensure the safe and sustainable development of nanoparticles that have already been approved for use in agriculture. This is necessary in order to guarantee the development of nanoparticles that are both secure and beneficial for the environment.

This is a deficiency of information that has been brought to light by the fact that there is a lack of comprehensive understanding. On the other hand, it is common knowledge that nano formulations play a key part in reducing the pace of degradation of the active component, improving the equilibrium of water solubility, and increasing the levels of active substances that are available to biological systems. To be more precise, in order to prevent the endemic infestation of pests, the harm that would be caused to plants, and the economic loss, would be the result of a drop in the number and quality of agricultural commodities and foods (Syafrudin et al. 2021). Soil leaching allows nanopesticides that are produced from runoff of agricultural and industrial wastewater to reach the water supply during the precipitation event. Consequently, this has results in reducing the quality of the water, increasing the amount of time that people are exposed to them, and raising worries about the ecological system. It has been discovered that nanoparticles have the potential to create toxicological consequences. This is because nanoparticles feature biomimetic properties including a high capacity for diffusion and bioaccumulation in soil, aquatic habitats, and foods. Additionally, nanoparticles have the ability to mimic biological processes. Mammals are a particularly good example of this (Hayles et al. 2017). It is possible for people to have both acute and chronic pathological symptoms as a consequence of the detrimental effects that are related with individual sensitivity and the length of time that they are exposed to nanoparticles. Some of the systems that may exhibit these symptoms are the respiratory system, the cardiovascular system, the lymphatic system, the autoimmune system, the nervous system, and numerous malignancies. These signs and symptoms may occur soon after exposure or a long time afterwards as a result of bioaccumulation (Khan et al. 2019) and the one-of-a-kind characteristics of nanoparticles (Ferreira et al. 2013). Modified-release nanopesticides may be categorized into two categories: chemically linked pesticides and physically incorporated formulations that exhibit rapid activity upon delivery in agricultural environments.

15.4 Innovations in Nanoencapsulation for Controlled Pesticide Delivery Systems

Encapsulation is the process of enclosing a biologically active component within capsule walls. This allows the core substance to be released into the environment around it under controlled circumstances over a specified period or when external factors cause the capsule walls to break, melt, or dissolve gradually. The formulation involves the encapsulation of the active ingredient within a nano capsule by either physical adsorption within a matrix or chemical bonding employing multiple methods. After that, the active component is liberated by either the breaking of chemical bonds or the diffusion of energy across the physical medium. Nanoencapsulation of pesticides serves various goals, including minimizing the reduction in efficiency caused by evaporation, degradation, and leaching, as well as boosting their efficacy through increased interaction with pathogens, insects, weeds, and other pests (Fig. 15.2). Nanoencapsulation of pesticides is a relatively new technique that has been widely used in recent years. Nevertheless, it is of the utmost importance to conduct a thorough analysis of the benefits and drawbacks that are connected to the use of pesticides in nonencapsulated systems. Novel materials systems are able to react to a wide variety of stimuli, including light, pH, temperature, enzymes, and more. This is in contrast to the existing practice of agrochemical

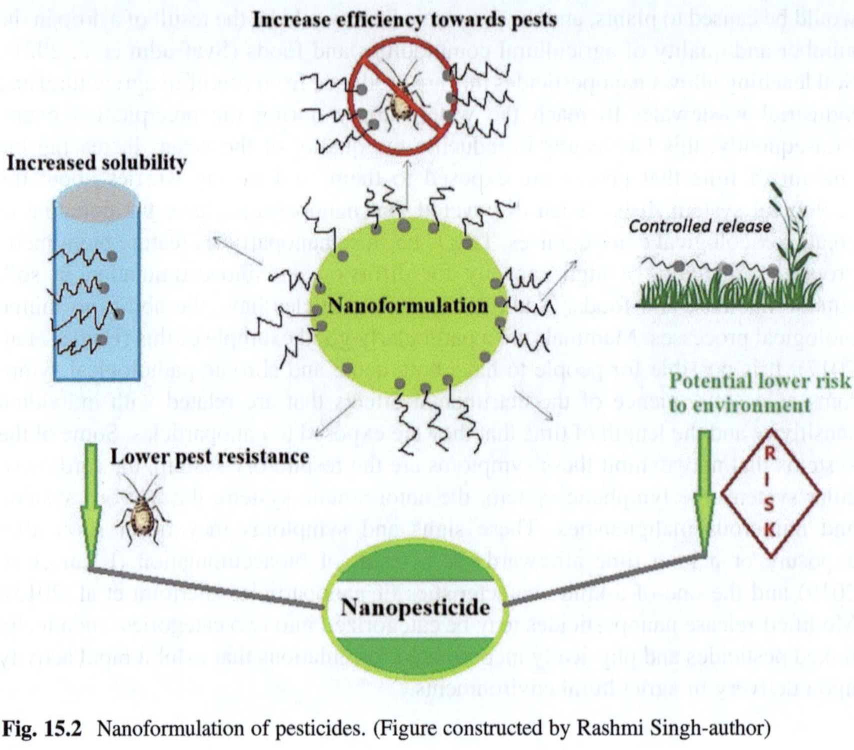

Fig. 15.2 Nanoformulation of pesticides. (Figure constructed by Rashmi Singh-author)

corporations, which emphasize crop laying regardless of whether or not the conditions are optimum. These systems have the ability to extend the release of agrochemicals, target delivery, reduce consumption, eliminate leaching and drift, and improve the efficiency with which pesticides are utilized (Zhang et al. 2019; Chen et al. 2018; Li et al. 2020). The depot systems are a sort of innovative controlled release formulation, which suggests that there is a lag time between the release (delivery) of the encapsulated nanopesticide to the crop and the beginning of the procedure with which agricultural chemicals are released. This is in contrast to typical formulations, which include the release of agricultural chemicals in a quick burst that takes place shortly after the formulation is applied to the crop. Consequently, the intended result will not be visible until the least effective concentration is achieved, regardless of whether it is for the depot system or an immediate alternative (Wang et al. 2017). In the context of a depot system, a loading rate that is much higher than average provides the benefits of controlled release over an extended period of time as well as a reduction in the frequency of dosing to once every 5 months. In addition to this, nanopesticides not only enhance the efficiency of the pesticides, but they also need much reduced application volumes when compared to the conventional formulation (Table 15.1) (Bahadir and Pfister 1990).

Through the employment of grafted target nanoparticle formulations in response to environmental stimuli, a substantial technical development has been made in the direction of assuring the safe application of pesticides and introducing innovative materials for nanopesticides. In the past, researchers have employed sensitive polymer materials in order to develop and functionalize nanostructured matrix systems. These systems include nano capsules, nanospheres, and nanovesicles, and their primary objective is to transport pesticides. For the purpose of reducing the toxicity of pyraclostrobin, a controlled release device that makes use of a humidity-sensitive

Table 15.1 Comparisons of nanofertilizers and conventional fertilizers

S. N.	Properties	Nanotechnology fertilizers	Conventional fertilizers
1.	Solubility and dispersion	Micronutrients engineered by nanotechnology may enhance bioavailability by diminishing soil absorption and fixation	Reduced solubility attributable to macroscopic characteristics
2.	Mode of nutrient release	Encapsulation allows for the regulation of both the rate and pattern of nutrient release in water-soluble fertilizers	The excessive application of fertilizers leads to toxicity and disrupts soil ecological equilibrium
3.	Duration of nutrient release	They may prolong the duration of nutrient availability from fertilizers in the soil for plant utilization	During delivery, they are utilized by plants, while the remainder persists as insoluble salts in the soil
4.	Uptake efficiency	The nano-dimensions may enhance nutrient absorption and fertilizer efficacy	It diminishes efficiency, and a substantial quantity is inaccessible to plant roots
5.	Rate of nutrient loss	They mitigate the depletion of fertilizer nutrients in the soil	They exhibit significant nutrient depletion resulting from leaching, runoff, and drift

submicron capsule of Seltima® was created. This method was developed with the particular intention of reducing any negative impacts on the aquatic environment, and it was specifically built for rice leaves as well as other crops. The thermosensitive and efficiency-controlled release time of pyraclostrobin nano capsule was obtained using emulsion polymerization using poly (N-isopropyl acrylamide-co-butyl methacrylate) (Sershen et al. 2000). This allowed for an increase in the stability of pesticides as well as the amount of time they took to take effect. The employment of a stimulus light-responsive controlled release system as a delivery mechanism in different agro-livestock application domains. This technology demonstrates ecologically benign properties in both field and greenhouse growth environments. To create the light-responsive co-polymer known as Poly (ethylene oxide-b-methacrylic acid) (PEOPMAA), light-sensitive groups were included into the synthesis process. It was possible to establish controlled release of the pesticide by using self-assembly and encapsulation, which led to a considerable residual activity of the formulation. A dual responsive system for pH and ion strength is an engineering endeavor that aims to reduce the amount of pesticides used in crops, implement innovative modified-release methods in complex environments, minimize the amount of pesticides that leach and decompose, and minimize the negative impacts that pesticides have on the environment. It is possible that a unique release mechanism that is intended for environmental stimuli has the potential to reduce the early development of complexes in agrochemicals, to impede the sulfidation process, and to display better efficacy in pest management (Gao et al. 2021). A dual responsive system for pH and ion strength is being implemented as part of an engineering project with the objective of lowering the number of pesticides that are used in agricultural production. In order to accomplish this objective, this system makes use of novel materials.

15.4.1 Physical and Chemical Systems for Agrochemical Delivery

Single-walled or multi-walled carbon nanotubes (MWCNT) have emerged as promising nanomaterials that have tremendous promise for a variety of applications in the agricultural sector. Their potential as biosensors for agricultural chemicals is one of these uses, along with their usage as slow-release systems for fertilizers and pesticides, other applications include their use as slow-release systems. An investigation has demonstrated that carbon nanotubes have a positive influence on the biomass of plants, the length of their roots and shoots, and the germination of seeds (Sarlak et al. 2014). Despite the fact that it has been observed that the mancozeb antifungal pesticide, when enclosed within a hybrid material consisting of MWCNT graft and poly-citric acid nanoparticles, demonstrates enhanced stability and efficacy in comparison to the bulk pesticide, as well as higher water solubility of agricultural chemicals, it is important to note that carbon nanotubes have been discovered to have an impact that is paradoxical. It is possible that the presence of carbon nanotubes might have negative effects on plants, leading to the death of cells and a decrease in the number of soil microorganisms as a result of the generation of reactive oxygen species (ROS) (Vithanage et al. 2017).

Proteins and peptides, such as those found in snail venom, hormones, and poisons for scorpions and other insects, have been demonstrated to possess multifunctional properties and insecticidal action. Nevertheless, the ease with which proteins or peptides are metabolized within the digestive system of insects is a significant barrier in the application of pesticide peptide-based chemicals. This is a serious limitation. The protease breakdown mechanism that takes place within the digestive system and the hemolymph, as well as the transit over the midgut ventriculus in insect pests, is the principal impediment that prevents pesticides from being effective. The capacity of insects to endure pyrethroids, organophosphates, organochlorines, and carbamates is hindered as a result of these impediments, which also have an impact on the production of food (Gahan et al. 2001). Innovative and intelligent nanopesticides have the goal of overcoming physical, chemical, and biological barriers in order to achieve their intended purpose. Insecticide action can be achieved through the creation of agricultural chemicals release systems. These systems utilize the conjugation of hydrophilic polymers with TMOF (trypsin modulating oostatic factor), which has been discovered to improve stability and reduce metabolic activity. The insecticide demonstrated enhanced bioaccumulation inside the gut lumen of the insect species *Spodoptera frugiperda*, which led to the inhibition of oocyte maturation and a reduction in the number of insects that reproduced (Shen et al. 2009; Patel et al. 2018). Ionotropic reaction-based particles have garnered attention as potential nanocarriers for pesticides due to their cost-effectiveness, ease of manufacture under moderate conditions, a Generally Recognised as Safe (GRAS) medium, size control, biocompatibility, non-toxicity, and the availability of several crosslinking agents. These characteristics have contributed to their popularity. In order to create loaded nanoparticles that contained cypermethrin and had a sustained-release effect, sodium alginate and hydro soluble calcium chloride were utilized in the manufacturing process. Cypermethrin molecules were discovered to be largely positioned on the inner side of the nanoparticles, as shown by the fact that the entrapment efficacy of cypermethrin was found to be 96% and the loading of cypermethrin was found to be 78%. Studies conducted in soil pots shown that the utilization of Ca-alginate nanoparticles loaded with cypermethrin produced superior results when compared to the direct administration of the chemical to plants (Patel et al. 2018). Because of its ability to transport nanopesticides, polyelectrolyte complexes have attracted the attention of both agricultural and agriculturally-based manufacturers. Because of their capacity to lower the necessary dose per hectare and to lessen the pace at which herbicides and agrochemicals drain into the soil, this is the reason why they are so effective. Clay-gelatine, which is used in the formulation of the Methylphenoxyacetic acid (MCPA) herbicide, has been discovered to reduce the dangers to the environment that are associated with water when compared to the conventional formulation of MCPA.K-salt solution in water (Parsons et al. 2018). There have been observations made on the usage of the polyelectrolyte complex system as a biosensor within the realm of electroanalytical chemistry for the purpose of detecting food poisoning. Without the presence of any metallic compounds, the electro-reductive determination of metribuzin, which is a pesticide, and metronidazole, which is an antibiotic, was carried out with the help of the chitosan-pectin polyelectrolyte complex (Ranganathan et al. 2019). In addition, the Interpol electrolyte complex (IPEC) system, is a cationic polyplex that has a high degree of solubility in water. Condensing genetic material to a colloidal size that falls anywhere between 50 and 500 nanometers is something that it is capable of doing. According to the findings of the release test, IPEC nano dispersions display characteristics that are similar to those of a pesticide reservoir. These nano dispersions demonstrate a regulated release rate that is governed by ionic exchange pH-based effects in physiological fluids (Alromeed et al. 2015). The introduction of double-stranded RNA (dsRNA) into insect crops for the goal of pest management has been made possible with the development of technology known as RNA interference (RNAi). Following the conversion of the dsRNA into siRNA effectors, the gene expression of *Spodoptera frugiperda* was inhibited by the dsRNA. Through the process of reversible addition-fragmentation chain transfer (α-RAFT) polymerization, a cationic polymer with biomimetic characteristics, such as pGMPA, was successfully produced. The development of an Interpol electrolyte complex was the end outcome of this procedure. This complex was successfully internalized by cells and acted as a catalyst for RNA interference (RNAi). There was research that demonstrated that this strategy was successful in deactivating the gene in an insect that was resistant to many drugs (Parsons et al. 2018). Metallic nanoparticles are produced using a single-phase reduction process that is carried out in an aqueous medium. This method involves the utilization of different metal precursors. These nanoparticles have a unique insecticidal property that can be attributed to the surface plasmon resonance characteristics that have been investigated extensively. Noble metal ions, including silver, gold, copper, and others, are chemically absorbed onto the surface of pathogens and insects by functional groups that are located on the cell wall of these organisms. Polysaccharides have been recognized as a possible substitute for conventional pesticides in the context of addressing the problem of bacterial resistance through the utilization of non-specific bacterial toxicity mechanisms (Sánchez-López et al. 2020).

The use of physical methods to provide medication is more ecologically friendly than the use of chemical methods. Nano encapsulating biocide and fertilizer over a longer period of time has been one of the applications of these technologies in the agricultural sector, with the goal of improving the efficacy, safety, and stability of agrochemical materials. The agrochemicals that are encased inside physical systems are referred to as "internal" or "filler material entrapped in nanoparticles," and these are the terminology that are used to describe them. Under these circumstances, it is generally accepted that the word "nano" should be utilized if the size of nanoparticles is more than 100 nanometers. Clay minerals (such as bentonite, smectite, chaolite, and montmorillonite), LDHs anionic clay (layered double hydroxides), lipids (such as waxes, triglycerides, fatty acids, and surfactants), inorganic porous materials (such as silica, ceramic, and polytriphenylamine), natural polymers (such as cellulose, starch, gelatin, albumin, and chitosan), and synthetic polymers (such as polyamidoamine, polyethylene oxide, polylactide, and cyanoacrylates) are some of the materials that have been utilized for nanoencapsulation in systems that are classified as physical (Alromeed et al. 2015). The agrochemicals that are encased inside physical systems are referred to as "internal" or "filler material entrapped in nanoparticles," and these are the terminology that are used to describe them. Under these circumstances, it is generally accepted that the word "nano" should be utilized if the size of nanoparticles is more than 100 nanometers. Nanoencapsulation in physical systems has been achieved using various materials such as clay minerals (e.g., bentonite, smectite, chaolite, montmorillonite), LDHs (anionic clay), lipids (e.g., waxes, triglycerides, fatty acids, surfactants), inorganic porous materials (e.g., silica, ceramic, polytriphenylamine), natural polymers (e.g., cellulose, starch, gelatine, albumin, chitosan), synthetic polymers (e.g., polyamidoamine, polyethylene oxide, polylactide, cyanoacrylates), and others. It has been established that the employment of nanoparticles based on LDHs and clay is effective in protecting pesticides against volatilization and photodegradation. It is well known that mineral clay nano formulations are environmentally safe. This is due to the fact that, in comparison to other carrier systems, they have lower drift and leaching rates. There is a connection between the chemical property of these nanocarriers and the reversible binding of the pesticide with the matrix. The numbers (Alromeed et al. 2015; Nuruzzaman et al. 2016).

For the purpose of constructing a light-responsive controlled-release pesticide nanoparticle, a porous system that was coated with amino silicon oil and grafted with azobenzene was successfully employed. Both attapulgite and biochar were the components that made up this nanoparticle. The employment of composites as an adsorbent for glyphosate is beneficial owing to the fact that they are ecologically friendly and possess a high capacity for adsorption. In order to investigate the release properties in water, an examination was carried out utilizing UV-VIS light (365–435 nm). UV-VIS light was responsible for regulating the core-shell nanostructures' high level of reactivity to stimuli, which was proved by the nanostructures. Because of their ability to cling to the leaves of weeds, there was a decrease in the amount of glyphosate that was lost, and there was a considerable improvement in the effectiveness of herbicides (Chen et al. 2017; Chen et al. 2018). When a dopant agent is introduced into a silica-based matrix, an organic molecule is able to become entrapped inside the inner porosity of the matrix. This is the moment that the sol-gel composite process is initiated. The sol-gel method is responsible for the flexibility of organic polymers as well as the characteristic tail shape, density, and surface features of organically modified silica nanoparticles. With these exceptional properties, a much bigger quantity of active material may be bonded together (up to 90% by weight), control can be exerted over the pace at which it is released over a period of many months, and stability can be improved even when exposed to extreme environmental conditions (Ciriminna et al. 2011). A new nanocarrier for the delivery of pesticides, the sol-gel composite is now being considered for use in this application. It possesses the capability to react to various stimuli. The purpose of this work was to evaluate the possibility of using carboxymethyl-chitosan to change the surface of mesoporous silica nanoparticles loaded with azoxystrobin as a sustainable fungicide for tomato plants. The results showed that there was an increase in the loading content, a pH-dependent controlled release, and an improvement in the fungicidal activity (Xu et al. 2018). The development of biopesticides as an alternative to chemical techniques of pest management has attracted a lot of interest due to the fact that they are accurate, they are able to target particular organisms, they are not poisonous, and they are able to overcome the obstacles that are provided by multidrug resistance occurring in illnesses. Through the utilization of sequence-specific endogenous RNA interference, the transmission of genetic material in a system that is not encapsulated is a unique and outstanding method for the management of pests and the protection of plants from insects. At the moment, it is utilized for the purpose of pest management, either with or without the activation of ds RNA (Fletcher et al. 2020).

15.4.2 Making Use of Lipid-Based Nanopesticides

Creating nano emulsion, micellar, and vesicular systems for the purpose of delivering pesticides to specific targets may be accomplished by the usage of certain lipids and surface modification chemical agents. These materials can be utilized to produce these systems. This strategy makes it possible to distribute pesticides in a controlled manner while maintaining sensitivity to the surrounding environment. For the goal of releasing agrochemicals, nano emulsified carriers are now the subject of a significant amount of research that is currently being accomplished. As a result of the potential for free energy, the broad surface area, the specific chemical characteristics, and the biocompatibility of the emulsified systems, a number of nanopesticide emulsion-based types have emerged in order to improve the utilization efficiency and manage the release profile of pesticides. Formulations that are based on emulsions are designed with the particular purpose of enhancing the bioavailability, stability, and wettability of active chemicals, as well as increasing their solubility properties. This results in an improvement in the management of weeds and pests, and it also adds to the advancement of technology in the delivery of agrochemicals to the agricultural sector (Hazra and Purkait 2019; Mustafa and Hussein 2020).

Colloidal systems that employ emulsion-based technologies as carriers for pesticides are depicted in the schematic examples. Such stimuli-responsive substances as heat, light, and pH are included into these systems through the process of doping. The spontaneous assembly of supramolecular nanoparticles is referred to as micellization, which is another name for the process of micelle creation. It is frequently seen as a collection of micelles made up of smaller particles that come together to form a sphere. Over the course of the synthesis process, auto-aggregation takes place spontaneously, which is credited with being the cause of this production (Zhang et al. 2019). After the creation of a composite consisting of amphiphilic polymers, the behavior of nanoscale mixed supramolecular cooperative assembly was investigated while pyrethrin was present. The formulation demonstrated a modified-release mode that was sensitive to temperature, which allowed for the regulation of the rate of pyrethrin administration. This was accomplished by the structural deformation of the copolymer that was induced by temperature. This approach revealed three key benefits, three of which are photostability, enhanced larvicide efficacy in natural situations, and prolonged release, as stated in reference (Zhang et al. 2019). Lipid colloidal dispersions that do not contain surfactants are referred to as Pickering emulsions (PE). Polymer particles or inorganic particles are responsible for their physical stability, and they have an exceptional capacity to be wetted. Because of their ability to respond to release cues, PEs have been used extensively in the production of pesticides. In response to variations in pH, temperature, and humidity, they are able to convert between stable and unstable forms (Tu and Lee 2014). This flexibility allows them to adapt to different conditions. Anisotropic silica nanoparticles alginate grafted has devised with the purpose of determining the influence of structure and pH responsiveness have on the capabilities of PE as a carrier for λ-cyhalothrin have discussed. The potential use of these nanoparticles in pH-triggered modified release systems were also discussed. Increasing the viscosity and surface charge of the nanoparticles led to an improvement in the stability of PE, according to the findings. At 240 min, the cumulative release of λ-cyhalothrin from pH 4 reached 87%, 28% (pH 2), and 53% (pH 9) at pH 4, demonstrating compatibility with sustained-release systems triggered by pH ranging from 2.0 to 9.0). In compared to conventional pesticides, nano emulsions that are based on lipids have the highest level of effectiveness. i.e. Neem oil (*Azadiractaindica*) and *citronella* oil (*Citronella nardus*), is widely recognized for its effectiveness in the control of pest diseases, however, the insufficient water solubility of these oils and other biopesticides makes it difficult to use them in pest control applications. Encapsulating the oils within the nanoemulsions system through the application of nanotechnology is one approach that may be taken to overcome this constraint. Through the utilization of the spontaneous emulsification method, the evaluation of the encapsulation and influence of neem oil and citronella oil on fungus was assessed. An important approach for administering hydrophobic pesticides to treat fungal-induced plant diseases was discovered as a result of the studies, which confirmed the efficiency of fungicides under laboratory conditions. Nevertheless, it is of the utmost importance to evaluate the environmental conditions of these nano emulsions that are produced by employing the oil itself (Osman Mohamed Ali et al. 2017). When it comes to the creation of pharmaceutical innovations, liquid crystals are a highly significant method for enhancing the water solubility, efficacy, stability, and safety of active agrochemical ingredients. Lyotropic liquid crystal is a sort of self-assembly nanostructure system that boasts particular properties as a result of its structural organization or the production of complicated three-dimensional network lamellas. It is possible for biological activity to occur at the molecular level because to these thermodynamic features (De Souza et al. 2020). As a result of the liquid crystal's ability to assist the penetration of the active agrochemical ingredients into the tissue and boost its attachment to the leaf, not only is the efficiency of the process improved, but the environmental damage is also decreased. Following the preparation of the lyotropic liquid crystal formulation of phytotantriol, an evaluation of its structural equilibrium was carried out. According to the Bisset et al. 2019,phytonatriol, when it is present in the form of an emulsion, demonstrates interactions with lyotropic liquid crystals that are combined with pesticides that are used as adjuvants. Based on this interaction, it appears that phytonatriol possesses the ability to transport hydrophobic herbicides in an unconstrained way (Bisset et al. 2019). In their capacity as nanocarriers, liposomes are matrix systems that resemble vesicles and are made up of materials that are both biodegradable and biocompatible. The modification of the release of active substances is made easier by these systems, which have been designed expressly for that purpose. As soon as they are manufactured, liposomes have the potential to serve as nanocarriers for insecticides. Certain compounds have the ability to modify the surface of liposomes, which can result in the development of characteristics such as lipophilicity, hydrophilicity, targeting, flexible wall, controlled release, and biodistribution. It is also feasible to alter the functionalization of the liposome surface in order to protect the active agricultural chemicals from photodegradation and thermal degradation (Hwang et al. 2011; Jampílek and Král'ová 2017). This may be done in order to keep the chemicals from being destroyed by the sun. For the purpose of acting as a nanocarrier for etofenprox and α-cypermethrin, a liposome was produced and then modified using chitosan. This alteration was made with the intention of modifying the intrinsic surface charge, increasing the thickness, establishing a depot system, and extending the amount of time that active agricultural chemicals were delivered (Bang et al. 2011).

15.5 Nanocarriers and Pesticide Toxicology

Chemicals that are toxic to nanocarriers and pesticides. Although nanotechnology has made significant advancements in plant science, our understanding of the interactions between nanomaterials and plants in the environment, such as bioavailability, bioaccumulation, drift, toxicity, and agro-ecosystem safety, is still inadequate for accurately assessing the advantages and disadvantages of agricultural chemicals nanotechnology in crops (Xu et al. 2022). In the present day, nanopesticides are being utilized in the process of modifying registered agricultural chemicals. The objective is to improve the performance of these chemicals in contrast to those of non-technological agricultural chemicals and to solve the shortcomings of the agrochemical products that are already available. In order to get an understanding of the plant's response to being exposed to tailored nanopesticides and to accurately evaluate the toxicity of the substance, it is necessary to conduct a complete inquiry that takes into account certain limits and aspects that have not yet been investigated. Nanocarriers have evolved as a significant way to meet the pressing modern agricultural challenges, notably those relating to the trade-off between production growth and environmental repercussions, as well as the requirement for reduced pesticide application rates (Kah and Hofmann 2014; Prasad et al. 2017). Nanocarriers have emerged as a prominent strategy to address these important challenges. The results of previous studies suggest that around 50% of the pesticides that are utilized are discovered in various microenvironments close to crops, which results in the pollution of the water table, rivers, and lakes. Leaching or air streams are the two channels via which this pollution takes place. In spite of this, it seems that accomplishing all agricultural goals without having a comprehensive grasp of the challenges will only serve to magnify the risks, particularly in situations when agricultural chemicals and nanoparticles offer a higher threat to both human health and the environment (Pérez-Iglesias et al. 2017). In order to successfully target pests, reduce environmental pollution, limit the negative impacts of bioaccumulation, and protect both ecological systems and human well-being, it is desirable to target and manage the release of nanopesticides based on triggers. This is because of the nanopesticides' capacity to efficiently target pests. Recently, there has been a rising interest in the subject of nanotechnology about the negative effects and risks associated with nanoparticles, anionic nanoparticles, surfactants, solvents, wetting agents, and agents that produce toxicity and genotoxicity in humans. This interest has been fueled by the fact that nanoparticles have been shown to have a number of adverse effects and hazards. The combination of atomic collisions and nanotechnology, on the other hand, appears to raise the potential toxicological dangers that are involved with accomplishing all of the objectives (Yadav et al. 2020; Mustafa and Hussein 2020). When it comes to the production of nanoparticles, the materials that are used have the potential to cause toxicity in a wide variety of creatures, including plants, people, and other vertebrates (Schnoor et al. 2018). It is standard practice to administer agrochemical ingredients through the use of lipid-based formulations such as liposomes, liquid crystals, and micelles. This is because these formulations have biomimetic qualities with the cellular wall. Typically, natural phospholipids and cholesterol are the components that make up these formulations. The low water solubility of agrochemical ingredients, on the other hand, makes it necessary to employ surfactants, which might lead to an increase in the toxicity of pesticides (Bang et al. 2011). The assessment and prediction of cytotoxicity and genotoxicity in nanopesticides provide substantial problems due to the complex nature of nanostructures, reactivity, size, form, and electric charge. These factors all play a role in the phenomenon. Consequently, the process of determining whether or not they are safe for biological use is a difficult endeavor. Concerns have been raised by manufacturers, workers, and others concerned about food security over the toxicity of glyphosate, which is a broad-spectrum organophosphate that does not exhibit selectivity. It has been determined that glyphosate and its metabolite, aminomethyl phosphonic acid, are the substances responsible for the negative effects that are experienced by living species as well as those that reside on land. Upon exposure to the human placental JEG3-cell line, it has been discovered that individuals have allergic reactions, difficulties in the cardiovascular and respiratory systems, and disturbances in their endocrine system. It has been demonstrated that the incorporation of albumin, surfactants, and solvents such 1,4-dioxane derivatives into nanoparticles can significantly increase their toxicity (Chen et al. 2018). Pesticides, such as the herbicide clomazone, have the ability to disseminate from crops in nearby regions after they have been applied. As a result, they pose major toxicity threats to both animals and people. It has been determined that this herbicide has an effect both when it is in its unbound condition and when it is contained inside polymeric mix nanoparticles that are ordered by ionic bonding (Patel et al. 2018; De Oliveira et al. 2016]. Tadpoles (*L. catesbeianus*) that were subjected to acute circumstances were used for the purpose of conducting an evaluation of the hepatic impact of nanoparticles and clomazone, in both of their forms, at doses that were considered to be sublethal. It was noted that the experimental groups that were exposed to clomazone had an increased incidence of macrophage aggregates (melano-macrophage centers), accumulation of eosinophils, and lipidosis in the liver (De Oliveira et al. 2016). This was based on the data that were obtained. According to the findings of the current experiment, the integration of polymeric nanoparticles did not result in an increase in the protective impact that the liver had on tadpoles. When placental cytotoxicity occurs, it almost always leads to difficulties in reproductive health. The method known as plant-mediated RNA interference (RNAi) is employed for the purpose of controlling insect pests. This method involves the targeting of certain genes and the utilization of double-stranded RNA (dsRNA) to offer protection against potential pests. In light of this, the careful selection of genes and the design of double-stranded RNA (dsRNA) fragments show tremendous potential in terms of lowering the danger of cross-toxicity with a target organism that is not selective. Nevertheless, despite the fact that it has received a lot of support, this concept has not been evaluated. It has been demonstrated that the particular sequence of the RNA endogenous interference pathway is efficient in inhibiting gene targets, and it also has the potential to make insecticides that have a half-life that is far longer. However, there are still concerns that have not been overcome that pertain to the delivery method, environmental stability, toxicity to insects that are not the target, the danger of insect resistance with extended exposure to RNAi plants, and the environmental destination (Fletcher et al. 2020). The products that are based on RNA interference (RNAi) have the potential to greatly revolutionize agricultural practice management and environmental protection in a timely and secure way, provided that these restrictions are overcome. It is widely acknowledged that atrazine, a herbicide that is widely used in the growth of maize and sugarcane, is a pesticide that is persistent in environmentally sensitive environments. Significant worry has been generated as a result of its adverse impacts on ecosystems that are found in soil and water. When it comes to hydrographic basins in Brazil, it is not uncommon to see levels that are nine times greater than the acceptable threshold of 2 g L^{-1} (Nogueira et al. 2012). The purpose of this study was to determine the toxicity of atrazine and the possible risks it poses to aquatic biodiversity. In order to do this, Pacu (*P. mesopotamicus*) fingerlings were tested for sensitivity using a dose of atrazine that was comparable to the lower control limit (LC50). A greater degree of damage was found to have occurred in the liver, according to the findings. Atrazine has been shown to generate cellular stress in the liver and kidney, despite the fact that there has been no discernible change in the frequency of micronuclei or nuclear abnormalities (De Paiva et al. 2017). In .the Asian clam (*C. flumineausing*), the purpose of the study was to determine whether or not atrazine may be harmful to the species. Based on the data, it was determined that atrazine had an effect on the process of biotransformation, which resulted in oxidative damage and DNA corruption. On the other hand, the antioxidant defenses of the Asian clam were not shown to have undergone any substantial changes (Dos Santos and Martinez 2014). In terms of the genotoxic effects seen on Asian clam (*C. flumineausing*) appears to be ascribed to the chemical interaction that takes place between DNA and guanine and adenine bases. This contact takes place through an intercalation process and results in the production of adducts (Dos Santos and Martinez 2014). Imazethapyr is an herbicide that is generated from nicotinic acid and belongs to the imidazolinones group. It is defined as a Class III herbicide that has a low level of toxicity. On the other hand, it has been established that this material is a cause of irritation for the respiratory system and the eyes, in addition to being a dangerous agent for the environment. However, earthworms show either a limited or nonexistent amount of sensitivity to imazethapyr, in contrast to the high level of sensitivity displayed by bees and annelids. For the purpose of determining the genotoxic effects of imazethapyr, the oxidative DNA damage technique was utilized (Kushwaha et al. 2011). The findings indicated that the administration of imazethapyr at concentrations ranging from 0.39 to 1.17 mg/L led to the development of nuclear abnormalities in tadpoles that were subjected to the substance for a period of 96 h. A significant rise in the genetic damage index (GDI) was seen over a period of 48 h. Evidence was presented by the researchers that indicated that imazethapyr has the potential to cause harm to animal species that are not the intended targets of the treatment when it is distributed through water or air (Pérez-Iglesias et al. 2017). Scientific publications have not found any study that addresses this dilemma through the utilization of nanostructured systems, despite the fact that imazethapyr plays an important role in the control of weeds and has a negative influence on the environment. Fipronil is a phenylpyrazole insecticide that has use in a variety of fields, including agriculture, veterinary medicine, and healthcare, among others. Through the process of attaching to -aminobutyric acid receptors in insects, fipronil is able to exercise its pharmacological actions, which ultimately results in a depolarizing reaction. It has been demonstrated that the level of toxicity of this chemical is greater in insects as compared to vertebrates. On the other hand, it has been reported to cause harmful consequences and to disturb things like the production of fat in the liver, the reproductive cycle, the integrity of DNA, and the death of spermatozoa. As a consequence of the findings of in vitro research, it has been demonstrated that fipronil has a negative influence on the development of mouse preimplantation embryos. This influence results in an increase in the rate of apoptosis and a decrease in the rate of cell proliferation. It has been established through in vivo trials that fipronil offers a risk to the reproduction of vertebrates, which lends credence to the findings which were obtained through in vitro research. Long-term exposure to potentially hazardous agrochemicals causes the accumulation of fatty acids in the cells of the liver, which in turn leads to the development of fibrosis and hepatitis (Wasef et al. 2021). Inflammation of the liver, which ultimately results in damage to liver tissue, is caused by the accumulation of reactive oxygen species (ROS), which plays an important function in the liver and is responsible for the inflammation.

15.6 Possible Beneficial Effects of Nanoparticles

Nanoparticles play an essential role in weed control, the use of agrochemical ingredients, which has a low absorption rate and a restricted bioavailability, creates environmental challenges because of the requirement for atypical quantities. In order to resolve the dose-effect dilemma and improve the safe threshold, polymeric nanoparticles that contain agrochemical ingredients have been evaluated for their potential in environmental protection and weed management. This evaluation was carried out with the intention of achieving positive results. An agrochemical ingredients model was utilized in order to regulate the release of atrazine through the utilization of a nanoparticulated system that was constructed with polylactic acid (PLGA). Because of this system, there was a decrease in the amount of environmental toxicity and an increase in the amount of bioavailability (Schnoor et al. 2018). Furthermore, the power of PLGA nanoparticles to increase the resistance of crops to atrazine is demonstrated in this study, which is the reason why this work is significant.

In therapeutic applications, cerium oxide nanoparticles have been utilized as an antioxidant for the purpose of scavenging reactive oxygen species (ROS) (Mittal et al. 2014). In addition, cerium oxide nanoparticles have been employed in order to evaluate the level of protection against hepatotoxicity and lipogenesis that are brought about by fipronil. Using in vivo testing, the purpose of this investigation was to determine whether or not nanoparticles composed of fipronil and cerium oxide had a beneficial impact on hepatic tissue. According to the findings, cerium oxide nanoparticles were able to successfully prevent inflammation in the blood of rats, which resulted in a reduction in the synthesis of fatty acids and a mitigation of liver damage (Fig. 15.3) (Wasef et al. 2021).

The manufacturing of metallic nanoparticles through the use of techniques such as GRAS is a method that is not only environmentally friendly but also does not require the utilization of reducing agents that are hazardous to the environment. The properties of metallic nanoparticles include a substantial surface area, a wide range of efficiency against infections (both bacterial and fungal), the absence of any risk of resistance development, and the cost-effectiveness of the treatment (Malandrakis et al. 2021). Unlike bulk metal homologues, metal nanoparticles demonstrate efficacy as a result of the antibacterial biochemical process that elucidates the activities of metallic nanoparticles (Santos et al. 2021). This process is responsible for the effectiveness of metal nanoparticles. As an additional point of interest, it has been shown that metallic nanoparticles, when utilized for rapid release in agricultural crops, provide effective control against bacterial and fungal infections (Keller et al. 2017). Nevertheless, Worthington et al. 2013 has demonstrated that the application of a copper nanoparticles (CuNPs) coating that contains polysaccharides and the exposure toCuNPs might result in inflammation of the lungsand a significant immune system response in living organisms (Zhang et al. 2012). An investigation was carried out to assess the dangers that are connected to six nanoparticles that are based on copper. It was the purpose of this study to shed light on the possible

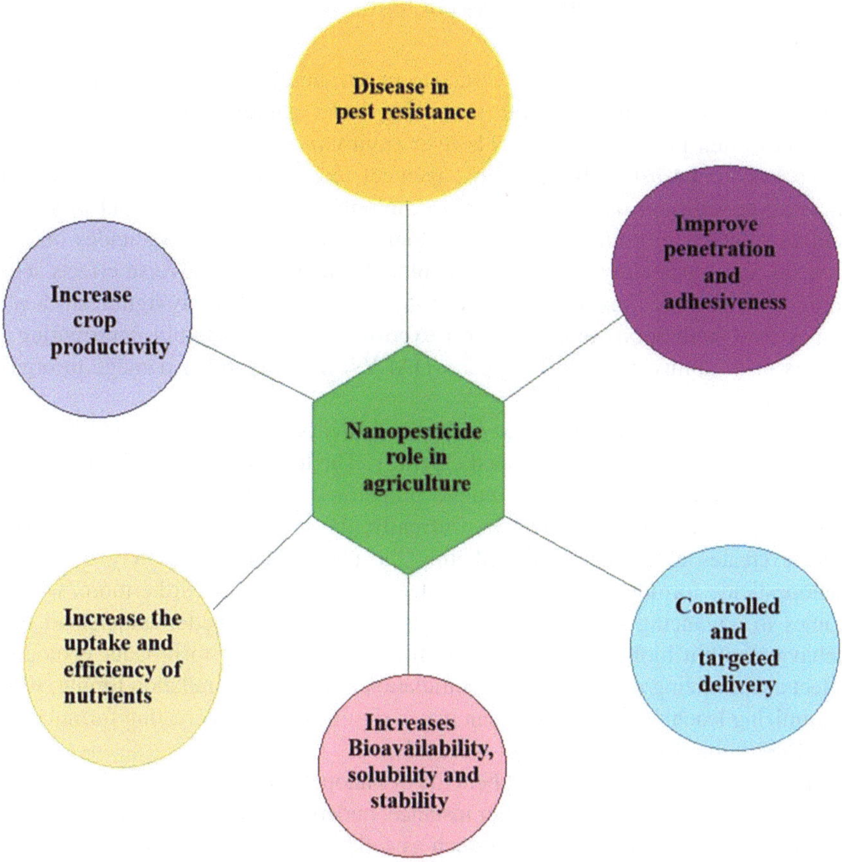

Fig. 15.3 Nanopesticide role in agriculture. (Figure constructed by Rashmi Singh-author)

dangers that may be posed by copper nanoparticles. The findings suggested that when silver nanoparticles (Ag-NPs) are put into the soil, it goes through a process of fast sulfation in settings that are reducing, such as silt from freshwater deposits. The sulfurized silver nanoparticles (sAg-NPs) experience a reduction in their bioavailability and solubility as a consequence of this process. As an additional point of interest, it was demonstrated that the cytotoxicity of sAg-NPs was reduced, which suggests that the process of sulfidation may result in a reduction in the environmental toxicity of Ag-NPs Based on the findings, it was determined that the production of reactive oxygen species (ROS) and the disturbance of nutritional balance that was brought about by the retention of anionic mineral components including phosphate, nitrate, and sulfate played a significant role in the mechanism of toxicity (Keller et al. 2017).

15.7 Conclusion and Prospects

The potential of nanopesticides to transform agricultural productivity while minimizing ecological impact is both promising and complex. Current agrochemical practices often prioritize crop yields over environmental conditions, which sometimes lead to unsustainable pest management techniques. A key challenge is to increase agricultural output while preserving environmental health and protecting human safety. Unfortunately, the toxicological impacts of nanopesticides on ecosystems and food safety remain underexplored. The potential adverse effects stemming from pesticide residues in food, water, and air can lead to significant health issues over both short- and long-term exposure periods, possibly contributing to deaths or disability-adjusted life years (DALYs) due to chronic damage to organs and tissues.

The interaction between nanopesticides and environmental systems adds layers of complexity that must be investigated carefully. The toxicity of these substances can vary significantly based on the unique properties of nanocarriers and the active agricultural chemicals they deliver. Currently, the limited data available on these effects creates a challenge in predicting the performance and safety of various nanopesticide formulations once they enter the environment. Furthermore, when it comes to advancing nanopesticide design, it is crucial to better understand their behavior within biological systems, from plant interactions to broader ecological effects. Optimizing reactivity, bioaccumulation rates, and degradation timing, while minimizing leaching and drifting, can play a significant role in creating formulations that meet both efficacy and safety standards.

To fully leverage the potential benefits of nanopesticides, researchers must focus on filling these knowledge gaps. For instance, understanding the interaction between nanocarriers and biological entities such as cell membranes, proteins, and enzymes will be instrumental in designing formulations that maximize safety and efficacy. Additionally, environmental conditions such as pH, temperature, and moisture levels can influence how nanopesticides perform, which underscores the importance of testing them under various field conditions.

Nanopesticides offer numerous advantages over traditional formulations, including enhanced stability, targeted delivery, and controlled release. These properties help address challenges typically encountered in pharmaceutical and agricultural science, such as poor solubility, degradation, low bioavailability, and the need for prolonged action. Advanced nanopesticide formulations can provide longer-lasting effects with fewer applications, which is a crucial benefit for reducing the overall volume of pesticides needed in agricultural settings.

Emerging perspectives in agricultural science suggest a pressing need to address issues such as the rise in resistance among pathogens to conventional pesticides. Nanopesticides, particularly those with multiple mechanisms of action, may help address this issue by delivering more effective, targeted interventions. For instance, metallic and bimetallic nanoparticles, or siRNA-encapsulated compounds, are being explored for their potential to enhance pest management by targeting specific biological pathways in pests. However, the introduction of such novel materials also brings new concerns regarding their environmental fate, particularly their impact on non-target organisms, such as pollinators essential for ecosystem functioning and food security. Pollinators, which contribute to approximately 80% of crop pollination, are crucial for maintaining biodiversity and supporting sustainable agriculture. As such, protecting these organisms while using nanopesticides will require careful assessment and design choices to avoid unintended ecological harm.

In response to these challenges, scientists are working on stimuli-responsive nanocarriers, which release pesticides in controlled amounts based on environmental conditions like light, moisture, pH, and temperature. This approach can improve the efficiency of pesticide utilization, reducing the risk of leaching and drift and, consequently, lowering environmental contamination. By matching pesticide release with the presence of pests or other specific conditions, these systems can offer precision that traditional pesticide applications lack. Ideally, this could result in a more sustainable pest management strategy with fewer negative effects on ecosystems.

The potential role of metallic nanoparticles in sustainable agriculture is another area of considerable interest, as these materials could facilitate controlled and effective pest management with minimal environmental impact. Going forward, integrating in silico models and predictive ecotoxicological frameworks will be essential for assessing nanopesticide risks. These computational tools can help simulate acute environmental exposures and predict potential toxicological impacts, thereby supporting more informed decision-making during nanopesticide development. Such an interdisciplinary approach, combining computational modeling with experimental validation, will improve our ability to create nanopesticide formulations that balance efficacy with safety.

Innovation in nanopesticide development will also benefit from contributions across fields. For instance, combining computational tools, chemical pest management techniques, and principles from engineering to design efficient nanocarriers can enhance targeted delivery systems. Such advancements will require collaboration between disciplines, allowing researchers to optimize nanopesticide efficiency, improve food safety, and minimize environmental risks. Additionally, automated systems that streamline the application of nanopesticides in the field may further reduce their environmental footprint by ensuring precise, measured doses and avoiding overspreading.

In summary, the future of nanopesticides in agriculture presents a unique opportunity to address longstanding challenges in pest control and crop protection. While these formulations have the potential to increase agricultural yields and reduce environmental pollution, they also come with unknowns that must be carefully studied. By prioritizing biosafety research, improving predictive modeling, and advancing interdisciplinary collaborations, the agricultural sector can create nanopesticides that meet the dual goals of productivity and environmental sustainability. Such an approach will ultimately enable safer, more effective pest control solutions that contribute to a resilient and eco-friendly agricultural system.

References

Agostini A, Mondragón L, Coll C, Aznar E, Marcos MD, Martínez-Máñez R, Sancenón F, Soto J, Pérez-Payá E, Amorós P (2012) Dual enzyme-triggered controlled release on capped nanometric silica mesoporous supports. Chem Open 1:17–20

Alromeed AA, Scrano L, Bufo SA, Undabeytia T (2015) Slow-release formulations of the herbicide MCPA by using clay-protein composites. Pest Manag Sci 71:1303–1310

Bahadir M, Pfister G (1990) Controlled release formulations of pesticides. In: Controlled release, biochemical effects of pesticides, inhibition of plant pathogenic fungi. Springer, Berlin/Heidelberg, pp 1–64

Bang SH, Hwang IC, Yu YM, Kwon HR, Kim DH, Park HJ (2011) Influence of chitosan coating on the liposomal surface on physicochemical properties and the release profile of nanocarrier systems. J Microencapsul 28:595–604

Bisset NB, Webster GR, Dong YD, Boyd BJ (2019) Understanding the kinetic mixing between liquid crystalline nanoparticles and agrochemical actives. Colloids Surf B Biointerfaces 175: 324–332

Bombo AB, Pereira AES, Lusa MG, De Medeiros Oliveira E, De Oliveira JL, Campos EVR, De Jesus MB, Oliveira HC, Fraceto LF, Mayer JLSA (2019) Mechanistic view of interactions of a nanoherbicide with target organism. J Agric Food Chem 67:4453–4462

Chaud M, Eliana B, Souto, Zielinska A, Severino P, Batain F, Oliveira-Junior J, Alves T (2021) Nanopesticides in agriculture: benefits and challenge in agricultural productivity, toxicological risks to human health and environment. Toxics 9(6):131

Chen K, Yu G, He F, Zhou Q, Xiao D, Li J, Feng Y (2017) A pH-responsive emulsion stabilized by alginate-grafted anisotropic silica and its application in the controlled release of λ-cyhalothrin. Carbohydr Polym 176:203–213

Chen C, Zhang G, Dai Z, Xiang Y, Liu B, Bian P, Zheng K, Wu Z, Cai D (2018) Fabrication of light-responsively controlled release herbicide using a nanocomposite. Chem Eng J 349:101–110

Ciriminna R, Sciortino M, Alonzo G, De Schrijver A, Pagliaro M (2011) From molecules to systems: sol-gel microencapsulation insilica-based materials. Chem Rev 111:765–789

De Oliveira CR, Fraceto LF, Rizzi GM, Salla RF, Abdalla FC, Costa MJ, Silva-Zacarin ECM (2016) Hepatic effects of the clomazone herbicide in both its free form and associated with chitosan-alginate nanoparticles in bullfrog tadpoles. Chemosphere 149:304–313

De Paiva PP, Delcorso MC, Matheus VA, do Nascimento de Queiroz SC, Collares-Buzato CB, Arana S (2017) Acute toxicity of commercial atrazine in Piaractus mesopotamicus: histopathological, ultrastructural, molecular, and genotoxic evaluation. Vet World 10:1008–1019

De Souza JF, da Silva Pontes K, Alves TFR, Torqueti de Barros C, Amaral VA, de Moura Crescencio KM, Rios AC, Batain F, Souto EB, Severino P et al (2020) Structural comparison, physicochemical properties, and in vitro release profile of curcumin-loaded lyotropic liquid crystalline nanoparticle: influence of hydrotrope as interface stabilizers. J Mol Liq 306:112861

Dos Santos KC, Martinez CBR (2014) Genotoxic and biochemical effects of atrazine and Roundup (®), alone and in combination, on the Asian clam Corbicula fluminea. Ecotoxicol Environ Saf 100:7–14

Ferreira AJ, Cemlyn-Jones J, Robalo Cordeiro C (2013) Nanoparticles, nanotechnology and pulmonary nanotoxicology. Rev Port Pneumol 19:28–37

Fletcher SJ, Reeves PT, Hoang BT, Mitter N (2020) A perspective on RNAi-based biopesticides. Front Plant Sci 11:51

Fraceto LF, Grillo R, de Medeiros GA, Scognamiglio V, Rea G, Bartolucci C (2016) Nanotechnology in agriculture: which innovation potential does it have? Front Environ Sci 4:1–5

Gahan LJ, Gould F, Heckel DG (2001) Identification of a gene associated with Bt resistance in Heliothis virescens. Science 293:857–860

Gao Y, Liu Y, Qin X, Guo Z, Li D, Li C, Wan H, Zhu F, Li J, Zhang Z et al (2021) Dual stimuli-responsive fungicide carrier based on hollow mesoporous silica/hydroxypropyl cellulose hybrid nanoparticles. J Hazard Mater 414:125513

Grillo R, dos Santos NZP, Maruyama CR, Rosa AH, de Lima R, Fraceto LF (2012) Poly (ε-caprolactone)nanocapsules as carrier systems for herbicides: physico-chemical characterization and genotoxicity evaluation. J Hazard Mater 231–232:1–9

Hayles J, Johnson L, Worthley C, Losic D (2017) Nanopesticides: a review of current research and perspectives. Elsevier, Amsterdam. ISBN 9780128042991

Hazra DK, Purkait (2019) A role of pesticide formulations for sustainable crop protection and environment management: a review. J Pharmacogn Phytochem 8:686–693

Hwang IC, Kim TH, Bang SH, Kim KS, Kwon HR, Seo MJ, Youn YN, Park HJ, Yasunaga-Aoki C, Yu YM (2011) Insecticidal effect of controlled release formulations of Etofenprox based on Nano-biotechnique. J Fac Agric Kyushu Univ 56:33–40

Jampílek J, Král'ová K (2017) Nanopesticides: preparation, targeting, and controlled release. Academic, Cambridge, MA. ISBN:9780128042991

Kah M, Hofmann T (2014) Nanopesticide research: current trends and future priorities. Environ Int 63:224–235

Keller AA, Adeleye AS, Conway JR, Garner KL, Zhao L, Cherr GN, Hong J, Gardea-Torresdey JL, Godwin HA, Hanna S et al (2017) Comparative environmental fate and toxicity of copper nanomaterials. Nano Impact 7:28–40

Khan I, Saeed K, Khan I (2019) Nanoparticles: properties, applications and toxicities. Arab J Chem 12:908–931

Kuhlbusch TAJ, Wijnhoven SWP, Haase A (2018) Nanomaterial exposures for worker, consumer and the general public. NanoImpact 10:11–25

Kushwaha S, Vikram A, Trivedi PP, Jena GB (2011) Alkaline, Endo III and FPG modified comet assay as biomarkers for thedetection of oxidative DNA damage in rats with experimentally induced diabetes. Mutat Res Genet Toxicol Environ Mutagen 726:242–250

Li GB, Wang J, Kong XP (2020) Coprecipitation-based synchronous pesticide encapsulation with chitosan for controlled spinosadrelease. Carbohyd Polym 249:116865

Malandrakis AA, Kavroulakis N, Chrysikopoulos CV (2021) Copper nanoparticles against benzimidazole-resistant Monilinia fructicola field isolates. Pestic Biochem Physiol 173:104796

Mgadi K, Ndaba B, Roopnarain A, Rama H, Adeleke R (2024) Nanoparticle applications in agriculture: overview and response of plant-associated microorganisms. Front Microbiol 6: 15–1354440

Mittal M, Siddiqui MR, Tran K, Reddy SP, Malik AB (2014) Reactive oxygen species in inflammation and tissue injury. Antioxid Redox Signal 20:1126–1167

Mustafa IF, Hussein MZ (2020) Synthesis and technology of nanoemulsion-based pesticide formulation. Nano 10:1608

Nogueira EN, Dores EFGC, Pinto AA, Amorim RSS, Ribeiro ML, Lourencetti C (2012) Currently used pesticides in water matrices in Central-Western Brazil. J Braz Chem Soc 23:1476–1487

Nuruzzaman M, Rahman MM, Liu Y, Naidu R (2016) Nanoencapsulation, Nano-guard for pesticides: a new window for safe application. J Agric Food Chem 64:1447–1483

Osman Mohamed Ali E, Shakil NA, Rana VS, Sarkar DJ, Majumder S, Kaushik P, Singh BB, Kumar J (2017) Antifungal activityofnano emulsions of neem and citronella oils against phytopathogenic fungi, *Rhizoctonia solani* and *Sclerotium rolfsii*. Ind Crop Prod 108:379–387

Osorio-Echavarría J, Osorio-Echavarría J, Ossa-Orozco CP, Gómez-Vanegas NA (2021) Synthesis of silver nanoparticles using white-rot fungus anamorphous Bjerkandera sp. R 1: influence of silver nitrate concentration and fungus growth time. Sci Rep 11:1–14

Parsons KH, Mondal MH, McCormick CL, Flynt AS (2018) Guanidinium-functionalized Interpol electrolyte complexes enabling RNAi in resistant insect pests. Biomacromolecules 19:1111–1117

Patel S, Bajpai J, Saini R, Bajpai AK, Acharya S (2018) Sustained release of pesticide (Cypermethrin) from nanocarriers: an effective technique for environmental and crop protection. Process Saf Environ Prot 117:315–325

Pérez-Iglesias JM, Ruiz de Arcaute C, Natale GS, Soloneski S, Larramendy ML (2017) Evaluation of imazethapyr-induced DNA oxidative damage by alkaline Endo III- and Fpg-modified single-cell gel electrophoresis assay in Hypsiboas pulchellus tadpoles (Anura, Hylidae). Ecotoxicol Environ Saf 142:503–508

Prasad R, Bhattacharyya A, Nguyen QD (2017) Nanotechnology in sustainable agriculture: recent developments, challenges, and perspectives. Front Microbiol 8:1–13

Ranganathan P, Mutharani B, Chen SM, Sireesha P (2019) Biocompatible chitosan-pectin polyelectrolyte complex for simultaneous electrochemical determination of metronidazole and metribuzin. Carbohydr Polym 214:317–327

Sánchez-López E, Gomes D, Esteruelas G, Bonilla L, Lopez-Machado AL, Galindo R, Cano A, Espina M, Ettcheto M, Camins A (2020) Metal-based nanoparticles as antimicrobial agents: an overview. Nanomaterials 10:292

Santos TS, Dos Passos EM, de Jesus Seabra MG, Souto EB, Severino P, da Costa MM (2021) Entomopathogenic fungibiomass production and extracellular biosynthesis of silver nanoparticles for bioinsecticide action. Appl Sci 11:2465

Sarlak N, Taherifar A, Salehi F (2014) Synthesis of nanopesticides by encapsulating pesticide nanoparticles using functionalized carbon nanotubes and application of new nanocomposite for plant disease treatment. J Agric Food Chem 62:4833–4838

Schnoor B, Elhendawy A, Joseph S, Putman M, Chacón-Cerdas R, Flores-Mora D, Bravo-Moraga-F, Gonzalez-Nilo F, Salvador-Morales C (2018) Engineering atrazine loaded poly (lactic-co-glycolic acid) nanoparticles to ameliorate environmental challenges. J Agric Food Chem 66:7889–7898

Sershen SR, Westcott SL, Halas NJ, West JL (2000) Temperature-sensitive polymer-nanoshell composites for photothermally modulated drug delivery. J Biomed Mater Res 51:293–298

Shen H, Brandt A, Witting-Bissinger BE, Gunnoe TB, Roe RM (2009) Novel insecticide polymer chemistry to reduce the enzymatic digestion of a protein pesticide, trypsin modulating oostatic factor (TMOF). Pestic Biochem Physiol 93:144–152

Syafrudin M, Kristanti RA, Yuniarto A, Hadibarata T, Rhee J, Al-onazi WA, Algarni TS, Almarri AH, Al-Mohaimeed AM (2021) Pesticides in drinking water—a review. Int J Environ Res Public Health 18:468

Tu F, Lee D (2014) Shape-changing and amphiphilicity-reversing Janus particles with pH-responsive surfactant properties. J Am Chem Soc 136:9999–10006

Vigneshwaran N, Kathe AA, Varadarajan PV, Nachane RP, Balasubramanya RH (2006) Biomimetics of silver nanoparticles by white rot fungus, *Phaenerochaete chrysosporium*. Colloids Surf B Biointerfaces 53:55–59

Vithanage M, Seneviratne M, Ahmad M, Sarkar B, Ok YS (2017) Contrasting effects of engineered carbon nanotubes on plants: a review. Environ Geochem Health 39:1421–1439

Wang N, Qi L, Wang Y, Li XG (2017) Preparation and performance of Thermo-sensitive Pyraclostrobin microcapsules. Chin J Pestic Sci 19:381–387

Wasef L, Nassar AMK, El-Sayed YS, Samak D, Noreldin A, Elshony N, Saleh H, Elewa YHA, Hassan SMA, Saati AA et al (2021) The potential ameliorative impacts of cerium oxide nanoparticles against fipronil-induced hepatic steatosis. The potential ameliorative impacts of cerium oxide nanoparticles against fipronil-induced hepatic steatosis Sci Rep 11:1–15

Worthington KLS, Adamcakova-Dodd A, Wongrakpanich A, Mudunkotuwa IA, Mapuskar KA, Joshi VB, Allan Guymon C, Spitz DR, Grassian VH, Thorne PS et al (2013) Chitosan coating of copper nanoparticles reduces in vitro toxicity and increases inflammation in the lung. Nanotechnology 24:395101

Xu C, Cao L, Zhao P, Zhou Z, Cao C, Li F, Huang Q (2018) Emulsion-based synchronous pesticide encapsulation and surface modification of mesoporous silica nanoparticles with carboxymethyl chitosan for controlled azoxystrobin release. Chem Eng J 348:244–254

Xu Z, Tang T, Lin Q et al (2022) Environmental risks and the potential benefits of nanopesticides: a review. Environ Chem Lett 20:2097–2108

Yadav RK, Singh NB, Singh A, Yadav V, Bano C, Khare S, Niharika (2020) Expanding the horizons of nanotechnology in agriculture: recent advances, challenges and future perspectives. Vegetos 33:203–221

Zainab R, Hasnain M, Ali F, Abideen Z, Siddiqui ZS, Jamil F, Hussain M, Park Y (2024) Prospects and challenges of nanopesticides in advancing pest management for sustainable agricultural and environmental service. Environ Res 261:119722

Zhang H, Ji Z, Xia T, Meng H, Low-Kam C, Liu R, Pokhrel S, Lin S, Wang X, Liao YP et al (2012) Use of metal oxide nanoparticle band gap to develop a predictive paradigm for oxidative stress and acute pulmonary inflammation. ACS Nano 6:4349–4368

Zhang Y, Chen W, Jing M, Liu S, Feng J, Wu H, Zhou Y, Zhang X, Ma Z (2019) Self-assembled mixed micelle loaded with natural pyrethrins as an intelligent nano-insecticide with a novel temperature-responsive release mode. Chem Eng J 361:1381–1391

Zielinska A, Costa B, Ferreira MV, Miguéis D, Louros JMS, Durazzo A, Lucarini M, Eder P, Chaud MV, Morsink M et al (2020) Nanotoxicology and nanosafety: safety-by-design and testing at a glance. Int J Environ Res Public Health 17:4657

Chapter 16
Sustainable Agriculture and the Role of Nanotechnology

Vidya Niranjan, Lavanya C, Anagha S Setlur, Chandrashekar Karunakaran, and Sinosh Skariyachan

Abstract Nanotechnology has emerged as a transformative tool across multiple domains, with agriculture and food industries being key beneficiaries. This book chapter presents a comprehensive overview of various nanoparticles, their synthesis, and applications, particularly in agricultural biotechnology. Major nanoparticles such as carbon nanotubes (CNTs), fullerenes, quantum dots, nanorods, and nanoencapsulations are discussed along with their unique properties and contributions to enhancing agricultural outputs. The chapter emphasizes green synthesis methods, highlighting the influence of parameters such as temperature, pH, salt concentration, and time on nanoparticle formation, which aligns with sustainable development goals. The need for nanotechnology in sustainable agriculture is underscored by examples like nanofertilizers and nanopesticides, which promote efficient nutrient delivery and pest control while minimizing environmental impact. Toxicological aspects, nanobiosensors, and nanoparticle delivery systems are also explored to address the balance between innovation and safety. The chapter further delves into the role of nanotechnology in the food industry, detailing applications in food processing, packaging, and labeling that improve product quality and shelf life. While the potential of nanoparticles is vast, the chapter also discusses associated risks, challenges, and regulatory concerns that may affect their adoption. The merits and limitations of this technology are critically evaluated, considering its role in advancing sustainable agriculture. Future perspectives provide insights into emerging trends and the need for further research to harness nanotechnology's full potential responsibly. The chapter concludes with a synthesis of findings, offering

V. Niranjan
Department of Biotechnology, RV College of Engineering, Bangalore, India

Research & Collaborations, MIT Vishwaprayag University, Solapur, India

Lavanya C · A. S. Setlur · C. Karunakaran
Department of Biotechnology, RV College of Engineering, Bangalore, India

S. Skariyachan (✉)
Department of Microbiology, St. Pius X College, Rajapuram, Kasaragod, Kerala, India
e-mail: sinoshmicro@stpius.ac.in

J. M. Al-Khayri et al. (eds.), *Nanotechnology Applications in Modern Agriculture*, Nanotechnology in Plant Sciences 3, https://doi.org/10.1007/978-3-031-90506-3_16

a roadmap for integrating nanotechnology into agricultural and food systems sustainably.

Keywords Green synthesis · Nano-biosensors · Nano-fertilizers · Nanotechnology · Nanotechnology-aided approaches · Sustainable agriculture

16.1 Introduction

An approach to farming that emphasizes on long-term environmental and economical balance whist tackling the social and economic aspects of food production is known as sustainable agriculture. It primarily aims to meet the present needs of agriculture without compromising the ability of the future generations to meet their requirements. One major challenge in sustainable agriculture is enhancing productivity and efficiency of resource while simultaneously minimizing the environmental adverse effects. Recently, nanotechnology has emerged as a promising arena with potential to revolutionize agricultural techniques. Nanotechnology is the manipulation of materials at a nano-scale (between 1 to 100 nm typically). It has potential benefits including food quality augmentation, enhancing safety, nutrient absorption into the soil at nanoscale levels, and diminishing agricultural inputs (Prasad et al. 2017). The primary application of nanomaterials in agriculture is to minimize chemical spread, nutrient losses and improve yield through sustainable management (Prasad et al. 2017).

The integration of nanotechnology into agriculture holds the promise of optimizing resource use, improving crop yields, and minimizing environmental impacts. Nanomaterials, such as nanoparticles and nanoscale delivery systems, have shown potential in enhancing the efficiency of nutrient delivery, pest control, and water management in agricultural systems. This intersection of sustainable agriculture and nanotechnology opens up new avenues for innovation and research. As the world grapples with the increasing demands for food production, coupled with the challenges of climate change and resource depletion, exploring the role of nanotechnology in sustainable agriculture becomes imperative. This chapter thus, comprehensively details the major applications and use of nanotechnology in sustainable agriculture.

16.2 Major Nanoparticles and Their Applications

In agriculture, optimum plant nutrition management is vital for sustainable productivity of the crops, and research in this field is now emerging. With regard to this, advancements in the area of crop nutrition have led to the realization of nano-nutrition for sustainable production (Ditta et al. 2015). Two primary sources of crop nutrition in terms of nano-particles (NPs) are currently being used: abiotic and biotic NPs. Biotic NPs are from organic sources that make them both environment friendly and bio-degradable, while abiotic NPS are from inorganic salts, but are non-biodegradable (Ditta et al. 2015). Increasing the nutrient uptake in plants is the focus of using NPs in sustainable agriculture. Some important NPs commonly being employed are provided below:

16.2.1 Metal Nanoparticles

Metal nanoparticles, such as silver (Ag), gold (Au), zinc (Zn), copper (Cu), and titanium (Ti), exhibit antimicrobial properties and are employed for crop protection and disease management (Raza et al. 2024). They also enhance nutrient uptake by plants and facilitate the delivery of nutrients to target tissues (Kah et al. 2018).

16.2.2 Metal Oxide Nanoparticles

Metal oxide nanoparticles like titanium dioxide (TiO2), zinc oxide (ZnO), and iron oxide (Fe2O3) have been utilized in soil remediation, nutrient delivery, and as carriers for pesticides and fertilizers (Dimkpa et al. 2012). Their applications extend to improving soil fertility and mitigating soil-borne diseases (Bandyopadhyay et al. 2020).

16.2.3 Silica Nanoparticles

Silica nanoparticles find applications in agriculture for controlled release of agro-chemicals, improving soil structure, and promoting plant growth (Naaz et al. 2023). They contribute to enhanced nutrient uptake efficiency and stress tolerance in plants (Gupta et al. 2020).

16.2.4 Carbon-Based Nanoparticles

Carbon nanoparticles, including carbon nanotubes (CNTs) and graphene, are explored for enhancing soil properties, water retention, and the targeted delivery of agrochemicals (Miralles et al. 2012). They offer potential benefits in terms of improving soil structure and enhancing nutrient availability (Mahajan et al. 2019).

16.2.5 Lipid-Based Nanoparticles

Lipid nanoparticles, such as liposomes and solid lipid nanoparticles (SLNs), are employed for targeted delivery of agrochemicals and enhancing their efficacy (Soni et al. 2020). They contribute to reduced environmental impact and improved crop productivity (Tripathi et al. 2023).

16.2.6 Polymeric Nanoparticles

Nanoparticles derived from biodegradable polymers like chitosan, poly (lactic-co-glycolic acid) (PLGA), and polyethylene glycol (PEG) are utilized for controlled release of agrochemicals and improving seed germination (Hu et al. 2014). They offer advantages such as sustained nutrient release and improved crop yield (Kaur et al. 2020).

16.2.7 Carbon Nanotubes (CNTs)

With a tensile strength of ~200 GPa, these are new forms of carbon, tantamount to 2D graphene sheets rolled in the form of a tube. Two types in CNTs are multi-walled and single-walled. CNTs comprise of graphitic layers wrapped seamlessly into cylinders, structurally (Isahak et al. 2023). Attributes such as electronic, thermal, mechanical, thermal, elastic, optical, are determined via geometrical dimensions by diameter, and the utilization of these properties will help open avenues to the development of several nanodevices that can aid in sustainable agriculture (Naaz et al. 2023). To decrease the amount of chemicals produced into the environment and/or degrade other plant tissues, targeting via CNT based delivery systems is a major application (Naaz et al. 2023; Hajirostamlo et al. 2015). Further, studies have previously reported that CNTs containing iron and Sulphur augmented water contaminant soaking efficiency, including pesticides, oil, fertilizers (Dehghani et al. 2024), thereby improving product yield.

16.2.8 Carbon Fullerenes

Carbon fullerenes, particularly C60, offer diverse applications in sustainable agriculture owing to their unique physicochemical properties. They serve as efficient carriers for nutrients and agrochemicals, enhancing their delivery to plants while protecting them from degradation, thus improving nutrient uptake efficiency (Gao et al. 2020). Moreover, carbon fullerenes demonstrate promise in soil remediation by binding with heavy metals and organic pollutants, contributing to soil health and fertility (Athanassiou et al. 2020). Additionally, fullerene-based nanomaterials have been studied for their ability to promote plant growth and development by acting as antioxidants, enhancing photosynthesis, and improving tolerance to abiotic stress factors such as drought and salinity (de Souza et al. 2025). They also show potential in pest and disease management through targeted delivery of pesticides and antimicrobial agents, reducing off-target effects and environmental contamination, while exhibiting antimicrobial properties themselves (Yin et al. 2021). Overall, the use of carbon fullerenes in agriculture aligns with principles of environmental sustainability, offering opportunities to enhance agricultural productivity while minimizing negative impacts on ecosystems and human health.

16.2.9 Quantum Dots (QD)

Semiconductor QDs are molar extinction co-efficient, with high quantum yield. QDs also have large excitation, resistance to photochemical degradation, and high resistance to photobleaching (Nabil and Megahed 2023). Due to these reasons, they are good materials for fluorescence, confinement of charge carrier's materials and have band energy that can be tunable easily (Androvitsaneas et al. 2016). Additionally, QDs also have novel spectral attributes when compared to traditional organic dyes, thereby, finding its application in biosensing and imaging (Hu et al. 2010). At a lower concentration, quantum dots were found to have no cyto-toxicity for the growth of seedlings and germination of seeds, therefore, allowing it to be used for imaging live in the plant root systems for the detection and identification of Phyto-physiological mechanisms (Hu et al. 2010; Das et al. 2015).

16.2.10 Nanorods

Nanorods are plasmonic materials that have good couple sensing phenomenon and adjustable energy regulation. Coupled with MEMS, nanorods can trigger specific field responses (Nabil and Megahed 2023). For instance, the gold nanorods were employed previously for detecting physiological modifications that occurred in watermelon plant (Bandi et al. 2023). Another study showed that nanorods were employed to assess its capability to transport 2,4-D, an auxin growth regulator in tobacco (Nima et al. 2014). Thus, nanorods hold promise in sustainable agriculture by facilitating efficient nutrient delivery, soil remediation, and pest control. Through their high aspect ratio and large surface area, nanorods enhance nutrient uptake by plants and adsorb pollutants from soil, contributing to improved crop growth and soil health. Additionally, they enable targeted delivery of pesticides and fertilizers, reducing chemical usage and environmental contamination.

16.2.11 Micro and Nano Encapsulations

The process where an object is embedded within a heterogenous or homogenous matrix/coating due to which the final formed capsule will be imbibed with useful properties, is called encapsulation. This process comes with several merits such as protection of the object, controlled release of substance and for precise targeting (Rodriguez et al. 2016; Ozdemir and Kemerli 2016). Micro and nano encapsulations give particles at a micro and nano scale. These capsules are systems in which the objects/substances are restricted to a cavity comprising of a liquid core surrounded by membranes made of polymers (Couvreur et al. 1995). Advancements in nanotechnology have made drug delivery better using nano and micro-NPs, thereby allowing for an increase in bioavailability in plant science (Ozdemir and Kemerli 2016). Studies have shown that nano encapsulation methods have been previously employed to deliver bio-actives to targeted tissues (Pohlmann, 2022). Thus, by encapsulating active compounds within protective matrices at the micro and nano scales, these technologies enhance the efficiency of nutrient utilization, reduce chemical leaching, and minimize environmental contamination. Microencapsulation ensures controlled release of nutrients and agrochemicals, promoting targeted delivery to plants while reducing the need for frequent applications. Nanoencapsulation, on the other hand, provides even finer control over release kinetics and enables tailored delivery to specific plant tissues or pests. These encapsulation approaches contribute to resource efficiency, soil health improvement, and reduced ecological impacts, thereby supporting sustainable practices.

16.2.12 Nano Emulsions

Formed by very minute emulsion nanoscale droplets (from oil/water systems), nano emulsions exhibit sizes lower than ~100 nm, with properties quite different from micro scale emulsions. Further, nano emulsions possess significantly larger surface area-to-volume ratios, Laplace pressure, and elastic moduli compared to ordinary emulsions, primarily due to the small size of droplets. Additionally, unlike conventional emulsions, most nano emulsions exhibit optical transparency, rendering them advantageous for incorporation into beverages. However, the formulation of nano emulsions demands high energy inputs, necessitating specialized equipment capable of generating extreme shear stress, such as high-pressure homogenizers or ultrasonic generators (Asua 2002; Gupta et al. 2016). Al-Hussaniy et al. (2023) introduced a "low-energy" method for nano emulsion formation, involving the contact of two liquid phases—a homogeneous mixture comprising a lipophilic phase, hydrophilic surfactant, and potentially a solvent, polymer, or drug, with an aqueous phase, including pure water. In this process, the hydrophilic components from the oily phase rapidly dissolve into the aqueous phase, leading to the remixing of oil into nano-droplets, promptly stabilized by amphiphiles (Anton and Vandamme 2010;

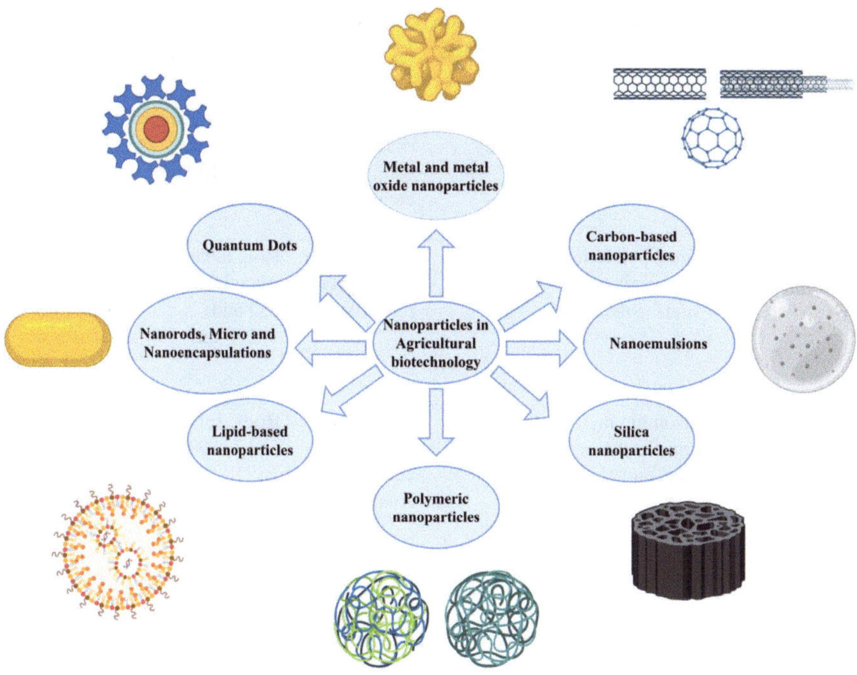

Fig. 16.1 Major types of nanoparticles used in agricultural biotechnology

Gupta et al. 2016). This method offers simplicity and does not require specialized high-energy devices. Figure 16.1 shows the major types of nanoparticles used in agricultural biotechnology, and Table 16.1 highlights the major salient features of nanoparticles used in agriculture suitability.

16.3 Green Synthesis of Nanoparticles

Green synthesis of nanoparticles for agriculture sustainability involves eco-friendly methods that utilize natural resources and environmentally benign approaches to produce nanoparticles. These methods typically involve the use of plant extracts, microorganisms, or agricultural waste materials as reducing agents and stabilizers in the synthesis process. Green synthesis offers several advantages over conventional chemical methods, including lower energy consumption, reduced toxicity, and biocompatibility (Kuppusamy et al. 2015).

These nanoparticles synthesized through green routes have shown promising applications in various aspects of agriculture, including crop enhancement, disease management, soil fertility improvement, and environmental remediation. For instance, silver nanoparticles synthesized using plant extracts have demonstrated potent antimicrobial activity against plant pathogens, helping to protect crops from

Table 16.1 Salient features of major nanoparticles used in agriculture suitability

Nanoparticles	Solubility	Toxicology	Optical properties	Structure and Size	References
Aluminum nanoparticles (AlNPs)	Soluble in ethanol and acetone but insoluble in H_2O	Low toxicity	Transparent or translucent; poorly conductive	Nano cubes, nanosphere and 1–100 nm	Wang et al. (2015), Yang et al. (2021), Altammar (2023)
Gold nanoparticles (AuNPs)	Great solubility in organic solvents (toluene)	Low toxicity	Gold color and highly reflective	Rod, hexagon, triangle, spherical and 1–100 nm	Yang et al. (2021), Guo et al. (2013), Altammar (2023), Khan et al. (2021)
Iron nanoparticles (FeNPs)	Insoluble in organic solutions and H2O	Low toxicity	Transparent or translucent; poorly conductive	Rods, spheres and 1–100 nm	Essajai et al. (2019), Yang et al. (2021), Khan et al. (2019), Altammar (2023)
Silver nanoparticles (AgNPs)	Colloidal stability is for long-term and very good water solubility	Low toxicity	Transparent or translucent; highly reflective	Triangle, rod, spherical, octagonal and 1–100 nm	Yang et al. (2021), Altammar (2023), Jana et al. (2007), Graf et al. (2003)
Copper nanoparticles (CuNPs)	At pH 9–11, minimal solubility of cu is observed	Low toxicity	Transparent or translucent; highly conductive	Tetrahedron, rods and cubes; 1–100 nm	Yang et al. (2021), Altammar (2023), Hortin et al. (2020), Khan et al. (2019)
Zinc oxide nanoparticles (ZnONPs)	Approximately 0.3–3.6 mg/L in aqueous medium	Low toxicity	Transparent or translucent; poorly conductive	Hexagonal and polycrystalline structure; 1–100 nm	Yang et al. (2021), Altammar (2023), Siddiqi and Husen (2016), Khan et al. (2019)
Carbon nanotubes (single and multi-walled carbon nanotubes)	Poorly soluble in polar solvents	Highly toxic	Not so transparent or translucent; moderately conductive	SWCNT – 0.2 to 5 μm; 0.5 to 2 nm MWCNT - 0.36 nm; 2 to 100 nm	Patel et al. (2020), Aacharya and Chhipa (2020)

diseases (Siddiqi et al. 2018). Similarly, copper nanoparticles derived from agricultural waste have been found effective in controlling fungal infections in plants (Dwivedi et al., 2012).

Furthermore, these green synthesized NPs may be used for delivery of nutrients and amendment of soil, thereby, contributing to improved plant yield and growth. Studies have previously shown that zinc oxide NPs synthesized via green techniques have demonstrated augmented nutrient uptake, in turn promoting crop growth (Ghorbanpour and hadian, 2017). Additionally, these nanoparticles can also be utilized in the formulation of nano-pesticides and nano-fertilizers, which offer targeted delivery and controlled release of active ingredients, thereby reducing the overall usage of agrochemicals and minimizing environmental pollution (Mehariya et al. 2021). Additionally, the green synthesis of nanoparticles aligns with the principles of sustainable agriculture by promoting eco-friendly practices, reducing dependability on chemical inputs and compounds, and also reducing the environmental impact. By utilizing renewable resources and adopting environmentally benign processes, green synthesis contributes to the conservation and preservation of natural resources.

16.3.1 General Procedure for Green Synthesis of NPs

The following are the steps involved in typical green synthesis of nanoparticles.

- Selection of eco-friendly materials: Identification of natural resources like plant extracts or microorganisms.
- Preparation of extracts or biomass: Extraction of desired components from plants or culture and harvesting the microorganisms.
- Mixing with precursor materials: Combining the extracts or biomass with metal salts or other compounds.
- Controlled synthesis: Subjecting the mixture to controlled conditions to facilitate formation of nanoparticles.
- Characterization: Analyses of nanoparticle properties using techniques like UV-Vis spectroscopy and electron microscopy.
- Optimization: Fine-tuning the synthesis parameters such as temperature and pH to enhance nanoparticle quality.
- Application testing: Evaluation of nanoparticle efficacy in agriculture through laboratory, greenhouse, or field studies

16.4 Factors Influencing Green Synthesis

Green synthesis of nanoparticles is influenced by several factors such as:

(a) **Synthesis source:** The choice of natural resources or eco-friendly materials, such as plant extracts, microorganisms, or agricultural waste, vastly influences the synthesis process. Different sources possess varying compositions of biomolecules that can act as reducing agents or stabilizers in nanoparticle formation (Kumar et al. 2018).
(b) **Temperature:** The temperature of the synthesis reaction plays a crucial role in controlling the rate of nanoparticle formation, size distribution, and crystallinity. Optimal temperatures can promote efficient reduction of metal ions and facilitate nucleation and growth of nanoparticles (Sathiyanarayanan et al. 2017).
(c) **Concentration of salts:** The concentration of metal salts or precursor materials in the reaction mixture influences the rate of nanoparticle formation and the final yield of nanoparticles. Higher salt concentrations can lead to faster nucleation and growth rates, affecting nanoparticle size and morphology (Singh and Singh, 2020).
(d) **pH levels:** pH levels of the reaction medium affect the stability of biomolecules, as well as the chemical reduction and nucleation processes involved in nanoparticle synthesis. Maintaining suitable pH conditions is essential for controlling nanoparticle size, shape, and surface charge (Jeevanandam et al. 2018).
(e) Time of synthesis: The duration of the synthesis reaction determines the kinetics of nanoparticle formation, affecting the size, shape, and crystallinity of the synthesized nanoparticles. Longer reaction times may result in larger nanoparticles or aggregation, while shorter times may yield smaller and more uniform nanoparticles (Gupta et al. 2019).

These factors should be optimized in order to achieve the required NP attributes for particular applications in agriculture. Figure 16.2 shows the overview of green synthesis of nanoparticles factors and applications.

16.5 Need for Nanotechnology in Sustainable Agriculture

The pressing challenges in relation to global production of food such as population growth, change in climate, dwindling natural resources, environmental degradation, triggers the advent of new technology and innovative solutions, that may be now possible through nanotechnology. Enhancement in agricultural productivity, improvement in resource efficiency and simultaneously minimizing the environmental impacts are the need of the hour. Nanotechnology holds immense promise in addressing these urgent challenges facing sustainable agriculture due to its unique capabilities to manipulate matter at the nanoscale. For instance, nanoscale sensors enable precise monitoring of crop health, soil moisture, and nutrient levels, facilitating targeted interventions and optimizing resource use. Nanoparticles, such as nano fertilizers and nano pesticides, enhance nutrient uptake by plants and deliver agrochemicals with precision, reducing input requirements and minimizing environmental pollution. Moreover, nanomaterials improve soil structure, water retention,

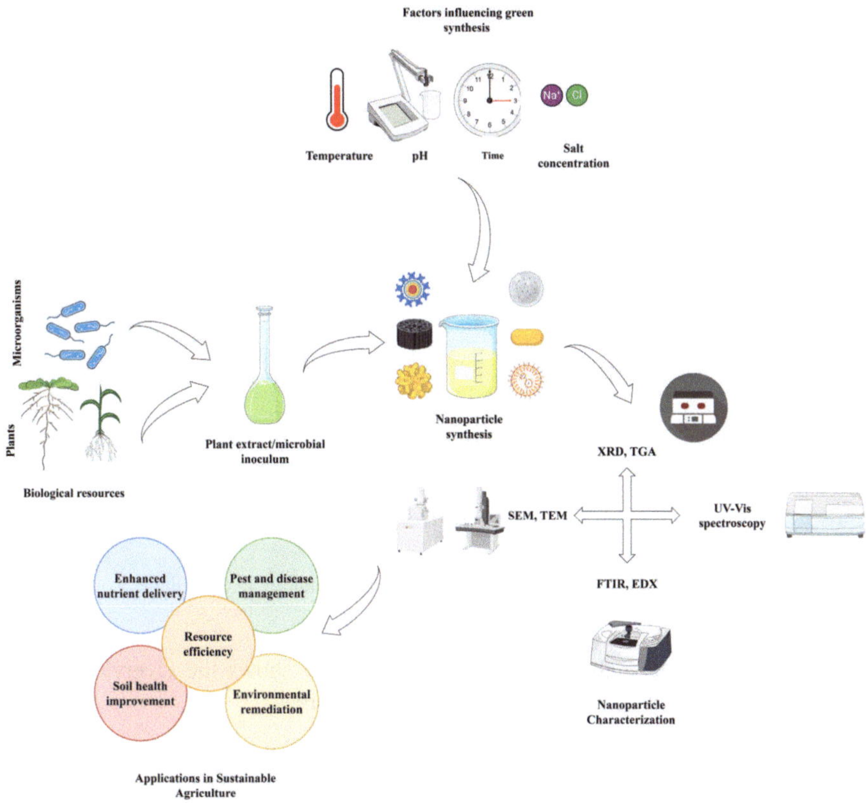

Fig. 16.2 The overview of green synthesis of nanoparticles, factors and applications in sustainable agriculture

and nutrient availability, promoting sustainable land use practices and ecosystem resilience.

Furthermore, nanotechnology contributes to sustainable agriculture by fostering resource efficiency, reducing energy consumption, and enhancing the resilience of agricultural systems to climate change and environmental stresses. By harnessing the unique properties of nanomaterials, such as high surface area-to-volume ratio, reactivity, and tunable properties, nanotechnology enables the development of advanced materials and technologies for agriculture. For example, nano-enabled materials for water purification and soil remediation help mitigate pollution and conserve water resources, essential for sustainable agricultural practices (Kah et al. 2015). Additionally, the integration of nanotechnology with biotechnology and bioinformatics offers novel approaches for crop improvement, disease resistance, and environmental monitoring. Despite the potential benefits, it is essential to address concerns related to the safety, regulatory aspects, and ethical considerations associated with nanotechnology applications in agriculture to ensure their

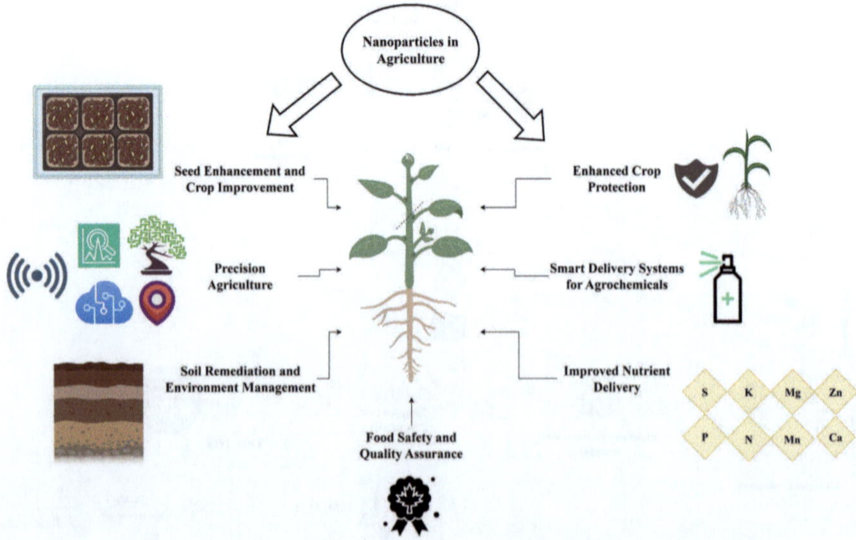

Fig. 16.3 Overview of the potential applications of nanotechnology in agriculture

responsible and sustainable deployment. Figure 16.3 shows the Overview of the potential applications of nanotechnology in agriculture.

16.6 Nanotechnology and Suitable Agricultural Developments

Nanotechnology holds great promise in revolutionizing sustainable farming practices, offering innovative tools for various aspects of agriculture. These include advanced solutions for pesticides, fertilizers, plant growth, and disease control, as well as precision farming techniques. The integration of nanotechnology has the potential to enhance crop yields and overall agricultural productivity (Saritha et al. 2022).

By leveraging nanotechnology, there is a possibility to decrease reliance on chemical inputs, reduce nutrient losses, and optimize pest and nutrient management strategies. This can result in more efficient and sustainable agricultural practices. Furthermore, the application of nanomaterials in plant growth, fertilizers, and pesticides has shown encouraging outcomes in improving both the quantity and quality of agricultural production (Yadav et al. 2023).

Nanotechnology brings valuable contributions to agriculture by improving crop management, reducing postharvest losses, and enhancing food quality and safety. Notable applications include the use of nanomaterials, like carbon nanoparticles, to boost crop productivity. In genetic modification, nanoparticles facilitate the transfer of genes or DNA in plants, leading to insect-resistant varieties and improved food processing (Panhwar et al. 2019). Additionally, nanotechnology is harnessed in pest and disease management through the creation of nano sensors for diagnostics, as well as minimizing the negative impacts of pesticides and fertilizers on crops. In essence, these advancements highlight how nanotechnology is playing a vital role in making agriculture more efficient, resilient, and sustainable (Kumari et al. 2023).

Nanotechnology brings valuable advantages to agriculture, promoting better crop productivity through the use of nanoparticles and nanostructures. These technologies enhance nutrient absorption, minimize water loss, and boost resistance to pests and diseases, ultimately improving overall crop yield (Muller et al. 2017). Furthermore, nanotechnology plays a role in environmental sustainability by reducing the environmental impact of farming practices. This is achieved by decreasing reliance on chemical inputs, mitigating nutrient losses, and enhancing soil quality, aligning with more eco-friendly agricultural practices. In the realm of postharvest management, nanotechnology is applied through the development of nano sensors for pest and disease diagnostics (Dar et al. 2020). Additionally, it contributes to minimizing postharvest losses by improving food packaging and storage methods, ensuring a more efficient and sustainable food supply chain. Moreover, nanotechnology enhances food quality and safety by detecting contaminants and pathogens, improving food processing techniques, and optimizing nutrient absorption from the soil. These advancements underscore the multifaceted positive impact of nanotechnology in agriculture, addressing key challenges while promoting sustainability and food security (Sarauskis et al. 2022; Si et al. 2018).

While nanotechnology holds promise for agriculture, it comes with its share of drawbacks. A major concern revolves around potential environmental and health risks linked to the use of nanomaterials. Insufficient research on the long-term impact of nanoparticles on the environment and human health raises uncertainties. Some studies even suggest adverse effects on plant growth and development due to nanoparticles (Shang et al. 2019). Another challenge lies in the high cost associated with nanotechnology research and development, posing accessibility issues for small-scale farmers. Despite these drawbacks, the potential benefits of nanotechnology in agriculture are substantial. Ongoing research endeavors are focused on addressing these concerns, aiming to ensure the safe and sustainable application of nanotechnology in agriculture (Panpatte et al. 2016).

16.6.1 Nano Fertilizers

In the ever-changing landscape of agriculture, the introduction of nano fertilizers emerges as a revolutionary stride, presenting a nuanced approach to managing nutrients in crops. Nano fertilizers, defined by their minuscule particle size, usher in a new era of efficiency in delivering nutrients to plant cells. This section thoroughly explores the multifaceted realm of nano fertilizers, delving into their merits, mechanisms of action, environmental impact, challenges, and the promising prospects they offer for sustainable agriculture.

Nano fertilizers carve a distinct niche in boosting nutrient absorption efficiency, departing from traditional fertilizers. The precision enabled by the minute size of these particles allows for targeted delivery to plant cells, ensuring that vital nutrients reach their intended destination. This precision not only fosters healthier plant growth but also minimizes wastage, marking a significant step toward sustainable farming practices (Mahapathra et al. 2022; Butt and Naseer, 2020). The allure of nano fertilizers extends to their controlled-release properties, providing a steady and regulated nutrient supply over an extended period. This not only caters to the temporal nutrient needs of plants but also addresses concerns surrounding nutrient runoff, thereby mitigating their environmental impact (Gogos et al. 2012).

The mechanism of action intrinsic to nano fertilizers is closely tied to their nano-sized particles. These particles facilitate a more efficient interaction with plant physiology, enhancing nutrient absorption. The intricate interplay between nanomaterials and plant cells underscores the precision and specificity that nano fertilizers bring to nutrient management. Grasping this mechanism becomes pivotal for unlocking the full potential of nano fertilizers and optimizing their application across diverse agricultural settings. Yet, as is typical with emerging technologies, nano fertilizers are not immune to challenges and concerns. Despite their potential benefits, uncertainties persist regarding the environmental and health risks associated with their use. A critical need for in-depth research arises to comprehend the long-term effects of nano fertilizers on both the environment and human health. This section acknowledges the existing gaps in knowledge, stressing the ongoing research imperative to ensure the safe and sustainable integration of nano fertilizers into mainstream agricultural practices (Rajamuthuramalingam et al. 2024).

The key knowledge gaps in nano fertilizers revolve around their long-term environmental and health impacts, including potential soil accumulation, toxicity to microorganisms, and bioaccumulation in plants, which could affect the food chain. Uncertainties also exist regarding their interactions with soil chemistry, ecological risks, and the possibility of nanoparticle resistance. To ensure sustainable integration, interdisciplinary research is needed to assess their environmental fate, develop eco-friendly formulations, and guide responsible use through regulatory frameworks. Long-term studies, education programs for farmers, and public awareness initiatives are essential to optimize their benefits while minimizing risks, ensuring nano fertilizers contribute effectively to sustainable agriculture.

Challenges of cost-effectiveness and scalability pose additional hurdles to the widespread adoption of nano fertilizers. The high costs tied to research and development may restrict accessibility, particularly for small-scale farmers. Addressing regulatory considerations and establishing standards for nano fertilizer use becomes imperative for fostering broader application (Brady et al. 2005). In practical terms, the application of nano fertilizers in crop management yields promising outcomes. Specific crops and soil types have been identified where nano fertilizers exhibit optimal effectiveness. Real-world case studies serve as tangible demonstrations, showcasing the adaptability of nano fertilizers in diverse agricultural scenarios. As these case studies unfold, they offer valuable insights into the practical implications of incorporating nano fertilizers into existing farming practices (Chinnamuthu & Boopathi, 2009; Colvin, 2003). By diminishing the environmental impact of traditional fertilization methods, nano fertilizers align with the principles of sustainability. Their role in championing resource efficiency and minimizing the ecological footprint of agriculture positions them as pivotal contributors to the pursuit of a more sustainable food production system (Allen et al. 1991).

As we gaze into the future, the symbiotic relationship between nanotechnology and agriculture holds immense promise, and the exploration of emerging trends signifies a dynamic landscape where innovation continually unfolds. The quest for sustainable and efficient farming practices propels researchers toward uncharted territories, where nano fertilizers are poised to play a pivotal role in shaping the future of agriculture (Barbarrick et al. 1990).

16.6.2 Nano Pesticides

Nano pesticides stand at the forefront of a modern agricultural revolution, introducing a paradigm shift in how we manage pests. These cutting-edge formulations, spanning nano capsules, nanosuspensions, nanoparticles, nano emulsions, and lipid-based nanocarriers, operate through sophisticated mechanisms. By enhancing penetration, absorption, and the controlled release of active ingredients, nano pesticides deliver heightened efficacy while minimizing environmental impact and enabling reduced application volumes (Pretty, 2018). Their advantages further extend to bolstering stability and prolonging shelf life, contributing significantly to sustainable and efficient crop management. Yet, the adoption of nano pesticides is not without its challenges and considerations. The paramount concerns revolve around environmental impact and safety, demanding a thorough examination of their potential effects on non-target organisms, ecotoxicological considerations, and assessments of persistence and bioaccumulation. The evolving regulatory landscape adds complexity, with existing regulations posing challenges and considerations in obtaining the necessary approvals. Nano pesticides seamlessly find application in precision farming, integrating with precision agriculture techniques to elevate overall crop management. Despite their promising attributes, stability and formulation challenges, along with cost considerations, and public perception and acceptance, present hurdles to their widespread adoption (Zhang et al. 2018). Looking ahead, the future of nano pesticides holds exciting prospects with emerging technologies and innovations poised for further research and development. A comparative analysis with conventional pesticides critically evaluates factors such as efficacy, performance, and environmental and health impacts. Addressing ethical considerations surrounding nano pesticides emphasizes the crucial need to strike a balance between the benefits they offer and potential risks (Bhattacharya 2015). The integration of nano pesticides with Integrated Pest Management (IPM) approaches underscores their role in synergizing with biological control methods. An exploration of commercialization and market trends offers a snapshot of their status and promising projections for the future. In conclusion, this journey into the realm of nano pesticides accentuates their transformative potential, reshaping pest management practices and presenting sustainable and efficient solutions for the future of agriculture (Wahab et al. 2024).

16.6.3 Toxicological Aspects of Nanoparticles in Agriculture

The incorporation of nanoparticles in agriculture has introduced a new era of innovation, promising advancements in crop yields and overall farming practices. However, this groundbreaking shift has also brought about significant concerns regarding the potential toxicological implications of nanoparticles within agricultural ecosystems. A comprehensive understanding of the risks and impacts is vital to ensure the safe and sustainable integration of these novel materials. Nanoparticles find entry into the agricultural environment through diverse pathways such as soil amendments, pesticides, and fertilizers. The direct and indirect exposure of crops, soil, water, and organisms necessitates a thorough examination of their toxicological effects. The uptake and accumulation of nanoparticles in plants, driven by mechanisms like root absorption and foliar uptake, involve intricate dynamics influenced by factors such as particle size, shape, and surface properties. These dynamics contribute to shaping the potential risks associated with nanoparticle exposure in agricultural settings. As nanoparticles interact with soil microorganisms, concerns emerge regarding their impact on microbial communities and overall soil health. Potential disruptions in nutrient cycling and alterations in microbial functions may pose challenges to the agroecosystem's soil fertility and sustainability (Lowry et al. 2019; Kah et al. 2014). Moreover, the toxicity of nanoparticles extends beyond plants to non-target organisms within the agricultural ecosystem, encompassing beneficial insects and earthworms. Understanding the ecological consequences and potential cascading effects is critical to maintaining the delicate balance of the agroecosystem. The growing concern over human exposure to nanoparticles in agriculture raises questions about potential pathways, including the ingestion of nanoparticle-exposed crops and inhalation during farming activities. The complex health implications of such exposures require in-depth investigation. Toxicological studies play a pivotal role in unravelling the intricate relationships between nanoparticle exposure and health risks, taking into consideration factors like particle size, concentration, and duration of exposure. The challenge lies in conducting comprehensive risk assessments that encompass the diverse facets of nanoparticle toxicity in agricultural contexts. Navigating the regulatory landscape for nanoparticles in agriculture entails understanding existing frameworks and standards. Current regulatory considerations aim to strike a balance between the potential benefits of nanoparticle use and the imperative to safeguard environmental and human health. Achieving this equilibrium necessitates the continuous refinement of regulations and guidelines to align with evolving scientific understanding and technological advancements.

Mitigating the toxicological impact of nanoparticles in agriculture requires the identification of strategies that minimize risks while preserving the benefits of nanotechnology. Sustainable practices, alternative technologies, and precision application methods present promising avenues for reducing nanoparticle-related risks (Yadav et al 2020). Moreover, incorporating risk mitigation strategies into agricultural practices demands collaboration among scientists, policymakers, and industry stakeholders to ensure effective implementation. Recognizing the existing research gaps and charting future directions in nanoparticle toxicology is crucial for advancing knowledge in this field. Identifying areas where information is lacking and prioritizing research initiatives will contribute to a more comprehensive understanding of the risks associated with nanoparticle use in agriculture. Additionally, examining case studies that highlight incidents or studies demonstrating toxicological effects in specific agricultural contexts provides valuable insights, enabling the refinement of best practices and guidelines for safer nanoparticle use. In conclusion, addressing the toxicological aspects of nanoparticles in agriculture demands meticulous attention and a multidisciplinary approach. As nanotechnology continues to reshape farming practices, it becomes imperative to navigate the delicate balance between innovation and environmental and human safety. By proactively addressing research gaps, implementing effective regulatory measures, and embracing sustainable practices, the agricultural community can harness the benefits of nanoparticles while minimizing potential risks, ensuring a safer and more sustainable future for agriculture (Bombo et al. 2019, Osorio et al. 2021).

16.6.4 Nano Biosensors

At the forefront of advanced technology, nano biosensors represent a revolutionary shift in sensing and detection applications. These tiny devices, encompassing diverse types such as surface-based and nanomaterial-based variants, operate on intricate principles that capitalize on the unique properties of nanomaterials. In the healthcare domain, nano biosensors have become indispensable for disease diagnostics and monitoring, ushering in a new era of personalized medicine. Their capability to identify biomarkers facilitates early detection and intervention rapidly and precisely. Additionally, in environmental monitoring, nano biosensors play a crucial role in detecting pollutants and contaminants, offering real-time monitoring, and contributing to early-warning systems for environmental threats (Vigneshwaran et al. 2006).

Within the realm of food safety and quality control, nano biosensors ensure swift and reliable detection of foodborne pathogens, thereby safeguarding the integrity and quality of the food supply chain. Their applications extend to defense, where they serve as essential instruments for the rapid identification of biological and chemical warfare agents, bolstering security and preparedness. Implantable nano biosensors represent another frontier, seamlessly integrating for in vivo monitoring and enabling continuous health monitoring and disease management.

While the potential of nano biosensors is immense, challenges persist, including issues of sensitivity, selectivity, and reproducibility that necessitate ongoing refinement. Ethical considerations, particularly in healthcare applications, prompt questions about privacy and data security. The critical choice of nanomaterials for biosensor construction adds complexity, with each material presenting distinct advantages and limitations that require careful consideration (Prasad et al. 2017).

Recent technological strides, such as the integration of artificial intelligence and machine learning in biosensor data analysis, highlight the rapid evolution of nano biosensors. With these advancements, commercialization is gaining momentum, driving significant market growth. However, ethical and regulatory frameworks will be crucial in ensuring their responsible development and use. A noteworthy case study is the application of nano biosensors in water quality monitoring, where real-time detection of contaminants like heavy metals and pathogens has improved public health outcomes by enabling swift interventions. These practical examples underscore the potential of nano biosensors to address complex challenges across industries. As the technology progresses, such case studies provide essential insights, guiding further innovation and helping avoid pitfalls. In conclusion, nano biosensors are poised to revolutionize healthcare, environmental monitoring, and other sectors, promising a future where precision sensing technologies address critical societal needs (Faceto et al. 2016).

16.7 Nanotechnology in Food Industry

The infusion of nanotechnology into the food industry has ignited a revolutionary transformation, reshaping key aspects of food production, processing, and safety. In the realm of food packaging, nanocomposites stand out as revolutionary, bolstering barrier properties and extending the shelf life of products. Intelligent packaging systems, equipped with nano sensors, offer real-time monitoring capabilities, actively contributing to the upkeep of optimal food quality (Zielinska et al. 2020). The spotlight on nanoencapsulation takes the lead in flavor and nutrient delivery, introducing controlled release mechanisms that elevate the bioavailability of essential nutrients. Furthermore, precision in quality control reaches new heights with nanoscale analytical techniques, where nano sensors play a pivotal role in detecting contaminants and pathogens, ensuring the safety and integrity of food products (Zielinska et al. 2020).

Innovations in food preservation are evident through nano emulsions and nano capsules, enabling the controlled release of preservatives. Antimicrobial nano coatings further enhance this preservation journey, extending the shelf life of a variety of food items. The landscape of functional foods undergoes a profound metamorphosis with nanotechnology paving the way for targeted nutrient delivery through nanoparticles, amplifying health benefits. In the additives arena, the introduction of nano-sized emulsifiers and stabilizers revolutionizes texture and stability. In beverage processing, the incorporation of nanoscale ingredients elevates both stability and flavor. Dairy and bakery products experience optimization in fortification and texture, thanks to the contributions of nanotechnology. Innovations in nanotechnology extend to meat and seafood processing, introducing advancements in packaging, tenderness, and flavor enhancement. Finally, the synergy of smart packaging and nanotechnology emerges as a potent force in reducing food waste through innovative strategies such as recycling and repurposing (Jamwal and Mittal 2024).

16.7.1 Food Processing

The infusion of nanotechnology into the food industry has ignited a wave of transformative progress, reshaping how we cultivate, process, and enjoy our food. Within food packaging, fore fronted by antimicrobial nanomaterials, we witness an extension of shelf life and the preservation of freshness. Nano-coatings, fortified with superior barrier properties, act as safeguards against contamination, and intelligent packaging systems leverage nanotechnology for real-time monitoring, ensuring optimal food quality. Innovations in food preservation unfold through nano-emulsions and nanoencapsulation, introducing controlled release mechanisms for preservatives, while edible nano films provide a shield against spoilage and microbial threats. The incorporation of nano sensors adds an additional layer of precision to detect and monitor instances of food spoilage, enhancing overall quality control (Muthukumar et al. 2023).

In the arena of food processing, the introduction of nanostructured materials elevates texture and flavor experiences, with nano emulsions and nano capsules facilitating the encapsulation of flavors and essential nutrients. Nanoscale enzymes and catalysts contribute to heightened processing efficiency, showcasing the multifaceted impact of nanotechnology. The sphere of food quality and safety reaps benefits from nano biosensors, swiftly identifying contaminants and pathogens, alongside nanomaterials aiding in toxin removal. The integration of nanotechnology ensures traceability, assuring the safety of the entire food supply chain (Pradhan et al. 2015).

Functional foods undergo a revolution with nanoparticle-enabled targeted nutrient delivery and improved bioavailability of essential elements. Nanocarriers play a pivotal role in encapsulating bioactive compounds, amplifying the health advantages of functional foods. Analytical techniques at the nanoscale bring about a transformation in food analysis, offering heightened sensitivity in testing and tools to combat food fraud. As nanotechnology permeates food additives, it introduces nano-sized emulsifiers and stabilizers, enhancing both texture and stability, thereby revolutionizing the functionality of these additives. In agriculture, the nanoscale delivery of fertilizers enhances nutrient uptake, while nano sensors monitor soil conditions and crop health, fostering sustainable and efficient food production. Navigating the evolving landscape of nanotechnology in food demands regulatory and safety considerations to ensure its responsible and ethical integration into our culinary practices (Saha et al. 2022).

The realm of food packaging has undergone a revolutionary metamorphosis through the integration of nanotechnology, marking a new era characterized by innovation and sustainability. Leading the charge are nanocomposites, propelling packaging materials to new heights with enhanced barrier properties, increased mechanical strength, and greater flexibility. The infusion of active and intelligent packaging systems, equipped with nano sensors, not only provides real-time monitoring capabilities but also imparts valuable information directly to consumers. Antimicrobial packaging, boasting nano coatings with controlled release properties, addresses critical challenges in food preservation by extending the shelf life of perishable goods. Nanoencapsulation takes a central role, introducing controlled release mechanisms for flavors and preservatives while simultaneously revolutionizing the precision of targeted nutrient delivery (Vanderrroost et al. 2014).

16.7.2 Food Packaging

The emergence of edible packaging enhanced with nanomaterials strengthens packaging durability while supporting environmental sustainability (Dainellu et al. 2008). Innovations such as active oxygen scavengers and antioxidant packaging, driven by nanotechnology, are crucial in preserving food quality and preventing oxidation (He et al. 2019). Additionally, nanostructured barrier coatings and films minimize gas and moisture permeation, significantly extending the shelf life of perishable items (Singh et al. 2023). The shift toward biodegradable and sustainable packaging is facilitated by nanomaterials, offering eco-friendly alternatives to mitigate plastic waste. Furthermore, nanostructured labels, smart barcodes, and advanced printing technologies enhance product traceability and enable real-time monitoring of food quality throughout the supply chain (He et al. 2019). This review underscores the broad applications and benefits of nanotechnology in food packaging, emphasizing the need for sustainable and safe solutions. These innovations not only optimize packaging performance but also pave the way for a more environmentally responsible and well-informed future for consumers (Singh et al. 2023).

16.7.3 Food Labelling

Navigating the intricate world of food labeling is essential for both consumers and producers alike, as it serves as a key interface for informed choices and responsible production practices. Nutritional labeling provides a comprehensive breakdown of macronutrients and micronutrients, aiding consumers in making health-conscious decisions. Allergen information, with its clear identification and disclosure, addresses the critical needs of individuals with sensitivities, fostering a safer food environment. The ingredient list, meticulously detailed and prioritized, promotes transparency and assists consumers in understanding product compositions. Country of Origin Labeling (COOL) not only aligns with international trade regulations but also satisfies the growing consumer demand for transparency in sourcing. Date labeling, an often-misunderstood aspect, requires educational efforts to minimize food waste and encourage accurate interpretation (Ganeson et al. 2023).

Genetically Modified Organism (GMO) labeling responds to consumer concerns about biotechnology, offering transparency and choice. Organic labeling, with its diverse categorizations, empowers consumers to align their choices with environmental and ethical considerations. Fair Trade labeling, a beacon for ethically sourced products, ensures fair compensation for producers, fostering a global sense of responsibility. Health claims, when regulated and substantiated, guide consumers towards healthier options. Non-GMO Project Verified labeling, driven by consumer demand, offers a voluntary standard for products avoiding genetic modification. Front-of-pack labels simplify nutritional information, aiding quick assessments for health-conscious consumers. Lastly, kosher and halal certifications cater to diverse dietary preferences, aligning food choices with religious beliefs (Mittal et al. 2020). Table 16.2 highlights the major applications of nanotechnology in the food industry.

Table 16.2 Major applications of nanotechnology in food industry

Application	Description	Specific nanoparticle	Reference
Nano sensors for food safety	Nanotechnology enables the development of highly sensitive sensors for detecting contaminants, pathogens, and spoilage indicators in food products.	Gold nanoparticles	Ramezani et al. (2020), Li and Sheng (2014)
Nanoencapsulation for nutrient delivery	Nanoencapsulation techniques improve the bioavailability and stability of nutrients, vitamins, and bioactive compounds, enhancing their incorporation into food products.	Lipid nanoparticles	Karunaratne et al. (2017), Nallamuthu et al. (2017)
Nanocomposites for packaging	Nanocomposite materials provide enhanced barrier properties, mechanical strength, and antimicrobial activity, extending the shelf life and safety of packaged foods.	Silver nanoparticles	Youssef (2013), De Azeredo (2009)
Nano emulsions for flavor delivery	Nano emulsion technology enables the creation of stable, transparent emulsions that improve the solubility and dispersion of flavors, leading to enhanced taste perception in food products.	Nanostructured lipid carriers	Saffarionpour (2019), Aswathanarayan and Vittal (2019)
Nanofiltration for water purification	Nanofiltration membranes with nanoscale pores effectively remove contaminants, pathogens, and impurities from water used in food processing, ensuring product safety and quality.	Carbon nanotubes	Das et al. (2014), Chen et al. (2016)

16.8 Merits and Demerits of Nanoparticle-Based Technology in the Future of Suitable Agriculture

The integration of nanoparticle-based technology holds great promise for the future of sustainable agriculture, ushering in a transformative era with multifaceted benefits. A standout advantage lies in the realm of precision farming, where the application of nanoparticles allows for the targeted delivery of fertilizers, pesticides, and agrochemicals directly to plant cells. This precise administration not only minimizes wastage but also contributes to a significant reduction in the environmental impact associated with chemical inputs, fostering heightened resource efficiency in agricultural practices. Nanoparticles further enhance nutrient management by optimizing nutrient uptake, minimizing losses, and enabling controlled-release mechanisms that finely tune crop nutrition. In the domain of disease and pest management, nanotechnology plays a pivotal role in the development of nano sensors for early diagnostics and treatment. This innovation not only improves the effectiveness of control measures but also aligns with environmentally friendly practices (Mohana et al. 2020).

Despite these promising merits, nanoparticle-based technology in agriculture is not without challenges and considerations. A significant concern revolves around the potential environmental and health risks linked to nanoparticle usage. The long-term impacts of these minute particles on soil ecosystems, water quality, and human health remain not fully understood, underscoring the imperative for comprehensive research to guarantee the safe and sustainable deployment of nanoparticle-based solutions. Additionally, the high costs associated with the research and development of nanotechnology applications pose potential barriers to accessibility, particularly for small-scale farmers in developing regions. Addressing these challenges requires the evolution of regulatory frameworks to effectively balance the innovative potential of nanotechnology with the essential need for stringent safety assessments. Navigating this evolving landscape demands a cautious and well-informed approach to unlock the full potential of nanoparticle-based technology, ensuring its positive impact on agriculture while mitigating potential risks to the environment and human health (Shelar et al. 2023; Wahab et al. 2023).

16.9 Conclusion and Future Prospects

The future of sustainable agriculture holds significant promise with the integration of nanotechnology, offering innovative solutions to address pressing challenges such as climate change, soil degradation, and resource scarcity. Nanotechnology enables precise manipulation at the nanoscale, facilitating the development of novel agrochemicals, smart delivery systems for nutrients and pesticides, and nano sensors for real-time monitoring of crop health and environmental conditions. By enhancing nutrient uptake efficiency, minimizing pesticide use, and optimizing resource utilization, nanotechnology has the potential to increase agricultural productivity while reducing environmental impacts.

Furthermore, nanomaterial-based approaches can contribute to the development of resilient crop varieties capable of withstanding biotic and abiotic stresses. However, as with any emerging technology, careful consideration of potential risks to human health, ecosystems, and ethical implications is essential to ensure the responsible and sustainable implementation of nanotechnology in agriculture. Collaboration among researchers, policymakers, industry stakeholders, and the public will be crucial to harnessing the full potential of nanotechnology in advancing sustainable agriculture for future generations.

In conclusion, the intersection of sustainable agriculture and nanotechnology presents a promising path toward addressing the complex challenges facing global food systems. Through the precision and efficiency offered by nanoscale interventions, such as smart delivery systems and nano sensors, agriculture can become more resource-efficient, environmentally friendly, and resilient to changing conditions. However, realizing the full potential of nanotechnology in sustainable agriculture requires careful consideration of potential risk factors. Collaborative efforts across disciplines and sectors will be essential to ensure responsible innovation and the equitable distribution of benefits. By leveraging the power of nanotechnology alongside sustainable practices, we can pave the way for a more resilient, productive, and sustainable agricultural future.

References

Aacharya R, Chhipa H (2020) Nanocarbon fertilizers: implications of carbon nanomaterials in sustainable agriculture production. In: Kamel A. Abd-Elsalam (Ed) Carbon nanomaterials for agri-food and environmental applications. Elsevier, pp 297–321. https://doi.org/10.1016/B978-0-12-819786-8.00015-3

Al-Hussaniy HA, Almajidi YQ, Oraibi AI, Alkarawi AH (2023) Nanoemulsions as medicinal components in insoluble medicines. Pharmacia (0428-0296) 70(3):537–547

Allen P, Van Dusen D, Lundy J, Gliessman S (1991) Integrating social, environmental, and economic issues in sustainable agriculture. Am J Altern Agric 6(1):34–39

Altammar KA (2023) A review on nanoparticles: characteristics, synthesis, applications, and challenges. Front Microbiol 14:1155622–1155622. https://doi.org/10.3389/fmicb.2023.1155622

Androvitsaneas P, Young AB, Schneider C, Maier S, Kamp M, Höfling S, Knauer S, Harbord E, Hu CY, Rarity JG, Oulton R (2016) Charged quantum dot micropillar system for deterministic light-matter interactions. Phys Rev B 93(24). https://doi.org/10.1103/physrevb.93.241409

Anton N, Vandamme TF (2010) Nano-emulsions and micro-emulsions: clarifications of the critical differences. Pharm Res 28(5):978–985. https://doi.org/10.1007/s11095-010-0309-1

Asua JM (2002) Miniemulsion polymerization. Prog Polym Sci 27(7):1283–1346. https://doi.org/10.1016/s0079-6700(02)00010-2

Aswathanarayan JB, Vittal RR (2019) Nanoemulsions and their potential applications in food industry. Front Sustain Food Syst 3. https://doi.org/10.3389/fsufs.2019.00095

Athanassiou A (ed) (2020) Sustainable food packaging technology. Wiley, USA ISBN (ePDF): 978-3-527-82008-5

Azeredo HMC (2009) Nanocomposites for food packaging applications. Food Res Int 42(9): 1240–1253. https://doi.org/10.1016/j.foodres.2009.03.019

Bandi R, Dadigala R, Alle M (2023) Emerging role of gold nanoparticles for healthier crop plants growth and enhanced yield. In: Azamal Husen (Ed) Engineered nanomaterials for sustainable agricultural production, soil improvement and stress management. Academic Press, pp 125–143. https://doi.org/10.1016/B978-0-323-91933-3.00014-3

Bandyopadhyay S, Novo LA, Pietrzykowski M, Maiti SK (2020) Assessment of forest ecosystem development in coal mine degraded land by using integrated mine soil quality index (IMSQI): the evidence from India. Forests 11(12):1310

Barbarick KA, Lai TM, Eberl DD (1990) Exchange fertilizer (phosphate rock plus ammonium-zeolite) effects on Sorghum-Sudangrass. Soil Sci Soc Am J 54(3):911–916

Bhattacharya SC (2015) Wood energy in India: status and prospects. Energy 85:310–316. https://doi.org/10.1016/j.energy.2015.03.043

Bombo AB, Pereira AES, Lusa MG, de Medeiros Oliveira E, de Oliveira JL, Campos EVR, de Jesus MB, Oliveira HC, Fraceto LF, Mayer JLS (2019) A mechanistic view of interactions of a nanoherbicide with target organism. J Agric Food Chem 67(16):4453–4462. https://doi.org/10.1021/acs.jafc.9b00806

Brady KU, Kruckeberg AR, Bradshaw HD Jr (2005) Evolutionary ecology of plant adaptation to serpentine soils. Annu Rev Ecol Evol Syst 36(1):243–266

Butt BZ, Naseer I (2020) Nanofertilizers. In: Javad, S. (eds) Nanoagronomy. Springer, Cham., pp 125–152. https://doi.org/10.1007/978-3-030-41275-3_8

Chen X, Qiu M, Ding H, Fu K, Fan Y (2016) A reduced graphene oxide nanofiltration membrane intercalated by well-dispersed carbon nanotubes for drinking water purification. Nanoscale 8(10):5696–5705. https://doi.org/10.1039/c5nr08697c

Chinnamuthu CR, Boopathi PM (2009) Nanotechnology and agroecosystem

Colvin VL (2003) The potential environmental impact of engineered nanomaterials. Nat Biotechnol 21(10):1166–1170. https://doi.org/10.1038/nbt875

Couvreur P, Puisieux F (1993) Nano-and microparticles for the delivery of polypeptides and proteins. Adv Drug Deliv Rev 10(2–3):141–162

Dainelli D, Gontard N, Spyropoulos D, Zondervan-van den Beuken E, Tobback P (2008) Active and intelligent food packaging: legal aspects and safety concerns. Trends Food Sci Technol 19: S103–S112. https://doi.org/10.1016/j.tifs.2008.09.011

Dar FA, Qazi G, Pirzadah TB (2020) Nano-biosensors: NextGen diagnostic tools in agriculture. In: Hakeem, K., Pirzadah, T. (eds) Nanobiotechnology in Agriculture. Nanotechnology in the Life Sciences. Springer, Cham. pp 129–144. https://doi.org/10.1007/978-3-030-39978-8_7

Das R, Ali ME, Hamid SBA, Ramakrishna S, Chowdhury ZZ (2014) Carbon nanotube membranes for water purification: a bright future in water desalination. Desalination 336:97–109. https://doi.org/10.1016/j.desal.2013.12.026

Das S, Wolfson BP, Tetard L, Tharkur J, Bazata J, Santra S (2015) Effect of N-acetyl cysteine coated CdS:Mn/ZnS quantum dots on seed germination and seedling growth of snow pea (Pisum sativum L.): imaging and spectroscopic studies. Environ Sci Nano 2(2):203–212. https://doi.org/10.1039/c4en00198b

Dehghani MH, Ahmadi S, Ghosh S, Khan MS, Othmani A, Khanday WA et al (2024) Sustainable remediation technologies for removal of pesticides as organic micro-pollutants from water environments: a review. Appl Surface Sci Adv 19:100558

de Souza NRD, Solberg MKM, Hu X, Cherubini F (2025) Environmental benefits and changes in ecosystem services of climate-smart agriculture relative to conventional agriculture in Norway. J Environ Manag 377:124633

Dimkpa CO, McLean JE, Britt DW, Anderson AJ (2012) Bioactivity and biomodification of Ag, ZnO, and CuO nanoparticles with relevance to plant performance in agriculture. Ind Biotechnol 8(6):344–357. https://doi.org/10.1089/ind.2012.0028

Ditta A, Arshad M, Ibrahim M (2015) Nanoparticles in sustainable agricultural crop production: applications and perspectives. In: Siddiqui, M., Al-Whaibi, M., Mohammad, F. (eds)

Nanotechnology and Plant Sciences. Springer, Cham, pp 55–75. https://doi.org/10.1007/978-3-319-14502-0_4

Dwivedi P, Narvi SS, Tewari RP (2012) Rudraksha assisted generation of silver nanoparticles for integrated application in the biomedical landscape. Int J Green Nanotechnol 4(3):248–261. https://doi.org/10.1080/19430892.2012.706116

Essajai R, Chafi FZ, Benhouria Y, Hassani A, Qjani M, Mzerd A, Hassanain N (2019) Structural and magnetic properties of iron nanoparticles: insights from Monte-Carlo and molecular-statics simulations. Mater Res Express 6(9):095097

Fraceto LF, Grillo R, de Medeiros GA, Scognamiglio V, Rea G, Bartolucci C (2016) Nanotechnology in agriculture: which innovation potential does it have? Front Environ Sci 4. https://doi.org/10.3389/fenvs.2016.00020

Ganeson K, Mouriya GK, Bhubalan K, Razifah MR, Jasmine R, Sowmiya S, Amirul A-AA, Vigneswari S, Ramakrishna S (2023) Smart packaging—a pragmatic solution to approach sustainable food waste management. Food Packag Shelf Life 36:101044. https://doi.org/10.1016/j.fpsl.2023.101044

Gao Y, Zhou L, Ouyang S, Sun J, Zhou Q (2024) Environmental applications and risks of engineered nanomaterials in removing petroleum oil in soil. Sci Total Environ 946:174165

Ghidan A Y, M. Al Antary T (2020) Applications of Nanotechnology in Agriculture. IntechOpen. https://doi.org/10.5772/intechopen.88390

Ghorbanpour M, Hadian J (2017) Engineered nanomaterials and their interactions with plant cells: injury indices and detoxification pathways. In: Ghorbanpour, M., Manika, K., Varma, A. (eds) Nanoscience and Plant–Soil Systems. Soil Biology, vol 48. Springer, Cham, pp 429–453. https://doi.org/10.1007/978-3-319-46835-8_16

Gogos A, Knauer K, Bucheli TD (2012) Nanomaterials in plant protection and fertilization: current state, foreseen applications, and research priorities. J Agric Food Chem 60(39):9781–9792. https://doi.org/10.1021/jf302154y

Graf C, Vossen DLJ, Imhof A, van Blaaderen A (2003) A general method to coat colloidal particles with silica. Langmuir 19(17):6693–6700. https://doi.org/10.1021/la0347859

Guo D, Xie G, Luo J (2013) Mechanical properties of nanoparticles: basics and applications. J Phys D Appl Phys 47(1):013001. https://doi.org/10.1088/0022-3727/47/1/013001

Gupta A, Eral HB, Hatton TA, Doyle PS (2016) Nanoemulsions: formation, properties and applications. Soft Matter 12(11):2826–2841. https://doi.org/10.1039/c5sm02958a

Gupta N, Yadav KK, Kumar V, Kumar S, Chadd RP, Kumar A (2019) Trace elements in soil-vegetables interface: translocation, bioaccumulation, toxicity and amelioration-a review. Sci Total Environ 651:2927–2942

Gupta S, Hengl T, Lehmann P, Bonetti S, Or D (2020) SoilKsatDB: global soil saturated hydraulic conductivity measurements for geoscience applications. Earth Syst Sci Data Discuss 2020:1–26

Hajirostamlo B, Mirsaeedghazi N, Arefnia M, Ali M, Shariati EAF (2015) The role of research and development in agriculture and its dependent concepts in agriculture. Sci Eng 4(1):79–81

He X, Deng H, Hwang H-M (2019) The current application of nanotechnology in food and agriculture. J Food Drug Anal 27(1):1–21. https://doi.org/10.1016/j.jfda.2018.12.002

Hortin JM, Anderson AJ, Britt DW, Jacobson AR, McLean JE (2020) Copper oxide nanoparticle dissolution at alkaline pH is controlled by dissolved organic matter: influence of soil-derived organic matter, wheat, bacteria, and nanoparticle coating. Environ Sci Nano 7(9):2618–2631. https://doi.org/10.1039/d0en00574f

Hou X, Liu H, Zhang Y, Jiang M, Yuan L, Shi J, Hou C (2020) Oxygen vacancies enhancing acetone-sensing performance. Mater Today Chem 18:100372. https://doi.org/10.1016/j.mtchem.2020.100372

Hu J (2018) Shape memory behavior of bio-based fibers and their applications. Nano Res Appl 04. https://doi.org/10.21767/2471-9838-c6-024

Hu Y, Li J, Ma L, Peng Q, Feng W, Zhang L, He S, Yang F, Huang J, Li L (2010) High efficiency transport of quantum dots into plant roots with the aid of silwet L-77. Plant Physiol Biochem 48(8):703–709. https://doi.org/10.1016/j.plaphy.2010.04.001

Hu YL, Wu FP, Zeng DH, Chang SX (2014) Wheat straw and its biochar had contrasting effects on soil C and N cycling two growing seasons after addition to a Black Chernozemic soil planted to barley. Biol Fertil Soils 50:1291–1299

Isahak WNRW, Shaker LM, Al-Amiery A (2023) Oxygenated hydrocarbons from catalytic hydrogenation of carbon dioxide. Catalysts 13(1):115

Jain SK, Chourasia MK, Masuriha R, Soni V, Jain A, Jain NK, Gupta Y (2005) Solid lipid nanoparticles bearing flurbiprofen for transdermal delivery. Drug Deliv 12(4):207–215. https://doi.org/10.1080/10717540590952591

Jamwal V, Mittal A (2024) Recent progresses in nanocomposite films for food-packaging applications: synthesis strategies, technological advancements, potential risks and challenges. Food Rev Intl 40:1–32

Jana NR, Earhart C, Ying JY (2007) Synthesis of water-soluble and functionalized nanoparticles by silica coating. Chem Mater 19(21):5074–5082. https://doi.org/10.1021/cm071368z

Jeevanandam J, Barhoum A, Chan YS, Dufresne A, Danquah MK (2018) Review on nanoparticles and nanostructured materials: history, sources, toxicity and regulations. Beilstein J Nanotechnol 9(1):1050–1074

Jones DC, Young JM, Lakshantha WJ, Singh S, Byers TA, Weathers DL, McDaniel FD, Rout B (2012) Redistribution of nickel ions embedded within indium phosphide via low energy dual ion implantations. J Nucl Phys Mat Sci Rad A 6(1):9–15. https://doi.org/10.15415/jnp.2018.61002

Kah M (2015) Nanopesticides and nanofertilizers: emerging contaminants or opportunities for risk mitigation? Front Chem 3:64–64. https://doi.org/10.3389/fchem.2015.00064

Kah M, Hofmann T (2014) Nanopesticide research: current trends and future priorities. Environ Int 63:224–235. https://doi.org/10.1016/j.envint.2013.11.015

Kah M, Beulke S, Tiede K, Hofmann T (2013) Nanopesticides: state of knowledge, environmental fate, and exposure modeling. Crit Rev Environ Sci Technol 43(16):1823–1867. https://doi.org/10.1080/10643389.2012.671750

Kah M, Kookana RS, Gogos A, Bucheli TD (2018) A critical evaluation of nanopesticides and nanofertilizers against their conventional analogues. Nat Nanotechnol 13(8):677–684. https://doi.org/10.1038/s41565-018-0131-1

Kaur G, Singh G, Motavalli PP, Nelson KA, Orlowski JM, Golden BR (2020) Impacts and management strategies for crop production in waterlogged or flooded soils: a review. Agron J 112(3):1475–1501

Karunaratne DN, Surandika Siriwardhana DA, Ariyarathna IR, Indunil Rajakaruna RMP, Banu FT, Karunaratne V (2017) Nutrient delivery through nanoencapsulation. In: Nutrient delivery. Elsevier, pp 653–680

Khan I, Saeed K, Khan I (2019) Nanoparticles: properties, applications and toxicities. Arab J Chem 12(7):908–931. https://doi.org/10.1016/j.arabjc.2017.05.011

Kumar S, Singh AK, Ghosh P (2018) Distribution of soil organic carbon and glomalin related soil protein in reclaimed coal mine-land chronosequence under tropical condition. Sci Total Environ 625:1341–1350

Kumari A, Rana V, Yadav SK, Kumar V (2023) Nanotechnology as a powerful tool in plant sciences: recent developments, challenges and perspectives. Plant Nano Biology 100046: 100046

Kuppusamy P, Ichwan SJA, Parine NR, Yusoff MM, Maniam GP, Govindan N (2015) Intracellular biosynthesis of Au and Ag nanoparticles using ethanolic extract of Brassica oleracea L. and studies on their physicochemical and biological properties. J Environ Sci 29:151–157. https://doi.org/10.1016/j.jes.2014.06.050

Lather V, Poonia N, Pandita D (2021) Mesoporous silica nanoparticles: Multifunctional Nanocarrier for Therapeutic Applications, In: Research anthology on synthesis, characterization, and applications of nanomaterials. IGI Global, pp 458–513 https://doi.org/10.4018/978-1-7998-8591-7

Li M, Chen Z, Yang L, Li J, Xu J, Chen C, Wu Q, Yang M, Liu T (2022) Antibacterial activity and mechanism of GO/Cu(2)O/ZnO coating on ultrafine glass fiber. Nanomaterials (Basel) 12(11): 1857. https://doi.org/10.3390/nano12111857

Li Z, Sheng C (2014) Nanosensors for food safety. J Nanosci Nanotechnol 14(1):905–912. https://doi.org/10.1166/jnn.2014.8743

Lowry GV, Avellan A, Gilbertson LM (2019) Opportunities and challenges for nanotechnology in the agri-tech revolution. Nat Nanotechnol 14(6):517–522. https://doi.org/10.1038/s41565-019-0461-7

Mahajan G, Mutti NK, Walsh M, Chauhan BS (2019) Effect of varied soil moisture regimes on the growth and reproduction of two Australian biotypes of junglerice (Echinochloa colona). Weed Sci 67(5):552–559

Mahapatra DM, Satapathy KC, Panda B (2022) Biofertilizers and nanofertilizers for sustainable agriculture: phycoprospects and challenges. Sci Total Environ 803:149990. https://doi.org/10.1016/j.scitotenv.2021.149990

Mehariya S, Kumar P, Marino T, Casella P, Iovine A, Verma P et al (2021) Aquatic weeds: a potential pollutant removing agent from wastewater and polluted soil and valuable biofuel feedstock. In: Bioremediation using weeds. Springer, Singapore, pp 59–77

Miralles DG, Van Den Berg MJ, Teuling AJ, De Jeu RAM (2012) Soil moisture-temperature coupling: a multiscale observational analysis. Geophys Res Lett 39(21):6

Mittal D, Kaur G, Singh P, Yadav K, Ali SA (2020) Nanoparticle-based sustainable agriculture and food science: recent advances and future outlook. Front Nanotechnol 2. https://doi.org/10.3389/fnano.2020.579954

Mohana NC, Mithun PR, Rao HY, Mahendra C, Satish S (2020) Nanoparticle applications in sustainable agriculture, poultry, and food: trends and perspective. In: Susai Rajendran, Anita Mukherjee, Tuan Anh Nguyen, Chandraiah Godugu, Ritesh K. Shukla (Eds) Nanotoxicity. Elsevier, pp 341–353. https://doi.org/10.1016/B978-0-12-819943-5.00016-6

Muller A, Schader C, El-Hage Scialabba N, Brüggemann J, Isensee A, Erb K-H, Smith P, Klocke P, Leiber F, Stolze M, Niggli U (2017) Strategies for feeding the world more sustainably with organic agriculture. Nat Commun 8(1):1290–1290. https://doi.org/10.1038/s41467-017-01410-w

Muthukumar J, Kumar S, Arunachalam N, Kumar L, Chidambaram R (2023) Quality assurance of packaged food using nanotechnology. In: Arun Sharma, P.S. Vijayakumar, Er. Pramod K Prabhakar, Ritesh Kumar (Eds) Nanotechnology applications for food safety and quality monitoring. Academic Press, pp 341–372. https://doi.org/10.1016/B978-0-323-85791-8.00019-7

Naaz H, Rawat K, Saffeullah P, Umar S (2023) Silica nanoparticles synthesis and applications in agriculture for plant fertilization and protection: a review. Environ Chem Lett 21(1):539–559

Nabil M, Megahed F (2023) Quantum dot nanomaterials: preparation, characterization, advanced bio-imaging and therapeutic applications. J Fluoresc:1–18

Nallamuthu I, Khanum F, Fathima SJ, Patil MM, Anand T (2017) Enhanced nutrient delivery through nanoencapsulation techniques: the current trend in food industry. In: Alexandru Mihai Grumezescu (ed) Nanotechnology in the Agri-Food Industry, Nutrient Delivery. Academic Press, pp 619–651. https://doi.org/10.1016/B978-0-12-804304-2.00016-0

Nima ZA, Lahiani MH, Watanabe F, Xu Y, Khodakovskaya MV, Biris AS (2014) Plasmonically active nanorods for delivery of bio-active agents and high-sensitivity SERS detection in planta. RSC Adv 4(110):64985–64993. https://doi.org/10.1039/c4ra10358k

Osorio-Echavarría J, Osorio-Echavarría J, Ossa-Orozco CP, Gómez-Vanegas NA (2021) Synthesis of silver nanoparticles using white-rot fungus anamorphous Bjerkandera sp. R1: influence of silver nitrate concentration and fungus growth time. Sci Rep 11(1):3842–3842. https://doi.org/10.1038/s41598-021-82514-8

Ozdemir M, Kemerli T (2016) Innovative applications of micro and nanoencapsulation in food packaging. In: Jamileh M. Lakkis (Ed) Encapsulation and controlled release technologies in food systems. 2nd edition, Wiley, pp 333–378. https://doi.org/10.1002/9781118946893.ch12

Pal G, Rai P, Pandey A (2019) Green synthesis of nanoparticles: a greener approach for a cleaner future. In: Ashutosh Kumar Shukla, Siavash Iravani (Eds) Green synthesis, characterization and applications of nanoparticles. Elsevier, pp 1–26. https://doi.org/10.1016/B978-0-08-102579-6.00001-0

Panhwar QA, Ali A, Naher UA, Memon MY (2019) Fertilizer management strategies for enhancing nutrient use efficiency and sustainable wheat production. In: Sarath Chandran, M.R. Unni, Sabu Thomas (Eds) Woodhead Publishing Series in Food Science, Technology and Nutrition, Organic Farming, Woodhead Publishing, pp 950 17–39. https://doi.org/10.1016/B978-0-12-813272-2.00002-1

Panpatte, D.G., Jhala, Y.K., Shelat, H.N. and Vyas, R.V. (2016) Nanoparticles: The Next Generation Technology for Sustainable Agriculture. In: Singh, D., Singh, H. and Prabha, R., (Eds.), Microbial Inoculants in Sustainable Agricultural Productivity, Springer, 289–300. https://doi.org/10.1007/978-81-322-2644-4_18

Patel DK, Kim H-B, Dutta SD, Ganguly K, Lim K-T (2020) Carbon nanotubes-based nanomaterials and their agricultural and biotechnological applications. Materials (Basel) 13(7):1679. https://doi.org/10.3390/ma13071679

Pohlmann MA (2022) Tracking the pulse of pyrogenic carbon and lithogenic solutes through surface soil of a western forest headwater catchment immediately following wildfire (Doctoral dissertation, The University of Arizona)

Pradhan N, Singh S, Ojha N, Shrivastava A, Barla A, Rai V, Bose S (2015) Facets of nanotechnology as seen in food processing, packaging, and preservation industry. Biomed Res Int 2015: 365672–365672. https://doi.org/10.1155/2015/365672

Prasad R, Bhattacharyya A, Nguyen QD (2017) Nanotechnology in sustainable agriculture: recent developments, challenges, and perspectives. Front Microbiol 8:1014. https://doi.org/10.3389/fmicb.2017.01014

Pretty J (2018) Intensification for redesigned and sustainable agricultural systems. Science 362(6417). https://doi.org/10.1126/science.aav0294

Rajamuthuramalingam T, da Silva W, Zuverza-Mena N, Dimkpa C, White JC (2024) Nano-sized metal oxide fertilizers for sustainable agriculture: balancing benefits, risks, and risk management strategies. Nanoscale

Ramezani M, Esmaelpourfarkhani M, Taghdisi SM, Abnous K, Alibolandi M (2020) Application of nanosensors for food safety. In: Baoguo Han, Vijay K. Tomer, Tuan Anh Nguyen, Ali Farmani, Pradeep Kumar Singh (Eds) Nanosensors for smart cities, Micro and Nano Technologies, Elsevier, pp 369–386. https://doi.org/10.1016/B978-0-12-819870-4.00021-9

Raza A, Khandelwal K, Pandit S, Singh M, Kumar S, Rustagi S et al (2024) Exploring the potential of metallic and metal oxide nanoparticles for reinforced disease management in agricultural systems: a comprehensive review. Environmental Nanotechnology, Monitoring & Management, 22: 100998 https://doi.org/10.1016/j.enmm.2024.100998

Rodríguez J, Martín MJ, Ruiz MA, Clares B (2016) Current encapsulation strategies for bioactive oils: from alimentary to pharmaceutical perspectives. Food Res Int 83:41–59. https://doi.org/10.1016/j.foodres.2016.01.032

Saffarionpour S (2019) Preparation of food flavor Nanoemulsions by high- and low-energy emulsification approaches. Food Eng Rev 11(4):259–289. https://doi.org/10.1007/s12393-019-09201-3

Saha NC (2022) Food packaging: concepts and its significance. In: Lecture notes in management and industrial engineering. Springer Nature Singapore, pp 1–45

Šarauskis E, Kazlauskas M, Naujokienė V, Bručienė I, Steponavičius D, Romaneckas K, Jasinskas A (2022) Variable rate seeding in precision agriculture: recent advances and future perspectives. Agriculture 12(2):305. https://doi.org/10.3390/agriculture12020305

Saritha GNG, Anju T, Kumar A (2022) Nanotechnology—big impact: how nanotechnology is changing the future of agriculture? J Agric Food Res 10:100457. https://doi.org/10.1016/j.jafr.2022.100457

Sathiyanarayanan G, Dineshkumar K, Yang YH (2017) Microbial exopolysaccharide-mediated synthesis and stabilization of metal nanoparticles. Crit Rev Microbiol 43(6):731–752

Servin AD, White JC (2016) Nanotechnology in agriculture: next steps for understanding engineered nanoparticle exposure and risk. NanoImpact 1:9–12. https://doi.org/10.1016/j.impact.2015.12.002

Shang Y, Hasan MK, Ahammed GJ, Li M, Yin H, Zhou J (2019) Applications of nanotechnology in plant growth and crop protection: a review. Molecules (Basel) 24(14):2558. https://doi.org/10.3390/molecules24142558

Shelar A, Nile SH, Singh AV, Rothenstein D, Bill J, Xiao J, Chaskar M, Kai G, Patil R (2023) Recent advances in nano-enabled seed treatment strategies for sustainable agriculture: challenges, risk assessment, and future perspectives. Nano-micro Letters 15(1):54–54. https://doi.org/10.1007/s40820-023-01025-5

Si Y, Liu J, Chen Y, Miao X, Ye F, Liu Z, Li J (2018) rGO/AuNPs/tetraphenylporphyrin nanoconjugate-based electrochemical sensor for highly sensitive detection of cadmium ions. Anal Methods 10(29):3631–3636. https://doi.org/10.1039/c8ay01020j

Siddiqi KS, Husen A (2016) Green synthesis, characterization and uses of palladium/platinum nanoparticles. Nanoscale Res Lett 11(1):482–482. https://doi.org/10.1186/s11671-016-1695-z

Siddiqi KS, Husen A, Rao RAK (2018) A review on biosynthesis of silver nanoparticles and their biocidal properties. J Nanobiotechnol 16(1):14–14. https://doi.org/10.1186/s12951-018-0334-5

Singh SP, Singh MK (2020) Soil pollution and human health. In: Plant responses to soil pollution. Springer, Singapore, pp 205–220

Singh, R., Dutt, S., Sharma, P., Sundramoorthy, A. K., Dubey, A., Singh, A., & Arya, S. (2023). Future of nanotechnology in food industry: Challenges in processing, packaging, and food safety. Global challenges (Hoboken, NJ), 7(4), 2200209–2200209. doi:https://doi.org/10.1002/gch2.202200209

Soni JK, Choudhary VK, Singh PK, Hota S (2020) Weed management in conservation agriculture, its issues and adoption: a review. J Crop Weed 16:09

Sood K, Kaur J, Singh H, Kumar Arya S, Khatri M (2019) Comparative toxicity evaluation of graphene oxide (GO) and zinc oxide (ZnO) nanoparticles on Drosophila melanogaster. Toxicol Rep 6:768–781. https://doi.org/10.1016/j.toxrep.2019.07.009

Tripathi S, Mahra S, Tiwari K, Rana S, Tripathi DK, Sharma S, Sahi S (2023) Recent advances and perspectives of nanomaterials in agricultural management and associated environmental risk: a review. Nano 13(10):1604

Vanderroost M, Ragaert P, Devlieghere F, De Meulenaer B (2014) Intelligent food packaging: the next generation. Trends Food Sci Technol 39(1):47–62. https://doi.org/10.1016/j.tifs.2014.06.009

Vigneshwaran N, Kathe AA, Varadarajan PV, Nachane RP, Balasubramanya RH (2006) Biomimetics of silver nanoparticles by white rot fungus, Phaenerochaete chrysosporium. Colloids Surf B: Biointerfaces 53(1):55–59. https://doi.org/10.1016/j.colsurfb.2006.07.014

Wahab A, Batool F, Muhammad M, Zaman W, Mikhlef RM, Naeem M (2023) Current knowledge, research Progress, and future prospects of phyto-synthesized nanoparticles interactions with food crops under induced drought stress. Sustain For 15(20):14792. https://doi.org/10.3390/su152014792

Wahab A, Muhammad M, Ullah S, Abdi G, Shah GM, Zaman W, Ayaz A (2024) Agriculture and environmental management through nanotechnology: eco-friendly nanomaterial synthesis for soil-plant systems, food safety, and sustainability. Sci Total Environ 171862:171862

Wang P, Menzies NW, Lombi E, Sekine R, Blamey FPC, Hernandez-Soriano MC, Cheng M, Kappen P, Peijnenburg WJGM, Tang C, Kopittke PM (2015) Silver sulfide nanoparticles (Ag₂S-NPs) are taken up by plants and are phytotoxic. Nanotoxicology 9(8): 1041–1049. https://doi.org/10.3109/17435390.2014.999139

Yadav RK, Singh NB, Singh A, Yadav V, Bano C, Khare S, Niharika. (2020) Expanding the horizons of nanotechnology in agriculture: recent advances, challenges and future perspectives. Vegetos 33(2):203–221. https://doi.org/10.1007/s42535-019-00090-9

Yang W, Wang L, Mettenbrink EM, DeAngelis PL, Wilhelm S (2021) Nanoparticle toxicology. Annu Rev Pharmacol Toxicol 61(1):269–289. https://doi.org/10.1146/annurev-pharmtox-032320-110338

Yadav A, Yadav K, Ahmad R, Abd-Elsalam KA (2023) Emerging frontiers in nanotechnology for precision agriculture: advancements, hurdles and prospects. Agrochemicals 2(2):220–256. https://doi.org/10.3390/agrochemicals2020016

Yin H, Cao Y, Marelli B, Zeng X, Mason AJ, Cao C (2021) Soil sensors and plant wearables for smart and precision agriculture. Adv Mater 33(20):2007764

Youssef AM (2013) Polymer nanocomposites as a new trend for packaging applications. Polym-Plast Technol Eng 52(7):635–660. https://doi.org/10.1080/03602559.2012.762673

Zhang H, Li Y, Zhu JK (2018) Developing naturally stress-resistant crops for a sustainable agriculture. Nat Plants 4(12):989–996

Zhao L, Lu L, Wang A, Zhang H, Huang M, Wu H, Xing B, Wang Z, Ji R (2020) Nano-biotechnology in agriculture: use of nanomaterials to promote plant growth and stress tolerance. J Agric Food Chem 68(7):1935–1947. https://doi.org/10.1021/acs.jafc.9b06615

Zielińska A, Costa B, Ferreira MV, Miguéis D, Louros JMS, Durazzo A, Lucarini M, Eder P, Chaud MV, Morsink M, Willemen N, Severino P, Santini A, Souto EB (2020) Nanotoxicology and nanosafety: safety-by-design and testing at a glance. Int J Environ Res Public Health 17(13): 4657. https://doi.org/10.3390/ijerph17134657

Chapter 17
Ethical and Environmental Considerations in Plant Nanotechnology

Jenifer Christy Yacob, Vidhya Ramachandran, Dhannia P. Narayanan, and Sumod George

Abstract The need for nanobiotechnological applications in plant and agriculture sciences has been growing rapidly in recent decades due to the exponential growth of the world's population and subsequent increase in demand for food supplies. Plant biotechnology has evolved as a highly promising approach in various agricultural sectors in the form of using nanoparticles as fertilizers, sensors, genetic modifiers, and so on. Though the researchers and environmental policy makers around the globe tend to cope with the process of fulfilling the demands, they equally work and constantly update the regulatory frameworks on the ethical and environmental impacts of using nanoparticles to ensure safe and right handling. But those are still in their infancy. Here we contextualized the existing ethical concerns, environmental considerations, and various regulations of using and handling nanotechnology in plant systems.

Keywords Aquatic organisms · Environment · Ethics · Metal oxides · Metal oxide nanoparticles · Nano-agrochemicals · Nanoparticles · Plant nanotechnology

---

J. C. Yacob
Peer Review Services Team (Wiley Publishing Company), Integra Software Services Pvt. Ltd, Pakkamudiyanpet, Puducherry, India

V. Ramachandran
Department of Biochemistry, Idhaya College of Arts and Science, Pakkamudayanpet, Pondicherry, India

D. P. Narayanan
Department of Biology, Mar Athanasius College (Autonomous), Kothamangalam, Kerala, India
e-mail: dhanniapn@macollege.in

S. George (✉)
Department of Chemistry, Nirmala College (Autonomous), Muvattupuzha, Kerala, India
e-mail: sumod@nirmalacollege.ac.in

J. M. Al-Khayri et al. (eds.), *Nanotechnology Applications in Modern Agriculture*, Nanotechnology in Plant Sciences 3, https://doi.org/10.1007/978-3-031-90506-3_17

17.1 Introduction

A new and exciting field of study called "plant nanobiotechnology" combines nanotechnology and plant science to create and use Engineered Nanomaterials (ENMs) for engineering and plant function research. (Zhang 2022; Arya et al. 2021). Although the nanotechnological field research activities were coined three decades ago, only recently the applications of nanotechnology were initiated imposing a significant impact on global economy (Haris et al. 2023; Usman et al. 2020). Interestingly nanotechnology was introduced in agriculture for meeting the global requirements of food deficiency. Owing to their unique physicochemical properties, engineered and tailor-made nanoparticles have the potential to double crop production and meet the global challenges (Fig. 17.1). Meeting these requirements is a task due to the different types of landscapes, high nutritional demands and lack of adequate technology. Adversely effecting agriculture by producing unhealthy flavoured agro-products, chemical fertilizers has also affected the environment on planet earth (Torado 2008). However, nanotechnology offers a well-organized, workable agriculture industry by reducing harmful chemicals and addressing many of the existing global challenges of food scarcities. Nanotechnology provides quick plant disease diagnosis, improved plant nutrient absorption and rare-earth nanoparticles as potent fertilizers. Despite their potent effects, researchers are unable to unlock their operation mechanism since nanoparticles (NPs) can cause cytotoxicity at higher concentration levels (Sukhanova 2018). Because of their tiny size

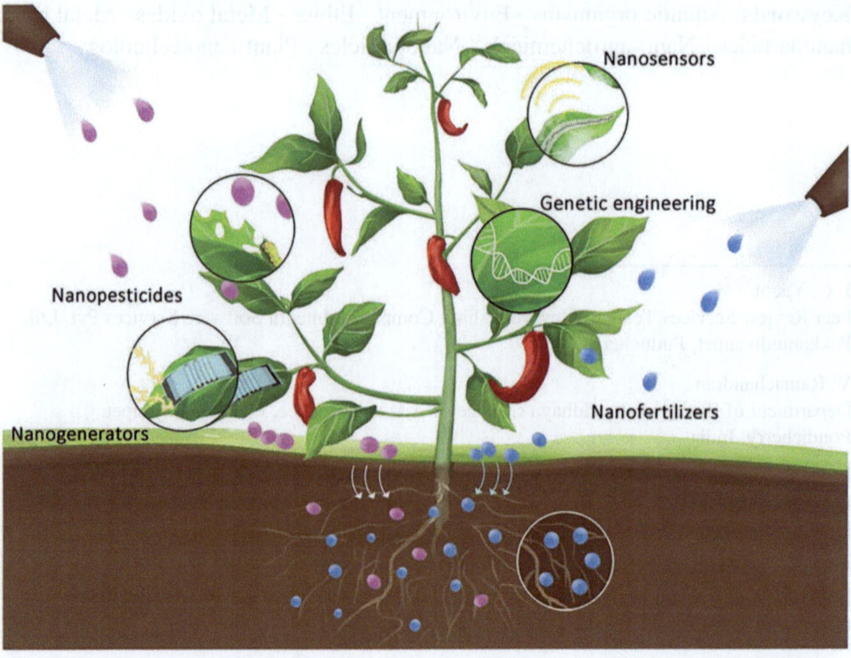

Fig. 17.1 Nanotechnological application in plants. (Adopted from Zhang et al. (2022))

nanoparticles are susceptible to penetrate cell membranes and form agglomerates with different microorganism and subsequently emerging with new bioavailability. The harmful consequences of metal oxide nanoparticles at different doses have been documented by researchers who have examined their effects and subsequently it is found that metal oxide nanoparticles had adverse effects if accumulates in the plant cells (Malhotra 2020). Further, the public is unaware of the toxic nature of nanotechnology and hence raises the concern of ethics as to whether it is appropriate to encourage the agriculture community, whose knowledge of nanotechnology is minimal, to apply nano-chemicals for better food productivity (Mwaanga 2018). This book chapter will highlight some of the most current developments in determining how nanoparticles affect the environment and ecosystem. In addition, this book chapter also examines the ethical issues concerning the usage of nanoparticles to enhance the crop enhancement in agriculture.

17.2 Plant Nanotechnology

Nanotechnology has influenced almost every essential sector of mankind with the aim of making products better and thereby achieving maximum productivity. Nanomaterials (NM) have unique physiochemical properties and has attracted plant science for the utilization of various activities like crop enhancements and better productivity. It is remarkable that nanomaterials are used as vehicles for the transportation of agrochemicals or biomolecules in plants which forms a part of the productivity enrichment in different crops (Khan et al. 2017). Plant nanotechnology includes areas such as nanoparticle-plant interaction and uptake, mobilization and accumulation of nanoparticles and its biological effects. Furthermore, there are other intriguing areas of plant nanoscience, such as the application of rare earth metal oxide nanomaterials in plants and the possible application of polymeric soft NMs, which offer a new and safer way to deliver biomolecules intelligently and ultimately enhance the defence or growth mechanisms of plants.

There are several nanomaterials used in plant sciences. Nanomaterials (NMs) are described by the American Society for Testing and Materials (ASTM) as manufactured or natural materials with a size between 1 and 100 nm (ASTM E2456-06, 2012). Interestingly, nanoparticles' high surface-to-volume ratio gives them exceptional physicochemical qualities when compared to their bulk counterparts (Roduner 2006). In plant science, metallic nanoparticles (NPs) like gold (Au) and silver (Ag) are often utilized to increase productivity. Nevertheless, their synthesis involves several hazardous chemicals and is not economical. Oxidised NMs like MgO, CaO, ZnO and TiO_2 are also used for crop enhancement due to their superior physicochemical properties (Jahan 2018). Another recently introduced nanomaterials for their versatility and flexibility are the polymeric nanomaterials. They can be synthesised cost-effectively and have high response to external stimuli (Baskar 2018). Because Core/Shell Nano Particles (NPs) are made from a range of material combinations, including inorganic/inorganic, inorganic/organic, organic/

inorganic, and organic/organic materials, they offer greater flexibility. The ability to use polymeric shells to increase NPs' biocompatibility is an extra benefit (Ghosh Chaudhuri 2012). Nanogels (NGs) are another intriguing nanomaterial utilized in plant nanotechnology. They are characterized as ionic and non-ionic hydrogels composed of natural or synthesized polymers at the nanoscale. It is noteworthy that NGs have a broad range of physicochemical characteristics, including high water content, high porosity and load capacity, colloidal stability, high encapsulation capacity, and more. Common NGs include chitosan, alginate, poly (vinyl alcohol), and poly (ethylene oxide) (Molina 2015).

Despite all of the previously mentioned enlightening uses, NMs' full potential in plant nanotechnology has not yet been realized, and there are still certain obstacles to be addressed. Primarily these include the need to create and synthesize what are generally thought to be safe NMs that do not impede plant growth and development (Sabo-Attwood 2012). Secondly a precise mechanism of NMs intake and immobilization in plants are not well understood. Therefore, it is realized there are insufficient interdisciplinary approaches and initiatives to bridge the knowledge gaps and address current issues.

17.3 Environmental Considerations in Plant Nanotechnology

Nanomaterials are used for enriching crop products and also cleaning pollutants from the environment and they include metal oxide nanoparticles as well. Consequently, majority of the organisms in the environment is exposed constantly to the crop enriching nanomaterials and there are adverse effects as well. According to studies, the accumulation of these metal oxide nanoparticles in the soil, especially copper oxide, lanthanum oxide, cerium oxide, and nickel oxide, can significantly slow down the pace at which plants in the soil transpire and perform photosynthesis (Vineeth Kumar 2022). Furthermore, several studies have demonstrated that crop-enhancing nanoparticles discharged into the environment as waste can harm marine life (Dube 2023). Although a comprehensive investigation is necessary for a deeper knowledge, a firm comprehension of the dangerous effect on the environment is still pending. The next section will cover some significant instances of harmful effects of nanomaterials' (NMs').

17.3.1 Impact of Copper Oxide Nanoparticles on the Environment

Copper oxide (CuO) NPs are used as agro-fertilizers because trace quantities of copper play a pivotal role in boosting photosynthesis and metabolism in many plants. CuO has also been found to enhance intake of nutrients and support root growth. In addition, they are available in large abundance and cost-effective which makes them further attractive as biofertilizers. But an excessive deposition of copper oxide NP in plants is a concern whether it will impact human health if consumed via food chain (Naz 2023; Rajput 2019). Furthermore, nothing is known about how Cu-based NPs affect soil organisms like nematodes, earthworms, mycorrhiza, fungi, and agriculturally significant microorganisms.

Owing to their tiny size, nanoparticles can interact with microorganisms known as plant growth-promoting rhizobacteria (PGPR) and cause environmental issues for ecosystems. The harmful impact of copper oxide nanoparticles on PGPR has lately been investigated by scientists (Tripathi 2024). Regardless of the physicochemical characteristics of the NPs, such as their solubility and larger hydrodynamic size, researchers examined the interaction of copper oxide nanoparticles with PGPR, Bacillius megaterium (Scholle 2003), in vitro and discovered that the toxicity of CuO NPs may occur even at a relatively lower dose and ultimately have a detrimental effect on the bacterium's function. Transmission Electron Microscopy (TEM) and Atomic Force Microscopy (AFM) photomicrographs showed abnormalities, distortion, and cellular structure loss, confirming cellular membrane leakage and nanoparticle uptake. Confocal microscopy revealed the production of reactive oxygen species (ROS), and flow cytometric analysis revealed the binding of propidium iodide solution (PI), which demonstrated cell death. In addition to serving as an indicator of the effect of nanoparticle stress on Bacillus megaterium, the reduction of Indole Acetic Acid (IAA) production verified the detrimental effect of CuO nanoparticles on the rhizobacteria's ability to promote plant growth. The current study also sheds light on the significance of the data regarding reduced concentrations of the agriculturally significant nanoparticles.

17.3.2 Toxicity of Rare Earth Nanoparticle Lanthanum Oxide (La_2O_3) on Agricultural Crops and Aquatic Organisms

Lanthanum is a Block F, period 6 element, likewise oxygen is a Block P, Period 2 element. Lathanum oxide solids are odorless white solids, insoluble in water while soluble in dilute acids. Lanthanum oxide nanoparticles have spherical morphology but are strongly hygroscopic and cause irritation to eyes, skin and is suffocating to respiratory systems. Plants are a significant component that act as a key conduit for the transport of lanthanum oxide nanoparticles (La_2O_3 NPs) and a means of accumulation in different organisms in the food chain since NPs are employed in agriculture to control plant development (Hanif 2024). Because lanthanum oxide is used extensively in many industries, it is unavoidably discharged into the environment and has negative effects on both plants and animals.

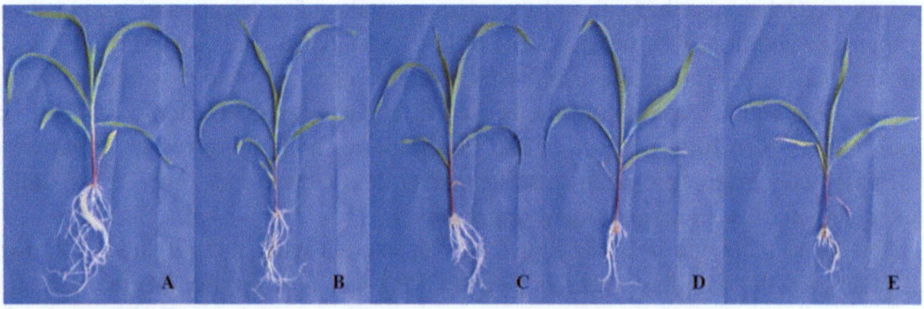

Fig. 17.2 Photomicrographs of maize plants exposed to different treatment of La_2O_3NPs. Image from left to right in each picture were: A. unexposed control plants, B to D exposed to 5, 10, 50, 100 mg/L La_2O_3NPs (adopted from the reference (Liu 2018) under the Creative Commons Attribution 3.0 licence, https://doi.org/10.1088/1755-1315/113/1/012020)

Lanthanum oxide nanoparticles' effects on maize (Zea mays L.) root elongation, biomass output, and chlorophyll content were examined by a team of researchers (Liu 2018). As shown in Fig. 17.2 below, maize plants were subjected to varying concentrations of La_2O_3 NPs, including 5, 10, 50, and 100 mg/L NPs. The morphology of maize roots was shown to be impacted; however the morphology of maize shoots was found to be less affected. Overall, the biomass of the roots and shoots was impacted by the La_2O_3 NPs. This study also looked into how La_2O_3 NPs affected the amount of chlorophyll. They found that the relative chlorophyll content (CCI) was not affected at low concentrations (5 mg/L), but it was significant when compared to higher concentrations, meaning that the amount of chlorophyll content decreased by 6.5%, 25.9% and 21.7% respectively for 10, 50 and 100 mg/L (Fig. 17.3).

A Turkish team of researchers looked into the potentially harmful effects of lanthanum oxide nanoparticles on aquatic life (Balusamy 2015). The researchers shared their findings after examining how lanthanum oxide affected two significant aqua species. Because of their crucial role in the aquatic food chain and sensitivity to a variety of contaminants, the crustacean Daphnia magna and the freshwater microalgae Chlorella sp. are regarded as remarkable biomonitoring aquatic species. To ascertain the smallest amount of lanthanum oxide nanoparticles required to cause toxicity in these organisms, their research included a growth inhibition assay. At a dosage of 1000 mg L-1, the growth inhibition assay investigations found that La_2O_3 NPs had no harmful effect on Chlorella sp. after 72 h of exposure, although D. magna was sensitive. At a very high concentration La_2O_3 NPs had a lethal effect on the two above species studied. To comprehend the relationship between La_2O_3 and aquatic life, they also employed photomicrographs. Overall, aquatic animals are toxically affected by very high concentrations of La_2O_3 NPs, making it difficult to maintain a clean environment when lanthanum NPs build up via the food chain.

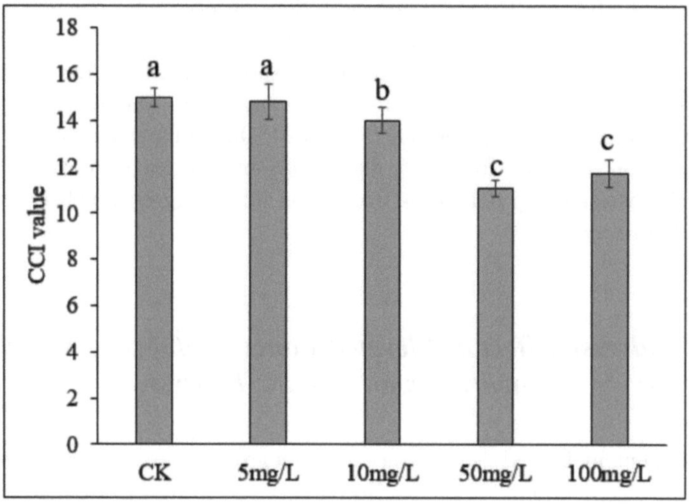

Fig. 17.3 The relative chlorophyll content (CCI value) of maize when exposed to La_2O_3 NPs at various concentrations. The values were represented as mean ± SD. (adopted from the reference (Liu 2018) under the Creative Commons Attribution 3.0 licence, https://doi.org/10.1088/1755-1315/113/1/012020)

17.3.3 Cerium Oxide Nanoparticles an Abiotic Stress Reducer with Toxic Nature at High Concentration

Abiotic stress and other environmental limitations have an impact on agricultural crops, which directly affects food production worldwide. In biological systems, nanoparticles like cerium oxide function as nanozymes to lessen the impacts of abiotic stress. However, because cerium oxide is produced on a massive scale each year—an estimated 10,000 tonnes—and is used so extensively in the modern world, it accumulates in the environment as waste products that pose deadly risks to aquatic life, there have been reports of its toxicity effect on theenvironment (Naiel 2022). With an atomic number of 58, cerium (Ce) is an inner transition element that is commonly found in monazite and bastnaesite ores. Cerium oxide is created by heating bastnaesite and then treating it with HCl (Ismael 2021). There are two oxidation states for cerium: Ce^{3+} and Ce^{4+}. When cerium is internalized, it is released as $Ce^{3+,}$ which may then be harmful to live cells.

Dedman and colleagues have used the ecologically significant marine *cyanobacterium Prochlorococcus* sp. MED4 (Dedman 2021) as a model to investigate the impact of CeO_2 on marine microbial species. The *cyanobacterium* underwent simulated natural and nutrient-rich growth and environmental circumstances while being exposed to $nCeO_2$ at a broad range of concentrations (1 μg L-1 to 100 mg L-1). They found that, in natural settings, $nCeO_2$ causes minimal toxicity to *Prochlorococcus* over an extended period of time, with the exception of higher concentrations (100 mg L-1). Additionally, they observed that microbial cultures with higher cyanobacterium cell densities cultivated in nutrient-rich media experienced greater CeO_2 toxicity after 240 hours of exposure due to increased interaction that caused the microbials and cerium NPs to aggregate. This hetero-aggregation between *Prochlorococcus* and $nCeO_2$ (Fig. 17.4). One of the main characteristics of *cyanobacterium* exposure to $nCeO_2$ that results in ecotoxicity as verified by the Dynamic Light Scattering (DLS) approach is the hetero-aggregation between $nCeO_2$ and *Prochlorococcus*.

17.3.4 Adverse Effect of Elevated Concentrations of Nickel Oxide Nanoparticles on Aquatic Ecosystem

A trace element, nickel (Ni) is necessary for plant growth and development as well as for increasing agricultural yields. It also plays a part in nitrogen fixation and urea breakdown. Nickel oxide nanoparticles at low concentration (50 mg/kg) promoted better yield, photosynthesis and nitrogen metabolism. Zhang et al. have conducted the role of Nickel NPs in soyabean recently (Zhang 2023). However, it has been found that nickel oxide nanoparticles (NPs) have negative impacts, including lowering biomass accumulation, root and shoot growth, and seed germination, as well as inhibiting physiological activities including photosynthesis and transpiration. We'll talk about a few of these impacts on the plant habitat subsequently.

Pseudokrichneriella subcapitata, an ecologically significant freshwater alga, was used to study the harmful environmental consequences of nickel oxide nanoparticles (Sousa 2018). In OECD medium, the NiO NP suspensions in algae were made. It's intriguing that the physicochemical characteristics of the possible risks—agglomeration, surface charges, stability (NP dissolution), and abiotic reactive oxygen species (ROS) production—were examined. They noticed that NiO NPs liberated Ni^{2+} into the medium and formed loose aggregates. The Ni^{2+} that the NPs released was the primary cause of the toxic effects seen in *P. subcapitata*. After being left for 72 h, the NiO NPs suspension had an EC50 of 1.6 mg L-1, which was then assessed using the algal growth inhibition assay and determined to be hazardous.

The team of scientists also observed that, in the absence of the plasma membrane, NiO NPs led to the intracellular buildup of reactive oxygen species (ROS), the bleaching of photosynthetic pigments, and the loss of metabolic esterase function. The photosynthetic process was also interfered with by the NiO NPs. It was also observed that the photosynthetic efficiency decreased along with the electron flow rate across the photosynthetic chain. One possible explanation for the buildup of ROS in algal cells is the electron leakage from the photosynthetic chains. An increase in cell volume, abnormal morphology, and the quenching of the cell cycle before cell division were the results of prolonged exposure to NiO NPs. Thus, this study emphasizes NiO's toxicological impacts.

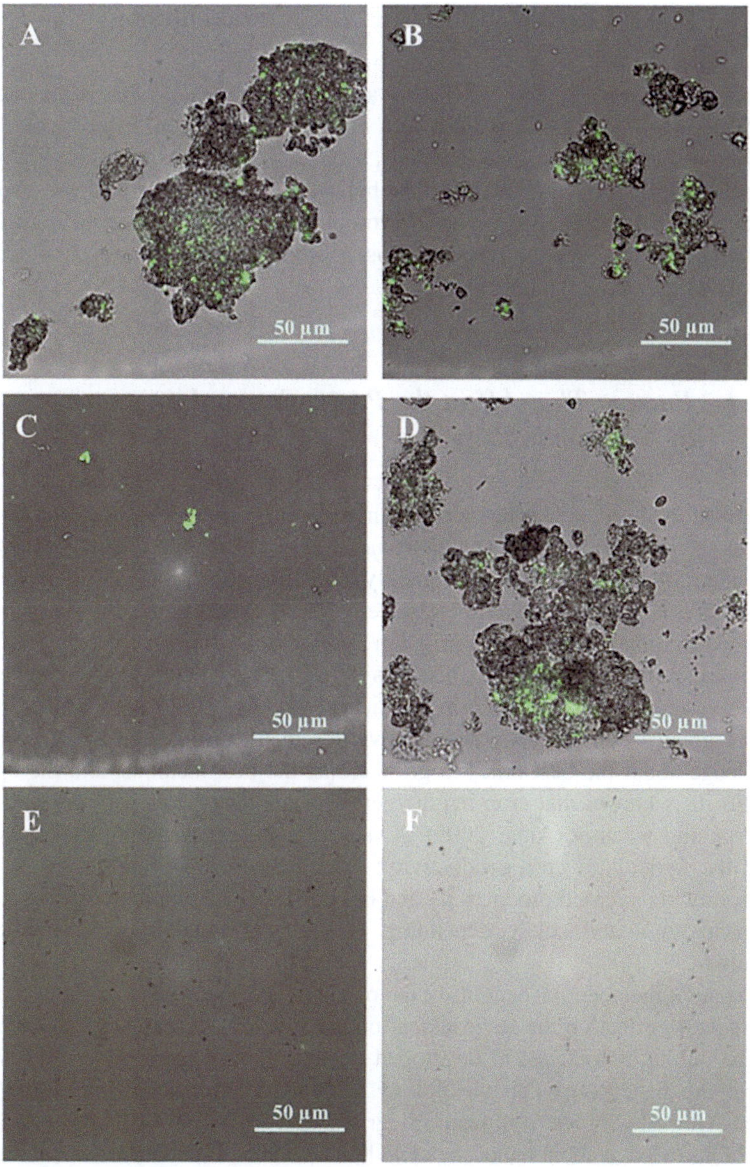

Fig. 17.4 Hetero-aggregation of *Prochlorococcus* sp. MED4 (green) and nCeO2 (10 and 100 mg L^{-1}) formed in natural oligotrophic seawater [(**a**); 10 mg L^{-1} and (**b**); 100 mg L^{-1}] and Pro99 media [(**c**); 10 mg L^{-1} and (**d**); 100 mg L^{-1})]. Photomicrographs were recorded using fluorescent microscopy under brightfield and GFP fluorescent and subsequently were merged. It is interesting to note that no such aggregates were visible in *Prochlorococcus* culture (**e**) or natural oligotrophic seawater (**f**) in the absence of nanoparticles. (Adopted from reference Dedman et al. (2021), under Creative Commons Attribution License (CC BY))

17.4 Ethical Consideration in Plant Nanotechnology

The collection of ideals and beliefs that address the issues of what is beneficial or detrimental to life on Earth is known as ethical consideration. Ethics looks at the motivations behind actions and inactions, such as accepting or disapproving behavior, believing or rejecting that certain behaviors are good or evil, or choosing between laws that are good or bad. Avoiding purposeful damage to a system is simply one aspect of ethical considerations; another is the need to prevent potentially detrimental unintentional effects.

17.4.1 Factors That Affect the Perception and Awareness of Nanotechnology

The global food production has seen an unprecedented growth in food production in the last decade and this is attributed to increased need for food materials owing to population growth as well as consequently implementation of advanced agricultural technologies in crop production (Joubert 2020). Nanotechnology is one such advanced technology that has contributed to this extraordinary food crop productivity. Introducing metal oxide nanoparticles (NPs), nanofertilizers, nanopesticides has enabled farmers to double the crop production (de Oliveira 2014). Farmers in poor nations greatly benefit from the creative discoveries since they lack technical knowledge about the best times to grow crops, the best methods, and the optimal dosages. It is known that crop requirements for agrochemicals vary depending on soil type and weather (Aktar 2009). A lack of this understanding may adversely affect the agriculture crop productivity; an enhanced use of fertilisers may not correspond to increased productivity and is a consequence of not embracing science and technology and solely depending on the traditional seasonal rainfall for agriculture.

In spite of the potential benefits of nanotechnology, the general public is unaware of its adverse effect on the environment on planet earth. Researchers have reported that the public in societies in developed countries like United States and Europe believe the benefits of nanotechnology outweighs the risks, while some other citizens viewed with less optimism (Siegrist 2008; Coles 2013). In this context the level of knowledge about nanoscience in less advanced countries and their societies could be far below the expectation. Consequently, the public's perception of and adoption of a new technology depends on how industries and scientists view the social and ethical issues surrounding the innovative technology when the public is less knowledgeable about it. Therefore, when the public lacks a basic knowledge about a new technology their belief about the safety and trust about nanotechnology will follow what the higher education institutions, scientists and industries perceive about nanotechnology.

The way the media covers technological issues is another factor that shapes public opinion. Compared to industrialized nations, impoverished nations receive less media attention about the uses of nanotechnology, and research and development knowledge is also lacking. Even in developed countries, where the research activities are at higher grades, the media coverage is only modest owing to dearth of expert media persons (Dudo 2011).

17.4.2 Ethical Aspects of Nanotechnology in the Context of Toxic Nanoparticles

There aren't many ethical problems with nanotechnology in the agricultural sector, according to observations and awareness. Is it morally acceptable to employ nanoparticles in agriculture given the current state of knowledge and the possible harm they could cause to the environment and food chain? When offered in the market, should agro-nano goods have specific labels? Is it ethical that the government institutions should fund the research and development in nanotechnology? The public should be made aware of the toxicity of that nanoparticles pose, which they do not have currently. Is it ethical to sell nanoparticles and nano-agrochemicals to the farmers who are less aware of their health consequences? Should there be regulatory rules for improving the public awareness? All these questions are to be addressed at an early stage so that the public as well as farmers are aware of the consequences of nanoparticles in the environment as well as public health.

17.5 Conclusion

Recently, in addition to being widely available in a variety of items like sunscreens, cosmetics, paints, and building materials, nanoparticles (NPs) have recently been integrated into a large number of traditional home and commercial appliances. Daily use of high concentrations of NPs will lead to massive waste discharge into the ecosystem and environmental pollution. As a result, aquatic and non-aquatic habitats are thought to be susceptible to various NP disposal, the effects of which have not yet been thoroughly identified. It has been noted that once NPs are in aquatic settings, transformations, agglomerations, and surface characteristics are important factors in influencing their bioavailability and toxicological characteristics. Therefore, the quantity of nanoparticles on the consumer products and their effects on the environment needs to be studied well and regulated by laws. The present book chapter has highlighted many of the adverse effects of nanoparticles and will provide guidelines while preparing nanoparticles for new applications and thus assist in meeting the challenges posed by these innovative materials to our environment. Value is added to the agricultural chain with the introduction of smart nano agrochemicals that are intended to improve physicochemical characteristics such as solubility, bioavailability, targeted distribution, and controlled release over an extended period. Therefore, by controlling fertilizer dosage, nanotechnology applied to plants and agriculture can lower chemical pollution and minimize contact between farmers and plants.

The public's lack of knowledge about nanotechnology and the general lack of publicity are two other concerning issues that have been raised regarding nanoparticles. Despite the advantages of nanotechnology, the general public will use heuristics like trust to gauge the risks and rewards when they are not well understood. Furthermore, the public is more likely to think that a certain product is safe if businesses and government regulatory agencies build strong relationships with the public by giving them accurate information. Furthermore, the media has a crucial influence in shaping public perceptions and attitudes toward novel and cutting-edge technologies. Since only less media persons are professionally trained to report appropriately on issues on nanotechnology, it is critical to follow the precise information. In addition, the labelling of nanoproducts should be encouraged so that the public is open to free choice for the selection of their desired products.

References

Aktar W, Sengupta D, Chowdhury A (2009) Impact of pesticides use in agriculture. Their benefits and hazards. Interdiscip Toxicol 2(1):1–12. https://doi.org/10.2478/v10102-009-0001-7

Arya SS, Tanwar N, Lenka SK (2021) Prospects of nano- and peptide-carriers to deliver CRISPR cargos in plants to edit across and beyond central dogma. Nanotechnol Environ Eng 6:22. https://doi.org/10.1007/s41204-021-00118-z

Balusamy, B., Sundaram, S., & Radhakrishnan, R. (2015). Influence of copper oxide nanoparticles on growth of beneficial soil bacterium Pseudomonas fluorescens. Ecotoxicology and Environmental Safety, 122, 336–344. https://doi.org/10.1016/j.ecoenv.2015.08.002

Baskar V, Meeran S, Shabeer STK, Sruthi S, Ali J (2018) Historic review on modern herbal nanogel formulation and delivery methods. Int J Pharm Pharm Sci 10:1–10. https://doi.org/10.22159/ijpps.2018v10i10.23071

Coles D, Frewer LJ (2013) Nanotechnology applied to European food production: a review of ethical and regulatory issues. Trends Food Sci Technol 34:32–43. https://doi.org/10.1016/j.tifs.2013.08.006

de Oliveira JL, Campos EVR, Bakshi M, Abilashi PC, Fraceto LF (2014) Applications of nanotechnology for encapsulation of botanical insecticides for sustainable agriculture and promises. Biotechnol Adv 23:1550–1561. https://doi.org/10.1016/j.biotechadv.2014.10.010

Dedman CJ, Rizk MMI, Christie-Oleza JA, Davies G-L (2021) Investigating the impact of cerium oxide nanoparticles upon the ecologically significant marine cyanobacterium *Prochlorococcus*. Front Mar Sci 8:668097. https://doi.org/10.3389/fmars.2021.668097

Dube E, Okuthe GE (2023) Engineered nanoparticles in aquatic systems: Toxicityand mechanism of toxicity in fish. Emerg Contam 9(2):100212. https://doi.org/10.1016/j.emcon.2023.100212

Dudo A, Choi D, Scheufele DA (2011) Food nanotechnology in the news. Coverage patterns and thematic emphases during the last decade. Appetite 56:78–79. https://doi.org/10.1016/j.appet.2010.11.143

Ghosh Chaudhuri R, Paria S (2012) Core/Shell nanoparticles: classes, properties, synthesis mechanisms, characterization and applications. Chem Rev 112:2373–2433. https://doi.org/10.1021/cr100449n

Hanif Z, Jabeen N, Anwaar S, Aftab A, Hussain SZ, Anwar T, Qureshi H, Munazir M, Zaman W, Soufan W (2024) Synthesis and characterization of lanthanum oxide nanoparticles using Citrus aurantium and their effects on Citrus Limon germination and Callogenesis. Sci Rep 14(1): 21737. https://doi.org/10.1038/s41598-024-73016-4

Haris M, Hussain T, Mohamed HI, Khan A, Ansari MS, Tauseef A et al (2023) Nanotechnology—a new frontier of nano-farming in agricultural and food production and its development. Sci Total Environ 857:159639. https://doi.org/10.1016/j.scitotenv.2022.159639

Ismael NE, Abdel El-hameed SA, Salama AM, Naiel MA, Abdel-Latif HM (2021) The effects of dietary clinoptilolite and chitosan nanoparticles on growth, body composition, haemato-biochemical parameters, immune responses, and antioxidative status of Nile tilapia exposed to imidacloprid. Environ Sci Pollut Res 1–16:29535–29550. https://doi.org/10.1007/s11356-021-12693-4

Jahan S, Alias YB, Bakar AFBA, Yusoff IB (2018) Toxicity evaluation of ZnO and TiO2 nanomaterials in hydrophobic red bean (Vigna angularis) plant: physiology, biochemistry and kinetic transport. J Eviron Sci 72:140–152. https://doi.org/10.1016/j.jes.2017.12.022

Joubert IA, Geppert M, Ess S, Nestelbacher R, Gadermaier G, Duschl A, Bathke AC, Himly M (2020) Public perception and knowledge on nanotechnology: a study based on a citizen science approach. NanoImpact 17(January):100201. https://doi.org/10.1016/j.impact.2019.100201

Khan MN, Mobin M, Abbas ZK, AlMutairi KA, Siddiqui ZH (2017) Role nanomaterials in plants under challenging environments. Plant Physiol Biochem 100:194–209. https://doi.org/10.1016/j.plaphy.2016.05.038

Liu Y, Xu L, Dai Y (2018) Phytotoxic effect of lanthanum oxide nanoparticles on maize (*Zea mays L.*). In: IOP Conf. Series: Earth and Environmental Science, vol 113. IOP Publishing, p 012020. https://doi.org/10.1088/1755-1315/113/1/012020

Malhotra N, Hsu H-S, Liang S-T, Roldan MJM, Lee J-S, Ger T-R, Hsiao C-D (2020) An updated review of toxicity effect of the rare earth elements (REEs) on aquatic organisms. Animals (Basel) 10(9):1663. https://doi.org/10.3390/ani10091663

Molina M, Asadian-Birjand M, Balach J, Bergueiro J, Miceli E, Calderon M (2015) Stimuli-responsive nanogel composites and their application in nanomedicine. Chem Soc Rev 44: 6161–6186. https://doi.org/10.1039/C5CS00199D

Mwaanga P (2018) Risks, uncertainities and ethics of nanotechnology in agriculture. New visions in plant science. Intech. https://doi.org/10.5772/intechopen.76590

Naiel MAE, Abdel-Latif HMR, Abd El-Hack ME, Khafaga AF, Elnesr SS, Dawood MAO, Alkazmi L, Elhady HA, Batiha GES, Alagawany M (2022) The applications of cerium oxide nanoform and its ecotoxicity in the aquatic environment: an updated insight. Aquat Living Resour 35:9. https://doi.org/10.1051/alr/2022008

Naz S, Gul A, Zia M, Javed R (2023) Synthesis, biomedical applications, and toxicity of CuO nanoparticles. Appl Microbiol Biotechnol 107:1039–1061. https://doi.org/10.11007/s00253-023-12364-z

Rajput V et al (2019) Interaction of copper-based nanoparticles to soil, terrestrial, and aquatic systems: critical review of the state of the science and future perspectives. In: de Voogt P (ed) Reviews of Environmental Contamination and Toxicology Volume 252. Reviews of Environmental Contamination and Toxicology, vol 252. Springer, Cham. https://doi.org/10.1007/398_2019_34

Roduner E (2006) Size matters: why nanomaterials are different. Chem Soc Rev 35:583–592. https://doi.org/10.1039/B502142C

Sabo-Atwood T, Unrine JM, Stone JM, Murphy CJ, Ghoshroy S, Blom D et al (2012) Uptake distribution and toxicity of gold nanoparticlces in tobacco (Nicotiana xanthi) seedlings. Nanotoxicology 6:353–360. https://doi.org/10.3109/17435390.2011.579631

Scholle MD, White CA, Kunnimalaiyaan M, Vary PS (2003) Sequencing and characterization of pBM400 from Bacillus megaterium QM B1551. Appl Environ Microbiol 69(2003):6888–6898. https://doi.org/10.1128/AEM.69.11.6888-6898.2003

Siegrist M, Stampfli N, Kastenholz H, Keller C (2008) Perceived risks and benefits of different nanotechnology tools and nanotechnology food packaging. Appetite 51:283–290. https://doi.org/10.1016/j.appet.2008.02.020

Sousa CA, Soares HMVM, Soares EV (2018) Toxic effects of nickel oxide (NiO) nanoparticles on the freshwater alga Pseudokirchneriella subcapitata. Aquat Toxicol 204:2018–2090. https://doi.org/10.1016/j.aquatox.2018.08.022

Sukhanova A, Bozrova S, Sokolov, Berestovoy PM, Karaulov A, Nabiev I (2018) Dependence of nanoparticle toxicity on their physical and chemical properties. Nanoscale Res Lett 13:44. https://doi.org/10.1186/s11671-018-2457-x

Tripathi, GD., Javed, Z., Dashora, K., (2024) Toxicity of copper oxide nanoparticles on agriculturally important soil rhizobacteria Bacillus megaterium, Emerging Contaminants 10(1): 100280, ISSN 2405-6650, https://doi.org/10.1016/j.emcon.2023.100280

Torado R, Englande AJ, Promakasikorn L, Novotny V (2008) Use of agrochemicals in thailand and its consequences for the environment. Greenpeace Research Laboratories Technical Note 03/2008

Usman M, Farooq M, Wakeel A, Nawaz A, Cheema SA, Ur Rehman H et al (2020) Nanotechnology in agriculture: current status, challenges and future opportunities. Sci Total Environ 721: 137778. https://doi.org/10.1016/j.scitotenv.2020.137778

Vineeth Kumar CM (2022) The impact of engineered nanomaterials on the environment: release mechanism, toxicity, transformation, and remediation. Environ Res 212:113202. https://doi.org/10.1016/j.envres.2022.113202

Zhang Q, Ying Y, Ping J (2022) Recent advances in plant nanoscience. Adv Sci 9(2):2103414. https://doi.org/10.1002/advs.202103414

Zhang P, Jiang Y, Adeel M, Shakoor N, Zhao W, Liu Y, Li Y, Li M, Azeem I, Rui Y, Tan Z, White JC, Guo Z, Lynch I, Zhang P (2023) Nickel oxide nanoparticles improve soyabean yields and enhance nitrogen assimilation. Environ Sci Technol 57(19):7547–7558. https://doi.org/10.1021/acs.est.3c00959

Chapter 18
Current Trends of Nanoparticles in Insect Pest Management

U. Pirithiraj, R. P. Soundararajan, M. Yuvaraj, Jaiby Cyriac, R. Sathya Priya, N. Jagathjothi, N. Suganthi, R. Sharmila, and M. Saranya

Abstract A wide range of new sciences and technologies were used to agriculture in order to boost productivity. Pest insects pose a substantial threat to crop productivity. According to recent data, arthropods are to blame for 18–20% of all agricultural losses worldwide, amounting to more than US$470 billion annually. We must make strides in developing pest management solutions that are safe, economical, and long-lasting. During the twentieth century, pesticides gained worldwide commercialization and increased importance for controlling pests of agriculture crops. After decades of practical application, researchers across the globe have studied and recorded the adverse effects of pesticides on non-targeted organisms in crop ecosystems. In order to overcome the limits of conventional pesticides, the twenty-first century is looking forward for alternative pest management approaches. Nanotechnology is a relatively recent science that, in the form of nano pesticides controlled release formulations, offers an alternative to conventional pesticides. Most synthetic chemical insecticides, considered outstanding for pest management during their discovery, were later recognized for their harmful environmental consequences. Similarly, nano pesticides have both beneficial and harmful consequences on the environment. This paper delves deeper into the significance of nano insecticides against numerous insect pests of crops, as well as their environmental consequences.

Keywords Agriculture · Controlled release · Formulations · Nanotechnology · Pest management · Pesticide · Pheromones

U. Pirithiraj · R. P. Soundararajan
Centre for Plant Protection Studies, Tamil Nadu Agricultural University, Coimbatore, India

M. Yuvaraj (✉)
Agricultural College and Research Institute, Vazhavachanure, Tamil Nadu, India

J. Cyriac
Department of Botany, Newman College, Thodupuzha, Kerala, India

R. S. Priya · N. Jagathjothi · N. Suganthi · R. Sharmila · M. Saranya
Tamil Nadu Agricultural University, Coimbatore, Tamil Nadu, India

J. M. Al-Khayri et al. (eds.), *Nanotechnology Applications in Modern Agriculture*, Nanotechnology in Plant Sciences 3, https://doi.org/10.1007/978-3-031-90506-3_18

18.1 Introduction

The term nanotechnology was coined by N. Taniguchi in 1974 during an International Conference on Industrial Production held in Tokyo, Japan (Srivastava et al. 2022). Nanotechnology dates back from 100 BC to 100 AD, when mankind produced cotton and silk materials and manufactured them into clothes with pore size ranging from 1 to 20 nanometers. As a result, they were classified as nanoporous materials (Tolochko 2009). Earlier, due to a lack of information and scientific contribution to the science of nanotechnology, the manufacture and exploitation of nanomaterials remained uncommon. The concept of nanotechnology was introduced in 1959 by physicist Richard Feynman in his address "There's Plenty of Room at the Bottom" (Feynman 1991; Mccray 2007). The industrial revolution in the middle of the eighteenth century and the green revolution in the 1960s preceded the current agricultural revolution, which is referred to as the nanotechnology revolution. Every new scientific finding has benefits and drawbacks for human life (Gulati et al. 2022). Nanotechnology's application in agriculture offers benefits and drawbacks. All across the world, nano pesticides were under commercialization for use in agriculture.

Pest insects inflict more crop damage than any other biotic agent. Insect's causes direct damage to fruits, vegetables, cereals, and so on, also cause indirect losses to food products by contaminating them with exuvia and insect excreta, which necessitates new management tactics. In recent years, nanotechnology has gained prominence in agriculture for improving crop protection measures such as controlled application of fertilizers and pesticides to boost crop productivity and protection (Ruparao 2020; Shekhar et al. 2021; Wang et al. 2022). This chapter thoroughly reviewed literatures pertaining to the impact of nano pesticides on insect pests, beneficial organisms and a way forward.

18.1.1 *Evolutionary Trend of Nanotechnology in Agriculture*

Agri-nanotechnology is a vast field that includes pest control, waste remediation, bioprocessing, genetics, and plant breeding, among other things (Khan et al. 2023). Food production losses cost the world over $2000 billion. Plant disease, insects, and weeds were estimated to be responsible for 13, 14, and 13% of food production losses, respectively (Pimentel 2009). During the early Second World War to the late 1950s, insecticides such as dichlorodiphenyltrichloroethane (DDT), organophosphates, and organochlorines were used indiscriminately to control crop loss caused by insect pests. Environmental impacts of these pesticides, such as adverse effects on non-targeted organisms and the development of resistant strains, were reported in the following decades (Beard 2003; Guillette 2012; Mei 2015; Naqqash 2016). Pesticides for agriculture must be target-specific, less hazardous to non-target organisms, and not breed any resistant strains of the targeted organisms. Due to the fact that

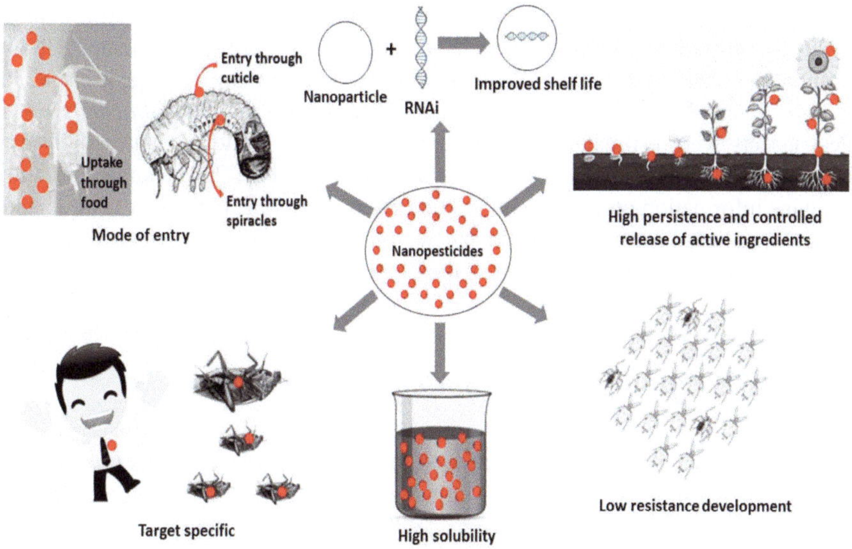

Fig. 18.1 Nanoparticle mediated pesticide delivery in the environment for pest management and their significance. (Figure constructed by U. Pirithiraj)

many of these pesticide formulations are unstable and quickly degrade in nature, development of nano formulations addresses these issues (Manna et al. 2023). The pesticides' nano formulations and encapsulation are precise, slow to release their active components, and stable (Li et al. 2023a). This opens the door for nano-based substances to be used as an alternative to traditional insecticides against insect pests (Fig. 18.1) (Anjali et al. 2012; Jerobin et al. 2012; Xu et al. 2010).

18.1.2 Importance of Insects in Agriculture

Like people, insects play a critical role in agriculture. They could be pests of crops, weed eaters, pollinators, parasitoids, predators, etc. Insects play two different roles in agriculture. Both as a pest that severely damages the crop resulting in yield loss and a pollinator that increases the pollination resulting in high yield. "Agriculture dies if beneficial insects die." The pest insects have become a source of financial concern since they drastically lower the value of the crops (Srishailam et al. 2022). A mature locust swarm of one square kilometer may consume enough food to feed 35,000 people for a whole day, according to estimations (Roychoudhury 2020). A fall armyworm outbreak in Kenya resulted in losses of nearly one million tons, or one-third, of the country's annual maize production signifying the importance of insects in agriculture (Groote 2020).

Each year, FAO member nations submit their requests for assistance with food and agriculture. Most member countries have approached FAO during the past

14 years with questions about outbreaks of insect pests. The invasive cypress aphid, *Cinara cupressivora*, is thought to have killed trees worth an estimated US$41 million per year in East and Southern Africa. Furthermore, Africa loses $2.25 million a year due to the two pine aphids, *Pineus boerneri* and *Eulachnus rileyi*. Pest insects were estimated to damage more than 52 million hectares of forests between 1980 and 2002 in 37 nations across the globe. The database currently contains information on more than 300 outbreaks involving 113 insect taxa, including 38 families, 79 genera, and 90 species (Allard 2003). After which there were record of number of outbreaks across the globe.

18.1.3 Ill Effects of Chemical Control on Environment

Persistent insecticides like DDT were used to manage a variety of insect species that are detrimental to different types of cultivated crops during the twentieth century. Unfortunately, these insecticides are made illegal in many countries throughout the world due to bioaccumulation and biomagnification. Along with being persistent, only 1% or fewer pesticides succeed in killing the intended target; the remaining 99% still affect organisms other than the intended target (Memarizadeh et al. 2014a, b). On the other hand, non-persistent pesticides break down more quickly and leave fewer residues in the biosphere. Because of this, they are somewhat ineffective against crop pests (Wurster 1972). It is necessary to take an alternative strategy to pest management because synthetic chemical insecticides have limitations. There were several environmentally friendly management strategies for reducing insect pest density such as using predators and parasitoids, growing cover crops, intercrops, mulching, using products derived from neem that weren't always feasible due to environmental factors. To control the growing pest population, an efficient substitute for conventional insecticides that is practical in all ecosystems is necessary (Zhou et al. 2023). A target-specific, highly toxic, environmentally safe, controlled-release formulation is required to safeguard the environment from synthetic chemical pesticides (Choudhury et al. 2012; Harish et al. 2022). The primary objective of nano pesticides is to facilitate the development of formulations that act on certain species more effectively than conventional pesticides (Atta et al. 2015; Villasenor and Ríos 2018).

18.2 Nanotechnology: Basic Concepts, a New Practice in Pest Management

Insect pests have been controlled using a variety of formulations, including emulsifiable concentrate (EC), dust (D), granules (G), wettable powder (WP), soluble powder (SP), etc. A substantial number of these compounds were environmentally unstable and biodegradable. Before they reach their target, they deteriorate. Contrarily, some pesticides such as DDT are environmentally stable and broad-spectrum and are transported up the food chain from one tropic to another tropic level, where they may accumulate over time and kill unintended creatures. Therefore, agriculture needs novel insecticidal formulations that are highly target-specific, gradually distributed in the environment, stable for a while, and degradable by microorganisms after release (Mubeen et al. 2023).

The use of polymers in nanotechnology enabled the manufacture of next-generation nano pesticides (Fang et al. 2023). These new formulations displayed host-specificity, regulated controlled release, and increased shelf-life. Some of them were covered here. Imidacloprid was released into water during a period ranging from 0.22 to 0.37 days in the studied formulations. In a controlled-release formulation, 50% of Imidacloprid took up to 9.31 days to release (Adak et al. 2012). Silica nanoparticles were reported to improve plant resistance to pests and also help in enhanced synthesis of volatile compounds thereby attracting natural enemies (Saw et al. 2023). The coconut mite (*Aceria guerreronis* Keifer) and rice weevil (*Sitophilus oryzae* Linnaeus) were discovered to be efficient against silica nanoparticles. Physical adsorption of nanoparticles on lipids in insect cuticles was thought to be the route of action of silica nanoparticles on insects (Barik et al. 2012). Several host crops such as cotton and vegetables were resistant to the *Spodoptera littoralis* Boisduval when treated with nano-silica particles as an antifeedant. It was discovered that the use of nano-silica particles tended to boost tomato plants' resilience to insects and other pests. Nanoparticles directly affect the biology of insects in addition to causing resistance in crop plants. Due to lower fecundity, adult longevity, etc., silica nanoparticle application reduced tomato pest damage (El-bendary and El-Helaly 2013).

Pesticides in the controlled release (CR) formulation performed better than those in commercial formulations. The pyrethroid pesticide Cyfluthrin's rate of administration can be altered by controlled release, perhaps leading to a higher rate of insect pest control (Loha et al. 2011). According to Pankaj et al. (2012), the carbofuran CR formulation proved effective in the field at suppressing the economically relevant tomato root-knot nematode. Thiamethoxam, a second-generation neonicotinoid, was demonstrated to be more efficacious in CR formulation than in commercial formulation, according to Sarkar et al. (2012).

18.2.1 Mode of Entry and Mode of Action of Nano Pesticides on Insects

Insecticides often enter an insect's body through feeding, physical contact, or spiracles that enter the trachea. Insecticides kill insects through a variety of mechanisms after entering the insect body, including acetylcholine inhibitors, GABA-gated chloride channel blockers, nicotinic acetylcholine receptors, and mitochondrial ATP synthase inhibitors. Nano pesticides work similarly to insecticides in that they enter an insect's body by contact, ingestion, and inhalation. Nano pesticides generally had similar mechanisms of action. Silver nanoparticles (AgNPs) have been reported to impact the Glutathione S transferase enzyme and interfere with melanin formation. Nanoparticle inhalation changed the activity of metabolic genes, lowering levels of fat, protein, and glucose as well as food intake, which finally caused an insect to die.

The neurological system and membrane potential of insects are also affected by inhalation. For instance, cholinergic esterase and carboxyl esterase are both binders of Zn nanoparticles and TiO_2 NPS, which both have an impact on their activity. Insects exposed to nanoparticles experience DNA damage as a result of free radical induction (Abd-Elsalam et al. 2018; Gustafson et al. 2015; Saini et al. 2020; Vinutha et al. 2013). Depending on the type of nanoparticle utilized, the mode of action of nano insecticides varies substantially among insects. According to Nair and Choi (2011), AgNPs control the expression and inhibition of the glutathione S transferase gene in *Chironomus riparius* Meigen. Green-produced Ag and zinc oxide nanoparticles affected the midgut epithelial cells of *Aedes aegypti* Linnaeus and *Anopheles stephensi* Liston. According to Abinaya et al. (2018), Banumathi et al. (2017), Ishwarya et al. (2018), Kalimuthu et al. (2017), Sundararajan and Kumari (2017) and treated insects also exhibited symptoms like shrinkage in the abdomen, altered thorax shape, damaged midgut, loss of setae on the tagma, and loss of anal gills.

Insect cell CYP450 enzyme activity was discovered to be inhibited by polystyrene nanoparticles (Frohlich et al. 2010). TiO_2 nanoparticles impacted the development and molting times as well as the 20-hydroxyecdysone production and cytochrome P450 genes (Li et al. 2014). Insects killed by SiO_2 nanoparticles in storage pests had altered cuticular lipids (Athanassiou et al. 2018; Barik et al. 2008; Debnath et al. 2011). *Drosophila melanogaster* Meigen died as a result of carbon black and multi-walled nanotubes simply adhering to the insect's body and interfering with its ability to move (Liu et al. 2009). According to Yasur and Usha-Rani (2015), oxidative stress was caused by Ag nanoparticles in the larval intestines of *Spodoptera litura* Fabricius and *Acheae janata* Linnaeus.

18.3 Polymer Based Nano Formulations of Pesticides

A formulation is the mixture of an active agent and suitable inert components. The formulation of a pesticide improves its storage, handling, safety, application, and effectiveness. Due to its extreme toxicity, which can even kill plants, the active ingredient cannot be used in the field as it is. Insecticide formulations currently in use cannot compete with nano formulations which have superior stability, chemical properties, effectiveness, or other aspects. Insecticidal efficacy of nanoparticles depends on sizes, shapes, and charges (Li et al. 2023b). Pesticide compounds have been incorporated in nanoparticles that can be synthesized into several

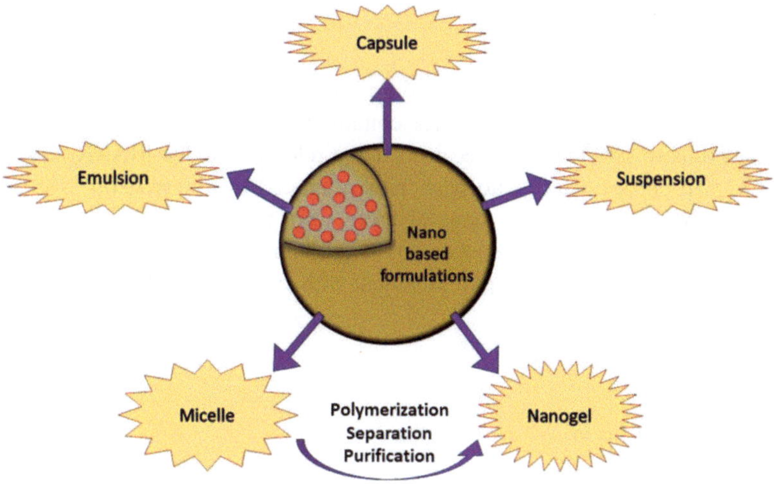

Fig. 18.2 Nano based formuations utilized for pesticide delivery system. (Figure constructed by U. Pirithiraj)

morphological shapes, such as nano emulsion, nanosuspension, nano capsules, and nanogels (Fig. 18.2) (Perlatti et al. 2013).

18.3.1 Nano Emulsion

The formulation known as a nano emulsion resembles the result of mixing water and oil. A liquid is disseminated into another liquid in a nano emulsion, although less surfactant is needed (Tamjidi et al. 2013). *Ephestia kuhniella* Zeller was tested against *Mentha longifolia* Linnaeus essential oils and its nano emulsions. Due to its high endurance, the nano emulsion demonstrated more toxicity against the insect (Louni et al. 2018).

18.3.2 Nano Suspension

By dispersing solid particles into liquid, a nano suspension formulation is developed. They serve as a delivery system, the most preferable method for nano emulsions and nano encapsulations (Godge 2020).

18.3.3 Nano Capsule

Nano capsule refers to the core-shell composition created by encapsulating a pesticide with a nanoparticle. The pesticide-containing core particle is surrounded with nanomaterials to provide regulated insecticide release. Subdue MAXX and Apron MAXX, two encapsulated nano pesticides produced by Bayer Crop Science and Syngenta, two of the pesticide manufacturers, were commercially accessible on the market (Kah et al. 2013; Huertas et al. 2010; Ezhilarasi et al. 2013). *Lipaphis erysimi* Kaltenbach, a mustard aphid, upon treatment with synthetic AgNPs recorded higher mortality (Kamili et al. 2017).

18.3.4 Nano Gels

Nano gels are strongly cross-linked three-dimensional hydrogels that contain water molecules without dissolving in aqueous media. In the drug delivery system, nanogels are becoming more significant (Kabanov and Vinogradov 2009; Ferreira et al. 2013; Soni and Yadav 2016). Due to their environmental friendliness and species-specificity, pheromones have emerged as a crucial part of integrated pest management. It was more effective since the evaporation of the pheromones was regulated by the nanogel composition. According to Herlekar and Ramaseshan (2014), nanogel-methyl eugenol attracts fruit flies, and its impact lasts around 30 weeks.

18.4 Efficacy of Nano Insecticides

Due to varying bodily structure and physiology, feeding habits and living habitats of different orders of insects, the efficacy of nano insecticide is not the same for all insects. However, the metabolism of insects might play a critical role in the efficacy of insects. The effectiveness of various nano insecticides on various groups of insect's viz., pests of crops, pests of stored products, foragers, and household pest's are discussed below.

18.4.1 Against Pests of Crops

Due to the harm that insects may do to plants, they are major crop pests. Recently, nano formulations have become more significant in treating insect infestations rather consisting environmental risks (Kannan et al. 2023). The Nano-acephate created by encasing acephate in polypolyethylene glycol (PEG) was successfully used to control the polyphagous pest *S. litura*. Acephate is a broad-spectrum organophosphate insecticide extremely hazardous to non-target creatures like fish, bees, and people. However, acephate nano formulation did not exhibit any harmful effects on people (Pradhan 2013). The polyphagous pest *Bemisia tabaci* Gennadius infests a wide variety of crops and serves as a vector for several plant pathogenic viruses. When *B. tabaci* was treated with nano formulations of neem-based pesticides enclosed in poly-b-hydroxybutyrate (PHB) and poly-e-caprolactone (PCL), the pest displayed higher mortality (Carvalho 2015).

The use of nano pyrifluquinazon was successful in managing the green peach aphid, *Myzus persicae* Sulzer, which feeds on several crops. Chitosan, a biodegradable polymer derived from chitin, was used as a carrier molecule to develop nano pyrifluquinazon (Kang et al. 2012). *S. littoralis* mortality from dehydration caused by amorphous silica nanoparticles (Ayoub et al. 2017). *Beauveria bassiana* Balsamo, an entomopathogenic fungus, was the source of the active ingredient utilized to create the nanoparticle insecticide chitosan-beauvericin. *S. litura*, often known as the cotton leaf worm or tobacco cutworm, demonstrated increased mortality after exposed to Chitosan-beauvericin nanoparticles (Bharani et al. 2014).

The majority of pests insects and microbes were mutually beneficial. Reduced intestinal microorganisms were recorded when AgNPs were employed to treat *Spodoptera furgiperda* Smith, the fall armyworm (Bharani et al. 2017). Citric acid polymer Titanium dioxide nanoparticles were used to treat larvae in their fifth instar of the lesser mulberry snout moth, *Glyphodes pyloalis* Walker, and they demonstrated better efficacy through controlled release (Memarizadeh et al. 2014a, b). When treated with fipronil nano formulations rather than commercial formulations, brown plant hoppers (*Nilaparvata lugens* Stal), a significant pest of rice, showed great mortality (Kumar et al. 2019).

18.4.2 Against Storage Pests

The rice weevil (*Sitophilous oryzae* Linnaeus), a significant pest of rice storage, was successfully controlled using a nano emulsion of jojoba (*Simmondsia chinensis* Link) seed oil. In comparison to bulk formulations, nano emulsions showed increased insect lethality (Sh et al. 2015). *Tribolium castaneum* Herbst, a secondary pest of grains in storage, was tested against a nano emulsion made of de-oiled karanja (*Pongamia glabra* Vent.) and jatropha (*Jatropha curcas* Linnaeus) aqueous filtrate in eucalyptus essential oils nano emulsion (77 nm), which demonstrated LC50 values of 0.1646 mg/L. While the LC50 value for the formulation without aqueous filtrate was 5.4872 mg/L. The results show that the toxicity of nano emulsions with aqueous filtrate is higher than without aqueous filtrate, even though higher LC values suggest decreased toxicity (Pant et al. 2014). Pesticides' activity was prolonged in the environment as a result of the controlled release, making them more effective against insects for a longer time. *Callosobruchus chinensis* Linnaeus was better controlled by the beta-cyfluthrin nano formulation than by the commercial formulation (Loha 2011).

18.4.3 Against Household Pests

The German cockroach, *Blattella germanica* Linnaeus (Dictyoptera: Blattellidae), is the most frequent pest encountered in homes. The pest is noticeable in homes and other public areas where people congregate. As vectors of bacteria, viruses, and nematodes among other pathogens, these cockroaches transmit several human diseases (Fotedar et al. 1991; Hamilton and Schal 1990; Pai et al. 2003). The nano formulation of bergamot (*Citrus reticulata* Linnaeus) and geranium (*Geranium* sp.) essential oils revealed unusual properties that could be utilized in effective pest management. According to Jorge Omar (2015), this nano formulation's features include contact toxicity, delayed and sustained release, and continual provision of active terpene form.

18.4.4 Against Vector Insects

According to the WHO (2014), Benelli and Mehlhorn (2016), Yakob and Walker (2016), and Vadivalagan et al. (2017) the three mosquito species *Anopheles* spp., *A. aegyptii*, and *Culex* spp. are the carriers of a range of diseases including filariasis, Japanese encephalitis, dengue fever, and yellow fever. According to Hemingway and Ranson (2000), Desneux et al. (2007) and Naqqash et al. (2016), the use of traditional insecticides to reduce mosquito populations resulted in resistance in mosquitoes, environmental pollution, toxicity to non-target organisms and other problems. The development of AgNPs from the Sundarbans mangroves was found to be efficient against various mosquito species (Alshehri et al. 2020). *A. aegypti*, a dengue vector, was managed by chitosan-pepper-rosmarin nanoparticles with a controlled release mode of action (Abreu et al. 2012).

18.4.5 Against Foragers

Nano pesticides are effective against several pest insects that are crucial to agriculture. Most pesticide formulations toxic to non-target species. Nanotoxicology studies should be done to examine the impact of nano pesticides on beneficial organisms that are not their intended targets. It has been proven that using nanoparticle cum pyrethrum extract was considerably safer for honey bees because it did not alter their digestive systems. Only pyrethrum extract and nanoparticles affected the digestive tract when administered. Honeybees are significantly safer when exposed to 1 ng of pyrethrum extract cum nanoparticles (Oliveira 2019). Chlorfenapyr was encapsulated through crosslinking nanoparticle (polysaccharide materials O-carboxymethyl chitosan and crosslinker-sodium tripolyphosphate) to form stable nano pesticide that exhibited low toxicity to worker bees (Hou et al. 2023).

18.4.6 Effect on Silkworm

AgNPs were toxic to animals and numerous insects when mixed with other pesticides. However, the effect of AgNPs on silkworms was studied, which opened up hitherto unexplored directions. *Bombyx moori* Linnaeus silkworms were exposed to varied quantities of AgNPs. A moderate dose of 400 mg/lit increased the weight of the *B. mori* cocoons, but a higher concentration of 800 mg/lit encouraged growth but resulted in the silkworm's death. The following are the leading causes of death (Meng 2017). Reduced resistance to stresses especially oxidative stress.

- Cell apoptosis got affected (Gurunathan et al. 2018).
- Changes in the metabolic pathways (Catto et al. 2019).
- Affected regulation of protein metabolism that caused cell necrosis (Karfa et al. 2020).

18.4.7 Effect on Other Beneficial Organisms

Natural enemies play a vital role in pest management under field conditions. Use of insecticides to manage pest population has been reported to have mere effect on population of predators and parasitoids (Sánchez-Bayo 2021). The parasitoid *Encarsia formosa* Gahan, a frequent parasitoid of whiteflies, *B. tabaci*, was not affected by the neem oil nano formulation when used as a carrier molecule (Carvalho et al. 2012). Volatile emitted from pest infested plants treated with silica nanoparticles significantly attracted more predators than untreated plants (Thabet et al. 2021). In contrast, silica nanoparticles application didn't attracted predators in case of sugar beet crop while managing *S. littoralis* (El-Samahy et al. 2015). The earthworm *Eisenia andrei* Bouche are litter-dwelling organisms, commonly known as the soil decomposer that replenish nutrients and improve plant growth. Zinc oxide nanoparticles have harmful effects on *E. andrei* (García-Gómez et al. 2019). A thiacloprid nano pesticide showed reduced toxicity to earthworms and highly toxic to aphids due to carrier delivery system specific to digestive tract of aphids but not to earthworms (Zhang et al. 2023). Even due to bioaccumulation nano pesticides have the tendency to carryover the food chain and may affect humans as shown in Fig. 18.3 (Da Silva Pontes et al. 2023).

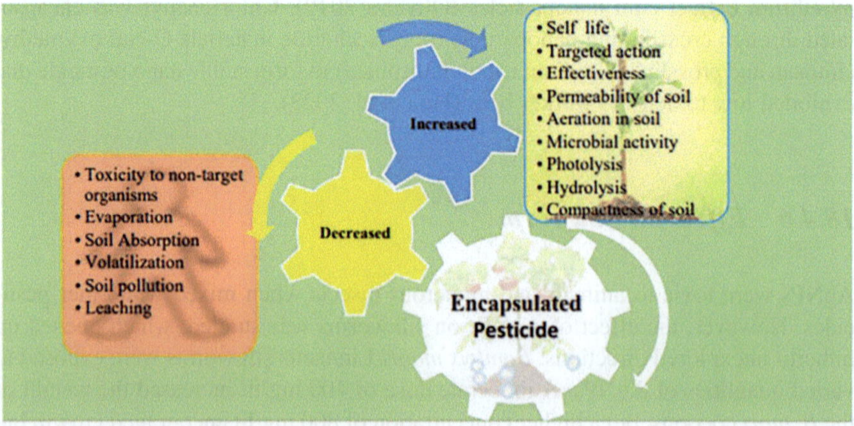

Fig. 18.3 Impacts of pesticides encapsulated in biopolymeric NPs on soil, organisms, and the environment. (Adopted from Kumar et al. (2022))

18.5 Nanoparticles Biocompatibility

The biocompatibility of nanoparticles with plant derivatives, entomopathogenic fungi, pheromones, and RNA interference (RNAi) for pest management is an evolving field that integrates nanotechnology with sustainable agricultural practices. Nanoparticles can enhance the efficacy of plant-derived pesticides by improving their solubility and bioavailability. Nanoparticles can be used as carriers for entomopathogenic fungi, improving their delivery and stability. Nanoparticles can serve as effective delivery systems for pheromones, allowing for controlled release and improved attraction of target pests. RNAi technology allows for specific targeting of pest genes, reducing the impact on non-target organisms. The compatibility reactions of nanoparticles are detailed hereunder.

18.5.1 Nanoparticles—Plant Derivatives

Because of their slower release, plant-derived nano pesticides expose higher toxicity to pest insects than bulk traditional pesticides (Fig. 18.4). Various neem-based products, including neem leaf extract, neem oil, and azharidactin, were capsulated by nanoparticles. These nano formulations were evaluated in maize plants against the fall armyworm, *S. furgiperda*, in both field and greenhouse conditions resulted in higher mortality percentages than convention ones. Giongo et al. (2016) found that nano formulations were less effective than organic neem oil. The third instar larvae of *Helicoverpa armigera* Hubner were treated with a AgNPs -fabricated leaf extract of *Aristolochia indica* Linnaeus, which demonstrated antifeedant efficacy against the pest (Siva 2015).

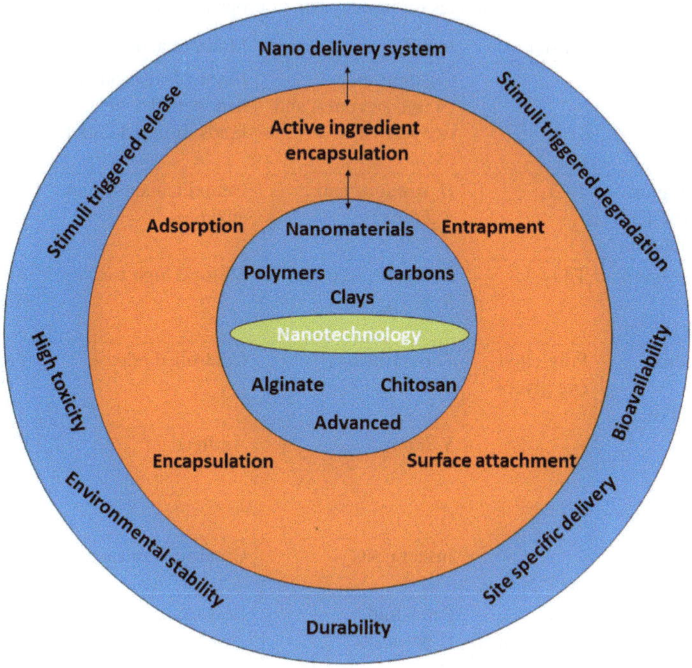

Fig. 18.4 Nanomaterials and different methods of encapsulation and effects up on delivery. (Figure constructed by U. Pirithiraj)

Nano formulations of neem bark gum extract (NBGE) were created using TiCl4 as a carrier and tested against *H. armigra* and *S. litura*. A nano formulation of NBGE influences the detoxifying enzyme in insect guts and acts as an antifeedant. The larva treated with NBGE nano formulation developed intermediate larval-pupal and pupal-adult forms (Table 18.1).

Apart from the nanoparticles characteristics, this NBGE nano formulation has been showed to be safer for non-target species (Kamaraj et al. 2018). Azharidactin and poly alpha-caprolactone nanoparticles were found to be highly effective against the larva of the diamondback moth, *Plutella xylostella* Linnaeus. Azharidactin was more stable in nanoparticles with a size of 245 nm (Forim et al. 2013). The tomato pinworm, *Tuta absoluta* Meyrick, was efficiently managed with the nanoparticles of essential oils isolated from mandarin, *Citrus reticulata* Blanco, which controlled 62% of the larval population (Compolo et al. 2017). A *Moringa oleifera* Linnaeus extract was used against the larvae of the red palm weevil, *Rhyncophorous ferrugenius* Olivier, using TiO_2 as a carrier with different dosages, including 50, 75, 100, 150, and 200 mg/lit, during the treatment. In comparison to previous dosages, the 75 mg/lit dose showed increased mortality on the second day of therapy and 100% death after 10 days (Al-Barly et al. 2015). In Brazil, the Nerium aphid, *Aphis nerri* Boyer, frequently affects attractive plants. Two brand-new nano pesticides (Ag and AgZn) were tested against the insect in the commercial formulation of imidacloprid. According to Rouhani et al. (2019), the death rates for imidacloprid,

Table 18.1 Effect of nano insecticides on insect pests of crops and their mode of action

Nano insecticide	Carrier	Insect	Mode of action	Reference
Silica nanoparticles	–	Coconut mite (*A. guerreronis*) and rice weevil (*S. oryzae*)	Physical adsorption of nanoparticles on lipids in insect cuticle	Barik et al. (2012)
Neem bark gum extract (NBGE) nano formulation	$TiCl_4$	*H. armigra* and *S. litura*	Detoxification of gut enzymes	Kamaraj et al. (2018)
Moringa oleifera extract nanoparticle	TiO_2	Red palm weevil, *R. ferrugenius*	Showed high toxicity	Al-Barly et al. (2015)
Garlic, *Allium sativum* L. essential oils	Poly ethylene glycol	*T. castaneum*	Controlled release	Yang et al. (2009)
Nano-silica particles	–	*S. littoralis*	Antifeedant	El-bendary and El-Helaly (2013)
CR formulation of β-cyfluthrin	–	Insect pests	Controlled release	Loha et al. (2011)
Neem oil nano formulation	Beta-cyclodextrin	Parasitoid, *E. formosa*	Less toxicity	Carvalho et al. (2012)
Nano pyrifluquinazon	Chitosan	Green peach aphid, *M. persicae*	Controlled release	Kang et al. (2012)
Nano-destruxin	–	*P. interpunctella*	Reduced fecundity	Sabbour (2014a, b); Sabbour (2015)
Silver nanoparticle	–	*H. armigera*	Antifeedant activity	Siva (2015)
Nano-ZnO and nano-ZnO–Al2O	–	Psyllids, *Agonoscena pistacie* Burckhardt & Lauterer	High toxicity	Samith et al. (2011)
Silver nanoparticles	–	*C. riparius*	Glutathion S transferase gene on/off	Nair and Choi (2011)
Nano emulsion of essential oils from lemongrass	–	Oat aphid, *R. padi*	Repellant	Pascual-Villalobos et al. (2017)

$TiCl_4$ Titanium tetrachloride, *TiO_2* Titanium dioxide

Ag, and Ag-Zn were 11.0% at 1 l/ml, 9.6% at 700 mg/ml, and 8.6% at 700 mg/ml, respectively. When tested against the oat aphid, *Rhopalosiphum padi* Linnaeus, the nano emulsion of essential oils from lemongrass and peppermint demonstrated a 68–100% repellent effect against winged females (Pascual-Villalobos et al. 2017). With a nano formulation mixture of ZnO + TiO_2 + Ag under laboratory conditions, higher mortality was seen in western flower thrips, *Frankliniella occidentalis* Pergande (Sinha et al. 2017).

In Iran, psyllids, *Agonoscena pistacie* Burelhardt and Lauerer are controlled by the insecticide amitraz. Nanoparticles of ZnO and ZnO-Al2O led to high mortality of this pest (Samith et al. 2011). The red spider mite, a polyphagous pest of more than 100 crops, was repelled by zein nanoparticles including citronella essential oils (De Oliveira et al. 2018). On the seventh day following treatment against the rice weevil, *S. oryzae*, aqueous *Euphorbia prostrata* Aiton leaf extract encapsulated in AgNPs demonstrated a maximum mortality of 100% (Zahir et al. 2012). Rice weevil, *S. oryzae* was resistant to the pesticidal effects of mangrove plant extract from *Avicennia marina* (Forssk.) encapsulated in Ag and lead nanoparticles (Sankar et al. 2015).

The red flour beetle, *T. castaneum*, is a global storage pest. Garlic essential oils, *Allium sativum* Linnaeus, encapsulated in nanoparticles using polyethylene glycol as a carrier, demonstrated lethal effects against it. Due to the delayed release, the effect of the nano formulation persisted for up to 5 months (Yang et al. 2009). It was discovered that the nano formulation of imidacloprid was more effective than the commercial version. Imidacloprid microcrystals were encapsulated with chitosan and sodium alginate, and this formulation was successful due to the synergistic gradual release of nanoparticles (Guan et al. 2008). The adult Mexican bean weevil, *Zabrotes subfasciatus* Boheman, demonstrated increased mortality when exposed to 1–5 nanometer-sized AZ-coated nano capsules. After 14 days of use, the nano capsules effects were apparent, and they showed excellent stability against UV radiation. According to Costa et al. (2014), the non-encapsulated compounds were unstable and deteriorated when exposed to UV.

18.5.2 Nanoparticles—Entomopathogens

Different forms of nano pesticides have been used against pests of agricultural importance for their effective management (Bhattacharyya et al. 2016; Rai and Ingle 2012). Different formulations of pesticides were used against insects, fungi, and bacteria. Their application is improved by nanotechnology, which also aids in controlled release and efficient management (Gajbhiye et al. 2009; Goswami et al. 2010). An improved substitute for synthetic insecticides is provided by the use of entomopathogenic fungus and applications of nanotechnology (Sabbour and Singer 2013; Sabbour and El Aziz 2015; Sahab et al. 2015).

Under controlled laboratory conditions, Multiwall carbon nanotubes (MWCNT) were used to treat *Isaria fumosorosea* Wize. The MWCNT doesn't appear to be poisonous to the fungus or its conidia. After the treatment of MWCNT, conidial germination remained unaltered. The entomopathogenic fungus encapsulated with nanoparticles increased the pathogenicity against the insects. In a lab setting, the entomopathogenic fungus *Metarhizium robertsii* Metchnikoff, formerly known as *M. anisopliae* demonstrated comparable effects on test insects (Gorczyca et al. 2015). Because they are not poisonous to fungi and boost the pathogenicity against the targeted insects, chitosan nanoparticles have enormous potential in entomopathogenic fungus applications (Yien et al. 2012). Entomopathogenic fungi, *Nomuraea rileyi* Farlow was tested on *S. litura* with three different formulations of chitosan nanoparticle coated fungal metabolite, an uncoated fungal metabolite and fungal spores. Results indicated the higher efficacy of chitosan-coated fungal metabolite than uncoated and fungal spores (Helan et al. 2013). A new nano insecticide of *Beauveria bassiana* Balsamo synthesized AgNPs showed 100% mortality against the first and second instar larva of *A. aegypti* (Banu and Balasubramanian 2014). Nanoparticles of destruxin from *M. robertsii* were tested against female grasshoppers *Hetiracris littoralis* Rambur and recorded a significant reduction in LC50 value. Nano-destruxin significantly reduced the fecundity of *Plodia interpunctella* Hubner under laboratory conditions (Sabbour 2014a, b).

18.5.3 Nanoparticles—Pheromones

The emission period of pheromones that can be used to successfully manage pests is continuously extended by the controlled release of pheromones using nanoparticles. Due to its gradual release action, *Scirpophaga incertulas* Walker nanosex pheromone had a longer duration of action than other pheromones and attracted a greater proportion of adults (Kannan 2018). When evaluated in an electro-antennogram, nano formulations containing pheromone and nano pheromone and insecticide had comparable effects on insects (Czarnobai et al. 2017; Kaur et al. 2021).

18.5.4 Nanoparticles—RNAi Technology

RNAi technology was employed to control a variety of significant pest insects. The use of nanoparticles as siRNA carriers against *Acyrthosiphon pisum* Harris, *Aphis glycines* Matsumura, and *Schizaphis graminum* Rondani resulted in higher efficacy of gene knockdown (Thairu et al. 2017). Fluorescent nanoparticles (Carrier) enhanced the siRNA delivery in to the cells of *H. armigera* that silences sterol carrier protein-2 thereby interacting in cholesterol absorption resulted in retarded growth and development (Geng et al. 2023). *S. furgiperda* chitin synthase (sfCHS), interfered by siRNA nanoparticle—cyromazine—chitosan complex resulted in higher mortality (Gong et al. 2023).

18.6 Regulations

In India, the registration of nano pesticides falls under the purview of the Central Insecticide Board and the Ministry of Agriculture and Farmers' Welfare. The regulatory framework is evolving to address the unique challenges posed by nanotechnology in agriculture, and ongoing research and guidelines are being developed to ensure the safe use of these products (Chakraborty et al. 2021; Kumari et al. 2023).

18.6.1 Registration of Nano Pesticides in India

There are no unanimously acceptable international guidelines for nano-agriproducts (NAPs). A few provisions that are in place globally for nanomaterials include Registration, Evaluation, Authorization and Restriction of Chemicals, US Environmental Protection Agency, Australian Pesticide and Veterinary Medicine Authority, Organization for Economic Co-operation and Development, and Food and Agricultural Organization/World Health Organization with certain specific guidelines for quality, safety and efficacy. However, innovations with alteration of functionality of nano systems make it difficult to apply a universal set of evaluation parameters for different nanoproducts with different applications.

The only region in the world with specific legal rules for dealing with nanoproducts is the European Union, together with Switzerland. In certain nations, there are no particular rules for nanoproducts and covered by the existing legislative and regulatory frameworks, often with modifications to take into account the unique characteristics of nanomaterials.

In India, there are different government bodies and provisions that regulate different agriproducts. Since different NAPs are considered in the guidelines, their evaluations were conducted as per the NAP type. Safety studies such as chemistry, bio-efficacy, residues, toxicity, packaging, and processing were done as per the guidelines of the Insecticides Act, of 1968. Nano pesticides for manufacture and import, registration should be done as per the regulatory aspect of section 9-(3) specified in the Insecticides Act, 1968 with additional criteria for inclusion of nano pesticides and plant growth regulators as per the guidelines of Central Insecticides Board and Registration Committee under the Ministry of Agriculture.

The Organization for Economic Co-operation and Development rules were followed for evaluating the created nanoproducts' safety. According to the regulations of the Insecticides Act, Fertilizer (Control) Order, Bureau of Indian Standards, and Food Safety and Standards Authority of India, the quality of the nanoproduct and other particular requirements were met. According to the criteria of the 1968 insecticide legislation, manufactured nano pesticides must adhere to their unique specifications (Anonymous 1). Hence frequent environmental risk assessment and regulatory guidelines has to be necessitated as shown in Fig. 18.5 (Kannan et al. 2023).

18.6.2 Biosafety Aspects

Any new science that is put into practice or enforced must take biosafety considerations into account. The commercialization of nano pesticides will have benefits and drawbacks, as was previously discussed. In numerous vertebrate and invertebrate animals, abnormalities, DNA damage, and interactions with enzyme synthesis have been linked to the use of AgNPs in insecticide formulations (Panacek et al. 2011;

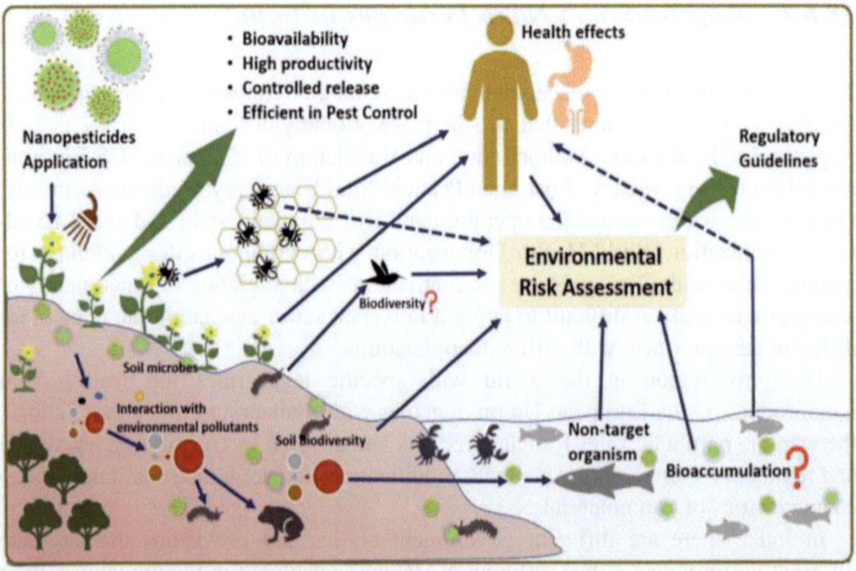

Fig. 18.5 Necessaries for environmental risk assessment and regulatory measure. (Adopted from Kannan et al. (2023))

Armstrong et al. 2013; Parveen et al. 2021). The denaturation of enzymes and cellular organelles occurs when the nanomaterials enter the intracellular environment and bind to the phosphorus in DNA (Jiang et al. 2015). Human lung inflammation is brought on by silica and carbon nanomaterials (Oberdorster et al. 2002; Warheit et al. 2007). Gold nanoparticles have been reported to have high toxicity to human lung epithelial cells (Egbuna et al. 2021). Dermal toxicity is caused by the use of nanoparticles like carbon, quantum dots, and TiO_2. Human kidney cells are impacted by SiO_2 nanoparticles (Mortensen et al. 2008; Wang et al. 2009). Therefore, after evaluating the risks to non-targeted creatures (both vertebrates and invertebrates), the use of nanoparticles must be increased.

18.7 Conclusion and Prospects

The concept of using nanoparticles for pest management is clearly discussed in this chapter. Nanoparticles may be the best component of integrated pest management, but they are not as effective as traditional host plant resistance and biological control, When evaluating the nano-formulations, it is important to consider the contentious ability of some nanoparticles used in pesticide formulations to directly affect skin toxicity of humans Although a benefit, the controlled release feature of nano insecticide raises concerns about high residual toxicity. Both conventional and nano pesticides have benefits and drawbacks. However, the controlled release of botanicals and pheromones was still encouraged because of its environment-friendly nature and high host specificity. Nano insecticides possess novel characteristics such as controlled release, target specificity, and environmental stability upon comparisons with conventional insecticides. Further comprehensive research on non-targeted organisms is required to make sure the safety of nanoparticles application. However, to the extent of current knowledge, the integration of nanoparticles as a component of integrated pest management would advance crop protection and productions.

References

Abd-Elsalam KA, Prasad R (eds) (2018) Nanobiotechnology applications in plant protection. Springer

Abinaya M, Vaseeharan B, Divya M, Sharmili A, Govindarajan M, Alharbi NS, Benelli G (2018) Bacterial exopolysaccharide (EPS)-coated ZnO nanoparticles showed high antibiofilm activity and larvicidal toxicity against malaria and Zika virus vectors. J Trace Elem Med Biol 45:93–103. https://doi.org/10.1016/j.jtemb.2017.10.002

Abreu FO, Oliveira EF, Paula HC, de Paula RC (2012) Chitosan/cashew gum nanogels for essential oil encapsulation. Carbohydr Polym 89(4):1277–1282. https://doi.org/10.1016/j.carbpol.2012.04.048

Adak T, Kumar J, Shakil NA, Walia S (2012) Development of controlled release formulations of imidacloprid employing novel nano-ranged amphiphilic polymers. J Environ Sci Health B 47(3):217–225. https://doi.org/10.1080/03601234.2012.634365

Al-Barly AMF, Hamza RZ (2015) Larvicidal, ani-oxidant activities and perturbation of transminases activities of titanium dioxide nanoparticles synthesized using *Moringa oleifera* leaves extract against the red palm weevil (*Rhynchophorus ferrugineus*). Eur J Pharm Med Res 2:49–54

Allard GB, Fortuna S, See LS, Novotny J, Baldini A, Courtinho T (2003) Global information on outbreaks and impact of major forest insect pests and diseases. XII World Forestry Congress, Quebec city

Alshehri MA, Trivedi S, Alanazi NA, Panneerselvam C, Baeshen R, Alatawi A (2020) One-step synthesis of ag nanoparticles using aqueous extracts from sundarbans mangroves revealed high toxicity on major mosquito vectors and microbial pathogens. J Clust Sci 31(1):177–184. https://doi.org/10.1007/s10876-019-01631-7

Anjali CH, Sharma Y, Mukherjee A, Chandrasekaran N (2012) Neem oil (*Azadiracta indica*) nanoemulsion-a potent larvicidal agent against *Culex quinquefasciatus*. Pest Manag Sci 68(2):158–163. https://doi.org/10.1002/ps.2233

Armstrong N, Ramamoorthy M, Lyon D, Jones K, Duttaroy A (2013) Mechanism of silver nanoparticles action on insect pigmentation reveals intervention of copper homeostasis. PLoS One 8(1):e53186. https://doi.org/10.1371/journal.pone.0053186

Athanassiou CG, Kavallieratos NG, Benelli G, Losic D, Rani PU, Desneux N (2018) Nanoparticles for pest control: current status and future perspectives. J Pest Sci 91(1):1–15. https://link.springer.com/article/10.1007/s10340-017-0898-0

Atta S, Bera M, Chattopadhyay T, Paul A, Ikbal M, Maiti MK, Singh NP (2015) Nano-pesticide formulation based on fluorescent organic photoresponsive nanoparticles: for controlled release of 2, 4-D and real time monitoring of morphological changes induced by 2, 4-D in plant systems. RSC Adv 5(106):86990–86996. https://doi.org/10.1039/C5RA17121K

Ayoub HA, Khairy M, Rashwan FA, Abdel-Hafez HF (2017) Synthesis and characterization of silica nanostructures for cotton leaf worm control. J Nanostructure Chem 7(2):91–100. https://doi.org/10.1007/s40097-017-0229-2

Banu AN, Balasubramanian C (2014) Myco-synthesis of silver nanoparticles using *Beauveria bassiana* against dengue vector, *Aedes aegypti* (Diptera: Culicidae). Parasitol Res 113(8): 2869–2877. https://doi.org/10.1007/s00436-014-3948-z

Banumathi B, Vaseeharan B, Ishwarya R, Govindarajan M, Alharbi NS, Kadaikunnan S, Benelli G (2017) Toxicity of herbal extracts used in ethno-veterinary medicine and green-encapsulated ZnO nanoparticles against *Aedes aegypti* and microbial pathogens. Parasitol Res 116(6): 1637–1651. https://doi.org/10.1007/s00436-017-5438-6

Barik TK, Sahu B, Swain V (2008) Nanosilica from medicine to pest control. Parasitol Res 103(2): 253–258. https://doi.org/10.1007/s00436-008-0975-7

Barik TK, Kamaraju R, Gowswami A (2012) Silica nanoparticle: a potential new insecticide for mosquito vector control. Parasitol Res 111(3):1075–1083. https://doi.org/10.1007/s00436-012-2934-6

Beard J, Sladden T, Morgan G, Berry G, Brooks L, McMichael A (2003) Health impacts of pesticide exposure in a cohort of outdoor workers. Environ Health Perspect 111(5):724–730. https://doi.org/10.1289/ehp.5885

Benelli G, Mehlhorn H (2016) Declining malaria, rising of dengue and Zika virus: insights for mosquito vector control. Parasitol Res 115(5):1747–1754. https://doi.org/10.1007/s00436-016-4971-z

Bharani RA, Namasivayam SKR, Shankar SS (2014) Biocompatible chitosan nanoparticles incorporated pesticidal protein beauvericin (CSNp-BV) preparation for the improved pesticidal activity against major groundnut defoliator *Spodoptera litura* (Fab.). (Lepidoptera; Noctuidae) Int Chem Tech Res 6(12):5007–5012

Bharani RA, Namasivayam SKR (2017) Biogenic silver nanoparticles mediated stress on developmental period and gut physiology of major lepidopteran pest Spodoptera litura (Fab.)(Lepidoptera: Noctuidae) An eco-friendly approach of insect pest control. Journal of environmental chemical engineering. 5(1), 453-467

Bhattacharya S, Samanta SK (2016) Soft-nanocomposites of nanoparticles and nanocarbons with supramolecular and polymer gels and their applications. Chem Rev 116(19):11967–12028. https://doi.org/10.1021/acs.chemrev.6b00221

Campolo O, Cherif A, Ricupero M, Siscaro G, Grissa-Lebdi K, Russo A, Biondi A (2017) Citrus peel essential oil nanoformulations to control the tomato borer, *Tuta absoluta*: chemical properties and biological activity. Sci Rep 7(1):1–10. https://doi.org/10.1038/s41598-017-13413-0

Carvalho SS, Vendramin JD, Pitra MM, Florin MR (2012) Efficacy of neem oil nanoformulations to *Bemisia tabaci* (Genn.) biotype B (Hemiptera: Aleyrodidae). Semina Cienc Agrar 33:193–202. https://doi.org/10.5433/1679-0359.2012v33n1p193

Carvalho SSD, Vendramim JD, Sa ICGD, Silva MFDGFD, Ribeiro LDP, Forim MR (2015) Systemic insecticidal effect of neem-based nanoformulations against Bemisia tabaci (Hemiptera: Aleyrodidae) biotype B in tomato. Bragantia. 74(3):298–306. https://doi.org/10.1590/1678-4499.0404

Catto C, Garuglieri E, Borruso L, Erba D, Casiraghi MC, Cappitelli F, Zanchi R (2019) Impacts of dietary silver nanoparticles and probiotic administration on the microbiota of an in-vitro gut model. Environ Pollut 245:754–763

Chakraborty R, Mukhopadhyay A, Sarkar S, Kundu A (2021) An insight into data requirements for registration of nanopesticides and nanofertilizers in India. Agric Food e-Newslett 3:453–455

Chowdhury M, Zaman A, Banik S, Uddin B, Moniruzzaman M, Karim N, Gan SH (2012) Organophosphorus and carbamate pesticide residues detected in water samples collected from paddy and vegetable fields of the Savar and Dhamrai Upazilas in Bangladesh. Int J Environ Res Public Health 9(9):3318–3329. https://doi.org/10.3390/ijerph9093318

Czarnobai De Jorge B, Bisotto-de-Oliveira R, Pereira CN, Santana J (2017) Novel nanoscale pheromone dispenser for more accurate evaluation of *Grapholita molesta* (Lepidoptera: Tortricidae) attract-and-kill strategies in the laboratory. Pest Manag Sci 73(9):1921–1926. https://doi.org/10.1002/ps.4558

Da Costa JT, Forim MR, Costa ES, De Souza JR, Mondego JM, Junior ALB (2014) Effects of different formulations of neem oil-based products on control *Zabrotes subfasciatus* (Boheman, 1833)(Coleoptera: Bruchidae) on beans. J Stored Prod Res 56:49–53. https://doi.org/10.1016/j.jspr.2013.10.004

Da Silva Pontes M, de Lima Forini MM, Caires ARL, Santiago EF, Tripathi DK, Kah M, Grillo R (2023) Nanopesticides modes of action and impacts. Nano-Enabled Sustainable and Precision Agriculture, vol 79–103. Elsevier. https://doi.org/10.1016/B978-0-323-91233-4.00001-6

Debnath N, Das S, Seth D, Chandra R, Bhattacharya SC, Goswami A (2011) Entomotoxic effect of silica nanoparticles against *Sitophilus oryzae* (L.). J Pest Sci 84(1):99–105. https://doi.org/10.1007/s10340-010-0332-3

Desneux N, Decourtye A, Delpuech JM (2007) The sublethal effects of pesticides on beneficial arthropods. Annu Rev Entomol 52:81–106. https://doi.org/10.1146/annurev.ento.52.110405.091440

Egbuna C, Parmar VK, Jeevanandam J, Ezzat SM, Patrick-Iwuanyanwu KC, Adetunji CO, Khan J, Onyeike EN, Uche CZ, Akram M, Ibrahim MS, el Mahdy NM, Awuchi CG, Saravanan K, Tijjani H, Odoh UE, Messaoudi M, Ifemeje JC, Olisah MC, Ibeabuchi CG (2021) Toxicity of nanoparticles in biomedical application: nanotoxicology. J Toxicol 9:9954443. https://doi.org/10.1155/2021/9954443

El-Bendary HM, El-Helaly AA (2013) First record nanotechnology in agricultural: Silica nanoparticles a potential new insecticide for pest control. App Sci Report 4(3):241–246. https://doi.org/10.1155/2021/9954443

El-Samahy MFM, Khafagy IF, El-Ghobary AMA (2015) Efficiency of silica nanoparticles, two bioinsecticides, peppermint extract and insecticide in controlling cotton leafworm, *Spodoptera littoralis* Boisd. and their effects on some associated natural enemies in sugar beet fields. J Plant Prot Pathol Mansoura Univ 6:1221–1230. https://doi.org/10.21608/jppp.2015.74934

Ezhilarasi PN, Karthik P, Chhanwal N, Anandharamakrishnan C (2013) Nanoencapsulation techniques for food bioactive components: a review. Food Bioprocess Technol 6(3):628–647. https://doi.org/10.1007/s11947-012-0944-0

Fang X, Zhang Q, Yu B, Wang B, Yang G, Wang B, Hua Z (2023) Plant-oil based polymeric emulsions as adhesive nanocarriers for enhancing the efficacy of nanopesticides. Ind Crop Prod 192:116020. https://doi.org/10.1016/j.indcrop.2022.116020

Ferreira SA, Gama FM, Vilanova M (2013) Polymeric nanogels as vaccine delivery systems. Nanomed: Nanotechnol Biol Med 9(2):159–173. https://doi.org/10.1016/j.nano.2012.06.001

Feynman R (1991) There's plenty of room at the bottom. Science 254(5036):1300–1301

Forim MR, Costa ES, Da Silva MFDGF, Fernandes JB, Mondego JM, Boiça Junior AL (2013) Development of a new method to prepare nano microparticles loaded with extracts of *Azadirachta indica*, their characterization and use in controlling *Plutella xylostella*. J Agric Food Chem 61(38):9131–9139. https://doi.org/10.1021/jf403187y

Fotedar R, Shriniwas UB, Verma A (1991) Cockroaches (*Blattella germanica*) as carriers of microorganisms of medical importance in hospitals. Epidemiol Infect 107(1):181–187. https://doi.org/10.1017/s0950268800048809

Fröhlich E, Kueznik T, Samberger C, Roblegg E, Wrighton C, Pieber TR (2010) Size-dependent effects of nanoparticles on the activity of cytochrome P450 isoenzymes. Toxicol Appl Pharmacol 242(3):326–332. https://doi.org/10.1016/j.taap.2009.11.002

Gajbhiye M, Kesharwani J, Ingle A, Gade A, Rai M (2009) Fungus-mediated synthesis of silver nanoparticles and their activity against pathogenic fungi in combination with fluconazole. Nanomed: Nanotechnol Biol Med 5(4):382–386. https://doi.org/10.1016/j.nano.2009.06.005

García-Gómez C, Babín M, Garcia S, Almendros P, Pérez RA, Fernandez MD (2019) Joint effects of zinc oxide nanoparticles and chlorpyrifos on the reproduction and cellular stress responses of the earthworm *Eisenia andrei*. Sci Total Environ 688:199–207. https://doi.org/10.1016/j.scitotenv.2019.06.083

Geng K, Zhang Y, Zhao X, Zhang W, Guo X, He L, Peng R (2023) Fluorescent nanoparticle-RNAi-mediated silencing of sterol carrier protein-2 gene expression suppresses the growth, development, and reproduction of *Helicoverpa armigera*. Nano 13(2):245. https://doi.org/10.3390/nano13020245

Giongo AMM, Vendramim JD, Forim MR (2016) Evaluation of neem-based nanoformulations as alternative to control fall armyworm. Cienciae Agrotecnologia 40(1):26–36. https://doi.org/10.1590/S1413-70542016000100002

Godge GR, Garje MA, Dode AB, Tarkase KN (2020) Nanosuspension technology for delivery of poorly soluble drugs and its applications: a review. Int J Pharm Sci Nanotechnol 13(4): 4965–4978. https://doi.org/10.37285/ijpsn.2020.13.4.1

Gong C, Hasnain A, Wang Q, Liu D, Xu Z, Zhan X, Wang X (2023) Eco-friendly deacetylated chitosan base siRNA biological-nanopesticide loading cyromazine for efficiently controlling *Spodoptera frugiperda*. Int J Biol Macromol 241:124575. https://doi.org/10.1016/j.ijbiomac.2023.124575

Gorczyca A, Kasprowicz MJ, Lemek T (2015) Bioactivity of MWCNT in conidia of entomopathogenic fungus *Isaria fumosorosea*. Water Air Soil Pollut 226:75–86. https://doi.org/10.1007/s11270-015-2350-5

Goswami A, Roy I, Sengupta S, Debnath N (2010) Novel applications of solid and liquid formulations of nanoparticles against insect pests and pathogens. Thin Solid Films 519(3): 1252–1257. https://doi.org/10.1016/j.tsf.2010.08.079

Guan HD, Chi J, Yu L, Li X (2008) A novel photodegradable insecticide: preparation, characterization and properties evaluation of nano-imidacloprid. Pestic Biochem Physiol 92:83–91. https://doi.org/10.1016/j.pestbp.2008.06.008

Guillette LJ, Iguchi T (2012) Life in a contaminated world. Science 337(6102):1614–1615. https://doi.org/10.1126/science.1226985

Groote H, Kimenju SC, Munyua B, Palmas S, Kassie M, Bruce A (2020) Spread and impact of fall armyworm (Spodoptera frugiperda JE Smith) in maize production areas of Kenya. Agriculture, ecosystems & environment 292:106804. https://doi.org/10.1016/j.agee.2019.106804

Gulati S, Kumar S, Jain S, Radhika, Sharma N, Batra K (2022) Toxicological perspectives and environmental risks of consumer nanoproducts. In: Handbook of consumer nanoproducts. Springer Nature Singapore, pp 1253–1275. https://doi.org/10.1007/978-981-16-8698-6-102

Gurunathan S, Qasim M, Park C, Yoo H, Choi DY, Song H, Hong K (2018) Cytotoxicity and transcriptomic analysis of silver nanoparticles in mouse embryonic fibroblast cells. Int J Mol Sci 19(11):3618

Gustafson HH, Holt-Casper D, Grainger DW, Ghandehari H (2015) Nanoparticle uptake: the phagocyte problem. Nano tod 10(4):487–510. https://doi.org/10.1016/j.nantod.2015.06.006

Hamilton RL, Schal C (1990) Sublethal effects of chlorpyrifos-methyl on reproduction in female German cockroaches (Dictyoptera: Blattellidae). J Econ Entomol 83(2):441–443. https://doi.org/10.1093/jee/83.2.441

Harish V, Tewari D, Gaur M, Yadav AB, Swaroop S, Bechelany M, Barhoum A (2022) Review on nanoparticles and nanostructured materials: bioimaging, biosensing, drug delivery, tissue engineering, antimicrobial, and agro-food applications. Nano 12:457. https://doi.org/10.3390/nano12030457

Helan JCh, Anand LFA, Karthick SRN, Bharani RSA (2013) Improved pesticidal activity of fungal metabolite from *Nomureae rileyi* with chitosan nanoparticles proceedings of the "International Conference on Advanced Nanomaterials and Emerging Engineering Technologies" (ICANMEET-20J3). New Delhi, India, 24–26 July, 387–390

Hemingway J, Ranson H (2000) Insecticide resistance in insect vectors of human disease. Annu Rev Entomol 45(1):371–391. https://doi.org/10.1146/annurev.ento.45.1.371

Huertas CE, Fessi H, Elaissari A (2010) Polymer-based nanocapsules for drug delivery. Int J Pharm 385(1–2)(142):113–142. https://doi.org/10.1016/j.ijpharm.2009.10.018

Herlekar I, Ramaseshan S (2014) Using nanotechnology to control pests: trapping fruit fly using pheromone gels. Curr Sci 106(1):14

Hou R, Li C, Tan Y, Wang Y, Huang S, Zhao C, Zhang Z (2023) Eco-friendly O-carboxymethyl chitosan base chlorfenapyr nanopesticide for effective pest control and reduced toxicity to honey bees. Int J Biol Macromol 224:972–983. https://doi.org/10.1016/j.ijbiomac.2022.10.182

Ishwarya R, Vaseeharan B, Kalyani S, Banumathi B, Govindarajan M, Alharbi NS, Benelli G (2018) Facile green synthesis of zinc oxide nanoparticles using Ulva lactuca seaweed extract and evaluation of their photocatalytic, antibiofilm and insecticidal activity. J Photochem Photobiol B Biol 178:249–258. https://doi.org/10.1016/j.jphotobiol.2017.11.006

Jerobin J, Sureshkumar RS, Anjali CH, Mukherjee A, Chandrasekaran N (2012) Biodegradable polymer based encapsulation of neem oil nanoemulsion for controlled release of Aza-A. Carbohydr Polym 90(4):1750–1756. https://doi.org/10.1016/j.carbpol.2012.07.064

Jiang X, Miclauş T, Wang L, Foldbjerg R, Sutherland DS, Autrup H (2015) Fast intracellular dissolution and persistent cellular uptake of silver nanoparticles in CHO-K1 cells: implication for cytotoxicity. Nanotoxicology 9:181–189. https://doi.org/10.3109/17435390.2014.907457

Kabanov AV, Vinogradov SV (2009) Nanogels as pharmaceutical carriers: finite networks of infinite capabilities. Angew Chem Int Ed Engl 48(30):5418–5429. https://doi.org/10.1002/anie.200900441

Kalimuthu K, Panneerselvam C, Chou C, Lin SM, Tseng LC, Tsai KH, Hwang JS (2017) Predatory efficiency of the copepod *Megacyclops formosanus* and toxic effect of the red alga *Gracilaria firma*-synthesized silver nanoparticles against the dengue vector *Aedes aegypti*. Hydrobiologia 785(1):359–372. https://doi.org/10.1007/s10750-016-2943-z

Kamaraj C, Gandhi PR, Elango G, Karthi S, Chung IM, Rajakumar G (2018) Novel and environmental friendly approach: impact of neem (*Azadirachta indica*) gum Nano formulation (NGNF) on *Helicoverpa armigera* (Hub.) and *Spodoptera litura* (Fab.). Int J Biol Macromol 107:58–69. https://doi.org/10.1016/j.ijbiomac.2017.08.145

Kamil D, Prameeladevi T, Ganesh S, Prabhakaran N, Nareshkumar R, Thomas SP (2017) Green synthesis of silver nanoparticles by entomopathogenic fungus *Beauveria bassiana* and their bioefficacy against mustard aphid (*Lipaphis erysimi* Kalt.). NISCAIR-CSIR India 555–561

Kang MA, Seo MJ, Hwang IC, Jang C, Park HJ, Yu YM, Youn YN (2012) Insecticidal activity and feeding behavior of the green peach aphid, *Myzus persicae*, after treatment with nano types of pyrifluquinazon. J Asia Pac Entomol 15(4):533–541. https://doi.org/10.1016/j.aspen.2012.05.015

Kannan M (2018) Electrospun nanofiber for controlled release of male attracting sex pheromone to monitor yellow stem borer, *Scirpophaga incertulas* in rice. In: International Conference on Biocontrol and Sustainable Insect Pest Management. ISBN: 978-93-81102-46-6

Kannan M, Bojan N, Swaminathan J, Zicarelli G, Hemalatha D, Zhang Y, Faggio C (2023) Nanopesticides in agricultural pest management and their environmental risks: a review. Int J Environ Sci Technol 20(9):10507–10532. https://doi.org/10.1007/s13762-023-04795-y

Karfa P, Majhi KC, Madhuri R (2020) Bringing awareness to the darker side of nanoparticles. The ELSI handbook of nanotechnology: risk, safety, ELSI and commercialization. Wiley, pp 135–163

Kaur K, Sharma S, Gupta R, Munikrishnappa VKT, Chandel M, Ahamed M, Shanmugam V (2021) Nanomaze lure: pheromone sandwich in graphene oxide interlayers for sustainable targeted pest control. ACS Appl Mater Interfaces 13(41):48349–48357

Khan N, Sudhakar K, Mamat R (2023) Seaweed farming: a perspectives of genetic engineering and nano-technology application. Heliyon 9(4):e15168. https://doi.org/10.1016/j.heliyon.2023.e15168

Kumar N, Kumar R, Shakil NA, Sarkar DJ, Chander S (2019) Evaluation of fipronil nanoformulations for effective management of brown plant hopper (*Nilaparvata lugens*) in rice. Int J Pest Manag 65(1):86–93. https://doi.org/10.1080/09670874.2018.1468046

Kumar R, Kumar N, Rajput VD, Mandzhieva S, Minkina T, Saharan BS, Kumar D, Sadh PK, Duhan JS (2022) Advances in biopolymeric nanopesticides: A new eco-friendly/eco-protective perspective in precision agriculture. Nanomaterials 10;12(22):3964

Kumari R, Suman K, Karmakar S, Mishra V, Lakra SG, Saurav GK, Mahto BK (2023) Regulation and safety measures for nanotechnology-based agri-products. Front Genome Ed 5:1200987

Li F, Gu Z, Wang B, Xie Y, Ma L, Xu K, Li B (2014) Effects of the biosynthesis and signaling pathway of ecdysterone on silkworm (*Bombyx mori*) following exposure to titanium dioxide nanoparticles. J Chem Ecol 40(8):913–922. https://doi.org/10.1007/s10886-014-0487-0

Li X, Chen Y, Xu J, Lynch I, Guo Z, Xie C, Zhang P (2023a) Advanced nanopesticides: advantage and action mechanisms. Plant Physiol Biochem 10:108051. https://doi.org/10.1016/j.plaphy.2023.108051

Li C, Han Y, Gao T, Zhang J, Xu DX, Wāng Y (2023b) Insecticidal activity of metallic nanopesticides synthesized from natural resources: a review. Environ Chem Lett 21(2):1141–1176. https://doi.org/10.1007/s10311-022-01548-0

Liu X, Vinson D, Abt D, Hurt RH, Rand DM (2009) Differential toxicity of carbon nanomaterials in Drosophila: larval dietary uptake is benign, but adult exposure causes locomotor impairment and mortality. Environ Sci Technol 43(16):6357–6363. https://doi.org/10.1021/es901079z

Loha KM, Shakil NA, Kumar J, Singh MK, Adak T, Jain S (2011) Release kinetics of β-Cyfluthrin from its encapsulated formulations in water. J Environ Sci Health B 46(3):201–206. https://doi.org/10.1080/03601234.2011.540200

Louni M, Shakarami J, Negahban M (2018) Insecticidal efficacy of nanoemulsion containing *Mentha longifolia* essential oil against *Ephestia kuehniella* (Lepidoptera: Pyralidae). J Crop Prot 7(2):171–182

Manna S, Roy S, Dolai A, Ravula AR, Perumal V, Das A (2023) Current and future prospects of "all-organic" nanoinsecticides for agricultural insect pest management. Front Nanotechnol 4:1082128. https://doi.org/10.3389/fnano.2022.1082128

McCray WP (2007) MBE deserves a place in the history books. Nat Nanotechnol 2(5):259–261. https://doi.org/10.1038/nnano.2007.121

Mei C, Chang CH, Tao L, Lu C (2015) Residential exposure to pesticide during childhood and childhood cancer: a meta-analysis. Pediatrics 136(4):719–729. https://doi.org/10.1542/peds.2015-0006

Memarizadeh N, Ghadamyari M, Adeli M, Talebi K (2014a) Preparation, characterization and efficiency of nanoencapsulated imidacloprid under laboratory conditions. Ecotoxicol Environ Saf 107:77–83. https://doi.org/10.1016/j.ecoenv.2014.05.009

Memarizadeh N, Ghadamyari M, Adeli M, Talebi K (2014b) Linear-dendritic copolymers/indoxacarb supramolecular systems: biodegradable and efficient nano-pesticides. Environ Sci Process Impacts 16(10):2380–2389. https://doi.org/10.1039/C4EM00321G

Meng X, Abdlli N, Wang N, Lü P, Nie Z, Dong X, Chen K (2017) Effects of Ag nanoparticles on growth and fat body proteins in silkworms (*Bombyx mori*). Biol Trace Elem Res 180(2):327–337. https://doi.org/10.1007/s12011-017-1001-7

Mortensen LJ, Oberdo�rster G, Pentland AP, DeLouise LA (2008) In vivo skin penetration of quantum dot nanoparticles in the murine model: the effect of UVR. Nano letters 10;8(9):2779-87

Mubeen I, Mfarrej MFB, Razaq Z, Iqbal S, Naqvi SAH, Hakim F, Mosa WFA, Moustafa M, Fang Y, Li B (2023) Nanopesticides in comparison with agrochemicals: outlook and future prospects for sustainable agriculture. Plant Physiol Biochem 198:107670. https://doi.org/10.1016/j.plaphy.2023.107670

Nair PMG, Choi J (2011) Identification, characterization and expression profiles of *Chironomus riparius* glutathione S-transferase (GST) genes in response to cadmium and silver nanoparticles exposure. Aquat Toxicol 101(3–4):550–560. https://doi.org/10.1016/j.aquatox.2010.12.006

Naqqash MN, Gökçe A, Bakhsh A, Salim M (2016) Insecticide resistance and its molecular basis in urban insect pests. Parasitol Res 115(4):1363–1373. https://doi.org/10.1007/s00436-015-4898-9

Oberdörster G, Utell MJ (2002) Ultrafine particles in the urban air: to the respiratory tract–and beyond. Environ Health Perspect 110(8):A440–A441. https://doi.org/10.1289/ehp.110-1240959

Oliveira JLD, Campos EV, Pereira AE, Pasquoto T, Lima R, Grillo R, Fraceto LF (2018) Zein nanoparticles as eco-friendly carrier systems for botanical repellents aiming sustainable agriculture. J Agric Food Chem 66(6):1330–1340. https://doi.org/10.1021/acs.jafc.7b05552

Oliveira CR, Domingues CE, de Melo NF, Roat TC, Malaspina O, Jones-Costa M, Fraceto LF (2019) Nanopesticide based on botanical insecticide pyrethrum and its potential effects on honeybees. Chemosphere 236:124282. https://doi.org/10.1016/j.chemosphere.2019.07.013

Pai HH, Ko YC, Chen ER (2003) Cockroaches (*Periplaneta americana* and *Blattella germanica*) as potential mechanical disseminators of *Entamoeba histolytica*. Acta Trop 87(3):355–359. https://doi.org/10.1016/s0001-706x(03)00140-2

Panacek A, Prucek R, Safarova D, Dittrich M, Richtrova J, Benickova K, Kvitek L (2011) Acute and chronic toxicity effects of silver nanoparticles (NPs) on *Drosophila melanogaster*. Environ Sci Technol 45(11):4974–4979. https://doi.org/10.1021/es104216b

Pankaj Shakil NA, Kumar J, Singh MK, Singh K (2012) Bioefficacy evaluation of controlled release formulations based on amphiphilic nano-polymer of carbofuran against *Meloidogyne incognita* infecting tomato. J Environ Sci Health B 47(6):520–528. https://doi.org/10.1080/03601234.2012.665667

Pant M, Dubey S, Patanjali PK, Naik SN, Sharma S (2014) Insecticidal activity of eucalyptus oil nanoemulsion with karanja and jatropha aqueous filtrates. Int Biodeterior Biodegrad 91:119–127

Parveen R, Maiti PK, Murmu N, Datta A (2021) Preparation of serum capped silver nanoparticles for selective killing of microbial cells sparing host cells. Sci Rep 11(1):11610. https://doi.org/10.1038/s41598-021-91031-7

Pascual-Villalobos MJ, Canio-Tejero M, Vallejo R, Guirao P, Rodriez-Rojo S, Cocero MJ (2017) Use of nanoemulsions of plant essential oils on aphid repellents. Ind Crop Prod 110:45–57. https://doi.org/10.1016/j.indcrop.2017.05.019

Perlatti B, de Souza Bergo PL, Fernandes JB, Forim MR (2013) Polymeric nanoparticle-based insecticides: a controlled release purpose for agrochemicals. IntechOpen, In Insecticides-Development of safer and more effective technologies. https://doi.org/10.5772/53355

Pimentel D (2009) Pesticides and pest control. In: Peshin R, Dhawan AK (eds) Integrated pest management: innovation-development process. Springer., Dordrecht 83–87. https://doi.org/10.1007/978-1-4020-8992-3

Pradhan S, Patra P, Das S, Chandra S, Mitra S, Dey KK, Akbar S, Palit P, Goswami A (2013) Photochemical modulation of biosafe manganese nanoparticles on Vigna radiata: a detailed molecular, biochemical, and biophysical study. Environmental Science & Technology 19;47 (22):13122-31

Rai M, Ingle A (2012) Role of nanotechnology in agriculture with special reference to management of insect pests. Appl Microbiol Biotechnol 94(2):287–293. https://doi.org/10.1007/s00253-012-3969-4

Rouhani M, Samih MA, Zarabi M, Beiki K, Gorji M, Aminizadeh MR (2019) Synthesis and entomotoxicity assay of zinc and silica nanoparticles against Sitophilus granarius (Coleoptera: Curculionidae). Journal of Plant Protection Research 26-31

Roychoudhury A (2020) 'Desert locust': a menace to Indian agriculture and economy. Young Scientist-Tomorrow's Science Begins Today 4(1):14–20

Sabbour MM (2014a) Evaluating toxicity of nano-extracted Destruxin from *Metarhizium anisopliae* against the grasshopper *Hetiracris littoralis* in Egypt. J Egypt Acad Soc Environ Dev 15(2):1–7. https://doi.org/10.13140/RG.2.1.2116.3365

Sabbour MM (2014b) Evaluating toxicity of extracted nano-Destruxin against the desert locust *Schistocerca gregaria* in Egypt. J Egypt Acad Soc Environ Dev 15(2):9–17. https://doi.org/10.13140/RG.2.1.4213.4887

Sabbour MM, El Aziz SEA (2015) Efficacy of some nano-Imidacloprid against red flour beetle *Tribolium castaneum* and confused flour beetle, *Tribolium confusum* (Coleoptera: Tenebrionidae) under laboratory and store conditions. Bull Env Pharmacol Life Sci 4(7):54–59. https://doi.org/10.13140/RG.2.1.4737.7766

Sabbour MM, Singer SM (2013) Efficacy of nano *Isaria fumosorosea* and *Metarhizium flavoviride* against corn pests under laboratory and field conditions in Egypt. Int J Sci Res 4(7):945–949. https://doi.org/10.13140/RG.2.1.1854.1927

Sahab AF, Waly AI, Sabbour MM, Nawar LS (2015) Synthesis, antifungal and insecticidal potential of chitosan (CS)-g-poly (acrylic acid) (PAA) nanoparticles against some seed borne fungi and insects of soybean. Int J Chem Tech Res 8(2):589–598

Saini RK, Patel S, Bajpai J, Bajpai AK (2020) Advanced controlled nanopesticide delivery systems for managing insect pests. In: K. R. R., Thomas S, Volova, T. K. J. (eds) Controlled release of pesticides for sustainable agriculture. Springer, Cham, pp 155–184. https://doi.org/10.1007/978-3-030-23396-9_7

Samith MA, Rouhani M, Aslani A, Beiki K (2011) Insecticidal properties of Amitraz, nano-Amitraz, nano ZnO and nano-ZnO-Al2O nanoparticles on *Agonoscena pistaciae* (Hemiptera: Aphelaridae). In: Proceedings of third international symposium on insect physiology, biochemistry and molecular biology, 2–5 July 2011, held at East China Normal University. Shanghai, p 131

Sánchez-Bayo F (2021) Indirect effect of pesticides on insects and other arthropods. Toxics 9(8). https://doi.org/10.3390/toxics9080177

Sankar MV, Abideen S (2015) Pesticidal effect of green synthesized silver and lead nanoparticles using *Avicennia marina* against grain storage pest *Sitophilus oryzae*. Int J Nanomater Biostruct 5(3):32–39

Sarkar DJ, Kumar J, Shakil NA, Walia S (2012) Release kinetics of controlled release formulations of thiamethoxam employing nano-ranged amphiphilic PEG and diacid based block polymers in soil. J Environ Sci Health A 47(11):1701–1712. https://doi.org/10.1080/10934529.2012.687294

Saw G, Nagdev P, Jeer M, Murali-Baskaran RK (2023) Silica nanoparticles mediated insect pest management. Pestic Biochem Physiol 105524:105524. https://doi.org/10.1016/j.pestbp.2023.105524

Sh A, Abdelrazeik AB, Rakha OM (2015) Nanoemulsion of jojoba oil, preparation, characterization and insecticidal activity against *Sitophilus oryzae* (coleoptera: Curculionidae) on wheat. Int J Agric Innov Res 4(1):72–75

Shekhar S, Sharma S, Kumar A, Taneja A, Sharma B (2021) The framework of nanopesticides: a paradigm in biodiversity. Mater Adv 2(20):6569–6588. https://doi.org/10.1039/D1MA00329A

Sinha K, Ghosh J, Sil PC (2017) New pesticides: a cutting-edge view of contributions from nanotechnology for the development of sustainable agricultural pest control. New Pesticides and Soil Sensors, vol 47–79. Elsevier. https://doi.org/10.1016/B978-0-12-804299-1.00003-5

Siva C, Kumar MS (2015) Pesticidal activities of ecofriendly synthesized silver particles using *Aristolochia indica* extract against *Helicoverpa armigera* Hubner (Lepidoptera: Noctuidae). Int J Adv Sci Tech Res 2:197–226

Soni G, Yadav KS (2016) Nanogels as potential nanomedicine carrier for treatment of cancer: a mini review of the state of the art. Saudi Pharm J 24(2):133–139. https://doi.org/10.1016/j.jsps.2014.04.001

Srishailam B, Sailaja V, Nikhitha A, Kiran PK (2022) Promoting startups in agriculture. An innovative approach for transforming agriculture to agri-business. Vigyan Varta 3(4):73–81

Srivastava S, Bhargava A, Srivastava S, Bhargava A (2022) Green nanotechnology: an overview. Green nanoparticles: the future of nanobiotechnology. Springer Singapore, pp 1–13

Sundararajan B, Kumari BR (2017) Novel synthesis of gold nanoparticles using *Artemisia vulgaris* L. leaf extract and their efficacy of larvicidal activity against dengue fever vector *Aedes aegypti* L. J Trace Elem Med Biol 43:187–196. https://doi.org/10.1016/j.jtemb.2017.03.008

Tamjidi F, Shahedi M, Varshosaz J, Nasirpour A (2013) Nanostructured lipid carriers (NLC): a potential delivery system for bioactive food molecules. Innov Food Sci Emerg Technol 19:29–43. https://doi.org/10.1016/j.ifset.2013.03.002

Thabet AF, Boraei HA, Galal OA, El-Samahy MF, Mousa KM, Zhang YZ, Nozaki T (2021) Silica nanoparticles as pesticide against insects of different feeding types and their non-target attraction of predators. Sci Rep 11(1):14484. https://doi.org/10.1038/s41598-021-93518-9

Thairu MW, Skidmore IH, Bansal R, Nováková E, Hansen TE, Li-Byarlay H, Hansen AK (2017) Efficacy of RNA interference knockdown using aerosolized short interfering RNAs bound to nanoparticles in three diverse aphid species. Insect Mol Biol 26(3):356–368. https://doi.org/10.1111/imb.12301

Tolochko NK (2009) History of nanotechnology. Encyclopedia of Life Support Systems (EOLSS)

Vadivalagan C, Karthika P, Murugan K, Panneerselvam C, Del Serrone P, Benelli G (2017) Exploring genetic variation in haplotypes of the filariasis vector *Culex quinquefasciatus* (Diptera: Culicidae) through DNA barcoding. Acta Trop 169:43–50. https://doi.org/10.1016/j.actatropica.2017.01.020

Villasenor MJ, Rios A (2018) Nanomaterials for water cleaning and desalination, energy production, disinfection, agriculture and green chemistry. Environ Chem Lett 16(1):11–34. https://doi.org/10.1007/s10311-017-0656-9

Vinutha JS, Bhagat D, Bakthavatsalam N (2013) Nanotechnology in the management of polyphagous pest *Helicoverpa armigera*. J Acad Indus Res 1(10):606–608

Wang D, Saleh NB, Byro A, Zepp R, Sahle-Demessie E, Luxton TP, Ho KT, Burgess RM, Flury M, White JC, Su C (2022) Nano-enabled pesticides for sustainable agriculture and global food security. Nat Nanotechnol 17(4):347–360. https://doi.org/10.1038/s41565-022-01082-8

Wang Z, Zhao J, Li F, Gao D, Xing B (2009) Adsorption and inhibition of acetylcholinesterase by different nanoparticles. Chemosphere 1;77(1):67-73

Warheit DB, Hoke RA, Finlay C, Donner EM, Reed KL, Sayes CM (2007) Development of a base set of toxicity tests using ultrafine TiO2 particles as a component of nanoparticle risk management. Toxicol Lett 171(3):99–110. https://doi.org/10.1016/j.toxlet.2007.04.008

World Health Organization (2014) Global status report on non-communicable diseases 2014 (No. WHO/NMH/NVI/15.1). World Health Organization

Wurster CF (1972) Effects of insecticides. In: Polunin N (ed) The environmental future. Macmillan, London, pp 293–310

Xu X, Lu P, Guo M, Fang M (2010) Cross-linked gelatin/nanoparticles composite coating on micro-arc oxidation film for corrosion and drug release. Appl Surf Sci 256(8):2367–2371. https://doi.org/10.1016/j.apsusc.2009.10.069

Yakob L, Walker T (2016) Zika virus outbreak in the Americas: the need for novel mosquito control methods. Lancet Glob Health 4(3):e148–e149. https://doi.org/10.1016/S2214-109X(16)00048-6

Yang FL, Li XG, Zhu F, Lei CL (2009) Structural characterization of nanoparticles loaded with garlic essential oil and their insecticidal activity against *Tribolium castaneum* (Herbst) (Coleoptera: Tenebrionidae). J Agric Food Chem 57(21):10156–10162. https://doi.org/10.1021/jf9023118

Yasur J, Rani PU (2015) Lepidopteran insect susceptibility to silver nanoparticles and measurement of changes in their growth, development and physiology. Chemosphere 124:92–102. https://doi.org/10.1016/j.chemosphere.2014.11.029

Yien L, Mohamad NZ, Sarwar A, Katas H (2012) Antifungal activity of chitosan nanoparticles and correlation with their physical properties. Int J Biomaterials 2012:1–9. https://doi.org/10.1155/2012/632698

Zahir AA, Rahuman AA (2012) Evaluation of different extracts and synthesised silver nanoparticles from leaves of Euphorbia prostrata against *Haemaphysalis bispinosa* and *Hippobosca maculata*. Vet Parasitol 187(3–4):511–520. https://doi.org/10.1016/j.vetpar.2012.02.001

Zhang X, He Y, Yuan Z, Shen G, Zhang Z, Niu J, Qian K (2023) A pH-and enzymatic-responsive nanopesticide to control pea aphids and reduce toxicity for earthworms. Sci Total Environ 861:160610. https://doi.org/10.1016/j.scitotenv.2022.160610

Zhou J, Liu G, Guo Z, Wang M, Qi C, Chen G, Huang X, Yan S, Xu D (2023) Stimuli-responsive pesticide carriers based on porous nanomaterials: a review. Chem Eng J 455:140167. https://doi.org/10.1016/j.cej.2022.140167

Chapter 19
Nanotechnology Solutions for Post-harvest Challenges

Emad Hamdy Khedr and Nagwa Khedr

Abstract Maintaining fruit quality is a critical goal for improving productivity and minimizing postharvest losses, which represents a significant challenge for the global agricultural sector. Fruits, being perishable commodities, encounter a range of issues from the moment they are harvested until they reach the consumer. These challenges include moisture loss, increased respiration rates, accelerated biochemical degradation, color changes, and vulnerability to diseases and pest infestations. Additionally, physical damage during handling and transportation further contributes to the deterioration of fruit quality. Such factors collectively result in substantial postharvest losses, which can negatively impact food availability, farmers' income, and consumer satisfaction. Addressing these complex postharvest issues requires innovative approaches, and one promising solution is the application of nanotechnology. This chapter delves into the cutting-edge advancements in the use of nanotechnology, particularly focusing on the role of nanoparticles in fruit coatings to combat moisture loss, maintain color, and reduce disease incidence. Furthermore, nano compounds are being utilized to regulate ethylene production, which is crucial for delaying fruit ripening and senescence. Nanosensors also offer significant advantages by enabling real-time monitoring of fruit quality throughout the supply chain. This chapter provides an in-depth exploration of how these nanotechnological innovations can effectively enhance postharvest management and improve fruit storability.

Keywords Antibacterial · Decay · Fruit quality · Food safety · Nano-coatings · Nanosensors · Shelf life

---

E. H. Khedr (✉) · N. Khedr
Department of Pomology, Faculty of Agriculture, Cairo University, Giza, Egypt
e-mail: emad.khedr@agr.cu.edu.eg; nagwa.h.khedr@agr.cu.edu.eg

J. M. Al-Khayri et al. (eds.), *Nanotechnology Applications in Modern Agriculture*, Nanotechnology in Plant Sciences 3, https://doi.org/10.1007/978-3-031-90506-3_19

19.1 Introduction

Fresh fruits and vegetables are a crucial source of fibers, vitamins, minerals, and bioactive compounds such as carotenoids, anthocyanins, and vitamin C, all of which are essential for human well-being. However, post-harvest losses pose significant challenges to global food security, leading to decreased fruit quality, nutritional degradation, and substantial economic losses (Arabpoor et al. 2021; Yang et al. 2021). These products are highly perishable, with a short shelf life, contributing to significant losses worldwide (Dargiri et al. 2024). Fruit quality is affected by numerous challenges that begin right from the cultivation stage. Selecting the appropriate soil type and ensuring optimal environmental conditions are crucial for each species and cultivar (Huang et al. 2022; Khedr 2022a). Equally important are agricultural practices such as fertilization, irrigation, and pruning, which are crucial in the estimating the final produce yield and quality (Castelo Branco Melo et al. 2018; Yang et al. 2021). Despite time, efforts, and costs invested in these processes, approximately 40–50% of fruits and vegetables are lost in the interval between harvest and usage. This substantial loss is largely by mishandling and poor post-harvest practices (Dargiri et al. 2024). Key challenges begin with harvesting at the incorrect maturity stage, essential to guarantee the best possible quality and shelf life (Al-Khayri and Khedr 2024a). The use of inappropriate tools and methods exacerbates the issue, causing physical damage such as bruises, wounds, and scratches (Rastegar et al. 2024). Furthermore, inadequate packaging materials and techniques compromise the integrity and appearance of the produce (Castelo Branco Melo et al. 2018). During transportation of produce presents additional risks, especially transported in non-cooled trucks that fail to maintain low temperatures (Al-Khayri and Khedr 2024b; Huang et al. 2022). The absence of proper pre-cooling protocols accelerates deterioration (Rastegar et al. 2024). Additionally, suboptimal storage conditions, such as improper temperature and humidity levels, contribute to further degradation of fruit quality (Khedr 2022a).

Handling in packaging facilities often lacks care, leading to increased respiration and ethylene production, hastening ripening, and spoilage. Another critical factor contributing to post-harvest losses is the improper storage of incompatible fruits together (Castelo Branco Melo et al. 2018). A few fruits emit ethylene gas, which can hasten how quickly delicately produce ripens and spoils (Khedr 2022b). Additionally, differing optimal humidity requirements among fruits can lead to dehydration or mold growth. These factors result in significant changes in texture, color, and overall marketability before the product even reaches consumers (de Oliveira et al. 2020). At the market level, inadequate shelf-life management and unfavorable conditions contribute to further deterioration until the produce reaches consumers (Khedr 2022a). Tackling these challenges demands an integrated approach that encompasses every stage of the supply chain, from production to market, to maintain the quality of fruits and vegetables while reducing losses. As a result, there is a growing need for advanced post-harvest technologies that prevent diseases and physiological disorders, slow down the aging process, and enhance both handling and quality preservation (Arabpoor et al. 2021).

Nanotechnology has emerged as a promising approach to overcome these challenges. The use of submicron-sized materials offers distinct advantages, including improved gas exchange, enhanced mechanical and optical properties, as well as boosted antioxidant and antimicrobial activities. Recently, nanotechnology has been utilized in post-harvest management of fruits and vegetables to develop advanced coatings like nanocomposites and nano emulsions with superior physical and antimicrobial characteristics (Khedr and Al-Khayri 2023; Khedr et al. 2024), and to create highly sensitive nanosensors capable of detecting volatile compounds such as ethylene, indicating the ripeness of produce (Khedr and Al-Khayri 2024). These innovations offer innovative solutions to improve the quality, safety, and shelf life of fruits during post-harvest storage. This chapter explores its applications in mitigating post-harvest issues such as spoilage, microbial contamination, and moisture loss. Key technologies discussed include edible nano-coatings, antimicrobial nanoparticles, smart packaging, and nanosensors for real-time quality monitoring (Khedr and Al-Khayri 2024). The chapter also addresses safety and regulatory considerations, ensuring that nanotechnology's implementation in the food industry adheres to health standards. Nanotechnology has the potential to revolutionize post-harvest management, reducing food waste and maintaining high-quality produce for longer periods, with promising innovations on the horizon. In recent years, nanotechnology has demonstrated remarkable efficacy in reducing microbial growth, preventing moisture loss, and providing barriers against gas exchange. Additionally, smart packaging and nanosensors enable real-time monitoring of food quality, allowing better management of storage conditions and timely interventions to prevent spoilage (EL-Beltagi and Khedr 2025). By integrating nanotechnology into post-harvest systems, we can reduce reliance on chemical preservatives, promote sustainable packaging solutions, and contribute to a more efficient and resilient food supply chain. As we explore these innovative approaches, we will also consider the potential risks, regulatory challenges, and prospects.

This chapter delves into the various applications of nanotechnology in addressing post-harvest challenges, focusing on its role in reducing losses, maintaining fruit quality, and ensuring food safety. The integration of nanotechnology into post-harvest systems has the potential to minimize the need for chemical preservatives, promote sustainable packaging solutions, and contribute to a more efficient and resilient food supply chain. As we explore these innovative approaches, we will also consider the potential risks, regulatory challenges, and the prospects of nanotechnology in post-harvest management.

19.2 Nanotechnology

Nanotechnology has introduced an innovative approach to developing edible coating components for the postharvest management of fresh produce. This involves reducing edible coating particles to a nanoscale, typically between 1 and 100 nanometers (He and Hwang 2016). Nanotechnology's application has proven highly effective in extending the shelf life of fresh fruits, as nanoparticles in edible coatings exhibit unique properties, such as antibacterial and antifungal activity (Liu et al. 2018). Recent research has explored adding nanoparticles to edible coatings, revealing that this approach not only boosts antimicrobial, antifungal, and antiviral effectiveness (Khedr and Khedr 2025) but also enhances the cost-efficiency of these coatings (Akintelu and Folorunso 2019). Liu et al. (2018) and Jamdagni et al. (2018), have compiled extensive data on functional coatings for tropical fruits and horticultural produce. These reviews emphasize safe nanoparticle concentrations and the use of commonly consumed plant extracts for fresh produce coatings. They detail the use of polysaccharides, proteins, and lipids as edible coatings for popular fruits, highlighting their potential to extend the postharvest shelf life of tropical fruits.

A wide array of nanomaterials have been developed to mitigate fruit damage in the food and agricultural industries, including those based on lipids, carbon, and inorganic metals and their oxides. Nanotechnology applications are invaluable in managing postharvest diseases, advancing packaging technologies with nano-bio films, reducing the effects of harmful gases and rays, and developing nano-biosensor chips for labeling fresh products.

Additionally, nanomaterials serve as antifungal agents for treating fungal infections in fruits. For example, an edible gelatin-based coating containing ethanol-based propolis extract to exhibit significant antifungal activity, effectively inhibiting mold growth, especially against *Penicillium digitatum* and *Botrytis cinerea* strains in raspberries (Nasirifar et al. 2018). A bioactive coating made of pectin, pullulan, and Vitis vinifera L. grape seed extract demonstrated potential in extending peanut shelf life by reducing lipid oxidation and delaying rancidity. This coating also had antibacterial properties against Escherichia coli and Listeria monocytogenes (Priyadarshi et al. 2022). Conversely, a pectin-based coating applied to tomatoes negatively affected their color, diminishing sensory quality (Chaturvedi et al. 2019).

19.3 Fruit Quality

Many kinds of physical deterioration, such as surface injuries, impact damage, and vibration bruising, often occurs in fruits and vegetables post-harvest, and these factors significantly contribute to the deterioration of fruit. Figure 19.1 presents main causes of fruit loss; biological (pest and disease), chemical (residual effect of pesticides, toxins produced by pathogens), mechanical (injuries, bruises, abrasions), physical (water loss, heating, freezing), physiological (chilling injury, shortage of calcium, browning) (Al-Khayri and Khedr 2024b; Khedr et al. 2023).

Fruit size and shape are essential quality traits that exhibit significant genetic variability (Khedr 2017). The increase in fruit volume results from the development of pericarp tissue, which accounts for more than two-thirds of the total fruit weight (Ho 1996). This tissue expansion not only influences fruit size but also impacts other quality aspects such as texture, composition, and overall consumer appeal. Texture plays a key role in determining the end-use value of fruits, whether intended for fresh

Fig. 19.1 Main causes of postharvest loss. (Constructed by Emad Khedr and Nagwa Khedr)

markets or industrial processing. It is a complex trait that includes various components such as firmness, and meltiness. These textural qualities are evaluated through sensory analysis or measured using objective instrumental methods, including mechanical testing, magnetic resonance imaging (MRI), and ultrasonic method. Among the mechanical methods, puncture methods are the most common (Barrett et al. 2010), offering precise assessments that complement the perceived texture by consumers.

The composition of fruit, particularly in terms of sugars and acids, is another critical factor that contributes to its overall flavor. Soluble sugars and organic acids are key compounds that accumulate in fruits and play a significant role in determining their sweetness and sourness (Davies and Hobson 1981). The balance between these compounds is vital for flavor development, with the structure and tissue distribution of fruit also influencing taste. For example, jelly tissues typically contain higher acid levels and lower sugar content than the pericarp. Research has documented the metabolic pathways involved in sugar and acid synthesis, as well as the enzymatic activities that regulate their accumulation (Etienne et al. 2013).

As fruits ripen, their color and appearance undergo noticeable changes, particularly due to the breakdown of chlorophyll. This pigment shift, which begins at the blossom end and moves toward the stem end, serves as a visual indicator of ripeness and enhances the fruit's marketability. The visual appeal created by this process is often one of the first factors to attract consumers, preceding the evaluation of texture, flavor, and nutritional content (Khedr and Khedr 2024a). In addition to these physical and sensory traits, polyphenols, a large family of phytochemicals, are important contributors to both plant health and human nutrition. In humans, these compounds are known for their health-promoting properties, including antioxidant, hepatoprotective, hypoglycemic, and antiviral activities (Khedr and Khedr 2024b).

Synthesized primarily from phenylalanine via the shikimic acid pathway, polyphenols contribute to both the fruit's nutritional value and its resistance to environmental stress (Al-Khayri et al. 2024).

Aroma and flavor are often the most subjective aspects of fruit quality but remain critical in shaping consumer preferences. These volatile compounds significantly influence human perception, with studies showing their impact on taste and preference (Klee and Tieman 2013). Together, the intricate balance of these components defines the overall eating experience, making aroma and flavor a central focus in both breeding and postharvest management.

19.4 Postharvest Challenges

Fruits face several challenges from harvest till they are consumed by the customer, many of which are influenced by environmental conditions, as depicted in Fig. 19.2. One of the most critical factors is temperature. Storing fruits outside their optimal temperature range often causes significant damage. Elevated temperatures accelerate respiration and metabolic breakdown, while temperatures that are too low can cause chilling injuries (Al-Khayri and Khedr 2024a; EL-Beltagi and Khedr 2025a; Rastegar and Khedr 2025). In addition to temperature, relative humidity is essential to the preservation of fruit. and closely linked to temperature effects. High humidity levels can promote the growth of diseases, particularly fungal infections,

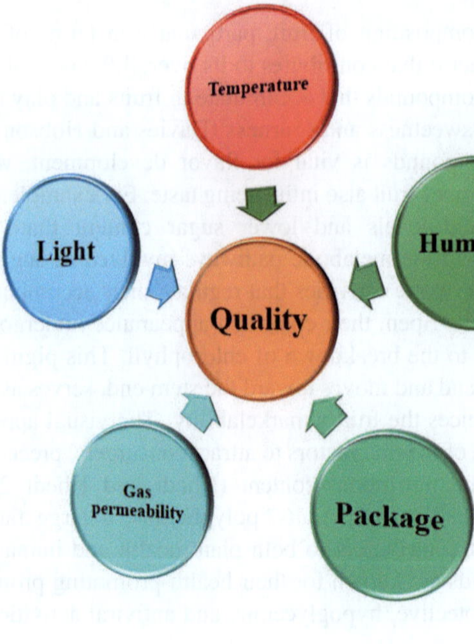

Fig. 19.2 Factors related to fruit storage life. (Constructed by Emad Khedr and Nagwa Khedr)

whereas low humidity can cause excessive moisture loss, leading to weight reduction, decreased nutritional value, and physical damage like shriveling and wilting.

Light exposure is another factor that can negatively affect fruits, as prolonged exposure can lead to the formation of toxic compounds, such as in potatoes. Furthermore, the composition and exchange of gases around the fruits are crucial to maintaining their quality. An improper gas balance can increase respiration rates or trigger anaerobic respiration, which damages the fruit. The type of packaging used plays a significant role in managing these factors. Effective packaging protects the fruit, controls gas exchange, and enhances its shelf life and marketability (Gholami et al. 2017). Table 19.1 summarizes how distinct nanoparticle applications influence fruit quality parameters.

Table 19.1 Effect of different nano particle treatments on fruit quality

Fruit type	Treatment	Main effects	Reference
Apple	SiOx film coating	Extending storage life and reducing weight loss	Liu et al. (2017)
Apple	ZnO Nanoparticles	Maintaining quality and inhibiting deterioration by maintaining total phenolic, and antimicrobial properties, and reducing, and fruit decay	Li et al. (2011)
Apple	Ag_2O packing bag	It has nanostructured elements in it that help keep cut apples from browning	Zhou et al. (2011)
Apple	A-Ag NPs	Serves as an antibacterial agent and is primarily applied to fruit postharvest treatments	Gudadhe et al. (2013)
Banana	ZnO Nanoparticles	Extending storage life and inhibiting deterioration by increasing: Titratable acidity, firmness, sensory and physiological attributes, and antifungal activity and reducing weight loss and total soluble solids	Li et al. (2019)
Bananas	Nano chitosan	Delayed ripening, resulting in decreased expression of the MaACS1 and MaACO genes in the coated bananas	Lustriane et al. (2018)
Bell pepper	Chitosan nanoparticles	Extending storage life, maintaining quality, and inhibiting deterioration	González-Saucedo et al. (2019)
Blueberry	Chitosan/Si nanoparticles	Extending storage life and enhancing antimicrobial influences can be achieved by increasing lightness and the blue component	Sami et al. (2021)
Blueberry	Chitosan coating based on SiO_2NP or TiO_2NP	Extending shelf life and maintaining quality can be achieved by increasing lightness, ascorbic acid, titratable acidity, and polyphenol oxidase activity, while reducing weight loss and peroxidase activity	Li et al. (2021a, b)

(continued)

Table 19.1 (continued)

Fruit type	Treatment	Main effects	Reference
Cantaloupe	Chitosan and transcinnamaldehyde	Maintaining quality can be achieved by increasing total carotenoids, TSS, ascorbic, antioxidant capacity, and structural integrity, while reducing weight loss, respiration rates, reactive oxygen species, polyphenol oxidase, pectinase, and peroxidase, as well as electrolyte leakage, lipid peroxidation degree, and color change	Carvalho et al. (2016)
Cherry	Chitosan nanoparticles	Extending storage life and reducing decay can be achieved by increasing firmness, while reducing weight loss, respiration rates, total bacterial count, color change, and titratable acidity	Arabpoor et al. (2021)
Cherry	Chitosan and SiOx nanoparticles	Maintaining quality can be achieved by increasing total soluble solids, soluble sugars, and proteins, while reducing weight loss, decay rate, titratable acidity, and pectin chain degradation	Xin et al. (2017)
Cherry	Chitosan nanoparticles	Maintain dehydroascorbatereductase, and reductase, dehydroascorbate and oxidized glutathione.	Ma et al. (2019)
Cherry	Si nanoparticles	Reduce hydrogenperoxide, superoxideanion, malondialdehyde, lipoxygenaseactivity	Meng et al. (2022)
Citrus	Nanoemulsion	Reducing decay and weight loss, respiration rate, color change, bruising index, electrolyte leakage, and spore germination of *Penicillium digitatum* is essential.	Yang et al. (2021)
Citrus	Soy-wax containing emulsion	Inhibiting deterioration Inhibiting deterioration can be achieved by reducing weight loss, firmness	Fei et al. (2021)
Citrus	Carnauba wax-nano-clay	Maintaining value can be achieved by increasing antioxidant activity and nutritional quality, while reducing weight loss and respiration rates.	Motamedi et al. (2018)
Cucumber	Polymeric coatings with CaO nanoparticles	Maintaining quality can be achieved by increasing firmness, and total phenolic, while reducing color change, dehydration, the ratio of TSS to titratable acidity, titratable acidity, and weight loss.	Cid-López et al. (2021)
Fresh-cut cucumber	LEO-A-P-NP coating	Maintaining quality can be achieved by increasing sensory attributes, chlorophyll, ascorbic acid, and firmness, while reducing weight loss, respiration rates, total soluble solids, pH, yellowing index, color change, activities of polyphenol oxidase, peroxidase, and pectin methylesterase	Zambrano-Zaragoza et al. (2021)

(continued)

Table 19.1 (continued)

Fruit type	Treatment	Main effects	Reference
Grape	Lemongrass nanoemulsions	Enhancing antimicrobial properties, firmness, total phenolic content, and DPPH radical scavenging activity, while reducing weight loss	Kim et al. (2014)
Grapes	Edible chitosan nanoparticle coatings	Maintaining quality can be achieved by increasing moisture retention, titratable acidity, anthocyanin content, and antimicrobial activity, while reducing weight loss, total soluble solids.	Castelo Branco Melo et al. (2018)
Guavas	Candeuba wax	Decrease in the permeability of O_2 and CO_2	Garcia-Betanzos et al. (2018)
Guavas	Chitosan, alginate and ZnO nanoparticles	Guavas' shelf life is extended to 3 weeks	Silva-Vera et al. (2018)
Litchi	Chitosan	Maintaining quality can be achieved by increasing sensory attributes, TSS, titratable acidity	Mei et al. (2022)
Litchi	Chitosan	To enhance quality maintenance and inhibit deterioration, while addressing issues such as fruit browning index and weight loss.	Su et al. (2023)
Litchi, strawberry	Chitosan	Reducing deterioration can be achieved by increasing sensory attributes, while reducing moldy area and weight loss	Huang et al. (2022)
Longan	$CaCO_3$ nanoparticles	For extending shelf life, improve sensory attributes while minimizing weight loss and loss and fruit browning index.	Fu et al. (2021)
Loquat	Si nanoparticles	Reducing browning and weight loss	Song et al. (2016)
Mango	HPMC-based nanoparticles	Minimizing the growth of *Colletotrichum gloeosporioides* and reducing color changes.	Klangmuang and Sothornvit (2018)
Mango	Chitosan/TiO_2NPs composite coatings	To maintain quality, increase firmness and total phenolic content, while reducing respiration rate, total soluble solids, weight loss	Xing et al. 2020
Mango	New chitosan/zein-cinnamaldehyde NCPs film	To maintain quality, increase superoxide dismutase activity while reducing color changes, respiration rate, weight loss, reactive oxygen species, malondialdehyde	Xiao et al. (2021)
Mushrooms	Coating of konjac glucomannan/carrageenan/SiO_2NPs	To extend shelf life, increase total soluble solids and firmness, while reducing weight loss, electrolyte leakage, and both surface and flesh color changes.	Zhang et al. (2019)

(continued)

Table 19.1 (continued)

Fruit type	Treatment	Main effects	Reference
Peach	ZnO nanoparticles	Reducing weight loss and TSS	Batool et al. (2022)
Plum	Chitosan nanoparticles	Decrease internal browning, weight loss, glycine betaine content, proline content, polyphenol oxidase activity, malondialdehyde, and hydrogen peroxide.	Mahmoudi et al. (2022)
Pomegranate	Chitosan nanoparticles	Reducing weight loss, yeast and mold counts, decay rate, and pH	Hasheminejad and Khodaiyan (2020)
Pomegranate arils	ZnONP + CMC coating	To extend shelf life, increase fruit juice content, total phenolic content, and anthocyanins	Saba and Amini (2017)
Shiitake mushrooms	Alginate-Ag nanoparticles	Reduce spoilage, weight loss, firmness loss, browning index, microbial counts, carbon dioxide production, electrolyte leakage, and total soluble solids.	Jiang et al. (2013)
Shiitake mushrooms	TiO_2 nanoparticles	To extend shelf life and maintain quality, increase electrolyte conductivity, levels of ATP and ADP, the activities of H + -ATPase, succinate ATPase, succinate dehydrogenase, cytochrome c oxidase, guanylic acid, glutamate, and proline. At the same time, reduce weight loss, firmness loss, color changes, AMP levels, total protease activity	Li et al. (2021b)
Spinach leaves	Nanofibrillated cellulose coating	To extend shelf life, enhance sensory physiological attributes and chlorophyll levels, while reducing respiration rate, color changes, and moisture content.	Pacaphol et al. (2019)
Strawberry	Nano-chitosan coating combined with $CaCl_2$	To extend storage life, improve total soluble solids, total phenolic content, ascorbic acid, and volatile compounds, while reducing weight loss, titratable acidity, malondialdehyde, and total anthocyanin content.	Nguyen et al. (2020)
Strawberry	Coating of carboxymethyl	Reducing deterioraation	Lin et al. (2020)
Strawberry	Chitosan	Increase total soluble solids, sensory physiological attributes, firmness, and pH, while reducing fungal decay, weight loss, and titratable acidity.	(Robledo et al. (2018)
Strawberry	Solid lipid nanoparticles (SLN)	Maintaining fruit quality	Zambrano-Zaragoza et al. (2020)

(continued)

Table 19.1 (continued)

Fruit type	Treatment	Main effects	Reference
Tomato	Edible coating incorporated with chitosan	Maintaining fruit quality	Cui et al. (2017)
Tomatoes	Ag NPs	Minimizing weight loss	Gao et al. (2017)
Tomatoes	Ag NPs	Effectively influences the fresh produce's sensory qualities and quality metrics	Chandirika et al. (2008)

19.4.1 Physiological Weight Loss

The primary source of unacceptability and food waste in fruits is their weight loss during storage (Khedr and Khedr 2023). This weight loss is mostly caused by the respiration and transpiration of the water content, which may be identified by adjusting the water vapor pressure between the fruits and their surroundings. By reducing the rate of respiration and water exchange through transpiration, edible coatings on tomatoes using Ag NPs (Gao et al. 2017), and halloysite NPs (Buendía-Moreno et al. 2020). Reducing fruits' physiological weight loss is the primary goal to preserve their postharvest quality characteristics and customer appeal (El-Beltagi et al. 2025b).

19.4.2 Respiration Rate and Ethylene Production

Fruit deterioration and shelf life during storage are mostly caused by high levels of ethylene and respiration rate (El-Beltagi et al. 2025c, d); these factors result in color browning, weight loss, deterioration, and off flavor (Khedr et al. 2023).

In sweet cherry fruit, the application of NO-releasing chitosan nanoparticles was demonstrated to effectively reduce ethylene production and respiration rates (Ma et al. 2019). In strawberries, the use of an edible nano-coating composed of alginate and limonene liposomes similarly decreased respiration rates (Dhital et al. 2018). For spinach leaves, a coating made from nano fibrillated cellulose demonstrated a notable reduction in respiration rates (Pacaphol et al. 2019). Additionally, in mangoes, a chitosan/zein-cinnamaldehyde nanocomposite film also contributed to a reduction in respiration (Xiao et al. 2021).

Furthermore, Tirgar et al. (2018) verified that it also offered a self-supporting way, indicating their greater prospective for use as packaging tools than the existing beads-in-pouch commercial products. Compared to other procedures, such ethylene scrubbing, these ethylene removal techniques are notably applicable, which suggests that they have a wide range of applications, especially in developing nations lacking access to cold storage facilities or sophisticated controlled atmospheres. Another intriguing use for nanomaterials has been the photocatalytic degradation process for the elimination of ethylene. TiO_2NPs have a strong and robust photocatalytic activity (Siripatrawan and Kaewklin 2018). TiO_2NPs were used as active packing into the chitosan film matrix by Siripatrawan and Kaewklin (2018) based on the photocatalyst.

Moreover, TiO_2NPs reduced the packaging film's optical transmittance (Siripatrawan and Kaewklin 2018). Furthermore, ethylene photocatalytic degradation by chitosan-TiO_2 nanocomposite films was triggered by ultraviolet (UV) light to produce essential reactive oxygen species (ROS), such as hydroxyl radical (·OH) and peroxynitrite anion ($O2^-$), having strong antimicrobial activity against gram-positive and gram-negative, as well as fungi (Aspergillus and Penicillium) (Siripawan and Kaewklin 2018).

19.4.3 Color Changes

Color plays a crucial role in the visual appeal of fruits and vegetables, as it is often the primary factor influencing consumer choice (Khedr et al. 2023). Typically, color browning occurs due to chlorophyll degradation, increased ethylene production, and a higher respiration rate (Khedr et al. 2023). Enzymatic activity is a key factor responsible for browning and the loss of color in fruits. Recent studies have shown that edible coatings, an eco-friendly solution, help preserve the color of fruits by minimizing enzymatic activity, preventing chlorophyll degradation, and reducing both respiration rates and ethylene biosynthesis.

In tomatoes, the combination of an active cardboard box and coatings of essential oils with cyclodextrins and/or halloysite nanotubes has been effective in extending shelf life and maintaining quality. This approach not only increases firmness but also reduces pH, titratable acidity (TA), color changes (ΔE), and decay incidence (Buendía-Moreno et al. 2020). Additionally, the use of edible nanolaminate coatings infused with extracts of *Flourensiacernua* has shown promise in prolonging shelf life and inhibiting decay, while also decreasing titratable acidity, ethylene production, and color changes (ΔE) (Salas-Méndez et al. 2019). For cucumbers, nano-coatings have been demonstrated to extend storage life and preserve quality by enhancing sensory attributes, maintaining chlorophyll levels, and minimizing color changes (ΔE), as well as reducing the activities of PPO and POD (Zambrano-Zaragoza et al. 2021). Moreover, mangoes coated with chitosan/zein-cinnamaldehyde nanocomposite films effectively reduce color changes (Xiao et al. 2021).

19.4.4 Firmness

Firmness is one of the most critical quality attributes of fruits, directly affecting their storability (Khedr 2022a). The firmness of any fruit can decline as insoluble protopectin breaks down into more soluble pectin and pectic acid (Khedr 2021).

Additionally, in non-climacteric fruits, this degradation process occurs much more slowly compared to climacteric fruits.

Strawberries coated with solid lipid nanoparticles (SLN) have demonstrated a significant decrease in weight loss and deterioration (Zambrano-Zaragoza et al. 2021). In tomatoes, the application of edible nanolaminate coatings enriched with extracts of *Flourensiacernua*, which increasing pH and firmness, and reducing titratable acidity, ethylene production, and color changes (ΔE) (Salas-Méndez et al. 2019). Additionally, strawberry nano-coatings made from quinoa protein/chitosan combined with thyme and neem extracts have been shown to extend storage life by enhancing total soluble solids, sensory physiological attributes, firmness, and pH, while also reducing fungal decay, weight loss (Robledo et al. 2018). Furthermore, a nano-chitosan combined with $CaCl_2$ has improved shelf life and sensory attributes, increasing firmness, TSS, antioxidant capacity, total phenolic content, ascorbic acid, and volatile compounds, while simultaneously reducing weight loss, titratable acidity, malondialdehyde, and total anthocyanin content (Nguyen et al. 2020).

19.4.5 Antimicrobial and Antifungal

Microbial spoilage is a major factor that diminishes the storage life and quality attributes of fruits by accelerating lipid oxidation, increasing ethylene production, and raising the respiration rate. Additionally, microbes can produce harmful mycotoxins, such as citrinin, patulin, and tremorgenic compounds, in fruits (Moraes Bazioli et al. 2019).

Antibacterial agents predominantly target t plasma membrane, which serves as a barrier between the extracellular environment and the cytosol. They do this by either increasing non-specific permeability or disrupting membrane integrity (Wang et al. 2017). Thyme essential oil and its active component, thymol, exhibit significant antibacterial effects at concentrations up to twice the minimum inhibitory concentration. The inactivation of bacteria attributed to these compounds' capacity to disrupt the cytoplasmic membrane, leading to polarization and functional loss (Kang et al. 2018). This disruption results in diminished intracellular ATP levels, altered metabolic activities, and reduced cell viability.

The antibacterial properties of chitosan are attributed to the relations between its positively charged amino and the negatively charged carboxylate on bacterial cell membranes (Andres et al. 2007). This electrostatic interaction disrupts the cell membrane, resulting in increased lipid peroxidation and protein leakage (Arroyo et al. 2020). Additionally, the long molecular chains of high molecular weight chitosan enhance their binding affinity to *E. coli*. The protonation of amino groups contributes to this antibacterial activity. Hydrolyzed chitosan products may also interact with microbial DNA, inhibiting protein and mRNA synthesis, which further supports its antimicrobial action (Qin et al. 2006).

In bacterial cells exposed to ethanol, membrane permeabilization results in the leakage of plasma components, including proteins (Hammer and Heel 2012). Various essential oils disrupt bacterial cell membranes, causing the leakage of nutrients and genetic material, leading to cell damage, irregular shapes, and shrinkage, as confirmed by scanning electron microscopy (Kang et al. 2018). While most essential oils primarily act as a membrane permeabilizers, certain compounds, such as cinnamon essential oil and cinnamaldehyde, are thought to inhibit metabolic enzymes, affecting energy production without significantly altering membrane permeability (Hammer and Heel 2012; Raybaudi-Massilia et al. 2008). Essential oils compromise eukaryotic cell membranes, including those of fungi, by depolarizing the mitochondrial membrane, reducing its potential, and initiating a cascade of changes. This mechanism resembles that seen in bacterial cells, where increased membrane permeability impacts proton pumps and ionic channels (Sánchez-González et al. 2011).

The essential oils of cinnamon, lemongrass, and cloves, along with their key active compounds, have been associated with the reduction of various pathogen virulence factors. For example, these oils have decreased motility, hemolytic activity, and proteinase production in Pseudomonas aeruginosa, as well as reduced cell surface hydrophobicity and hyphal formation in *Candida spp.* (Khan and Ahmad 2013). Thyme essential oil has reduced biofilm production in Bacillus cereus, an important virulence factor related to disease transmission (Kang et al. 2018). Cinnamon essential oil has also been effective in decreasing extracellular materials linked to biofilm formation in *Pseudomonas aeruginosa* (Kalia et al. 2015).

Metal ions can inflict biological damage by generating reactive species such as superoxide anions. For instance, copper ions produce hydrogen peroxide radicals, which, when combined with superoxide dismutase anions generated by TiO_2 under sunlight, lead to oxidative damage in bacterial cells (Manzl et al. 2004). The stimulation of zinc oxide (ZnO) by UV and visible light produces electron-hole pairs that initiate reactions converting dissolved oxygen into superoxide radical anions. Heinlaan et al. (2008) noted that water molecules also contribute to this process, yielding radicals that react with electrons to form H_2O_2.

The absorption of silver ions disrupts DNA replication and ATP synthesis, contributing to antimicrobial effects. The proposed mechanisms of action include essential transport activities (Carvalho et al. 2014). Microbial sensitivity to chitosan varies based on differences in cell membrane electrochemistry, composition, thickness, and structure. Notably, the antibacterial efficacy of films containing nanoparticles (NPs) primarily arises from NP activity. For instance, silver nanoparticles combined with chitosan interact with negatively charged bacterial membranes, increasing permeability and forming pores that lead to cell death (Davoodbasha et al. 2016). The incorporation of nanoparticles (NPs) such as silver (AgNPs), silica (SiOxNPs), zinc oxide (ZnONPs), and titanium dioxide (TiO_2NPs) has been shown to enhance polymer materials with antimicrobial properties against a diverse range of microorganisms (Fernández et al. 2010).

19.4.6 Nano-Sensors

Nano sensors offer real time data for immediate pathogen recognition and localization. Also used to monitor environments, contaminants, pesticides, and assess fruit quality (Khedr et al. 2024; Khedr and Al-Khayri 2024). This information aids in boosting agricultural productivity by optimizing nutrient application and protecting crops from viruses, diseases, and pests (Li et al. 2022). Nano-sensors enhance selectivity and sensitivity compared to traditional methods, with broad applications in agriculture and food production. By utilizing nanomaterials, nano-sensors enable precision farming with minimal effort, cost, and environmental impact. They assist in monitoring crop health, soil nutrients, water conditions, disease outbreaks, and environmental risks from agrochemicals (Cheng et al. 2016). Nanomaterials can amplify biological organisms' natural sensory capabilities, improving their sensitivity to various signals (Dubey et al. 2019). According to El Beyrouthya and El Azzi (2014), nano-sensor networks can wirelessly transmit real-time field data throughout the seasons, allowing remote monitoring of agricultural conditions. In addition to tracking soil moisture and automating irrigation, nano-sensors can also detect potential pests and diseases, including insects and microbial infections. For example, carbon NPs are being used to identify pesticide residues in plants. Fluorescent silica NP probes are used for the early detection of pathological decay in oranges, freezing damage in sweet lemons and oranges caused by the plant pathogen *Xanthomonas axonopodis*, which causes the bacterial spot disease in plants of the Solanaceae family (Yao et al. 2009).

19.4.7 Nano-Bubbles

Micro/nano-bubble technology is an innovative postharvest management approach has several advantages over traditional soaking and fumigation methods. This technique utilizes bubbles with diameters of 50 micrometers (μm) or 1 nanometer (nm). It not only enhances the solubility of gases in water but also increases their stability (Patel et al. 2021). Additionally, this non-chemical cleaning method minimizes mechanical loss to fresh fruits during postharvest handling, as it does not rely on detergents, thus helping to mitigate environmental pollution. The technology can alter gas or solvent permeability, reduce pathogen spread, and boost the antioxidant function of fruits (Wang et al. 2020). When a bubble with such significant potential bursts, it generates a shock wave that can promote spontaneous water oxidation and produce hydroxyl radicals, leading to the generation of ROS that help prevent pathogen infections (Xiao et al. 2019).

19.4.8 Nano-Packaging

Safeguarding fruits from mechanical damage and quality loss is a crucial task in postharvest management. Effective packaging must be resilient to physical impacts, cost-effective, disposable and safe. The type and composition of packaging materials significantly affect food quality. Incorporating nanomaterials into packaging can create lightweight options with enhanced strength, and heat resistance (Kumar et al. 2018). Nanoparticles (NPs) in food packaging has several advantages, including antibacterial properties, improved oxygen transfer, enzyme mobilization, and indicators of degradation exposure. Nano-packaging materials provide enhanced sensing capabilities compared to traditional packaging options (Sharma et al. 2017).

19.4.9 Antioxidant Effects

Growing research highlights the beneficial effects of nanomaterial packaging in enhancing the antioxidant capacity of fresh fruits, thereby extending their shelf life (El-Beltagi et al. 2025e). Nanocomposite packaging effectively regulates the antioxidant systems of fresh fruit during storage. Reactive oxygen species such as H_2O_2, superoxide, and hydroxyl radicals (OH) are key contributors to oxidative reactions that damage membrane fatty acids and overall cellular structures, leading to senescence in postharvest (Tabassum et al. 2023).

Nanotechnology offers a promising pathway to reduce post-harvest losses, providing innovative strategies for enhancing produce quality. Research has shown that nanotechnology-based edible coatings—such as nanocomposite coatings and nano emulsions—exhibit superior physical and antimicrobial properties compared to traditional coatings. These advanced materials play a crucial role in improving the physicochemical, microbiological, and physiological attributes of fresh produce (Riseh et al. 2023).

Incorporating nanoparticles into polymeric packaging marks a significant advancement in active food packaging. The addition of nanomaterials like silver (Ag), titanium dioxide (TiO_2), copper, zinc oxide (ZnO), and silica (SiO_2) have been demonstrated to enhance the physical, mechanical, optical, and antimicrobial characteristics of packaging (Li et al. 2021a, b; Xing et al. 2020). This innovation prolongs the quality and shelf life of fruits and vegetables. However, more research is needed to assess the potential migration of these nanoparticles under various storage conditions to ensure compliance with safety regulations.

The deployment of nanosensors in packaging represents one of the most exciting applications of nanotechnology for monitoring and preserving the quality of fruits and vegetables throughout the logistic process. The positive results observed in several studies underscore the appeal of this technology. Continued research is essential to refine nano sensor applications, particularly in developing user-friendly tools that allow consumers to accurately interpret color changes in colorimetric nanosensors using smartphone cameras, thereby minimizing misinterpretations (Sharma et al. 2023).

Safety remains a critical concern regarding the use of nanomaterials. Certain nanoparticles may pose health risks due to their capacity to penetrate cells, leading to adverse reactions in human and animal tissues as well as in plants. Future research must prioritize addressing these safety concerns by exploring greener synthesis methods and creating efficient, cost-effective protocols for the degradation and removal of nanomaterials from contaminated environments (Cheng et al. 2024).

Sustainable nano-application emphasizes the environmentally responsible use of nanomaterials in the postharvest management of fresh produce. Green nanotechnology should focus on synthesizing nano-products from natural materials, such as plant secondary metabolites, which are eco-friendly, cost-effective, biodegradable, and highly bioactive. Strategies for utilizing these nano-products should include recycling residual waste materials and producing new batches of nanomaterials. In summary, the future of nanotechnology in postharvest management looks bright, but it is crucial to balance innovation with safety and sustainability to fully realize its potential in the agricultural sector (Riseh et al. 2023).

19.5 Conclusion and Prospects

Nanotechnology presents a groundbreaking opportunity to tackle post-harvest challenges, offering innovative solutions that enhance fruit quality, prolong shelf life, and reduce spoilage. Through applications such as nano-coatings, antimicrobial nanoparticles, and smart packaging, nanotechnology not only addresses the pressing issues of food waste but also contributes to more sustainable practices in agriculture and food processing. The continued development of nanosensors and nano-based materials promises even greater advancements in real-time quality monitoring and preservation techniques.

However, for these technologies to achieve their full potential, it is crucial to address the challenges of cost, scalability, and regulation. As research and innovation in this field progress, nanotechnology could become a vital component of global food security strategies, ensuring that fruits and vegetables maintain their nutritional and sensory qualities from harvest to consumption. The future of post-harvest management lies in the hands of nanotechnology, with the promise of reducing food losses, enhancing sustainability, and creating a more resilient food supply chain.

The anticipated advancements in nanotechnology applications for enhancing fruit quality are significant. These developments encompass various innovations, including the creation of nano-smart packaging, the formulation of advanced antioxidants, and the evolution of diverse coating types and techniques to improve their effectiveness and applicability. Furthermore, a variety of nanosensors, particularly e-nose technologies, are expected to increased implementation, facilitating real-time monitoring of fruit quality. It is essential that these approaches work in synergy by emphasizing the importance of rigorous testing and research to evaluate safety and residual effects on both fruit quality and consumer health.

References

Akintelu SA, Folorunso AS (2019) Characterization and antimicrobial investigation of synthesized silver nanoparticles from *Annona muricata* leaf extracts. Nanotechnol Nanomed Nanobiotechnol 6:22

Al-Khayri JM, Khedr EH (2024a) Augmenting apricot fruit quality and prolonging storage life via modulating ethylene metabolism and antioxidants with 1-(3-phenyl-propyl) cyclopropene and psyllium treatments. J Hortic Sci Biotech 1–17:696–712. https://doi.org/10.1080/14620316.2024.2338130

Al-Khayri JM, Khedr EH (2024b) Enhancing peach quality and extending storability by regulating ethylene metabolism and enzymes using 1-(3-phenyl-propyl) cyclopropene and salicylic acid. J Hortic Sci Biotech:1–16

Al-Khayri JM, Abdel-Haleem M, Khedr EH (2024) Harnessing GABA Pathways to Improve Plant Resilience Against Salt Stress. Horticulturae 10(12):1296. https://doi.org/10.3390/horticulturae10121296

Andres Y, Giraud L, Gerente C, Cloirec PL (2007) Antibacterial effects of chitosan powder: mechanisms of action. Environ Technol 28:1357–1363. https://doi.org/10.1080/09593332808618957

Arabpoor B, Yousefi S, Weisany W, Ghasemlou M (2021) Multifunctional coating composed of *Eryngium campestre* L. essential oil encapsulated in nano-chitosan to prolong the shelf-life of fresh cherry fruits. Food Hydrocoll 111:106394. https://doi.org/10.1016/j.foodhyd.2020.106394

Arroyo BJ, Bezerra AC, Oliveira LL, Arroyo SJ, Melo EA, Santos AMP (2020) Antimicrobial active edible coating of alginate and chitosan add ZnO nanoparticles applied in guavas (*Psidium guajava* L.). Food Chem 309:125566. https://doi.org/10.1016/j.foodchem.2019.125566

Barrett DM, Beaulieu JC, Shewfelt R (2010) Color, flavor, texture, and nutritional quality of fresh-cut fruits and vegetables: desirable levels, instrumental and sensory measurement, and the effects of processing. Crit Rev Food Sci Nutr 50(5):369–389

Batool R, Kazmi SAR, Khurshid S, Saeed M, Ali S, Adnan A, Altaf F, Hameed A, Batool F, Fatima N (2022) Postharvest shelf life enhancement of peach fruit treated with glucose oxidase immobilized on ZnO nanoparticles. Food Chemistry 366, p.130591

Buendía-Moreno L, Sánchez-Martínez MJ, Antolinos V, Ros-Chumillas M, Navarro-Segura L, Soto-Jover S, Martinez-Hernandez GB, López-Gómez A (2020) Active cardboard box with a coating including essential oils entrapped within cyclodextrins and/or halloysite nanotubes: A case study for fresh tomato storage. Food Control 107:106763. https://doi.org/10.1016/j.foodcont.2019.106763

Carvalho P, Sampaio P, Azevedo S, Vaz C, Espinós JP, Teixeira V, Carneiro JO (2014) Influence of thickness and coatings morphology in the antimicrobial performance of zinc oxide coatings. Appl Surf Sci 307:548–557. https://doi.org/10.1016/j.apsusc.2014.05.051

Carvalho RL, Cabral MF, Germano TA, de Carvalho WM, Brasil IM, Gallão MI, Moura CFH, Lopes MMA, de Miranda MRA (2016) Chitosan coating with trans-cinnamaldehyde improves structural integrity and antioxidant metabolism of fresh-cut melon. Postharvest Biol Technol 113:29–39. https://doi.org/10.1016/j.postharvbio.2015.11.004

Castelo Branco Melo NF, de Mendonça Soares BL, Marques Diniz K, Ferreira Leal C, Canto D, Flores MAP, Henrique da Costa Tavares-Filho J, Galembeck A, Montenegro Stamford TL, Montenegro Stamford-Arnaud T, Montenegro Stamford TC (2018) Effects of fungal chitosan nanoparticles as eco-friendly edible coatings on the quality of postharvest table grapes. Postharvest Biol Technol 139:56–66. https://doi.org/10.1016/j.postharvbio.2018.01.014

Chandirika JU, Selvi ST, Annadurai G (2018) Synthesis and characterization of silver nanoparticle using Melia azedarach for vegetable coating and antibacterial activity. Journal of Innovations in Pharmaceutical and Biological Sciences 5(2):38-42

Chaturvedi K, Sharma N, Yadav SK (2019) Composite edible coatings from commercial pectin, corn flour and beetroot powder minimize post-harvest decay, reduces ripening and improves sensory liking of tomatoes. Int J Biol Macromol 133:284–293

Cheng JH, Lan L, Sun DW (2024) Innovative strategy for fabricating nanomaterials to ensure food safety by cold plasma: principles, mechanisms, and applications. Trends Food Sci Technol 104644:104644

Cheng HN, Klasson KT, Asakura T, Wu Q (2016) Nanotechnology in agriculture. In: Nanotechnology: delivering on the promise. American Chemical Society, pp 233–242. https://doi.org/10.1021/bk-2016-1224.ch012

Cid-López ML, Soriano-Melgar LAA, García-González A, Cortéz-Mazatán G, Mendoza-Mendoza E, Rivera-Cabrera F, Peralta-Rodríguez RD (2021) The benefits of adding calcium oxide nanoparticles to biocompatible polymeric coatings during cucumber fruits postharvest storage. Sci Hortic 287:110285. https://doi.org/10.1016/j.scienta.2021.110285

Cui H, Yuan L, Li W, & Lin L (2017) Edible film incorporated with chitosan and Artemisia annua oil nanoliposomes for inactivation of Escherichia coli O157: H7 on cherry tomato. International Journal of Food Science and Technology 52(3):687-698

Dargiri SA, Rastegar S, Khedr EH (2024) Multilayer edible coatings technology for fresh fruits and vegetables. In: Sustainable postharvest technologies for fruits and vegetables. CRC Press, pp 386–397

Davies JN, Hobson GE (1981) The constituents of tomato fruit: the influence of environment, nutrition, and genotype. Crit Rev Food Sci Nutr 15:205–280. https://doi.org/10.1080/10408398109527317

Davoodbasha M, Kim S-C, Lee S-Y, Kim J-W (2016) The facile synthesis of chitosan-based silver nano-biocomposites via a solution plasma process and their potential antimicrobial efficacy. Arch Biochem Biophys 605:49–58. https://doi.org/10.1016/j.abb.2016.04.001

de Oliveira LIG, de Oliveira KÁR, de Medeiros ES, Batista AUD, Madruga MS, Lima MDS, de Souza EL, Magnani M (2020) Characterization and efficacy of a composite coating containing chitosan and lemongrass essential oil on postharvest quality of guava. Innov Food Sci Emerg Technol 66:102506

Dhital R, Mora NB, Watson DG, Kohli P, Choudhary R (2018) Efficacy of limonene nano coatings on post-harvest shelf life of strawberries. LWT—Food Sci Technol 97:124–134

Dubey PK, Shukla RN, Srivastava G, Mishra AA, Pandey A (2019) Study on quality parameters and storage stability of mango coated with developed nanocomposite edible film. Int J Curr Microbiol App Sci 8(4):2899–2935. https://doi.org/10.20546/ijcmas.2019.804.339

EL-Beltagi, HS, Khedr EH (2025) Biosensors and nanosensors: Innovative tools for precision fruit quality assessment. Journal of Food Composition and Analysis 146:107957. https://doi.org/10.1016/j.jfca.2025.107957

El-Beltagi, H. S., Abdel-Haleem, M., Al Saikhan, M. S., & Khedr, E. H. (2025a). Optimization of progesterone treatment via response surface methodology to maintain quality and mitigate chilling injury in Prunus persica L. New Zealand Journal of Crop and Horticultural Science,1–19. https://doi.org/10.1080/01140671.2025.2458819

El-Beltagi HS, Abdel-Haleem M, Adel, AR, Khedr EH (2025b) Progesterone as a plant physiochemical regulator: Mechanisms and efficacy in alleviating abiotic stress. Journal of Crop Health 77:60. https://doi.org/10.1007/s10343-025-01125-9

El-Beltagi, H. S., Abdel-Haleem, M., Al Saikhan, M. S., Shalaby, T. A., Mohamed, A. A., & Khedr, E. H. (2025c). Antibacterial coatings: pioneering solutions for maintaining fruit quality and prolonging storability. The Journal of Horticultural Science and Biotechnology, 1–25. https://doi.org/10.1080/14620316.2025.2476660

EL-Beltagi, H. S.; Abdel-Haleem, M.; Khedr, N.; Shalaby, T. A.; El-Mogy, M.M.; and Khedr, E.H. (2025d). Millets as Climate-Resilient Crops: Emerging Technologies for Sustainable Postharvest Management, Journal of Food Composition and Analysis, 147, 107999. https://doi.org/10.1016/j.jfca.2025.107999

El-Beltagi HS, Gad M, Abdel-Haleem M, Shalaby TA, Rezk AA, Khedr EH (2025e) Triacontanol: a Multifunctional Growth Regulator for Enhancing Stress Tolerance. Journal of Crop Health 77: 78. https://doi.org/10.1007/s10343-025-01144-6

El Beyrouthya M, El Azzi D (2014) Nanotechnologies: novel solutions for sustainable agriculture. Adv Crop Sci Technol 2(03):8863. https://doi.org/10.4172/2329-8863.1000e118

Etienne A, Génard M, Lobit P, Mbeguié-A-Mbéguié D, Bugaud C (2013) What controls fleshy fruit acidity? A review of malate and citrate accumulation in fruit cells. J Exp Bot 64(6):1451–1469. https://doi.org/10.1093/jxb/ert035

Fei T, Leyva-Gutierrez FMA, Wan Z, Wang T (2021) New insights into nano-polymeric-based edible coatings in fresh fruits and vegetables preservation: advances in the last decade. Food Packag Shelf Life 30:100745

Fernández A, Picouet P, Lloret E (2010) Cellulose-silver nanoparticle hybrid materials to control spoilage-related microflora in absorbent pads located in trays of fresh-cut melon. Int J Food Microbiol 142:222–228. https://doi.org/10.1016/j.ijfoodmicro.2010.07.007

Fu B, Mei S, Su X, Chen H, Zhu J, Zheng Z, Lin H, Dai C, Luque R, Yang DP (2021) Integrating waste fish scale-derived gelatin and chitosan into edible nanocomposite film for perishable fruits. Int J Biol Macromol 191:1164–1174. https://doi.org/10.1016/j.ijbiomac.2021.09.171

Gao L, Wang X, Zeng H, Zheng Y (2017) Silver nanoparticles biologically synthesized using tea leaf extracts and their use for extension of fruit shelf life. IET Nanobiotechnol 11(6):637–643. https://doi.org/10.1049/iet-nbt.2017.0010

Garcia-Betanzos CI, Hernández-Sánchez H, Quintanar-Guerrero D, Galindo-Pérez MJ, Zambrano-Zaragoza ML (2018) Influence of solid lipid nanoparticle/xanthan gum coatings on compositional and enzymatic changes in guava (*Psidium guajava* L.) during ripening. Acta Hortic 1194: 289–295. https://doi.org/10.17660/ActaHortic.2018.1194.42

Gholami R, Ahmadi E, Farris S (2017) Shelf life extension of white mushrooms (Agaricus bisporus) by low temperatures conditioning, modified atmosphere, and nanocomposite packaging material. Food Packag Shelf Life 14:88–95. https://doi.org/10.1016/j.fpsl.2017.09.001

González-Saucedo A, Barrera-Necha LL, Ventura-Aguilar RI, Correa-Pacheco ZN, Bautista-Baños S, Hernández-López M (2019) Extension of the postharvest quality of bell pepper by applying nanostructured coatings of chitosan with Byrsonimacrassifolia extract (L.) Kunth. Postharvest Biol Technol 149:74–82

Gudadhe JA, Patil SB, Bhosale RS (2013) Preparation of an agar-silver nanoparticles (A-AgNp) film for increasing the shelf-life of fruits. IET Nanobiotechnol 8(4):190–195. https://doi.org/10.1049/iet-nbt.2012.0014

Hammer KA, Heel KA (2012) Use of multiparameter flow cytometry to determine the effects of monoterpenoids and phenylpropanoids on membrane polarity and permeability in staphylococci and enterococci. Int J Antimicrob Agents 40(3):239–245. https://doi.org/10.1016/j.ijantimicag.2012.04.008

Hasheminejad N, Khodaiyan F (2020) The effect of clove essential oil loaded chitosan nanoparticles on the shelf life and quality of pomegranate arils. Food Chemistry 309:125520.

He X, Hwang HM (2016) Nanotechnology in food science: functionality, applicability, and safety assessment. J Food Drug Anal 24(4):671–681. https://doi.org/10.1016/j.jfda.2016.06.001

Heinlaan M, Ivask A, Blinova I, Dubourguier H-C, Kahru A (2008) Toxicity of nanosized and bulk ZnO, CuO and TiO2 to bacteria Vibrio fischeri and crustaceans Daphnia magna and Thamnocephalusplatyurus. Chemosphere 71:1308–1316. https://doi.org/10.1016/j.chemosphere.2007.11.048

Ho LC (1996) Tomato. Marcel Dekker Inc., New York

Huang G, Huang L, Geng C, Lan T, Huang X, Xu S, Shen Y, Bian H (2022) Green and multifunctional chitosan-based conformal coating as a controlled release platform for fruit preservation. Int J Biol Macromol 219:767–778. https://doi.org/10.1016/j.ijbiomac.2022.08.038

Jamdagni P, Khatri P, Rana JS (2018) Green synthesis of zinc oxide nanoparticles using flower extract of *Nyctanthes arbor-tristis* and their antifungal activity. J King Saud Univ-Sci 30(2): 168–175. https://doi.org/10.1016/j.jksus.2016.10.002

Jiang T, Feng L, Wang Y (2013) Effect of alginate/nano-Ag coating on microbial and physicochemical characteristics of shiitake mushroom (*Lentinus edodes*) during cold storage. Food Chem 141:954–960. https://doi.org/10.1016/j.foodchem.2013.03.093

Kalia M, Yadav VK, Singh PK, Sharma D, Pandey H, Narvi SS, Agarwal V (2015) Effect of cinnamon oil on quorum sensing-controlled virulence factors and biofilm formation in Pseudomonas aeruginosa. PloS One 10(8):e0135495. https://doi.org/10.1371/journal.pone.0135495

Kang J, Liu L, Wu X, Sun Y, Liu Z (2018) Effect of thyme essential oil against Bacillus cereus planktonic growth and biofilm formation. Appl Microbiol Biotechnol 102:10209–10218. https://doi.org/10.1007/s00253-018-9340-3

Khan MSA, Ahmad I (2013) In vitro inhibition of growth and virulence factors production in azole-resistant strains of non-albicans Candida by Cinnamomum verum, Cymbopogon citratus, Cymbopogon martini and Syzygium aromaticum essential oils. J Biol Act Prod Nat 3(2): 139–153. https://doi.org/10.1080/22311991.2013.11009533

Khedr EH (2021) Does the quality of pomegranate (Punica granatum L., cv. 116) disparity in response to different coatings during prolonged storage? Plant Cell Biotechnol Mol Biol 22(53–54):135–155

Khedr EH (2022a) Application of different coating treatments to enhance storability and fruit quality of pomegranate (Punica granatum L., cv. Wonderful) during prolonged storage. Rev Bras Frutic 44:e-855

Khedr EH (2022b) Improving the quality of 'Keitt' mango using pre-harvest calcium lactate treatment and post-harvest applications of kojic and ascorbic acids. J Hortic Sci Biotech 97(6):757–772

Khedr EH, Al-Khayri JM (2023) Synergistic effects of tragacanth and anti-ethylene treatments on postharvest quality maintenance of mango (Mangifera indica L.). Plan Theory 12(9):1887

Khedr, E. H. (2017). Does the quality of Valencia orange vary in response to different coatings during cold storage? Egyptian Journal of Horticulture, 44(2), 219–233. https://doi.org/10.21608/ejoh.2018.2220.1031

Khedr, E.H., Al-Khayri, J.M. (2024). Biosensors and Nanosensors for Determination of Harvest Quality Parameters and Fruit Handling Processes Monitoring. In: Al-Khayri, J.M., Alnaddaf, L.M., Jain, S.M., Penna, S. (eds) Innovative Methods in Horticultural Crop Improvement. Advances in Plant Breeding Strategies, vol 2. Springer, Cham. pp. 143-190. https://doi.org/10.1007/978-3-031-61095-0_7

Khedr EH, Khedr N (2023) Optimization of postharvest progesterone treatment to alleviate chilling injury in mango fruit, maintaining intracellular energy, cell wall stability, and antioxidant activity. Postharvest Biol Technol 206:112572

Khedr EH, Khedr N (2024a) Unraveling the mechanisms behind salt stress alleviation and enhanced fruit storability in pomegranate through triacontanol treatment. Sci Hortic 325:112634

Khedr EH, Khedr N (2024b) Enhancing productivity, modifying biochemical parameters, and regulating the phenylpropanoid pathway in 'Le-Conte' pears through optimal protocatechuic acid treatments. BMC Plant Biol 24(1):50

Khedr EH, Khedr N, Abdel-Haleem M (2023) Harnessing the metabolic modulatory and antioxidant power of 1-(3-phenyl-propyl) cyclopropane and melatonin in maintaining mango fruit quality and prolongation storage life. BMC Plant Biol 23(1):464

Khedr, E.H., Al-Qahtani, W.H., Khedr, N. (2024). Biosensors and Nanosensors for Determination of Fruit Safety. In: Al-Khayri, J.M., Alnaddaf, L.M., Jain, S.M., Penna, S. (eds) Innovative Methods in Horticultural Crop Improvement. Advances in Plant Breeding Strategies, vol 2. Springer, Cham. Pp. 191-234. https://doi.org/10.1007/978-3-031-61095-0_8

Khedr EH, Khedr N (2025) Nanotechnology in Edible Coatings: A Novel Strategy for Fruit Quality Preservation. In: Al-Khayri, J.M., Yatoo, A.M., Jain, S.M., Penna, S. (eds) Handbook of Agricultural Technologies. Springer, Singapore. https://doi.org/10.1007/978-981-99-0862-2_63-1

Kim IH, Oh YA, Lee H, Song KB, Min SC (2014) Grape berry coatings of lemongrass oil-incorporating nanoemulsion. LWT—Food Sci Technol 58:1–10. https://doi.org/10.1016/j.lwt.2014.03.018

Klangmuang P, Sothornvit R (2018) Active coating from hydroxypropyl methylcellulose-based nanocomposite incorporated with Thai essential oils on mango (cv. Namdokmai Sithong). Food Biosci 23:9–15. https://doi.org/10.1016/j.fbio.2018.02.012

Klee HJ, Tieman DM (2013) Genetic challenges of flavour improvement in tomato. Trends Genet 29(4):257–262. https://doi.org/10.1016/j.tig.2012.12.003

Kumar S, Shukla A, Baul PP, Mitra A, Halder D (2018) Biodegradable hybrid nanocomposites of chitosan/gelatin and silver nanoparticles for active food packaging applications. Food Packag Shelf Life 16:178–184. https://doi.org/10.1016/j.fpsl.2018.03.008

Li D, Wang D, Fang Y, Li L, Lin X, Xu Y, Chen H, Zhu M, Luo Z (2021a) A novel phase change coolant-assisted chitosan/gelatin nanocomposite coating with Nano-ZnO particles maintains strawberry quality under cold storage. Food Packag Shelf Life 28:100636. https://doi.org/10.1016/j.fpsl.2021.100636

Li J, Sun Q, Sun Y, Chen B, Wu X, Le T (2019) Improvement of banana postharvest quality using a novel soybean protein isolate/cinnamaldehyde/zinc oxide bionanocomposite coating strategy. Sci Hortic 258:108786

Li XH, Li WL, Xing YG, Jiang YH, Ding YL, Zhang PP (2011) Effects of nano-ZnO power-coated PVC film on the physiological properties and microbiological changes of fresh-cut "Fuji" apple. Adv Mat Res 152:450–453

Li Y, Rokayya S, Jia F, Nie X, Xu J, Elhakem A, Almatrafi M, Benajiba N, Helal M (2021b) Shelf-life, quality, and safety evaluations of blueberry fruits coated with chitosan nano-material films. Sci Rep 11:55

Li Z, Zhu L, Zhao F, Li J, Zhang X, Kong X, Zhang Z (2022) Plant salinity stress response and nano-enabled plant salt tolerance. Front Plant Sci 13:843994. https://doi.org/10.3389/fpls.2022.843994

Lin M, Fang S, Zhao X, Liang X, Wu D (2020) Natamycin-loaded zein nanoparticles stabilized by carboxymethyl chitosan: evaluation of colloidal/chemical performance and application in postharvest treatments. Food Hydrocoll 106:105871. https://doi.org/10.1016/j.foodhyd.2020.105871

Liu R, Liu D, Liu Y, Song Y, Wu T, Zhang M (2017) Using soy protein SiOx nanocomposite film coating to extend the shelf life of apple fruit. Int J Food Sci Technol 52:2018–2030

Liu X, Han W, Zhu Y, Xuan H, Ren J, Zhang J, Ge L (2018) Anti-oxidative and antibacterial self-healing edible polyelectrolyte multilayer film in fresh-cut fruits. J Nanosci Nanotechnol 18: 2592–2600

Lustriane C, Dwivany FM, Suendo V, Reza M (2018) Silver nanoparticles as a preservative agent in fruit coating films. J Plant Biotechnol 45(1):36–42. https://doi.org/10.5010/JPB.2018.45.1.036

Ma Y, Fu L, Hussain Z, Huang D, Zhu S (2019) Enhancement of storability and antioxidant systems of sweet cherry fruit by nitric oxide-releasing chitosan nanoparticles (GSNO-CS NPs). Food Chem 285:10–21. https://doi.org/10.1016/j.foodchem.2019.01.156

Mahmoudi R, Razavi F, Rabiei V, Gohari G, Palou L (2022) Application of Glycine betaine coated chitosan nanoparticles alleviate chilling injury and maintain quality of plum (Prunus domestica L.) fruit. International Journal of Biological Macromolecules 207:965-977

Manzl C, Enrich J, Ebner H, Dallinger R, Krumschnabel G (2004) Copper-induced formation of reactive oxygen species causes cell death and disruption of calcium homeostasis in trout hepatocytes. Toxicology 196:57–64. https://doi.org/10.1016/j.tox.2004.06.009

Mei S, Fu B, Su X, Chen H, Lin H, Zheng Z, Dai C, Yang DP (2022) Developing silk sericin-based and carbon dots reinforced bio-nanocomposite films and potential application to litchi fruit. LWT−food. Sci Technol 164:113630. https://doi.org/10.1016/j.lwt.2022.113630

Meng X, Chen C, Song T, Xu J, Zhang X, Wang J, Pan Z, Zhang H, Zhang HJ (2022) Effect of nano-silica coating combined with pressurized Ar treatment on postharvest quality and reactive oxygen species metabolism in sweet cherry fruit. Food Chem 374:131715. https://doi.org/10.1016/j.foodchem.2021.131715

Moraes Bazioli J, Belinato JR, Costa JH, Akiyama DY, Pontes JGDM, Kupper KC et al (2019) Biological control of citrus postharvest phytopathogens. Toxins 11(8):460. https://doi.org/10.3390/toxins11080460

Motamedi E, Nasiri J, Malidarreh TR, Kalantari S, Naghavi MR, Safari M (2018) Performance of carnauba wax-nanoclay emulsion coatings on postharvest quality of 'Valencia' orange fruit. Sci Hortic 240:170–178. https://doi.org/10.1016/j.scienta.2018.06.002

Nasirifar SZ, Maghsoudlou Y, Oliyaei N (2018) Effect of active lipid-based coating incorporated with nanoclay and orange peel essential oil on physicochemical properties of *Citrus sinensis*. Food Sci Nutr 6:1508–1518

Nguyen VTB, Nguyen DHH, Nguyen HVH (2020) Combination effects of calcium chloride and nano-chitosan on the postharvest quality of strawberry (*Fragaria* × *ananassa* Duch.). Postharvest Biol Technol 162:111103. https://doi.org/10.1016/j.postharvbio.2019.111103

on cherry tomato. (n.d.) *International Journal of Food Science and Technology*, 52(3), 687–698. doi:10.1111/ijfs.13322

Pacaphol K, Seraypheap K, Aht-Ong D (2019) Development and application of nanofibrillated cellulose coating for shelf life extension of fresh-cut vegetable during postharvest storage. Carbohydrate Polymers 224:115167

Patel AK, Singhania RR, Chen CW, Tseng YS, Kuo CH, Wu CH, Dong CD (2021) Advances in micro- and nano bubbles technology for application in biochemical processes. Environ Technol Innov 23:101729. https://doi.org/10.1016/j.eti.2021.101729

Priyadarshi R, Riahi Z, Rhim JW (2022) Antioxidant pectin/pullulan edible coating incorporated with *Vitis vinifera* grape seed extract for extending the shelf life of peanuts. Postharvest Biol Technol 183:111740

Qin C, Li H, Xiao Q, Liu Y, Zhu J, Du Y (2006) Water-solubility of chitosan and its antimicrobial activity. Carbohydr Polym 63:367–374. https://doi.org/10.1016/j.carbpol.2005.09.002

Rastegar S, Dargiri SA, Khedr EH (2024) Antagonistic microorganism technology for horticultural produce. In: Sustainable postharvest technologies for fruits and vegetables. CRC Press, pp 431–441

Rastegar S, Khedr EH (2025) GABA in relation to cold and chilling stress. In Singh S, Tripathi DK, Singh VP, editors. GABA in plants. Plant Development, and Food Security 243-262, Hoboken, NJ: John Wiley & Sons. https://doi.org/10.1002/9781394217786.ch14

Raybaudi-Massilia RM, Rojas-Graü MA, Mosqueda-Melgar J, Martín-Belloso O (2008) Comparative study on essential oils incorporated into an alginate-based edible coating to assure the safety and quality of fresh-cut Fuji apples. J Food Prot 71(6):1150–1161. https://doi.org/10.4315/0362-028X-71.6.1150

Riseh RS, Vatankhah M, Hassanisaadi M, Kennedy JF (2023) Chitosan-based nanocomposites as coatings and packaging materials for the postharvest improvement of agricultural product: a review. Carbohydr Polym 309:120666

Robledo N, López L, Bunger A, Tapia C, Abugoch L (2018) Effects of antimicrobial edible coating of thymol nanoemulsion/quinoa protein/chitosan on the safety, sensorial properties, and quality of refrigerated strawberries (*Fragaria* × *ananassa*) under commercial storage environment. Food Bioproc Tech 11:1566–1574. https://doi.org/10.1007/s11947-018-2124-3

Saba MK, Amini R (2017) Nano-ZnO/carboxymethyl cellulose-based active coating impact on ready-to-use pomegranate during cold storage. Food Chemistry 232:721-726

Salas-Méndez EJ, Vicente A, Pinheiro AC, Ballesteros LF, Silva P, Rodríguez-García R, Hernández-Castillo FD, Díaz-Jiménez M d LV, Flores-López ML, Villarreal-Quintanilla JÁ, Peña-Ramos FM, Carrillo-Lomelí DA, Jasso de Rodríguez D (2019) Application of edible nanolaminate coatings with antimicrobial extract of *Flourensiacernua* to extend the shelf-life of tomato (*Solanum lycopersicum* L.) fruit. Postharvest Biol Technol 150:19–27. https://doi.org/10.1016/j.postharvbio.2018.12.008

Sami R, Soltane S, Helal M (2021) Microscopic image segmentation and morphological characterization of novel chitosan/silica nanoparticle/nisin films using antimicrobial technique for blueberry preservation. Membranes 11(5):303

Sánchez-González, L., Vargas, M., González-Martínez, C., Chiralt, A., & Cháfer, M. (2011). Use of essential oils in bioactive edible coatings: a review. Food Engineering Reviews, 3, 1-16.

Sharma C, Dhiman R, Rokana N, Panwar H (2017) Nanotechnology: an untapped resource for food packaging. Front Microbiol 8:1735. https://doi.org/10.3389/fmicb.2017.01735

Sharma A, Ranjit R, Kumar N, Kumar M, Giri BS (2023) Nanoparticlesbasednanosensors: principles and theirapplicationsin activepackagingforfoodquality and safety detection. Biochem Eng J 193:108861

Silva-Vera W, Zamorano-Riquelme M, Rocco-Orellana C, Vega-Viveros R, Gimenez-Castillo B, Silva-Weiss A, Osorio-Lira F (2018) Food bioprocess technology. Food Bioprocess Technology

Siripatrawan U, Kaewklin P (2018) Fabrication and characterization of chitosan-titanium dioxide nanocomposite film as ethylene scavenging and antimicrobial active food packaging. Food Hydrocoll 84:125–134. https://doi.org/10.1016/j.foodhyd.2018.04.049

Song H, Yuan W, Jin P, Wang W, Wang X, Yang L, Zhang Y (2016) Effects of chitosan/nano-silica on postharvest quality and antioxidant capacity of loquat fruit during cold storage. Postharvest Biol Technol 119:41–48. https://doi.org/10.1016/j.postharvbio.2016.04.015

Su X, Lin H, Fu B, Mei S, Lin M, Chen H, Zheng Z, Bo H, Yang DP, Lin Y (2023) Egg-yolk-derived carbon dots@albumin bio-nanocomposite as multifunctional coating and its application in quality maintenance of fresh litchi fruit during storage. Food Chem 405:134813. https://doi.org/10.1016/j.foodchem.2022.134813

Tabassum Z, Mohan A, Mamidi N, Khosla A, Kumar A, Solanki PR et al (2023) Recent trends in nanocomposite packaging films utilising waste generated biopolymers: industrial symbiosis and its implication in sustainability. IET Nanobiotechnol 17(3):127–153

Tirgar A, Han D, Steckl AJ (2018) Absorption of ethylene on membranes containing potassium permanganate loaded into alumina-nanoparticle-incorporated alumina/carbon nanofibers. J Agric Food Chem 66:5635–5643. https://doi.org/10.1021/acs.jafc.7b05037

Wang LH, Zhang ZH, Zeng XA, Gong DM, Wang MS (2017) Combination of microbiological, spectroscopic and molecular docking techniques to study the antibacterial mechanism of thymol against Staphylococcus aureus: membrane damage and genomic DNA binding. Anal Bioanal Chem 409:1615–1625. https://doi.org/10.1007/s00216-017-0394-6

Wang L, Shao S, Madebo MP, Hou Y, Zheng Y, Jin P (2020) Effect of nano-SiO2 packing on postharvest quality and antioxidant capacity of loquat fruit under ambient temperature storage. Food Chem 315:126295. https://doi.org/10.1016/j.foodchem.2020.126295

Xiao J, Gu C, Zhu D, Huang Y, Luo Y, Zhou Q (2021) Development and characterization of an edible chitosan/zein-cinnamaldehyde nano-cellulose composite film and its effects on mango quality during storage. LWT-Food Sci Technol 140:110809. https://doi.org/10.1016/j.lwt.2020.110809

Xiao Z, Aftab TB, Li D (2019) Applications of micro-nano bubble technology in environmental pollution control. Micro and Nano Letters 14:782–787. https://doi.org/10.1049/mnl.2018.5710

Xin Y, Chen F, Lai S, Yang H (2017) Influence of chitosan-based coatings on the physicochemical properties and pectin nanostructure of Chinese cherry. Postharvest Biol Technol 133:64–71. https://doi.org/10.1016/j.postharvbio.2017.06.010

Xing Y, Yang H, Guo X, Bi X, Liu X, Xu Q, Wang Q, Li W, Li X, Shui Y, Chen C, Zheng Y (2020) Effect of chitosan/Nano-TiO_2 composite coatings on the postharvest quality and physicochemical characteristics of mango fruits. Sci Hortic 263:109135. https://doi.org/10.1016/j.scienta.2019.109135

Yao KS, Li SJ, Tzeng KC, Cheng TC, Chang CY, Chiu CY et al (2009) Fluorescence silica nanoprobe as a biomarker for rapid detection of plant pathogens. Adv Mat Res 79:513–516

Yang L T, Pan JF, Hu NJ, Chen HH, Jiang HX, Lu YB, Chen LS (2021) Citrus physiological and molecular response to boron stresses. Plants 11(1):40

Zambrano-Zaragoza ML, Quintanar-Guerrero D, Gonzalez-Reza RM, Cornejo-Villegas MA, Leyva-Gomez G, Urban-Morlan Z (2021) Effects of UV-C and edible nano-coating as a combined strategy to preserve fresh-cut cucumber. Polymers 13:3705. https://doi.org/10.3390/polym13213705

Zambrano-Zaragoza ML, Quintanar-Guerrero D, Del Real A, González-Reza RM, Cornejo-Villegas MA, Gutiérrez-Cortez E (2020) Effect of nano-edible coating based on beeswax solid lipid nanoparticles on strawberry's preservation. Coatings 10(3):253

Zhang R, Wang X, Li L, Cheng M, Zhang L (2019) Optimization of konjac glucomannan/carrageenan/nano-SiO2 coatings for extending the shelf-life of *Agaricus bisporus*. Int J Biol Macromol 122:857–865. https://doi.org/10.1016/j.ijbiomac.2018.10.165

Zhou L, Wang J, Zhang H, Chen Y (2011) Effect of PE/Ag_2O nano-packaging on the quality of apple slices. J Food Qual 34(3):171–176. https://doi.org/10.1111/j.1745-4557.2011.00406.x

Chapter 20
Nanomaterials for Enhancing Fruit Yield and Quality

Emad Hamdy Khedr and Nagwa Khedr

Abstract To meet the increasing demand for higher fruit yield and quality, various advanced agricultural techniques have been developed and refined. Among these, nanotechnology stands out as one of the most promising and transformative solutions in modern agriculture, particularly within the field of fruit production. Nanotechnology's potential to revolutionize this sector is especially apparent in the development of nano fertilizers, which have demonstrated remarkable success in stimulating both vegetative and reproductive growth. By enhancing flowering and fruit set, nano fertilizers contribute to significant increases in crop yield and improve fruit quality by boosting nutrient uptake and optimizing metabolic processes within the plant. Beyond boosting productivity, nanotechnology also plays a crucial role in extending the shelf life of fruits and minimizing post-harvest losses, addressing key concerns in food security and supply chains. However, despite these benefits, the precise mechanisms through which nano fertilizers influence fruit trees and their physiological processes remain underexplored. This underscores the need for further research to fully understand their long-term effects and potential environmental impacts. This chapter delves into the expansive applications of nanotechnology in fruit production, emphasizing its roles in enhancing growth, increasing yield, improving plant tolerance to abiotic stress, and supporting pest control. It also highlights the future potential of nanosensors in precision agriculture, especially in fruit quality monitoring and sustainable farming practices.

Keywords Abiotic stress · Nano fertilizers · Nano sensors · Nanotechnology · Quality

---

E. H. Khedr (✉) · N. Khedr
Department of Pomology, Faculty of Agriculture, Cairo University, Giza, Egypt
e-mail: emad.khedr@agr.cu.edu.eg; nagwa.h.khedr@agr.cu.edu.eg

J. M. Al-Khayri et al. (eds.), *Nanotechnology Applications in Modern Agriculture*, Nanotechnology in Plant Sciences 3, https://doi.org/10.1007/978-3-031-90506-3_20

20.1 Introduction

Nanotechnology, a rapidly expanding area of modern science, is attracting interest in various fields, including physics, chemistry, and particularly agriculture. This discipline focuses on manipulating materials at the atomic or molecular level, sizes smaller than 100 nanometers. Often referred to as micro-material technology or microtechnology, nanotechnology focuses on the precise modification of molecules or atoms to engineer new materials with enhanced properties (Ilaria et al. 2022). This technique allows scientists to capture, manipulate, and reconfigure nanoparticles, which can then combine with atoms from other substances to form crystalline structures with improved performance characteristics (Huang et al. 2015).

Nanotechnology has already started to revolutionize key areas, including plant production, and protection, nano fertilizers, precision farming, soil enhancement, water retention and purification, pollutant remediation, nanosensors, plant breeding, genetic modification, and the food industry (Fig. 20.1). Due to the exceptional physicochemical properties of nanomaterials, such as plasmonic, magnetic, electronic, and catalytic capabilities nanostructured materials have become essential components of modern agricultural practices (Nasrollahzadeh et al. 2019). These materials, which can vary in size, shape, and composition, include metal oxides, metals, carbon-based materials (e.g., graphene, nanodiamonds, fullerenes, carbon nanotubes), and both organic and inorganic composites. Their dimensions typically range from 1 to 100 nanometers, and they can take on diverse shapes (Zahedi et al. 2020).

Nano fertilizers, have emerged as a promising innovation in fruit crop nutrition. Due to their smaller size, nano fertilizers can be easily absorbed and translocated within plants, offering efficient nutrient delivery with minimal waste. They can be applied as soil-based or spray-based treatments, providing more precise and effective nutrient distribution assessed with conventional fertilizers (Zahedi et al. 2019a).

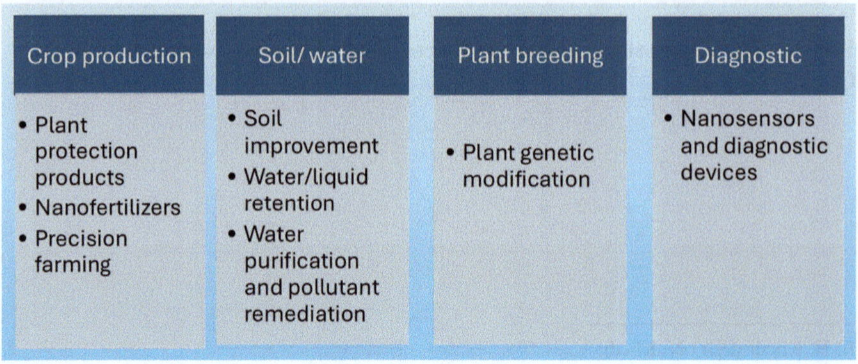

Fig. 20.1 Key efforts of nanotechnology in agriculture. (Constructed by Emad Khedr and Nagwa Khedr)

The integration of nanotechnology in plant production, through nano fertilizers, nano pesticides, and nano sensors, has ushered in a new era of "smart agriculture." This technology reduces the excessive use of chemicals, enables targeted delivery of agricultural inputs, and minimizes harm to non-target organisms (Zahedi et al. 2019b). However, while nanotechnology offers several benefits, concerns over its potential toxicity to plants and ecosystems have also been raised (Dargiri et al. 2024). For example, specific concentrations of metal-based nanoparticles enhance plant growth, seed germination, and pest control, although there are environmental risks associated with their use (Chhipa 2019). The prospects of nanotechnology in agriculture are broadening with advancements in polymer-based nanoparticles and green-synthesized nanomaterials, which offer promising applications. In this chapter, we investigate the diverse applications of nanotechnology in agriculture, including metal nanoparticles, metal oxide nanoparticles, polymer-based nanomaterials, and green-synthesized nanomaterials. We analyze their impacts on plant growth, pest management, pathogen control, and the development of nanosensors for monitoring soil nutrients and ensuring food safety.

20.2 Nano Fertilizers

Fruit yield and quality are influenced by varied factors (Al-Khayri and Khedr 2024a, b), each playing a critical role in determining overall productivity (Fig. 20.2). Genetic factors, such as the plant's genetic makeup, variety, and breeding history, are fundamental to both yield potential and the intrinsic quality of the fruit (Khedr 2021; Khedr 2022a, b). These traits directly influence attributes like size, flavor, and resistance to environmental stress. Environmental conditions, including temperature, light, and rainfall, also play a significant role in plant growth, flowering, fruit set, and overall yield. Soil properties, such as composition, pH, nutrient availability, and water retention capacity, are crucial for healthy plant growth, nutrient uptake, and proper fruit development (Barrios et al. 2016; Rai et al. 2015). Water management is another essential factor; adequate and consistent water supply is vital for optimal fruit growth, while poor irrigation practices can negatively impact both yield and quality. Nutrient management, particularly through balanced fertilization and the application of nano fertilizers, enhances nutrient absorption, leading to improved fruit quantity and quality. Effective pest and disease control, including the use of nano pesticides, minimizes damage from pests and pathogens, contributing to better yield outcomes (Zhu et al. 2023, 2024). Additionally, proper post-harvest practices, such as harvesting, handling, storage, and transportation are necessary to maintain fruit quality and extend shelf life. Finally, the plant's ability to withstand abiotic stressors like drought, salinity, and extreme temperatures, often enhanced through technological interventions, is crucial for sustaining both high yield and fruit quality.

Nanoparticles can penetrate plant cells, as well as through mechanisms like endocytosis (Rico et al. 2011). Their smaller size compared to conventional

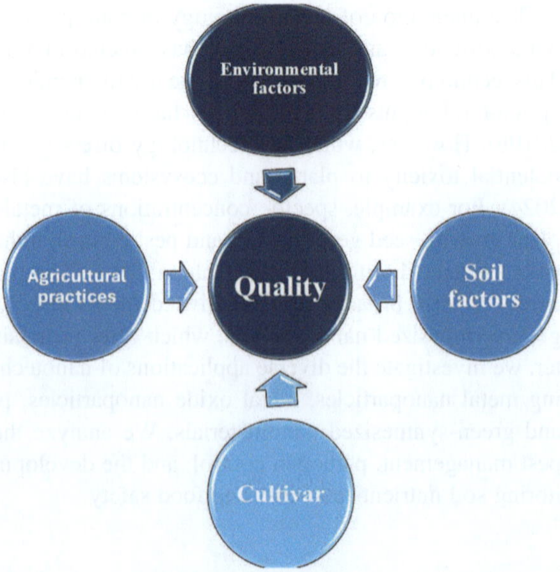

Fig. 20.2 Factors affecting fruit yield and quality. (By Emad Khedr and Nagwa Khedr)

materials, combined with unique shapes, higher penetrability, and a larger surface area-to-weight ratio, enables nanoparticles to have a more profound influence on plant development and development (Khedr and Khedr 2025). They can also enter leaf tissues directly through stomata, enhancing nutrient delivery (Barrios et al. 2016).

The impacts of nanoparticles on trees depend on the dosage and timing of application. While low doses of nanoparticles can promote growth, high quantities may have detrimental effects due to their small size, which increases their solubility and reactivity within plant tissues. To mitigate these risks, nanoparticles are applied in low concentrations. Encapsulation or combining nanoparticles with other beneficial materials has reduced potential harm and enhance yield, as demonstrated in date palm studies (Sabir et al. 2014).

The optimization of nano fertilizer formulations is done with encapsulation techniques to maximize nutrient absorption and minimize waste. There are three main methods for encapsulating nano fertilizers: (1) using NP-porous materials, (2) coating with thin polymer films, and (3) delivering nutrients as particles or emulsions of nanoparticles (Rai et al. 2015). Additionally, surface-modified nanoparticles, such as nanocomposites containing macronutrients like N, P, and K, as well as micronutrients, can prevent nutrient losses by avoiding interactions with soil, microorganisms, water, and air during nutrient uptake by crops like tomatoes (Barrios et al. 2016). This enhances nutrient efficiency and reduces environmental waste.

Nutrients are essential for plants to sustain vital physiological activities such as growth and development, are required in large quantities (Khedr et al. 2023). Nanoparticles are increasingly used alongside these key nutrients to enhance distribution efficiency and improve overall plant performance. However, the overuse of traditional organic and mineral fertilizers has severely impacted soil and water quality worldwide. For instance, the excessive application of conventional nitrogen (N) fertilizers has resulted in significant nitrogen losses to the environment, leading to problems such as eutrophication of water bodies, soil acidification, and reduced biodiversity. Moreover, synthetic nitrogen fertilizers contribute to the emission of N_2O, a potent greenhouse gas exacerbating global warming. Despite various efforts to address these issues, nitrogen use efficiency (NUE) in agriculture remains low, with over 50% of applied nitrogen lost to the environment globally (Govindasamy et al. 2023).

The challenge facing modern agriculture is to sustainably increase food production for a growing global population (Khedr and Al-Khayri 2023; Khedr and Khedr 2023). Improving NUE is crucial, and this can be achieved by adopting eco-friendly agronomic practices, enhancing agricultural system efficiencies, and leveraging innovative technologies like nano fertilizers. Nitrogen-based nano fertilizers show promise in minimizing nitrogen losses while optimizing N delivery to plants, thus contributing to more sustainable agriculture. Studies showed positive impacts of nano fertilizers on crop performance. For example, when blueberries were treated with Nano calcium under saline stress, increased vegetative growth and higher chlorophyll content in the leaves (Sabir et al. 2014). Similarly, the use of Nano boron on mango trees improved overall yield and enhanced leaf nutrient content, including nitrogen, phosphorus, potassium, manganese, magnesium, boron, zinc, and iron (Akl et al. 2019). Boron-based nano fertilizers have also been shown to enhance olive tree growth, resulting in larger fruits, higher yields, and increased oil content, alongside improved concentrations of essential nutrients (Abd-Elall 2018). Likewise, the application of nano zinc on mango trees increased both fruit weight and quantity, further boosting yield, chlorophyll content, and carotene levels in the leaves (Zagzog and Gad 2017).

Research on pomegranate trees treated with nano zinc and nano boron before full flowering revealed improvements in fruit quality, increased fruit number per tree, enhanced phenols, and overall fruit yield (Davarpanah et al. 2016). In palm trees, the use of a nano Zn-Fe-Mn-B solution increased leaf area, total chlorophyll content, and nutrient concentrations, resulting in greater fruit production by reducing precipitation rates and increasing fruit weight (El-Sayed 2018). Nano silicon application to loquat fruits demonstrated its effectiveness in reducing fruit weight loss, maintaining the total soluble solids (TSS) ratio, increasing glucose and fructose content, and enhancing resistance to cold stress. These benefits collectively extended the storage life of loquat fruits in refrigerated conditions, while preserving fruit quality (Song et al. 2016). These findings illustrate the potential of nano fertilizers to enhance crop yield, nutrient content, and postharvest performance, offering promising solutions for sustainable agricultural practices.

Due to their ability to penetrate plant tissues via stomata, nanoparticles can be applied to the branches and leaves of fruit trees, can significantly enhance absorption rates, leading to improvements in both physiological functions and productivity (Zagzog et al. 2017). For instance, foliar sprays of nanoparticles boost nutrient uptake substantially. In one study, the application of boron (B) and zinc (Zn) nanoparticles increased nutrient levels by 19.9% and 60.9% in pomegranate trees (Davarpanah et al. 2016). Similarly, selenium (Se) nanoparticles improved the nutritional status of pomegranate trees, further demonstrating the efficacy of nanoparticle-based fertilization strategies (Zahedi et al. 2019c).

Additionally, advanced nanotechnology has enabled the development of highly effective products like super iron-chelating nanoparticles, that are combined with zinc, iron, and manganese. These nanoparticles contain Fe, Zn, and Mn and characterized by their high absorption efficiency and reduced risk of damage at low concentrations (Kamiab and Zamanibahramabadi 2016). Such studieshighlight the potential of nano fertilizers to revolutionize nutrient delivery in agriculture, optimizing plant health and boosting productivity while minimizing environmental impact.

20.3 The Role of Nanomaterials in Stimulating Growth

Nano fertilizers, compared to conventional bulk fertilizers, offer unique advantages due to their nano-level nutrients, which can easily bond with soil ions and release nutrients in a controlled, slow-release manner when coated with polymers (Shang et al. 2019). These nanoparticles are potent growth stimulators in fruit trees, enhancing plant development (Subramanian et al. 2015). However, research on the innovative applications of nano fertilizer technology in fruit trees remains limited.

One promising nanomaterial is chitosan, a polymer derived from seafood waste. Chitosan has various agricultural applications, including boosting crop yields and improving the quality of agricultural products (Malerba and Cerana 2018; Hosseini et al. 2018). When utilized at 10 mg/L, chitosan nanoparticles produced using a nano-spray drier significantly enhanced seedling growth, photosynthesis rates, chlorophyll content, and nutrient absorption in coffee plants compared to controls (Van et al. 2013). Similarly, chitosan nanoparticles promoted vegetative growth in mango branches and leaves (Zagzog et al. 2017).

In date palms, the application of Zn-Fe-Mn-B nanoparticles at a 0.04% concentration significantly improved various growth parameters, including photosynthetic pigments and nutrient content. The treated trees exhibited higher chlorophyll and carotenoid levels than those treated with conventional ZnFeMnB fertilizers at 0.05% (El-Sayed 2018). Additionally, applying nano-boron (nano-B) at concentrations of 0.1–0.25% and wheat seed sprout extract (0.5–2.0%) positively impacted the size, number of leaves, and growth of pinnae in date palms.

Nanomaterials also show potential in mitigating environmental stress. For example, while heavy metals are toxic to mango, Si nanoparticles have demonstrated protective effects under drought stress (Li et al. 2012). In strawberries, the application of nano-silica (Si-NPs) at 5 mM reduced transpiration rates and specific leaf area while enhancing leaf number (Dehghanipoodeh et al. 2016). This suggests that nanomaterials can significantly enhance growth.

20.4 Enhancing the Yield of Fruit Trees Through Nanomaterials

The application of nanoparticles (NPs) has shown promising effects on the blooming and yield stages of fruit trees (Álvarez-Paino et al. 2017). Although limited studies exist, research suggests that zinc nanoparticles (Zn-NPs) positively influence the blooming process in fruit crops. For example, grapevines treated with calcite nano fertilizers exhibited earlier blooming and better-quality compared to untreated vines, with full bloom occurring a day earlier in treated vines (24.1 days in the control versus 23.1 days in the NF-treated group). Similarly, spraying mango trees with Zn-NPs resulted in an increase in both flower panicles, and ripe fruit production (Zagzog et al. 2017).

Nanoparticles such as silicon (Si-NPs) are effective at mitigating environmental stressors (Cui et al. 2017). In strawberries, the application of nanosilica (SiO_2) enhanced flowering. This treatment accelerated floral organ differentiation, especially during the transition of crown buds from vegetative growth to blooming (Dehghanipoodeh et al. 2018). Furthermore, research on various fruit trees has demonstrated improved fertilization and fruit yield following the application of nanocomposites containing essential macronutrients and micronutrients (Kamiab and Zamanibahramabadi 2016).

In almond trees, a foliar application of iron chelate and ZnFeMnB (ZFM) nanocomposite (2 $g\,L^{-1}$) on three cultivars (Shokufeh, Monagha, and Sahand) resulted in significant yield increases, with treated branches yielding approximately 350 g compared to 120 g in the control group. Also, it reduced fruit abscission by 25% (Kamiab and Zamanibahramabadi 2016). In pomegranates, a foliar spray of Zn-NPs and boron nanoparticles (B-NPs) (6.5 mg L^{-1}) was more effective at increasing fruit yield and fruit count per branch than higher concentrations of the same nutrients. B-NPs had a particularly strong impact on these parameters compared to Zn-NPs and untreated controls (Davarpanah et al. 2016).

Boron shows a vital role in pollen tube elongation, fruit set, sugar translocation, and hormone regulation, which directly impacts fruit yield (Marschner 2012). In pomegranates, spraying selenium nanoparticles increased the number of fruits, fruit diameter, and peel thickness, with significant improvements over the control (Hasan 2015). Similarly, in mango trees, foliar application of a nano fertilizer containing nitrogen, phosphorus, potassium, and magnesium (NF-NPKMg) boosted both fruit production and fruit quality, including increased flesh content and higher vitamin C levels compared to macronutrient sprays alone (Saied 2018).

Nitrogen-containing nano fertilizers have also proven highly effective in boosting fruit yield. For instance, spraying pomegranate branches and leaves with nitrogen nanoparticles resulted in higher fruit yields compared to conventional urea treatments (Davarpanah et al. 2017). The treatment of N-NPs during the full bloom stage and a month later increased leaf nitrogen concentration by 2%, playing a key role in the physiological and metabolic processes necessary for fruit set and flowering (Hasan 2015). The application of nano fertilizers has demonstrated superior results over conventional fertilizers in enhancing the growth, yield, and quality of various fruit crops. For example, when compared to control trees, the use of Zn-NPs on mango shoots and leaves prior to flowering reduced malformation, increased fruit weight, and boosted yield (Refaai 2014). Additionally, chitosan nanoparticles (5 $mL\,L^{-1}$) applied to mango trees increased fruit number per tree by 35.28% and improved fruit quality, both physically and chemically, compared to non-nano fertilizer treatments.

Chitosan nanoparticles also promote enzymatic activity crucial for fruit development (El-Beltagi et al. 2025a, b). For instance, turmeric plants sprayed with chitosan nanoparticles exhibited high protease inhibitors, glucanases, and peroxidase (Anusuya and Sathiyabama 2016). In date palm (cultivar Zaghloul), the use of boron nanoparticles combined with wheat seed sprout extract led to significant improvements in fruit weight pulp percentage. At the same time, undesirable attributes such as acidity, fiber, tannin content, and seed percentage were significantly reduced (Li et al. 2012). This demonstrates how nanoparticles can enhance desirable fruit qualities while minimizing undesirable ones.

Spraying a date palm cultivar (Sakkoti) with a nanocomposite containing zinc, iron, manganese, and boron (ZnFeMnB) also significantly improved both fruit quality and yield, outperforming traditional fertilizers (Jędrszczy and Ambroszczyk 2016). In pomegranates, the application of Zn-NPs and B-NPs led to improved fruit quality, including increased TSS, higher total phenolic content, and decreased titratable acidity. These changes resulted in an enhanced maturity index (TSS/TA), further demonstrating the effect of nanoparticles on fruit quality (Davarpanah et al. 2016).

20.5 Effect of Nanoparticles on Abiotic Stress Resistance of Plants

Plants exhibit multiple defense mechanisms to manage abiotic stress conditions like drought, salinity, and nutrient deficiency (Al-Khayri et al. 2024; Khedr and Khedr 2024a, b). However, these defenses often activate too late, leading to irreversible damage (Rastegar et al. 2024). To mitigate this, nano-stimulators, such as chemical compounds, plant extracts, or nanocomposites, can be applied to enhance plant resilience and productivity in stressful environments (Brown and Saa 2015; Jędrszczy and Ambroszczyk 2016).

Nanotechnology-based solutions, particularly nanocomposites or nano-metals, have been shown to enhance abiotic stress resistance (El-Beltagi et al. 2025c, d) by improving nutrient uptake efficiency (Rastegar and Khedr 2025). Nanocomposites made of macronutrients and micronutrients, such as calcium carbonate ($CaCO_3$), silicon dioxide (SiO_2), and iron oxide (Fe_2O_3), provide essential elements that are more readily available for plant roots. These nano fertilizers reduce nutrient losses to soil, water, and air and promote efficient absorption, thus enhancing plant growth even under stressful conditions (Barrios et al. 2016). For instance, in the grape cultivar *Narince*, grown in calcareous orchards under alkaline stress, the application of nano fertilizers (0.5 $g\,L^{-1}$) containing $Cacao_3$, SiO_2, MgO, and Foo_3 shortened growth stages. Nano fertilizer-treated vines had better performance in terms of fruit weight, and yield. This treatment led to earlier version (26.07 vs. 30.07 in the control) and maturity (25.08 vs. 29.08 in the control), enhancing overall yield and quality (Sabir et al. 2014).

Zinc's role in enhancing stress tolerance is crucial, as it contributes to auxin production by promoting the production of tryptophan, a precursor of indole-3-acetic acid. This in turn promotes better pollen production, berry set, and shoot growth in grapevines, resulting in higher yields and uniformity in berry bunches (Hanafy et al. 2012; Sabir et al. 2014). The application of zinc-based nano fertilizers has been shown to increase the concentration of endogenous zinc, which enhances IAA synthesis, leading to more vigorous shoot and berry growth, reduced undeveloped berries, and improved tolerance to stress factors (Arrobas et al. 2014).

In strawberries, foliar treatment of selenium nanoparticles improved the plants' salinity tolerance by protecting photosynthetic pigments. This treatment resulted in a 12.19% and 13.63% increase in chlorophyll a and b, respectively, and elevated free proline levels compared to non-nanoparticle treatments, contributing to better plant performance under stress (Zahedi et al. 2019a). In pomegranates, Se-NP treatments (1 and 2 μM) significantly reduced fruit cracking caused by drought stress, with 2 μM showing a greater reduction (4.0%) compared to 1 μM (7.2%) and the non-treated control (8.03%) (Hasan 2015).

20.6 Effects of Nanotechnology in Pest Management

Nanotechnology offers promising solutions for pest management by enhancing the effectiveness and sustainability of pest control methods while minimizing environmental impact. One such innovation is the use of nano-gelled pheromones, which provide an eco-friendly alternative for controlling pests like *Bactrocera dorsalis*, a destructive pest of fruits like guava. A nano-gel containing immobilized pheromone can be placed above water in traps, luring fruit flies into plastic bottles where they drown, reducing the need for harmful chemicals. This method also improves the transportation and storage of pheromones without the need for freezing (Vargas et al. 2009; Bhaga et al. 2013). Nanotechnology has also been applied to disease management through the development of nano-fungicides. These are seen as greener alternatives to traditional synthetic fungicides, offering better target specificity and reduced environmental harm. Studies by Malandrakis et al. (2019) demonstrated the antifungal characteristics of metal nanoparticles such as zinc oxide, silver, and copper against significant plant pathogens. Cu-NPs were found to be highly effective in reducing mycelial growth. Ag-NPs showed strong reduction against *Botrytis cinerea*, while CuO-NPs were less effective against fungal species. Nanoparticles exhibited greater toxicity to fungal spores than hyphae, suggesting their potential use as antifungal agents.

Traditional fungicides, including dicarboximides, cupric salts, dithiocarbamates, thiabendazole, and thiocarbamates, often suffer from poor target specificity. Nanotechnology helps address this issue by enhancing the precision of these fungicides. Kumar et al. (2016) developed a nano formulation containing 50% tebuconazole and 25% trifloxystrobin to combat *Macrophominaphaseolina*, demonstrating the potential of nanotechnology to improve fungicide performance and specificity. Nano silver colloidal solutions have shown particular promise in controlling powdery mildew (*Sphaerothecapannosa*) on roses. A 5000 ppm nano silver solution effectively managed this common fungal disease, making it a viable, cost-effective alternative to conventional fungicides. Furthermore, a low-cost copper-zinc nanoparticle complex was developed by Antonoglou et al. (2018) to manage fungal infections such as *Saccharomyces cerevisiae* that impact crops like tomatoes by damaging photosystem II.

Copper nanoparticles and copper oxide nanoparticles are proposed as effective as nano fungicides against a variety of mycotoxigenic fungi. These formulations not only combat fungal pathogens but also provide essential copper for biological processes such as chlorophyll formation, seed development, and rooting. Mali et al. (2020) showed that copper nanoparticles significantly inhibited *Fusarium oxysporum*, demonstrating their potential in managing soil-borne pathogens while promoting plant health.

In agricultural production, using nematodes to control pests is a complicated task. Traditional nematodes used in agriculture include aldicarb, fosthiazate, and oxamyl compounds. These nematodes pose a threat to both humans and the environment due to the rapid release of their toxic active components. Innovative nanotechnology methods have made it possible to address these problems successfully. An outstanding example of this technique is the management of *Malologyne incognita* and *M. graminis* with silver nanoparticles. In addition, green silver nanoparticles have demonstrated greater efficacy in suppressing the nematode of the mango (*M. javanica*) than conventional chemical nematicides (Dzięgielewska et al. 2023).

In any event, there is hope that nano pesticides can overcome these disadvantages. In the presence of the proper NMs, controlled release and gradual degradation of active ingredients can result in long-term efficacy in pest control. Because they may lessen the need for synthetic chemicals and the risks that come with them, the use of nano pesticides is therefore essential for effective and environmentally friendly pest management. Nano pesticides are therefore more successful than conventional ones. The transportation of the dissolved and colloidal nanoparticles may differ from that of regular solutes (Tang et al. 2023).

The use of nematodes to manage agricultural pests has long been a double-edged sword. Traditional nematicides, such as aldicarb, fosthiazate, and oxamyl, effectively control harmful nematodes like *Meloidogyne incognita* and *M. graminis*, but they pose significant threats to both human health and the environment due to the rapid release of toxic compounds. Nanotechnology offers a more sustainable and efficient alternative, addressing these challenges with innovative methods. Silver nanoparticles have emerged as a potent tool in managing nematodes (Khan et al. 2023). They have demonstrated high efficacy against *M. incognita* and *M. graminis*, and green silver nanoparticles have been especially effective in suppressing *M. javanica*, a nematode that attacks mango trees. These green nanoparticles outperform traditional chemical nematicides, offering a safer, eco-friendly alternative (Khan et al. 2023).

However, challenges such as photolysis, hydrolysis, and environmental drift caused by wind or rain remain obstacles to the widespread use of bio-nano nematicides. To counter these issues, the development of nano-encapsulated nematicides is being promoted. These nano-capsulated packages allow for the controlled release of the active ingredients, melting and decomposing within the plant to maintain stability and consistency. This method reduces the number of chemical inputs, shortens application time, and enhances the precision of nematode control. Research into natural nematicides has also shown promise. Extracts from *Paeonia rockii* and *Camellia oleifera* have demonstrated potent nematocidal activity against *M. incognita* in vitro. The stem extract from *P. rockii* and cake extract from *C. oleifera* effectively inhibited egg hatching and showed toxicity against the second juvenile stage (J2) nematodes. The *P. rockii* extract, in particular, showed superior results, inactivating J2 nematodes at a concentration of 5 mg/mL within 7 days (Usman et al. 2020).

While biopesticides provide a more environmentally friendly alternative to synthetic pesticides. Nanopesticides address these shortcomings by offering controlled release and gradual degradation of active ingredients, leading to prolonged pest control efficacy. Nanoparticles enhance the solubility of active compounds, improving their absorption by soil-dwelling microorganisms and reducing the need for frequent applications. This results in significant savings in time and resources. Despite the advantages of nano pesticides, there are concerns regarding their potential health impacts.

20.7 Nanosensors

Nano-sensors are revolutionizing agriculture by providing real-time data on various critical parameters, including pathogen detection, pesticide levels, microbiological conditions, pollutants, and food freshness. By offering unprecedented sensitivity and selectivity, these advanced tools play a crucial role in enhancing crop protection, optimizing nutrient application, and improving agricultural yields (Khedr et al. 2024; Khedr and Al-Khayri 2024).

20.7.1 Applications for Nano-Sensors in Agriculture

1. **Early Detection of Pathogens and Pests**: Nano-sensors can identify the presence of pests, diseases, and viruses at an early stage, preventing large-scale infestations and minimizing the need for chemical interventions. For instance, carbon nanoparticles can detect pesticide residues in plants, providing crucial information to reduce harmful chemical use while maintaining crop health.
2. **Monitoring Crop and Soil Health**: Nano-sensors offer real-time monitoring of crop health, soil nutrient levels, and water conditions. This enables precise application of fertilizers, irrigation, and pest control measures, minimizing waste and environmental damage. As Cheng et al. (2016) highlighted, nanosensors provide a detailed overview of agrochemical risks and environmental conditions, allowing for more informed decision-making in farming practices (EL-Beltagi and Khedr 2025).
3. **Detection of Plant Pathogens**: Nano-sensors have been successfully employed in detecting specific plant pathogens. For example, the bacterial spot disease caused by *Xanthomonas axonopodis* in crops like tomatoes and peppers can be monitored using nano-sensing technology. Additionally, fluorescent silica nanoparticles (NPs) have been used to detect early signs of pathological decay in oranges, particularly in response to damage caused by freezing, ensuring timely interventions to prevent crop loss (Malandrakis et al. 2019).

20.7.2 Precision Farming with Nano-Sensors

Nano-sensors contribute significantly to precision farming by enabling more accurate and timely interventions. They help farmers optimize resource usage, such as water and nutrients, by providing precise data on soil moisture levels and nutrient content. This technology reduces the need for manual monitoring, enhancing efficiency while minimizing labor and costs. Moreover, nano-sensors help lower the environmental impact of farming by reducing agrochemical overuse and preventing pollution (Tong et al. 2023).

20.7.3 Wireless Communication and Remote Monitoring

Networks of nano-sensors can wirelessly transmit data on field environments, enabling remote monitoring of crops throughout the year. This capability facilitates decision-making even in hard-to-reach or large-scale farming areas, further optimizing farm management by allowing real-time responses to changing conditions in the field (El Beyrouthya and El Azzi (2014).

20.8 Conclusion and Prospects

Nanotechnology has shown tremendous promise in enhancing the growth, yield, and overall health of fruit crops, with nano fertilizers emerging as one of the most promising applications. The use of nanomaterials in fruit trees like pomegranate, mango, grapes, and strawberry has led to significant improvements in fruit quality and production, even at low concentrations. These nano fertilizers optimize nutrient uptake and contribute to resistance against environmental and biological stresses, though the long-term effects and mechanisms behind these benefits are still not fully understood. Despite these advancements, there are risks associated with higher concentrations of nanoparticles, which could lead to toxicity, as evidenced in isolated cases. Moreover, the variability in species and cultivar responses to nanoparticles highlights the need for more targeted research to understand their full potential and limitations.

Future research should be focused on several key areas to maximize the benefits of nanotechnology in fruit production. More studies are required to assess the effects of nanomaterials across a broader range of fruit species and cultivars to account for their diverse responses. This will help identify the most effective nano fertilizer formulations and application methods for specific crops. Additionally, research should aim to uncover the mechanisms through which nanoparticles enhance nutrient uptake and resistance to stresses, both environmental and biological. Furthermore, long-term studies are necessary to evaluate the safety and potential toxicity of nanoparticles, particularly at higher concentrations, to ensure sustainable agricultural practices. As the field progresses, innovations like nano-based biocontrol agents and nanosensors will likely play a significant role in improving fruit quality, yield, and resilience, contributing to more sustainable and efficient agricultural systems.

References

Abd-Elall EH (2018) Effect of macro nutrients and Nano-boron foliar application on vegetative growth, yield and fruit quality of Manzanillo olive. Alex Sci Exch J 39:394–400

Akl AMMA, Ahmed YM, Mohamed AZ (2019) Response of keitte mango trees to spray boron prepared by nanotechnology technique. New York Sci J 12(6):48–55

Al-Khayri JM, Khedr EH (2024a) Augmenting apricot fruit quality and prolonging storage life via modulating ethylene metabolism and antioxidants with 1-(3-phenyl-propyl) cyclopropene and psyllium treatments. J Hortic Sci Biotechnol:1–17. https://doi.org/10.1080/14620316.2024.2338130

Al-Khayri JM, Khedr EH (2024b) Enhancing peach quality and extending storability by regulating ethylene metabolism and enzymes using 1-(3-phenyl-propyl) cyclopropene and salicylic acid. J Hortic Sci Biotechnol:1–16

Al-Khayri JM, Abdel-Haleem M, Khedr EH (2024) Harnessing GABA Pathways to Improve Plant Resilience Against Salt Stress. Hortic 10(12):1296. https://doi.org/10.3390/horticulturae10121296

Álvarez-Paino M, Muñoz-Bonilla A, Fernández-García M (2017) Antimicrobial polymers in the nano-world. Review. Nanomaterials 7(2):1–44

Antonoglou O, Moustaka J, Adamakis IDS, Sperdouli I, Pantazaki AA, Moustakas M, Dendrinou-Samara C (2018) Nanobrass CuZn nanoparticles as foliar spray nonphytotoxic fungicides. ACS Appl Mater Interfaces 10(5):4450–4461

Anusuya S, Sathiyabama M (2016) Effect of chitosan on growth, yield and curcumin content in turmeric under field condition. Biocat Agric Biotechnol 6:102–106

Arrobas M, Ferreira IQ, Freitas S, Verdial J, Rodrigues MÂ (2014) Guidelines for fertilizer use in vineyards based on nutrient content of grapevine parts. Sci Hortic 172:191–198

Barrios AC, Rico CM, Trujillo-Reyes J, Medina-Velo IA, Peralta-Videa JR, Gardea-Torresdey JL (2016) Effects of uncoated and citric acid coated cerium oxide nanoparticles, bulk cerium oxide, cerium acetate, and citric acid on tomato plants. Sci Total Environ 563–564:956–964

Bhaga D, Samanta SK, Bhattacharya S (2013) Efficient management of fruit pests by pheromone nanogels. Sci Rep 3:1294

Brown P, Saa S (2015) Biostimulants in agriculture. Front Plant Sci 6:671

Cheng L, Zhang K, Zhou CC, Weir MD, Zhou XD, Xu HH (2016) One-year water-ageing of calcium phosphate composite containing nano-silver and quaternary ammonium to inhibit biofilms. Int J Oral Sci 8(3):172–181

Chhipa H (2019) Applications of nanotechnology in agriculture. In: Methods in microbiology, vol 46. Academic, pp 115–142

Cui J, Liu T, Li F, Yi J, Liu C, Yu H (2017) Silica nanoparticles alleviate cadmium toxicity in rice cells: mechanisms and size effects. Environ Pollut 228:363–369

Dargiri SA, Rastegar S, Khedr EH (2024) Antagonistic microorganism technology for horticultural produce. In: Sustainable postharvest technologies for fruits and vegetables. CRC Press, pp 431–441

Davarpanah S, Tehranifar A, Davarynejad G, Abadía J, Khorassani R (2016) Effects of foliar applications of zinc and boron nano-fertilizers on pomegranate (*Punica granatum* cv. Ardestani) fruit yield and quality. Sci Hortic 210:57–64

Davarpanah S, Tehranifar A, Davarynejad G, Aran M, Abadía J, Khorassani R (2017) Effects of foliar nano-nitrogen and urea fertilizers on the physical and chemical properties of pomegranate (*Punica granatum* cv. Ardestani) fruits. HortScience 52:288–294

Dehghanipoodeh S, Ghobadi C, Baninasab B, Gheysari M, Bidabadi SS (2016) Effects of potassium silicate and nanosilica on quantitative and qualitative characteristics of a commercial strawberry (*Fragaria* × *ananassa* cv. 'Camarosa'). J Plant Nutr 39:502–507

Dehghanipoodeh S, Ghobadi C, Baninasab B, Gheysari M, Bidabadi SS (2018) Effect of silicon on growth and development of strawberry under water deficit conditions. Hortic Plant J 4:226–232

Dzięgielewska M, Skwiercz A, Wesołowska A, Kozacki D, Przewodowski W, Kulpa D (2023) Effects of silver, gold, and platinum nanoparticles on selected nematode trophic groups. J Hortic Res 31(2):23–34

EL-Beltagi HS, Khedr EH (2025) Biosensors and nanosensors: Innovative tools for precision fruit quality assessment. J Food Compos Anal 146:107957. https://doi.org/10.1016/j.jfca.2025.107957

El-Beltagi HS, Abdel-Haleem M, Al Saikhan MS, Khedr EH (2025a) Optimization of progesterone treatment via response surface methodology to maintain quality and mitigate chilling injury in Prunus persica L. NZ J Crop Hort Sci 1–19. https://doi.org/10.1080/01140671.2025.2458819

El-Beltagi HS, Abdel-Haleem M, Adel AR, Khedr EH (2025b) Progesterone as a plant physiochemical regulator: Mechanisms and efficacy in alleviating abiotic stress. J Crop Health 77:60. https://doi.org/10.1007/s10343-025-01125-9

El-Beltagi HS, Abdel-Haleem M, Al Saikhan MS, Shalaby TA, Mohamed AA, Khedr EH (2025c) Antibacterial coatings: pioneering solutions for maintaining fruit quality and prolonging storability. J Hortic Sci Biotechnol 1–25. https://doi.org/10.1080/14620316.2025.2476660

EL-Beltagi HS, Abdel-Haleem M, Khedr N, Shalaby TA, El-Mogy MM, Khedr EH (2025d) Millets as climate-resilient crops: Emerging technologies for sustainable postharvest management. J Food Compos Anal 147:107999. https://doi.org/10.1016/j.jfca.2025.107999

El Beyrouthya M, El Azzi D (2014) Nanotechnologies: novel solutions for sustainable agriculture. Adv Crop Sci Technol 2(03):8863

El-Sayed EM (2018) Effect of spraying some micronutrients via normal versus nano technology on fruiting of Sakkoti date palms. Research 10:39–43

Govindasamy P, Muthusamy SK, Bagavathiannan M, Mowrer J, Jagannadham PTK, Maity A et al (2023) Nitrogen use efficiency—a key to enhance crop productivity under a changing climate. Front Plant Sci 14:1121073

Hanafy Ahmed AH, Khalil MK, Abd El-Rahman AM, Nadia AMH (2012) Effect of zinc, tryptophan and indole acetic acid on growth, yield and chemical composition of Valencia orange trees. J Appl Sci Res 8:901–914

Hasan S (2015) A review on nanoparticles: their synthesis and types. Res J Recent Sci 4:1–3

Hosseini MS, Zahedi SM, Abadía J, Karimi M (2018) Effects of postharvest treatments with chitosan and putrescine to maintain quality and extend shelf-life of two banana cultivars. Food Sci Nutr 6:1328–1337

Huang, S., Wang, L., Liu, L., Hou, Y., & Li, L. (2015) Nanotechnology in agriculture, livestock, and aquaculture in China. A review. Agronomy for Sustainable Development, 35, 369-400

Ilaria C, Iole V, Francesco T, Carlo P (2022) Environmental safety of nanotechnologies: the eco-design of manufactured nanomaterials for environmental remediation

Jędrszczy E, Ambroszczyk A (2016) The influence of NANO-GRO® organic stimulator on the yielding and fruit quality of field tomato (*Lycopersicon esculentum* Mill.). Folia Hortic 28:87–94

Kamiab F, Zamanibahramabadi E (2016) The effect of foliar application of nano-chelate super plus ZFM on fruit set and some quantitative and qualitative traits of almond commercial cultivars. J Nuts 7:9–20

Khan MR, Haroun SA, Rizvi TF (2023) Novel nanomaterials and nanoformulations for nematode management in agricultural crops. In: Novel biological and biotechnological applications in plant nematode management. Singapore, Springer Nature Singapore, pp 227–243

Khedr EH (2021) Does the quality of pomegranate (*Punica granatum* L., cv. 116) disparity in response to different coatings during prolonged storage? Plant Cell Biotechnol Mol Biol 22(53–54):135–155

Khedr EH (2022a) Application of different coating treatments to enhance storability and fruit quality of pomegranate (*Punica granatum* L., cv. Wonderful) during prolonged storage. Rev Bras Frutic 44:e-855

Khedr EH (2022b) Improving the quality of 'Keitt'Mango using pre-harvest calcium lactate treatment and post-harvest applications of kojic and ascorbic acids. J Hortic Sci Biotechnol 97(6):757–772

Khedr EH, Al-Khayri JM (2023) Synergistic effects of tragacanth and anti-ethylene treatments on postharvest quality maintenance of mango (Mangifera indica L.). Plan Theory 12(9):1887

Khedr EH, Al-Khayri JM (2024) Biosensors and nanosensors for determination of harvest quality parameters and fruit handling processes monitoring. In: Innovative methods in horticultural crop improvement. Springer Press

Khedr EH, Khedr N (2023) Optimization of postharvest progesterone treatment to alleviate chilling injury in mango fruit, maintaining intracellular energy, cell wall stability, and antioxidant activity. Postharvest Biol Technol 206:112572

Khedr EH, Khedr N (2024a) Unraveling the mechanisms behind salt stress alleviation and enhanced fruit storability in pomegranate through triacontanol treatment. Sci Hortic 325:112634

Khedr EH, Khedr N (2024b) Enhancing productivity, modifying biochemical parameters, and regulating the phenylpropanoid pathway in 'Le-Conte' pears through optimal protocatechuic acid treatments. BMC Plant Biol 24(1):50

Khedr EH, Khedr N (2025) Nanotechnology in Edible Coatings: A Novel Strategy for Fruit Quality Preservation. In: Al-Khayri JM, Yatoo AM, Jain SM, Penna S (eds) Handbook of agricultural technologies. Springer, Singapore. https://doi.org/10.1007/978-981-99-0862-2_63-1

Khedr EH, Khedr N, Abdel-Haleem M (2023) Harnessing the metabolic modulatory and antioxidant power of 1-(3-Phenyl-Propyl) cyclopropane and melatonin in maintaining mango fruit quality and prolongation storage life. BMC Plant Biol 23(1):464

Khedr EH, Al-Qahtani WH, Khedr N (2024) Biosensors and nanosensors for determination of fruit safety. In: Innovative methods in horticultural crop improvement. Springer Press

Kumar GD, Natarajan N, Nakkeeran S (2016) Antifungal activity of nanofungicideTrifloxystrobin 25%+ Tebuconazole 50% against Macrophominaphaseolina. Afr J Microbiol Res 10(4): 100–105

Li C, Adamcik J, Mezzenga R (2012) Biodegradable nanocomposites of amyloid fibrils and graphene with shape-memory and enzyme-sensing properties. Nat Nanotechnol 7:421–427

Malandrakis AA, Kavroulakis N, Chrysikopoulos CV (2019) Use of copper, silver and zinc nanoparticles against foliar and soil-borne plant pathogens. Sci Total Environ 670:292–299

Malerba M, Cerana R (2018) Recent advances of chitosan applications in plants. Polymers 118:1–10

Mali SC, Dhaka A, Githala CK, Trivedi R (2020) Green synthesis of copper nanoparticles using Celastrus paniculatus Willd. Leaf extract and their photocatalytic and antifungal properties. Biotechnol Rep 27:e00518

Marschner H (2012) Mineral nutrition of higher plants. Academic Press Limited, Harcourt Brace and Company, pp 347–364

Nasrollahzadeh M, Sajadi MS, Atarod M, Sajjadi M, Isaabadi Z (2019) Types of nanostructures. In: An introduction to green nanotechnology, interface science and technology. Elsevier Science Publishers, Amsterdam, pp 29–80

Rai M, Ribeiro C, Mattoso L, Duran N (2015) Nanotechnologies in food and agriculture. Springer International Publishing, p 347

Rastegar S, Khedr EH (2025) GABA in relation to cold and chilling stress. In: Singh S, Tripathi DK, Singh VP (eds) GABA in plants. Plant development, and food security. Hoboken, NJ: John Wiley & Sons, pp 243–262. https://doi.org/10.1002/9781394217786.ch14

Rastegar S, Dargiri SA, Khedr EH (2024) Multilayer edible coatings technology for fresh fruits and vegetables. In: Sustainable postharvest technologies for fruits and vegetables. CRC Press, pp 386–397

Refaai MM (2014) Response of Zaghloul date palms grown under Minia region conditions to spraying wheat seed sprout extract and nano-boron. Stem Cell 5:22–28

Rico CM, Majumdar S, Duarte-Gardea M, Peralta-Videa JR, Gardea-Torresdey JL (2011) Interaction of nanoparticles with edible plants and their possible implications in the food chain. J Agric Food Chem 59:3485–3498

Sabir A, Yazar K, Sabir F, Kara Z, Yazsci MA, Goksu N (2014) Vine growth, yield, berry quality attributes and leaf nutrient content of grapevines as influenced by seaweed extract (*Ascophyllum nodosum*) and nanosize fertilizer pulverizations. Sci Hortic 175:1–8

Saied HHM (2018) Response of keitte mango trees to spraying nano NPK Mg fertilizers. Researcher 10:1–5

Shang Y, Hasan MK, Ahammed GJ, Li M, Yin H, Zhou J (2019) Applications of nanotechnology in plant growth and crop protection: a review. Molecules 24(13):2558

Song HW, Yuan WM, Jin P, Wang W, Wang XF, Yang LM et al (2016) Effects of chitosan/nano-silica on postharvest quality and antioxidant capacity of loquat fruit during cold storage. Postharvest Biol Biotechnol 119:41–48

Subramanian KS, Manikandan A, Thirunavukkarasu M, Sharmila Rahale C (2015) Nano fertilizers for balanced crop nutrition. In: Rai M, Ribeiro C, Mattoso L, Duran N (eds) Nanotechnologies in food and agriculture. Springer, pp 69–80

Tang Y, Zhao W, Zhu G, Tan Z, Huang L, Zhang P et al (2023) Nano-pesticides and fertilizers: solutions for global food security. Nanomaterials 14(1):90

Tong SW, Goh WP, Jiang C (2023) Recent advances in nanosensors for precision agriculture. J Electrochem Soc 170(12):127507

Usman M, Farooq M, Wakeel A, Nawaz A, Cheema SA, ur Rehman H et al (2020) Nanotechnology in agriculture: current status, challenges and future opportunities. Sci Total Environ 721:137778

Van SN, Minh HD, Anh DN (2013) Study on chitosan nanoparticles on biophysical characteristics and growth of robusta coffee in green house. Biocat Agric Biotechnol 2:289–294

Vargas RI, Piñero JC, Mau RFL, Stark JD, Hertlein M, Mafra-Neto A (2009) Attraction and mortality of oriental fruit flies to SPLAT-MAT-methyl eugenol with spinosad. Entomol Exp Appl 131:286–293

Zagzog OA, Gad MM (2017) Improving growth, flowering, fruiting and resistance of malformation of mango trees using nano-zinc. Middle East J Agric Res 6:673–681

Zagzog OA, Gad MM, Hafez NK (2017) Effect of nano-chitosan on vegetative growth, fruiting and resistance of malformation of mango. Trends Hortic Res 7:11–18

Zahedi SM, Abdelrahman M, Hosseini MS, Hoveizeh NF, Tran LSP (2019a) Alleviation of the effect of salinity on growth and yield of strawberry by foliar spray of selenium-nanoparticles. Environ Pollut In Press, Hoveizeh NF and Tran LSP

Zahedi SM, Hosseini MS, Daneshvar Hakimi Meybodi N, Teixeira da Silva JA (2019b) Foliar application of selenium and nano-selenium affects pomegranate (*Punica granatum* cv. MalaseSaveh) fruit yield and quality. Afr J Bot 124:350–358

Zahedi SM, Karimi M, Teixeira da Silva JA (2020) The use of nanotechnology to increase quality and yield of fruit crops. J Sci Food Agric 100:25–31. https://doi.org/10.1002/jsfa.10004

Zahedi SM, Karimi M, Teixeira da Silva JA (2019c) The use of nanotechnology to increase quality and yield of fruit crops. J Sci Food Agric.https://doi.org/10.1002/jsfa.10004

Zhu M, Ma W, Deng J, Liu J, Kang L, Yu J (2023) Effect of nano-calcium on carotenoid and anthocyanin contents of nectarine fruit. Plant Physiol Biochem 204:108088

Zhu M, Li J, Liu Y et al (2024) Preharvest nano-calcium reduces the table grape berry abscission by regulating ethylene production during storage. J Plant Growth Regul 43:1400–1409. https://doi.org/10.1007/s00344-023-11192-9

Chapter 21
Nano Fertilizers Influence on Plant Physiological Activities

Ghulam Abbas, Muneeba, Muhammad Asif Jamal, Kaleemul Din, Ilkay Yavas, Safdar Ali, Asif Ameen, Saddam Hussain, and Mustafa Surmen

Abstract The indiscriminate and imbalanced application of chemical fertilizers leads to significant environmental and ecological consequences with a damage to topsoil profile. Furthermore, the anticipated population growth necessitates alternative solutions to ensure continuous food supply. Nanotechnology presents a capable approach to enhance crop yields while reducing conventional chemical fertilizer use. The properties of nano-materials (such as size, charge and morphology) and plant physiology are key factors influencing plant growth. Extensive scientific research has demonstrated the potential of nanoparticles to improve plant development by modulating physiological processes and metabolic pathways. These nanoparticles have been shown to regulate photosynthesis, nutrient concentrations, and enzyme activities under normal and stress conditions. Additionally, nano-fertilizers can facilitate the slow and sustained release of nutrients, thereby improving nutrient use efficiency in crop plants. However, the technical advancements in

G. Abbas
Department of Agronomy, University of Agriculture, Faisalabad, Pakistan

Agronomic Research Institute, Ayub Agricultural Research Institute, Faisalabad, Pakistan
e-mail: ghulam.abbas@aari.punjab.gov.pk

Muneeba · M. A. Jamal · S. Ali · S. Hussain
Department of Agronomy, University of Agriculture, Faisalabad, Pakistan

K. Din
Department of Botany, University of Agriculture, Faisalabad, Pakistan

I. Yavas (✉)
Koçarlı Vocational School, Department of Crop and Animal Production, University of Aydın Adnan Menderes, Aydın, Turkey
e-mail: iyavas@adu.edu.tr

A. Ameen
Agronomic Research Institute, Ayub Agricultural Research Institute, Faisalabad, Pakistan

M. Surmen
Department of Field Crops, Faculty of Agriculture, University of Aydın Adnan Menderes, Aydın, Turkey
e-mail: mustafa.surmen@adu.edu.tr

J. M. Al-Khayri et al. (eds.), *Nanotechnology Applications in Modern Agriculture*, Nanotechnology in Plant Sciences 3, https://doi.org/10.1007/978-3-031-90506-3_21

nanotechnology must be accompanied by scientific updates for regulations and legal outlines to facilitate their broader commercialization. While nano-fertilizers offer a promising strategy to improve plant nutrition, growth and development, their application in sustainable agriculture remains a significant challenge. This chapter highlights the comprehensive review of nanofertilizers influence on plant physiological activities (including photosynthesis, plant water relations, nutrients uptake, translocation, enzymatic activity and carbohydrate metabolism), plant defense mechanisms, stress responsive gene expression and protecting integrity of cell membranes / cellular organelles against multiple abiotic stresses (e.g. salt stress, drought stress and heavy metal stress) with aemphasis on current progress in nanofertilizers research and explores their future prospects.

Keywords Food security · Nanofertilizer · Nutrient use efficiency · Plant physiology · Stress conditions

21.1 Introduction

The rapid increase in population is raising the concerns regarding food security. The extensive use of the conventional fertilisers to increase the agricultural productivity is disturbing the natural ecosystem by contaminating the soil, water, and environment. One of the healthier alternatives to ascertain the food security-fulfilling the food requirement of the future generations without compromising the ecosystem health- is the use of the nanofertilizers (Channab et al. 2024). Nano fertilizers boast distinctive physicochemical characteristics, rendering them superior to conventional chemical fertilizers. Nanofertilizers (NFs) encompass nutrients encased or coated within nanomaterial, facilitating controlled release and gradual diffusion of nutrients into the soil. With particle dimensions typically between 1 and 100 nanometers, nanofertilizers possess the capability to infiltrate plant systems effectively, whether applied as foliar spray or at the base (Seleiman et al. 2020, 2021). Due to their extremely minute size, NFs exhibit a notable increase in surface-to-volume ratio and surface area, thereby enhancing nutrient retention, absorption and utilisation efficiency in contrast to conventional bulkier chemical fertilizers (Hussain et al. 2022). This enables NFs to release nutrients progressively in accordance with crop requirements, without any environmental hazard (Siddiqi and Husen 2017). This chapter endeavours to reveal potential advantages of nanoparticles by highlighting its impacts on different plant physiological processes. Although, a significant number of studies related to the use of nanofertilizers in variety of crop species have been conducted; yet, response of plant physiological attributes to nanofertilizers application is inadequate and needs to be explored. By thoroughly scrutinizing how nanofertilizers affect essential physiological processes including photosynthesis, nutrient absorption, water regulation, growth, and stress management, readers will acquire valuable insights into the intricate interplay between nanoparticles and plant biology. Ultimately, this chapter aims to highlight the multifaceted connection between nanofertilizers and plant physiological processes to enhance our comprehension of how nanofertilizers can be utilized to enhance plant vitality, increase crop yields, ensuring sustainability and promote agricultural sustainability amidst ongoing global climate change in future.

21.2 Nanofertilizers and Plant Physiological Attributes

21.2.1 Role in Photosynthesis

Nanofertilizers have considerable potential to enhance the photosynthesis in crop plants and boosting crop productivity by various means either by protecting the photosynthetic apparatus, improving photosynthetic pigments, increased carbon dioxide fixation and balanced plant metabolism (Fig. 21.1). Iron, one of the essential micro nutrients, owing to its oxidation- reduction properties regulate many physiological and biochemical pathways in plants. Thus, iron significantly influences the DNA synthesis, respiration, photosynthesis ultimately final yield of plant (Finazzi et al. 2015; Kobayashi et al. 2018).

Iron chelated nano fertilizers present a potent and dependable source of divalent iron for plants owing to their exceptional stability and controlled iron release across a broad pH spectrum. One notable benefit of this nanofertilizer is its exclusion of ethylene ($H_2C{=}CH_2$) compounds in its composition. Ethylene facilitates the growth and mitigates the appearance of chlorosis symptoms in leaves (Monsef-Afshar et al. 2013). Iron plays a fundamental role in chlorophyll synthesis, photosynthesis and various enzyme systems essential for plant respiration. Furthermore, a prominent

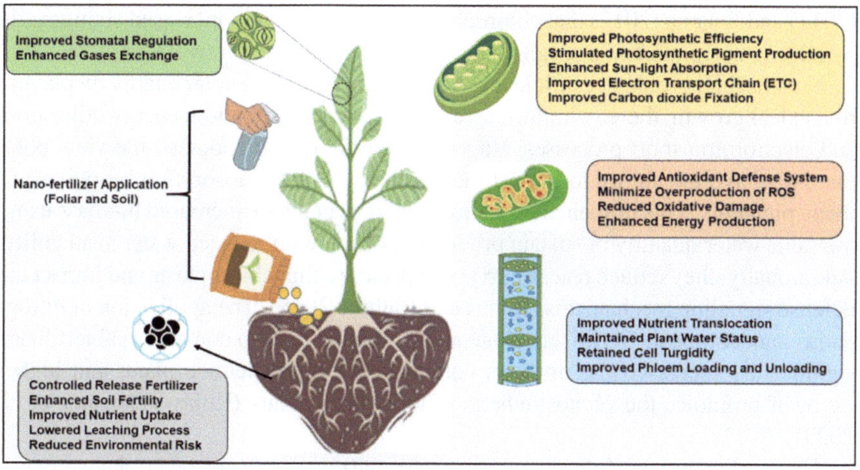

Fig. 21.1 Plant physiological attributes as influenced by nanoparticles. Key: ETC = Step involved in the conversion of glucose into energy, ROS = Unstable and highly reactive molecules carrying oxygen. (Figure prepared by Ghulam Abbas)

manifestation of iron deficiency is the reduction in photosynthetic pigment content, causing an imbalance between carotenoids and chlorophylls, resulting in the characteristic yellowing observed in chlorotic leaves (Pirzad 2010). Various nanohematite/nanoferrihydrite formulations significantly enhance the iron concetration- reducing the iron deficiency- lead to elevated leaf chlorophyll contents, plant photosynthetic activities, its growth and in turn the final yield (Thiruvengadam et al. 2024). Nadi et al. (2013) affirmed through his study that applying nano-iron fertilizer can improve iron concentration in grains, leaf chlorophyll, grain protein (%) as iron is pivotal in chlorophyll synthesis and ultimately photosynthesis due to its essentiality in enzyme systems and pathways involved in plants respiration.

Dey et al. (2023) investigated the positive role of foliar applied green synthesised bimetallic nanoparticles; and revealed that the NFs significantly improve the germination, growth, photosynthetic rate and overall yield of rice crop by maintaining the structural integrity without any adverse effect on environment. Foliar application of NFs with different concentrations of Calcium (Ca) and Magnesium (Mg) significantly increases the nodulations in groundnut plants (Singh et al. 2023). There was a notable disparity in yield between plots treated with nano-potassium and muriate of potash (MOP), with the nano-fertilizer resulting in an 8% yield increase over MOP at a rate of 20 kg potassium oxide per hectare (Rezaei and Abbasi 2014). Mir et al. (2015) found that nanofertilizers and manure in combination had significant positive effect on chlorophyll contents (a & b) carotenoids and carbohydrates. Nano-fertilizers positively affected the chlorophyll concentration as compared to control treatment where no NFs were applied, witnessed in broad bean, where sulfur and zinc applications showed significantly promoting effect on the chlorophyll content (Ghidan et al. 2020). Both micronutrients are also necessary in enzymes activation for chlorophyll formation existent in photosynthetic tissues. The same findings were also described by Karthick and Chitrakala (2011) and Salama (2012) that chlorophyll concentration was increased significantly by silver -nanoparticles in sorghum and common bean, respectively.

Nanofertilizers enhance efficiency of plants to capture the solar energy by playing role in leaf growth; thereby improving chloroplast carbon fixation, energy utilisation, and electron transport processes. Nayeri et al. (2023) also endorsed the view point that NFs have the ability to impact the efficiency of light absorption by photosynthetic pigments. This influence may allow plants to generate increased biomass using the same water quantity or sustain productivity even with reduced water availability. Additionally, they reduce reactive oxygen species within chloroplasts and impact the defense signaling mechanisms (Khatri and Rathore 2018). The application of nitrogenous nanocomposite fertilizers alone and in combination to conventional fertilizers significantly increased the nitrogen concentration in the lettuce plant, and higher nitrogen promotes the photosynthetic efficiency of plants (Rojas-Velázquez et al. 2023).

The application of ZnO nanofertilizer to maize and cluster bean plants significantly enhanced the photosynthetic pigments (chlorophyll a, b, carotenoids) in both soil and foliar applications (Azam et al. 2022; Raliya and Tarafdar 2013). An increase of 276.2% in chlorophyll contents was observed for cluster bean at application rate of 10 mg/L versus control. Zinc play a vital role in plant metabolism, transcription and translation factors (Shaw et al. 2020), as zinc-dependent enzymes are present in DNA and RNA polymerases, as well as chloroplast and mitochondrial RNA editing enzymes (Eide 2020). Furthermore, zinc also acts as cofactor in chloroplast for carbonic anhydrase to uplift carbon dioxide levels (Saini et al. 2021). So, deficiency of zinc hampers the photosynthetic process in maize plant.

Nanoparticles, when applied through foliar application, can cause a higher inclusion of encapsulated nutrients through tissues (Monreal et al. 2015). These nutrients are used in protein and/or other enzymes synthesis intervening Calvin cycle (Elsheery et al. 2020), enhancing carboxylation efficiency (Morales-Díaz et al. 2017), pigments concentrations (Raliya et al. 2017; Wang et al. 2020). Furthermore, by the deposition of nanoparticles a dense layer is formed on cuticle in many species and act as barrier reducing water loss between plants surface and the atmosphere and results in water use efficiency (WUE) improvement and net photosynthetic rate. Foliar fertilization of SiO_2-NFs in *Theobroma cacao* L. increased leaf gas exchange attributes and photochemical activity of photosystem-II (PSII) by improving electron transport, photosynthetic rate, photochemical extinction coefficient, total soluble proteins (TSPs) concentration, and leaf N-P-K-Si concentrations (Gómez-Vera et al. 2021).

NFs are more efficient as compared to conventional fertilizers in terms of photosynthetic efficiency of crops, if applied in proper dose and concentration. In near future, with the optimized use of nano-fertilizers in different crops, the agriculture production system will prove to be more economical, eco-friendly, climate smart, and sustainable.

21.2.2 Role in Plant Water Relations

Under the climate change scenerio, water stress resulting from water scarcity signifies a decrease in turgor pressure within plant cells, leading to inhibited growth. Under water stress, an elevation in leaf canopy temperature is observed due to stomata closure at less negative water potential within plant. This can hinder photosynthesis by restricting the CO_2availability within leaves. Application of nano-fertilizers increases the root growth, enhances the water use efficiency, regulating the temperature within plant system, which ultimately maintains the respiration and photosynthetic rate of crop plants under drought stress conditions (Tambussi et al. 2007).

The decline in chlorophyll levels is regarded as an adaptive mechanism in plants cultivated under water scarcity. Chlorophyll Content Index (CCI) notably reduced under moderate to severe water deficits. This reduction in CCI may be attributed to increased chlorophyllase activity, leading to chloroplast structural degradation and unstability of the pigment protein complex. Plants treated with nano-SiO_2 exhibited the highest CCI, followed by those treated with nano-Mn_2O_3, possibly due to the elevated leaf water content in plants treated with nano-fertilizers (Ghassemi-Golezani and Afkhami 2018).

The percentage of green cover in various crops such as safflower, soybean, and chickpea declined with longer irrigation intervals. This reduced ground-green cover percentage due to drought stress can be credited to plant competition for nutrients and water. Leaf area of crop plants in negatively affected under drought stress conditions due to reduction in leaf numbers and less turgor pressure. Diminished turgor potential in leaves and photosynthetic activities under drought stress hinder cell growth and its expansion, resulting in reduced leaf area and green cover (Anjum et al. 2011). This reduction in leaf area on the one side serves as an adaptive mechanism to mitigate higher transpiration rates by minimizeing the radiation absorption due to water deficiency and on otherside decreases the potential yield of crop (Hayatu et al. 2014). Enhancing green canopy cover through foliar application of nanofertilizers has the potential to improve light interception, photosynthesis, plant height, branches and capitols per plant in safflower and ultimately crop yield by lessening the competition among sinks like root and shoot for partitioning of assimilates (Ghassemi-Golezani and Afkhami 2018).

On a molecular scale, nanoparticles (NPs) have the capacity to engage with genetic makeup of plant cells, potentially influencing the gene expression associated with responses to water stress (Al-Khayri et al. 2023). For example, the transport of hormones or genetic material via nanoparticles can assist the crop plants in preserving osmotic balance and maintaining turgor pressure of plant cells at optimal levels, thereby improving their resilience to different stresses like salinity or drought. This mechanism operates by regulating stomatal closing and opening patterns, thus optimizing water utilization. So, the above quoted examples provide evidence that nanofertilizers are supportive in regulating the water balance between plants and their environment.

21.2.3 Role in Nutrients Uptake and Translocation

A significant portion of conventional chemical fertilizers depleted due to runoff, leaching, evaporation, soil microorganisms' degradation and hydrolysis. Consequently, slow-release nano-fertilizers present a viable substitute to mineral fertilizers for enabling plants to uptake a greater portion of nutrients without significant losses due to NPs small size and increased surface area which ultimately improves the crop growth and development. Nano-nutrients can efficiently penetrate the plant leaf (through cuticle, epidermis and stomata) (Wang et al. 2023), activate nutrient pathways (either symplastic or apoplastic) (Avellan et al. 2021; Hong et al. 2021; Su et al. 2019) and achieve high nutrient use efficiency (NUE). This viewpoint was witnessed when nano-Fe treated plants and mineral fertilizers treated plants were compared and showed a notable increase in translocation of nutrients from roots to aerial plant parts (stems, leaves and fruits). Application of nano-Fe significantly raised the concentrations of nutrients such as Mg, Cu, Zn, Na, and Mo in the roots.

Among the plant parts, nutrients concentration shows a decreasing trend from higher in leaves, then followed by stems, then roots and lowest in fruits. Particularly, nano-Fe treatment in tomato demonstrated a higher translocation of nutrients from roots to fruits, with the translocation order being nano iron> chelated Fe > control. Evaluating the percentage nutrients translocation in tomatoes, the nano iron treatment showed a greater increase (16.7% and 3% respectively) in nutrient availability compared to control and chelated iron treatments. This enhancement could be attributed to the indirect outcomes of increased synthesis of leaf chlorophyll (a and b), consequently leading to greater nutrient translocation in tomatoes (Arifur Rahman et al. 2024).

The application of a nano-engineered composite comprising of nutrients (macro and micro), amino acids and mannose promotes the uptake and crop nutrients use (Jinghua 2004). Moreover, synchronization feature of nano-fertilizers in terms of release and the uptake of NPK fertilizer suppresses undesirable losses of these nutrients (DeRosa et al. 2010). Findings of Hemdan et al. (2020) are also in agreement, stating that foliar spray of nano-fertilizers affected the macro nutrients (nitrogen "N", phosphorus "P", potassium "K", calcium "Ca" and magnesium "Mg"), micro nutrients (iron "Fe", zinc "Zn", copper "Cu" and manganese "Mn") uptake in lupine plant at growth stage in comparison to soil applied and mineral foliar NPK treatments. Llanderal et al. (2020) stated that nutrients content in the petiole cells extract is a function of the ion demand level for plant physiological processes, like absorption, assimilation and nutrients storage and it is significantly affected by the application of nano-fertilizers.

Applying phosphatic nano fertilizers to pearl millet crop improve phosphorus use efficiency by reducing P losses possibly due to increase concentration and uptake. Controlled, slow and steady release of nanophosphatic fertilizer may balance and enhance the nitrogen uptake too which ultimately play role in increased protein content and potassium content in grain. It might be due to greater concentration in reactive cell areas which improve the nitrogen translocation that ultimately led to high chlorophyll contents, moisture percentage in leaves, crude protein, plant nutrients and carbohydrates contents (Dhansil et al. 2018).

The cementing and coating of nano-composites have the potential to control the nutrients release from fertilizers. Additionally, nano particles possessing both negatively and positively charged binding sites can adsorb available nitrogen in the soil, thereby reducing various types of losses. This results in enhanced nitrogen uptake by crops and improved nutrient transport to plants (Kaviani et al. 2016). Increased metabolic activity due to nanoparticles absorption in plant cells can provoke root exudation resulting in soil acidification which can cause phosphate desorption through a ligand exchange reaction, altered the adsorption desorption equilibrium and release of readily available phosphorus in soil solution. These results were also witnessed by Soliman et al. (2016) stating that nanophosphatic fertilizer application may expose NPK nutrients to greater surface area and improves their fixation, reduces losses and releasing at later crop growth stages when plants need.

It was also observed that NPK nanofertilizers even at application rates below the recommended doses of traditional chemical fertilizers (NPK) to faba bean resulted in elevated nutrient uptake which in turn enhances dry weight and crop yield (Abou-yuoseff et al. 2022). Nano fertilization promotes agricultural output and bio-source utilization due to enhanced photosynthetic activity, plant growth, metabolism and nutrient transport. Nano-fertilizers have the ability to release their contents in a controlled way, either slowly or rapidly, in response to different stimuli such as heat, drought etc. Studies have indicated that nano-fertilizers can be employed to regulate the gradual nutrients release of from fertilizer granules, thereby enhancing nutrient use efficiency and preventing nutrient ions from being lost or fixed in the medium (Abou-yuoseff et al. 2022).

The interaction between plants and water encompasses the absorption, movement, and water utilization within plants, being vital processes for their growth, development, photosynthesis, and final productivity (Rasheed et al. 2022). NPs have the capability to transport and release nutrients in a regulated fashion, guaranteeing a consistent provision of vital nutrients to plants without excessive supply of water (El-Saadony et al. 2022). Nano fertilizers, comprising essential nutrients encapsulated within nanoparticles, have the ability to augment the accessibility and absorption of nutrients by plants. Owing to their diminutive size, expansive surface area and capability to transport and release nutrients in regulated fashion, NPs have the potential to enhance nutrient transport to plants (Elemike et al. 2019). Nanoparticles have the capacity to modulate plant physiological functions, resulting in improved uptake of nutrients. For example, research indicates that specific nano-fertilizers can stimulate the expression of nutrient transporters, thereby enhancing nutrient absorption particularly in challenging environmental conditions (Junedi et al. 2023; Tan et al. 2017). The application of Selenium NFs has been documented to elevate the selenium concentration in plants, consequently augmenting their nutritional profile (Bano et al. 2021; Thabet and Alqudah 2023). Furthermore, nanoparticles have the capability to stimulate the bioactive compounds and secondary metabolites production, thereby enhancing the qualitative attributes of crops.

Although, uptake and translocation of nanofertilizers in different crop plants has not been properly studied but it is evident now that, with the use of nanofertilizers in agriculture, nutrient uptake, translocation from source to sink, availability, and nutrient use efficiency will be increased subject to optimal doses of various nanofertilizers applied.

21.2.4 Role in Enzymatic Activity

NFs are promising approach to get well balanced plant enzymatic activities. Phosphatase activity (both in alkaline and acidic phases) and dehydrogenase activity increased significantly by the application of mineral based nutrients source (IMBP) in tandem with nanofertilizers (e.g. diammonium phosphate (DAP) and potassium

(K) based) in comparison with recommended doses of phosphorus and potassium alone (Jayara et al. 2023). They also reported an increase in organic carbon and available nutrients after harvest (nitrogen, phosphorus, potassium, sulphur and silicon) of wheat plants by IMBP + nano fertilizers application. NFs can also regulate the enzymatic activities engaged in nitrogen metabolism like nitrate reductase enzyme and glutamate dehydrogenase enzyme (Ghosh and Bera 2021).

Foliar application of nano fertilizers can induce the release of various enzymes by the plant into the rhizosphere, resulting in a significant increase in release of beneficial enzymes such as acid phosphatase, esterase, phytase, alkaline phosphatase, nitrate reductase, aryl sulphatase, urease, lignase, cellulose and hemi cellulase. Nanoparticles enter the plant through shoots, roots and leaves. Once inside, nanoparticles move through the cell sap and trigger various enzyme systems, with many particles potentially agglomerating to form larger particles along the pathway, primarily depositing in the vacuole. Moreover, nanofertilizers have the capacity to boost nutrient utilization efficiency and augment soil fertility in an environmentally friendly manner by regulating the enzymatic activity. They can also prolong the release of nutrients, thereby extending the period of effectiveness of the fertilizer (Tarafdar et al. 2015).

21.2.5 Modulating Carbohydrate Metabolism

Sugar molecules serve not only as metabolic intermediates but also as signaling molecules and cosmo-protectants, aiding plants in adapting to stressful conditions (Sharma et al. 2019). Hence, modulation of sugar metabolism holds promise for enhancing crop plants resilience in challenging environments. Certain nanoparticles, upon application to plants, can participate in pathways associated in sugar metabolism by interacting directly with enzymes and some intermediates. Additionally, NPs have the ability to influence the gene expressions associated with carbohydrates metabolism. This gene-level modulation can either strengthen or attenuate the key enzymes synthesis crucial for sugar metabolism, enabling plants to regulate their metabolic pathways in response to environmental signals (Selvakesavan et al. 2023). Various nano-fertilizers have been explored for their potential to modulate plant sugar metabolic activities. For example, exposure to silver-based nanoparticles (AgNPs) has been shown to augment sugar concentration in certain plants (Kim et al. 2024a). This phenomenon may arise from the oxidative stress induced by silver nanoparticles, prompting plants for sugars accumulation as a defensive mechanism (Abasi et al. 2022). Silicon, recognized for its advantageous effects on plants, when administered in nanoparticles form, can enhance sugar molecules metabolism, particularly under drought conditions. It can elevate enzymes activity and expedite improved sugar metablism in crop plants (Irfan et al. 2023).

21.3 Nanofertilizers and Plant Defense Mechanisms

21.3.1 Nanofertilizers in Salt Stress

Soil salinity poses a significant hazard to the world's major food crops by inhibiting plants' water uptake abilities and impairing overall plant performance. Various negative effects imposed by salinity stress are nutrient imbalances, inhibition in enzymatic activity and restricted growth in the plants. Plants frequently elevate the synthesis of phenolic compounds, secondary metabolites renowned for their pivotal roles in stress management owing to their UV-absorbing, antioxidant, and signalling attributes to cope with salt stress (Lin et al. 2016). Nanoparticles (NPs) have been observed to augment the accumulation of phenols under stress conditions, thus bolstering plant stress tolerance. NPs possess the capability to stimulate phenolic compounds biosynthesis (Rasheed et al. 2022). Recent investigations have highlighted that the administration of silver (AgNPs) and zinc oxide (ZnONPs) can amplify the phenolic compounds production in plants subjected to various abiotic stresses (mostly under drought and salinity) (Kim et al. 2024b; Kumari et al. 2022). Nanoparticles exert regulatory control over the metabolic pathways associated with phenolic compounds (Bontempo et al. 2023). This modulation is accomplished through the manipulation of the pivotal enzymes activities in these pathways, such as phenyl-alanine-ammonia-lyase (PAL) enzyme, which catalyze the initial step in biosynthesis of phenolic compounds. Experimentation have illustrated that NPs like nano-TiO_2 can boost PAL activity, thereby facilitating phenolic compund's metabolic processes (Shah et al. 2021). Moreover, NPs can enhance the storage ability and required mobilization of phenolic compounds, a critical aspect under stress exposure (Geremew et al. 2023). For example, SiO_2NPshave been presented to optimize the phenolic compounds compartmentalization within vacuoles, safeguarding plant cells against potential toxicity and ensuring their availability as needed (Yin et al. 2012). A study carried out on wheat crop under saline conditions revealed that applying zinc and iron nano oxide fertilizers improved the chlorophyll contents with a comparison to control, along with increase in proline, total soluble sugars, anti-oxidant enzymes (catalase, peroxidase and polyphenol oxidase), relative water content (RWC) and ultimately efficiency of PSII (Babaei et al. 2017). Weisany et al. (2012) also documented that the protective role of nano zinc was attributed to its ability to preserve the integrity of plasma membranes, thereby regulating the uptake of toxic ions under salinity stress.

NPs exert influence on stomatal behaviour, a critical determinant in regulating water loss via transpiration. They modulate stomatal aperture dynamics, curbing transpiration rates and preserving water particularly under drought or salinity stresses (Seleiman et al. 2021). This regulatory mechanism sustains plants at optimal hydration levels and strengthens the slogan more crop per drop (Cominelli et al. 2010).

21.3.2 Nanofertilizers and Plant Drought Tolerance

Water is crucial for nutrients transport in plants, and its deficit can lead to drought stress (temporarily or permanently), causing physiological, morphological, and biochemical changes that reduce crop yields. Plants experience drought stress, when rate of transpiration is more than water availability to the roots, commonly in arid tosemi-arid regions worldwide. Under drought conditions, the critical trait for assessing the performance of agricultural crops is water use efficiency (WUE). Drought-induced reductions in intercellular CO_2 (Ci) due to stomatal closure lead to decreased net photosynthesis (Pn). This, combined with reduced pigmentation, ribulose-1,5-bisphosphate carboxylase oxygenase (Rubisco) synthesis and its activity, and diminished electron transport to PS-II, results in photo inhibition (Pagter et al. 2005). Dalal and Tripathy (2012) found that drought also decreases the production of chlorophyll intermediates, such as δ-aminolevulinic acid (δ-ALA). However, the nano-compost application has been effective in enhancing the δ-ALA content and enhancing water content of tissues through increased osmolyte synthesis, ultimately improving chlorophyll levels and photosynthetic performance in tomato plants, thus mitigating the adverse effects of drought (Ahanger et al. 2021).

One of the well-documented functions of abscisic acid (ABA), during periods of drought, involves regulating stomatal opening and closing to mitigate water shortage. Research suggests that silver (Ag) nanoparticles have the capability to induce stomatal closure, potentially through elevating ABA levels. Upon absorption by roots or uptake through leaves, AgNP scan trigger the plant's signaling mechanisms, leading to heightened abscisic acid synthesis and subsequent reduction in transpiration via stomatal closure (Bharath et al. 2021). Increased levels of reactive oxygen species serve as indicators of stress conditions, prompting plants to enhance the antioxidants production and signalling molecules e.g. ABA to alleviate stress (Li et al. 2022). Consequently, silver NPs may indirectly elevate ABA concentrations by promoting reactive oxygen species generation (Abasi et al. 2022).

ZnO nano-fertilizers when applied to maize plants showed positive significant effect on the non-enzymatic antioxidants (ascorbic acid, total flavonoids, total phenolic and anthocyanins) in both soil and foliar applications, which maximizes the plants' ability to cope with oxidative stress-induced risks (Azam et al. 2022). Foliar fertilization of SiO_2-NF when applied to water deficit criollo cocoa clones' results in an increase of photochemical efficiency, while transpiration rate, relative water content (RWC) and stomatal conductance decreased (Zanetti et al. 2016). So, nanofertilizers are applied for successful crop production, fostering plant growth, development and alleviating negative effects of drought stress.

21.3.3 Nanofertilizers Application Under Heavy Metal Toxicity

Heavy metals (HM) are injurious to plants disturbing their physiology, morphology and biochemistry. They diminish chlorophyll content and growth; causes oxidative stress in plants, increased malondialdehyde levels, lesser glutathione content and antioxidants (enzymatic & non enzymatic) activity. To mitigate HM toxicity in plants, NPs employed various strategies: by reducing HM availability in the soil, enhance antioxidant defense system abilities, regulate the expression of HM transport genes, improved physiological activities, and promote the synthesis of defensive substances e.g. root exudates, organic acids, and phytochelatins. The findings of Jalil et al. (2023) underscore the potential of nano zinc (nZn) in alleviating the adverse effects of arsenic (As) accumulation in *Oryza sativa* L. plants. The utilization of nZn effectively mitigated As toxicity, leading to a notable decrease in arsenic contamination in rice seedlings (shoots and roots). Moreover, nZn exhibited the capability to neutralize ROS such asmalondialdehyde(MDA), superoxide anion (O^{-} ^{2}) and hydrogen peroxide (H_2O_2) because unwarranted production of hydrogen peroxide in guard cells provoked the stomata closing in plants, while significantly enhancing chlorophyll content (depicted by SPAD value, which is relative chlorophyll concentration in a leaf and estimated by SPAD meter), gas exchange parameters, antioxidant enzyme activities and fluorescence efficiency (Fv/m) under arsenic induced stress conditions. Adding nZn to plant leaves experiencing arsenic stress resulted in improved chloroplast development, including enhanced starch granules, grana thylakoids, and osmophilic plastoglobuli, along with reduced damage to the nucleolus in mesophyll cells indicating that nano-Zn maintained the chloroplasts structural integrity. Stomata assist as vital conduits for water and gas exchange between plants and environment. Similarly, alterations in the guard cells morphology under arsenic stress might result from the inhibition of particular metabolic pathways essential for sustaining turgor pressure in these cells. nZn effectively mitigated the adverse influence of As stress on stomatal aperture in rice shoots and roots, as evidenced by a notable increase in stomatal opening observed in treated rice plants. The application of nano zinc (nZn) amendments resulted in increased crop biomass and alleviated cadmium (Cd) toxicity by boosting superoxide dismutase, peroxidase and GAT activities, thereby inhibiting reactive oxygen species activities in leaves (Ramzan et al. 2022). So, nanoparticles can play a substantial role in mitigating stress effects more effectively than conventional fertilizers.

21.3.4 Role in Stress-Responsive Gene Expression

Nanoparticles have emerged as remarkable contributor to the enhancement of stress-responsive gene expression like DDF2, AREB and SOS2, particularly those implicated in salt stress responses in plants (Sarkar and Kalita 2023; Sheteiwy et al. 2021;

Spanò et al. 2020). In *Arabidopsis thaliana,* cerium dioxide (CeO_2) based nanoparticles are documented to upregulate gene expression of NHX1 under salinity conditions (Wu et al. 2018) and also in rice (Zhou et al. 2021), thus alleviating the detrimental effects of salinity stress. Likewise, ceria nanoparticles (Zhou et al. 2021) and silver-based nanoparticles (Almutairi 2016) up-regulate the SOS2 gene associated with salt overly sensitive(SOS) pathway for maintaining ion homeostasis. These genes involved in stress responses encoding proteins engaged in diverse processes, encompassing ion transport, signaling cascades, and ROS detoxification. Transcription factors, pivotal proteins governing the expression of other genes, play a crucial role in orchestrating plant responses to stress. NPs possess the ability to modulate the activity of these transcription factors, thereby exerting influence over the expression of stress-responsive genes (Trono and Pecchioni 2022).

Nanoparticles application up-regulates the expression of drought stress responsive genes including GmDREB2, GmRD20A, GmFDL19, GmMYB174, GmNAC11, GmERD1, GmWRKY27 and GmMYB118 (Kandhol et al. 2022). Same findings were witnessed by the application of ZnO-NPs and CuO-NPs in wheat sown under drought stress (Yang et al. 2018). Similarly, chitosan nanoparticles play a vital role in up-regulating the gene expression associated with antioxidant production and biosynthesis of alkaloids in *Catharanthus roseus* grown in drought conditions (Ali et al. 2021). Different stress responsive genes linked with increased alkaloid concentration are strictosidine synthase (STR), deacetylvindoline-4-Oacetyltransferase (DAT), geissoschizine synthase (GS) and peroxidase 1 (PRX1) (Vakilian 2019). Silved based nanoparticles increased the gene expression linked with RD22 proteins, 9-cis carotenoid dioxygenase (NCED3) and IAA while decrease the ACC oxidase 2 and ACC 7 synthase (ACS7) genes, when applied by Pérez-Labrada et al. (2020) on Arabidopsis under drought stress. In kidney beans, proteins responsible for carbohydrates metabolism and nutrients storage are down-regulated while stress responsive protein expression is up-regulated, when applied with ceria-based nanoparticles (Majumdar et al. 2015). Nanoparticles regulate the biosynthesis of abscisic acid (ABA) as well as stomata functioning under drought stress by the increased expression of GmWRKY27 gene (Linh et al. 2020). Expression of GmMYB174, GmWRKY27 and GmMYB118 genes was up-regulated under drought conditions by the application of nanofertilizers which in turn increased the secondary metabolites synthesis, signaling of hormones, lignin content and seed germination (Rushton et al. 2010). Similar findings were observed by Linh et al. (2020) when iron and cobalt nanoparticles was applied under drought stress and they have significant positive effect on GmRD20A gene expression which is responsible for calcium binding. Similarly, application of zinc nanoparticles in wheat crop under drought stress significantly up-regulates the genes (ZIP1, HMA2 and WRKY1) expression associated with transcription factors causing a considerable increase in drought mitigation (RaeesiSadati et al. 2021).

21.3.5 Role of Nanofertilizersin Protecting Integrity of Cell Membranes/Cellular Organelles

Nanoparticles play a pivotal role in safeguarding the cell membranes integrity amidst environmental changes by instigating ultra-structural alterations at both cellular and sub-cellular levels. Through interaction with the lipid bilayer, NPs effectively mitigate membrane integrity loss, diminish membrane permeability, and bolster overall membrane structural stability (Wang et al. 2017). This shielding mechanism preserves homeostasis at cellular levels and impedes the ingress of detrimental ions, thereby heightening stress tolerance (Bora et al. 2022).

The NPs exhibit the capability to safeguard sub-cellular organelles (i.e. chloroplasts and mitochondria) against stressinduced damage by mitigating oxidative stress, preserving structural integrity, and augmenting photosynthetic efficiency. These organelles portray an indispensable role in efficient energy generation, metabolic pathways, and stress response mechanisms (Alam et al. 2021; Kumar et al. 2024; Wahab et al. 2023). Furthermore, nanoparticles can instigate alterations at the subcellular level, fostering the development and functionality of peroxisomes and endoplasmic reticulum. These organelles are pivotal in stress-associated processes, including reactive oxygen species (ROS) detoxification and protein synthesis, thereby enhancing the plant's resilience to abiotic stresses like drought and salinity (Mariyam et al. 2024; Sharma et al. 2012). Such alterations contribute significantly to the enhancement of cellular integrity, regulation of water balance, protection of organelles, and optimization of sub-cellular functionalities (Table 21.1).

21.4 Safety and Sustainability

Nanoparticles can cause phytotoxicity (Jiang et al. 2022; Mazari et al. 2021; Saritha et al. 2022) owing to different factors like inappropriate concentration, composition, size, application method, surface area, exposure time, plantspecies, physicochemical properties, plant age, NPs stability, bioavailability and poresize (Li et al. 2022; Liu et al. 2022; Ma et al. 2010; Sun et al. 2022). It is worth mentioning that nanomaterials have the ability to invade human skin, gastrointestinal tract and lungs, enabling their translocation into tissues. Similarly, NPs may alter the microbiota composition in human gut and may cause severe health risks (Utembe et al. 2022). However, there should be ethical guidelines with nanotechnology that must be considered to minimize risks and to foster overall benefit.

Table 21.1 Overview of NPs functions in various crops under different stress conditions

NPs used	Function performed	Crop	Stress imposed	References
Iron oxide	Improve water use efficiency and chlorophyll content	*Triticum aestivum*	Drought stress	Manzoor et al. (2023)
Zinc oxide	Enhanced antioxidant enzyme activity	*Hordeum vulgare*	Salinity stress	Singh et al. (2022)
	Mitigate the harmful effects of reactive oxygen species (ROS) and enhance the activities of antioxidant enzymes such as superoxide dismutase (SOD), catalase (CAT) and peroxidase (POD). Developing salt tolerance and regulating ion homeostasis	*Zea mays* seedlings	Drought and salinity stresses	Rehman et al. (2023) Faizan et al. (2021)
	Improve WUE by regulating transpiration rate and stomatal conductance which could help plants maintain their water balance	*Solanum tuberosum*	Drought conditions	Seleiman et al. (2023)
	Minimizing lipid peroxidation	*Lupinus*	Salinity stress	Abdel Latef et al. (2017)
Silver nanoparticles (AgNPs)	Improve antioxidant activity, reducing oxidative damage caused by the increased ROS production, enhancing plant growth	*Triticum aestivum*	Drought stress	El-Saadony et al. (2022) Al-Khayri et al. (2023) Khan et al. (2020, 2022)
	Improve plant water use efficiency and modify stomatal conductance to reduce water loss	–	Drought stress	Yan and Chen (2019)
	Modulating ionic imbalance by the uptake and accumulation of ions	–	Salinity stress	Abasi et al. (2022)
	Induction of metallothioneins expression; proteins responsible for binding and sequestering heavy metals, thus minimizing their toxicity	Food crops	Heavy metals stress	Rai et al. (2023)
	Improving photosynthesis by enchancing chlorophyll content	*Triticum aestivum* seedlings	Drought and salinity stress	Vannini et al. (2014)
	Affecting the levels of abscisic acid (ABA) in plants that can play a vital role in plants tress tolerance. ABA being regulator of seed dormancy and seed germination may influence seed	–	Drought and salinity stress	Alabdallah and Hasan (2021) Nadarajah (2020) Noga et al. (2023)

(continued)

Table 21.1 (continued)

NPs used	Function performed	Crop	Stress imposed	References
	germination and overall plant growth			
TiO_2	Increasing chlorophyll content, higher photosynthetic efficiency and antioxidant enzymes activation	*Helianthus annuus*	Drought stress	Ramadan et al. (2022)
	Elevated antioxidant activity resulting in protection of plant cells from oxidative damage caused by ROS and and maintains membrane stability	*Solanum lycopersicum* seedlings	Drought stress	Cevik (2023) Samanta et al. (2024) Al-Khayri et al. (2023)
NPK	Nanofertilizers containing NPK or micronutrients can improve plant nutrition	–	Abiotic stress conditions	Nongbet et al. (2022)
SiO_2	Mitigate the ionic imbalance thereby increasing nutrient uptake	*Triticum aestivum*	Salinity stress	Etesami et al. (2021) Tripathi et al. (2017)
	Improve the stability of photosystem II and enhance photosynthetic activity	*Solanum lycopersicum*	Mild heat stress Drought stress	Qi et al. (2013) Rasheed et al. (2022)
Biochar NP's	Change soil properties, increase nutrient retention, soil water holding capacity thus enhancing nutrient availability	–	Drought and salinity stress	Liu and Lal (2015)
CeO_2 and TiO_2	Protecting photosynthetic apparatus (pigments such as chlorophyll, chloroplasts and photosystems) from oxidative damage. They can act as ROS scavengers too.	*Oryza sativa* *Helianthus annuus* *Glycine max*	Heavy metals stresses	Chen et al. (2015) Kumar et al. (2023) Singh and Lee (2016) Choudhary et al. (2024)
Zinc oxide and TiO_2	Can absorb or scatter harmful UV rays (working as UV filters) thus protecting the photosynthetic machinery from potential UV-induced damage	–	–	Liang et al. (2022)
Manganese NP's	Protecting photosynthetic apparatus from oxidative damage caused by ROS through SOD production (a natural antioxidant enzyme in plants)	–	–	Padmanaban et al. (2023)
Hydrogel NP's	Absorb water and release it slowly, enabling plants to access water for longer periods,	–	Drought stress	Patra et al. (2022)

(continued)

Table 21.1 (continued)

NPs used	Function performed	Crop	Stress imposed	References
	so that photosynthesis process may continue without being disrupted and results in increased water use efficiency in plants			
Carbon based NP's	Can modulate and stimulate the activity and expression of genes responsible for chloroplast development and photosynthesis, thereby improving the photosynthetic efficiency	–	–	Chandrashekar et al. (2023) Rasheed et al. (2022)
NP's in general	Thermo protection ensures the integrity of the photosynthetic apparatus enabling the plants to survive in high temperature	–	Heat stress conditions	Karumannil et al. (2023)

Key: WUE = Water use efficiency is the total biomass produced per unit of water, ROS = Unstable and highly reactive molecules carrying oxygen, NPK = Nitrogen Phosphorus Potassium. (Tablegenerated by Ghulam Abbas)

21.5 Conclusion and Future Prospects

The consideration of nanoparticles (NPs) in agriculture, as delineated in this chapter, presents an innovative avenue for bolstering plant resilience amidst stressors. The distinct attributes of nanoparticles (size, precision in delivery, and reactivity) opens avenues for mitigating stresses under the changing climate scenario. From enhancing nutrient assimilation to eliciting stress response mechanisms, nanoparticles exhibit substantial role in enriching plants against a spectrum of abiotic stresses like salinity, drought and metal toxicity by down-regulating or up-regulating the expressions of stress responsive genes. Nanoparticles play a considerable role in plant physiological attributes by improving photosynthesis, increased carbon fixation, enhanced solar interception, maintaining optimum plant water balance, minimizing nutrients losses, better nutrients uptake and enzymatic activities in different metabolic pathways. Nonetheless, while the potential benefits loom large, it is imperative to consider the environmental repercussions and potential health hazards linked to nanoparticles adoption in agriculture specifically determination of their toxic levels. Rigorous comprehension and vigilant oversight in the integration of nanotechnology in agriculture are crucial to ensure its safety and sustainability such as biosynthesized NP-based fertilizers can be an emerging technology. Further inquiries, specifically at the genomic, proteomic and metabolomic levels, are imperative to unravel the intricate mechanisms governing the interplay between nanoparticles and plant systems under diverse environmental stressors. Interdisciplinary synergies melding nanotechnology, crop plant biology and agricultural sciences are pivotal for unlocking the full potential of NPs in enhancing plant resilience.

References

Abasi F, Raja NI, Mashwani ZU et al (2022) Biogenic silver nanoparticles as a stress alleviator in plants: a mechanistic overview. Molecules 27:3378

Abdel Latef AA, Abu Alhmad MF, Abdelfattah KE (2017) The possible roles of priming with ZnO nanoparticles in mitigation of salinity stress in lupine (*Lupinustermis*) plants. J Plant Growth Regul 36:60–70

Abou-yuoseff AM, El-Khair A, El-Mohtasem MO et al (2022) Impacts of nano-fertilizers and chemical fertilizers on plant growth and nutrient uptake by faba bean (*Viciafaba* L.) plant. Al-Azhar J Agric Res 47(1):216–229

Ahanger MA, Qi M, Huang Z et al (2021) Improving growth and photosynthetic performance of drought stressed tomato by application of nano-organic fertilizer involves up-regulation of nitrogen, antioxidant and osmolyte metabolism. Ecotoxicol Environ Safety 216:112195

Alabdallah NM, Hasan MM (2021) Plant-based green synthesis of silver nanoparticles and its effective role in abiotic stress tolerance in crop plants. Saudi J Biol Sci 28(10):5631–5639

Alam P, Balawi TH, Altalayan FH et al (2021) Silicon attenuates the negative effects of chromium stress in tomato plants by modifying antioxidant enzyme activities, ascorbate–glutathione cycle and glyoxalase system. Acta Physiol Plant 43(7):110

Ali EF, El-Shehawi AM, Ibrahim OH et al (2021) A vital role of chitosan nanoparticles in improvisation the drought stress tolerance in *Catharanthus roseus* (L.) through biochemical and gene expression modulation. Plant Physiol Biochem 161:166–175. https://doi.org/10.1016/j.plaphy.2021.02.008

Al-Khayri JM, Rashmi R, Surya Ulhas R et al (2023) The role of nanoparticles in response of plants to abiotic stress at physiological, biochemical, and molecular levels. Plants (Basel) 12(2):292

Almutairi ZM (2016) Effect of nano-silicon application on the expression of salt tolerance genes in germinating tomato (*Solanum lycopersicum* L.) seedlings under salt stress. Plant Omics 9(1): 106–114

Anjum SA, Xie XY, Wang LC et al (2011) Morphological, physiological and biochemical responses of plants to drought stress. Afric J Agric Res 6:2026–2032

Arifur Rahman M, Harker T, Lewis W et al (2024) Chelated- and nano iron fertilization affects nutrient uptake and translocation in fresh market tomatoes. J Plant Nutri 47:1152–1174. https://doi.org/10.1080/01904167.2024.2304166

Avellan A, Yun J, Morais BP et al (2021) Critical review: role of inorganic nanoparticle properties on their foliar uptake and in planta translocation. Environ Sci Technol 55:13417–13431

Azam M, Bhatti HN, Khan A et al (2022) Zinc oxide nano-fertilizer application (foliar and soil) effect on the growth, photosynthetic pigments and antioxidant system of maize cultivar. Biocatal Agric Biotechnol 42:102343

Babaei K, SeyedSharifi R, Pirzad A et al (2017) Effects of bio fertilizer and nano Zn-Fe oxide on physiological traits, antioxidant enzymes activity and yield of wheat (*Triticumaestivum* L.) under salinity stress. J Plant Interact 12(1):381–389

Bano I, Skalickova S, Sajjad H et al (2021) Uses of selenium nanoparticles in the plant production. Agronomy 11(11):2229

Bharath P, Gahir S, Raghavendra AS (2021) Abscisic acidinduced stomatal closure: an important component of plant defense against abiotic and biotic stress. Front Plant Sci 12:615114

Bontempo P, De Masi L, Rigano D (2023) Functional properties of natural products and human health. Nutrients 15(13):2961

Bora KA, Hashmi S, Zulfiqar F et al (2022) Recent progress in bio-mediated synthesis and applications of engineered nanomaterials for sustainable agriculture. Front Plant Sci 13:999505

Cevik S (2023) TiO_2 nanoparticles alleviates the effects of drought stress in tomato seedlings. Bragantia 82:e20220203

Chandrashekar HK, Singh G, Kaniyassery A et al (2023) Nanoparticle-mediated amelioration of drought stress in plants: a systematic review. 3 Biotech 13(10):336

Channab BE, Idrissi AE, Ammar A et al (2024) Recent advances in nano-fertilisers: synthesis, crop yield impact, and economic analsysis. Nanoscale 16:4484–4513

Chen J, Liu X, Wang C et al (2015) Nitric oxide ameliorates zinc oxide nanoparticles-induced phytotoxicity in rice seedlings. J Hazard Material 297:173–182

Choudhary R, Rajput VD, Ghodake G et al (2024) Comprehensive journey from past to present to future about seed priming with hydrogen peroxide and hydrogen sulfide concerning drought, temperature, UV and ozone stresses- a review. Plant Soil 500:1–23

Cominelli E, Galbiati M, Tonelli C (2010) Transcription factors controlling stomatal movements and drought tolerance. Transcription 1(1):41–45

Dalal VK, Tripathy BC (2012) Modulation of chlorophyll biosynthesis by water stress in rice seedlings during chloroplast biogenesis. Plant Cell Environ 35:1685–1703

DeRosa MC, Monreal C, Schnitzer M et al (2010) Nanotechnology in fertilizers. Nat Nanotechnol J 5:91

Dey S, Nath S, Ansari TA et al (2023) Application of green synthesized bimetallic nZVI-Cu nanoparticle as a sustainable alternative to chemical fertilizers to enhance growth and photosynthetic efficiency of rice seedlings. Plant Physiol Biochem 201:107837

Dhansil A, Zalawadia N, Prajapat BS et al (2018) Effect of nanophosphatic fertilizer on nutrient content and uptake by pearl millet (*Pennisetumglaucum* L.) crop. Int J Curr Microbiol Appl Sci 7:2327–2337

Eide DJ (2020) Transcription factors and transporters in zinc homeostasis: lessons learned from fungi. Crit Rev Biochem Mol Biol 55:88–110

Elemike EE, Uzoh IM, Onwudiwe DC et al (2019) The role of nanotechnology in the fortification of plant nutrients and improvement of crop production. Appl Sci 9(3):499

El-Saadony MT, Saad AM, Soliman SM et al (2022) Role of nanoparticles in enhancing crop tolerance to abiotic stress: a comprehensive review. Front Plant Sci 13:946717

Elsheery NI, Sunoj VS, Wen Y et al (2020) Foliar application of nanoparticles mitigates the chilling effect on photosynthesis and photoprotection in sugarcane. Plant Physiol Biochem 149:50–60

Etesami H, Fatemi H, Rizwan M (2021) Interactions of nanoparticles and salinity stress at physiological, biochemical and molecular levels in plants: a review. Ecotoxicol Environ Saf 225:112769

Faizan M, Bhat JA, Chen C et al (2021) Zinc oxide nanoparticles (ZnO-NPs) induce salt tolerance by improving the antioxidant system and photosynthetic machinery in tomato. Plant Physiol Biochem 161:122–130

Finazzi G, Petroutsos D, Tomizioli M et al (2015) Ions channels/transporters and chloroplast regulation. Cell Calcium 58:86–97

Geremew A, Carson L, Woldesenbet S et al (2023) Effect of zinc oxide nanoparticles synthesized from *Caryaillinoinensis* leaf extract on growth and antioxidant properties of mustard (*Brassica juncea*). Front Plant Sci 14:1108186

Ghassemi-Golezani K, Afkhami N (2018) Changes in some morpho-physiological traits of safflower in response to water deficit and nano-fertilizers. J Biodiversity Environ Sci 12(3): 391–398

Ghidan A, Kahlel A, Al-Antary TA et al (2020) Efficacy of nanotechnology liquid fertilizers on weight and chlorophyll of broad bean (*Viciafaba* L.). Fresen Environ Bull 29(6):4789–4793

Ghosh SK, Bera T (2021) Molecular mechanism of nano-fertilizer in plant growth and development: a recent account. In: Advances in nano-fertilizers and nano-pesticides in agriculture. Woodhead Publishing, pp 535–560

Gómez-Vera P, Blanco-Flores H, Francisco AM et al (2021) Silicon dioxide nanofertilizers improve photosynthetic capacity of two *Criollo cocoa* clones (*Theobroma cacao* L.). Exp Agric 57(2):85–102

Hayatu M, Muhammad SY, Abdu HU (2014) Effect of water stress on the leaf relative water content and yield of some cowpea (*VignaUnguiculata* (L) Walp.) genotype. Int J Sci Technol Res 3(7)

Hemdan NA, Hussein MM, El-Ashry S (2020) Conventional and nano fertilizers and its effect on growth and nutrient status of lupine plants. Curr Sci Int 9(4):698–705

Hong J, Wang C, Wagner DC et al (2021) Foliar application of nanoparticles: mechanisms of absorption, transfer, and multiple impacts. Environ Sci Nano 8(5):1196–1210

Hussain N, Bilal M, Iqbal HM (2022) Carbon-based nanomaterials with multipurpose attributes for water treatment: greening the 21st century nanostructure materials deployment. Biomaterials Polymers Horizon 1(1):48–58

Irfan M, Maqsood MA, Rehman HU et al (2023) Silicon nutrition in plants under water-deficit conditions: overview and prospects. Water 15(4):739

Jalil S, Alghanem SM, Al-Huqail AA et al (2023) Zinc oxide nanoparticles mitigated the arsenic induced oxidative stress through modulation of physio-biochemical aspects and nutritional ions homeostasis in rice (*Oryza sativa* L.). Chemosphere 338:139566

Jayara AS, Kumar R, Singh AV et al (2023) Assessing the impact of mineral based nutrient sources and nano-fertilizers on root growth, soil health, and nutrient availability in wheat crop. https://doi.org/10.21203/rs.3.rs-3444003/v1

Jiang Y, Zhou P, Zhang P et al (2022) Green synthe-sis of metal-based nanoparticles for sustainable agriculture. Environ Pollut 309:119755

Jinghua G (2004) Synchrotron radiation, soft X-ray spectroscopy and nano-materials. J Nanotechnol 1:193–225

Junedi MA, Mukhopadhyay R, Manjari KS (2023) Alleviating salinity stress in crop plants using new engineered nanoparticles (ENPs). Plant Stress 9:100184

Kandhol N, Jain M, Tripathi DK (2022) Nanoparticles as potential hallmarks of drought stress tolerance in plants. Physiol Plant 174(2):e13665. https://doi.org/10.1111/ppl.13665

Karthick S, Chitrakala K (2011) Ecotoxico-logical effect of Lecani cilium Lecanii (Asco-mycota: Hypocereales) based silver nanoparti-cles on growth parameters of economically im-portant plants. J Biopest 4(1):97–101

Karumannil S, Khan TA, Kappachery S et al (2023) Impact of exogenous melatonin application on photosynthetic machinery under abiotic stress conditions. Plan Theory 12(16):2948

Kaviani B, Ghaziani MV, Negahdar N (2016) Effect of application methods and different concentrations of biologic nano-fertilizer, especial for ornamental plants on some morphological, physiological and proliferation traits and enhancing the quality of buxushyrcanapojark. Int J Adv Sci Eng Tech Spl Issue 2:206–212

Khan I, Raza MA, Awan SA et al (2020) Amelioration of salt induced toxicity in pearl millet by seed priming with silver nanoparticles (AgNPs): the oxidative damage, antioxidant enzymes and ions uptake are major determinants of salt tolerant capacity. Plant Physiol Biochem 156:221–232

Khan I, Awan SA, Rizwan M et al (2022) Nanoparticle's uptake and translocation mechanisms in plants via seed priming, foliar treatment, and root exposure: a review. Environ Sci Pollut Res 29(60):89823–89833

Khatri K, Rathore MS (2018) Plant nanobionics and its applications for developing plants with improved photosynthetic capacity. In: Photosynthesis: from its evolution to future improvements in photosynthetic efficiency using nanomaterials, vol 95. InTech, p 1016

Kim DY, Patel SK, Rasool K et al (2024a) Bioinspired silver nanoparticle-based nanocomposites for effective control of plant pathogens: a review. Sci Total Environ 908:168318

Kim DY, Yang T, Srivastava P et al (2024b) Alginic acid-functionalized silver nanoparticles: a rapid monitoring tool for detecting the technology-critical element tellurium. J Hazard Mater 465:133161

Kobayashi T, Nozoye T, Nishizawa NK (2018) Iron transport and its regulation in plants. Free Radic Biol Med 133:11–20. https://doi.org/10.1016/j.freeradbiomed.2018.10.439

Kumar D, Dhankher OP, Tripathi RD et al (2023) Titanium dioxide nanoparticles potentially regulate the mechanism(s) for photosynthetic attributes, genotoxicity, antioxidants defense machinery, and phytochelatins synthesis in relation to hexavalent chromium toxicity in *Helianthus annuus* L. J Hazard Mater 454:131418

Kumar D, Singh R, Upadhyay SK et al (2024) Review on interactions between nanomaterials and phytohormones: novel perspectives and opportunities for mitigating environmental challenges. Plant Sci 340:111964

Kumari S, Khanna RR, Nazir F et al (2022) Bio-synthesized nanoparticles in developing plant abiotic stress resilience: a new boon for sustainable approach. Int J Mol Sci 23(8):4452

Li J, Wang C, Yue L et al (2022) Nano-QSAR modeling for predicting the cytotoxicity of metallic and metaloxide nanoparticles: a review. Ecotoxicol Environ Saf 243:113955

Liang Y, Simaiti A, Xu M et al (2022) Antagonistic skin toxicity of co-exposure to physical sunscreen ingredients zinc oxide and titanium dioxide nanoparticles. Nano 12(16):2769

Lin D, Xiao M, Zhao J et al (2016) An overview of plant phenolic compounds and their importance in human nutrition and management of type 2 diabetes. Molecules 21(10):1374

Linh TM, Mai NC, Hoe PT et al (2020) Metal-based nanoparticles enhance drought tolerance in soybean. J Nanomater 2020:4056563. https://doi.org/10.1155/2020/4056563

Liu R, Lal R (2015) Potentials of engineered nanoparticles as fertilizers for increasing agronomic productions. Sci Total Environ 514:131–139

Liu Z, Xu Z, Xu L et al (2022) Modified biochar: synthesis and mechanism for removal of environmental heavy metals. Carbon Res 1:8

Llanderal A, García-Caparrós P, Pérez-Alonso J et al (2020) Approach to petiole sap nutritional diagnosis method by empirical model based on climatic and growth parameters. Agronomy 10(2):188

Ma X, Geiser-Lee J, Deng Y, Kolmakov A (2010) Interactions betweenengineered nanoparticles (ENPs) and plants: phytotoxicity, uptakeandaccumulation. Sci Total Environ 408:3053–3061

Majumdar S, Almeida IC, Arigi EA et al (2015) Environmental effects of nanoceria on seed production of common bean (*Phaseolus vulgaris*): a proteomic analysis. Environ Sci Technol 49:13283–13293. https://doi.org/10.1021/acs.est.5b03452

Manzoor N, Ali L, Al-Huqail AA et al (2023) Comparative efficacy of silicon and iron oxide nanoparticles towards improving the plant growth and mitigating arsenic toxicity in wheat (*Triticumaestivum* L.). Ecotoxicol Environ Saf 264:115382

Mariyam S, Upadhyay SK, Chakraborty K et al (2024) Nanotechnology, a frontier in agricultural science, a novel approach in abiotic stress management and convergence with new age medicine-a review. Sci Total Environ 912:169097

Mazari SA, Ali E, Abro R et al (2021) Nanoma-terials: applications, waste-handling, environmental toxicities, andfuture challenges –a review. J Environ Chem Eng 9:105028

Mir S, Sirousmehr A, Shirmohammadi E (2015) Effect of nano and biological fertilizers on carbohydrate and chlorophyll content of forage sorghum (speedfeed hybrid). Int J Biosci 6: 157–164

Monreal CM, DeRosa M, Mallubhotla SC et al (2015) Nanotechnologies for increasing the crop use efficiency of fertilizer-micronutrients. Biol Fertil Soils 52:423–437

Monsef-Afshar R, Hadi H, Pirzad A (2013) Effect of nano-iron on the yield and yield components of cowpea under end season water deficit. Int J Agric 3(1):27–34

Morales-Díaz AB, Ortega-Ortíz H, Juárez-Maldonado A et al (2017) Application of nanoelements in plant nutrition and its impact in ecosystems. Adv Natural Sci: Nanosci Nanotechnol 8:1–13

Nadarajah KK (2020) ROS homeostasis in abiotic stress tolerance in plants. Int J Mol Sci 21(15): 5208

Nadi E, Aynehband A, Mojaddam M (2013) Effect of nano-iron chelate fertilizer on grain yield, protein percent and chlorophyll content of Faba bean (*Viciafaba* L.). Int J Biosci 3(9):267–272

Nayeri S, Dolatyari M, Mouladoost N et al (2023) Ag/ZnO core–shell NPs boost photosynthesis and growth rate in wheat seedlings under simulated full sun spectrum. Sci Rep 13(1):14385

Noga M, Milan J, Frydrych A et al (2023) Toxicological aspects, safety assessment, and green toxicology of silver nanoparticles (AgNPs)-criticalreview: state of the art. Int J Mol Sci 24(6): 5133

Nongbet A, Mishra AK, Mohanta YK et al (2022) Nanofertilizers: a smart and sustainable attribute to modern agriculture. Plants 11(19):2587

Padmanaban S, Pully D, Samrot AV et al (2023) Rising influence of nanotechnology in addressing oxidative stress-related liver disorders. Antioxidants 12(7):1405

Pagter M, Bragato C, Brix H (2005) Tolerance and physiological responses of *Phragmites australis* to water deficit. Aquat Bot 81:285–299

Patra SK, Poddar R, Brestic M et al (2022) Prospects of hydrogels in agriculture for enhancing crop and water productivity under water deficit condition. Int J Polym Sci 1:4914836

Pérez-Labrada F, Hernández-Hernández H, López-Pérez MC et al (2020) Nanoparticles in plants: morphophysiological, biochemical, and molecular responses. In: Plant life under changing environment. Elsevier, pp 289–322. https://doi.org/10.1016/B978-0-12-818204-8.00016-3

Pirzad A (2010) The effect of iron, zinc and manganese on quality and quantity of soybean seed. J Phyto 2(11):73–79

Qi M, Liu Y, Li T (2013) Nano-TiO_2 improve the photosynthesis of tomato leaves under mild heat stress. Biol Trace Elem Res 156(1–3):323–328. https://doi.org/10.1007/s12011-013-9833-2

RaeesiSadati SY, JahanbakhshGodehkahriz S, Ebadi A (2021) Study of expression pattern of some transcription factors in wheat under drought stress and zinc nanoparticles. Plant Genet Res 7: 135–144. https://doi.org/10.52547/pgr.7.2.11

Rai PK, Song H, Kim KH (2023) Nanoparticles modulate heavy-metal and arsenic stress in food crops: hormesis for food security/safety and public health. Sci Total Environ 902:166064

Raliya R, Tarafdar JC (2013) ZnO nanoparticle biosynthesis and its effect on phosphorous-mobilizing enzyme secretion and gum contents in Clusterbean (*Cyamopsistetragonoloba* L.). Agric Res 2:48–57

Raliya R, Saharan V, Dimkpa C et al (2017) Nanofertilizer for precision and sustainable agriculture: current state and future perspectives. J Agric Food Chem 66:6487–6503

Ramadan T, Sayed SA, Abd-Elaal AK et al (2022) The combined effect of water deficit stress and TiO2 nanoparticles on cell membrane and antioxidant enzymes in *Helianthus annuus* L. Physiol Mol Biol Plants 28(2):391–409

Ramzan M, Ayub F, Shah AA et al (2022) Synergistic effect of zinc oxide nanoparticles and moringaoleifera leaf extract alleviates cadmium toxicity in *Linumusitatissimum*: antioxidants and physiochemical studies. Front Plant Sci 13:900347

Rasheed A, Li H, Tahir MM et al (2022) The role of nanoparticles in plant biochemical, physiological, and molecular responses under drought stress: a review. Front Plant Sci 13:976179

Rehman A, Weng J, Li P et al (2023) Green synthesized zinc oxide nanoparticles confer drought tolerance in melon (*Cucumismelo* L.). Environ Exp Bot 212:105384

Rezaei M, Abbasi H (2014) Foliar application of nanochelate and non-nanochelate of zinc on plant resistance physiological processes in cotton (*Gossipiumhirsutum* L.). Iran J Plant Physiol 4(4): 1137–1144

Rojas-Velázquez ÁN, Guillén-Castillo OI, Alcalá-Jauregui JA et al (2023) Effect of a nitrogenous nanocomposite on leaching and N content in lettuce in soil columns. Discover Nano 18:98

Rushton PJ, Somssich IE, Ringler P (2010) WRKY transcription factors. Trends Plant Sci 15:247–258. https://doi.org/10.1016/j.tplants.2010.02.006

Saini P, Nagpal S, Saini P et al (2021) Microbial mediated zinc solubilization in legumes for sustainable agriculture. In: Phytomicrobiome interactions and sustainable agriculture, pp 254–276

Salama H (2012) Effects of silver nanoparti-cles in some crop plants, common bean (*Phaseolus vulgaris* L.) and corn (*Zea mays* L.). Int Res J Biotech 3(10):190–197

Samanta S, Seth CS, Roychoudhury A (2024) The molecular paradigm of reactive oxygen species (ROS) and reactive nitrogen species (RNS) with different phytohormone signaling pathways during drought stress in plants. Plant Physiol Biochem 206:108259

Saritha GNG, Anju T, Kumar A (2022) Nanotechnology -big impact: how nanotechnology is changing the future of agriculture? J Agric Food Res 10:100457

Sarkar RD, Kalita MC (2023) Alleviation of salt stress complications in plants by nanoparticles and the associated mechanisms: an overview. Plant Stress 7:100134

Seleiman MF, Almutairi KF, Alotaibi M et al (2020) Nano-fertilization as an emerging fertilization technique: why can modern agriculture benefit from its use? Plan Theory 10(1):2

Seleiman MF, Al-Suhaibani N, Ali N et al (2021) Drought stress impacts on plants and different approaches to alleviate its adverse effects. Plants (Basel) 10(2):259

Seleiman MF, Al-Selwey WA, Ibrahim AA et al (2023) Foliar applications of ZnO and SiO_2 nanoparticles mitigate water deficit and enhance potato yield and quality traits. Agronomy 13(2):466

Selvakesavan RK, Kruszka D, Shakya P et al (2023) Impact of nanomaterials on plant secondary metabolism. In: Al-Khayri JM, Alnaddaf LM, Jain SM (eds) Nanomaterial interactions with plant cellular mechanisms and macromolecules and agricultural implications. Springer International Publishing, Cham, pp 133–170

Shah T, Latif S, Saeed F et al (2021) Seed priming with titanium dioxide nanoparticles enhances seed vigor, leaf water status, and antioxidant enzyme activities in maize (*Zea mays* L.) under salinity stress. J King Saud Univ – Sci 33(1):101207

Sharma P, Jha AB, Dubey RSP et al (2012) Reactive oxygen species, oxidative damage, and antioxidative defense mechanism in plants under stressful conditions. J Bot 2012:217037

Sharma A, Shahzad B, Kumar V et al (2019) Phytohormones regulate accumulation of osmolytes under abiotic stress. Biomolecules 9(7):285

Shaw D, Hermoso A, Lluch-Senar M, Shaw D (2020) Comparative gene essentiality across the bacterial domain. bioRxiv. https://doi.org/10.1101/2020.02.28.969238

Sheteiwy MS, Shaghaleh H, Hamoud YA et al (2021) Zinc oxide nanoparticles: potential effects on soil properties, crop production, food processing, and food quality. Environ Sci Pollut Res 28(28):36942–36966

Siddiqi KS, Husen A (2017) Plant response to engineered metal oxide nanoparticles. Nanoscale Res Lett 12:1–18

Singh J, Lee BK (2016) Influence of nano-TiO_2 particles on the bioaccumulation of Cd in soybean plants (*Glycine max*): a possible mechanism for the removal of Cd from the contaminated soil. J Environ Manag 170:88–96

Singh A, Sengar RS, Rajput VD et al (2022) Zinc oxide nanoparticles improve salt tolerance in rice seedlings by improving physiological and biochemical indices. Agriculture 12(7):1014

Singh A, Pankaczi F, Rana D et al (2023) Coated hematite nanoparticles alleviate iron deficiency in cucumber in acidic nutrient solution and as foliar spray. Plants 12(17):3104

Soliman AS, Hassan M, Abou-Elella F et al (2016) Effect of nano and molecular phosphorus fertilizers on growth and chemical composition of baobab (*Adansoniadigitata* L.). J Plant Sci 11(4):52–60

Spanò C, Bottega S, Bellani L et al (2020) Effect of zinc priming on salt response of wheat seedlings: relieving or worsening? Plants 9(11):1514

Su Y, Ashworth V, Kim C et al (2019) Delivery, uptake, fate, and transport of engineered nanoparticles in plants: a critical review and data analysis. Environ Sci Nano 6(8):2311–2331

Sun Y, Zhu G, Zhao W et al (2022) Engineered nanomaterials for improving the nutritional quality of agricultural products: a review. Nanomaterials (Basel) 12:4219

Tambussi EA, Bort J, Guiamet JJ et al (2007) The photosynthetic role of ears in C_3 cereals: metabolism, water use efficiency and contribution to grain yield. Crit Rev Plant Sci 26:1–16

Tan W, Du W, Barrios AC et al (2017) Surface coating changes the physiological and biochemical impacts of nano-TiO_2 in basil (*Ocimumbasilicum*) plants. Environ Pollut 222:64–72

Tarafdar JC, Rathore I, Thomas E (2015) Enhancing nutrient use efficiency through nano technological interventions. Indian J Fertil 11(12):46–51

Thabet SG, Alqudah AM (2023) New genetic insights into improving barley cope with salt stress via regulating mineral accumulation, cellular ion homeostasis, and membrane trafficking. Environ Exp Bot 208:105252

Thiruvengadam M, Chi HY, Kim SH (2024) Impact of nanopollution on plant growth, photosynthesis, toxicity, and metabolism in the agricultural sector: an updated review. Plant Physiol Biochem 207:108370

Tripathi DK, Singh S, Singh VP et al (2017) Silicon nanoparticles more effectively alleviated UV-B stress than silicon in wheat (*Triticumaestivum*) seedlings. Plant Physiol Biochem 110:70–81
Trono D, Pecchioni N (2022) Candidate genes associated with abiotic stress response in plants as tools to engineer tolerance to drought, salinity and extreme temperatures in wheat: an overview. Plants 11(23):3358
Utembe W, Tlotleng N, Kamng'ona AW (2022) A systematic review on the effects of nanomaterials on gut microbiota. Curr Res Microb Sci 3:100118
Vakilian KA (2019) Gold nanoparticles-based biosensor can detect drought stress in tomato by ultrasensitive and specific determination of miRNAs. Plant Physiol Biochem 145:195–204. https://doi.org/10.1016/j.plaphy.2019.10.042
Vannini C, Domingo G, Onelli E et al (2014) Phytotoxic and genotoxic effects of silver nanoparticles exposure on germinating wheat seedlings. J Plant Physiol 171(13):1142–1148
Wahab A, Batool F, Muhammad M et al (2023) Current knowledge, research progress, and future prospects of phyto-synthesized nanoparticles interactions with food crops under induced drought. Stress 15(20):14792
Wang L, Hu C, Shao L (2017) The antimicrobial activity of nanoparticles: present situation and prospects for the future. Int J Nanomedicine 12:1227–1249
Wang A, Jin Q, Xu X et al (2020) High-throughput screening for engineered nanoparticles that enhance photosynthesis using mesophyll protoplasts. J Agric Food Chem 68:3382–3389
Wang X, Xie H, Wang P, Yin H (2023) Nanoparticles in plants: uptake, transport and physiological activity in leaf and root. Materials (Basel) 16(8):3097. https://doi.org/10.3390/ma16083097
Weisany W, Sohrabi Y, Heidari G et al (2012) Changes in antioxidant enzymes activity and plant performance by salinity stress and zinc application in soybean (*Glycine max* L.). Plant Omic J 5: 60–67
Wu H, Shabala L, Shabala S, Giraldo JP (2018) Hydroxyl radical scavenging by cerium oxide nanoparticles improves Arabidopsis salinity tolerance by enhancing leaf mesophyll potassium retention. Environ Sci Nano 5(7):1567–1583
Yan A, Chen Z (2019) Impacts of silver nanoparticles on plants: a focus on the phytotoxicity and underlying mechanism. Int J Mol Sci 20(5):1003
Yang KY, Doxey S, McLean JE et al (2018) Remodeling of root morphology by CuO and ZnO nanoparticles: effects on drought tolerance for plants colonized by a beneficial pseudomonad. Botany 96:175–186. https://doi.org/10.1139/cjb-2017-0124
Yin L, Colman BP, McGill BM et al (2012) Effects of silver nanoparticle exposure on germination and early growth of eleven wetland plants. PLoS One 10(7):e47674
Zanetti LV, Milanez CR, Gama VN et al (2016) Leaf application of silicon in young cacao plants subjected to water deficit. Pesqui Agropecu Bras 51:215–223
Zhou H, Wu H, Zhang F et al (2021) Molecular basis of cerium oxide nanoparticle enhancement of rice salt tolerance and yield. Environ Sci Nano 8(11):3294–3311

Chapter 22
Future Trends and Challenges in Agricultural Nanotechnology

Viji Mary Varghese, O. Kavitha, Gigi Poulose, and Nithya Joseph

Abstract Nanotechnology, a rapidly rising field, has laid its footprint in a multitude of sectors including textiles, electronics, healthcare, environment remediation, biomaterials and agriculture, influencing everyday life. It has shown promising results in the agricultural sector through the application of nano fertilizers, nano pesticides, and nano sensors. Agricultural nanotechnology provides innovative approaches to tackle the issue of increased food demand due to the rising population while consuming minimum resources. It has enormous potential to impact agriculture by increasing soil quality, controlling nutrients, monitoring pesticide use, and water quality. It also contributes to sustainable development by improving resource management and soil fertility. This chapter provides an overview of the recent advances and future trends in nanotechnological applications in agriculture. Advancements in nanotechnology such as nano priming, nano aptamers, nanocarriers for gene delivery, and integration with other platforms have the capability to completely transform farming and agriculture. Despite its many advantages, agricultural nanotechnology also has a number of drawbacks. Therefore, this chapter also focuses on the various challenges and constraints pertaining to the application of nanomaterials in agriculture. The long-term environmental fate of nanoparticles is poorly understood which necessitates further research to assess their ecological risk. Regulatory guidelines and safety standards are being increasingly implemented. Public attitude and perception of the usage of nanomaterials in agriculture is another element that needs to be taken into consideration.

V. M. Varghese (✉) · O. Kavitha
Department of Biotechnology, St. Joseph's College (Autonomous), Irinjalakuda, Thrissur, Kerala, India
e-mail: vijimary@stjosephs.edu.in; kavithao@stjosephs.edu.in

G. Poulose
Department of Zoology, St. Joseph's College (Autonomous), Irinjalakuda, Thrissur, Kerala, India
e-mail: gigip@stjosephs.edu.in

N. Joseph
Bioscience and Water Filtration Division, Parker Hannifin Corporation, Oxnard, CA, USA
e-mail: nithya.joseph@parker.com

J. M. Al-Khayri et al. (eds.), *Nanotechnology Applications in Modern Agriculture*, Nanotechnology in Plant Sciences 3, https://doi.org/10.1007/978-3-031-90506-3_22

Keywords Agriculture · Challenges · Future trends · Nanocarriers · Nanotechnology · Public perception

22.1 Introduction

More than being a major livelihood of the people in developing countries, agriculture is essential for the perpetuation of life on earth. It is essential for ensuring food security, and research has been done worldwide concerning nutrition and food safety from the perspective of agricultural land systems, with an emphasis on increasing food production to satisfy the needs of an expanding population. The yield in agriculture is highly affected by the changes in environment and climate such as deficiencies in micro and macronutrients, erratic weather patterns, rise in global temperature, depletion in water sources and presence of invasive species and pests (Liliane and Charles 2020). Nevertheless, the rapidly growing world population escalates the demand for increased agricultural production. Hence, the farm output must rise significantly to guarantee an adequate supply of raw materials for food, feed, and fiber. Efficient utilization of land, water, energy, and inputs such as fertilizers and pesticides is essential to minimize environmental impacts, conserve natural resources, and promote sustainable agriculture. Sustainable farming practices aim to meet present-day food requirements while safeguarding resources for future generations. Policies and plans are implemented internationally to achieve food security by using novel technologies for increasing crop output through rigorous and widespread farming. The utilization of advanced technology, from mechanizing tilling and harvesting to employing genetic engineering to enhance crop yields (Sehgal Foundation 2023), contributes significantly to increasing production and reducing food waste. Technological advancements like satellite and GPS technologies, sensors, smart irrigation, drones and blockchain technologies assist farmers to attain the goal of high output yield along with effective resource utilization (Hegde 2023). Recently, the potential of nanotechnology has been intensely explored in tackling the challenges and issues in agriculture (Fraceto et al. 2016). Agricultural nanotechnology offers innovative solutions to meet the challenges of producing more food while minimizing resource usage. It supports the Sustainable Development Goals, especially Goal 2 (Zero hunger) and Goal 12 (Responsible consumption and production) (Gonzalez et al. 2024; Mukherjee et al. 2019; Viana et al. 2021).

Nanotechnology, the manipulation of matter at atomic and molecular scale, is a rapidly evolving field which aids in scientific advancement in many sectors such as healthcare, biomaterials, environmental protection, textiles and electronics (Singh et al. 2020a). Norio Taniguchi of Tokyo University of Science first used the word "nanotechnology" in 1974 (Khan and Rizvi 2014). Nanotechnology limits the material size to 1–100 nm (Dazon et al. 2020; Shang et al. 2019), making it superior to many other technologies. It is regarded as a pioneering approach in developmental research concerning the exploration and manipulation of materials at atomic, molecular, or macromolecular scales. At these levels, the properties of substances undergo significant alterations compared to their behavior at larger scales. The newly acquired properties of nanomaterials(NMs),such as their small size, large surface area, and high reactivity, could be used for manufacturing novel products and tools that could operate with enhanced efficiency (Neme et al. 2021). Their small size gives them unique physicochemical properties (Wang et al. 2016a, b), and they are easily conjugable (Hu et al. 2020). Other distinctive qualities include their specificity to the target, self-regulation and programmable nature (War et al. 2020). Since the size of the majority of nanoparticles (NPs) is less than the dimension of cells, these particles are able to interact and attach with biological entities without altering their functions (Nam and Luong 2019).

Photochemical reactions, natural processes, simple abrasion, wildfires, and living things, including microorganisms, can produce nanoparticles in a natural way. Based on the shape, size, characteristics and sources, NPs can be classified as carbon-based, metal, ceramics, lipid-based, semiconductor and polymeric (Altammar 2023). In the field of agriculture, two categories of NPs are mainly used: (1) metal-based and (2) carbon-based (Gupta et al. 2023a). Frequently used metal-based NPs are silver, gold, and copper (Khan et al. 2021a, b).

The use of carbon nanotubes, one of the most common forms of carbon-based NMs, with unique physiochemical and mechanical characteristics was reported in rice (*Oryza sativa* L), tomato *(Solanum lycopersicum* L), and Bt cotton (Bhattacharjee et al. 2022; Singh et al. 2017). Metal based NMs are in the form of oxides. Oxides of silicon, zinc, titanium, magnesium and copper have already been proven effective. Silicon dioxide (SiO_2) NPs are reported to have plant germination-enhancing capacity (Disfani et al. 2017), lateral root proliferating ability in seedlings (Baz et al. 2020), insecticidal influence on the grain (Saranya et al. 2020) and also the ability to promote the biosynthesis of defense compounds (Sheng et al. 2018). Reports indicate that the growth and development of soybean, peanut, and wheat can be enhanced through the application of Zinc oxide (ZnO) NPs (Prasad et al. 2012; Ramesh et al. 2014; Sedghi et al. 2013). Titanium dioxide (TiO_2)NPs have demonstrated insecticidal and antibacterial properties (Marimuthu et al. 2013), as well as the ability to enhance nitrogen metabolism (Kamal and Mogazy 2023) and accelerate growth (Ali et al. 2021a). Magnesium oxide (MgO) NPs have shown to provide resistance against diseases in crops (Imada et al. 2016; Ogunyemi et al. 2020). In a waterlogged paddy field, copper oxide (CuO) nanoparticles decreased the soil microbiota and the activity of enzymes such as urease and phosphatase and total phospholipid fatty acids (Xu et al. 2015). These NPs can influence microbial composition and diversity (Frenk et al. 2013; Xu et al. 2015) and root shoot development (Song et al. 2021). In plants, CuO, SiO2, and TiO2 NPs can also serve as delivery vehicles (Jahan et al. 2018).

Simple metals are also found to be effective NPs with various applications in agriculture. Improved rate of seed germination, increased yield of seeds and enhanced plant growth are some of the reported effects of gold (Au) nanoparticles (Alagoz et al. 2022; Avellan et al. 2017; Barrena et al. 2009; Saeed et al. 2021; Savithramma et al. 2012). It has been shown that silver (Ag) NPs improve the biochemical characteristics and yields of brown mustard (*Brassica juncea* L) and maize (*Zea mays* L) in addition to increasing their ability for seed germination

(Savithramma et al. 2012). They also have strong antimicrobial activity ((El-Sheekh and El-Kassas 2016) and are superior fungicides (Gupta et al. 2023a, b).

Numerous techniques are utilized for synthesizing nanomaterials, broadly categorized into chemical, physical, and biological methods (Gupta et al. 2023a, b). Physical and chemical approaches have several drawbacks, including low production yield, excessive energy use, excessive heat generation, and the requirement for expensive equipment (Soni et al. 2018). Green synthesis is often considered more beneficial since it prevents the formation of undesirable by-products. This technique makes use of microorganisms and their enzymes, plants, degradable polymers, and polysaccharides (Roychoudhury 2020). Bacteria, fungi, yeast, actinomycetes, algae and plants can be effectively used for the synthesis of nanoparticles. Compared to traditional NPs made by physical and chemical processes, these NPs often have superior qualities (Gupta et al. 2023a, b).

The demand for fast, reliable, and cost-effective systems in agriculture has led to the exploration of green nanotechnology since they offer a safer, energy-efficient, and environmentally friendly alternative to chemically synthesized counterparts (Prasad 2014; Prasad et al. 2014, 2016; Sagadevan and Periasamy 2014; Vidotti et al. 2011). Green nanotechnology, with its focus on sustainability and reduced environmental impact, presents promising advancements in the field, although challenges remain in ensuring its long-term environmental sustainability (Kandasamy and Prema 2015).

22.2 Recent Advances in Agricultural Nanotechnology

By improving soil quality, managing nutrients, tracking pesticide use, enhancing the efficiency of nutrient and water use and monitoring water quality, nanotechnology has the potential to change agriculture completely (Gruère 2012; Mukhopadhyay 2014; Prasad et al. 2014; Shukla et al. 2024). The significance of nanotechnology for sustainable development is highlighted by its prospective applications in agriculture, including drug delivery, resource management, and soil fertility maintenance (Sabir et al. 2014). Additionally, it supports appropriate food processing, risk assessment, and biomass utilization in agriculture (Floros et al. 2010). Nano sensors, including biosensors and optical sensors, are increasingly utilized for analysing soil and water contamination, particularly for detecting heavy metals at trace levels. Nano bioremediation, employing NMs to enhance microbial degradation of waste and toxins, offers sustainable solution for removing heavy metals from agricultural soil and water, aligning with the principles of sustainable remediation technologies (Dixit et al. 2015).

22.2.1 Nano Fertilizers

The use of fertilizers in agriculture is always required because of the rising global need for food. To maximize yields with limited land and water resources, conventional agricultural methods often rely on chemical fertilizers. Although agricultural productivity has increased, chemical fertilizers negatively impact the environment, soil, and human health. A significant portion of these chemicals can seep into the soil or be carried by water into water bodies, harming the environment and ecosystem (Mejias et al. 2021; Nongbet et al. 2022). Nanotechnology enables the creation and utilization of nanostructured materials as "smart fertilizers," which offer various benefits including effective nutrient absorption, soil fertility restoration, and reduced environmental pollution (Soni et al. 2024). These materials consist of NPs that contain macro- and microelements. They provide nutrients to crops at a controlled rate, which enhances nutrient uptake efficiency, reduces the amount of fertilizer required, and minimizes the risk of nutrient leaching and runoff (Anil et al. 2024; Kopittke et al. 2019). Nano fertilizers influence rhizosphere bacteria by promoting secondary metabolite formation (Nongbet et al. 2022; Panichikkal et al. 2019). They can increase the crop yield by triggering seed germination, nitrogen metabolism, protein and carbohydrate synthesis, and by improving stress tolerance (Rautela et al. 2021). Many nano fertilizers can increase photosynthesis rate and stimulate antioxidant enzyme activity (Rana et al. 2024). NMs utilized for the manufacturing of fertilizers include graphene, quantum dots and carbon nanotubes (Yadav et al. 2023).

In order to generate precise nano fertilizers, NMs are being combined with macronutrients, including nitrogen, phosphorous, potassium, calcium, magnesium, and sulphur. This approach addresses issues like poor nutrient uptake efficiency and the negative environmental effects of conventional fertilizers. Advanced nano fertilizers, for instance, urea-coated hydroxyapatite nanoparticles for nitrogen and phosphorus delivery, offer enhanced efficiency and slow-release mechanisms, opening the door to environmentally friendly substitutes for conventional fertilizers (Nongbet et al. 2022). Micronutrients such as zinc, boron, iron, manganese, and silica are crucial for plant growth and development, with nanomaterial-enhanced delivery showing promising results in improving crop yield and quality (Dimkpa and Bindraban 2016). Additionally, NMs like TiO_2 and carbon-based nanomaterials exhibit beneficial effects on various crops, enhancing germination rates, biomass, photosynthetic activity, and nutrient content, thereby offering potential solutions to tackle challenges such as salinity, drought, and pest management in agriculture (Mittal et al. 2020). Nanofertilizers can help tackle micronutrient deficiency in the global population by increasing the essential mineral content in various crops (Santos et al. 2024). Huang et al. (2024) proved that the foliar application of manganese ferrite nanoparticles could increase the concentration of manganese and iron in wheat grains.

Nano biofertilizers, formed by combining nanoparticles with beneficial microorganisms, stimulate the symbiotic relationship between plants and microorganisms, and optimize soil productivity through the consistent release of nutrients, nitrogen fixation and phosphorus solubilization (Rezaei and Khanmirzaei 2024; Saurabh et al. 2024; Yadav et al. 2023). Nano biofertilizers are expected to exhibit the advantages of both NPs and biofertilizers. Nevertheless, their commercialization is not widely undertaken due to concerns regarding the interaction between nano biofertilizers and plants (Akhtar et al. 2022).

22.2.2 Nano Delivery of Phytohormones

Phytohormones regulate plant growth and development. For the purpose of applying these hormones to the plant, either by leaf spraying or root absorption—nanocarriers are utilized (Liu et al. 2024). The positive impact of nanoparticles and growth regulators on the germination of Maize seeds was reported by Yousif et al. (2023). The utilization of nano-formulated hormones, particularly Nano-indole acetic acid (IAA), demonstrated significant enhancements in coffee survival rates and eggplant germination rates in acidic soil, along with increased root emergence in Bougainvillea (Fernando et al. 2022). The physiological characteristics (root length, shoot length, and biomass) of chickpea plants were improved by CuO nanomaterials, which acted as nano-carriers of IAA (Hanif et al. 2023).

22.2.3 Nano Pesticides

Another promising application of nanotechnology in agriculture is nano pesticides. One of the primary objectives of agrochemicals is to protect crops from pests, including pathogens, harmful insects and parasitic weeds, which drastically reduce agricultural productivity. However, the excessive use of pesticides has negatively affected the environment by diminishing the quality of soil, inducing pathogen resistance, increasing bioaccumulation of pesticides and eliminating beneficial symbiotic organisms (Chaud et al. 2021). Nanopesticides are plant protection products, where either the active ingredient or the carrier molecule is synthesized through nanotechnology (Kookana et al. 2014; Rajna et al. 2019). Some nanomaterials with specific antibacterial or insecticidal properties include, but are not limited to, silver (Kumari et al. 2017), MgO (Imada et al. 2016), TiO_2 (Rajna et al. 2019),CuO, and calcium oxide (Ayoub et al. 2018). Nano pesticides can also be developed using innovative formulations of nanocarriers such as nanoemulsions, nanosuspension, nanoencapsulation, nanogels, nanospheres, nanotubes, polymeric nanoparticles and nanofibers (Rajna et al. 2019). Nanocarriers are mostly based on materials like silica, lipids, polymers, copolymers, ceramic, metal, carbon and others (Chaud et al. 2021). Nanocarriers not only enable quick and efficient targeted delivery of pesticides but also ensure environmental safety (Singh et al. 2024). Nano pesticides enhance precision farming by facilitating more effective "smart field management." (Pradhan and Mailapalli 2020). These nano pesticides can be customized specifically to target agricultural pests with enhanced efficiency, offering improved durability and bioavailability without introducing any environmental hazards (Alam et al. 2024; Choudhury et al. 2012; Chowdhury et al. 2017). Nano-encapsulation technique is employed to improve the efficiency of pesticides (Scrinis and Lyons 2007). In the nano-encapsulation approach, a thin-walled shell acts as a protective layer over the nano-sized active pesticide component. The primary aim of the nanoencapsulation approach is to control the release of the active compound. This method enhances the effectiveness of the active substance while reducing pesticide usage, thereby protecting the environment (Ali et al. 2014).

Long-term usage of the same herbicide usually causes weeds to become resistant to it. To address this, herbicide-loaded nanoparticles have been developed to penetrate weed roots, translocate to cells, stop metabolic processes like glycolysis, and eventually kill the weeds (Ali et al. 2014). Nano herbicides, designed for target-specific delivery in weed roots, can be highly beneficial in eliminating weeds from crops in an environmentally friendly way (Perez-de-Luque and Rubiales 2009).

22.2.4 Nano Remediation

Nano remediation of soil offers an interesting approach for amplifying agricultural productivity. Soil quality is continuously depleted due to the presence of heavy metals, pesticides, persistent organic pollutants and excess nutrients. This depletion poses significant environmental concerns, affecting not just food security but also human well-being (Bakshi and Abilash 2020). The application of nanomaterials for the remediation or revitalization of soil has been investigated using a variety of methods, including adsorption, redox reactions, precipitation, co-precipitation, heterocatalysis, photodegradation, electro-nanobioremediation and soil immobilization (Mukhopadhyay et al. 2022; Rajput et al. 2022). Heavy metals such as Lead, Chromium, Copper, Cadmium and Mercury can be removed from soil through the applications of nanotechnology (Kristanti et al. 2021).

The distinctive characteristics of NPs, such as large specific surface area, microporous channels, network structure, and relatively high levels of adsorption capacity (Zahedifar 2021), contribute to enhanced efficiency in removing pollutants from vast areas *in situ*. Nano remediation for soil contamination has primarily utilized metallic and magnetic nanoparticles, carbon nanotubes, nanoscale zero-valent iron(nZVI), polymeric nanoparticles and nanoparticles supported on inert materials (Del Prado-Audelo et al. 2021; Adetutu et al. 2024). Among the nanomaterials, nZVI is commonly used for removing heavy metal pollutants due to its high efficiency in converting contaminants into less harmful compounds (Fajardo et al. 2012).

Kristanti et al. (2021) explored the efficacy of reduction reactions, particularly with nZVI nanoparticles, in removing heavy metals and organic compounds from contaminated soil and water sources. nZVI's widespread application is attributed to its small size and high surface area, facilitating direct contact with contaminants, and its ability to convert toxic compounds into less harmful forms. Moreover, the combination of biochar and nZVI enhances reduction reaction capacity, increasing removal efficiency, with notable success observed in reducing Chromium (VI) content in soil samples (Qiu et al. 2020). The combination of Carboxy methyl cellulose and CMC-stabilized nZVI showed significant capacity in converting Chromium (VI) contaminants into less bioavailable forms and efficiently removed organic contaminants such as dichlorodiphenyltrichloroethane (DDT), polychlorinated biphenyls (PCB), and pesticides from soil columns (El-Temsah et al. 2016). The dichlorination of herbicide atrazine in soil and water has been studied using nZVI, ensuring no toxic residues are left in the soil or environment (Satapanajaru et al. 2008).

Nanotechnology also plays a role in immobilizing and remediating soil contamination. Nanomaterials, such as carbon NMs, are crucial components of this technique due to their high adsorption abilities and structural configurations (Guerra et al. 2018; Zhang et al. 2019). Specifically, carbon nanotubes demonstrate significant potential for reducing organic contaminants like polycyclic aromatic hydrocarbons (PAHs) by hindering their movement and decreasing their bioavailability in soil environments (Guerra et al. 2018; Qian et al. 2020).

Metal oxide nanomaterials, including Iron (III) oxide and TiO_2, are effective in removing contaminants through surface complexation. Iron (III) oxide showcases a notable adsorption capacity, efficiently eliminating heavy metals like Cadmium (Qian et al. 2020), whereas TiO_2 demonstrates significant sorption rates, particularly in acidic soil conditions, thereby enhancing the surface complexation process for the efficient removal of Thorium (IV) (Tan et al. 2007). These findings emphasize the potential of metal oxide nanomaterials in implementing cost-effective and environmentally friendly soil remediation strategies.

Nano bioremediation, an alternative approach that employs microorganisms along with NMs to break down pollutant compounds and improve soil quality, is a cost-effective method which accelerates traditional bioremediation techniques (Singh et al. 2020a, b; Singh and Saxena 2022). By promoting the growth of microorganisms or immobilizing pollutants through the synthesis of microbial enzymes, nanomaterials help in bioremediation (Bhatt et al. 2022).

22.2.5 Nano Sensors and Precision Farming

Precision farming is another area where nanotechnology can be used to improve the efficiency of agricultural practices. This methodology seeks to maximize crop yields by applying precise amounts of various inputs like fertilizers, water and pesticides at the appropriate time, facilitated by measuring highly localized environmental variables (Manjunatha et al. 2019). Smaller detection limits, quick response time, high surface-to-volume ratio, and more reliable results make nano sensors an integral part of precision farming (Beegum and Das 2022). A nano sensor is capable of transmitting data and information about the properties and behavior of nanoscale particles to the macroscopic scale. Based on applications, nano sensors can be classified as chemical nano sensor, biosensor, electrometers, and deployable nano sensor (Pooja et al. 2021). Nano biosensors can be extremely useful in agriculture since they can detect and monitor several characteristics pertaining to plant health and environmental conditions (Kashyap et al. 2019; Shawon et al. 2020). Early detection of diseases, pests, and pathogens by these small sensors enables rapid intervention to reduce crop losses (Li et al. 2020). They provide accurate and real-time data on soil conditions such as nutrient levels, metabolites, pesticides, soil moisture, and temperature, allowing farmers to optimize the application of pesticides, water and other resources effectively (Chhipa 2019; El-Chaghaby and Rashad 2024; Shawon et al. 2020; Zain et al. 2024). Since they enable targeted and precise application of agricultural inputs, nano sensors help in reducing wastage and environmental impact (Garg et al. 2024). Nano sensors can also monitor environmental pollution, contributing to sustainable agriculture and environmental systems.

22.3 Future Trends in Agricultural Nanotechnology

Future prospects in agricultural nanotechnology point to continued advancements in improving crop yields, nutrition, and sustainability with the potential to revolutionize agriculture. Green technology innovations and other emerging technologies like Artificial Intelligence (AI) must be integrated with agricultural nanotechnology to create long-lasting and effective farming practices for the future. The anticipated progress in agricultural nanotechnology should be able to address global food security challenges by making agriculture more efficient, sustainable, and resilient to climate change (Fig. 22.1).

22.3.1 Aptamer Based Nanotechnology

Aptamers are single-stranded oligonucleotides or peptide molecules that serve as bioreceptors due to their high stability in various pH and temperatures and low development costs. They are usually 15 to 40 bases long and can form three-dimensional structures. Non-covalent interactions like van der Waals forces, hydrogen bonding, and electrostatic interactions enable them to be specific in binding to their targets (Mondal et al. 2022). The combination of aptamers and nanotechnology has accelerated advancements in various fields like disease diagnosis, drug delivery, and cancer chemotherapy (Liu et al. 2022). The potential of aptamers in agricultural nanotechnology has not yet been fully investigated. Aptasensor-based detection of pesticides has been developed by Talari et al. (2021). Nucleic acid aptamers are being used to construct aptasensors for the detection of pesticides, toxins, biomarkers of crops, and plant growth regulators (Hu et al. 2022). Further research and development of advanced nano sensors based on aptamers could lead to a significant breakthrough in accurately determining soil quality, thereby enabling

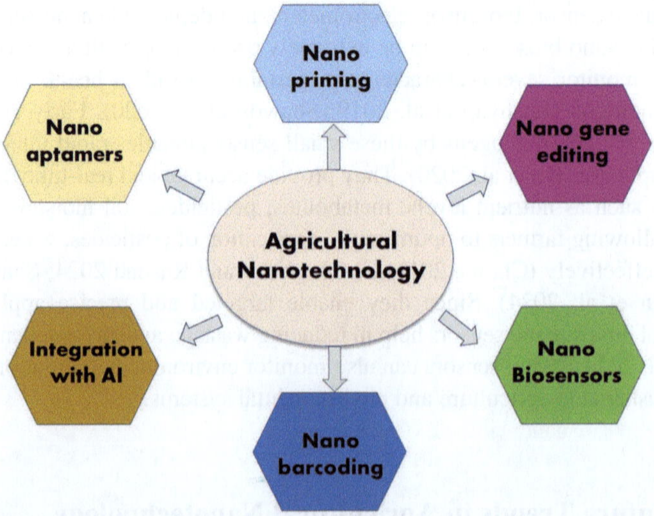

Fig. 22.1 Representation of future trends in agricultural nanotechnology. (Figure constructed by Dr. Viji Mary Varghese)

more informed and effective agricultural practices. Reports indicated that nano aptamers can potentially improve nutrient absorption by plants from the soil (Yadav et al. 2023).

22.3.2 *Nano Priming*

The purpose of seed priming is to accelerate germination while reducing the germination period. Nanotechnology-based seed priming is a novel way to increase germination, viability and yield. Seed primed with ZnONPs has been shown to mitigate both abiotic and biotic stresses, acting as a bio stimulant that enhances germination rate, seedling and plant growth, overall fresh weight, biomass, and photosynthetic efficiency. This beneficial effect is mediated through various mechanisms, including improved seed metabolic rate, vigor index, and seedling characteristics. ZnO NPs enhance nutrient uptake by altering seed metabolism, increasing α-amylase activity via gibberellic acid, and promoting the breakdown of stored starch and release of growth regulators. Furthermore, NPs can enhance stress tolerance and reduce the negative impacts of heavy metals on plant growth by modulating physiological and biochemical parameters, leading to decreased toxic metal accumulation and improved antioxidant enzyme activity (Donia and Carbone 2023). Graphene oxide nanoparticles, for example, can improve seed germination and protect against cadmium's adverse impact on seed germination and seedling growth (Shelar et al. 2023; Yin et al. 2018). Compared to the conventional method of hydropriming, nano priming has demonstrated a positive impact on biomass production in bitter gourds (Yeasmin et al. 2024). Watermelon seeds primed with iron nanoparticles exhibited more germination and seedling development (Kasote et al. 2019). Nano priming causes hydroxyl radicals to develop, which weaken cell walls and accelerate the hydrolysis of starch. It also activates the antioxidant and reactive oxygen species (ROS) systems in seeds (Nile et al. 2022). A further benefit of nano priming is that it increases disease resistance and decreases the need for fertilizers and pesticides.

22.3.3 Nanoparticle Mediated Gene Editing and Transformation

Advancements in nanotechnology have made it an efficient tool for plant genetic engineering, offering several advantages over conventional methods. Particles in the nanoscale are efficient carriers capable of delivering genetic materials into the target cells, enabling efficient gene transfer and stable genetic modifications (Ahmar et al. 2021). Conventional gene transfer methods often cause tissue damage and have low transformation efficiency. Nanotechnology-aided gene transfer has gained significant attention due to its remarkable traits and ability to mitigate the challenges posed by conventional methodologies (Saritha et al. 2022). Many nanomaterials are being investigated for their possible application in plant genetic engineering, including silica NPs, metal NPs, carbon-based nanotubes, polymer-based NPs, and nanomaterials with magnetic properties (Jat et al. 2020).

Nanoparticles can be designed with specific characteristics to optimize gene delivery. For instance, effective DNA delivery into the cells of both dicots (*Nicotiana tabacum* L) and monocots (*Oryza sativa* L) is possible with Au NPs (Saritha et al. 2022). According to Demirer et al. (2019) nanotubes made of carbon are capable of loading a substantial amount of plasmid DNA when the aspect ratio is increased. The transfer of foreign DNA into the pollen grains of Oil palm (*Elaeis guineensis* Jacq.) utilizing single-walled carbon nanotubes with imidazolium moieties (IM-SWCNTs) has been reported (Ahmar et al. 2021; Saritha et al. 2022). This method proves especially beneficial for economically significant crops such as palm, sorghum, and cotton, where plant regeneration via tissue culture has posed obstacles (Saritha et al. 2022). The precision and efficiency of crop improvement have been significantly enhanced by integrating nanotechnology with plant genetic engineering. It will also promote crop breeding. On further advancements in nanotechnology research, nanotechnology is anticipated to play an increasingly vital role in meeting the rising demand for food, medicine, and materials (Avnesh et al. 2023).

Nanoparticles with magnetic properties are being used for delivering nucleic acids to cotton, resulting in a stable genetic transformation (Zhao et al. 2017). Nanocarriers like Polyamidoamine (PAMAM) dendrimers and surface functionalized mesoporous silica nanoparticles (MSN) can also effectively deliver target genes into plant cells. PAMAM dendrimers, when complexed with DNA, can penetrate plant cell walls intact, enabling the expression of delivered genes (Pasupathy et al. 2008; Yashveer et al. 2014). MSN, with its porous structure, can encapsulate chemicals and gene fragments, allowing the delivery of multiple types of molecules into plant cells (Saritha et al. 2022).

Due to their ability to load and transfer plasmid DNA, mRNA, and RNPs, nanoparticles have become a promising delivery system for plant genome editing. According to reports, RNA interference (RNAi) molecules, small interfering RNA (siRNA), and plasmid DNA might be delivered through cell walls using a variety of nanoparticles, including clay nanosheets, DNA nanostructures, carbon nanotubes, and SiO2, without the need for mechanical assistance (Javaid et al. 2024). The CRISPR/Cas9 system, a third-generation genome editing tool, has been successfully utilized for crop improvement and protection, leveraging the advancements of nanotechnology (Ahmar et al. 2021). A notable example of this application is the delivery of SDN-1, a CRISPR tool, to wheat (*Triticum aestivum*L.) to confer resistance against powdery mildew (Gao 2018). The precise delivery of nucleic acid and protein into target cells, the crucial aspect of CRISPR/Cas9, is facilitated by nanotechnology (Saritha et al. 2022). It has been investigated whether direct cytoplasmic delivery of the Cas9 protein with the single guide RNA (sgRNA) complex can result in more effective gene editing (Wu et al. 2023).

NPs can be used for gene silencing by delivering dsRNA and is effective against pests and pathogens, disrupting pathogen cell walls, and protecting RNAi from degradation. According to studies, RNAi techniques based on nanocarriers are efficient against a variety of pests and pathogens, such as plant viruses and mosquitoes, and they hold significant promise in modern agriculture (Ghosh et al. 2023). There is much potential for managing pests and controlling diseases by spray-induced gene silencing employing nanoparticles. Nanovesicles can stop RNA from degrading quickly in the environment (Chen et al. 2023). Focusing on this field could lead to environmentally friendly agricultural pests and disease management systems.

Innovative plant genome editing techniques have opened novel avenues for speeding up crop enhancement and developing more sustainable agricultural practices. Specifically, the CRISPR platform, which originated from prokaryotes, enables targeted gene editing, facilitating the development of high-yield, stress-resistant crop varieties. Nanotechnology could further enhance these advancements by offering a rapid, universal delivery method for genome editing, applicable across various plant species (Sharma and Lew 2022).

22.3.4 Plant Disease Detection and Management Using Nanotechnology

Plant pathogenic diseases brought on by various microorganisms can cause a big implication on global economy due to catastrophic crop loss. Therefore, it is crucial to detect and manage these diseases at their very onset (Verinder et al. 2024). Living organisms and their components from an analyte can be detected using biosensor with the help of a transducer. Nano biosensors which are in nano scale size, can be used to detect very small levels of analytes. With its non-destructive, minimally invasive, cost-effective, and user-friendly systems that offer improved detection limits, sensitivity, specificity, and on-site plant pathogen detection, nano biosensor-based technology opens up new possibilities for plant disease diagnostic systems (Kashyap et al. 2019). An AuNP-based immunosensor was used to detect Karnal bunt disease in wheat plants, while a quantum-based biosensor was employed to identify the pathogen causing lime plant's Witche's broom disease (Alghuthaymi et al. 2021). Apart from nano biosensors, nano barcode technology is also emerging as a significant tool for pathological detection (Saritha et al. 2022). Plant disease diagnostics might undergo an unprecedented shift with the use of quantum dots nano barcode microbead systems (Eastman et al. 2006) and DL-DNA fluorescence nanobarcodes in multiplexed detections (Um et al. 2006).

22.3.5 Integration with Other Emerging Technologies

Many obstacles, including population growth, deteriorating soil quality, altered climate etc. are presenting risk to global food security. In order to find sustainable solutions, it is necessary to integrate knowledge from a variety of disciplines. More recently, drone technology, AI, machine learning, and Internet of Things (IoT) are being adopted in agricultural practices in order to mark a significant evolution.

Integration of agricultural nanotechnology with AI is a promising avenue for sustainable food production. AI can be applied to improve focus, absorption, distribution, and nutrient acquisition. Nanoscale agrochemicals can also be designed using AI, which has an optimal safety and functionality profile (Zhang et al. 2021). Nano informatics, utilizing AI and machine learning, plays a crucial role in relating nano structural features to functional properties (Burello and Worth 2011; Karatzas et al. 2020; Winkler 2020). While adopting agricultural nanotechnology, it is essential to ensure environmental and human safety, manage risks, and optimize the development of the technique. In nanotechnology, AI and machine learning have significantly improved computational risk assessment (Cohen et al. 2013), classification of nanomaterials (Lamon et al. 2018), safety-oriented design of nanomaterials (Varsou et al. 2019) and prediction of their interactions in various environments (Afantitis et al. 2018; Ban et al. 2020; Duan et al. 2020; Findlay et al. 2018). In agriculture, AI can predict nanomaterial association with soil, microbial ecosystem, and plant uptake, enabling precision farming (McManus et al. 2018; Zhang et al. 2019, 2020). Integration of AI and existing models can optimize nanomaterial use in agriculture, balancing safety and functionality for sustainable practices (Zhang et al. 2021). AI can also forecast weather patterns, pest infestations and potential crop diseases. For instance, Intel's OPENVINO toolkit, with its capacity to apply deep learning models for visual data processing, can be utilized for the identification of pathological diseases in plants.

Research is being conducted extensively on the use of machine learning in nanomaterial risk assessment and the creation of safe and eco-friendly nanomaterials. Emerging nano Quantitative Structure-Activity Relationship (QSAR) models connect specific properties of nanomaterials to their uptake and effects on cells or organisms. Additionally, these models can determine surface functionalization strategies aimed at enhancing or reducing protein binding and cellular association, which is vital for precision nano-agriculture (Zhang et al. 2021). Deep learning using automated data partitioning and assessment metrics is being developed to predict optimal pharmaceutical formulations and doses, which can be extended to precision nano agriculture for understanding factors controlling release rates and transport of nanomaterial carriers (Zhang et al. 2021). Even though NMs are promising in sustainable and precision agriculture, safety issues and hazards have to be resolved before they are widely adopted (Zhang et al. 2021). In silico computer-based chemical modelling technologies have the potential to facilitate the risk assessment of nano-enabled pesticides, but an evaluation of their suitability is needed (Zhang et al. 2021). Overall, AI and machine learning are critical for driving progress in nano-enabled agriculture, enabling the design of safe and sustainable nanomaterials for precision farming.

Incorporating AI in agricultural nanotechnology involves leveraging AI techniques like Artificial Neural Networks (ANNs) to enhance precision and sustainability in agriculture. By utilizing AI, particularly ANNs, in nanotechnology applications, several key benefits can be realized. In addition to providing useful insights for designing environmentally friendly nano enabled pesticides and fertilizers, Deng et al.'s (2023) machine learning models have been found to be effective in data interpretation and optimization, predicting plant responses to nanomaterials, uptake, and transport. Additionally, these models can facilitate more targeted and efficient agricultural practices by enabling the effective quantification and understanding of the collective effects of various physical and chemical factors, as well as the exposure dose.

By integrating AI into agricultural nanotechnology, researchers can enhance precision, sustainability, and productivity in agricultural methods. This will eventually lead to addressing global food security challenges and promoting environmentally responsible agriculture.

The uncertainty in nano-enabled agriculture is influenced by factors like the categories of nanoparticles, plant varieties, and ecological and climatic factors (Buriak et al. 2022). Traditional methods, such as QSARs and molecular dynamics simulations, have limitations in predicting systemic toxicity responses and NP translocation and accumulation at large scales (Geitner et al. 2017; Wang et al. 2017). To address these limitations, data-driven methods such as machine learning models are suggested to uncover intricate relationships and predict future trends (Yan et al. 2020; Yu et al. 2021).

Plant and nanoparticle interaction studies have been made possible by the introduction of machine learning models, revealing hidden relationships along with crucial roles of nanoparticle features and climatic/environmental factors (Deng et al. 2023). Nano informatics holds significant potential in the realm of nano-enabled farming, a field which remains largely unexplored. This includes predicting the interactions and impacts of nanomaterials with rhizosphere secretions, understanding nanomaterial changes prior to and at the time of uptake and transport, assessing the effects of nanomaterials on microbes present in the soil, and forecasting plant absorption after foliar treatment. Emerging experimental data are beginning to shed light on all these aspects (McManus et al. 2018; Zhang et al. 2019, 2020).

22.4 Challenges and Limitations of Agricultural Nanotechnology

Although nanotechnology is at the forefront of scientific advancement, offering nanoparticles that have the potential to benefit agriculture, it is also important to carefully consider their negative impacts. Multiple challenges and limitations need to be taken into account (Majumdar and Keller 2021; Shukla et al. 2024) (Fig. 22.2).

22.4.1 Environmental Challenges

The application of nanotechnology in agriculture raises concerns about its potential environmental impact. The introduction of nanoparticles into air, water, and soil ecosystems could lead to water pollution, soil degradation, harmful effects on

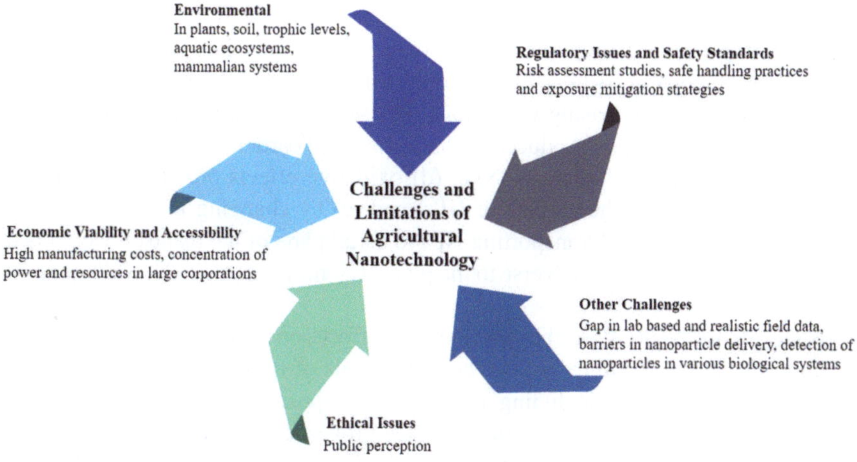

Fig. 22.2 Representation of challenges and limitations of agricultural nanotechnology. (Figure constructed by Dr. Gigi Poulose)

non-target organisms, and contributions to climate change. The long-term persistence and dynamics of nanoparticles in the environment are not well understood, making it challenging to accurately assess their ecological risks. The potential accumulation of nanomaterials in soil, water, and plants, along with their possible incorporation into the food web, adds further complexity. The toxic effects of nanoformulations can affect both living and non-living components of the ecosystem, causing detrimental impacts on various forms of life. Upon releasing into the environment, nanoparticles may interact with existing pollutants, potentially causing poorly understood biological cascade effects (Deng et al. 2017; Jan et al. 2022; Nile et al. 2020; Sadeghi et al. 2017; Siddiqui et al. 2022).

Nanoparticles are being introduced into agriculture by different sources such as nano pesticides, nano herbicides, genetic engineering, agrochemical delivery, soil condition sensors, pathogen detection, and nano fertilizers (Ameen et al. 2021). NPs can be efficiently stored and delivered under controlled conditions through nanoscale carriers with the principles of encapsulation, entrapment, polymers and dendrimers, surface ionic attachments, and weak bond attachments (Ditta 2012; Mukherjee et al. 2019). A variety of nanomaterials are being developed to enhance food safety and agricultural productivity, stimulate plant growth, and promote biofortification techniques. These include nanometals, nanorods, nanofilms, nanotubes, nanofibers, nanolayers, nano-emulsions, nano-suspensions, nanospheres, nanoencapsulations, nanogels, nanodiamonds, nanofoams, nanofoils, and nanosheets (Ansari 2023; Gupta et al. 2023a, b; Kapoor et al. 2022; Mwaanga 2018). Accumulating nano formulations have an adverse effect on the ecosystem as well as the flora and fauna that inhabit it (Saritha et al. 2022). Nano phytotoxicity poses a significant risk and could endanger the agricultural industry's future.

22.4.1.1 Concern in Plants

In plants, exposure to nanoparticles produces a range of physiological and biochemical effects. Nanoparticles interact with living things in five different ways (Dietz and Herth 2011): (a) by releasing metal ions into solution evoking a chemical effect; (b) the mechanical effects produced by hard spherical particles and their aggregates; (c) the surface catalytic effect of NPs; (d) oxidative effects produced by NPs' or binding of proteins to the surface of NPs; and (e) by changing the pH. Plants are capable of absorbing and transporting NPs to aerial parts of the plants, the effects of which may be beneficial or adverse to the plants (Gorczyca et al. 2021; Siddiqui et al. 2015).

Nano phytotoxicity describes the harmful effects on plants or the environment resulting from the exposure to toxic concentrations of nanomaterials (Ahmar et al. 2021). Various factors, including unsuitable composition, dosage, particle size, application technique, exposure time, surface properties, chemical attributes, plant type, growth stage, nanoparticle stability, pore dimensions, and bioavailability, can negatively influence the effectiveness of NPs in agriculture (Li et al. 2022; Ma et al. 2010). These factors are essential in figuring out how NPs affect plants (Shukla et al.

2024). NPs in plants can disrupt structural and functional stability of cellular and sub-cellular organelles, modify the composition of proteins, lipids, and nucleic acids through the generation of hydroxyl radicals (Rajput et al. 2020).

The positive and negative impacts of nanoparticles on the plant life cycle have been enumerated by Shukla et al. (2024). The negative effects include disruption of the growth of the reproductive system, decreased rate of germination of seeds, inhibited elongation of the roots, increased production of reactive oxygen species and oxidative damage to lipids, decreased content of chlorophyll, necrosis, delayed flowering, disrupted seed development, reduced photosynthetic efficiency, impaired chlorophyll function, decreased biomass production, lowered transpiration rates, mitochondrial malfunction, and other related issues (Gao et al. 2023; Juárez-Maldonado et al. 2018; Majumdar and Keller 2021; Prasad et al. 2020; Shukla et al. 2024; Wang et al. 2012; Yin et al. 2011). Accumulation of NPs can affect nutrient uptake, stomatal conductance, chloroplast microstructure, and secondary metabolic processes, protein synthesis, repression and expression. They can also enhance chromatin compaction and regulate gene expression, leading to DNA disorganization causing its modification/ damage. Research indicates that ZnO and CuO nanoparticles can traverse different routes within plant tissues, influencing biological nitrogen fixation and causing damage to plant cells (Rajput et al. 2020). Further research is needed on the application, deposition, distribution, bioaccumulation, and biotransformation of nanoparticles within plant cells (Ahmar et al. 2021; Mwaanga 2018).

22.4.1.2 Concern in Soil

Of all the inorganic pollutants that potentially accumulate in soil, NMs are transforming into major ones by undergoing bio-geotransformation processes such as clumping, disintegration, adherence, corrosion, reduction, and assimilation (Grün et al. 2019; Rajput et al. 2023).

Soil microflora plays a vital role in ecosystem health and soil fertility and the quality of microorganisms inhabiting it is influenced by their nature (Nkongolo and Narendrula-Kotha 2020). Nanoparticles, spread through industrial discharges, waste sites, and wastewater, can undergo bio-geotransformation causing toxic effects on soil microbiota. The toxicity of nanoparticles to beneficial soil bacteria strains is influenced by several factors, such as the size, shape, chemical composition, concentration, and exposure duration of the nanoparticles (Ali et al. 2021b). Nitrogen-fixing bacteria and soil-decomposing microorganisms are vulnerable to the effects of nanoparticles, as marked by the drop in Arbuscular mycorrhiza growth in soybeans (Burke et al. 2015). ZnO nanoparticles and cerium dioxide nanoparticles (CeO2) have been shown to adversely affect nitrogen-fixing microorganisms, decomposer bacteria, and methane-oxidizing microbes. ZnO NPs reduced enzyme activity in *Pseudomonas aeruginosa* and *Bacillus subtilis*, while CeO2 NPs inhibited growth and altered protein structure in *Sinorhizobium meliloti* (Dimkpa 2014). The adverse effects arise from the interaction of nanoparticles with microbial cell biomolecules, leading to disruptions in metabolism, protein expression, and cellular stability (Achari and Kowshik 2018).

The aggregation rate and stability of NPs are influenced by soil properties, such as water content, organic matter, and pH (Benoit et al. 2013). Studies on Ag NPs' effects on soil microbial communities discovered that they had a notable impact (Johansen et al. 2008). The accumulation of soil nanoparticles like CuO- and ZnO-NPs can affect the nutrient cycle negatively. The activity of enzymes in important soil microbiota such as *Rhizobiales*, *Azotobacter*, and *Sphingomonas* can also be negatively affected (Rajput et al. 2020). Nanoparticles disrupt the shape of cells, reduce soil microbial viability, and block some molecules necessary for soil fertility (Ali et al. 2021b). The phytohormone IAA, produced by symbiotic and nitrogen-fixing bacteria, plays a vital role in plant development and governs numerous cellular functions (Enders and Strader 2015; Spaepen et al. 2007). As cited by Ahmed et al. (2020) the exposure to ZnO NPs and Ag NPs inhibited IAA secretion as a function of dose, and complete inhibition achieved at a dose of 1000 μgmL^{-1}. Nanoparticles of Ag and ZnO also possess the ability to disrupt bacteria's surface adhering ability and subsequent inhibition of growth. There are various factors which determine the toxicity of nanoparticles towards bacterial cells. These factors include the species of the bacteria, type of metal oxides, and the dose of nanoparticles used (Ameen et al. 2021; Kaweeteerawat et al. 2015). Nanoparticles can harm non-target soil microflora, inhibit target organism growth, and cause toxicity to non-target microbes.

Research on nanotechnology's potential to enhance the rhizosphere microbiome is limited due to limited experiments under realistic conditions. Understanding the involvement of nanomaterials in enriching the microbial community is vital for enhancing soil characteristics and addressing soil stresses, including heat, drought, salinity, nutrient deficiency, and toxic elements. A thorough understanding of the ecological behaviors of nanomaterials, in conjunction with soil microorganisms, is essential for the safe use of nano-enabled products in agriculture. Thorough examination of different-level interactions between plants and nanomaterials is also necessary to minimize phytotoxic effects and maintain soil health (Rajput et al. 2023).

22.4.1.3 Concern in Trophic Levels

Nanoparticles can be transferred across various trophic levels when organisms consume plants treated with nanoparticles, thereby moving via the food chain. Although this aspect of nanotechnology remains underexplored, the accumulation of nanoparticles at different levels could adversely impact progeny and metabolic processes (Cedervall et al. 2012; Juárez-Maldonado et al. 2018). While data on the biomagnification of nanoparticles through the food chain is limited, some notable examples include the transfer of quantum dots (QDs) from algae to *Ceriodaphnia dubia* and from ciliated protozoans to rotifers. Additionally, TiO2 nanoparticles were observed to transfer from daphnia to zebrafish (Deng et al. 2014).

Bullfrogs (*Rana catesbeina*) were found to be more accessible to nano Au via trophic transfers, such as earthworms to bullfrogs, than when bullfrogs were directly exposed to soil contaminated with nano Au (Unrine et al. 2010, 2012). These findings foster worries about the possibility of trophic exposure and biomagnification from engineered nanomaterials (ENMs) in agriculture (Saritha et al. 2022). Common metal oxide nanoparticles like CuO, ZnO, TiO2, stannic oxide (SnO2), CeO2, and ferric oxide (Fe2O3) have been linked to various adverse effects on organisms (Matranga and Corsi 2012; Mwaanga 2018; Mwaanga et al. 2017; Pérez et al. 2009). Nanomaterials have been shown to impact the gut microbiota of various organisms in both *in vivo* and *in vitro* studies, with effects varying based on factors such as size, dose, duration, and functional groups. Increasing evidence points to dose-dependent adverse effects (Utembe et al. 2022).

Nanoparticle-infused pesticides (NBPs) have been observed to negatively affect bees, which are essential pollinators. Bees may come into contact with NBPs during pollen collection, leading to increased toxicity in beehives. Additionally, the accumulation of NBPs in the atmosphere could pose a lethal threat to bee populations, potentially disrupting pollination processes (Hooven et al. 2019).

22.4.1.4 Concern in Aquatic Ecosystems

The increasing concentrations of nanoparticles have the potential to infiltrate aquatic ecosystems, potentially driving toxicity in these environments. Although studies are limited, engineered nanoparticles (ENPs) have been observed to affect three levels of the trophic hierarchy within aquatic food webs. The biomagnification ratio of various nanoparticles (such as QD, Au, CeO_2 and TiO_2) exceeds one across two trophic levels, indicating accumulation through the food chain. ENPs have been found to impact the physiological processes of a wide range of organisms, including both flora and fauna within aquatic habitats (Krishna and Sachan 2021).

22.4.1.5 Concern in Mammalian Systems

The entry of NPs to the human body and its harmful impacts on humans is gaining serious concern (Verma et al. 2019). It has been identified that there are situations of direct entry into life sustaining organs like the brain, heart, liver, blood, and spleen. The significant surface area compared to volume of nanoparticles allows them to easily enter the nucleus, potentially damaging DNA (Diengngan 2019). The resemblance of nanomaterials with cell organelles enhances their binding capability and cellular engagement. The understanding of NP-cellular interactions remains limited, with research still in its early stages (Sadeghi et al. 2017). Certain studies highlight the adverse effects of NPs on gut microbiota, including *E. coli* within the digestive system (Ali et al. 2021b; Radziwill-Bienkowska et al. 2018; Utembe et al. 2022; Waller et al. 2017; Xuan et al. 2023). Studies have investigated the impact of nanoparticles on cell cultures, mice, and advanced models such as 3D organoids to explore their carcinogenic potential and underlying mechanisms. Research on the impact of NPs on different human systems—such as the respiratory, digestive, nervous, endocrine, immune, and reproductive systems—has identified outcomes like ROS buildup, mitochondrial dysfunction, inflammation, DNA damage, cell cycle disruption, and changes in epigenetic regulation. Nanoparticles have been shown to induce brain inflammation (caused by Fe2O3 NPs), disrupt signaling pathways and neurotransmitter balance (caused by TiO2 NPs), and trigger oxidative stress along with mitochondrial dysfunction (caused by Ag NPs) in the human body (Ali et al. 2021b; Aschner et al. 2023).

Studies reveal the negative effects of nanoparticles in humans, environment and ecosystem organisms. NPs affect all organisms, including essential components for human survival. Future research should focus on mitigating adverse effects and remediating and managing NPs released into the environment (Saritha et al. 2022).

22.4.2 Other Challenges

A major obstacle is the substantial disparity between the interpretation of lab-based results and actual field data (Ali et al. 2021b). The barriers encountered in nanoparticle delivery to plants have been analyzed and it has emphasized the need for innovative approaches (Sembada and Lenggoro 2024). The complex biological structure and small size of plant cells can hinder the detection of nanoparticles (Laborda et al. 2016; Praetorius et al. 2017). Since nanoparticles are comparable in size to plant cells, visualizing and accurately quantifying their presence becomes challenging (Ma et al. 2010; Nima et al. 2014). When NPs are mixed with biological fluids, physicochemical characterization is crucial to determine the effects on protein and the subsequent physiological changes (Ali et al. 2021b; Ray et al. 2009). High spatial resolution is also difficult, especially for *in vivo* imaging techniques, and it may involve the disintegration of the sample for analysis (Nečemer et al. 2008).

Even though hundreds of research studies have been conducted over the past decade, the approaches in each study vary significantly in terms of mechanisms of action, results, insufficient controls, and the lack of standardized analytical methods. When evaluating the toxicity of nanoformulations, it is essential to include surface-area dosimetry in the methodology. Majumdar and Keller (2021) highlight the importance of utilizing multiomics to identify biomarkers upregulated in response to ENMs, providing mechanistic insights and aiding in the design of safer agricultural products.

However, there are no studies that show the impact of NPs after a continuous study of an experimental area or its surroundings for more than a decade. Exposure to NPs over a couple of decades can result in changes in ecosystems, including humans. Traditional animal tests are costly and time-consuming, and the data from these tests may have inherent uncertainty (Mwaanga 2018). Toxicity evaluation needs to begin on formulations of nanoparticles, and it should be evaluated involving at least five species from different trophic levels to extract predicted no-effect concentrations (Krishna and Sachan 2021).

The toxicity caused by nanoparticles needs to be studied in detail to reduce their adverse effects so that they can be used to solve major issues like climate change and the food crisis. A comprehensive analysis of how different plant species respond to specific nanoparticles emphasizes the need to establish tailored application conditions (Ahmad et al. 2022).

22.4.3 Regulatory Issues and Safety Standards

Nanomaterials pose potential risks to human health and the environment due to their small size and high surface area. Exposure to certain NPs has been linked to adverse health effects, including respiratory problems, inflammation, and genotoxicity. The potential accumulation in food crops raises concerns about food safety and consumer exposure. To address these toxicity concerns, rigorous risk assessment studies, safe handling practices and exposure mitigation strategies are required.

Tracking the fate of nanomaterials and their potential adverse effects on ecosystems is of utmost interest to nanotoxicologists as these may lead to the dysfunction of agrocenosis components (Kookana et al. 2014; Navarro et al. 2008; Rana and Kalaichelvan 2013). However, the task is not simple, and public doubt about the fate of nanoparticles has led to concerns about their direct impact on the environment (Gorczyca et al. 2021). If toxicity is detected, scientists can modify nanomaterials to make them non-toxic, which will accelerate approval for commercial production and ensure market compatibility. The "Safe-by-Design" concept encourages agri-food companies to invest more in R&D for nanotechnology, leading to the development of safer and more advanced nano products. Rigorous safety testing of these new nano products will promote collaboration with professional associations and social experts, thereby increasing public acceptance of the products (Ali et al. 2021b). The challenges related to human safety, the environment, and the ecosystem remain unresolved, particularly in relation to nanomaterials and the agri-food chain, which can have harmful effects on human health and the environment due to their non-target areas (Mukhopadhyay 2014).

Nanotechnology requires a regulatory system to monitor safety standards. Global agencies like UNEP (United Nations Environmental Program), EUSCA (European Union Scientific Committees and Agencies), JMPR (Joint FAO/ WHO Meeting on Pesticide Residues), FDA (United States Food and Drug Administration), ISO (International Standard Organization), OECD (Organization for Economic Cooperation and Development), REACH (Registration, Evaluation, Authorization and Restriction of Chemicals) and UNITAR (United Nations for Training and Research) provide guidelines and policies for safe nanoproduct usage. CLP (Classification, Labelling and Packaging) serves as a complementary tool for REACH and OECD (Chávez-Hernández et al. 2024; Saritha et al. 2022). As the nano-Agri sector grows, new approaches like self-regulation, co-regulation, and meta regulation are being recognized (Abbot 2012). Self-regulation involves private agencies, industries, and academic institutions assessing their own activities, while co-regulation involves state agencies and industry firms collaborating. Meta regulation involves regulatory agencies like government bodies, Non-Governmental Organizations (NGOs), private agencies, research institutions, and third-party companies, regulated by regulators (Saritha et al. 2022). The main motive behind meta regulation is to instill social values like occupational health and consumer safety, known as 'responsibilisation'. Consequently, every country has its own frameworks, regulations and policies to adopt methods regarding nanotechnology and nanoproducts (He et al. 2019; Kumari et al. 2023; Saritha et al. 2022; Soltani and Pouypouy 2019).

Regulation in nanotechnology involves two forms: mandatory and voluntary, with voluntary regulation being mainly used as a reference in emerging technologies like nanotechnology. Regulation can be horizontal or vertical, with horizontal regulation affecting two or more sectors and vertical regulation affecting a particular sector. Due to concerns about commercialization, health and environmental impact, voluntary regulation is essential for industries to demonstrate consumer responsibility in using nanotechnology products. There are various types of voluntary regulations, including registers, labeling, codes of conduct, risk management systems, guidelines, and technical standards. Since the properties of NMs may change throughout their life cycle, continuous assessment is essential for effective regulation. Regulatory agencies worldwide should collaborate to advance research and implementation in nanotoxicology. By promoting actions to establish regulations, frameworks, and legislation for NMs, we can ensure sustainable development and enhance the quality of life for all people (Chávez-Hernández et al. 2024).

The European Union (EU) has started to adopt regulations for ENMs in food and agriculture, but the complex nature of agricultural systems makes this difficult. To address safety concerns and inappropriate generalization, ongoing toxicity testing projects should fill the regulatory gap. EU and European Nano Safety Cluster jointly produced a document identifying four major areas for nano safety research during 2015–2025 which could greatly benefit our understanding of ENM features, exposure, hazard mechanisms, risk assessment, and management (Iavicoli et al. 2017; Lowry et al. 2019; Savolainen et al. 2013).

Current risk assessment approaches for ENMs in food and agriculture are suitable, but workers and animals may be exposed to these materials through oral routes, inhalation, and skin contact. The number of published risk assessments is growing slowly, but it is difficult to determine the true number of substances assessed and specific nanomaterial-related issues. Particle risk assessment is a new field, and it needs to be done on a case-by-case basis, considering factors like size, shape, chemical composition, surface area, and surface charge (FAO/WHO Technical Paper 2013; Park et al. 2010).

Global efforts are underway to regulate nanotechnologies through advice, supervision, and legislation. However, no single law is fully committed to these directives. Existing laws are considered adequate, but modifications have been recommended by various governments and NGOs (Pandey and Jain 2020). Building consensus on risks and benefits is crucial for effective regulatory policy and public trust (Sylvester et al. 2009). A One Health approach for managing nanotechnologies in agriculture can provide indirect benefits (Lombi et al. 2019). Concerns about toxicity and inconsistent global surveillance policies hinder the development and promotion of nanomaterials in farming (Gupta et al. 2023a, b). Prioritizing risk assessment and risk management is essential for efficient regulatory measures (Mishra et al. 2017).

22.4.4 *Ethical Issues*

Nanotechnology has numerous benefits, but it also presents potential risks and ethical issues (Ferreira and Filipe 2022). Some nanoparticles could be toxic due to their nanoscale behavior, which could cross the blood-brain barrier. When nanoscale systems are designed, it is essential to inform all stakeholders about the short-term and long-term advantages as well as the limitations and hazards of nanotechnology (Khan 2015). Ethical concerns in agri-food nanotechnology include non-malfeasance, food and process safety, the need for effective risk assessment, transparency, consumer benefits, animal welfare, and environmental protection. The European Commission and other regulatory bodies recognize that existing regulatory instruments and risk assessment processes may need to be adapted to accommodate nanotechnology applications (Coles and Frewer 2013). Ethical issues may also arise regarding the public's perception and awareness of nanotechnology in agriculture. Factors influencing public perception include whether to continue using nano systems, mandate labelling of nano-enabled agricultural products, funding for nanotechnology development, and the amount of information the public should be made aware of the nanotechnology in the agricultural value chain. If these issues are not fully addressed, the negative perception and reluctance to accept nanotechnology in agriculture may be experienced again (Mwaanga 2018).

22.4.4.1 Public Perception

Research indicates that public awareness and interest in nanotechnology are low, despite its exponential growth in commercial application over the past decade. Nanotechnology's unique physicochemical properties can optimize agricultural systems with precision farming. However, there are moral concerns about nano pollution and nanoparticle-based hazards. Public opinion surveys can help understand the community's perception of nanotechnology and its risks. Risk perception varies among individuals based on their personality and socio-cultural background. Researchers analyze the findings of the previous studies on nanotechnology by meta-analysis to reach sensible conclusions (Satterfield et al. 2009). To have a change in certain cultural/ psycho-social beliefs and attitudes, accurate information about NPs' potential and risks, open access research reports, open discussions, debates, and inclusion in schools and educational curriculums should be ensured (Mukherjee et al. 2019). Societal stigma must also be addressed to improve public acceptance of nanotechnology. The widespread implementation of nanotechnology in agriculture could potentially displace traditional farming practices and lead to a loss of employment in the agricultural sector (Shukla et al. 2024). Concerns about potential health impacts, side effects, long-term effects, and overall safety are also raised. Scientific literacy remains low worldwide, and personal beliefs and perceptions about science and technology influence public responses. To increase consumer acceptance, informative and educational programs should be implemented to communicate aspects related to nanotechnology and its uses in all sectors (Ali et al. 2021b; Chávez-Hernández et al. 2024; Giles et al. 2015; Saritha et al. 2022; Sylvester et al. 2009).

22.4.5 *Economic Viability & Accessibility*

The adoption of agricultural nanotechnology has socioeconomic implications, especially for small-scale farmers and developing countries. The high cost of nanotechnology-based input may exacerbate existing inequalities in access to agricultural resources and technologies, widening the gap between large commercial farms and small holder producers. Reliance on nanotechnology could lead to dependency on external inputs and undermine traditional farming practices. The scalability and cost-effectiveness of nanotechnology applications in agriculture may limit widespread adoption in resource-limited agricultural systems (Ali et al. 2021b). Higher regulatory limitations can cause financial investments to decline, and big players in the agri-food sector may lose confidence in nanotechnology's power. Intellectual property rights (IPRs) play a crucial role in supporting innovative ideas, but they increase prices and create a gap at the economic level (Chugh and Jahangir Dar 2021). The main obstacles to the commercialization of nanotechnology include high production costs, challenges in the research and development of nanoproducts, difficulties in scaling up production, and low public awareness about the safety of nanoproducts (Gondal and Tayyiba 2022). While nanotechnology has the potential to make a positive impact on agriculture, there is a risk that production could become concentrated in the hands of large companies, potentially leading to economic imbalances and increased inequalities (Shukla et al. 2024).

22.5 Conclusion and Prospects

Agriculture plays a critical role in meeting global food demand, ensuring food security, supporting economies and promoting sustainable development. Progress in agriculture, characterized by economic viability, environmental sustainability and efficiency, is crucial for addressing the limitations of natural resources and increasing global population. Throughout the world, farmers are compelled to concentrate on applying cutting-edge discoveries and technological advancements to increase agricultural yields. Agricultural nanotechnology, the application of nanotechnology in agriculture, has gained significant momentum in recent years by offering creative solutions to the most pressing challenges. Because of their distinct physiochemical and biological characteristics, nanomaterials—which are mostly measured in nanometers—are well-suited for a variety of agricultural applications. Current advances in agricultural nanotechnology include the use of nano fertilizers, nano pesticides, nanocarriers and nano sensors, which enable crop improvement, precision farming, nano remediation and soil nutrient replenishment. The advancements in nanotechnology also make it possible for agriculture to become digital. Nanosensors facilitate the development of high-tech agricultural farms by providing real-time analytical data on soil conditions and plant health, as well as optimizing and automating irrigation systems. Another promising application of nanotechnology is the use of nanobiosensors for early-stage pathogen detection, enabling effective disease management. Recent research has shown the potential of nanotechnology to improve seed germination and tolerance to abiotic stresses through nano priming. Moreover, nanotechnology can also play a remarkable role in crop improvement by using nanocarriers for genetic manipulation. Although a number of obstacles must be addressed, including off-target editing, low efficiency, and nanostructure optimization, direct germline editing and plastid and mitochondrial genome modification may be possible through the use of several nano-delivery techniques. Integration of agricultural nanotechnology with other emerging technologies like AI and machine learning will ultimately enhance the precision and productivity in farming practices.

Agricultural nanotechnology is emerging as a transformative force in modern farming; however, it must ensure the preservation of human health, soil microbiota, and soil fertility over long-term use. The emphasis should be on sustainable practices and scalable applications that address the unique needs of diverse farming systems. Before the widespread adoption of nanotechnology in agriculture, it is crucial to understand the long-term effects of nanomaterials on the environment and human health. Therefore, conducting comprehensive risk assessment studies and implementing safety regulations is essential. The molecular mechanisms and signaling cascades activated in plant cells by the application of nanomaterials, particularly during nanopriming and gene editing using nanocarriers, need to be extensively studied. It is vital to carefully evaluate the benefits and drawbacks of nanotechnology, considering the challenges and constraints associated with its applications. Governments should allocate funds to establish proper infrastructure and recruit qualified personnel for innovative research and risk assessment of nanomaterials in agriculture. Collaborations among governments, industry, and academia will play a critical role in driving the adoption and innovation of agricultural nanotechnology.

References

Abbot C (2012) Bridging the gap – non-state actors and the challenges of regulating new technology. J Law Soc 39(3):329–358. https://doi.org/10.1111/J.1467-6478.2012.00588.X

Achari GA, Kowshik M (2018) Recent developments on nanotechnology in agriculture: plant mineral nutrition, health, and interactions with soil microflora. J Agric Food Chem 66(33): 8647–8661

Adetutu BO, Buba AB, Akinhanmi F et al (2024) Relevance of nanotechnology in agriculture. In: Adetunji CO, Egbuna C, Ficai A et al (eds) Handbook of agricultural biotechnology. Scrivener Publishing LLC. https://doi.org/10.1002/9781394234769.ch72

Afantitis A, Melagraki G, Tsoumanis A et al (2018) Nano informatics decision support tool for the virtual screening of gold nanoparticle cellular association using protein corona fingerprints. Nanotoxicology 12:1148–1165

Ahmad A, Hashmi SS, Palma JM et al (2022) Influence of metallic, metallic oxide, and organic nanoparticles on plant physiology. Chemosphere 290:133329. https://doi.org/10.1016/J.CHEMOSPHERE.2021.133329

Ahmar S, Mahmood T, Fiaz S et al (2021) Advantage of nanotechnology-based genome editing system and its application in crop improvement. Front Plant Sci 12:663849. https://doi.org/10.3389/FPLS.2021.663849/BIBTEX

Ahmed B, Ameen F, Rizvi A et al (2020) Destruction of cell topography, morphology, membrane, inhibition of respiration, biofilm formation, and bioactive molecule production by nanoparticles of Ag, ZnO, CuO, TiO2, and Al2O3 toward beneficial soil bacteria. ACS Omega 5(14): 7861–7876. https://doi.org/10.1021/ACSOMEGA.9B04084/SUPPL_FILE/AO9B04084_SI_001.PDF

Akhtar N, Ilyas N, Meraj TA et al (2022) Improvement of plant responses by nano biofertilizer: a step towards sustainable agriculture. Nanomaterials (Basel) 12(6):965. https://doi.org/10.3390/nano12060965

Alagoz SM, Arvanaghi HR, Dolatabadi N et al (2022) Impact on nutritional status of plants treated with nanoparticle. In: Toxicity of nanoparticles in plants: an evaluation of cyto/morpho-physiological, biochemical and molecular responses. Academic press, pp 333–358. https://doi.org/10.1016/B978-0-323-90774-3.00005-2

Alam MW, Junaid PM, Gulzar Y et al (2024) Advancing agriculture with functional NM: "pathways to sustainable and smart farming technologies". Discov Nano 19:197. https://doi.org/10.1186/s11671-024-04144-z

Alghuthaymi MA et al (2021) Nanohybrid antifungals for control of plant diseases: current status and future perspectives. J Fungi (Basel) 7(1):48. https://doi.org/10.3390/jof7010048

Ali MA, Rehman I, Iqbal A et al (2014) Nanotechnology, a new frontier in agriculture. Adv Life Sci 1(3):129–138

Ali S, Mehmood A, Khan N (2021a) Uptake, translocation, and consequences of nanomaterials on plant growth and stress adaptation. J Nanomater 2:1–17

Ali SS, Al-Tohamy R, Koutra E et al (2021b) Nanobiotechnological advancements in agriculture and food industry: applications, nanotoxicity, and future perspectives. Sci Total Environ 792: 148359. https://doi.org/10.1016/J.SCITOTENV.2021.148359

Altammar KA (2023) A review on nanoparticles: characteristics, synthesis, applications, and challenges. Front Microbiol 14:1155622. https://doi.org/10.3389/fmicb.2023.1155622

Ameen F, Alsamhary K, Alabdullatif JA et al (2021) A review on metal-based nanoparticles and their toxicity to beneficial soil bacteria and fungi. Ecotoxicol Environ Saf 213:112027. https://doi.org/10.1016/J.ECOENV.2021.112027

Anil BC, Savita A, Shubhan S et al (2024) Application of nanotechnology in agricultural systems: improving efficiency and sustainability. IJISAE 12(10s):7–12. https://ijisae.org/index.php/IJISAE/article/view/4343

Ansari MA (2023) Nanotechnology in food and plant science: challenges and future prospects. Plants (Basel, Switzerland) 12(13). https://doi.org/10.3390/PLANTS12132565

Aschner M, Skalny AV, Santamaria A et al (2023) From mechanisms to implications: understanding the molecular neurotoxicity of titanium dioxide nanoparticles. FBL 28(9):204. https://doi.org/10.31083/J.FBL2809204/2768-6698-28-9-204/FIG1.JPG

Avellan A, Schwab F, Masion A et al (2017) Nanoparticle uptake in plants: gold nanomaterial localized in roots of Arabidopsis thaliana by X-ray computed nanotomography and hyperspectral imaging. Environ Sci Technol 51:8682–8691. https://doi.org/10.1021/acs.est.7b01133

Avnesh K, Varnika R, Sudesh KY et al (2023) Nanotechnology as a powerful tool in plant sciences: Recent developments, challenges and perspectives. Plant Nano Biol (5). https://doi.org/10.1016/j.plana.2023.100046

Ayoub HA, Khairy M, Elsaid S et al (2018) Pesticidal activity of nanostructured metal oxides for generation of alternative pesticide formulations. J Agric Food Chem 66:5491–5498

Bakshi M, Abilash PC (2020) Nanotechnology for soil remediation: revitalizing the tarnished resource. In: Singh P, Borthakur A, Mishra PK, Tiwary D (eds) Nano-materials as photocatalysts for degradation of environmental pollutants. Elsevier. https://doi.org/10.1016/B978-0-12-818598-8.00017-1

Ban Z, Yuan P, Yu F et al (2020) Machine learning predicts the functional composition of the protein corona and the cellular recognition of nanoparticles. PNAS 117:10492–10499

Barrena R, Casals E, Colón J et al (2009) Evaluation of the ecotoxicity of model nanoparticles. Chemosphere 75(7):850–857

Baz H, Creech M, Chen J et al (2020) Water-soluble carbon nanoparticles improve seed germination and post-germination growth of lettuce under salinity stress. Agronomy. https://doi.org/10.3390/agronomy10081192

Beegum S, Das S (2022) Nanosensors in agriculture. In: Ghosh S, Thongmee S, Kumar A (eds) Agricultural nanobiotechnology. Woodhead publishing, pp 465–478

Benoit R, Wilkinson KJ, Sauvé S (2013) Partitioning of silver and chemical speciation of free Ag in soils amended with nanoparticles. Chem Cent J 7(1):1–7. https://doi.org/10.1186/1752-153X-7-75/FIGURES/5

Bhatt P, Pandey SC, Joshi S et al (2022) Nanobioremediation: a sustainable approach for the removal of toxic pollutants from the environment. J Hazard Mater 427:12803. https://doi.org/10.1016/j.jhazmat.2021.128033

Bhattacharjee R, Kumar L, Mukerjee N et al (2022) The emergence of metal oxide nanoparticles (NPs) as a phytomedicine: a two-facet role in plant growth, nano-toxicity and anti-phyto-microbial activity. Biomed Pharmacother. https://doi.org/10.1016/j.biopha.2022.113658

Burello E, Worth A (2011) A theoretical framework for predicting the oxidative stress potential of 624 oxide nanoparticles. Nanotoxicology 5:228–235

Buriak JM, Liz-Marzán LM, Parak WJ et al (2022) Nano and Plants. ACS Nano 16:1681–1684

Burke DJ, Pietrasiak N, Situ SF et al (2015) Iron oxide and titanium dioxide nanoparticle effects on plant performance and root associated microbes. Int J Mol Sci 16(10):23630–23650. https://doi.org/10.3390/IJMS161023630

Cedervall T, Hansson LA, Lard M et al (2012) Food chain transport of nanoparticles affects behaviour and fat metabolism in fish. PLoS One 7(2):e32254. https://doi.org/10.1371/JOURNAL.PONE.0032254

Chaud M, Souto EB, Zielinska A et al (2021) Nanopesticides in agriculture: benefits and challenge in agricultural productivity, toxicological risks to human health and environment. Toxics 9(6):131. https://doi.org/10.3390/toxics9060131

Chávez-Hernández JA, Velarde-Salcedo AJ, Navarro-Tovar G et al (2024) Safe nanomaterials: from their use, application, and disposal to regulations. Nanoscale Adv 6(6):1583–1610. https://doi.org/10.1039/D3NA01097J

Chen A, Halilovic L, Shay JH et al (2023) Improving RNA-based crop protection through nanotechnology and insights from cross-kingdom RNA trafficking. Curr Opin Plant Biol. https://doi.org/10.1016/j.pbi.2023.102441

Chhipa H (2019) Applications of nanotechnology in agriculture. Methods Microbiol 46:115–142. https://doi.org/10.1016/bs.mim.2019.01.002

Choudhury SR, Pradhan S, Goswami A (2012) Preparation and characterisation of acephatenanoencapsulated complex. Nanosci Methods 1:9–15. https://doi.org/10.1080/17458080.2010.533443

Chowdhury P, Gogoi M, Borchetia S et al (2017) Nanotechnology applications and intellectual property rights in agriculture. Environ Chem Lett 15:413–419. https://doi.org/10.1007/s10311-017-0632-4

Chugh P, Jahangir Dar S (2021) Nanotechnology: applications, opportunities, and constraints in agriculture. Nanotechnology 125–148. https://doi.org/10.1201/9781003120261-4

Cohen Y, Rallo R, Liu R et al (2013) In silico analysis of nanomaterials hazard and risk. Acc Chem Res 46:802–812

Coles D, Frewer LJ (2013) Nanotechnology applied to European food production – A review of ethical and regulatory issues. https://philpapers.org/rec/COLNAT-5

Dazon C, Fierro V, Celzard A et al (2020) Identification of nanomaterials by the volume specific surface area (VSSA) criterion: applications to powder mixes. Nanoscale Adv 2:4908–4917. https://doi.org/10.1039/D0NA00395F

Del Prado-Audelo ML, García Kerdan I, Escutia-Guadarrama L et al (2021) Nanoremediation: nanomaterials and nanotechnologies for environmental cleanup. Front Environ Sci. https://doi.org/10.3389/fenvs.2021.793765

Demirer GS, Zhang H, Goh NS et al (2019) Carbon nanotube-mediated DNA delivery without transgene integration in intact plants. Nat Protoc 14:2954–2971

Deng YQ, White JC, Xing BS (2014) Interactions between engineered nanomaterials and agricultural crops: implications for food safety. JZUSA 15(8):552–572. https://doi.org/10.1631/JZUS.A1400165/FIGURES/2

Deng R, Lin D, Zhu L et al (2017) Nanoparticle interactions with co-existing contaminants: joint toxicity, bioaccumulation and risk. Nanotoxicology 11(5):591–612. https://doi.org/10.1080/17435390.2017.1343404

Deng P, Gao Y, Mu L et al (2023) Development potential of nanoenabled agriculture projected using machine learning. PNAS. https://doi.org/10.1073/pnas.2301885120

Diengngan S (2019) Prospects of nanotechnology for crop improvement and production. Acta Hortic 1241:247–253. https://doi.org/10.17660/ACTAHORTIC.2019.1241.34

Dietz KJ, Herth S (2011) Plant nanotoxicology. Trends Plant Sci 16(11):582–589. https://doi.org/10.1016/J.TPLANTS.2011.08.003

Dimkpa CO (2014) Can nanotechnology deliver the promised benefits without negatively impacting soil microbial life? J Basic Microbiol 54(9):889–904

Dimkpa CO, Bindraban PS (2016) Fortification of micronutrients for efficient agronomic production: a review. Agron Sustain Dev 36:7. https://doi.org/10.1007/s13593-015-0346-6

Disfani MN, Mikhak A, Kassaee MZ (2017) Effects of nano Fe/SiO2 fertilizers on germination and growth of barley and maize. Arch Agron Soil Sci 63(6):817–826

Ditta A (2012) How helpful is nanotechnology in agriculture? ANSN 3(3):033002. https://doi.org/10.1088/2043-6262/3/3/033002

Dixit R, Malaviya WD, Pandiyan K et al (2015) Bioremediation of heavy metals from soil and aquatic environment: an overview of principles and criteria of fundamental processes. Sustainability 7:2189–2212. https://doi.org/10.3390/su7022189

Donia DT, Carbone M (2023) Seed priming with zinc oxide nanoparticles to enhance crop tolerance to environmental stresses. Int J Mol Sci. https://doi.org/10.3390/ijms242417612. PMID: 38139445; PMCID: PMC10744145

Duan Y, Coreas R, Liu Y et al (2020) Prediction of protein corona on nanomaterials by machine learning using novel descriptors. Nano Impact. https://doi.org/10.1016/j.impact.2020.100207

Eastman PS, Ruan W, Doctolero M et al (2006) Qdot nanobarcodes for multiplexed gene expression analysis. Nano Lett 6(5):1059–1064. https://doi.org/10.1021/nl060795t

El-Chaghaby GA, Rashad S (2024) Nanosensors in agriculture: applications, prospects, and challenges. In: Ali GAM, Chong KF, Makhlouf ASH (eds) Handbook of nanosensors. Springer, Cham. https://doi.org/10.1007/978-3-031-16338-8_52-1

El-Sheekh MM, El-Kassas HY (2016) Algal production of nano-silver and gold: their antimicrobial and cytotoxic activities: a review. J Genet Eng Biotechnol 14:299–310

El-Temsah YS, Sevcu A, Bobcikova K et al (2016) DDT degradation efficiency and exotoxicological effects of two types of nano-sized zero-valent iron (nZVI) in water and soil. Chemosphere 144:2221–2228. https://doi.org/10.1016/j.chemosphere.2015.10.122

Enders TA, Strader LC (2015) Auxin activity: past, present, and future. Am J Bot 102(2):180–196. https://doi.org/10.3732/AJB.1400285

Fajardo C, Ortiz LT, Rodríguez-Membibre ML et al (2012) Assessing the impact of zero-valent iron (ZVI) nanotechnology on soil microbial structure and functionality: a molecular approach. Chemosphere 86(8):802–808. https://doi.org/10.1016/j.chemosphere.2011.11.041

FAO/WHO Technical Paper (2013) State of the art on the initiatives and activities relevant to risk assessment and risk management of nanotechnologies in the food and agriculture sectors: FAO/WHO technical paper. (n.d.). from https://iris.who.int/handle/10665/87458

Fernando LM, Parami JMK, Bautista HNF et al (2022) Effectiveness of nanoformulated plant growth regulators in the production and propagation of high-value crops and ornamentals. Philipp J Sci 153(2):585–595

Ferreira MMA, Filipe JA (2022) Ethical considerations on nanotechnology. arXiv preprint. http://arxiv.org/abs/2202.01063

Findlay MR, Freitas DN, Mobed-Miremadi M et al (2018) Machine learning provides predictive analysis into silver nanoparticle protein corona formation from physicochemical properties. Environ Sci Nano 5:64–71

Floros JD, Newsome R, Fisher W et al (2010) Feeding the world today and tomorrow: the importance of food science and technology. CRFSFS 9:572–599. https://doi.org/10.1111/j.1541-4337.2010.00127.x

Fraceto LF, Grillo R, de Medeiros GA et al (2016) Nanotechnology in agriculture: which innovation potential does it have? Front Environ Sci 4:20. https://doi.org/10.3389/fenvs.2016.00020

Frenk S, Ben-Moshe T, Dror I et al (2013) Effect of metal oxide nanoparticles on microbial community structure and function in two different soil types. PLoS One. https://doi.org/10.1371/journal.pone.0084441

Gao C (2018) The future of CRISPR technologies in agriculture. Nat Rev Mol Cell Biol 19:275–276. https://doi.org/10.1038/nrm.2018.2

Gao M, Chang J, Wang Z et al (2023) Advances in transport and toxicity of nanoparticles in plants. J Nanobiotechnol 21(1):75. https://doi.org/10.1186/S12951-023-01830-5

Garg S, Rumjit NP, Roy S et al (2024) Smart agriculture and nanotechnology: technology, challenges, and new perspective. Adv Agrochem 3(2):115–125. https://doi.org/10.1016/j.aac.2023.11.001

Geitner NK, Zhao W, Ding F et al (2017) Mechanistic insights from discrete molecular dynamics simulations of pesticide–nanoparticle interactions. Environ Sci Technol 51:8396–8404

Ghosh S, Patra S, Ray S (2023) A combinatorial nanobased spray-induced gene silencing TECHNIQUE for crop protection and improvement. ACS Omega 8(25):22345–22351. https://doi.org/10.1021/acsomega.3c01968

Giles EL, Kuznesof S, Clark B et al (2015) Consumer acceptance of and willingness to pay for food nanotechnology: a systematic review. J Nanopart Res 17(12):1–26. https://doi.org/10.1007/S11051-015-3270-4/TABLES/5

Gondal AH, Tayyiba L (2022) Prospects of using nanotechnology in agricultural growth, environment and industrial food products. Rev Agric Sci 10:68–81. https://doi.org/10.7831/RAS.10.0_68

Gonzalez C, Antonio J, Chchávez-Hernández C et al (2024) Safe nanomaterials: from their use, application, and disposal to regulations. https://doi.org/10.1039/d3na01097j

Gorczyca A, Pociecha E, Matras E (2021) Nanotechnology in agriculture, the food sector, and remediation: prospects, relations, and constraints. pp 1–34. https://doi.org/10.1007/978-981-15-5499-5_1

Gruère GP (2012) Implications of nanotechnology growth in food and agriculture in OECD countries. Food Policy 37:191–198. https://doi.org/10.1016/j.jhazmat.2014.05.079

Grün AL, Manz W, Kohl YL et al (2019) Impact of silver nanoparticles (AgNP) on soil microbial community depending on functionalization, concentration, exposure time, and soil texture. Environ Sci Eur 31(1):1–22. https://doi.org/10.1186/S12302-019-0196-Y/TABLES/4

Guerra FD, Attia MF, Whitehead DC et al (2018) Nanotechnology for environmental remediation: materials and applications. Molecules. https://doi.org/10.3390/molecules23071760

Gupta A, Rayeen F, Mishra R et al (2023a) Nanotechnology applications in sustainable agriculture: an emerging eco-friendly approach. Plant Nano Biol. https://doi.org/10.1016/j.plana.2023.100033

Gupta P, Sharma M, Sharma S et al (2023b) Nanotechnology in agriculture: prospect, opportunities and challenges. Emer Life Sci Res 9(1):164–176. https://doi.org/10.31783/elsr.2023.91164176

Hanif S, Javed R, Khan A et al (2023) IAA-decorated CuO nanocarriers significantly improve Chickpea growth by increasing antioxidative activities. 3 Biotech 13(3). https://doi.org/10.1007/s13205-023-03516-z

He X, Deng H, Aker WG et al (2019) Regulation and safety of nanotechnology in the food and agriculture industry. In: Molina G, Inamuddin, Pelissari FM et al (eds) Food applications of nanotechnology, 1st edn. CRC press, pp 525–536. https://doi.org/10.1201/9780429297038-23

Hegde P (2023) New technologies in agriculture: for 2023 and beyond. https://www.cropin.com/blogs/new-technologies-in-agriculture. Accessed 08 April 2024

Hooven LA, Chakrabarti P, Harper BJ et al (2019) Potential risk to pollinators from nanotechnology-based pesticides. Molecules (Basel, Switzerland) 24(24). https://doi.org/10.3390/MOLECULES24244458

Hu P, An J, Faulkner MM et al (2020) Nanoparticle charge and size control foliar delivery efficiency to plant cells and organelles. ACS Nano 14:7970–7986. https://doi.org/10.1021/acsnano.9b09178

Hu R, Shi J, Tian C et al (2022) Nucleic acid aptamers for pesticides, toxins, and biomarkers in agriculture. ChemPlusChem 87(11). https://doi.org/10.1002/cplu.202200230

Huang X, Xin W, Xingxing L et al (2024) Nanotechnology in agriculture: manganese ferrite nanoparticles as a micronutrient fertilizer for wheat. Plants 13(10):1395. https://doi.org/10.3390/plants13101395

Iavicoli I, Leso V, Beezhold DH et al (2017) Nanotechnology in agriculture: opportunities, toxicological implications, and occupational risks. Toxicol Appl Pharmacol 329:96. https://doi.org/10.1016/J.TAAP.2017.05.025

Imada K, Sakai S, Kajihara H et al (2016) Magnesium oxide nanoparticles induce systemic resistance in tomato against bacterial wilt disease. Plant Pathol 65:551–560

Jahan S, Alias YB, Bakar AFBA et al (2018) Toxicity evaluation of ZnO and TiO2 nanomaterials in hydroponic red bean (*Vigna angularis*) plant: physiology, biochemistry and kinetic transport. J Environ Sci 72:140–152

Jan N, Majeed N, Ahmad M et al (2022) Nano-pollution: why it should worry us. Chemosphere 302:134746. https://doi.org/10.1016/J.CHEMOSPHERE.2022.134746

Jat SK, Bhattacharya J, Sharma MK (2020) Nanomaterial based gene delivery: a promising method for plant genome engineering. J Mater Chem B 8:4165–4175

Javaid A, Hameed S, Li L et al (2024) Can nanotechnology and genomics innovations trigger agricultural revolution and sustainable development? Funct Integr Genomics 24:216. https://doi.org/10.1007/s10142-024-01485-x

Johansen A, Pedersen AL, Jensen KA et al (2008) Effects of C60 fullerene nanoparticles on soil bacteria and protozoans. Environ Toxicol Chem 27(9):1895–1903. https://doi.org/10.1897/07-375.1

Juárez-Maldonado A, González-Morales S, Cabrera-De la Fuente M et al (2018) Nanometals as promoters of nutraceutical quality in crop plants. In: Grumezescu AM, Holban AM (eds) Impact of nanoscience in the food industry. Academic Press, pp 277–310. https://doi.org/10.1016/B978-0-12-811441-4.00010-8

Kamal R, Mogazy AM (2023) Effect of doping on TiO2 nanoparticles characteristics: studying of fertilizing effect on cowpea plant growth and yield. J Soil Sci Plant Nutr 23:325–337

Kandasamy S, Prema RS (2015) Methods of synthesis of nano particles and its applications. J Chem Pharm Res 7:278–285

Kapoor P, Dhaka RK, Sihag P et al (2022) Nanotechnology-enabled biofortification strategies for micronutrients enrichment of food crops: current understanding and future scope. NanoImpact 26. https://doi.org/10.1016/J.IMPACT.2022.100407

Karatzas P, Melagraki G, Ellis LJA et al (2020) Development of deep learning models for predicting the effects of exposure to engineered nanomaterials on Daphnia Magna. Small. https://doi.org/10.1002/smll.202001080

Kashyap PL, Kumar S, Jasrotia P et al (2019) Nanosensors for plant disease diagnosis: current understanding and future perspectives. In: Pudake R, Chauhan N, Kole C (eds) Nanoscience for sustainable agriculture. Springer, Cham. https://doi.org/10.1007/978-3-319-97852-9_9

Kasote DM, Lee JHJ, Jayaprakasha GK et al (2019) Seed priming with iron oxide nanoparticles modulate antioxidant potential and defense-Linked hormones in watermelon seedlings. ACS Sustain Chem Eng 7:5142–5151

Kaweeteerawat C, Ivask A, Liu R et al (2015) Toxicity of metal oxide nanoparticles in Escherichia coli correlates with conduction band and hydration energies. Environ Sci Technol 49(2): 1105–1112. https://doi.org/10.1021/ES504259S/SUPPL_FILE/ES504259S_SI_001.PDF

Khan A (2015) Ethical and social implications of nanotechnology. In: QScience Proceedings (Engineering Leaders Conference 2014), p 57. https://doi.org/10.5339/qproc.2015.elc2014.57

Khan MR, Rizvi TF (2014) Nanotechnology: scope and application in plant disease management. Plant Pathol J 13:214–231

Khan M, Khan MSA, Borah KK et al (2021a) The potential exposure and hazards of metal-based nanoparticles on plants and environment, with special emphasis on ZnO NPs, TiO2 NPs, and AgNPs: a review. Environ Adv. https://doi.org/10.1016/j.envadv.2021.100128

Khan ST, Adil SF, Shaik MR et al (2021b) Engineered nanomaterials in soil: their impact on soil microbiome and plant health. Plan Theory. https://doi.org/10.3390/plants11010109

Kookana RS, Boxall ABA, Reeves PT et al (2014) Nanopesticides: guiding principles for regulatory evaluation of environmental risks. J Agric Food Chem 62(19):4227–4240

Kopittke PM, Lombi E, Wang P et al (2019) Nanomaterials as fertilizers for improving plant mineral nutrition and environmental outcomes. Environ Sci 6:3513–3524. https://doi.org/10.3390/biology10111123

Krishna D, Sachan HK (2021) Nano-toxicity and aquatic food chain. In: Singh P, Singh R, Verma P et al (eds) Plant-microbes-engineered nano-particles (PM-ENPs)Nexus in agro-ecosystems. Advances in science, technology and innovation. Springer, pp 189–198. https://doi.org/10.1007/978-3-030-66956-0_13

Kristanti RA, Liong RMY, Hadibarata T (2021) Soil remediation applications of nanotechnology. Trop Aquat Soil Pollut 1(1):35–45

Kumari M, Pandey S, Bhattacharya A et al (2017) Protective role of biosynthesized silver nanoparticles against early blight disease in Solanum lycopersicum. Plant Physiol Biochem 121:216–225

Kumari R, Suman K, Karmakar S et al (2023) Regulation and safety measures for nanotechnology-based agri-products. Front Genome Ed 5:1200987. https://doi.org/10.3389/FGEED.2023.1200987/BIBTEX

Laborda F, Bolea E, Cepriá G et al (2016) Detection, characterization and quantification of inorganic engineered nanomaterials: a review of techniques and methodological approaches for the analysis of complex samples. Anal Chim Acta 904:10–32. https://doi.org/10.1016/J.ACA.2015.11.008

Lamon L, Richarz A, Joossens E et al (2018) Grouping of nanomaterials to read-across hazard endpoints: from data collection to assessment of the grouping hypothesis by application of chemoinformatic techniques. Part Fibre Toxicol 15:37. https://doi.org/10.1186/s12989-018-0273-1

Li Z, Yu T, Paul R et al (2020) Agricultural nanodiagnostics for plant diseases: recent advances and challenges. Nanoscale Adv 2:3083–3094. https://doi.org/10.1039/C9NA00724E

Li J, Wang C, Yue L et al (2022) Nano-QSAR modeling for predicting the cytotoxicity of metallic and metal oxide nanoparticles: a review. Ecotoxicol Environ Saf 243:113955. https://doi.org/10.1016/J.ECOENV.2022.113955

Liliane TN, Charles MS (2020) Factors affecting yield of crops. In: Amanullah (ed) Agronomy-climate change and food security. IntechOpen. https://doi.org/10.5772/intechopen.90672

Liu M, Wang L, Lo Y et al (2022) Aptamer-enabled nanomaterials for therapeutics, drug targeting and imaging. Cells 11(1):159. https://doi.org/10.3390/cells11010159

Liu Z, Wen F, Cheng X et al (2024) Nano-controlled release of phytohormones will broaden its application on plant protection. Adv Agrochem 3(1):39–42

Lombi E, Donner E, Dusinska M et al (2019) A One Health approach to managing the applications and implications of nanotechnologies in agriculture. Nat Nanotechnol 14(6):523–531. https://doi.org/10.1038/s41565-019-0460-8

Lowry GV, Avellan A, Gilbertson LM (2019) Opportunities and challenges for nanotechnology in the agri-tech revolution. Nat Nanotechnol 14(6):517–522. https://doi.org/10.1038/S41565-019-0461-7

Ma X, Geiser-Lee J, Deng Y et al (2010) Interactions between engineered nanoparticles (ENPs) and plants: phytotoxicity, uptake and accumulation. Sci Total Environ 408(16):3053–3061. https://doi.org/10.1016/J.SCITOTENV.2010.03.031

Majumdar S, Keller AA (2021) Omics to address the opportunities and challenges of nanotechnology in agriculture. Crit Rev Environ Sci Technol 51(22):2595–2636. https://doi.org/10.1080/10643389.2020.1785264

Manjunatha RL, Naik D, Usharani KV (2019) Nanotechnology application in agriculture: a review. J Pharmacogn Phytochem 8(3):1073–1083

Marimuthu S, Rahuman AA, Jayaseelan C et al (2013) Acaricidal activity of synthesized titanium dioxide nanoparticles using *Calotropis gigantea* against *Rhipicephalus microplus* and *Haemaphysalisbispinosa*. Asian Pac J Trop Med 6(9):682–688

Matranga V, Corsi I (2012) Toxic effects of engineered nanoparticles in the marine environment: model organisms and molecular approaches. Mar Environ Res 76:32–40. https://doi.org/10.1016/J.MARENVRES.2012.01.006

McManus P, Hortin J, Anderson AJ et al (2018) Rhizosphere interactions between copper oxide nanoparticles and wheat root exudates in a sand matrix: influences on copper bioavailability and uptake. Environ Toxicol Chem 37(10):2619–2632

Mejias JH, Salazar F, Pérez Amaro L et al (2021) Nanofertilizers: a cutting-edge approach to increase nitrogen use efficiency in grasslands. Front Environ Sci 9:635114. https://doi.org/10.3389/fenvs.2021.635114

Mishra S, Keswani C, Abhilash PC et al (2017) Integrated approach of agri-nanotechnology: challenges and future trends. Front Plant Sci 8:254477. https://doi.org/10.3389/FPLS.2017.00471/BIBTEX

Mittal D, Kaur G, Singh P et al (2020) Nanoparticle based sustainable agriculture and food science: recent advances and future outlook. Front Nanotechnol 2:579954. https://doi.org/10.3389/fnano.2020.579954

Mondal R, Dam P, Chakraborty J et al (2022) Potential of nanobiosensor in sustainable agriculture: the state-of-art. Heliyon 8(12). https://doi.org/10.1016/j.heliyon.2022.e12207

Mukherjee A, Maity A, Pramanik P et al (2019) Public perception about use of nanotechnology in agriculture. In: Ghorbanpour M, Wani SH (eds) Advances in phytonanotechnology. Academic Press, pp 405–418. https://doi.org/10.1016/B978-0-12-815322-2.00019-5

Mukhopadhyay SS (2014) Nanotechnology in agriculture: prospects and constraints. Nanotechnol Sci Appl 7(2):63–71. https://doi.org/10.2147/NSA.S39409
Mukhopadhyay R, Sarkar B, Khan E et al (2022) Nanomaterials for sustainable remediation of chemical contaminants in water and soil. Crit Rev Environ Sci Technol 52(15):2611–2660
Mwaanga P (2018) Risks, uncertainties, and ethics of nanotechnology in agriculture. In: Çelik Ö (ed) New visions in plant science. Intechopen. https://doi.org/10.5772/INTECHOPEN.76590
Mwaanga P, Mbulwe S, Shumbula P et al (2017) Investigating the toxicity of Cu, CuO and ZnO nanoparticles on earthworms in urban soils. J Pollut Eff Cont 5(3):2–9
Nam NH, Luong NH (2019) Nanoparticles: synthesis and applications. In: Grumezescu V, Grumezescu AM (eds) Materials for Biomedical Engineering. Elsevier, pp 211–240. https://doi.org/10.1016/B978-0-08-102814-8.00008-1
Navarro E, Baun A, Behra R et al (2008) Environmental behavior and ecotoxicity of engineered nanoparticles to algae, plants, and fungi. Ecotoxicology 17(5):372–386. https://doi.org/10.1007/S10646-008-0214-0
Nečemer M, Kump P, Ščančar J et al (2008) Application of X-ray fluorescence analytical techniques in phytoremediation and plant biology studies. Spectrochim Acta Part B Atomic Spectrosc 63(11):1240–1247. https://doi.org/10.1016/J.SAB.2008.07.006
Neme K, Nafady A, Uddin S et al (2021) Application of nanotechnology in agriculture, postharvest loss reduction and food processing: food security implication and challenges. Heliyon 7(12): e08539
Nile SH, Baskar V, Selvaraj D et al (2020) Nanotechnologies in food science: applications, recent trends, and future perspectives. Nano-Micro Lett 12:1–34. https://doi.org/10.1007/S40820-020-0383-9
Nile SH, Thiruvengadam M, Wang Y et al (2022) Nano-priming as emerging seed priming technology for sustainable agriculture—recent developments and future perspectives. J Nanobiotechnol 20:254. https://doi.org/10.1186/s12951-022-01423-8
Nima ZA, Lahiani MH, Watanabe F et al (2014) Plasmonically active nanorods for delivery of bio-active agents and high-sensitivity SERS detection in planta. RSC Adv 4(110): 64985–64993. https://doi.org/10.1039/C4RA10358K
Nkongolo KK, Narendrula-Kotha R (2020) Advances in monitoring soil microbial community dynamic and function. J Appl Genet 61(2):249–263. https://doi.org/10.1007/S13353-020-00549-5
Nongbet A, Mishra AK, Mohanta YK et al (2022) Nanofertilizers: a smart and sustainable attribute to modern agriculture. Plants 11:2587. https://doi.org/10.3390/plants11192587
Ogunyemi SO, Zhang M, Abdallah Y et al (2020) The bio-synthesis of three metal oxide nanoparticles (ZnO, MnO2, and MgO) and their antibacterial activity against the bacterial leaf blight pathogen. Front Microbiol 11:588326. https://doi.org/10.3389/fmicb.2020.588326
Pandey G, Jain P (2020) Assessing the nanotechnology on the grounds of costs, benefits, and risks. Beni-Suef Univ J Basic Appl Sci 9(1):1–10. https://doi.org/10.1186/S43088-020-00085-5
Panichikkal J, Thomas R, John JC et al (2019) Biogenic gold nanoparticle supplementation to plant beneficial Pseudomonas monteilii was found to enhance its plant probiotic effect. Curr Microbiol 76:503–509
Park EJ, Bae E, Yi J et al (2010) Repeated-dose toxicity and inflammatory responses in mice by oral administration of silver nanoparticles. Environ Toxicol Pharmacol 30(2):162–168. https://doi.org/10.1016/J.ETAP.2010.05.004
Pasupathy K, Lin S, Hu Q et al (2008) Direct plant gene delivery with a poly(amidoamine) dendrimer. Biotechnol J 3(8):1078–1082. https://doi.org/10.1002/biot.200800021
Pérez S, la Farré M, Barceló D (2009) Analysis, behavior and ecotoxicity of carbon-based nanomaterials in the aquatic environment. TrAC Trends Anal Chem 28(6):820–832. https://doi.org/10.1016/J.TRAC.2009.04.001
Perez-de-Luque A, Rubiales D (2009) Nanotechnology for parasitic plant control. Pest Manag Sci 65(5):540–545

Pooja, Rana M, Chowdhury P (2021) Modern applications of quantum dots: environmentally hazardous metal ion sensing and medical imaging. In: Hussain CM, Kailasa SK (eds) Micro and nano technologies. Handbook of nanomaterials for sensing applications, Elsevier, pp 465–503

Pradhan S, Mailapalli DR (2020) Nanopesticides for pest control. In: Lichtfouse E (ed) Sustainable agriculture reviews 40. Springer, Cham, pp 43–74. https://doi.org/10.1007/978-3-030-33281-5_2

Praetorius A, Gundlach-Graham A, Goldberg E (2017) Single-particle multi-element fingerprinting (spMEF) using inductively-coupled plasma time-of-flight mass spectrometry (ICP-TOFMS) to identify engineered nanoparticles against the elevated natural background in soils. Environ Sci Nano 4(2):307–314. https://doi.org/10.1039/C6EN00455E

Prasad R (2014) Synthesis of silver nanoparticles in photosynthetic plants. J Nanoparticles. https://doi.org/10.1155/2014/963961

Prasad TNVKV, Sudhakar P, Sreenivasulu Y et al (2012) Effect of nanoscale zinc oxide particles on the germination, growth and yield of peanut. J Plant Nutr 35(6):905–927

Prasad R, Kumar V, Prasad KS (2014) Nanotechnology in sustainable agriculture: present concerns and future aspects. Afr J Biotechnol 13:705–713. https://doi.org/10.5897/AJBX2013.13554

Prasad R, Pandey R, Barman I (2016) Engineering tailored nanoparticles with microbes: quo vadis? Wiley Interdiscip Rev: Nanomed Nanobiotechnol 8:316–330. https://doi.org/10.1002/wnan.1363

Prasad M, Gupta A, Ranjan R (2020) Nano-contaminants: sources and impact on agriculture. In: Naeem M, Ansari A et al (eds) Contaminants in agriculture. Springer, Cham. https://doi.org/10.1007/978-3-030-41552-5_8

Qian Y, Qin C, Chen M et al (2020) Nanotechnology in soil remediation – applications vs. implications. Ecotoxicol Environ Saf 201:110815. https://doi.org/10.1016/j.ecoenv.2020.110815

Qiu Y, Zhang Q, Gao B et al (2020) Removal mechanisms of Cr(VI) and Cr(III) by biochar supported nanosized zero-valent iron: Synergy of adsorption, reduction and transformation. Environ Pollut 265. https://doi.org/10.1016/j.envpol.2020.115018

Radziwill-Bienkowska JM, Talbot P, Kamphuis JBJ et al (2018) Toxicity of Food-Grade TiO2 to commensal intestinal and transient food-borne bacteria: new insights using Nano-SIMS and synchrotron UV fluorescence imaging. Front Microbiol 9:360789. https://doi.org/10.3389/FMICB.2018.00794/BIBTEX

Rajna S, Paschapur AU, Raghavendra KV (2019) Nanopesticides: its scope and utility in pest management. Indian Farmer 6(1):17–21

Rajput V, Minkina T, Mazarji M et al (2020) Accumulation of nanoparticles in the soil-plant systems and their effects on human health. Ann Agric Sci 65(2):137–143. https://doi.org/10.1016/J.AOAS.2020.08.001

Rajput VD, Minkina T, Upadhyay SK et al (2022) Nanotechnology in the restoration of polluted soil. Nanomaterials (Basel) 12(5):769

Rajput VD, Kumari A, Upadhyay SK et al (2023) Can nanomaterials improve the soil microbiome and crop productivity? Agriculture 13(2):231. https://doi.org/10.3390/AGRICULTURE13020231

Ramesh M, Palanisamy K, Babu K et al (2014) Effects of bulk & nano-titanium dioxide and zinc oxide on physio-morphological changes in *Triticum aestivum* Linn. J Glob Biosci 3(2):415–422

Rana S, Kalaichelvan PT (2013) Ecotoxicity of nanoparticles. ISRN toxicology. pp 1–11. https://doi.org/10.1155/2013/574648

Rana L, Kumar M, Rajput J et al (2024) Nexus between nanotechnology and agricultural production systems: challenges and future prospects. Discov Appl Sci 6:555. https://doi.org/10.1007/s42452-024-06265-7

Rautela I, Dheer P, Thapliyal P et al (2021) Current scenario and future perspectives of nanotechnology in sustainable agriculture and food production. Plant Cell Biotechnol Mol Biol 22:99–121

Ray PC, Yu H, Fu PP (2009) Toxicity and environmental risks of nanomaterials: challenges and future needs. J Environ Sci Health A C27(1):1–35. https://doi.org/10.1080/10590500802708267

Rezaei S, Khanmirzaei A (2024) Nano-biofertilizers: a new area for enhancing plant nutrition. In: Plant holobiome engineering for climate-smart agriculture. Springer Nat. Singapore, Singapore, pp 617–632. https://doi.org/10.1007/978-981-99-9388-8_32

Roychoudhury A (2020) Yeast-mediated green synthesis of nanoparticles for biological applications. Indian J Pharm Biol Res 8(03):26–31

Sabir S, Arshad M, Chaudhari SK (2014) Zinc oxide nanoparticles for revolutionizing agriculture: synthesis and applications. Sci World J. https://doi.org/10.1155/2014/925494

Sadeghi R, Rodriguez RJ, Yao Y et al (2017) Advances in nanotechnology as they pertain to food and agriculture: benefits and risks. Annu Rev Food Sci Technol 8:467–492. https://doi.org/10.1146/ANNUREV-FOOD-041715-033338

Saeed A, Qureshi S, Bhutta WM (2021) Applications of nanobiotechnology in plant sciences. Biomed J Sci Tech Res 35(1):27236–27240

Sagadevan S, Periasamy M (2014) Recent trends in nanobiosensors and their applications – a review. Rev Adv Mater Sci 36:62–69

Santos PA, Biraku X, Nielsen E et al (2024) Agricultural nanotechnology for a safe and sustainable future: current status, challenges, and beyond. J Sci Food Agric. https://doi.org/10.1002/jsfa.13922

Saranya S, Selvi A, Babujanarthanam R et al (2020) Insecticidal activity of nanoparticles and mechanism of action. In: Siddhardha B, Dyavaiah M, Kasinathan K (eds) Model organisms to study biological activities and toxicity of nanoparticles. Springer, Singapore, p 12

Saritha GNG, Anju T, Kumar A (2022) Nanotechnology – big impact: how nanotechnology is changing the future of agriculture? J Agric Food Res. https://doi.org/10.1016/j.jafr.2022.100457

Satapanajaru T, Anurakpongsatorn P, Pengthamkeerati P et al (2008) Remediation of atrazine-contaminated soil and water by nano zerovalent iron. Water Air Soil Pollut 192:349–359

Satterfield T, Kandlikar M, Beaudrie CEH et al (2009) Anticipating the perceived risk of nanotechnologies. Nat Nanotechnol 4(11):752–758. https://doi.org/10.1038/nnano.2009.265

Saurabh K, Prakash V, Dubey AK et al (2024) Enhancing sustainability in agriculture with nanofertilizers. Discov Appl Sci 6:559. https://doi.org/10.1007/s42452-024-06267-5

Savithramma N, Ankanna S, Bhumi G (2012) Effect of nanoparticles on seed germination and seedling growth of *Boswellia ovalifoliolata* an endemic and endangered medicinal tree taxon. Nano Vis 2:61–68

Savolainen K, Backman U, Brouwer D et al (2013) Nanosafety in Europe 2015-2025: towards safe and sustainable nanomaterials and nanotechnology innovations. www.ttl.fi/en/publications/electronic_publications/pages/default.aspx

Scrinis G, Lyons K (2007) The emerging nano-corporate paradigm: nanotechnology and the transformation of nature, food and agri-food systems. Int J Sociol Food Agric 15:22–44

Sedghi M, Hadi M, Toluie SG (2013) Effect of nano zinc oxide on the germination parameters of soybean seeds under drought stress. AWUT-SerBio 16(2):73–78

Sehgal Foundation (2023) Role of modern technology in agriculture. https://www.smsfoundation.org/role-of-modern-technology-in-agriculture/. Accessed 09 April 2024

Sembada AA, Lenggoro IW (2024) Transport of nanoparticles into plants and their detection methods. Nanomaterials 14(2):131. https://doi.org/10.3390/NANO14020131

Shang Y, Hasan MK, Ahammed GJ (2019) Applications of nanotechnology in plant growth and crop protection: a review. Molecules 24(14):2558. https://doi.org/10.3390/molecules24142558

Sharma P, Lew TTS (2022) Principles of nanoparticle design for genome editing in plants. Front Genome Ed 4:846624. https://doi.org/10.3389/fgeed.2022.846624

Shawon ZBZ, Hoque ME, Chowdhury SR (2020) Nanosensors and nanobiosensors: agricultural and food technology aspects. In: Kaushik P, Gomes F (eds) Micro and nano technologies. Nanofabrication for smart nanosensor applications. Elsevier, pp 135–161. https://doi.org/10.1016/b978-0-12-820702-4.00006-4

Shelar A, Nile SH, Singh AV et al (2023) Recent advances in nano-enabled seed treatment strategies for sustainable agriculture: challenges, risk assessment, and future perspectives. Nano-Micro Lett 15:54. https://doi.org/10.1007/s40820-023-01025-5

Sheng H, Ma J, Pu J et al (2018) Cell wall-bound silicon optimizes ammonium uptake and metabolism in rice cells. Ann Bot 122(2):303–313

Shukla K, Mishra V, Singh J et al (2024) Nanotechnology in sustainable agriculture: a double-edged sword. J Sci Food Agric. https://doi.org/10.1002/JSFA.13342

Siddiqui MH, Al-Whaibi MH, Firoz M et al (2015) Role of nanoparticles in plants. In: Siddiqui M, Al-Whaibi M, Mohammad F (eds) Nanotechnology and plant sciences. Springer, Cham. https://doi.org/10.1007/978-3-319-14502-0_2

Siddiqui MH, Kalaji HM, Zhang Z et al (2022) Nanoparticles in environment and plant system: a boon or bane. Chemosphere 308:136320. https://doi.org/10.1016/J.CHEMOSPHERE.2022.136320

Singh Y, Saxena MK (2022) Insights into the recent advances in nanobioremediation of pesticides from the contaminated soil. Front Microbiol 13:982611. https://doi.org/10.3389/fmicb.2022.982611

Singh A, Singh NB, Hussain I (2017) Effect of biologically synthesized copper oxide nanoparticles on metabolism and antioxidant activity to the crop plants *Solanum lycopersicum* and *Brassica oleracea* var. *botrytis*. J Biotechnol 262:11–27

Singh A, Suki M, Sharma R et al (2020a) Applications of nanotechnology: a review. Int J Adv Res Chem Sci 7(2):16–32

Singh R, Behera M, Kumar S (2020b) Nano-bioremediation: an innovative remediation technology for treatment and management of contaminated sites. In: Bharagava R, Saxena G (eds) Bioremediation of industrial waste for environmental safety. Springer, Singapore. https://doi.org/10.1007/978-981-13-3426-9_7

Singh A, Rawat S, Rajput VD et al (2024) Nanotechnology products in agriculture and environmental protection: advances and challenges. Egypt J Soil Sci 64(4):1355–1378

Soltani AM, Pouypouy H (2019) Standardization and regulations of nanotechnology and recent government policies across the world on nanomaterials. In: Ghorbanpour M, Wani SH (eds) Advances in phytonanotechnology: from synthesis to application. Academic Press, pp 419–446. https://doi.org/10.1016/B978-0-12-815322-2.00020-1

Song Y, Wang BQ, Qiu DY et al (2021) Melatonin enhances metallic oxide nanoparticle stress tolerance in rice via inducing tetrapyrrole biosynthesis and amino acid metabolism. Environ Sci Nano 8:2310–2323

Soni M, Mehta P, Soni A et al (2018) Green nanoparticles: synthesis and applications IOSR. J Biotechnol Biochem 4:78–83

Soni SK, Dogra S, Sharma A et al (2024) Nanotechnology in agriculture: enhancing crop productivity with sustainable nano-fertilizers and nano-biofertilizers. J Soil Sci Plant Nutr. https://doi.org/10.1007/s42729-024-01988-3

Spaepen S, Vanderleyden J, Remans R (2007) Indole-3-acetic acid in microbial and microorganism-plant signalling FEMS. Microbiol Rev 31(4):425–448. https://doi.org/10.1111/J.1574-6976.2007.00072.X

Sylvester DJ, Abbott KW, Marchant GE (2009) Not again! Public perception, regulation, and nanotechnology. Regul Gov 3(2):165–185. https://doi.org/10.1111/J.1748-5991.2009.01049.X

Talari FF, Bozorg A, Faridbod F et al (2021) A novel sensitive aptamer-based nanosensor using rGQDs and MWCNTs for rapid detection of diazinon pesticide. J Environ Chem Eng. https://doi.org/10.1016/j.jece.2020.104878

Tan X, Wang X, Chen C et al (2007) Effect of soil humic and fulvic acids, pH and ionic strength on Th(IV) sorption to TiO2 nanoparticles. Appl Radiat Isot 65(4):375–381. https://doi.org/10.1016/j.apradiso.2006.10.014

Um S, Lee J, Kwon S et al (2006) Dendrimer-like DNA-based fluorescence nanobarcodes. Nat Protoc 1:995–1000. https://doi.org/10.1038/nprot.2006.141

Unrine JM, Hunyadi SE, Tsyusko OV et al (2010) Evidence for bioavailability of Au nanoparticles from soil and biodistribution within earthworms (*Eiseniafetida*). Environ Sci Technol 44(21): 8308–8313. https://doi.org/10.1021/ES101885W/SUPPL_FILE/ES101885W_SI_001.PDF
Unrine JM, Shoults-Wilson WA, Zhurbich O et al (2012) Trophic transfer of Au nanoparticles from soil along a simulated terrestrial food chain. Environ Sci Technol 46(17):9753–9760. https://doi.org/10.1021/ES3025325
Utembe W, Tlotleng N, Kamng'ona AW (2022) A systematic review on the effects of nanomaterials on gut microbiota. Curr Res Microb Sci 3:100118. https://doi.org/10.1016/J.CRMICR.2022.100118
Varsou DD, Afantitis A, Tsoumanis A et al (2019) A safe-by-design tool for functionalised nanomaterials through the Enalos Nanoinformatics Cloud platform. Nanoscale Adv 1:706–718. https://doi.org/10.1039/C8NA00142A
Verinder V, Himani D, Khushbu T et al (2024) Amelioration in nanobiosensors for the control of plant diseases: current status and future challenges. Front Nanotechnol. https://doi.org/10.3389/fnano.2024.1310165
Verma SK, Das AK, Gantait S et al (2019) Applications of carbon nanomaterials in the plant system: a perspective view on the pros and cons. Sci Total Environ 667:485–499. https://doi.org/10.1016/J.SCITOTENV.2019.02.409
Viana CM, Freire D, Abrantes P et al (2021) Agricultural land systems importance for supporting food security and sustainable development goals: a systematic review. Sci Total Environ 806(3): 150718. https://doi.org/10.1016/j.scitotenv.2021.150718
Vidotti M, Carvalhal RF, Mendes RK et al (2011) Biosensors based on gold nanostructures. J Braz Chem Soc 22:3–20. https://doi.org/10.1590/S0103-50532011000100001
Waller T, Chen C, Walker SL (2017) Food and industrial grade titanium dioxide impacts gut microbiota. Environ Eng Sci 34(8):537–550. https://doi.org/10.1089/ees.2016.0364
Wang Z, Xie X, Zhao J et al (2012) Xylem- and phloem-based transport of CuO nanoparticles in maize (Zea mays L.). Environ Sci Technol 46(8):4434–4441. https://doi.org/10.1021/ES204212Z/SUPPL_FILE/ES204212Z_SI_001.PDF
Wang X, Yang X, Chen S et al (2016a) Zinc oxide nanoparticles affect biomass accumulation and photosynthesis in *Arabidopsis*. Front Plant Sci 12(6):1243. https://doi.org/10.3389/fpls.2015.01243
Wang Z, Xu L, Zhao J et al (2016b) CuO nanoparticle interaction with Arabidopsis thaliana: toxicity, parent progeny transfer, and gene expression. Environ Sci Technol 50:6008–6016
Wang W, Sedykh A, Sun H et al (2017) Predicting nano-bio interactions by integrating nanoparticle libraries and quantitative nanostructure activity relationship modeling. ACS Nano 11(12): 12641–12649. https://doi.org/10.1021/acsnano.7b07093
War JM, Fazili MA, Mushtaque W et al (2020) Role of nanotechnology in crop improvement. In: Hakeem K, Pirzadah T (eds) Nanotechnology in agriculture, Nanotechnology in the Life Sciences. Springer, Cham, pp 63–97
Winkler DA (2020) Role of artificial intelligence and machine learning in nanosafety. Small. https://doi.org/10.1002/smll.202001883
Wu K, Xu C, Li T et al (2023) Application of nanotechnology in plant genetic engineering. Int J Mol Sci 24(19):14836. https://doi.org/10.3390/ijms241914836
Xu C, Peng C, Sun L et al (2015) Distinctive effects of TiO2 and CuO nanoparticles on soil microbes and their community structures in flooded paddy soil. Soil Biol Biochem 86:24–33
Xuan L, Ju Z, Skonieczna M et al (2023) Nanoparticles-induced potential toxicity on human health: applications, toxicity mechanisms, and evaluation models. MedComm 4(4). https://doi.org/10.1002/MCO2.327
Yadav A, Yadav K, Abd-Elsalam KA (2023) Nanofertilizers: types, delivery and advantages in agricultural sustainability. Agrochemicals 2:296–336
Yan X, Sedykh A, Wang W et al (2020) Construction of a web-based nanomaterial database by big data curation and modeling friendly nanostructure annotations. Nat Commun. https://doi.org/10.1038/s41467-020-16413-3

Yashveer S, Singh V, Kaswan V (2014) Green biotechnology, nanotechnology and bio-fortification: perspectives on novel environment-friendly crop improvement strategies. J Sustain Agric:113–126. https://doi.org/10.1080/02648725.2014.992622

Yeasmin S, Dipto AR, Zakir AB et al (2024) Nanopriming and AI for sustainable agriculture: boosting seed germination and seedling growth with engineered nanomaterials, and smart monitoring through deep learning. ACS Appl Nano Mater 7(8):8703–8715

Yin L, Cheng Y, Espinasse B et al (2011) More than the ions: the effects of silver nanoparticles on *Loliummultiflorum*. Environ Sci Technol 45(6):2360–2367. https://doi.org/10.1021/ES103995X/SUPPL_FILE/ES103995X_SI_001.PDF

Yin L, Wang Z, Wang S et al (2018) Effects of graphene oxide and/or Cd2+ on seed germination, seedling growth, and uptake to Cd2+ in solution culture. Water Air Soil Pollut 229(5):151. https://doi.org/10.1007/s11270-018-3809-y

Yousif AM, Ahmed SH, Nuaman RS et al (2023) The effect of nanoparticles and plant growth regulators on germination of maize seeds: a comparative study. J Med Chem Sci 6:1246–1253

Yu F, Wei C, Deng P et al (2021) Deep exploration of random forest model boosts the interpretability of machine learning studies of complicated immune responses and lung burden of nanoparticles. Sci Adv. https://doi.org/10.1126/sciadv.abf4130

Zahedifar M (2021) Nanomaterials in soil remediation: an introduction. In: Amrane A, Mohan D, Assadi AA et al (eds) Nanomaterials for Soil Remediation. Elsevier, pp 3–13

Zain M, Ma H, Rahman SU et al (2024) Nanotechnology in precision agriculture: advancing towards sustainable crop production. Plant Physiol Biochem 206:108244. https://doi.org/10.1016/j.plaphy.2023.108244

Zhang P, Ma Y, Xie C et al (2019) Plant species-dependent transformation and translocation of ceria nanoparticles. Environ Sci: Nano (6):60–67

Zhang P, Guo Z, Zhang Z et al (2020) Nanomaterial transformation in the soil–plant system: implications for food safety and application in agriculture. Small. https://doi.org/10.1002/smll.202000705

Zhang P, Guo Z, Ullah S et al (2021) Nanotechnology and artificial intelligence to enable sustainable and precision agriculture. Nat Plants 7(7):864–876. https://doi.org/10.1038/s41477-021-00946-6

Zhao X, Meng Z, Wang Y et al (2017) Pollen magnetofection for genetic modification with magnetic nanoparticles as gene carriers. Nat Plants 3(12):956–964

Index

A
Abiotic stress, 37, 38, 62, 90, 92, 107, 119, 120, 138, 156, 166–169, 171, 204, 214–222, 224, 245, 261, 292, 312, 354, 411, 429, 443–444, 512–513, 532, 536, 538, 539
Abiotic stress tolerance, 203–226
Absorption, 5, 18, 37, 53, 54, 57–60, 68, 70, 82, 85, 86, 94, 108, 109, 140, 158, 159, 165, 168, 169, 185, 186, 190, 192, 193, 205, 216, 218, 220, 224, 225, 242, 247, 260, 261, 279, 287, 309, 314, 329, 330, 334–340, 342, 353, 387, 398, 408, 419–422, 438, 466, 492, 507, 508, 510, 513, 515, 523, 526, 528–530, 551, 552, 556, 559, 561
Aceria guerreronis, 455, 464
Acetylcholine inhibitors, 455
Acheae janata, 456
Adenosine Triphosphate (ATP), 92, 105, 117, 121, 188, 456, 488, 491, 492
Advantage of nano pesticides, 6, 248, 308, 400, 515
Aeroponics, 327–342
Agglomerates, 20, 338, 439
Agricultural biotechnology, 199, 413
Agricultural chemicals, 11, 382, 383, 387–389, 394, 395, 400
Agricultural inputs, 2, 51–74, 155, 171, 238, 258, 308, 348, 372, 408, 507, 555
Agricultural productivity, 4, 60, 82, 154, 239, 311, 328, 342, 349, 355, 364, 367, 369, 400, 411, 416, 418, 429, 493, 524, 552, 562
Agriculture, 2, 27, 52, 82, 104, 134, 154, 180, 204, 237, 272, 308, 328, 348, 382, 408, 438, 452, 493, 506, 530, 548
Agri-nanotechnology, 452
Agrochemical ingredients (AgI), 393, 395, 398
Agrochemicals, 27, 36, 52–57, 61, 63, 82, 84, 110, 205, 257, 258, 272, 284, 286, 296, 299, 314, 320, 328, 331, 342, 350, 351, 355, 360–362, 370, 382–384, 386, 388–392, 394, 397, 400, 409, 410, 412, 415, 416, 428, 439, 446, 447, 493, 516, 552, 559, 562
Agro-livestock, 388
Algae, 23–24, 39, 89, 91, 122, 186, 212–213, 218, 444, 550, 564
Amphiphilic polymers, 392
Antifungal, 20, 21, 24, 60, 222, 253, 254, 278, 296, 355, 388, 482, 485, 491–492, 514
Antimicrobial, 19, 20, 22, 23, 25, 27, 36, 39, 57, 85–89, 91–93, 139, 161, 170, 212, 263, 274, 276, 296, 298, 348, 352, 353, 363, 372, 409, 411, 413, 424–427, 481, 482, 485, 487, 490–492, 494, 495, 550
Antioxidant activity, 39, 160, 161, 215, 223, 291, 336, 486, 537, 538
Antioxidant enzymes, 32–34, 37, 38, 138, 140, 162, 167, 216, 221, 223, 290, 354, 534, 537, 538, 556
Antioxidants, 21, 33, 35, 38, 65, 92, 93, 138, 139, 159–160, 167, 169, 204, 215, 217, 219, 222, 223, 241, 244, 280, 282, 289, 292, 298, 397, 398, 411, 426, 481, 483, 486, 493–495, 532–535, 557
Apoplastic/symplastic transport, 338

J. M. Al-Khayri et al. (eds.), *Nanotechnology Applications in Modern Agriculture*, Nanotechnology in Plant Sciences 3, https://doi.org/10.1007/978-3-031-90506-3

Aquatic organisms, 441–443
Arabidopsis thaliana, 29, 93, 107, 122, 217, 253, 289, 312, 316, 337, 535
Atomic Force Microscopy (AFM), 441

B
Bacteria, 11, 19, 21–23, 25–27, 36, 39, 71, 85, 87, 89, 90, 110, 111, 115, 139, 161, 209, 214, 219, 222, 260, 273, 282, 313, 364, 460, 465, 491, 550, 551, 563, 564
Beauveria bassiana, 459, 466
Bemisia tabaci, 459, 461
Bioaccumulation, 272, 313, 368, 369, 385, 389, 394, 395, 400, 420, 421, 454, 461, 552, 563
Bioactive compounds, 7, 39, 164, 165, 284, 287, 350, 355, 363, 425, 427, 480, 530
Bioavailability, 5, 18, 39, 52–57, 61, 71, 157, 209, 241, 245, 283, 285, 297, 360, 362, 363, 384, 387, 398, 400, 412, 424, 425, 427, 439, 447, 462, 536, 554, 562
Biochemical processes, 118, 120–121, 204, 218, 219, 224, 398
Biocompatibility, 25, 64, 87, 107, 209, 252, 276, 389, 392, 413, 440, 462–466
Biomagnification, 454, 564, 565
Biomass accumulation, 196, 444
Biomimetic properties, 385
Bio/nano-sensors (bNSRs), 256, 257
Biopolymers, 88, 160, 297, 298, 313, 462
Bio-reductant, 21
Biosafety aspects, 467–468
Biosensors, 83, 106, 107, 111, 112, 115, 123, 124, 256, 257, 363–365, 388, 390, 424, 550, 555, 559
Biosphere pollution, 4
Biotic, 18, 27, 30, 82, 92, 107, 119, 120, 122, 156, 171, 204, 225, 312, 408, 429, 452, 552
Biotic stress, 108, 120, 154, 156, 170, 171, 203–226, 429, 556
Blattella germanica, 282, 460
Bombyx moori, 461
Bottom-up, 18, 86, 87, 180–182, 206, 207, 210

C
Cadmium (Cd^{2+}), 25, 37, 106, 166, 167, 216, 242, 244, 260, 534, 553, 554, 556
Cadmium toxicity, 38
Callosobruchus chinensis, 459
Carbohydrates, 34, 36, 111, 163, 165, 197, 290, 336, 526, 529, 531, 535, 551
Carbon-based, 84, 157, 309, 349, 409, 506, 549, 557
Carbon-based nano-fertilizers, 157
Carbon-based nanomaterials, 83, 87, 156, 182, 247, 287, 288, 351, 549, 551
Carbon fullerenes, 410–411
Carbon nanomaterial, 18, 239, 314, 351, 468, 554
Carbon nanotubes (CNTs), 32, 34, 37, 107, 118, 119, 121–123, 137, 157, 169, 180, 182, 186, 204, 211, 247, 253, 277, 287, 298, 313, 315, 318, 349, 351, 353, 388, 409, 410, 414, 427, 465, 549, 551, 553, 554, 557, 558
Carotenoids, 34, 35, 136–138, 164, 243, 291, 292, 319, 480, 486, 510, 526
Carrier molecule, 459, 461, 552
Cas protein, 316, 317
Cerium oxide (CeO_2), 86, 125, 137, 168, 216, 224, 287, 312, 319, 336, 354, 398, 440, 443–444, 535, 538, 563, 565
Chelated, 70, 157, 525, 529
Chemical methods, 18, 19, 88, 208, 390, 413
Chemiluminescence (CL), 257
Chitosan (CS), 8, 28, 61, 88, 112, 135, 159, 213, 253, 276, 311, 331, 391, 410, 440, 459, 485, 510, 535
Chitosan-beauvericin nanoparticles, 459
Chitosan-pepper-rosmarin nanoparticles, 460
Chlorophyll Content Index (CCI), 442, 527
Classification of nanopesticides, 276–277
Climate change, 27, 82, 140, 154, 166, 180, 238, 258, 272, 297, 348, 408, 417, 428, 525, 527, 555, 562, 567
Clustered regularly interspaced short palindromic repeats (CRISPR), 260, 316–318, 558
Coating, 18, 27, 30, 31, 59, 65, 87, 119, 137, 156, 180, 186, 245, 247, 251, 296, 320, 331, 351, 353, 398, 412, 426, 481, 482, 485–491, 494, 495, 508, 529
Cold storage, 489
Commercial hydroponic systems, 334
Composite edible coatings, 494
Composites, 27, 118, 124, 287, 293, 391, 392, 487, 506, 529
Controlled pesticide delivery systems, 386–394
Controlled release, 28, 31, 36, 52, 53, 55–57, 59, 61, 62, 64, 66, 68, 70, 71, 73, 74, 83, 84, 89, 137, 153, 156, 157, 159, 160, 170, 240, 248, 249, 253, 276, 285, 293, 308, 311, 330, 331, 342, 350–353, 355, 362, 370, 383, 387, 388, 391, 392, 394, 400, 409, 412, 415, 420, 421, 424–426, 428, 448, 455, 459, 460, 462, 464–466, 468, 469, 514, 515, 524

Controlled-release formulation, 249, 387, 454
Copper (Cu^{2+}), 4, 11, 18, 27, 35, 36, 85, 88, 92, 106, 157, 163, 219, 243, 258, 260, 276, 335–337, 390, 398, 409, 440, 492, 494, 514, 529, 549
Copper nanoparticles (CuNPs), 21, 88, 92, 118, 163, 164, 216, 219, 243, 339, 398, 399, 414, 415, 514
Copper oxide (CuO), 30, 35, 60, 86, 90, 135, 140, 160, 162, 163, 170, 211, 213, 319, 514, 549, 552, 563, 565
Copper oxide nanoparticles (CuO-NPs), 30, 35, 60, 135, 140, 160, 170, 440–441, 514, 535, 549, 552, 563
Core-shell, 30, 56, 350, 391, 458
Cost-effectiveness, 5, 28, 72, 239, 256, 276, 297, 389, 398, 420, 570
CRISPR-Cas, 316–318, 321
Crop enriching, 440
Crop productivity, 36, 37, 39, 59, 63, 70, 155–157, 159, 160, 168, 238, 240, 246, 255, 258, 262, 274, 295, 308, 309, 311, 312, 318, 342, 348, 350, 418, 446, 452, 525
Crop protection, 2, 7, 13, 60, 85, 240, 241, 257, 271–299, 313–314, 349, 361, 362, 401, 409, 452, 469, 515
Crop yield improvement, 154, 160, 216, 244, 255, 290, 299, 309, 351, 355, 356, 408, 410, 551
Cross-linking, 8, 286, 314, 389, 461
Culture media, 338
Cutting-edge technologies, 2, 55, 205, 448
Cyanobacterium Prochlorococcus, 443
CYP450 enzyme, 456
Cytometric analysis, 441

D

Degradation, 9, 52, 54, 57, 58, 60, 61, 66, 83, 140, 154, 157, 169, 211, 212, 251, 272, 275, 283, 295–298, 308, 311, 317, 342, 357, 360–362, 367, 384, 386, 394, 400, 410, 411, 416, 428, 480, 486, 489–491, 494, 495, 514, 515, 527, 528, 550, 558, 561
Dendrimers, 18, 26, 239, 252, 277, 287, 292, 328, 384, 557, 562
Deoxy ribo nucleic acid (DNA), 84, 91, 94, 104, 106, 107, 115, 165, 218, 244, 245, 251, 257, 280–282, 288, 315–317, 340, 355, 397, 419, 456, 467, 468, 491, 492, 525, 527, 557–559, 563, 565, 566
Dichlorodiphenyltrichloroethane (DDT), 452, 454, 455, 554
Disability-adjusted life years (DALY), 385, 400
DNA damage, 92, 244, 282, 397, 456, 467, 566
DNA synthesis, 525
Drought stress, 38, 168, 215, 218, 225, 243, 354, 510, 513, 527, 528, 533, 535, 537, 538
Drought tolerance, 243, 354, 533
Drug delivery, 84, 180, 412, 458, 550, 555
Dynamic Light Scattering (DLS), 444

E

Eco-friendly, 5, 22, 88, 180, 313, 370, 413, 415, 416, 419, 420, 426, 490, 495, 509, 513, 515, 527, 560
Eco-friendly agricultural system, 401
Economical, 19, 88, 256, 439, 527
Ecosystem, 52, 58, 67, 69, 71, 73, 82, 104, 133, 159, 213, 214, 255, 256, 258, 262, 272, 320, 330, 357, 362, 364, 367, 369, 370, 372, 382, 396, 401, 411, 417, 422, 428, 429, 439, 441, 444–445, 447, 454, 507, 524, 551, 559, 561–563, 565–567
Ecotoxicological framework, 401
Edible coating, 481, 482, 489, 490, 494
Effective pest control, 401
Efficiency, 2, 18, 52, 120, 157, 184, 216, 239, 273, 309, 330, 348, 383, 408, 444, 482, 508, 526, 549
Electrochemical sensors, 107, 111, 116, 119, 124
Electron transport, 10, 35, 120, 121, 280, 288, 315, 352, 526, 533
Empowering plant growth with nanomaterials, 287
Encarsia formosa, 461, 464
Engineered nanomaterials (ENMs), 157, 203–226, 287, 288, 315, 438, 565, 566, 568
Enhanced crop production, 171, 257
Environment, 8, 18, 57, 82, 102, 141, 160, 183, 207, 238, 281, 308, 330, 349, 382, 408, 438, 453, 491, 509, 526, 548
Environmental contamination, 18, 283, 294, 309, 330, 360–362, 401, 411, 412
Environmental impacts, 3, 52, 54, 55, 61, 64, 67, 69, 71, 73, 83, 124, 154, 171, 240, 245, 246, 249, 258, 261–263, 276, 277, 295, 298, 299, 308, 348, 349, 354, 356, 358, 360, 362, 372, 381–401, 408, 410, 420, 421, 428, 429, 452, 493, 510, 516, 548, 550, 555, 561, 568

Environmental pollution, 10, 156, 170, 248, 249, 254, 258, 272, 295, 349, 352, 362, 364, 367, 368, 395, 401, 415, 447, 460, 493, 551, 555
Environmental safety, 2, 281, 552
Environmental stress, 102, 166, 169, 224, 242, 298, 333, 348, 352, 355, 417, 507, 510, 539
Environmental sustainability, 74, 258, 259, 349, 357, 364, 419, 426, 550, 570
Environment protection, 71, 155, 156, 171, 396, 398, 548, 569
Ephestia kuhniella, 457
Essential oils nano emulsion, 459
Ethics, 439, 446
Evaporation, 7, 64, 65, 108, 109, 205, 208, 252, 286, 330, 352, 386, 458, 528

F
Fertilizers, 3, 18, 52, 82, 137, 154, 205, 239, 272, 309, 328, 350, 383, 409, 438, 452, 506, 524, 548
Foliage, 68, 154, 284, 335, 339, 361
Foliar application, 29, 34, 68, 74, 140, 158, 159, 163, 164, 170, 220, 222, 242, 335, 354, 511, 526–528, 531, 533, 551
Food chain, 71, 94, 257, 259, 308, 341, 368, 441, 442, 447, 455, 461, 564, 565
Food labelling, 426–427
Food packaging, 298, 424–426, 494
Food packaging and storage, 298–299, 419
Food processing, 257, 298–299, 419, 425–427, 495, 550
Food security, 2, 13, 39, 61, 73, 125, 126, 154, 160, 171, 204, 240, 297–299, 308, 318, 348, 349, 352, 362, 364, 367, 372, 373, 395, 401, 480, 495, 524, 553, 559, 560
Formulation, 6, 36, 60, 94, 137, 248, 274, 329, 351, 384, 412, 453, 495, 508, 526, 552
Formulation optimization, 508
Formulations of nanopesticides, 384
Frankliniella occidentalis, 464
Free radical induction, 456
Fruit, 19, 62, 107, 136, 161, 242, 275, 332, 355, 452, 480, 506, 528
Fruit extracts, 22
Fruit flies, 458, 513
Fruit quality, 480, 482–485, 488, 489, 495, 496, 507, 509, 511, 512, 517
Fruit quality improvement, 509, 512, 517
Fruit safety, 481, 496
Fruit yield enhancement, 506–517
Fullerenes, 27, 204, 247, 287, 289, 410–411, 506
Fungi, 19, 24–25, 34, 39, 85, 89, 90, 110, 111, 113, 123, 186, 209, 213–214, 222, 224, 253, 273, 278, 313, 351, 364, 441, 462, 465, 466, 490, 492, 514, 550

G
GABA-gated chloride channel blockers, 455
Gas exchange attributes, 527
Gastrointestinal tract, 536
Genetically modified organisms (GMOs), 238, 260, 262, 318, 321, 427
Genome editing, 316, 317, 558
Genotoxic, 94, 276, 383, 384, 395, 397, 567
Germination index, 336
Global warming, 154, 166, 171, 509
Gold, 18, 23–25, 27, 32, 35, 36, 57, 86, 90, 91, 112, 113, 123, 124, 136, 206, 257, 259, 287, 289, 318, 337, 349, 390, 409, 411, 414, 439, 549
Gold nanoparticles (AuNPs), 19, 21–23, 25, 26, 33, 34, 54, 87, 92, 106, 107, 112, 113, 115, 116, 123, 125, 135, 137, 138, 217, 241, 256, 257, 260, 282, 283, 289, 290, 337, 414, 427, 468, 559
Grafted target nanoparticle formulations, 387
Greenhouse gas emission, 213
Greenhouse growth environments, 388
Green nanoparticles, 89, 515
Green nanotechnology, 88, 89, 550
Green synthesis of nanoparticles, 19–26, 88–89, 413–415, 417
Groundwater contamination, 11

H
Heat stress, 38, 169, 221–222, 290, 538, 539
Heavy metal remediation, 259
Heavy metals, 11, 23, 37, 38, 83, 90, 105, 106, 166–167, 216, 219–221, 243, 258–260, 329, 352, 357, 366–368, 371, 411, 424, 510, 534, 537, 538, 550, 553, 554, 556
Heavy metal stress, 166–167, 219
Helicoverpa armigera, 60, 252, 462, 464, 466
Host-specificity, 455, 469
Human life, 452
Hydroponics, 36, 154, 291, 328–342

I

Impact of nano pesticides on the human, 281
Indian Council for Agricultural Research (ICAR), 247
Innovation, 28, 36, 40, 56, 57, 72, 73, 155, 254, 257, 298, 311, 348, 373, 386–394, 401, 408, 421–426, 428, 429, 467, 481, 494, 495, 506, 513, 517, 555, 571
Inorganic nanoparticles, 53, 247, 287, 289
Insecticidal agents, 382
Insecticides, 5, 7, 54, 55, 60, 61, 68, 85, 89, 123, 205, 215, 240, 248, 251, 252, 262, 273, 275, 276, 296, 297, 311, 351, 354, 382, 384, 389, 394, 396, 397, 452–456, 458–461, 464–469
Integrated pest management (IPM), 279, 421, 458, 468, 469
Interfacial deposition, 7
Interpol electrolyte complex (IPEC), 390
In-vitro, 38–39
In vivo methods, 335
Ionic imbalance, 168, 537, 538
Iron nanoparticles, 25, 168, 225, 290, 291, 339, 414, 557
Iron oxide nanoparticles (Fe_2O_3NPs), 20, 21, 25, 160, 216, 218, 225, 244, 537, 566

L

Lab-on-a-chip, 8, 107
Lanthanum oxide (La_2O_3), 440–443
Layer by layer deposition, 286
Leaching, 3, 52, 57, 58, 69, 70, 154, 157–160, 244, 252, 309, 330, 350, 351, 361, 362, 369, 385–387, 391, 395, 400, 401, 412, 528, 551
Lead (Pb), 11, 25, 32, 106, 167, 219, 220, 258, 260, 337, 465, 553
Leaf abscission, 336
Lipid-based edible coatings, 482
Lipid-based nanomaterials encapsulation, 287
Lipid nanoparticles, 384, 410, 427
Lipophilic pathway, 339
Longevity, 85, 294, 335, 358, 455

M

Macronutrient-based nano fertilizers, 157
Management of insect-pests, 295–297
Managing food security, 297–298
Manganese nanoparticles (MnNPs), 169, 243–244
Marketing, 126
Maturity indices, 512
Media, 21, 190, 197, 208, 338, 339, 444, 445, 447, 448, 458
Membrane structural stability, 536
Mesoporous silica nanoparticles (MSNs), 28, 29, 35, 247, 277, 278, 314, 392, 557, 558
Metabolic pathways, 165
Metallic nanoparticles, 36, 253, 289, 390, 398, 401, 439
Metalloids, 18, 27, 239, 259
Metal nanoparticles (NPs), 18, 33, 34, 53, 71, 86–88, 209, 296, 328, 331, 334, 337–339, 398, 409, 507, 514
Metal oxide nanoparticles, 291, 312, 409, 439, 440, 446, 507, 565
Metal oxides, 27, 28, 124, 140, 280, 287, 296, 349, 356, 506, 554, 564
Micelle, 6, 7, 56, 251, 277, 283, 285, 286, 313, 314, 392, 395
Microbes, 36, 71, 83, 85, 89, 90, 92, 123, 139, 185, 205, 213, 262, 263, 278, 313, 459, 491, 561, 563, 564
Microbiological flora, 328
Microemulsions, 283–284
Microencapsulation, 412
Micronutrient-based nano fertilizers, 157
Micronutrients, 4, 70, 83, 88, 116, 137, 157, 163, 164, 213, 224, 225, 248, 261, 298, 342, 358, 387, 426, 508, 511, 513, 526, 538, 551
Micronutrient supplements, 4
Microorganisms, 5, 8, 9, 11, 36, 88, 89, 94, 113, 156, 159, 170, 209, 256, 272, 273, 298, 308, 310, 369, 389, 413, 415, 416, 420, 422, 439, 441, 455, 459, 492, 508, 515, 528, 549–551, 554, 558, 563, 564
Milligram/litre (mg/L), 118, 160, 163, 164, 198, 222, 242–244, 297, 397, 414, 442, 459, 461, 463, 510, 527
Millimolar (mM), 25, 118, 119, 214, 220, 243, 511
Mineral perlite, 334
Mitochondrial ATP synthase inhibitors, 121, 456
Mode of action against insects, 282, 455–456
Modified nanoparticles, 27
Morphology, 111, 137, 169, 186, 190, 194, 210, 215, 244, 416, 441, 442, 444, 534
Mosquito species, 282, 460
Multi-walled carbon nanotubes (MWCNT), 34, 112, 113, 136–138, 222, 287–289, 316, 336, 388, 414, 465
Myzus persicae, 459, 464

N
Nano, 4, 18, 52, 82, 104, 136, 156, 180, 205, 243, 274, 308, 328, 350, 382, 408, 439, 453, 485, 509, 526, 550
Nano-agrochemicals, 382, 447
Nano-agrotechnology, 275
Nano-based pesticide formulations, 60, 274, 275, 277, 283–287, 400
Nano-bionics, 180–182, 189, 199, 311
Nano-biopesticides, 5, 6
Nano-biosensors, 8, 83, 106, 123, 256, 257, 273, 308, 309, 312, 423–425, 482, 555, 559
Nano capsules, 7–8, 36, 52, 56–57, 62–65, 67, 73, 250, 251, 259, 277, 286, 287, 297, 313, 350, 386, 387, 421, 425, 457, 458, 465
Nanocarriers, 31, 53, 57, 60–67, 69, 72, 84, 85, 157, 240, 245, 246, 260, 273, 275–277, 279, 286, 299, 314, 315, 317, 321, 360–363, 370, 371, 384, 389, 391, 394–397, 400, 401, 421, 425, 552, 557, 558, 571
Nano-chemicals, 439
Nanocoating, 6, 137, 248
Nano-delivery systems, 245
Nano-destruxin, 464, 466
Nano dispersion, 250, 283, 285, 390
Nano emulsion, 53–54, 156, 249–250, 253, 283–285, 297, 311, 349, 354, 384, 392, 393, 412–413, 421, 424, 425, 427, 457, 459, 464, 481, 486, 487, 552, 562
Nanoencapsulation, 4, 52–74, 248, 251, 252, 275, 283, 285–287, 298, 311, 351, 355, 386–394, 412, 424–427, 457, 552, 553, 562
Nano fertilizers, 4, 18, 59, 83, 154, 204, 240, 287, 309, 329, 350, 387, 415, 446, 506, 524, 551
Nano formulations, 7, 67, 94, 245, 248, 249, 253, 274, 275, 283, 295, 297, 299, 311, 352, 354, 355, 385, 386, 391, 453, 456–466, 468, 514, 562, 566
Nano fungicide, 27, 36, 253, 276, 513, 514
Nanogel-methyl eugenol, 458
Nanogels, 55–56, 250, 251, 286, 297, 313, 314, 440, 457, 458, 513, 552, 562
Nano herbicides, 6–7, 61, 63, 252, 253, 311, 553, 562
Nanomaterials, 5, 18, 55, 83, 111, 134, 156, 180, 205, 239, 273, 308, 328, 348, 383, 408, 439, 452, 482, 506, 524, 549
Nanomaterials for post-harvest management, 294–299
Nanometer (nm), 5, 18, 19, 21–23, 25, 53, 55, 56, 59, 64, 82, 84, 102, 106, 122, 137, 138, 155, 156, 160, 184, 186, 190, 192–194, 198, 199, 204, 207, 208, 213, 220, 246, 283, 285, 292, 315, 328, 331, 337, 339, 349, 357, 361, 390, 391, 408, 412, 414, 439, 452, 459, 463, 465, 481, 493, 506, 524, 548, 571
Nanoparticle-based technology, 428
Nanoparticle-mediated stress resistance, 512–513
Nanoparticles (NPs), 3, 18, 52, 83, 106, 134, 154, 180, 205, 239, 275, 308, 328, 348, 382, 408, 438, 453, 481, 506, 524, 549
Nanoparticles for insect control, 281–283
Nanoparticles in agriculture, 39, 40, 54, 239, 242, 422–423, 447
Nano pesticides, 6, 60, 85, 139, 155, 245, 274, 308, 354, 383, 415, 446, 452, 507, 552
Nano pyrifluquinazon, 459, 464
Nano remediation, 9–11, 167, 349, 366–368, 553–554, 571
Nanorobotics, 13
Nanorods, 23, 107, 318, 411, 562
Nanoscale, 18, 20, 36, 52, 56, 60, 68, 83, 86, 93, 102, 118, 121, 156, 157, 208, 240, 245, 247, 251, 255, 259, 285–287, 294, 296, 298, 309, 328–342, 348, 349, 361–363, 365, 372, 392, 408, 412, 416, 424, 425, 427–429, 440, 481, 554, 557, 559, 569
Nano scale guardians, 275–299
Nanoscale materials, 83, 296, 342
Nanoscience, 257, 272, 363, 439, 446
Nano sensors, 7, 60, 82, 102, 155, 241, 273, 315, 348, 419, 481, 506, 550
Nanosensors in precision agriculture, 12, 355–362
Nano-silver, 94, 139, 238, 280, 336, 514
Nanospheres, 6, 250, 251, 259, 277, 286, 296, 313, 387, 414, 552, 562
Nano suspension, 250, 252, 283, 285, 421, 457, 552, 562
Nanotechnology, 2, 29, 52, 82, 102, 133, 155, 204, 238, 272, 308, 328, 348, 382, 408, 438, 452, 481, 506, 536, 548
Nanotechnology applications in agriculture, 2–13, 246, 311, 348–373, 417, 452, 570
Nanotechnology in crop production, 312, 349, 369
Nanotechnology in pest control, 248, 273–275, 311, 383–384, 513
Nanotechnology in sustainable farming, 418
Nanotoxicology, 155, 460, 567, 568

Nanovesicles, 387, 558
Neurotoxicity, 252
Nickel (Ni), 27, 90, 113, 116, 163, 260, 444
Nickel oxide (NiO), 135, 287, 440, 444
Nilaparvata lugens, 459
Nitrate ion, 118, 335
Nitric oxide, 120, 122, 288
Nitrosoglutathione-containing alginate/chitosan nanoparticles (GSNO), 253
Non-metals, 18, 27, 114, 239
Non-target DNA, 317
Non-targeted organisms, 452, 469
Nutrient bioavailability, 157, 424
Nutrient deficiencies, 102, 157, 159, 217, 224, 245, 257–258, 279, 364, 512, 551, 564
Nutrient deficits, 2, 243
Nutrient delivery, 3–5, 13, 54, 156, 157, 205, 258, 263, 274, 328–342, 348, 349, 408, 409, 411, 424, 426, 427, 506, 508, 510
Nutrient enrichment, 171
Nutrient monitoring, 114–118
Nutrient uptake, 3, 4, 34, 37, 39, 64, 137, 154, 157, 158, 163, 239, 244, 245, 252, 261, 274, 279, 318, 329, 348, 350, 353, 372, 409–411, 415, 416, 425, 428, 507, 508, 513, 530, 538, 551, 556, 563
Nutritional values, 59, 263, 287, 294, 311, 364, 484, 485

O

Optical sensors, 109, 114, 115, 123–125, 550
Oxidative stress, 30, 33–35, 37–39, 71, 91, 93, 139, 167–169, 215–220, 222, 280, 281, 283, 456, 461, 531, 534, 536, 566
Oxide, 18, 54, 83, 104, 135, 160, 204, 239, 280, 312, 328, 349, 391, 409, 439, 456, 482, 506, 526, 549
Oxygen, 62, 109, 121–122, 208, 298, 441, 492, 494, 525, 539

P

Paraquat (PQ), 253
Parts per million (ppm), 21, 93, 118, 123, 125, 164, 165, 243, 289, 291, 336, 338, 514
Pathogen detection, 241, 515, 559, 562, 571
Percentage (%), 37–39, 108, 160, 194, 213, 316, 462, 512, 528, 529
Performance enhancement, 315
Pest control, 52, 54, 61, 245, 248, 249, 255, 273–276, 279, 294, 296, 311, 348, 354, 359, 372, 383–384, 401, 408, 411, 452, 455, 507, 513, 514, 516
Pesticide detection, 105, 123–125, 255, 256
Pesticides, 2, 28, 52, 82, 102, 140, 155, 239, 272, 308, 333, 348, 382, 409, 452, 482, 507, 548
Pest management, 40, 60, 82, 84, 155, 241, 255, 263, 274, 281, 287, 296, 308, 348, 358, 360, 388, 390, 392, 400, 401, 422, 428, 451–469, 507, 513–515, 551
Pest management in fruit production, 513
Phenolic compounds, 22, 532
pH impact on nanoparticle synthesis, 282
Phosphate ion, 117, 119, 335
Photocatalyst, 20, 490
Photomicrographs, 441, 442, 445
Photosynthesis, 33–35, 38, 39, 108, 120, 121, 136, 138, 159–161, 163, 167, 180, 188, 189, 197, 216, 217, 219, 220, 224, 247, 252, 275, 279, 352–353, 358, 411, 440, 444, 510, 524–528, 530, 533, 537, 539, 551
Photosystem-II (PSII), 34, 35, 167, 169, 514, 527, 532, 538
Physical, 18, 19, 30, 32, 53, 54, 68, 82, 84, 88, 93, 102, 103, 168, 206–209, 218, 244, 256, 274, 281, 286, 295, 314, 321, 334, 337, 340, 357, 366, 386, 388–393, 455, 464, 480–483, 485, 494, 550, 560
Physical methods, 18, 88, 208, 390
Physicochemical properties, 87, 93, 204, 259, 342, 410, 438, 439, 506, 536, 549, 569
Phyto-molecules, 18
Phytotoxicity, 18, 27, 220, 536, 562
Pickering emulsions (PE), 393
Pigments, 138, 161, 162, 188, 191, 194, 197, 216, 221, 223, 291, 292, 444, 483, 510, 513, 525–527, 538
Plant analytes, 121–123
Plant disease, 27, 36, 60, 63, 81–94, 110–115, 170, 222, 263, 273, 329, 356, 362–366, 393, 438, 452, 558–559
Plant extracts, 18, 19, 89, 91, 121, 180, 206, 209, 211, 212, 351, 413, 416, 465, 482, 512
Plant growth, 3, 23, 54, 93, 102, 137, 154, 190, 205, 239, 275, 312, 329, 351, 409, 440, 461, 507, 530, 549
Plant growth enhancement, 5, 65, 198, 332, 537
Plant growth promotion, 351
Plant growth regulation, 64
Plant hormones, 62–65, 82, 93, 105, 120–121, 158, 260, 352

Plant metabolism, 118–123, 125, 309, 525, 527
Plant morphogenesis, 120
Plant nanotechnology, 437–448
Plant pathogens, 36, 40, 82, 92, 106–107, 111, 139, 161, 170, 241, 245, 255–257, 262, 274, 363–365, 372, 413, 459, 493, 514, 516, 558, 559
Plant resilience to stress, 245, 352
Plants' resilience, 245, 352, 531, 539
Polymer-based nanoencapsulation materials, 286
Polymer-based nano fertilizers, 157
Polymeric nanoparticles, 55, 64, 156, 259, 286, 330, 396, 398, 410, 552, 553
Polysaccharides, 24, 84, 390, 398, 461, 482, 550
Polyvinyl pyrrole, 6
Postharvest, 273, 275, 294, 295, 353, 355, 372, 418, 419, 479–496, 507, 509
Post-harvest management, 275, 294–299, 356, 419, 481, 484, 493–495
Potassium ion, 117, 119, 168, 329
Potential of hydrogen (pH), 8, 21, 23, 26–29, 55, 63, 65–67, 72, 117, 134, 136, 137, 160, 212–214, 221, 248, 251, 260, 282, 285, 292, 294, 296, 329, 335, 356–358, 361, 362, 386, 388, 390, 392, 393, 400, 401, 414–416, 486, 488, 490, 491, 507, 525, 555, 562, 564
Precise irrigation practices, 10
Precision agriculture, 12, 52, 63, 110, 153–171, 238, 245, 246, 255, 318, 348, 349, 355–362, 372, 421, 560
Precision farming, 2, 5, 8, 9, 12–13, 28, 37, 82, 83, 240, 258, 371, 418, 428, 493, 506, 516, 552, 554–555, 559, 560, 569
Preparation of nano pesticides, 277
Protein-based edible coatings, 482
Protospacer, 317
Protospacer Adjacent Motif (PAM), 293, 316, 317
Pseudokrichneriella subcapitata, 444
Pseudomonas, 25, 26, 90, 107, 139, 222, 278, 492, 563
Public awareness, 73, 370, 447, 569, 570

Q

Quantum dots (QDs), 18, 27, 83, 84, 106, 107, 124, 125, 180, 182, 239, 256, 314, 318, 328, 357, 411, 468, 551, 559, 564, 565

R

Rare-earth nanoparticles, 438, 441–443
Reactive oxygen species (ROS), 10, 33, 88, 92–94, 105, 109, 139, 162, 165, 167, 169, 170, 214, 215, 219, 222, 223, 276, 280, 282, 292, 312, 315, 336, 352, 358, 389, 397–399, 441, 444, 486, 487, 490, 493, 494, 525, 526, 533–539, 557, 563, 566
Reduced fecundity, 464
Reducing agent, 18, 19, 21, 25, 86, 91, 212, 398, 413, 416
Remote sensing (RS), 9, 102, 155
Reproductive growth promotion, 563
Residual effect, 60, 482, 496
Resistance, 6, 28, 39, 53, 54, 62, 87, 88, 92, 93, 105, 109, 110, 117, 122, 140, 161, 168–170, 214–224, 251–253, 261–263, 272, 283, 284, 295, 329, 348, 353, 355, 362, 364, 370, 372, 383, 390, 392, 396, 398, 400, 411, 417, 419, 420, 455, 460, 461, 468, 484, 494, 507, 509, 512, 513, 517, 549, 552, 557
Reversible addition-fragmentation chain transfer (α-RAFT), 390
Ribo nucleic acid (RNA), 92, 245, 316, 317, 349, 353, 392, 396, 527, 558
RNAi technology, 462, 466
Role in food processing, 298–299
Runoff, 11, 58, 69, 116, 158, 330, 350, 362, 369, 372, 385, 387, 420, 528, 551

S

Saccharide fluxes, 122
Safety and sustainability, 369–370, 495, 536–539
Salinity, 37, 39, 62, 108, 138, 166, 168, 169, 216, 241, 244, 245, 258, 261, 335, 352, 354, 357, 358, 411, 507, 512, 528, 532, 535, 536, 539, 551, 564
Salinity stress, 38, 213, 215, 216, 532, 535, 537, 538
Salinity tolerance, 169, 513
Salt concentration effect on synthesis, 416
Scalability, 56, 72, 261, 420, 495, 570
Seed germination, 18, 20, 21, 33, 34, 36, 63, 68–70, 83, 122, 137, 138, 159, 161, 169, 211, 217, 219, 241–243, 288–293, 311–313, 336, 337, 410, 444, 507, 537, 549, 551, 556, 571
Seed nano priming, 158, 161–163

Selenium nanoparticles (SeNPs), 39, 163, 164, 169, 242–243, 290, 511, 513
Shelf-life, 65, 69, 85, 86, 249, 262, 263, 275, 295, 298, 309, 311, 329, 353, 421, 424–427, 455, 480–482, 485, 487–491, 494, 495
Shelf life extension in fruits, 482, 507
Silica nanoparticles (SiNPs), 63, 64, 106, 218–220, 242, 251, 298, 338, 391, 393, 409, 455, 459, 461, 464
Silicon dioxide (SiO_2), 32, 35, 37, 38, 112, 161, 222, 223, 279, 280, 292, 313, 328, 336, 513, 538, 549, 558
Silver (Ag), 18, 57, 85, 116, 136, 211, 257, 276, 336, 349, 409, 439, 492, 514, 531, 549
Silver nanoparticles (AgNPs), 19, 54, 86, 118, 134, 168, 216, 242, 282, 336, 399, 413, 456, 492, 514, 526
Single-walled carbon nanotubes (SWCNTs), 112, 120–122, 138, 287–289, 315, 316, 388, 410, 414, 557
SiO_2 nanoparticles, 32, 167, 292, 337, 456, 468, 485, 487, 532, 549
Sitophilous oryzae, 282, 296, 297, 455, 459, 464, 465
Slow-release, 28, 140, 293, 309, 311, 313, 318, 388, 510, 528, 551
Smart agriculture, 507
Smart delivery systems, 36, 40, 82, 171, 328, 372, 428, 429
Smart nano fertilizers, 157
Smart packaging, 481, 495
Sodium (Na), 29, 37, 105, 139, 168, 169, 216, 219, 224, 260, 279, 389, 465, 528
Sodium chloride (NaCl), 217, 224, 243, 292
Soil application, 67, 69, 70, 158, 163, 320, 335
Soil-borne pest infestations, 334
Soil contamination, 10, 61, 258, 553
Soil health, 2, 9–10, 13, 52, 55, 69, 71, 82, 134, 155–157, 213, 348, 357, 358, 362, 364, 366–369, 371, 411, 412, 422, 516, 564
Soil remediation, 246, 259–260, 308, 351, 367, 409–411, 417, 554
Soil treatment, 87, 158
Solid lipid nanoparticles (SLNs), 53, 55, 57, 286, 287, 314, 410, 488, 491
SPAD value, 534
Spodoptera furgiperda, 459, 462, 466
Spodoptera littoralis, 252, 455, 459, 461, 464
Spraying, 5, 68, 158, 243, 291, 330, 331, 335, 337, 382, 511, 512, 552
Spreading, 108
Stability, 5, 18, 21, 24–26, 28, 35, 52, 53, 55–57, 61–67, 69, 73, 74, 84, 109, 110, 122, 157, 180, 205, 209, 215, 223, 248, 249, 255, 275, 280, 283–285, 295, 297, 309, 337, 351, 357, 360, 362–364, 383, 388–393, 396, 400, 414, 416, 421, 423–425, 427, 440, 444, 456, 462, 465, 469, 493, 515, 525, 536, 538, 555, 562–564
Stimuli, 13, 25, 28, 55, 56, 63, 72, 157, 252, 342, 361, 362, 386–388, 391, 392, 401, 439, 530
Stress-responsive gene expression, 534–535
Stress tolerance, 33, 39, 62, 83, 118, 139, 170, 180, 204–226, 245, 261, 349, 352, 354, 355, 409, 513, 532, 536, 551
Sulfidation process, 388, 399
Surface-enhanced Raman scattering (SERS), 105, 257, 259, 365
Sustainability, 40, 52, 55, 56, 61, 67, 70, 73, 74, 82, 155, 204, 205, 239, 240, 257–259, 262, 263, 272, 308, 309, 348, 349, 352, 355, 357, 360–362, 364, 369–370, 372, 373, 413, 419, 422, 426, 495, 513, 525, 539, 550, 555, 560, 570
Sustainable agriculture, 10, 56, 61, 62, 83, 125, 156, 169, 258, 274, 299, 329, 372, 401, 408–429, 509, 548, 555
Sustainable development, 312, 385, 550, 568, 570
Sustainable farming, 13, 38
Sustainable farming practices, 4, 52, 261, 275, 372, 418, 548

T

Targeted nutrient delivery, 156, 258, 424, 426
Thylakoids, 138, 188, 534
Titanium dioxide nanoparticles (TiO_2NPs), 21, 32, 33, 37, 54, 85, 136, 163, 167, 169, 244, 291, 296, 409, 456, 459, 485, 487, 490, 492, 549, 566
Top-down, 18, 86, 87, 180–182, 206–208, 211
Toxic, 18, 88, 116, 244, 260, 275, 279–281, 311, 313, 329, 340, 369, 383, 394, 414, 439, 443–444, 447, 454, 460, 461, 485, 510, 514, 515, 532, 539, 553, 554, 556, 562–564, 569
Toxic mechanism of nano pesticides, 279–280
Toxicological aspects of nanoparticles, 422–423
Toxicological characteristics, 447

Toxins, 104, 107, 245, 256, 257, 425, 482, 491, 550, 555
Traditional agricultural practices, 52, 334
Transmission Electron Microscopy (TEM), 29, 30, 190, 194, 338, 441
Transpiration, 62, 108, 111, 330, 339, 358, 444, 489, 511, 528, 532, 533, 537, 563
Tribolium castaneum, 283, 296, 297, 459, 464, 465
Trypsin modulating oostatic factor (TMOF), 389
Turgor pressure, 527, 528, 534

U
US Food and Drug Administration (FDA), 342, 567

V
Volatile organic compounds (VOCs), 109, 111–113, 115, 122–123, 367

W
Wastewater treatment, 11
Water and humidity, 104–106
Water management, 10–11, 13, 82, 155, 275, 352, 372, 507
Water management using nanotechnology, 10, 155
Water stress, 107–110, 220, 358, 527, 528
Water use efficiency (WUE), 10, 62, 162, 527, 533, 537, 539
Waxing, 68, 138, 280, 282, 390, 391, 486, 487
Wireless sensor network, 493

X
X-ray absorption spectroscopy technique, 338

Y
Yeast, 90, 213, 214, 295, 488, 550
Yield Per Unit Input, 4

Z
Zinc (Zn), 4, 21, 59, 86, 116, 136, 157, 213, 247, 276, 336, 353, 409, 509, 526, 549
Zinc oxide nanoparticles (ZnONPs), 20–22, 28, 32, 33, 35, 54, 85, 136, 138, 164, 216, 219, 222, 225, 243, 253, 291, 292, 296–298, 318, 331, 336, 353, 409, 414, 415, 456, 461, 488, 492, 532, 535, 537, 549, 556